Commonly used statistical symbols and abbreviations†

Name	Symbol / abbreviation: Sample statistic	Population parameter	Other
Analysis of variance			ANOVA
Chi-square statistic	χ^2		
Coefficient of determination	R^2		
Coefficient of multiple determination	R^2_{mult}		
Coefficient of variation	CV		
Correlation coefficient	r	ρ	
Degrees of freedom			df
F-statistic	F		
Fisher's least significant difference	LSD		
Mean	$\bar{x}$	μ	
Mean squared error	MSE		
Probability			p, P
Probability of a Type 1 error			α
Probability of a Type 2 error			β
Regression coefficient(s)	b_1 (b_2, b_3,...)	β_1 (β_2, β_3,...)	
Regression constant (intercept)	a or b_0	α or β_0	
Sample size	n	N	
Significant at α = 0.05, 0.01, 0.001, respectively	*, **, ***		
Standard deviation or root mean square error	s, SD, RMSE	σ	
Standard error‡	SE		
Standard error of the mean	$s_{\bar{x}}$, SEM	$\sigma_{\bar{x}}$	
Student's *t* statistic	t		
Variance	s^2	σ^2	

†Adapted from Table 4-1, Publications Handbook and Style Manual, Alliance of Crop, Soil, and Environmental Science Societies. Available at https://dl.sciencesocieties.org/publications/style/

‡The standard error (general) is defined as an estimate of the standard deviation of a sampling distribution. While this is commonly a sampling distribution of the mean (in which case SE = SEM), any statistic from regression coefficients to residual variance can be the sampling distribution of interest.

Applied Statistics
in Agricultural, Biological, and Environmental Sciences

Barry Glaz and Kathleen M. Yeater, editors

American Society of Agronomy
Soil Science Society of America
Crop Science Society of America

5585 Guilford Road, Madison, WI 53711-58011 USA
agronomy.org · soils.org · crops.org
dl.sciencesocieties.org
SocietyStore.org

ISBN: 978-0-89118-359-4 (print)
ISBN: 978-0-89118-360-0 (online)
doi: 10.2134/appliedstatistics
Library of Congress Control Number: 2017949064

Cover design: Karen Brey
Cover photo: USDA NRCS Texas

CONTENTS

Foreword vii

Introduction ix

Acknowledgments xi

A Statistical Fable. The Story of a Grumpy Ox xiii
Barry Glaz

CHAPTER 1. Errors in Statistical Decision-Making 1
Kimberly Garland-Campbell

CHAPTER 2. Analysis of Variance and Hypothesis Testing 19
Marla S. McIntosh

CHAPTER 3. Blocking Principles for Biological Experiments 53
Michael D. Casler

CHAPTER 4. Power and Replication—Designing Powerful Experiments 73
Michael D. Casler

CHAPTER 5. Multiple Comparison Procedures: The Ins and Outs 85
David J. Saville

CHAPTER 6. Linear Regression Techniques 107
Christel Richter and Hans-Peter Piepho

CHAPTER 7. Analysis and Interpretation of Interactions of Fixed and Random Effects 177
Mateo Vargas, Barry Glaz, Jose Crossa, and Alex Morgounov

CHAPTER 8. The Analysis of Combined Experiments 201
Philip M. Dixon, Kenneth J. Moore, and Edzard van Santen

CHAPTER 9. Analysis of Covariance 235
Kevin S. McCarter

CHAPTER 10. Analysis of Repeated Measures for the Biological and Agricultural Sciences 279
Salvador A. Gezan and Melissa Carvalho

CHAPTER 11. The Design and Analysis of Long-term Rotation Experiments 299
Roger William Payne

CHAPTER 12. Spatial Analysis of Field Experiments 319
Juan Burgueño

CHAPTER 13. Augmented Designs—Experimental Designs in which All Treatments are Not Replicated **345**
Juan Burgueño, Jose Crossa, and Kathleen M. Yeater

CHAPTER 14. Multivariate Methods for Agricultural Research **371**
Kathleen M. Yeater and María B. Villamil

CHAPTER 15. Nonlinear Regression Models and Applications **401**
Fernando Miguez, Sotirios Archontoulis, and Hamze Dokoohaki

CHAPTER 16. Analysis of Non-Gaussian Data **449**
Walter W. Stroup

APPENDIX A.1. Author Supplements **A1**

APPENDIX A.2. Author Supplements **Online only**

APPENDIX B. Genstat Translation **Online only**
Vanessa Cave

APPENDIX C. R Translation **Online only**
Jon Baldock and Kimberly Garland-Campbell

Supplemental materials are available in Appendix A and online at https://dl.sciencesocieties.org/publications/additional-appendices-app-stats. These materials include answers to review questions and exercises, datasets, software code in SAS (SAS Institute, Cary, NC), R (CRAN, www.r-project.org [verified 23 Apr. 2018], and Genstat (VSN International LTD., Hemel Hempstead, UK) along with output from these software products. Most chapters utilize SAS software code to illustrate the examples within the chapter text, with R and Genstat also utilized, depending on the authors' preference. Supplemental Appendices are available online for R and Genstat translations where applicable. A great voluntary effort was undertaken to provide software code for all examples in the three major languages used in this text.

FOREWORD

In the early days of agronomy, researchers were perplexed by the natural variation in the fields they studied. They knew intuitively that whatever treatment they applied to a crop, some of the response they measured was due in part to the plot of land on which it was planted. Louie Smith in the very first volume of what later became ***Agronomy Journal*** reported on a uniformity trial conducted in central Illinois[1]. It was a relatively large trial comprised of 120 plots each 1/10 acre in size. The field, which had previously been in pasture, was planted to corn in three consecutive years. As most agronomists today would anticipate, there was a great amount of variation in yield among the plots even though the field appeared to be uniform to the eye. Not only that, the variation seemed to change along with the prevailing weather from one growing season to the next. Based on these observations he concluded:

"The difficulty lies in the extremely complicated sets of factors involved. When we consider all of the physical, chemical, and biological processes of the soil, their inter-relations, and their dependence upon climatic conditions, the problem of their control becomes well-nigh overwhelming."

That trial which so clearly demonstrated the challenges of spatial and temporal variation in agronomic research was conducted in the late 1800's and was described in one of several papers published in that first volume of our journal that addressed the issue of field-plot or what we would later call experimental design.

It was not until the 1920s and the publication of R. A. Fisher's book *Statistical Methods for Research Workers,* that experiments began to be routinely designed in ways to account for and remove uncontrollable sources of variation related to soil and weather. The first citation of this work in ***Agronomy Journal*** appeared in a 1926 paper by William Sealy Gossett under the alias 'Student'[2]. His paper, *Mathematics and Agronomy,* clearly laid out the premise behind the analysis of variance and its application to agronomy. Many of the principles and practices he discussed are still relevant to agronomic research today.

Having served us well, the analysis of variance remains an important paradigm that informs both how we design experiments and analyze the data collected from them. Refinements to the analysis of data have been made along the way, particularly since computers began shattering computational limits. The issues of correlated errors over time and space can now be easily addressed with modern statistical software. Transforming data to an approximately normal (Gaussian) distribution is no longer necessary because it is now possible to use alternative distributions. Advances in computing have thus enabled us to overcome issues that we otherwise had no practical way to address. We long realized that we had these issues but we chose to either ignore or address them through ancillary means.

Of course there are other statistical tools beyond the analysis of variance that have great value both in how we think about research problems and construe the meaning of the data we collect. As powerful a tool that it is for studying treatment effects, the analysis of variance intentionally isolates and ignores the variation inherent in agronomic systems. The world outside of designed experiments is functionally far more complex and variable than we can ever distill with a few treatment factors and a linear model. Multivariate methods recognize this complexity and can provide greater insight into systems that are unapproachable using other methods. They hold great promise for untangling the complex interactions that occur when the experiment moves beyond the field to a landscape, region, or any experimental situation where variation in the response is impacted, often marginally, by a large number of uncontrolled variables.

Statistical methods are tools we use as researchers to make inferences from the experiments we conduct. In a very real way, we are searching for truth in the data we collect. We want to avoid misreading the meaning of our results and then making inaccurate or faulty conclusions. The statistical methods we use are not the end, but our means of achieving it. Natural curiosity and a desire to understand what we study remain the most important characteristics that drive us as researchers. Statistical methods are the instruments that we play – they are not the music. If nothing else, statistics teaches us that we all make mistakes (probably at a rate higher than 5%) and that is OK as long as we learn from them. However, learning and knowing when and how to use increasingly complex statistical tools is critical to becoming a successful researcher. It is important not to become overwhelmed by the sheer amount of tools available and the disagreements and controversies surrounding their use. The chapters in this volume were written by leading researchers in our sciences who know the tools and how to use them. They have provided a wealth of information on the latest statistical methods as well as examples for using them. Most of the examples use real-world datasets and provide solutions using SAS, R, and Genstat to help readers conduct proper analyses with the software they already use.

This book represents the shared vision of Barry Glaz and Kathleen Yeater who wanted to create a volume that would apply directly to the needs of researchers in the agricultural and environmental sciences and function as a text for graduate students. That they have richly succeeded in doing so is a testament to their vision and persistence in pursuing it.

Ken Moore

Charles F. Curtiss Distinguished Professor in Agriculture and Life Sciences
Pioneer Hi-Bred Professor of Agronomy
Department of Agronomy
Iowa State University

Endnotes

[1]Smith, Louie H. 1909. Plot arrangement for variety experiments with corn. Proceedings of the American Society of Agronomy 1:84-89.

[2]Student. 1926. Mathematics and agronomy. Agronomy Journal 18:703-719.

INTRODUCTION

Agricultural and biological scientists conduct research to answer questions about what is happening in a farmer's field or in nature. The process includes conceiving a research idea, designing an experiment, conducting the designed research, analyzing the data, and interpreting the results of the analysis. The goal of this book is to help scientists apply statistical concepts and methodologies as they make decisions throughout each step of the research process. The more often the scientist makes decisions throughout the research process that are properly based on statistical principles, the more career-long impact he or she will have on a given field of science and on the beneficiaries of the research. Making informed statistical decisions while designing, conducting, analyzing, and interpreting one's research can be the difference between making and missing crucial scientific advances.

This book has two primary purposes. One is to serve as a text in the second or third statistics class for students majoring in the agricultural or biological sciences. The other purpose is to provide a textbook for practicing scientists. The editors and authors of this book worked together to develop the subject matters (chapters) they felt were most needed in a book of this nature. All authors felt it was important to put together a book that would be practical; in most cases the chapters show scientists how to analyze data and interpret results by providing examples with real datasets and code using popular software products. In addition, each chapter has a list of Key Learning Points that help the reader identify what is important about that chapter and serve as easy-to-find reference points throughout the book to help readers locate important information as they read.

Scientists today have abundant resources for conducting statistical analyses compared with what was available when I began my career in 1977. As software has become more sophisticated, scientists can substantially improve their statistical analyses and obtain results faster, but there is a price. Our tools, usually software packages, are constantly improving and expanding in capabilities. It is time consuming and difficult to stay current with these new capabilities and to learn how to use them properly. While some scientists admirably handle their statistics-related issues from experiment conception through data analysis and interpretation, it is an understatement that many of us have severe problems properly using modern statistical software and/or understanding how to interpret results of analyses, even when they are conducted properly. Although I did not realize it at first, it soon became clear that the reason for conceiving this book was to develop a resource that I wish I had decades earlier in my career as a Research Agronomist.

This Introduction concludes with a fable. Two statistical problems that inhibit the impact of many agricultural and biological scientists are their misunderstanding of the relative importance of α and β errors and their misuse of multiple comparison procedures. Another major problem is that many scientists ignore interactions in their research. The main objective of the fable is to use a silly story to help readers recognize the importance of these three problems with the aim that they then use this book even more enthusiastically. The fable concludes by providing a quick explanation of each chapter in the book. I hope the fable encourages you to use this book to help you better handle the problems like those encountered by its main characters, as well as illustrates that learning the concepts and methods in each chapter can improve your research and its impact.

The fable is not meant to demean or ridicule any person or group. There was no intention to diminish the good work of anybody who developed or contributed to the development of any statistical principle and methodology or agronomic research that by chance is similar in any way to those mentioned in the fable. The fable does not describe any actual person, place, or thing.

Barry Glaz

ACKNOWLEDGMENTS

I wish to thank Nicole Sandler, Lisa Al-Amoodi, Brett Holte, and Danielle Lynch; each a staff member of ACCESS who helped substantially with this book. All four of these individuals were often called on to develop creative solutions to thorny problems. Clearly, without them, this book would not have been published.

It was not an easy path from book conception to completion. Challenges arose throughout the process. My coeditor, Dr. Kathleen Yeater, contributed often and skillfully to resolve many of these challenges. In addition to all of her help resolving issues along the way, Kathy was also the lead author on one chapter, coauthor on another, and an anonymous reviewer of the other 14 chapters. Without Kathy, we would not have this completed book.

In addition to the hardest task of all, writing the chapters, many authors also reviewed chapters and often volunteered in nontechnical ways to help ensure that the book would move forward.

Each author of this book wrote an important and timely chapter that reflects his or her opinions and understanding of the subject matter. However, for many of these chapters, reaching the final version involved robust give and take between authors and reviewers. Every review was courteous and respectful and technically outstanding. The author–reviewer interaction was so productive that I am confident that many authors would agree that the reviewers contributed substantially to this book. The contributions of the reviewers were of such great importance to this book, that with their permission, I felt it necessary to list each reviewer's name and number of chapters reviewed. I hope that I have listed the name of each reviewer who granted permission to do so. If any are missing, I apologize.

CONTRIBUTORS

Reviewer	No. Of Chapters Reviewed
Jon Baldock	2
Juan Burgueño	1
Kimberly Garland Campbell	2
Michael Casler	5
Jose Clavijo	1
Philip Dixon	2
Vasilia Fasoula	1
Edward Gbur	1
Salvador Gezan	7
Alexander Lipka	1
Raúl Macchiavelli	1
Jack Martin	1
Marla McIntosh	1
Stanislaw Mejza	1
Fernando Miguez	1
Kenneth Moore	4
Roger Payne	1
Hans-Peter Piepho	4
Christel Richter	5
David Saville	2
Maninder Singh	1
Haile Tewolde	2
Edzard van Santen	1
Maria Villamil	3
Kathleen Yeater	14
Mark West	2
Yang Zhao	1

THE STORY OF A GRUMPY OX:

PART 1—A NON-GAUSSIAN STATISTICAL TAIL

Barry Glaz

1.1 Rho, Your Fable Narrator

I am Rho, here to narrate an unlikely fable about Alpha and Beta, who are destined to become famous statistical spouses as they erroneously evaluate (or choose not to evaluate) oxen, snakes, the oils of snakes, and interactions hidden in plain sight.

1.2 Qualitative and Quantitative Treatments

Long ago, in a faraway and wondrous land, Alpha and Beta were brand new Apprentices (Apps) working toward their Agronomist licenses. Sigma, their advisor, was famous for calculations that helped people know how far they were deviating from their means.

Alpha and Beta planned to jointly conduct two three-year experiments aimed at improving yields. Their presentations on the results of both experiments needed to pass peer review for them to obtain their Agronomist licenses. One experiment compared three rates (1, 2, and 3 L ha^{-1}) of snakes' oil collected from two snake types, slippery and slimy. The second experiment compared three ox-plowing methods:

- The current practice, a carefree ox skips along merrily while pulling the plow (Control).
- The farmer reviews statistics with a studious ox while plowing (Statistical Plowing).
- A purposeful ox jogs and monitors its heart rate while plowing (Aerobic Plowing).

Looking three years ahead, Sigma taught Alpha and Beta about the intricacies of testing for significant treatment differences. The Lowest Seeming Difference (LSD) was used near and far to separate means. Sigma liked the LSD for qualitative treatments like ox-plowing methods, but was concerned that there was not a good method available for quantitative treatments like rates of snakes' oil. To resolve this, Sigma funded a four-month Post App named Delta.

Using Sigma's approach to calculate deviations from one's means, Delta calculated the straight line that would deviate least from the mean responses of treatment rates. Sigma helped Delta extend this concept for squiggly curves that went up-down or down-up (later known as quadratic), and up-down-up or down-up-down (later

---

Silly Abbreviations: App, Apprentice; LSD, Lowest Seeming Difference; SSREG, Straight–Squiggly Response Estimate Graph; SI and SI Units, Squiggly Ink.

known as cubic). Sigma named the new method the Straight-Squiggly Response Estimate Graph (SSREG).

1.3 Snake Experiment Data Analysis

After three years, it was time to analyze the snake data. The only significant main effect in the ANOVA was Snakes' Oil (p = 0.0498). Alpha and Beta sadly showed Sigma that Snake Type was not significant (p = 0.0501). With a big agronomic smile (which meant important teaching moment), Sigma pointed out to the confused Apps that they had ignored the significant Snakes' Oil × Snake Type interaction (p = 0.0109). Sigma further explained that the significance of each of the two main effects was not very important; the key issue for their analysis was that the interaction was significant. Using SSREG to understand why this interaction was significant, they found a positive linear increase in yield to oil from slippery snakes whereas the linear response was negative to oil from slimy snakes (Fig. 1). Our Apps had learned their lesson; indeed, it is crucial to explore significant interactions.

1.4 Presentations and Errors

We are now at the Annual Near and Far Meeting where Alpha has been speaking for about 10 min on the ox study. Let's listen in on Alpha speaking in a scholarly monotone.

Results: Ox-Plowing Experiment

"This slide shows that Aerobic Plowing caused reduced yields compared with the Control (Fig. 2). Perhaps monitoring their heart rates while jogging and plowing was asking too much of the oxen. We are still optimistic about this approach and plan to look at jogging and heart-rate monitoring as separate treatments in future research."

(Sigma whispers proudly to a colleague that Alpha and Beta will soon be famous agronomists. Alpha points the remote and advances to the next slide.)

"Here we see that Statistical Plowing improved yields compared with the Control. There are no surprises here for Beta, my coauthor, and me. Epsilon et al. have shown that yields increase as oxen improve their knowledge of math. Therefore, it is logical that plowing with statistically savvy oxen employees should improve yields."

The Decider accepted Alpha's presentation. Also, the presentation had major impact as all farmers in The Wondrous Land wanted to prove for themselves that Statistical Plowing would improve yields and planted identical halves of each field using the Control or Statistical Plowing. Alpha and Beta were off to a great start!

Snake Results

Next, Beta presented the snake results. Beta had an excellent slide with a nice graph summarizing SSREG results showing that yield responses to slippery and slimy snakes crossed over as rates of their oils went from lowest to highest (Fig. 1). Excited about oil from slippery snakes, the audience gave Beta a standing ovation, which was unheard of for a mere App! But alas, the Decider rejected Beta's presentation. The reviewers were not comfortable with SSREG although they did not explain their problem with it. Feeling a tinge of sympathy, the Decider gave Beta and Alpha the option to present their snake research again at the next annual meeting using the

LSD instead of SSREG. So with only one of two presentations approved, Alpha and Beta needed to wait at least one more year for their Agronomist licenses.

The crop harvest was now beginning and farmers near and far were finding out, to their dismay, that studying statistics with their oxen did not improve yields. Instead, yields of Statistical Plowing and the Control were equal. Thus, Alpha's presented results were not correct, and the decision to accept Alpha's presentation was reversed. Worse yet, Alpha now had a namesake error. An Alpha error would heretofore and forevermore be known as declaring one mean superior to (or worse than) another mean, when in fact, the two means were equal.

Another big surprise of the ox-plowing research came when Wondrous Land U reported that 95% of the oxen assigned to Statistical Plowing scored ≥ 95% on their statistics' test. As the news of this non-Gaussian statistical tail spread near and far, some agronomists in the Wondrous Land began to ponder about potential analytical challenges related to these strange tails.

It was now time for Beta's second presentation at the next Annual Meeting. Alpha and Beta wanted to have at least one accepted presentation so they decided to use the LSD even though they knew SSREG was the better method. Beta reported that there were no significant LSD differences that coherently explained the interaction between snake types and oils. At a coffee break, several agronomists pretended that Beta's SSREG results from the previous year accurately described the population of responses to these treatments. If we accept this make-believe as real they continued (and this is legal in a

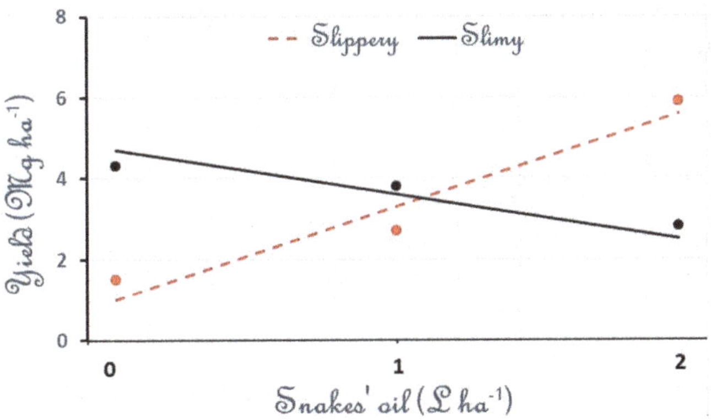

Fig. 1. Response of yield to three rates of snakes' oil from slippery and slimy snakes.

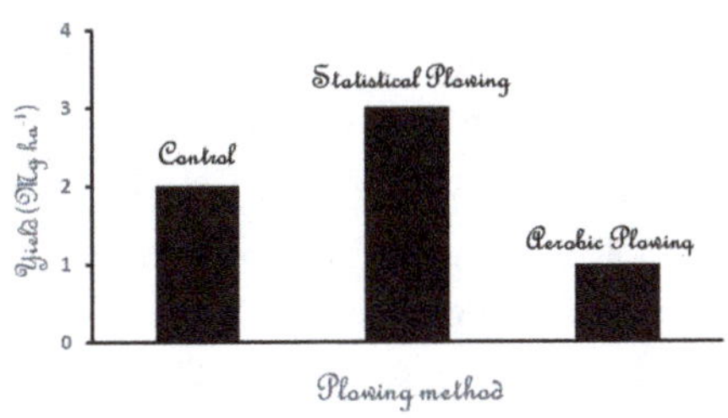

Fig. 2. Yield response to plowing by traditional method in The Wondrous Land (Control) and two new methods, statistical plowing (ox studies statistics while plowing) and aerobic plowing (ox jogs and monitors heart rate while plowing).

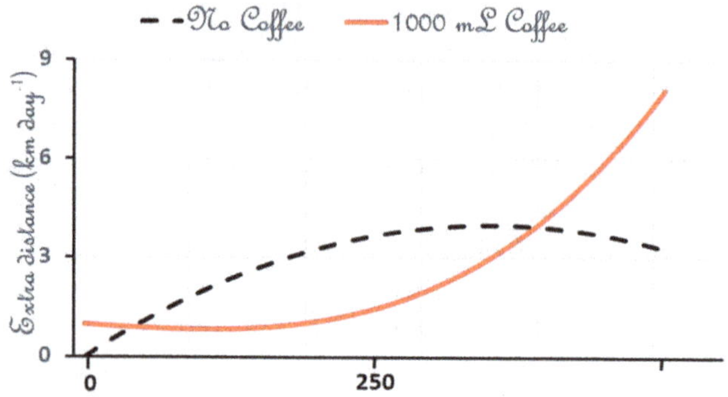

Fig. 3. Response of extra oxcart pulling distance to donut rates with and without 1000 mL coffee. Extra distance is based on improvement compared with previous morning snack.

fable), then many comparisons that failed to detect significant differences using the LSD were failing to identify true differences between means. And forevermore, not depicting a true difference between means was known as a Beta error. Now Beta also had a namesake error, but at least the Decider accepted Beta's presentation!

Later that year, Delta's presentation describing SSREG went so well that attendees and reviewers decided it would be a crime to evaluate responses to quantitative treatments with the LSD instead of SSREG. And in a cruel twist of fable fate, Alpha and Beta were sentenced to 1 yr in Agronomy Prison for the crime of analyzing quantitative treatments with the LSD instead of SSREG. This also triggered an automatic reject decision on Beta's presentation. So now, after more than 4 yr as Apps, all Alpha and Beta had to show for their hard work were two namesake errors, no approved presentations, and prison records.

After leaving prison, Alpha and Beta learned that disappointed farmers were no longer providing grant funding to Sigma due to Alpha's error. This was causing agronomists near and far to obsess over Alpha errors. Due to Beta's error, farmers were not using oil from slippery snakes and were therefore unknowingly losing profits. Maybe worse, snakes' oil will never be tested again because the very words "snake oil" are now synonymous with "worthless". However, agronomists lost no sleep over Beta errors because they would go unnoticed, and anyway, many had no idea how to calculate the probability of a Beta error...Are you wondering what happened to Delta?

PART 2 — CORPORATE RESEARCH

2.1 Delta becomes a Fortunate Entrepreneur

Delta opened a consulting firm called Squiggly Ink (SI) with the idea of doing the same consulting work on SSREG as a Post App but getting paid way more. The only downside for Delta as Big Boss and only employee of SI was the frequent travel on The Wondrous Land's Fortunate 0.05 Hundred Oxcart Transport Company.

Delta unexpectedly changed the acronym name SI to SI Units, and strangely, business boomed. To this day, the secret of this success remains unknown. It was as if agronomists thought they would never get a manuscript published unless they used SI Units. Now rich,

Delta purchased the ox transport company and named it Delta Oxlines. Sadly, Delta soon found that bossing around hundreds of humans and oxen was not fun. Delta thought, "I need some help. Do I have any friends who used to be highly-rated Apps or App advisors, and due to some bad luck could really use a break?" You guessed it! Delta hired Alpha, Beta, and Sigma as Associate Big Bosses of SI Units and Delta Oxlines.

2.2 Improved Oxcart Pulling Efficiency

Sigma entered private industry happily but soon began having problems with regression. At first it was just simple and linear, but later Sigma became consumed with multiple regressions and exponential responses.

Meanwhile, Alpha and Beta began corporate research that looked at effects of ox diet on oxcart pulling efficiency. One experiment to optimize the oxen's morning snack tested five rates of coffee (0, 250, 500, 750, and 1000 mL) and donut (0, 125, 250, 375, and 500 g) per ox. Their ANOVA showed that Coffee was not significant, but the oxcart-pulling response to Donut was quadratic with a maximum at 350 g donut. So Alpha and Beta changed the oxen's morning snack to no coffee and 350 g donut. Taking away their coffee made the oxen grumpier than ever, but with their 350-g donuts, they were pulling their carts 4 km further per day!

2.3 Rho Wraps Up

In a fable bombshell, the oxen revealed that they liked being bossed around by Delta, but not by Alpha and Beta. Delta helped them appreciate the need for change in the workplace, while Alpha and Beta took away their coffee and then had the nerve to tell them to stop being so grumpy.

Eternally perplexed that agronomists were fixated on Alpha and ignored Beta, Alpha and Beta considered writing a book on statistics. However, they insisted the book be written in stone, and since all the publishers were no longer doing hard copies, they did not write the book.

Perhaps this was for the best, because years later, while reviewing the analyses of the ox-diet research, Sigma unleashed a horrible agronomic scream heard near and far and beyond the borders of The Wondrous Land: "**Those δαρυ Apps**!" And Sigma continued: "They did it again, this time they ignored the significant Donut × Coffee interaction! Why didn't they ask Rho for some help?" ☺ Had they asked me, I would have told them that this interaction was significant because an oxcart-pulling response of 8 extra km per day occurred at 1000 mL coffee and 500 g donut (Fig. 3). This was substantially higher than the extra 4 km per day for oxen on 350 g donut without coffee. This slip-up cost Delta Oxlines millions in lost profits. Exasperated, Sigma said, "We might as well make this a namesake error too." And theretofore and forevermore, it would be an Alpha × Beta (Type 4) Error to evaluate research based on main effects while ignoring significant interactions.

Irate, Delta demoted Alpha and Beta to Insignificant Big Bosses. Later, a sassy Post App named Are advised Delta that this position title was an ox moron because an Insignificant Big Boss's job was significantly ($p \leq 0.0001$) not significant. Delta hired Are. Alpha and Beta became jealous of this new employee with a name sounding like a letter from a different alphabet. Worse yet, Alpha and Beta were confused by Are's analyses of what seemed to them like mixed up models with fixed and random ideas. Not able to cope with these complex concepts, Alpha and Beta relapsed and sought comfort by illegally using the LSD to conceal and avoid dealing with the realities of what could have been high-impact research results.

Later, Sigma told Delta, "I should have imposed a stricter standard; deviations facing Alpha and Beta were too large. Had we used more replications and/or locations and tried other experimental designs, we might have reduced the chances of those Alpha and Beta errors." Delta added, "Also, shortly before we began the ox diet study, cigarettes were introduced into The Wondrous Land and all the oxen were smoking between 1 and 10 cigarettes per day. We should have used daily cigarettes smoked per ox as a covariate."

Sigma eventually grew proud of the first two namesake errors as they became important statistical concepts. However, Sigma knew that many agronomists were missing out on major discoveries throughout their careers due to a poor understanding of the relative importance of Alpha and Beta errors plus a perplexing propensity to commit Alpha × Beta errors and to use the LSD to separate means of quantitative treatments. With a long sigh, Sigma said, "Wow Delta, Alpha and Beta were finally right. We really need a book on statistics that's written to meet the needs of agricultural and biological scientists."

(After devouring a 500-g donut, Rho delicately sips a 1000-mL latte. While ruminating over an e-cigarette, Rho suddenly realizes the narration is not finished.)

Oh, by the way,

The Moral of this Story is:

Use each chapter in this book wisely and diligently so your experimental designs and data analyses remain efficient and optimized heretofore and forevermore.

Key Moral Support

- Chapter 1 provides a comprehensive explanation of Type 1 and Type 2 errors.
- To learn what ANOVA can do for you, check out Chapter 2.
- For strategies that can help you maximize correct decisions, have a look at Chapters 3 and 4.
- Read Chapter 5 to learn when it is appropriate to use a multiple comparison procedure and why that procedure should be the unrestricted LSD.
- If your treatments are quantitative, then do not use the LSD. Instead, read Chapters 6 and 14 to learn about linear and nonlinear regression. These chapters will be well worth reading even if you deny having problems with regression.
- If you need to analyze an experiment with more than one effect or location, read Chapter 7 on interactions and Chapter 8 on analyzing combined experiments.
- Thanks to Delta realizing that Alpha and Beta should have used cigarettes smoked per day as a covariate, this book includes Chapter 9 which explains the proper use and potential benefits of Analysis of Covariance.
- In the coffee and donut experiment, if Alpha and Beta had measured oxcart distance at 4 monthly intervals, they could have studied Chapter 10 and used repeated measures analysis.
- Although not relevant to the fable, Sigma had a long-term rotation experiment. This special case is discussed in Chapter 11.
- Although not revealed in the fable, the soil Alpha and Beta used in the

plowing-methods experiment was highly variable. Before beginning this experiment, they should have read Chapter 12 on spatial statistics.

- Alpha, Beta, and Sigma were not breeders. However, for those of you who are breeders, you should study Chapter 13 on augmented designs to learn how to best handle your early selection stages when you cannot replicate your genotypes.
- What a shame Alpha and Beta did not measure ox weight, heart rate, and cholesterol as responses in the morning-snack experiment. Had they done so, they could have studied multivariate analysis in Chapter 15 and their results would have been envied near and far in The Wondrous Land.
- Read Chapter 16 to learn how to find out if your statistical distribution (or tail) is Gaussian, and what to do if it is not.

Review Questions

1. Alpha and Beta rejected a H_0 based on results with $p = 0.0498$ and did not reject when $p = 0.0501$. Were these good decisions?

   a. Yes, we live and die by 0.05.

   b. They should have rejected the H_0 in both cases.

   c. They should have accepted the H_0 in both cases.

   d. Had they also considered the effects of a Beta error, it is extremely likely that they would have either rejected or accepted the H_0 in both cases.

2. Rho did not understand statistics. You should ignore a significant interaction if at least one main effect is significant.

   a. True

   b. False

3. Alpha and Beta's research on maximizing the morning snack of the oxen employees of Delta Oxlines should follow up with higher rates of coffee and donut.

   a. True

   b. False

4. What is δαρν in English?

   a. The equation of a complex mixed model.

   b. A Type 5 error.

   c. The most famous ox fraternity in The Wondrous Land.

   d. darn.

CHAPTER 1: ERRORS IN STATISTICAL DECISION MAKING

Kimberly Garland-Campbell

"Every judgment teeters on the brink of error."— Frank Herbert, *Children of Dune*, 1976

Agronomic and environmental research experiments result in data that are analyzed using statistical methods. These data are unavoidably accompanied by uncertainty. Decisions about hypotheses, based on statistical analyses of these data are therefore subject to error. This error is of three types: Type 1 (α) is a false positive, Type 2 (β) is a false negative, and Type 3 (γ) is directional. Type 1 and Type 2 errors are both prevalent, but often only the α error is controlled when experiments are designed. Statistical decisions are therefore subject to more error than is usually acknowledged. The goal of this chapter is to explore the consequences of that error and uncertainty in the context of null hypothesis testing and the use of the normal distribution, methods that are usually taught in introductory statistics courses. Concepts to be discussed include effect size, noncentral distributions, and power analysis. Two examples are analyzed. One example illustrates methods to obtain the minimum average error for an experimental design, and the other illustrates the effect of a split plot design and linear field trends on estimates of error variance. The following best practices are recommended: (i) design experiments that answer the research question, without unnecessary complications; (ii) determine a relevant effect size for the treatments to be tested prior to conducting the experiment; (iii) benchmark against similar experiments; determine a range of suitable ratios of α and β error; (iv) define costs of each variable in the experimental design; (v) determine the optimal error of an experiment under various experimental design scenarios; and (vi) conduct a power analysis.

Statistics is the science of predicting outcomes on the basis of data. These predictions require that uncertainty in the data be measured, controlled, and communicated (Davidian and Lewis, 2012). In agronomic and environmental research, data may be measured on a plant or plot basis as, for example, height, weight, or concentration, or it may be a visual rating on an ordinal scale (Beres et al., 2016; Tackenberg et al., 2003). Data may be colorimetric or fluorescent measurements that are calibrated to a

---

Abbreviations: H_A, alternative hypothesis; H_0:, null hypothesis; λ, noncentrality parameter; MST, mean square for treatment.

USDA-ARS, Wheat Health, Genetics and Quality Research Unit, 209 Johnson Hall, Washington State University, Pullman WA. 99164 (kim.garland-campbell@ars.usda.gov).

doi:10.2134/appliedstatistics.2016.0007

Applied Statistics in Agricultural, Biological, and Environmental Sciences
Barry Glaz and Kathleen M. Yeater, editors

biological event, or data may be the counting and classification of objects (Hernández and Kubota, 2016; Smiley et al., 2016). Repeated measurements of the same phenomena will vary due to measurement error or environmental influences, and all data are accompanied by uncertainty. The data collected are only a sample of the possible measurements that we could take to better estimate the true value. The existence of uncertainty in data stymied early mathematicians until Legendre explained that the method of least squares could be used to distribute errors equally among observations in an experiment (Stigler, 1981). Statistical theory then advanced based on the assumption that the average of the errors distributed over the dataset was equal to zero, and that this error could be minimized and randomly distributed over the experimental units within each experiment or set of data. The goal of this chapter is to explore the consequences of that error and uncertainty in the context of well-known statistical methods like null hypothesis testing and the use of the normal distribution, methods usually taught in introductory statistics courses. We will discuss the important concepts of effect size, noncentral distributions, and power analysis, which are not always covered in introductory statistics classes.

Conventional statistics uses local control, randomization, and replication to estimate truth in the face of uncertainty. Local control incorporates systematic error into experimental design (e.g., through blocking, spatial experimental designs, or covariance analysis), randomization reduces the influence of systematic error, and replication is used to estimate the magnitude of the remaining unexplained error (Kirk, 2009).

The basic statistical concepts of the mean and variance of a set of data, the central limit theorem, and the normal distribution have been widely used to model the degree of uncertainty associated with data. Other distributions such as the β, binomial, exponential, and Poisson, often fit the data better than the normal distribution, and recent advances in generalized linear modeling software tools allow the use of these distributions in statistical inference (Gbur et al., 2012). For this chapter, however, we are going to focus on the normal distribution as our foundation for hypothesis testing because it is still the most widely used distribution and fits many applications in agronomic and environmental research. The central limit theorem states that "whatever the shape of the frequency distribution of the original population of data, the frequency distribution of the means of repeated random samples of size n tends to become normal as n increases" (Snedecor and Cochran, 1967). The central limit theorem permits the use of the normal distribution to estimate the probability (p) of observing data at least as extreme as that observed. Therefore, the central limit theorem is the foundation for the null hypothesis test.

Statistical Decisions in Agronomic and Environmental Research Are Often Framed as Null Hypothesis Tests

Fisher wrote "Every experiment may be said to exist only to give the facts a chance of disproving the null hypothesis" (Fisher, 1951). The null hypothesis (H_0) declares that an effect is not different from zero (Gill, 1999). The H_0 has multiple specific meanings depending on the experiment and the type of effect that is being evaluated. The H_0 can be defined as no difference between treatments ($T_1 = T_2 = T_3 = T_n$), or as no difference between a control and treatment ($C = T_n$). The H_0 can also be defined as a lack of correlation, $r = 0$, where r is the correlation coefficient. In a regression, the H_0 is defined as no association, $R^2 = 0$, where R^2 is the coefficient of determination for the total regression model, or as $b_1 = 0$, where b = the partial regression coefficient. The decision to accept or reject the H_0 is quantified by a p value that is calculated based on the distribution of the underlying test statistic (e.g., t, F, or X^2). If the p value is sufficiently small, then the H_0 is rejected and the effect is considered to be different from zero.

Our overemphasis on null hypothesis testing has been criticized recently (Gill, 1999; Anderson et al., 2000; Fidler et al., 2006; Nakagawa and Cuthill, 2007). We have become comfortable with the idea of making a false positive mistake no more than once every 20 times and $p \leq 0.05$ has been adopted as *the* critical point to decide to reject the H_0 and conclude that an effect is significantly different from 0. Therefore, published reports of experiments have overused $p \leq 0.05$ to detect "significance." Even so, the null hypothesis test is the dominant method of statistical decision making in published agronomic and environmental research because the approach fits so many research questions. Therefore, it is important that users understand the meaning of decisions made on the basis of the null hypothesis test.

Most published research reports the results from experiments where the H_0 is rejected. At this point, we can't resist asking the question: "What now?" The null hypothesis test was modified by Neyman and Pearson (1928a,b) to include an alternative hypothesis, H_A. This modification is what is commonly used in practice in conventional statistics. Experiments are conducted to determine if there is an effect, where the H_0 is defined as effect = 0, versus an alternative hypothesis, H_A, that there is a relevant non-zero effect that is defined by the research problem.

Effect Size

The critical effect size is the minimum difference between H_0 and H_A that is of functional significance (Cohen, 1992). A critical effect size can be expressed on the scale of the data, but in practice, to facilitate comparison between studies, the critical effect size is often expressed as a proportion of the experimental variation. Standard measures of effect size can be classified into two general categories: those that measure differences between groups and those that measure association between variables (Sullivan and Feinn, 2012). Common examples of the first category of critical effect sizes are the difference between two means as a proportion of their standard error (Student's t) or of their standard deviation (Cohen's d). Examples of critical effect sizes for the second category include the strength of the correlation between variables (the Pearson product–moment correlation coefficient r), the ratio of the proportion of variation explained by a multiple regression model to that remaining unexplained (the partial coefficient of regression r^2, Cohen's f^2 or η^2) and others (Nakagawa and Cuthill, 2007; Ellis, 2010; Selya et al., 2012). Biologically relevant critical effect sizes should be determined before the experiment is conducted based on prior knowledge about the system and research questions that are being studied (Mudge, 2013).

In an analysis of variance (ANOVA) or regression, the significance of an effect for a specific treatment is determined based on its variance component relative to the variance component for unexplained experimental error. The ratio of the mean square for a specific treatment (MST) to the mean square for the error (MSE) of the experiment is used to calculate an F test to determine this significance. The ratio is compared to the appropriate central F distribution as reference. To explain further, when the effect size is 0, the F statistic follows a central F distribution (Kutner et al., 2005), and the exact shape of this central F distribution depends on the experimental design, namely the number of degrees of freedom (df) for the treatment and the df for the experimental error. A critical F statistic can be calculated, representing the cutoff where the probability that the effect size fits the central F distribution is less than a given amount; usually $p \leq 0.05$.

The Noncentral Distribution

And now for something somewhat new—when the H_0 is not zero, the H_A follows a noncentral F distribution. The exact shape of the noncentral F distribution for the H_A depends on the numerator and denominator df as above and on the noncentrality parameter (λ), which depends on the effect size of the H_A. The λ is defined as the ratio of treatment variation to the error variation for the population being tested. Because we are working with samples from true populations, we correct for the sample size and the λ is defined as $(n - 1)$*MST/MSE, where n = the sample size or number of treatments, MST is the treatment mean square, and MSE is the mean square error, as described previously. If results are available from a previous similar experiment, λ is equal to the treatment df multiplied by the MST/MSE ratio (or its algebraic equivalent, the F statistic for treatment) (Kirk, 2013). The noncentral F distribution becomes more stretched out to the right as the effect size increases (Fig. 1), or as the sample size increases, or as the variance due to unexplained error decreases (Gbur et al., 2012). When the F distribution becomes more stretched out to the right, the overlap between the central and noncentral F distributions lessens and our ability to declare that the H_A is more probable than the H_0 increases.

Biologically relevant critical effect sizes can be estimated based on similar experiments. It is common practice in agronomy and ecological experiments to report a desired least significant difference (LSD) or coefficient of variation (CV). We can use these reports from previous experiments to calculate a range of estimates for effect size and error variance for our proposed experiments. These data can be manipulated to determine a range of expected critical F values and corresponding λ so that power analysis can be conducted (Gbur et al., 2012; Welham et al., 2014) (see below and Chapter 4, Casler).

Errors in Statistical Decision Making

Uncertainty accompanies data even in well designed and executed experiments. Errors are also made in the interpretation of data. These errors exist in multiple forms, commonly defined as Type 1 (α), Type 2 (β), and Type 3 (γ). Type 1 error is a false positive, the rejection of H_0 when it is actually true. Type 2 error is a false negative, the failure to reject H_0 when a real difference does exist. Type 3 error is a directional error, the rejection of the H_0 but for the wrong reason, such as the acceptance of a difference opposite in sign to the true difference. The focus our attention in this chapter will be on α and β errors because Type 3 errors are rare, although they can be significant for small effect sizes

The goal of scientific research is to investigate phenomena and interpret the results of experiments to make things better. We want to design and conduct

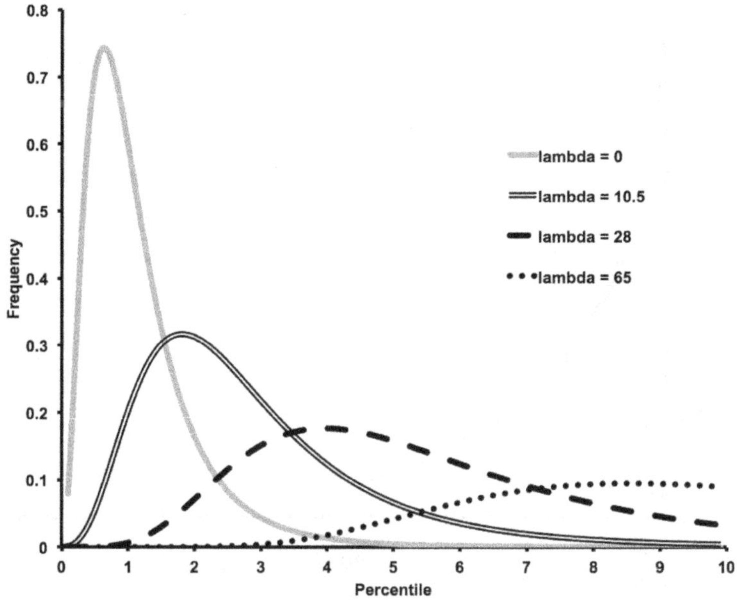

FIG. 1. Probability density functions (pdf) of central and noncentral *F* distributions as influenced by the noncentrality parameter (λ). The *x* axis represents the percentile of the pdf. Values for the probability density functions were derived from analysis of a simulated experiment with 8 entries, 3 replications, a grand mean of 5000 kg and a standard deviation of 500 kg (CV = 10). The central distribution (solid gray line), where λ = 0 is equivalent to an effect size of 0 or failure to reject the H_0. Values for λ were simulated as a small effect (double line) equal to 10% of the mean (λ = 10.05); a moderate effect (dashed line) equal to 16% of the mean (λ = 28) and a large effect (dotted line) equal to 25% of the mean (λ = 65). The central and noncentral distributions overlap for the 10% difference scenario, indicating that the power of the experiment to detect those differences will be lower than for the larger effects. The numerator df were equal to 7 and the denominator df were equal to 14 in these analyses.

experiments and interpret results to make a correct decision. Recognizing that α and β error are inevitable, this goal is best reached if their average is minimized. False positive and false negative errors are inversely related, all other experimental parameters being equal, but their relationship is not linear. The curve of their relationship depends on λ (Fig. 2) and as detailed in the example below. Therefore, an optimal minimum average of α and β error exists for every combination of treatment groups, replications, effect size, and relative emphasis on the two types of error (Fig. 3) (Mudge et al., 2012a). In practice, minimizing average statistical error for all effects tested, and especially for interactions, is often not feasible due to cost, time, and space constraints. Good experimental design must consider the tradeoffs between the consequences and costs of statistical error types, effect sizes, and these other constraints.

Example

Keeping the previous thoughts in mind, we want to design an experiment to detect cultivar yield differences of 10%. Based on previous experience, the mean yield is typically 5000 kg ha^{-1} and we would like to detect differences of 10% or 500 kg ha^{-1} among treatments. We have decided that the consequences of α and β error are of equal importance in this case. A conventional CV for crop yield trials is 10% or less. From the CV, we can estimate a standard deviation for the experiment of 500 kg ha^{-1},

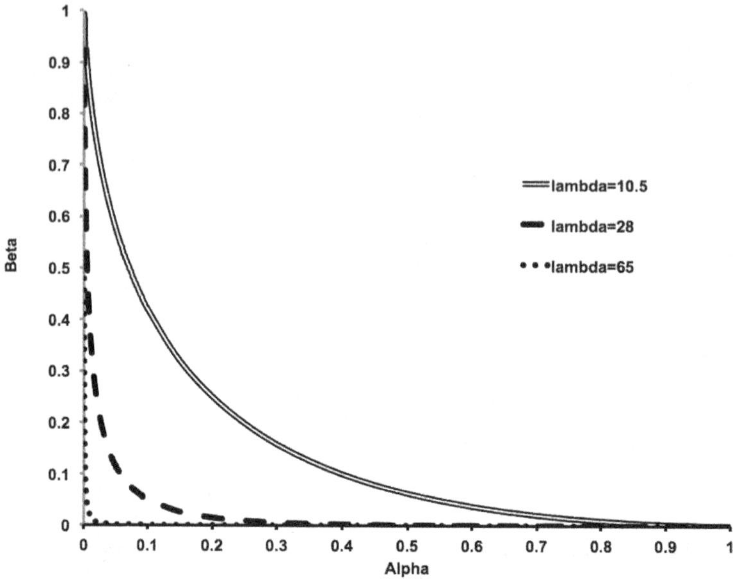

FIG. 2. A comparison of α and β error for the simulated experiment described in the text and in Fig. 1 with three increasing values for λ. Even for the smaller effect size (λ = 10.5, double line) the relationship is not linear. Instead the relationship between the two types of errors is negative and monotonic, becoming increasingly positively skewed as λ increases. The inverse relationship of α and β error is evident.

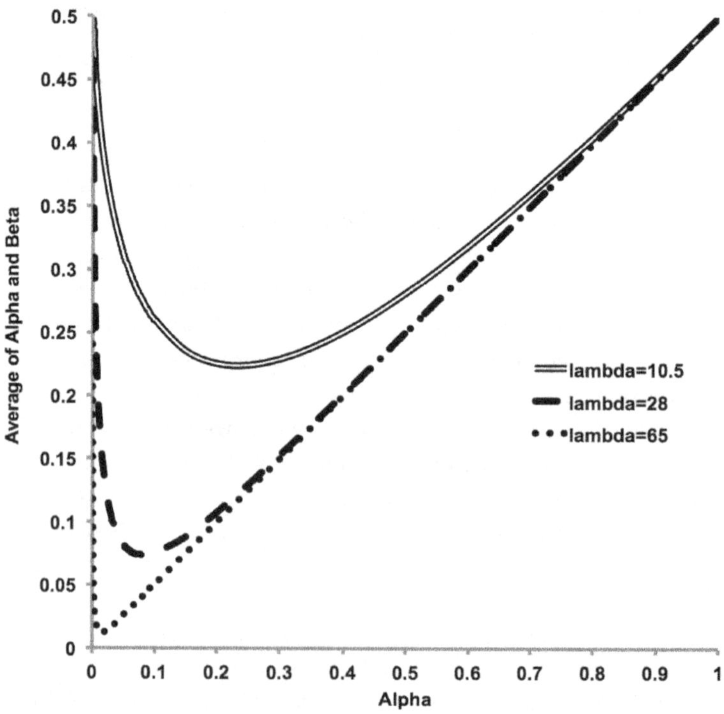

FIG. 3. The relationship between the α error rate and the average of α and β error for the simulated experiment described in the text and in Fig. 1 and 2 comprising 8 treatments and 3 replications. The emphasis on α error can be adjusted to obtain a minimum average error for the entire experiment. The minimum average error depends on the effect size that is desired to be detected.

with a corresponding error variance of 250,000 kg^2 ha^{-2}. We have eight cultivars to evaluate and plan to use three replications. The standard error of a difference (SED) between two means for this experiment is the square root of (2 × 250,000/3) = 408 kg ha^{-1}. The lower limit of the mean difference that we want to detect can be thought of as 10%, or 500 kg ha^{-1}. If we compared just two treatments with each other, we might use the t statistic and the t value for this difference is calculated as 500/408 = 1.22. This is equivalent to an F value of 1.50 where $F = t^2$.

These same calculations can be used to calculate our power to detect a 10%, a 16%, and a 25% difference between the treatment means in our planned experiment (Table 1). The λ for the smaller effect is 10.5, calculated as $\lambda = 1.5 \times 7$, where 7 is the treatment df in this experiment. The λ values for the other two effect sizes are calculated similarly. We examined our proposed experiments over a range of α levels from $0.001 < \alpha < 0.99$ in approximately 0.01 increments for the three effect sizes (see SAS and R code in the supplemental material; the variable names in the code are included in parentheses here). The critical F values (defined as FCRIT) were calculated with 7 degrees of freedom for the numerator (NUMDF) and 14 degrees of freedom for the denominator (DENDF). Those critical values were used to compute power curves (POWER) based on the λ values (NCP). The β error probabilities (BETA) were calculated as the inverse of power. The average of the α and β error (AVEERROR) was then computed across the range of α values for each effect size.

This analysis allows us to discover the α and β levels corresponding to the minimum average error (Fig. 3). A power of 0.8 or higher is generally recommended for good experimental design (Cohen, 1992; Welham et al., 2014). We will be able to achieve a power ≥ 0.8 for all three effect sizes if we use the α and β levels associated with the minimum average error (Table 1). For the small effect size, however, maintaining α at 0.05 severely compromises power, and we will need to relax the α to

TABLE 1. Experimental design and power analysis for simulated experiment.†

Source	No.	df	Mean square	Grand mean	CV	SD	SED		
Entries	8	7	–	5000	10	500	408		
Replications	3	2	–						
Error		14	250,000						
Percent difference to be detected	Difference	T value for difference‡	F value for difference §	λ for H_a ¶	Min. avg. of α and β error#	α at minimum	β at minimum	Power at minimum	Power at $\alpha = 0.05$
	kg								
10%	500	1.22	1.5	10.5	0.22	0.25	0.20	0.80	0.42
16%	816	2.00	4.0	28	0.07	0.08	0.06	0.94	0.89
25%	1244	3.05	9.3	65	0.01	0.02	0.01	0.99	1.00

† Simulated experiment is for 8 entries with 3 replications. Previous experiments had a grand mean of 5000 kg ha^{-1} and a CV of 10%. Therefore the SD is calculated at 500 kg. The SED is calculated as sqrt((2*250,000)/3).

‡ The t value for the difference between two entries is calculated as their difference/SED.

§ The F value for the difference between two entries is calculated as t^2.

¶ λ is calculated as the F value × df for entries.

Other data points in the table were calculated based on code for power analysis for simulated experiment.

0.25 to obtain an optimal power greater than 0.8. It's clear that if we want to detect an effect size of 10%, we should be prepared to replicate more extensively, find ways to reduce experimental error, or be content with a higher probably of false positives. These scenarios can be tested in a similar manner as above by using different estimates for the number of replications, the SED, and for α. And here is another new thought: *All three choices are valid responses, depending on the constraints and the specific goals of the experiment.* This exercise demonstrates that taking β error into account will cause us to rethink our devotion to $p = 0.05$ for α alone. When β is ignored, the true error of our experiments is almost always greater than we assume.

Rethinking Standard Operating Procedure:

Common practice in agronomic and environmental research (and most other research) has been to control α error, the probability of a false positive. Type 2 error probabilities are rarely reported in agricultural research. Instead, many experiments use the same statistical decision making threshold of $\alpha = 0.05$, regardless of the hypotheses and rationale for the research in the first place. This practice wasn't recommended by Fisher, nor by Neyman and Pearson, nor is it recommended currently (Fisher, 1951; Neyman and Pearson, 1933; Mudge et al., 2012b). The objectives and goals of experiments are not all the same.

If the α error rate is maintained at 0.05 or less for all experiments and for all effects within experiments, the β error rate is determined by sample size, by the realized rather than the desired effect size, and by the variability and uncertainty in the data of the experiment. Critical effect sizes are not determined a priori by the relevant biological research question but are instead determined by the experimental design (Mudge et al., 2012b). The LSD values are calculated after the experiment is conducted and then used as the defacto critical effect sizes for treatments. Mea culpa—we have all done this. And to be honest, the LSD from a similar experiment conducted previously is a good reference point for a future decision on a reasonable effect size. We can assume that we want an LSD that is the same or less than the current one.

If we don't take the time to learn from previous experiments and figure out what biologically relevant critical effect size we want when we design experiments, our interpretation of the results from our experiments is muddled, and the power of our experiments is reduced. Even if correct, the failure to prove an effect and reject the H_0 does not prove that there is NO effect (Parkhurst, 2001). Too often, if an effect cannot be detected at $p \leq 0.05$, it is assumed to be irrelevant, even if it could be detected at $p \leq 0.06$ or even $p \leq 0.10$, and it is not reported. Unfortunately, these practices have been typical of agronomic and environmental science reporting, although they are beginning to change (Anderson et al., 2000; Fidler et al., 2006; Begley and Ellis, 2012; Bosker et al., 2013).

Power, as defined by Casler in Chapter 4, is the likelihood of "getting it right" and making the correct decision based on the data collected in an experiment (Casler, 2016). Since it is the aim of many scientific experiments to evaluate the effect of a specific treatment or combination of treatments, the power of an experiment has been defined as the probability of correctly rejecting the H_0 when it is false. Another way of saying this is that the power of an experiment is the probability of not making a false negative error $(1 - \beta)$. Chapter 4 provides methods of power analysis to assist with this process of experimental design at $\alpha = 0.05$, similar to those used for the example above. Various α levels can be substituted into the calculations introduced in Chapter 4 to explore other scenarios for improving power. Power can be improved by increasing the size of the experiment, by decreasing the uncertainty associated with data measurements, or by decreasing the emphasis on α error.

Relative Importance of α and β Errors

In agronomic and environmental research, we rarely specify the relative consequences of making either type of error. Our major focus is on quantifying the α error rate (often at 5%), and we rarely consider or try to quantify β error. This practice implies a higher cost to false positives than to false negatives because it completely ignores false negatives. This is a mistake because the relative importance of false positives and false negatives depends on the consequences of each type of error for the research problem (Carmer, 1976; Campbell and Lipps, 1998). As an example, the differences in the relative consequences of α and β errors for typical judicial, medical, and agronomic decisions can be substantial (Table 2). The ideal situation is when the average error is low and the probability of both a false positive and a false negative is low, but this ideal situation is often not possible due to the cost, time, space, and logistical constraints mentioned earlier. Therefore, decisions based on statistics are usually made with either a greater α or greater β errors than what would be optimal. We need to take the time to determine the costs and consequences of greater than optimal errors for each type of decision that we make. Once we engage in this process, we will make better decisions going forward.

In many cases, the more critical problem is the false negative (Table 2). In exploratory science or environmental impact studies, false negatives maintain the status quo (Mudge et al., 2012b). For example, a false negative error would occur when an experiment failed to show that a fertilizer application increased grain yield, a spectral reflectance index was correlated with grain yield, application of biochar improved soil properties, and so forth, when all of these things truly occur (see Carmer, 1976, for additional examples from variety testing trials). When a new treatment or new germplasm is being evaluated for disease resistance, a false negative might result in that germplasm being discarded as susceptible while a false positive would likely result in additional evaluation. Additional testing often occurs after α errors are made. This additional testing means we have another chance to identify false positives. In contrast, additional testing is unlikely when β errors are made in biological experiments. Therefore false negatives should be weighted more strongly when experiments are exploratory and additional confirmatory research is planned. When the probability and cost ratios of α and β errors are equal, the optimal relative probabilities of the two types of error are determined by the effect size and by residual error in the experiment. The proposed experiment analyzed above considered the importance of α and β error to be equivalent. If β error is determined to be of more consequence than α error, their average can be weighted by the relative consequences. R-script to simplify these calculations, given an effect size, estimated experimental error, and relative cost of error types, has been published by Mudge et al. (2012a).

TABLE 2. Type 1 (a) and Type 2 (b) errors contrasted, with different scenarios for varying emphases of Type 1 (a) and Type 2 (b) errors.

Type 1 (α) and Type 2 (β) error contrasted		
	Low probability of false negative ↓ $\beta < 0.05$ Power high ↑	High probability of false negative ↑ $\beta > 0.05$ Power low ↓
Low probability of false positive $\alpha < 0.05$ ↓	Ideal situation Frequently not achievable, but can be done with large effect sizes	Typical of many published experiments in agronomy
High probability of false positive $\alpha > 0.05$ ↑	Exploratory research Positive effects are often re-evaluated.	Not an ideal situation Precision may be improved by increasing sample size or reducing experimental complexity.
Scenario for varying emphases of Type 1 (α) and Type 2 (β) error—Law		
Murder trial	β ↓ $(1 - \beta)$ ↑	β ↑ $(1 - \beta)$ ↓
α ↓	Correct decision to jail	Crime spree
α ↑	Jail for an innocent person	Hung jury
Scenario for varying emphases of Type 1 (α) and Type 2 (β) error—Medicine		
Test for cancer	β ↓ $(1 - \beta)$	β ↑ $(1 - \beta)$ ↓
α ↓	Correct decision to treat cancer	Disease untreated
α ↑	Additional testing, healthy person given treatment	Suffering, possible death
Scenario for varying emphases of Type 1 (α) and Type 2 (β) error—Agronomy		
Variety trials	β ↓ $(1 - \beta)$ ↑	β ↑ $(1 - \beta)$ ↓
α ↓	Growers adopt new variety and are happy.	Growers don't adopt new variety. Cost is associated with unrealized potential profit from increased yields of new variety.
α ↑	Growers adopt new variety, but it doesn't perform better than old variety. Cost differential depends on relative cost of seed. Or Type 3 error occurs, and new variety actually performs worse than old variety.	Experiment is too variable to make a decision. Resources for trialing wasted.
Scenario for varying emphases of Type 1 (α) and Type 2 (β) error—Soil science		
Is profit from no-till equal to or better than conventional till?	β ↓ $(1 - \beta)$ ↑	β ↑ $(1 - \beta)$ ↓
α ↓	H_0 not rejected; no difference between conventional till and no-till detected. Farmers adopt no-till. Less soil erosion.	H_0 not rejected, but no-till not as productive as conventional till. Some farmers switch to no-till and are not as profitable.
α ↑	H_0 not rejected; farmers don't adopt no-till. Conventional till and soil erosion continue.	System is too variable; time and effort to do experiment is not conclusive.

Reasons for Lack of Significance Not Reported

The problems with our tendency to ignore β errors becomes critical when β errors are *really* more of a cost than false positives. Often, the result of an experiment will be that the H_0 is not rejected. While these negative results are common, they are rarely published in the literature. The current interest in meta-analysis enables the correction of statistical errors and better decision making on the basis of combined datasets. Meta-analysis requires information on estimated effect sizes and experimental variation be included in the research report (Ellis, 2010). Too often, in published research, the research community isn't given enough information to verify the results of specific experiments so that they can be compared with other similar research. These problems led the National Institutes of Health and editors of leading science journals to develop "Principles and Guidelines for Reporting Preclinical Research" including standardized procedures for experimental design and data reporting (McNutt, 2014; National Institutes of Health, 2016). While these guidelines are focused on medical research, many of the same principles apply to agronomic and ecological research design and reporting.

When the H_0 is *not* rejected at $p \leq 0.05$, the reasons may be several: (i) no treatment effect; (ii) the variability in the experiment and lack of adequate replication obscured a small effect size; (iii) the subplots of a split plot design were adequately replicated and measured with precision, but the main plots suffered from inadequate or pseudo-replication; (iv) a systematic environmental trend obscured the true treatment effect and blocking was ineffective; or (v) another correlated variable influenced the variable that was measured (these points are further discussed by Casler in Chapters 3 and 4 and by McCarter in Chapter 9). When experiments are designed without attention to both α and β error, we don't know which of these reasons is true. Often, the conclusion is that no treatment effect exists, especially for the main plots when one of the other reasons listed above is the actual cause. Good experimental design will help to account for possibilities (ii) through (v), so that the correct conclusion can more often be made for the correct reason.

Complicated Experimental Designs with Multiple Possible Hypotheses Are Frequently Under-Powered

Most research in agronomy and environmental science is conducted to evaluate response to multiple treatment factors or multiple independent effects and analyzed using analysis of variance or multiple regression procedures. Complicated experimental designs, several levels of multiple treatments, and factorial arrangements of interactions are common. When multiple treatment factors are tested in a single experiment, factor interactions can be discovered. Most experiments aren't designed with enough power to detect these interactions, however. Usually four to eight times the number of replications are required for tests of interaction effects than for main effects to achieve similar power (Schauber and Edge, 1999; Tempelman, 2008). Multiple H_A are possible in factorial experiments. Different critical effect sizes can be expected for different contrasts among treatments and treatment interactions within the same experiment with different consequences for the relative importance of α and β error (Kirk, 2013).

Experiments in the agronomic and environmental sciences are frequently designed and analyzed based on tradition rather than statistical best practices. A common type of experiment reported in journals is a factorial design of two or three treatment factors arranged as a split plot, with three or four replications of the main plot effects. Split plot experiments have two measures of standard error, one for the main plots and another for the subplots. The subplot error is measured based on more samples than the main plot error and is therefore a more accurate estimate of uncertainty in the data (Yates, 1935).

Errors in Split Plot Experiments

For example, in the agridat package of R compiled by Wright (2012), several datasets provide examples of experiments in agriculture including 'yates.oat'. In 1935, Yates reported an experiment evaluating three oat cultivars and four levels of manure. (Yates, 1935). The experiment was arranged as a split plot with six blocks of each of the three cultivars as main plots. Each main plot was subdivided into four subplots to evaluate the manure levels for a total of 72 plots (Fig. 4). The experimental units were unique for the time when the experiment was conducted and are difficult to convert accurately to standard units. The original units are maintained here because the point of this exercise is to evaluate experimental design decisions rather than to make predictions about oat production. The three oat cultivars evaluated were named 'Golden Promise', 'Marvelous', and 'Victory'. The fertility treatments were 0, 0.2, 0.4, and 0.6 hundredweight of manure per acre. Grain yields were recorded in units of 0.25 lb per subplot (0.0125 acre). Yields averaged over the whole experiment were 104 of these units, which is the equivalent of approximately 2330 kg ha^{-1}. Because of the split plot structure of the experiment, the main effect of cultivar was evaluated with six replications and the error term to test the cultivar main effect was the block by cultivar interaction. This means that the MSE for the main plot was equal to 601.33. The manure treatment subplot effect was evaluated with 18 replications and the error term to test the subplot effect was the residual error (the mean squared error) from the experiment, which is equivalent of the ((manure*block) w/in cultivar) interaction (MSE = 177.08). The main effect in this split plot experiment was not evaluated with as much precision as the subplot effect, although the level of replication was high by modern standards in agronomic and environmental research. Yates noted that use of a design where all treatment combinations were randomized equally would result in an MSE equal to 254.22 for this experiment (Yates, 1935). In addition, as pointed out by Wright in the notes for the agridat package, the blocking for the experiment was inadequate to a correct for a yield trend occurring across the field (Fig. 4). When the columns in the design were added to the model as a covariate the mean squared error for the subplots was reduced to 111.18, but the main plot error was not affected because the larger main plots covered multiple columns. Although the maximum and minimum cultivar means differed by more than 10%, the cultivar effect was not considered to be different from 0 in this experiment. The $LSD_{0.05}$ for differences in yield among cultivars was 31.54 units (see above for unit description). In contrast, the manure effect was considered to be different from 0 with an $LSD_{0.05}$ equal to 7.08 units. If these two LSD values are used as minimum effect sizes for the experiment and α is maintained at 0.05, then the β error for the

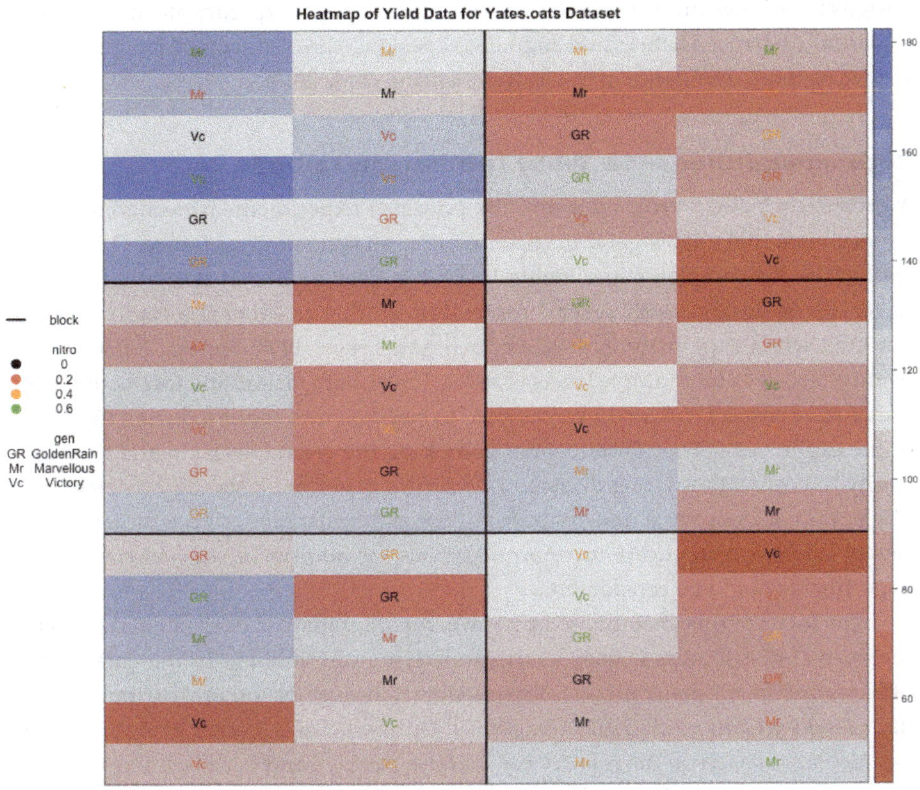

FIG. 4. The layout for the split plot oat trial described by Yates (1935) and plotted using the desplot function in R. The yield trend across the field is evident with higher yields represented by the blue highlighting on the left side of the field and lower yields represented by the red highlighting on the right side of the field. Units in the figure are those published in the original paper. The names of the cultivars (called GEN in the dataset in the agridat package) are represented by abbreviations and the levels of manure (called NITRO in the dataset in the agridat package) are represented by colors in the plots on the figure. Plot first published in agridat.pdf (Wright, 2012).

manure effect was very low at $\beta < 0.001$, but the β error for the cultivar effect was 0.75, and power for the main cultivar effect was = 25% (where power = $(1-\beta)*100$). The CV of the whole experiment, calculated using the MSE after accounting for linear trend, was 10%, which is generally considered to be adequate for yield trials. The CV for the main plots, however, was 24% when calculated based on the main plot error. The split plot arrangement and inadequate blocking made detection of any critical effects for cultivar very unlikely. The inability to detect the main plot effect may be due to a true lack of effect, but is likely due to poor experimental design.

The experiment was conducted in 1930, so we aren't actually trying to improve research per se. But, what can we learn from this experiment? First, the ability to detect main plot treatment effects is normally quite a bit less than that of subplot treatment effects in a split plot arrangement. Second, field trends should have been minimized with suitable blocking. The need for corrections for spatial trends should be explored and included in the data analysis. Third, β error can vary for effects within the same experiment due to experimental design and variation for that effect. Fourth, it's a waste of time and effort to evaluate effects in severely under-powered experiments. Yates (1935) recognized the inadequacies and recommended use of Latin square or lattice designs to improve this type of experiment. Other types of experimental designs, such as partially balanced lattices and augmented designs, can also account for spatial error to give us more confidence in the data (Moehring et al., 2014).

Recommendations—So What Is a Person to Do?

A greater focus on β error rates and the power of experiments is needed to design experiments that enable good decision making in spite of uncertainty. This is especially true for complex experiments involving split plots and multiple treatment factorials or designs with several regressors, so that positive associations can be detected when they truly exist (Campbell and Lipps, 1998; Stroup, 2002). The relevant effect sizes for multiple levels of several treatment factors and interactions plus the relative costs of α, β, and γ errors need to be taken into account during the design of the experiment. The consequences of making the wrong decision and the opportunity to correct the wrong decision in a future experiment should be determined. The option of improving and simplifying the experimental design with fewer treatments, additional replications, more locations, or additional time to conduct the experiment should be considered.

The focus of this chapter has been on errors in statistical decision making when the normal distribution is used as the underlying distribution for the data. Design of experiments for generalized linear models that may use other distributions was discussed by Gbur et al. (2012, Chapter 7). Optimal α and β error probabilities can be calculated based on the relative cost of false positive and false negative errors for various types of experiments for given experimental and effect sizes.

Step-by-Step Recommendations

- Design experiments appropriately, so all treatment and interaction effects can be detected if they exist.
- Determine a relevant effect size for the treatments and interactions to be tested.
- Benchmark against other similar experiments with good experimental design.
- Determine a range of suitable ratios of type α and β error.
- Define costs of each variable under your control in the experimental design.
- Define the costs of α and β error for each effect to be tested.
- Determine the optimal error for your experiment under various experimental design scenarios.
- Conduct a power analysis.
- Determine if treatments might be tested again in a later experiment, providing the opportunity to correct α errors.
- Revise plans according to the results above.

Summary

In summary, while the use of the normal distribution, the central *F* distribution, and control of false positives are well understood, the use of the noncentral *F* distribution, the identification of target effect sizes, and methods to calculate the rate of false negatives are not well understood by those of us who conduct agricultural and environmental research. As a result, error rates are greater than calculated, and too many experiments are under powered. Several tools are now available, including the software code included in the supplement. While every judgement teeters on the brink of error, good experimental design can keep it from teetering over into the deep.

Key Learning Points

- All data are collected with uncertainty.
- Uncertainty can be quantified as standard error, or variance about a mean.
- Relevant effect sizes for experimental treatments should be determined while the experiment is being designed.
- All types of error need to be considered during experimental design.
- The importance of false positive (α) and false negative (β) error will change depending on the desired outcomes of the experiment.
- The amount of α and β error can and should be estimated for various experimental designs during the planning phase of the experiment.
- Devotion to $p = 0.05$ should not be dogma.

Review Questions

True or False:

1. The central *F* distribution is calculated based on the numerator and error degrees of freedom.
2. Type 1 error should always be controlled below 5% whenever possible.
3. The noncentrality parameter is associated with the effect size.
4. Experiments should always be designed to obtain the minimum average error.
5. Effect sizes can be divided into those that measure differences and those that measure association.
6. When spatial variation is discovered after the experiment is conducted, it will need to be included in the unexplained error for the experiment.
7. An experiment with a good deal of power will be associated with a lower probability of false positives.
8. The null hypothesis test is a valid approach to agronomic and environmental research.

Exercises

For the following exercise, we will be working with the yates.oats.csv dataset described previously (Yates, 1935; R agridat package).

1. Instead of using a split plot design, investigate other designs, such as an RCB. What is the pooled error term if the design is analyzed as an RCB? What does use

of a pooled error term do to the significance of the main effects in the experiment?

2. Investigate whether incorporating a linear trend along the columns of the plot layout reduces the error in the RCB design.

3. Investigate whether incorporating a linear trend along the rows of the plot layout reduces the error in the RCB design.

4. Calculate the treatment and error variances for these scenarios and determine the average α and β errors if α is held to 0.05.

   (Note that due to the split plot structure, the actual number of plots per cultivar is 24, and the actual number of plots per manure level is 18 for this design).

   Calculate α and β errors for a 10% and 25% difference in the cultivar means in each case.

   Calculate α and β errors for a 10% and 25% difference in the nitrogen means.

   Calculate α and β errors to detect a difference of 10% and 25% for the interaction term.

References

Anderson, D.R., K.P. Burnham, and W.L. Thompson. 2000. Null hypothesis testing: Problems, prevalence, and an alternative. J. Wildl. Manage. 64:912–923. doi:10.2307/3803199

Begley, C.G., and L.M. Ellis. 2012. Drug development: Raise standards for preclinical cancer research. 483:531–533. doi:10.1038/483531a

Beres, B.L., T.K. Turkington, H.R. Kutcher, B. Irvine, E.N. Johnson, J.T. O'Donovan, K.N. Harker, C.B. Holzapfel, R. Mohr, G. Peng, and D.M. Spaner. 2016. Winter wheat cropping system response to seed treatments, seed size, and sowing density. Agron. J. 108:1101–1111. doi:10.2134/agronj2015.0497

Bosker, T., J.F. Mudge, and K.R. Munkittrick. 2013. Statistical reporting deficiencies in environmental toxicology. Environ. Toxicol. Chem. 32:1737–1739. doi:10.1002/etc.2226

Carmer, S.G. 1976. Optimal significance levels for application of the least significant difference in crop performance trials. Crop Sci. 16:95–99. doi:10.2135/cropsci1976.0011183X001600010024x

Campbell, K.A.G., and P.E. Lipps. 1998. Allocation of resources: Sources of variation in fusarium head blight screening nurseries. Phytopathology 88:1078–1086. doi:10.1094/PHYTO.1998.88.10.1078

Cohen, J. 1992. Statistical power analysis. Curr. Dir. Psychol. Sci. 1:98–101. doi:10.1111/1467-8721.ep10768783

Davidian, M., and T. Lewis. 2012. Why statistics. Science 336:12. doi:10.1126/science.1218685

Ellis, P.D. 2010. The essential guide to effect sizes: Statistical power, meta-analysis, and the interpretation of research results. Cambridge Univ. Press, New York. doi:10.1017/CBO9780511761676

Fidler, F., M.A. Burgman, G. Cumming, R. Buttrose, and N. Thomason. 2006. Impact of criticism of null-hypothesis significance testing on statistical reporting practices in conservation biology. Conserv. Biol. 20:1539–1544. doi:10.1111/j.1523-1739.2006.00525.x

Fisher, R.A. 1951. The design of experiments. Oliver and Boyd, Edinburgh, Scotland.

Gill, J. 1999. The insignificance of null hypothesis significance testing. Polit. Res. Q. 52:647–674. doi:10.1177/106591299905200309

Gbur, E.E., W.W. Stroup, K.S. McCarter, S. Durham, L.J. Young, M. Christman, M. West, and M. Kramer. 2012. Designing experiments. In: Analysis of generalized linear mixed models in the agricultural and natural resources sciences. ASA, SSSA, and CSSA, Madison, WI. p. 237–270, doi:10.2134/2012.generalized-linear-mixed-models.c7.

Hernández, R., and C. Kubota. 2016. Physiological responses of cucumber seedlings under different blue and red photon flux ratios using LEDs. Environ. Exp. Bot. 121:66–74. doi:10.1016/j.envexpbot.2015.04.001

Kirk, R.E. 2009. Experimental Design. In: R.E. Millsap and A. Maydeu-Olivares, editors, The SAGE handbook of quantitative methods in psychology. SAGE, London, p. 23–45, doi:10.4135/9780857020994.n2.

Kirk, R.E. 2013. Experimental design: Procedures for the behavioral sciences. 4th ed. Sage, Thousand Oaks, CA. doi:10.4135/9781483384733

Kutner, M.H., C.J. Nachtsheim, J. Neter, and W. Li. 2005. Applied linear statistical models. 5th ed. MGraw-Hill, Irwin, NY.

McNutt, M. 2014. Journals unite for reproducibility. Science 346:679. doi:10.1126/science.aaa1724

Moehring, J., E.R. Williams, and H.P. Piepho. 2014. Efficiency of augmented p-rep designs in multi-environmental trials. Theor. Appl. Genet. 127:1049–1060. doi:10.1007/s00122-014-2278-y

Mudge, J.F. 2013. Explicit consideration of critical effect sizes and costs of errors can improve decision-making in plant science. New Phytol. 199:876–878. doi:10.1111/nph.12410

Mudge, J.F., L.F. Baker, C.B. Edge, and J.E. Houlahan. 2012a. Setting an optimal a that minimizes errors in null hypothesis significance tests. PLoS One 7:e32734. doi:10.1371/journal.pone.0032734

Mudge, J.F., T.J. Barrett, K.R. Munkittrick, and J.E. Houlahan. 2012b. Negative consequences of using alpha = 0.05 for environmental monitoring decisions: A case study from a decade of Canada's Environmental Effects Monitoring Program. Environ. Sci. Technol. 46(17):9249–9255. doi:10.1021/es301320n

Nakagawa, S., and I.C. Cuthill. 2007. Effect size, confidence interval and statistical significance: A practical guide for biologists. Biol. Rev. Camb. Philos. Soc. 82:591–605. doi:10.1111/j.1469-185X.2007.00027.x

National Institutes of Health. 2016. Principles and guidelines for reporting preclinical research. http://www.nih.gov/research-training/rigor-reproducibility/principles-guidelines-reporting-preclinical-research (accessed 30 June 2016).

Neyman, J., and E.S. Pearson. 1928a. On the use and interpretation of certain test criteria for purposes of statistical inference: Part I. Biometrika 20A:175–240. doi:10.2307/2331945

Neyman, J., and E.S. Pearson. 1928b. On the use and interpretation of certain test criteria for purposes of statistical inference: Part II. Biometrika 20A:263–294. doi:10.2307/2332112

Neyman, J., and E.S. Pearson. 1933. On the problem of the most efficient tests of statistical hypotheses. Philos. Trans. R. Soc. London A. 231:289–337. doi:10.1098/rsta.1933.0009

Parkhurst, D.F. 2001. Statistical significance tests: Equivalence and reverse tests should reduce misinterpretation. 51:1051–1057. doi:10.1641/0006-3568(2001)051{$[$}1051:SSTEAR{$]$}2.0.CO;2

Schauber, E.M., and W.D. Edge. 1999. Statistical power to detect main and interactive effects on the attributes of small-mammal populations. Can. J. Zool. 77:68–73. doi:10.1139/z98-182

Selya, A.S., J.S. Rose, L.C. Dierker, D. Hedeker, and R.J. Mermelstein. 2012. A practical guide to calculating Cohen's f^2, a measure of local effect size, from PROC MIXED. Front. Psychol. 3:111. doi:10.3389/fpsyg.2012.00111

Smiley, R.W., S. Machado, K.E.L. Rhinhart, C.L. Reardon, and S.B. Wuest. 2016. Rapid quantification of soilborne pathogen communities in wheat-based long-term field experiments. Plant Dis. doi:10.1094/PDIS-09-15-1020-RE

Snedecor, G.W., and W.G. Cochran. 1967. Statistical methods. 6th ed. Iowa State Univ. Press, Ames.

Stigler, S.M. 1981. Gauss and the invention of least squares. Ann. Stat. 9:465–474. doi:10.1214/aos/1176345451

Stroup, W.W. 2002. Power analysis based on spatial effects mixed models: A tool for comparing design and analysis strategies in the presence of spatial variability. J. Agric. Biol. Environ. Stat. 7:491–511. doi:10.1198/108571102780

Sullivan, G.M., and R. Feinn. 2012. Using effect size—Or why the *P* value is not enough. J. Grad. Med. Educ. 4:279–282. doi:10.4300/JGME-D-12-00156.1

Tackenberg, O., P. Poschlod, and S. Bonn. 2003. Assessment of wind dispersal potential in plant species. Ecol. Monogr. 73:191–205. doi:10.1890/0012-9615(2003)073[0191:AOWDPI]2.0.CO;2

Tempelman, R.J. 2008. Statistical analysis of efficient unbalanced factorial designs for two-color microarray experiments. Int. J. Plant Genomics 584360. doi:10.1155/2008/584360

Welham, S.J., S.A. Gezan, S.J. Clark, and A. Mead. 2014. Replication and power In: Statistical methods in biology: Design and analysis of experiments and regression. CRC Press, Boca Raton, FL. p. 241–256.

Wright, K. 2012. agridat: Agricultural Datasets. R package version 1.4. https://cran.r-project.org/web/packages/agridat/index.html (accessed 30 June 2016).

Yates, F. 1935. Complex experiments. J. R. Stat. Soc. Suppl. 2:181–247. doi:10.2307/2983638

CHAPTER 2: ANALYSIS OF VARIANCE AND HYPOTHESIS TESTING

Marla S. McIntosh*

Abstract

This introductory chapter offers a summary review and practical guidance for using analysis of variance (ANOVA) to test hypotheses of fixed effects. The target audience is agricultural, biological, or environmental researchers, educators, and students already familiar with ANOVA. Descriptions of ANOVA, experimental design, linear models, random and fixed effects, and ANOVA components are presented along with discussions of their roles in scientific research. A case study with data provided that is balanced and normally-distributed is used to illustrate ANOVA concepts and offer hands-on ANOVA experience. The case study involves a factorial experiment conducted at three locations analyzed using a mixed model approach. Analysis of variance results are explained and used to construct ANOVA tables that effectively communicate key details to verify that the statistical design, analysis, and conclusions are appropriate and valid. The goal of this chapter is to: i) empower readers to make informed decisions to choose appropriate experimental designs, statistical estimates, and tests of significance to address the research objectives; and ii) provide advice on presenting statistical details in research papers to ensure that the experimental design, analysis, and interpretation were valid and reliable.

Research in agricultural, biological, and environmental science disciplines is a major contributor to our basic scientific understanding that can lead to new and improved technologies, safer and healthier food supply, and best practices for sustaining our ecosystems. And it can be argued that we owe much of the overall success of research in these disciplines to the widespread use of analysis of variance (ANOVA) that began the modern era of applied statistics.

Analysis of variance was initially conceived for agricultural researchers conducting field experiments to determine whether differences between treatments were significant (i.e., reliable and repeatable) rather than an artifact of variable environmental conditions (Fisher, 1926). Subsequently, the classical ANOVA concepts

---

Abbreviations: ANOVA, analysis of variance.

M.S. McIntosh, Professor Emerita, Department of Plant Science and Landscape Architecture, University of Maryland, College Park, Maryland, 20742. *Corresponding author (mmcintos@umd.edu)

doi:10.2134/appliedstatistics.2016.0009

Applied Statistics in Agricultural, Biological, and Environmental Sciences
Barry Glaz and Kathleen M. Yeater, editors

have been built on, refined, and adapted across disciplines leading to new ANOVA methodologies that have substantially extended the potential scope and application of ANOVA. In this chapter, we begin with classical ANOVA concepts and terminology to serve as the foundation and give context for understanding the contemporary ANOVA developed for mixed models. A case study provides a real-world example of ANOVA based on an RCBD factorial experiment conducted at three locations. The case study analyzes a balanced data set that meets the classical ANOVA assumptions. The ANOVA results are described and readers are encouraged to replicate the ANOVA and/or devise their own analyses. A broad overview and practical guidance on effective use of ANOVA is intended for scientists, educators, and students who have a basic understanding of statistics and experimental design.

The ANOVA Process

The experimental process begins with a researcher interested in testing a hypothesis that, according to the scientific literature and the researcher's opinion, is reasonable and justifiable. Based on the experimental objectives and scope of the inferences desired, the research team determines appropriate treatments to be applied and dependent variables to be measured. Subsequently, the experimental design, including the number and size of experimental and sampling units, is planned (often with the help of a statistician). The experimental design should provide the desired level of precision and power balanced by the perceived limits on resources, time, effort, and expense [see Chapters 1, 3, and 4 (Garland-Campbell, 2018; Casler, 2018a,b]. A linear model is then proposed to explain the variation observed in the dependent variable and a suitable statistical method is chosen to conduct the analysis.

Once the plans for the experiment are complete and thoroughly reviewed, the experiment is conducted and data carefully recorded along with field or laboratory notes. After the data are collected and checked for errors and outliers, the statistical analysis is conducted. The statistical results are often found within seemingly endless computer output filled with numbers that appear to be so precise that they are presented with ten digits. The output includes all kinds of statistics, parameter estimates, and probabilities and it is up to the research team to use the appropriate statistics to make sense of the data and arrive at valid conclusions.

The experimental process as described illustrates that statistics are not something that happens after an experiment is completed, but a component in every step of the process. Also, the decisions related to the design and analysis of the experiment can be complex and interdependent, requiring a scientific and practical understanding of the populations being investigated as well as a working knowledge of ANOVA concepts and procedures. As too many researchers have learned by experience, data analysis mistakes can be corrected, but poor experimental design choices can be irreversible and require repeating all or part of an experiment.

The point is that statistics matter and can determine whether your research accomplishes its objectives and your conclusions will stand the test of time or even sees the light of day. Obviously, the impact of the research is greatly magnified if published in a peer-reviewed, high-quality scientific journal. These journals have their own standards, which invariably include scientific merit. Well-designed and statistically sound experiments are integral to scientific quality and integrity. Thus, scientific journals often require that papers describe the statistical design and analyses in order to verify they are correct and appropriate for the stated objectives and conclusions. Meta-analysis has shown that statistical significance often determines if a study is publishable and sets a path for future research (Mervis, 2014). This can have negative consequences. Depending on the effect tested, knowing that an explanatory effect is not significant can be an important discovery (e.g., chemical *Y* does not significantly affect beneficial insect counts) or can discourage further research on ineffective treatments (e.g., chemical *Y* does not significantly affect targeted pest insect counts). On the other hand, an effect deemed not significant as a result of a Type 2 error should not be published, but the Type 2 error rate is rarely known and can be surprisingly large. Thus, if a paper with nonsignificant results is published, it should include a justification for the adequacy of the experimental design and power of the tests.

History of Analysis of Variance

Analysis of variance was introduced at a time when agronomists were desperately seeking a statistically sound technique to improve the credibility of field researchers who were (and still are) plagued by inherent variations caused by uncontrolled factors that can mask or exaggerate treatment effects. Agricultural researchers needed a scientific method to design and analyze data that would provide a measure of confidence that the measured differences in yield between treated plots and their controls were reliable and repeatable. Agronomic researchers were all too aware of the vagaries of field variation that often overshadowed the actual effects of treatments and R.A. Fisher developed ANOVA as a statistical method to address this problem. Fisher was hired by the Rothamsted Research Station to statistically analyze data and observations collected from continuous research on wheat that was conducted for over 70 yr to determine the causes of variation in yields (Fisher, 1921, Fisher and Mackenzie, 1923). Making use of the extensive long-term field and weather data and notes, Fisher developed ANOVA to estimate and compare the relative contributions of various factors (e.g., weather, soil fertility, weed cover, and manure treatments) to the observed variations in yield. Over the next few years, Fisher further developed the principles and practices for ANOVA and wrote the groundbreaking classical statistics treatise, "Statistical Methods for Research Workers" (Fisher, 1925). This book, written for agronomic researchers, educators, and students rather than statisticians, set the statistical standards for the design and analysis of field experiments. Subsequently, ANOVA has been adapted for statistical testing of experimental data throughout most scientific disciplines. As its use has grown, so has its theory and applications.

Fisher used probability distributions to determine if treatment means were "significantly" different or different by random chance. The ANOVA technique estimated the probability that the means were not different, assuming the population(s) of means were normally and independently distributed with equal error variation. Early agronomic researchers routinely replicated treatment plots to measure plot-to-plot variability and calculate a standard deviation (SD) for each treatment mean. However, researchers typically designed experiments where control plots were replicated to be compared with adjacent treated plots to minimize error variation. Fisher successfully demonstrated that this practice led to inherently biased estimates of error variation and argued that ANOVA requires that treatments and control plots must both be replicated and randomized. This new understanding of the importance of randomization and replication on estimating error variance was arguably the most revolutionary aspect of ANOVA that advanced applied statistics by exploiting the power of experimental design. Analysis of variance brought a new age of field experimentation that fostered the creation of experimental designs to reduce random error and gain power by increasing the number of replications and/or samples.

In the basic ANOVA process, sources of variation are identified, estimated, and compared to test null hypotheses based on statistical probability. The ANOVA addressed the most pressing concern of agricultural researchers by providing a scientifically sound and systematic process that could be taught to researchers so they could: i) design experiments to incorporate the ANOVA principles of treatment replication and randomization; ii) identify the sources (factors) that contribute to the total variation of the data; iii) analyze data using a simple step-by-step procedure to calculate the variation contributed by different sources; and iv) use the ratio of the variation among treatments to the variation within treatments to test the significance of a treatment effect.

Unfortunately, the mathematical simplicity that makes classical ANOVA practical and appealing to researchers comes with limitations. The theoretical probabilities used to test the significance of the differences among means are based on assumptions that the means being compared are from normally-distributed populations and their error variances are independent and equal. However, Box (1976) pointed out that "in nature there never was a normal distribution, there never was a straight line, yet with normal and linear assumptions, known to be false, he can often derive results which match, to a useful approximation, those found in the real world". Thus, in practice, the assumptions of normality and homogeneity of variance are intended as approximate. Fortunately, based on the Central Limit Theorem, if a dependent variable is not normally distributed, the distribution of its means will be normal given a sufficiently large sample size. Also, there is substantial evidence that minor deviations from the basic assumptions have little impact on ANOVA results (Acutis et al., 2012). However, deviations from these assumptions that are not trivial can lead to invalid probabilities for tests of significance and improper inferences.

In response to the shortcomings of classical ANOVA, newer approaches to ANOVA for fitting models, estimating variance and/or covariance parameters, and for testing significance of effects have been developed that are not conditional on the classical ANOVA assumptions. Contemporary ANOVA methods that employ computer-intensive maximum likelihood estimation rather than the classical method of moments have been designed for analysis of mixed models containing both fixed and random effects. More recently, generalized linear mixed model analysis has been developed for mixed model analysis of response variables that are not necessarily normally-distributed. The newer statistical methods have expanded capabilities and applications that allow ANOVA to address the modern-day challenges posed by "big data" and understanding complex systems (Gbur et al., 2012). Statistical software is widely available for scientists and educators to utilize generalized linear mixed model analysis for ANOVA of response variables of various distributions (e.g., normal, binomial, Poisson, lognormal), explanatory effects that are continuous or categorical, and models that are fixed, random, or mixed. However, the flexibility and power of generalized model fitting also requires a level of statistical skill and expertise far above the classical methods to effectively and correctly perform and interpret the ANOVA.

In this chapter, the statistical analyses are limited to continuous and normally distributed response variables that are fitted to a linear mixed model as implemented using PROC MIXED (SAS 9.4). The analysis of non-Gaussian (non-normal) and categorical data is best analyzed using a generalized linear model approach, which is the focus of Chapter 16 by Stroup (2018).

Planning an Experiment

A successful experiment, one that provides statistically sound evidence to reach correct conclusions about hypotheses, requires careful and skillful planning. The strength and correctness of an experiment is greatly influenced by the many decisions made during the planning process. Thus, before walking or running (but never jumping) to conclusions, well-defined experimental goals and objectives should be identified to enable the researcher to make informed choices to optimize the experimental design and analysis. These choices require considerations of experimental space and time, number and arrangement of experimental and sampling units, precision and suitability of laboratory and/or field techniques, equipment and personnel availability and costs, and the available and future funding levels. Keeping the research goals in mind, the researcher strives for a desired level of power for ANOVA tests in view of the practical considerations of resources, time, effort, and expense. The final plan should ensure that the results of the data analysis lead to statistically valid and convincing evidence to support the research conclusions. Table 1 contains a checklist of points useful for evaluating the quality and validity of the experimental design and analysis.

Even when collaborating with a statistician, researchers should take an active role in determining the best experimental design and analysis. Choices determining the quality of both the data and the statistical tests are made at each stage of designing and analyzing an experiment. These choices require both statistical expertise and a scientific and practical understanding of the populations being investigated. Regardless of who designs and analyzes the experiments, the primary researchers and/or authors are responsible for the integrity of the research. In other words, the statistical aspects of research should not be viewed as a separate endeavor to be handed over to a statistician but rather as a collaborative and highly interactive effort. To ensure a successful outcome, joint consultation beginning at the planning stages is essential.

The Linear Model

The linear model is the mathematical framework for ANOVA and provides a succinct description of the effects (known and postulated) that contribute to the total variation of the observations. A linear equation is used to specify all of the terms in the model. The simplest linear model outlines an experiment with treatments replicated in a completely randomized design (CRD), where the value of an observation is equal to the overall mean of the observations plus the effect of its treatment mean and its random error (Eq. [1]).

TABLE 1. Checklist for evaluating the quality and validity of ANOVA.

Experimental and treatment design	
1.	Description of experimental and treatment designs include sufficient detail to be judged or repeated.
2.	Experimental units of the treatments are randomized and replicated.
3.	Replications of experimental units and sampling units are adequate.
4.	Blocks, sites, years, sampling units considered to be random effects have adequate replication.
5.	Designs such as split-plot and sampling designs that result in multiple error terms are mentioned.
6.	Factorial treatment designs are used effectively to test relationships of treatments and add power to tests.
Statistical analyses	
1.	Statistical analyses are described in sufficient detail to be evaluated or repeated.
2.	Quality and quantity of data is evident.
3.	Statistical methods are appropriate for data and objectives.
4.	Theoretical assumptions are validated or justified.
5.	P-values for tests of significance are used correctly.
6.	Power of tests is sufficient to meet research objectives.
Hypothesis testing	
1.	Tests of significance are meaningful and preferably not confounded.
2.	Effects are classified as fixed or random to determine whether the inference space (narrow, intermediate, or broad) and parameters of interest are appropriate for research objectives/questions.
3.	Inferences for fixed effects are limited to the range of levels of the effect in the experiment. The parameters of interest are means.
4.	Inferences for random effects are broad and refer to the population of all possible levels of the effect. The parameters of interest are variances.
5.	Random effects are estimated with at least five degrees of freedom if used as error terms or test of significance will lack power, resulting in a high probability of a Type 2 error.
6.	For factorial treatment designs, tests of significance (fixed effects) or variance estimates (random effects) are conducted for main effects and interactions.
7.	Treatment means that are not structured are compared or rated using an appropriate multiple comparison procedure, usually a least significant difference (LSD) which is equivalent to multiple t-tests.

$$Y = \text{overall mean} + \text{treatment effect} + \text{random error} \qquad [1]$$

The linear model of an experiment becomes more complicated as experimental and treatment design factors are added. And the more complicated the model, the more helpful the linear model is for describing an experiment. The format of a linear model is flexible and can be adapted for specific uses and users such as: i) statistical textbooks on ANOVA and its mathematical foundation useful for teachers and students (Eq. [2]); ii) papers on ANOVA theory and application in journals intended for statisticians and mathematicians (Eq. [3]); iii) papers on research analyzed using ANOVA techniques in journals intended for scientists (Eq. [4]).

Examples of different formats of linear models are found throughout this book. Here, a linear model for an RCBD factorial experiment with two treatment factors is shown in three different formats Eqs. [2–4]. This linear model includes effects associated with the experimental design (block and random error) and the factorial treatment design (*A* and *B* main effects and $A \times B$ interaction).

$$Y = \text{overall mean} + \text{block effect} + A \text{ main effect} + B \text{ main effect} + A \times B \text{ interaction} + \text{random error} \quad [2]$$

$$Y_{ijk} = \mu + \rho_i + A_j + B_k + AB_{jk} + \varepsilon_{ijk:} \quad [3]$$

Y_{ijk} = the observation at the *i*th block, *j*th level of *A*, and *k*th level of *B*; μ = the overall mean; ρ_i = the effect of the *i*th block, A_j = the effect of the *j*th level of *A*, B_k = the effect of the *k*th level of *B*, AB_{jk} = the interaction effect of the *j*th level of *A* and *k*th level of *B* and ε_{ijk} = the residual error of the *i*th block, *j*th level of *A*, and *k*th level of *B*.

$$Y = \text{Block} + A + B + A{\times}B + \text{Error} \quad [4]$$

Equation [2] uses general descriptive statistical terms to explain that the dependent observation is equal to the overall mean plus the effects of the block, treatment factors, and experimental error. Although lacking detail, Eq. [2] defines the effects tested for significance (*A*, *B*, and $A \times B$) as well as the "nuisance" effects (Block and Error) associated with the experimental design.

Equation [3] uses a format common to mathematicians and statisticians in which a Greek or Latin letter represents each effect and subscript notation characterizes the effect level. This format provides a level of detail, clarity, and flexibility that is appropriate for graduate level statistical textbooks, papers in statistical journals, and experiments with complex or unorthodox linear models. However, this format may contain a level of detail that can overwhelm and confuse introductory-level students and non-statisticians. The major drawback, even for statisticians, is that the symbols are not standard. Thus, the symbol and subscript for each effect in the equation must be defined within each publication. For complicated models with many terms this can be tedious and time-consuming to follow especially when comparing models between publications.

Equation [4] is based on the syntax of the MODEL statement used by PROC GLM except that it includes the experimental (residual) error. This format is the most parsimonious and does not include the overall mean (μ) which is a constant and does not contribute to the variation of the dependent variable.

Fixed and Random Effects

Each effect in the linear model is considered to be either fixed or random. Determining whether an effect should be fixed or random is crucial because fixed and random effects satisfy different experimental objectives, have different scopes of inference, and lead to different statistical estimators and test statistics. Fixed effects are treatment factors that include all levels or types of the effect and are used to make narrow inferences limited to the treatments in the experiment and tests of their significance. Fixed effects are usually the treatment (explanatory) effects being investigated, such as fertility treatments, fertility rates, cultivars, and temperature levels. For fixed effects, the estimates of the treatment means and tests of significance of effects and differences between means are of primary interest. In contrast, the levels or types of a random effect represent random samples of all possible levels and types of the effect. Random effects are used to make broad inferences about the general variation caused by the random effect that is not limited to the levels of the effect included in the experiment. Effects that are associated with years, locations, or genotypes are often considered random if they constitute a sufficiently large random sample of the defined population [See Vargas et al. (2018) Chapter 7]. For random effects, inferences about their variation are of primary interest rather than differences between means. Effects associated with design factors and error variation such as blocks, experimental and sampling variation should always be random. Although it is usually readily apparent whether an effect is fixed or random, sometimes the effect does not neatly or exclusively fit the criteria for either random or fixed effects. Categorizing years and environments as fixed or random effects is particularly problematic and controversial for analyses of multi-environment experiments. Rationale and significance for categorizing years and locations for combined analysis has been explained by Moore and Dixon (2015) and in Chapter 8 by Dixon et al. (2018). Also, a comprehensive and practical review of mixed models by Yang (2010) provides criteria and other useful information for categorizing fixed and random effects. As with many statistical decisions, deciding whether effects are fixed or random may be subjective and require sound judgement based on scientific expertise and expectations of the researcher in addition to statistical considerations. When uncertain, expert advice and assistance of a statistician may be sought to ensure that the fixed and random effects are analyzed and interpreted correctly.

Fixed, Mixed, and Random Models

A linear model is categorized as fixed, random, or mixed based on the types of effects (other than the residual error) contained in the model. Fixed and random models contain all fixed or all random effects, respectively. A mixed model contains both fixed and random effects. Classical ANOVA procedures were developed for fixed and random models that are calculated using the mathematically simple method of moments and least squares estimation. Using classical least squares methods, the GLM procedure (SAS) was developed for fixed models to test significance of fixed effects and the VARCOMP procedure was developed for random models to estimate variance components. More recently, PROC MIXED was developed to properly analyze mixed models of normally-distributed data and PROC GLIMMIX to analyze generalized mixed models of data that are not necessarily normally-distributed. The mixed model approach treats random and fixed effects differently and can incorporate covariance structures for fitting models (Littell et al., 2006). PROC MIXED is flexible and appropriate for fitting normally-distributed dependent variables to mixed, fixed, or random models and designs with correlated errors such as repeated measures and split plots. For ANOVA of balanced, normally-distributed fixed models, the PROC GLM and PROC MIXED outputs look different but the statistical results are identical. However, PROC GLM does not automatically use the correct error term for *F*-tests of models with random designs or treatment effects so the MIXED procedure is recommended for most ANOVA analyses (Yang, 2010).

ANOVA COMPONENTS

Despite the new tools added to the ANOVA toolbox, the original ANOVA components are still useful for understanding ANOVA practice and procedures in a modern-day context. These components are: sources of variation (SOV), degrees of freedom (df), sums of squares (SS), mean squares (MS), *F*-ratios, and the probability of a greater *F*-value ($P > F$). Each component is associated with a step in the least squares calculations used to estimate variances and test significance of effects. The following is a brief explanation of each ANOVA component and its informational value.

Sources of Variation

Sources of variation provide the foundation for the ANOVA process and each SOV represents a term the linear model. Therefore, rather than presenting the linear model of a factorial RCBD ANOVA as an equation (Eqs. [2–4]), the same effects can be shown as SOV in an ANOVA table. By including all effects (random and fixed; treatment and design) in an ANOVA table, readers and reviewers can easily and unambiguously understand and judge the validity and appropriateness of the linear model. Correct accounting of the linear model effects is the most important component of ANOVA since every step of the ANOVA procedure depends on it. The SOV summarize the essential ingredients of an experiment including the experimental, sampling, and treatment designs.

Degrees of Freedom

After the SOV are identified, their df are calculated. The df is the number of independent differences used to calculate the ANOVA components for each SOV. The df for a SOV conveys the size and scope of an experiment and further clarifies the experimental, sampling, and treatment designs deduced from the SOV. The df in an ANOVA table are especially helpful when describing experiments with complex designs and analyses because df can be easily translated into the numbers of replications, factors, experimental units, and sampling units of the experiment, as well as the number of levels of each factor. Moreover, df provide insight into the adequacy of the statistical tests of the data. In general, error terms with few df indicate low power, poor repeatability of the *F*-test, and a high probability of a Type 2 error [See Chapter 1 by Garland-Campbell (2018) for a discussion of statistical errors].

In designing experiments, it is often preferable but not always practical or possible to have an equal number of observations in each cell. For experiments with unbalanced data, the df of the random SOV are not independent and should be adjusted using the Satterthwaite or Kenward-Rogers methods (Spilke et al., 2005). In cases with insufficient data for valid hypothesis testing, researchers should consider confidence interval estimation and data visualization techniques (Majumder et al., 2013) or conduct additional experiments to obtain adequate data needed to meet their research objectives.

Sums of Squares

The SS partition the total variation into the variation contributed by each SOV. In general, the difference between each effect mean and the overall mean is squared and then totaled. Given that the SOV are independent of each other, the SS of each

SOV sum to the total SS. The SS are used to determine the percent of the variation accounted for by individual terms in the model and play a primary role for fitting models and determining goodness of fit. For testing significance of effects, the SS are an intermediary step for calculating mean squares.

Mean Squares

The average of a SS (SS/df) is the MS, which estimates the average variation associated with a SOV. A MS is used to construct *F*-ratios for testing significance of effects and to estimate the variance of a random effect. Both usages are based on the concept that a MS is a sample statistic that estimates a population parameter known as an expected mean square (EMS). In turn, an EMS is a linear function that contains one or more variance components, which are population parameters of interest. Later in the chapter, examples will be used to demonstrate the role of EMS for estimating variances and significance testing.

The MS of the error (MSE) and its derivatives play a fundamental role in descriptive statistics as measures of variation within normally-distributed populations (s_e^2, experimental error, residual variation). The MSE is in squared units and can be difficult to interpret in the context of the observations. Instead, the standard deviation (SD), calculated as $\sqrt{\mathrm{MSE}}$, is often the preferred statistic since it is in the same units as the observations. Another common statistic, the standard error (SE), also referred to as the standard error of the mean (SEM), is similar to and often confused with the SD. The key difference between these two statistics is that the SD describes the population distribution of individual samples, whereas the SE describes the population distribution of sample means. The relationship between these statistics is SE = SD/$\sqrt{r} = \sqrt{\mathrm{MSE/r}}$, where r = the number of replications estimating the mean. The SE is used to estimate the precision and confidence intervals of means and commonly is presented with the mean as $\overline{Y} \pm \mathrm{SE}$. Another related statistic is the standard error of the difference (SED) = $\sqrt{2 \times \mathrm{MSE}/r}$, which is used for *t* tests and multiple comparison procedures such as the LSD (least signficant difference) to test for significant differences between pairs of means. An LSD procedure is the same as multiple *t*-tests to determine whether the difference between any pair of means is significant at a chosen α level. The LSD(α) = t_α(SED). If the difference between two means is greater than or equal to the LSD value, then those two means are considered to be significantly different. Note that the LSD and multiple *t*-tests use the SED based on the MSE from the ANOVA.

As seen in their calculations, the SE and SED decrease as the number of replications increase. This improves the precision of mean estimation, the power of the test, and reduces confidence intervals. Although increasing the number of replications does not necessarily increase or decrease the MSE or SD, sufficient replication is critical for an MS to be a reliable and unbiased estimate of variation. As previously noted, based on the Central Limit Theorem, even if the population of individuals is not normally-distributed, the population of means will be normal if the mean sample size is sufficiently large. Consequently, it is tempting to use a large number of replications but it is not always wise. In planning an experiment, it is important to determine, or at least estimate, the number of replications needed for a valid estimate of MSE and to detect meaningful differences, recognizing the real-world and practical constraints of time and resource limitations.

F-values

An *F*-statistic (also termed *F*-ratio or *F*-value) is a ratio of MS's used to compare variances. To test the null hypothesis that the treatment variance is not significantly different from zero, an *F*-statistic is constructed that compares the Treatment MS to the Error MS. This *F*-statistic is analogous to a signal-to-noise ratio, where the signal is the effect and the random variation is the noise. The Treatment MS, calculated from the deviations of the treatment means from the overall mean, contains variations due to both the signal and the noise. The Error MS, calculated from the deviations of the observations from their treatment mean, contains only variation due to the noise. Thus, the basic form for the $F = \frac{\text{signal + noise}}{\text{noise}}$. If the null hypothesis is true, the signal (effect) is zero and *F* reduces to one. The differences among the means are found to be significant when they are larger than would be expected from random variation. The MS's that comprise the numerator and denominator of the *F*-statistic are based on the variance components contained in their EMS (Table 2). Here, the *F*-statistic is the $\frac{\text{Treatment MS}}{\text{Error MS}}$ which is an estimate of $\frac{\text{Var(Residual)} + r\,\text{Var(Treatment)}}{\text{Var(Residual)}}$.

The probability density (frequency) distribution of *F*-statistics, known as the *F*-distribution, is a function of the df of the numerator (ndf), the denominator (ddf), and the hypothesis effect size. If the null hypothesis is true, the effect size, by definition, is equal to zero and the *F*-statistics follow a central *F*-distribution. For alternate hypotheses that propose an effect size of interest, the *F*-statistics follow a non-central *F*-distribution that is a function of the ndf, ddf, and the chosen effect size. The effect size is often given in relative terms as the non-centrality parameter (λ) calculated as the Treatment SS/Error MS.

The critical *F*-value is at the point on the central *F*-distribution where the $P>F$ is the chosen significance level. Prior to computers, researchers relied on *F*-tables containing critical *F*-values for *F*-distributions of varying ndf and ddf for common significance levels (e.g., 0.10, 0.05, 0.01). *F*-tables are relics of the past but are still useful for understanding the relationships between df, *F*-values, and *p*-values. For example, Table 3

TABLE 2. Expected mean squares of RCBD experiment with factors *A* and *B* from SAS (PROC MIXED Method=Type3).

	Fixed Model A-Fixed, B-Fixed		Mixed Model A-Random, B-Fixed	
Source	Expected Mean Square	Error Term	Expected Mean Square	Error Term
Blk	Var(Residual) + tVar(Blk)†	MS(Residual)	Var(Residual) + tVar(Blk)	MS(Residual)
A	Var(Residual) + Q(A,A×B)	MS(Residual)	Var(Residual) + rVar(A×B) + rb Var(A)	MS(A×B)
B	Var(Residual) + Q(B,A×B)	MS(Residual)	Var(Residual) + rVar(A×B) + Q(B)	MS(A×B)
A×B	Var(Residual) + Q(A×B)	MS(Residual)	Var(Residual) + rVar(A×B)	MS(Residual)
Residual	Var(Residual)		Var(Residual)	

† Var and Q are variances of random and quadratic functions of fixed effects, respectively. *r*=number of replications, *t*=number of treatments, and *b*=number of levels of *B*.

TABLE 3. Critical *F*-values for different degrees of freedom for *p*-values ≤ 0.01 and ≤ 0.05.

Critical *F*-values *P* ≤ 0.01								Critical *F*-values *P* ≤ 0.05							
	Numerator df								Numerator df						
Den df	1	2	3	4	5	10	20	Den df	1	2	3	4	5	10	20
2	99.0	-	-	-	-	-	-	2	19.0	-	-	-	-	-	-
3	34.1	30.8	-	-	-	-	-	3	10.1	9.5	-	-	-	-	-
4	21.2	18.0	16.7	-	-	-	-	4	7.7	6.9	6.6	-	-	-	-
5	16.3	13.3	12.1	11.4	-	-	-	5	6.6	5.8	5.4	5.2	-	-	-
10	10.0	7.6	6.5	6.0	5.6	-	-	10	5.0	4.1	3.7	3.5	3.3	-	-
20	8.1	5.8	4.9	4.4	4.1	3.4	-	20	4.3	3.5	3.1	2.9	2.7	2.3	-
120	6.8	4.8	3.9	3.5	3.2	2.5	2.0	120	3.9	3.1	2.7	2.5	2.3	1.9	1.3

shows the impact that ndf and ddf have on the critical *F*-values. Critical *F*-values with only 1,2 df are very large. However, critical *F*-values decrease exponentially as the ddf increase becoming relatively stable and reliable at five ddf for $\alpha = 0.05$ and ten ddf for $\alpha = 0.01$. When interpreting tests of significance, it is important to recognize that a MS estimated with few df tends to be inaccurate and imprecise, resulting in large critical *F*-values and *F*-tests lacking repeatability and power.

It is especially important to recognize that some tests of effects within experiments with multiple error terms, such as split plot experiments and experiments combined over a limited number of environments or years, often have few ddf and a higher risk of a Type 2 error. There are related concerns about the precision of estimates of variance components of random effects, and it is recommended to analyze random effects having less than 10 levels as fixed effects (Piepho et al., 2003, Yang, 2010) [See Chapter 7 (Vargas et. al, 2018) for a detailed discussion on levels of random effects].

On the other hand, *F*-values based on sufficient df are robust and minor deviations from normality or homogeneity of variance have a negligible impact on *F*-distributions and do not invalidate *F*-tests (Acutis et al., 2012). For experiments with heterogeneous error variances, mixed model approaches based on maximum likelihood estimates that can detect and fit different covariance structures are recommended.

The influence of the ndf and ddf on the shape and spread of the *F*-distribution is illustrated in Fig. 1, where the central and non-central *F*-distributions are shown for a small (ndf = 1, ddf = 4), a medium (ndf = 5, ddf = 20), and a large (ndf = 20, ddf = 100) experiments. For the small experiment (ddf = 4), the *F*-distribution is highly skewed towards 0 and displays extreme kurtosis. As the df increase as seen for the medium (ddf = 20) and large (ddf = 100) experiments, the central *F*-distribution becomes less dispersed with a higher density of *F*-values near the true *F*=1. Similarly, the non-central *F*-distribution also has a higher density and is less dispersed as the df increase. When both the central and non-central distribution curves are less dispersed, their overlap is also less. This translates into a more powerful *F*-test. The overlap between the *F*-distrubtions of the null and alternate hypotheses also depends on the effect size. In general, as the effect size of the alternate hypothesis increases, the non-central *F*-distribution shifts to the right, becomes more dispersed, and the overlap between the null and alternate *F*-distributions decreases (not shown). The relevance of the overlap between the central and non-central *F*-distributions to hypothesis testing is discussed in Chapter 1 (Garland-Campbell, 2018) and later in this chapter.

Although not common practice, *F*-values can be used to assess and compare the relative magnitude of fixed effects (McIntosh, 2015). If an effect is null, the variance component of the effect is 0 and the numerator and denominator estimate the same expected mean square value and the theoretical *F* ("true" *F*) is 1. Therefore, (*F* minus 1, *F*-1) estimates the magnitude of the variance due solely to the effect variance relative to its error variance. Since (*F*-1) increases as the effect size and variance component increases, the *F*-value calculated from MS with adequate df should increase as the size of the "true" effect increases. Thus, a ratio of *F*-values can be used as a simple and quick tool to compare the magnitudes of effects and to indicate their relative importance. As an informal and exploratory statistic, this "ratio of *F*-ratios" provides a rudimentary quantitative assessment of the relative magnitudes of effects to augment the qualitative tests of effect significance. The "ratio of *F*-ratios" for comparing a main effect to an interaction, calculated as *F*(main effect)/*F*(interaction), can help the researcher to decide whether to conduct post-planned comparisons between main effect or interaction means. This convenient ratio does not have the drawbacks noted by Saville (2015) from relying on the significance of the interaction to determine whether to conduct an LSD to compare means.

P-value

A *p*-value is the probability of a test statistic (*F*, *t*, or χ^2) equal to or greater than the calculated value, given that the null hypothesis is true. For ANOVA, *p*-values for the calculated *F*-statistics are determined using the appropriate central *F*-distribution. An appropriately small *p*-value is chosen to guard against incorrectly rejecting a null hypothesis and to ensure scientific credibility when claiming differences between means. During the early development of ANOVA, Fisher recommended using $p \leq 0.05$ as a reasonable and convenient *p*-value to determine statistical significance, which has since become a rote practice. Also by common convention, *p*-values of 0.05, 0.01, and 0.001 are designated as

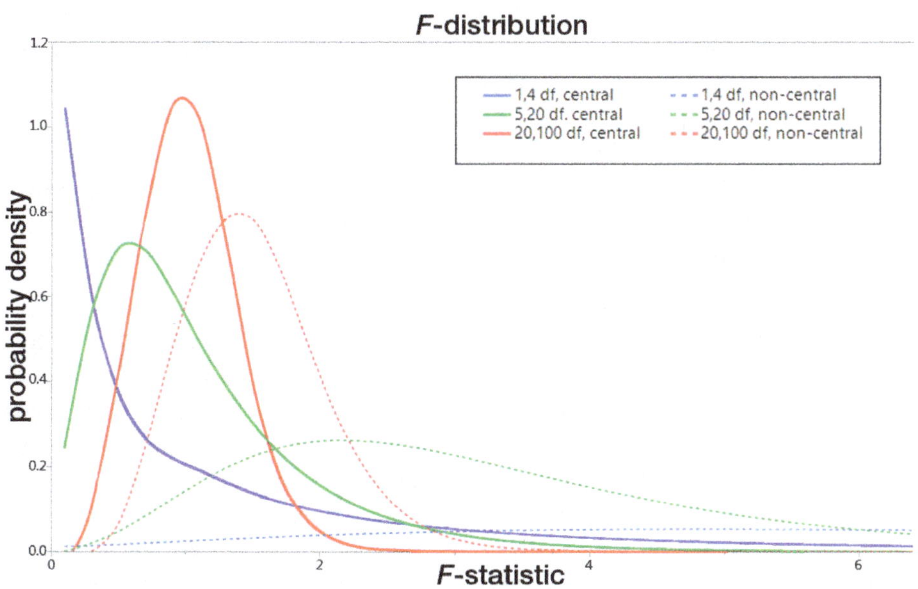

FIG. 1. Central (solid lines) and non-central (dashed lines) *F*-distributions (λ=10) for selected numerator and denominator degrees of freedom.

significant (*), very significant (**), and highly significant (***), respectively. An alternate approach to significance testing, championed by Neyman, considers two types of statistical errors (Neyman and Tokarska, 1936). Incorrectly rejecting a null hypothesis and falsely declaring a significant difference is a Type 1 error. A second type of error, a Type 2 error occurs when a false null hypothesis is accepted and the effect is not found to be significant. The test of significance uses a *p*-value from the central *F*-distribution to place a fixed limit on the Type 1 error rate (α). Whereas the Type 2 error rate (β) is based on the cumulative probability from the non-central *F*-distribution of the alternate hypothesis, which is the area to the left of the critical *F*-value. Thus, α and β are inversely related and reducing the Type 1 error rate (decreasing the *p*-value used to determine significance) will increase the Type 2 error rate. It is important that researchers choose a Type 1 error rate (or significance level) that also balances the relative risks of Type 2 errors. However, since β is unknown and changes with sample size, error variance, and effect size, the *de facto* significance level for α is the *p*-value of 0.05, not coincidentally the same as suggested by Fisher (Lehmann, 1993). In Chapter 1, Garland-Campbell (2018) provides a thorough discussion on this topic.

Fisher's *p*-values and Neyman's α-levels represent rival statistical philosophies regarding significance testing. Fisher's focus was on scientific inquiry and using significance tests to help the researcher draw conclusions to understand and learn from the experimental data. In contrast, Neyman focused on making correct decisions about rejecting or accepting the null hypothesis in relation to the relative seriousness of Type 1 and Type 2 errors. These two conceptual views of significance, similar yet different, are commonly conflated and used interchangeably. The consequence has contributed to misconceptions and misapplications of *p*-values (Nuzzo, 2014).

Regardless of their shortcomings, *p*-values are ubiquitous throughout the scientific literature. They are used as the universal statistic to convey confidence in the conclusions about the experimental results. Scientific journals have grown to rely on *p*-values as the deciding factor that separates scientific evidence from anecdote. Meanwhile, statisticians have become increasingly concerned about the impact that frequent misuse and misinterpretation of *p*-values have on scientific integrity and progress. The longstanding debate over the proper role of *p*-values has become increasingly heated, prompting the American Statistical Association to issue a policy statement with supplemental commentaries to address major concerns associated with the uses of *p*-values (Wasserstein and Lazar, 2016).

The following is a list of fundamental characteristics of *p*-values that are often overlooked.

1. A *p*-value is an inferential statistic that estimates the parameter *P* (the true probability). Just like treatment means are calculated from samples of populations to estimate the population mean, *p*-values are estimated using the distribution of the samples that represent the population distribution. In fact, *p*-values are subject to variation that can be surprisingly and disappointingly large. Based on simulations of typical data situations, standard errors of mean *p*-values between 0.00001 and 0.10 typically range from 10–50% of the mean and only the magnitude of a *p*-value is well-determined (Boos and Stefanski, 2011). Thus, using a strict cutoff such as 0.05 as a dividing line for significance is problematic since a *p*-value is not necessarily replicable (Amrhein et al., 2017).

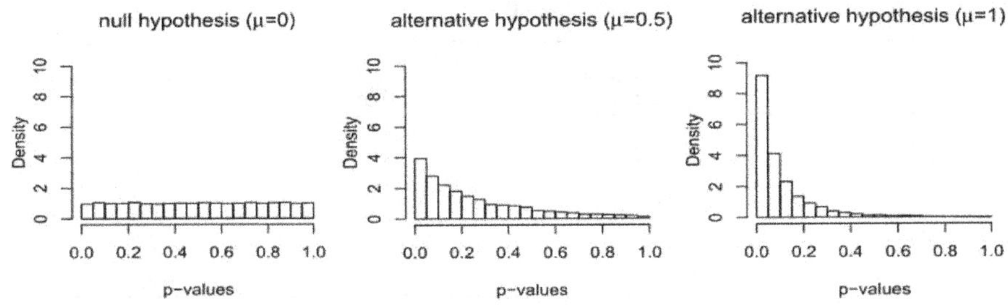

FIG. 2. Distributions of *p*-values for null hypothesis (μ=0) and alternate hypotheses (μ=0.05 and μ=0.1) based on 10,000 simulated *t*-tests (Murdoch, et al., 2008).

2. *P*-values are random variables and the distribution of *p*-values for the null and alternate hypotheses depends on the true values of the treatment means (μ). This is illustrated (Fig. 2) with histograms of simulated *p*-values generated from *t* tests where the true means are equal (μ = 0), differ by half the SD (μ = 0.5), and differ by one SD (μ = 1) (Murdoch et al., 2008). If the null hypothesis is true (treatment means are equal), the distribution of *p*-values is flat and evenly distributed from 0 to 1. In contrast, if the alternate hypothesis is true (the treatment means are not equal), the distribution of *p*-values is skewed toward 0. As the difference between the means increases, the *p*-values cluster nearer to 0, showing that the power of the *t* test increases as the difference between means increases.

3. Small *p*-values, those *p*-values that confer significance are the least reliable. *F*-distributions when the null hypothesis is true are highly skewed resulting in *p*-values being quite insensitive at the tail end of the *F*-distribution. This is illustrated in the nonlinear exponential relationship between *F* and *p*-values shown in Fig. 3 for *F*-values with 1,20 df. This is also demonstrated by comparing the differences between the critical *F*-values at different *p*-values in Table 4. For example, the difference between the critical *F*-values at the 0.10 and 0.05 significance levels is small (1.38) compared with the large difference (10.87) between the critical *F*-values for significance at the 0.0010 and 0.0001 significance levels.

4. The *p*-value for a given *F*-value decreases as sample size increases (Fig. 4). If sample size is small, large differences may not be statistically significant producing a Type 2 error. Conversely, if the sample size is very large, even trivial effects can be statistically significant, which can be misinterpreted to infer biological significance.

5. A small *p*-value is used as a criteria to reject (disprove) a null hypothesis. If a *p*-value $\leq$ 0.05

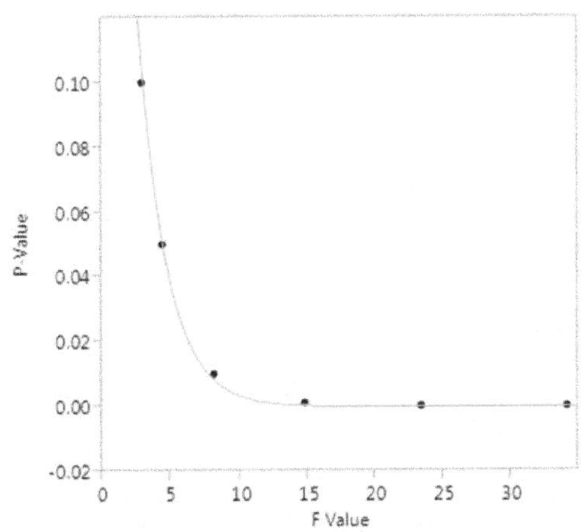

FIG. 3. *P*-value vs. *F*-value for 1,20 degrees of freedom.

TABLE 4. Critical *F*-values at *p*-values ranging from 0.1 to 0.0001.

P-value	Critical *F* value		
	1,2 df	1,20 df	24,96 df
0.1	8.5	3.0	1.5
0.05	18.5	4.3	1.6
0.01	98.5	8.1	2.0
0.001	998.5	14.8	2.5
0.0001	9998.5	23.4	2.9

is chosen as the criteria to reject the null hypothesis and the p-value > 0.05, the null hypothesis is accepted (not rejected). However, a p-value > 0.05 is too often misinterpreted as the probability that the null hypothesis is true or that the p-value is the probability that the alternative hypothesis is true.

6. *P*-values are quantitative statistics often transformed into a binary measure of significance. Although this has been defended as a safeguard against bias and subjectivity, it can create a cascade of bad decisions because statistical significance often determines if a study is publishable and sets a path for future research (Mervis, 2014). In fact, authors often do not write up nonsignificant findings creating a publication bias that increases the probability of experimentwise Type 1 errors and inflates effect sizes (Franco et al., 2014). According to the American Statistical Association, "The widespread use of "statistical significance" (generally interpreted as "$p \leq 0.05$") as a license for making a claim of a scientific finding (or implied truth) leads to considerable distortion of the scientific process" (Wasserstein and Lazar, 2016).

7. *P*-values are essentially irreproducible (Amrhein et al., 2017). There are countless possible experiments that could be designed to test a hypothesis, understand a phenomenon, or determine and predict effects of treatments. Therefore, there are numerous ways to statistically analyze experimental data. The best experimental design and analysis is always an unknown and can only be proposed based on the existing evidence, experimental objectives, and practical constraints. With so many possible permutations, there is no universal solution. Instead, we must still rely on researchers and statisticians

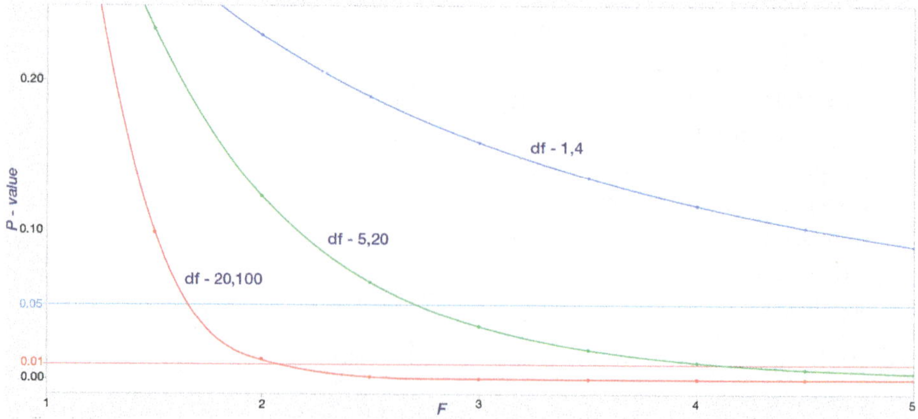

FIG. 4. *P*-value vs. *F* for varying numerator and denominator degrees of freedom.

to use informed judgement when designing and conducting an experiment and recognize that scientific advancement is usually the product of the accumulation of experimental evidence rather than the result of a single experiment.

8. *P*-values are sometimes used inappropriately to conduct meta-analyses across experiments. *P*-values should not be used to compare treatment effects across experiments because using significance levels rather than direct comparisons leads to flawed conclusions caused by ignoring the effect size and the power of the test. (Gelman and Stern, 2006).

Contrasts and Multiple Comparison Procedures

In addition to using ANOVA to test the significance of effects in the linear model, ANOVA can also test more in-depth and narrowly-focused hypotheses of interest. Contrasts and multiple comparison procedures can perform additional tests of significance using an error term from the ANOVA. Contrasts are constructed to test specific differences between and among means, groups of means, or polynomial relationships among means. Contrasts are usually planned comparisons that are conceived as part of the experimental planning process to investigate treatment effects at a fine level. As an alternative to contrasts, multiple comparison procedures are used to test for significance of differences between multiple pairs of treatments means. These procedures are most suitable for qualitative and unstructured treatments (cultivars, chemical formulations, soil types, etc.) and used to determine the best treatment(s) and rank treatments into clusters. Saville (2018) in Chapter 5 provides a critique of multiple comparison procedures. He recommends that if a multiple comparison procedure is justified, the best choice is an unrestricted LSD, which is equivalent to conducting *t*-tests for all possible treatment pairs.

Case Study: The Story of Statbean: From Discovery to Field Testing

Introduction

This case study provides the context for an example for readers to practice conducting and interpreting an ANOVA (the data, the SAS, the R code for the analyses of all dependent variables, and SAS outputs for the example are in the supplemental documentation). Hopefully, readers will share some of Dr. Y's enthusiasm for research. And by working with her data, you will appreciate how ANOVA can help organize and summarize the observations into useful information. The purpose of the example is not to provide a recipe for ANOVA but to illustrate a thought process and rationale that drives the ANOVA based on both contemporary and classic ANOVA theory. The example is intentionally framed to be of general interest and void of statistical complications. The analysis and results presented are also meant to serve as a platform for discussion. Readers who perform different analyses with the sample data can compare their own results to those from the analysis of the example. The purpose of the example is not how, nor what, but why.

Dr. Y was beginning her career as an agronomist in search of developing alternative crops to improve sustainability through genetic, economic, and commodity diversification. When she read that the indigenous people of Lala Land used a native herb to enhance performance for their traditional equation-solving competition, she was intrigued. Did this plant really have bioactive properties that improved brain function? Could this species be cultivated as a medicinal plant? If so, can this species become a new crop and open new opportunities for farmers and gardeners to grow and market as a herbal supplement? Her curiosity about this plant was intense as she envisioned herself leading multidisciplinary research needed to develop a new and valuable alternative crop.

Y was awarded a "seed" grant from WOMEN (World Organization of Mathematicians Exploring Nutraceuticals) to go to Lala Land to investigate and obtain samples of this promising medicinal herb. She traveled to Lala Land where leaders from the local population taught her how to grow plants to be formulated into a tonic. They gave her seed and tonic from this plant species, she named statbean (*Plantus statisticus).* In exchange, she taught them how to design and statistically analyze field trials using ANOVA and promised them a share of the future profits from the production or germplasm development of statbean.

Upon returning home to Reality, Y conducted efficacy trials with volunteers from college statistics classes and found that the tonic significantly increased student ability to solve equations for up to three hours. Eureka! Using data demonstrating the benefit of statbean tonic, Y received funds from the State of Reality Experiment Station to investigate the potential to cultivate statbean in Reality.

Research Objectives

A field study was conducted to determine the effects of soil calcium and mulch on the establishment of statbean in the State of Reality. This project consisted of an experiment that was replicated at three locations.

The objectives and challenges of these three statbean field experiments were similar to those encountered by the early agronomists conducting yield trials on crops to advise farmers on the effects of fertilizers and manures on crop yields. When Y conducted her experiments, she was able to benefit from a century of advancements in statistics to design and analyze her research to be confident that her results were repeatable, reliable, and valid.

Experimental Description and Design– Randomization and Replication of Treatments

Y established statbean research plots at three locations, the Western, Central, and Eastern Reality Research and Education Centers. These three research centers were chosen to represent the growing conditions of eastern, central, and western regions of Reality. At each location, 10 treatments were replicated in a RCBD with three blocks. The treatment design was a 5 × 2 factorial consisting of five Ca treatments (Ca_Trt: control (0), lime 1X (L1X), lime 2X (L2X), gypsum 1X (G1X), gypsum 2X (G2X) and two mulch treatments (with and without mulch). The lime ($CaCO_3$) and gypsum ($CaSO_4$) treatments both added Ca to the soil at two equivalent rates (1X and 2X). Lime also increases soil pH. Thus, the effect of lime on plant establishment confounds the effects of Ca with pH. Because gypsum does not increase soil pH, the gypsum treatments were used to separate soil Ca and soil pH effects on plant establishment. The location and treatment factors and interactions were considered fixed. A summary of the experiment is shown in Table 5.

The field plot plan for the 2 × 5 factorial randomized in three blocks was generated for the three locations using PROC PLAN of SAS (Table 6). The Ca and mulch treatments (labeled 1–10) were applied to the field plots (experimental units), which were then seeded with 100 statbeans per plot.

Sampling Description and Design–Measuring Dependent Variables

The primary objective was to investigate the effects of selected soil treatments on statbean production in three locations. Plant establishment (Ptotal) was the dependent variable used as an indirect measure of production. The Ptotal was calculated as the number of plants per 100 seeds sown per plot.

Soil pH and Ca concentration were also measured as independent variables. The lime Ca treatments were used to increase both soil pH and soil Ca, while the gypsum Ca treatments were used to increase soil Ca without changing soil pH. Thus, the ANOVAs of soil pH and soil Ca were conducted to verify and quantify the direct Ca treatment effects and interactions on the soil. Composite soil samples of 6 cores per plot were analyzed for pH and Ca concentration.

Preparing, Correcting, and Knowing the Data

Before using the ANOVA results, the data were scrutinized to verify that the correct data and model were being analyzed. "Garbage in, garbage out" is a familiar warning that you can trust the computer program to perform the math but the result will be garbage if the data or programming is incorrect. Regardless of what tools or whose assistance is used to perform the ANOVA, it is the researcher who is responsible for the integrity of the results. It is also a good idea to spend time to learn about and from your data before conducting an ANOVA. To do this, simple descriptive statistics and diagnostic plots can be useful. By assessing the data, the researcher can identify and resolve data issues and preview the means to be compared.

Descriptive Summary of the Data

Summary tables, plots, and histograms of the soil pH measurements were used to learn the pH range and distribution, discover patterns in the data, and even guestimate whether means were significantly different. The summary table (Table 7) shows that there were no missing (N) or out-of-range values (Min, Max). The pH means were

TABLE 5. Summary of the experiment at each location.

Linear Model - Y = Blk + Ca_Trt + Mulch + Ca_Trt × Mulch + Error
Experimental design – RCBD, 3 blocks
Treatment Design — 2 × 5 factorial, 10 treatments
Factors –Calcium Treatments (Ca_Trt) and Mulch Treatments (Mulch)
Ca_Trt levels – Control, 1X Lime, 2X Lime, 1X Gypsum, 2X Gypsum
Mulch levels – no mulch, mulch
Experimental Unit – 3 m × 3 m field plot planted with 100 seed
Dependent variables – Soil pH, Soil Ca, plant establishment (PTotal)
PTotal Sampling Unit - plant count/100
Soil pH and soil Ca Sampling Unit — composite of 6 soil samples/plot

TABLE 6. SAS code and output of plot plan for a 2×5 factorial RCBD randomized at three locations.†

```
Proc plan seed=101420171;
Factors Loc=3 ordered Blk=3 ordered trt_no=10;
Run;
```

† Seed number used to allow plot plan to be duplicated.

The PLAN Procedure

Factor	Select	Levels	Order
Loc	3	3	Ordered
Blk	3	3	Ordered
trt_no	10	10	Random

Loc	Blk	trt_no									
1	1	5	1	6	10	2	8	9	7	4	3
	2	6	9	10	5	1	2	8	7	4	3
	3	7	10	1	3	9	5	6	2	4	8
2	1	10	7	1	6	4	3	8	5	2	9
	2	9	7	2	3	8	5	4	1	6	10
	3	3	10	2	6	5	9	8	4	1	7
3	1	7	6	2	3	4	8	1	9	5	10
	2	6	4	1	7	10	2	8	9	5	3
	3	8	10	7	5	9	6	2	4	1	3

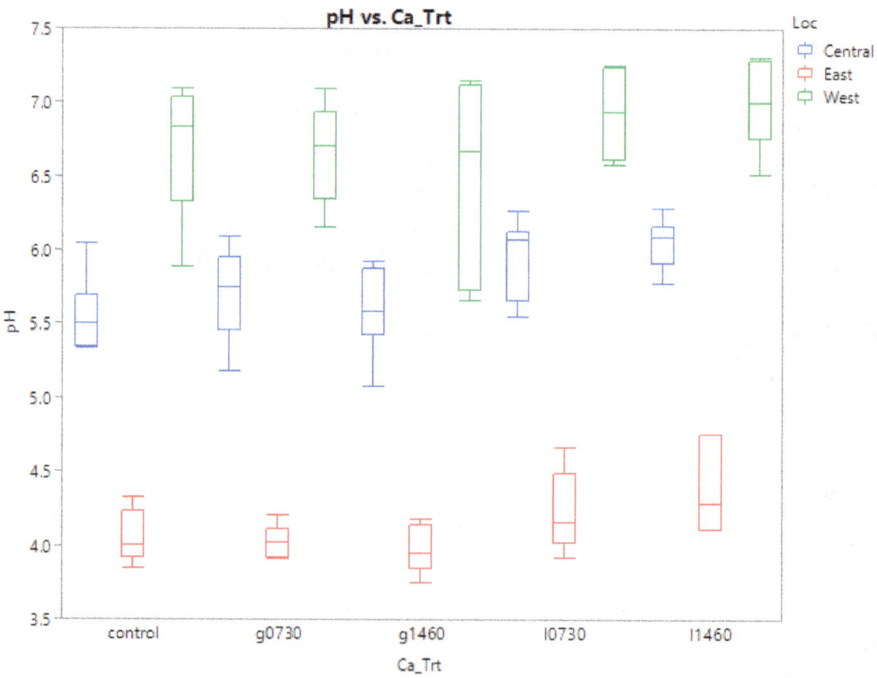

FIG. 5. Box plots of pH at Central, East, and West locations using JMP 11.1 Pro.

slightly higher for the lime than the control or gypsum treatments, while the mulch did not appear to affect soil pH. However, the standard deviations in the summary table were based on too few observations (N = 3) to be used to determine if the differences between means were due to the treatments or mere random variation. As previously noted, an estimate of error variance based on fewer than five replications is not reliable. Box plots provide a visual summary of the pH treatment means and distribution at each location (Fig. 5). The differences among pH treatment means within locations appear small and their distributions overlap, whereas the pH means show large differences between locations and the pH distributions do not overlap. Just as a picture is worth a thousand words, box plots offer an intuitive understanding of the pH data. In contrast, summary tables provide values and require more time to assess but are more precise and can be used for rigorous statistical analysis.

(Note to readers: In this chapter, JMP 11 was used to explore the data for errors and outliers. To see an example of data exploration using R software packages, refer to Chapter 14 on Multivariate Methods by Yeater and Villamil, 2018).

ANOVA by Location

The ANOVA was conducted first for each location as a separate experiment and then as one experiment combined over locations. The ANOVA by location is used to make inferences about the mulch and Ca treatment effects and interactions. A separate ANOVA at each location avoids issues related to heterogeneity of error variances across locations that occur with using a pooled residual error averaged over locations. However, the ANOVA combined over locations has an expanded model that includes location effects and interactions. For our example, the analyses of soil pH at the Central location and combined over locations are shown using the linear models

TABLE 7. Summary table of soil pH data at three locations.

		Location															
		Central					East					West					
Mulch	Ca_Trt	N	Mean	Std Dev	Min	Max	N	Mean	Std Dev	Min	Max	N	Mean	Std Dev	Min	Max
no	control	3	5.4	0.1	5.3	5.6	3	4.0	0.2	3.9	4.2	3	6.6	0.6	5.9	7.1
	G1X	3	5.5	0.4	5.2	5.9	3	4.0	0.1	3.9	4.1	3	6.6	0.4	6.2	6.9
	G2X	3	5.7	0.2	5.6	5.9	3	4.0	0.2	3.9	4.2	3	6.5	0.8	5.7	7.1
	L1X	3	6.0	0.4	5.6	6.3	3	4.2	0.3	3.9	4.4	3	6.9	0.3	6.6	7.2
	L2X	3	6.1	0.1	6.0	6.1	3	4.4	0.3	4.1	4.8	3	6.9	0.4	6.5	7.3
yes	control	3	5.7	0.3	5.5	6.1	3	4.1	0.2	4.0	4.3	3	6.8	0.3	6.5	7.0
	G1X	3	5.9	0.2	5.7	6.1	3	4.0	0.2	3.9	4.2	3	6.7	0.3	6.4	7.1
	G2X	3	5.5	0.4	5.1	5.9	3	3.9	0.2	3.8	4.1	3	6.5	0.7	5.8	7.2
	L1X	3	6.0	0.2	5.7	6.1	3	4.3	0.3	4.1	4.7	3	6.9	0.3	6.6	7.3
	L2X	3	6.1	0.3	5.8	6.3	3	4.3	0.4	4.1	4.8	3	7.1	0.2	6.8	7.3

TABLE 8. Linear model effects for one location and combined over locations.

	One Location	Combined Locations
Linear Model	pH = Blk+ Ca_Trt + Mulch + Ca_Trt×Mulch + Error	pH = Location + Blk(Location) + Ca_Trt + Mulch + Ca_Trt ×Mulch + Location×Ca_Trt + Location×Mulch + Location×Ca_Trt ×Mulch + Error
Fixed Effects	Ca_Trt, Mulch, Ca_Trt ×Mulch	Location, Ca_Trt, Mulch, Ca_Trt ×Mulch, Location× Ca_Trt, Location×Mulch, Location× Ca_Trt ×Mulch
Random Effects	Blk, Error	Blk(Location), Error

TABLE 9. SAS Code – PROC MIXED for pH at each location.

```
data anova.statbean; set anova.statbean;
proc sort; by loc;
run;
Title 'Statbean Data';
proc print;
run;
Title 'Mixed pH ANOVA by location';
proc mixed data=anova.statbean plots=residualpanel method=type3; by loc;
class Blk Mulch Ca_Trt;
model pH=Mulch Ca_Trt Mulch*Ca_Trt;
random Blk;
lsmeans Mulch Ca_Trt Mulch*Ca_Trt;
run;
```

in Table 8. All effects in both models except for Blk(Loc) were fixed and inferences were limited to the locations and treatments included in the experiments.

ANOVA by Location SAS code for PROC MIXED

The PROC MIXED statements to perform an ANOVA to test the significance of the effects of mulch, Ca treatments (Ca_Trt), and Ca_Trt × Mulch for soil pH at each location (Table 9) are: i) PROC statement to invoke the Mixed procedure with options a) plots = residualpanel to request plots of residuals panels, and b) method = type3 to print a comprehensive ANOVA table with EMS; ii) BY statement to request a separate analysis for each location; iii) CLASS statement to identify the classification (qualitative) model effects; iv) MODEL statement to define the fixed effects to be tested for significance for one dependent variable; v) RANDOM statement to identify the random model effects other than the residual (random error) effect; and vi) LSMEANS statement to request least squares means and their standard erorrs.

Residual Plots

The pH data have been checked and found to be free of typing errors and outliers. We also need to determine whether the classical ANOVA assumptions are justified to choose the most appropriate ANOVA analyses. These assumptions can be assessed using statistical tests and/or visual interpretation of graphs. Although statistical tests may seem to be more objective and definitive, plots of residual values are more powerful and useful for identifying the cause of an assumption violation (Kozak and Piepho, 2017, Loy et al., 2016). As part of the PROC MIXED analysis, diagnostic plots of the residuals (observed-predicted) were requested to check for normality, homogeneity, and independence of error. Diagnostic plots of the residual values for pH at the Central location requested using the "plot = residualpanel" option are shown in Fig. 6.

The upper left residual panel is a scatterplot of residual vs. predicted values with a reference line for residual values at zero. If the residual values are not randomly scattered across the range of predicted values, this can indicate outliers or a violation of the classical assumption that the residuals are independent and the variation is homogeneous. If one or a few residuals are unusually distant from the reference line and most points are tightly clustered near the reference line, the distant points should be investigated to determine if they are outliers and need to be removed from the data. If the residuals form a pattern, often a cone shape, the means and variances are correlated, indicating the variances may not be independent or homogeneous. Another reason to further investigate the data prior to analysis.

The upper right panel is a histogram of the residuals with a line referencing a normal distribution. The residuals show a nearly normal distribution. The minor deviations from the curve are expected based on the small sample size ($n = 30$) for the analysis of a single location.

The lower left panel is a normal probability or Q-Q plot, a widely used and powerful tool for checking for normality. The residuals are plotted against the quantiles of the normal distribution with a reference line for the normal distribution. Deviations from the line indicate deviations from the normal distribution. The Q-Q plot also shows that the distribution is approximately normal. The very minor deviations from normality at the tails of the distribution would have at most a nominal effect on the ANOVA *F*-tests.

The lower right panel reports the simple statistics that help spot problems in the data as follows: i) number of observations- missing values, ii) minimum and maximum residuals- outliers, iii) standard deviation–compare error variances between analyses. The fit statistics are used to compare models and identify covariance structures, which we are assuming to be unstructured. In this example, the statistics do not reveal any outliers and the classical ANOVA assumptions appear justified.

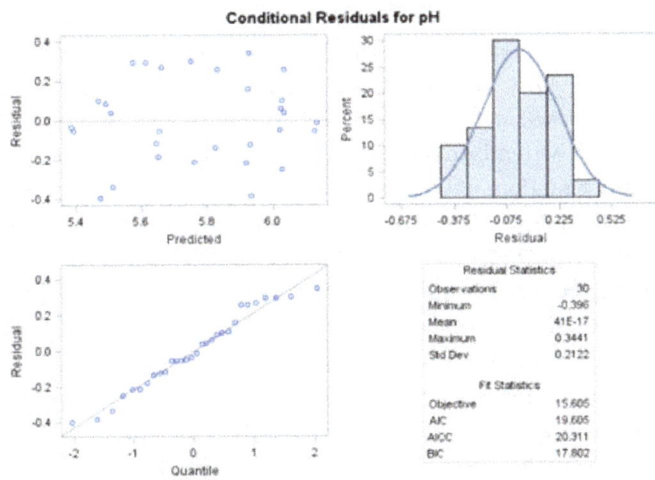

FIG. 6. Residual panels for pH at Central location.

TABLE 10. PROC MIXED output for pH at the Central location using Type III estimation method.

Model Information	
Data Set	ANOVA.STATBEAN
Dependent Variable	pH
Covariance Structure	Variance Components
Estimation Method	Type 3
Residual Variance Method	Factor
Fixed Effects SE Method	Model-Based
Degrees of Freedom Method	Containment

Class Level Information		
Class	Levels	Values
Blk	3	1 2 3
Mulch	2	0 1
Ca_Trt	5	0 G1X G2X L1X L2X

Covariance Parameter Estimates	
Cov Parm	Estimate
Blk	0.006969
Residual	0.06877

Type 3 Tests of Fixed Effects				
Effect	Num DF	Den DF	F Value	Pr > F
Mulch	1	18	0.65	0.4299
Ca_Trt	4	18	4.35	0.0123
Mulch*Ca_Trt	4	18	0.94	0.4618

Type 3 Analysis of Variance								
Source	DF	Sum of Squares	Mean Square	Expected Mean Square	Error Term	Error DF	F Value	Pr > F
Mulch	1	0.044853	0.044853	Var(Residual) + Q(Mulch,Mulch*Ca_Trt)	MS(Residual)	18	0.65	0.4299
Ca_Trt	4	1.196980	0.299245	Var(Residual) + Q(Ca_Trt,Mulch*Ca_Trt)	MS(Residual)	18	4.35	0.0123
Mulch*Ca_Trt	4	0.259447	0.064862	Var(Residual) + Q(Mulch*Ca_Trt)	MS(Residual)	18	0.94	0.4618
Blk	2	0.276927	0.138463	Var(Residual) + 10 Var(Blk)	MS(Residual)	18	2.01	0.1625
Residual	18	1.237940	0.068774	Var(Residual)				

Results

The PROC MIXED output using the Method=Type3 option for soil pH at the Central location is shown in Table 10. The Model Information provides important details about the statistical methods used for the mixed model analysis. The Class Levels are useful for checking the class level values and their order sorted. For example, the standard Type III least squares method (Method = Type3 option), rather than the default REML method, was used to estimate the variances of random effects in order to obtain a comprehensive ANOVA table. The example data are balanced, so the Type III and REML estimates of variance components are the same. Also, note in the Model Information that the Covariance Structure used is Variance Components. Thus, the Covariance Parameter Estimates in the PROC MIXED output are variance component estimates of VAR(Blk) and VAR(Residual). These variance components comprise the EMS for the Blk and Residual SOVs. The Blk MS (0.138463) estimates Var(Residual) +10 Var(Blk) = (0.06877+ 10(0.0006969) and the Residual MS estimates VAR(Residual) = 0.06877.

It is common with mixed model ANOVAs that tests of significance are important for fixed but not random effects, especially random design effects such as blocking factors. Thus, the PROC MIXED default output does not include a standard ANOVA table. Instead, a table of *F*-tests of fixed effects is given showing the ndf and ddf for the *F*-test, the calculated *F*-value, and probability of a greater *F*-value ($P > F$). The comprehensive ANOVA table presents the same results for tests of fixed effects along with additional ANOVA components that provide an overview of the experimental design and enough information to assess the appropriateness and power of the analysis.

Whether testing hypotheses by choosing a Type 1 error rate ($\alpha = 0.05$) as a significance level or using the $P > F$ (*p*-value ≤ 0.05) to reject the null hypothesis, the Ca_Trt main effect was significant for soil pH but the Mulch main effect and the Mulch × Ca_Trt interaction were not significant. Thus, we infer that the Ca treatments but not the mulch treatments significantly affected the soil pH and that the Ca_Trt effect was the same for the mulched and non-mulched plots.

Planned and Multiple Pairwise Comparisons

The ANOVA tests of the significance of the linear model effects did not include tests of all hypotheses of interest for the case study or test for significant differences between individual means. Regardless of whether the Ca_Trt effect was signficant, planned comparisons (contrasts) can be used to partition the Ca_Trt main effect and the Ca_Trt × Mulch interaction to delve deeper into the effects and interactions of lime and gypsum treatments. Remember that the Ca treatments added Ca to the soil in different forms (lime or gypsum) and that both lime and gypsum were expected to increase soil Ca concentration. However, the lime but not the gypsum was also expected to increase soil pH. The test of the Ca_Trt effect confirmed that the Ca treatments significantly affected soil Ca concentration (ANOVA in the online supplement) and soil pH (Table 10) at the Central location. While contrasts were used to test whether the effects of lime and gypsum were significantly different; determine if the response to rates of lime or gypsum is linear or quadratic; and test for their significant interactions with mulch. The SAS code and output for these contrasts are given in Table 11 for soil Ca and soil pH at the Central location.

These contrasts help us interpret the variation in soil Ca and soil pH means in the context of the error variation. The contrasts confirmed that: lime significantly linearly increased both soil Ca and soil pH; gypsum did not significantly affect soil pH or soil Ca; and the effects of lime and gypsum on soil pH, but not soil Ca, were significantly different. Even though gypsum adds Ca to the soil, the increase in soil Ca concentration was not significant. This is probably due to a Type 2 error because of a lack of power of the test.

In addition to understanding the basic soil responses to the lime and gypsum treatments, the treatment means of the soil pH and Ca concentration are themselves important. PROC MIXED automatically gives the least squares means and standard errors for fixed effects. However, since only the Ca_Trt main effect was significant, the Ca_Trt means and LSD(0.05) are shown in a bar graph (Fig. 7). The LSD bar, centered around each mean serves as a confidence interval. Thus, if the LSD bars of two means do not overlap, the means are significantly different at the 0.05 level.

For the untreated plots at the Central location, the mean soil pH was 5.6 and mean soil Ca concentration was 622 mg kg^{-1}. It can be seen from the graphs that the pH of the lime treatments were significantly higher than the control but the difference between the 1X and 2X rates was not significant. As expected, the mean pH's of the 1X and the 2X gypsum treatments did not differ significantly from the control or from each other. Similar to the soil pH, the mean soil Ca concentrations of the lime treatments were significantly higher than the control but not from each other, while the differences due to gypsum treatments were not significant.

Although the LSD procedure is often used to assign letters to means to indicate significantly differences between means, the LSD bars can be more informative because they offer a quantitative measure of precision associated with the differences between means. For example, the soil Ca graph shows an LSD value of 400 mg kg^{-1} Ca,

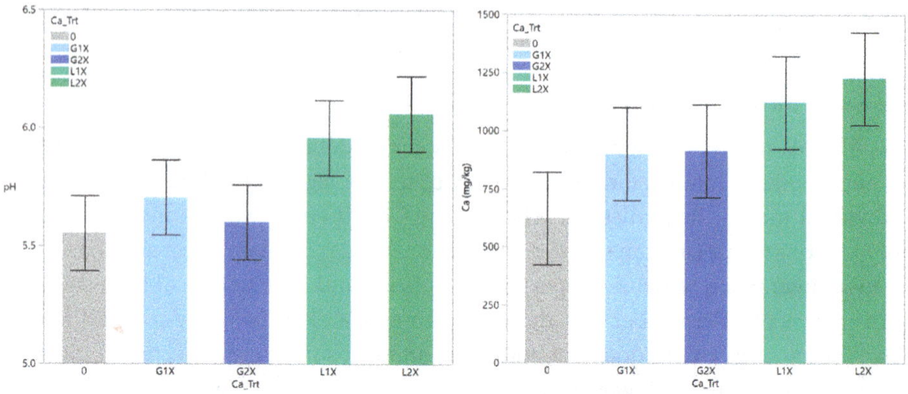

FIG. 7. Bar graph of Ca_Trt means and LSD(0.05) for soil pH and soil Ca concentrations at the Central Location. The LSD bar is centered around each mean and the means are significantly different if their LSD bars do not overlap. The LSD(0.05) for pH = 0.32; LSD(0.05) for Ca concentration = 400 mg kg^{-1}.

TABLE 11. SAS code and output for contrasts testing the effects of lime and gypsum Ca treatments (Ca_Trt) on soil pH and soil Ca at the Central location.

```
contrast "Main Effect-Gypsum Linear" Ca_Trt -1 0 1 0 0/E;
contrast "Main Effect-Gypsum Quadratic" Ca_Trt 1 -2 1 0 0/E;
contrast "Main Effect-Lime Linear" Ca_Trt -1 0 0 0 1/E;
contrast "Main Effect-Lime Quadratic" Ca_Trt 1 0 0 -2 1/E;
contrast "Main Effect-Lime vs Gypsum" Ca_Trt 0 1 1 -1 -1/E;
contrast "Interaction-Gypsum Linear*Mulch" Ca_Trt*Mulch -1 0 1 0 0 1 0 -1 0 0/E;
contrast "Interaction-Gypsum Quadratic* Mulch" Ca_Trt*Mulch 1 -2 1 0 0 -1 2 -1 0 0/E;
contrast "Interaction-Lime Linear*Mulch" Ca_Trt*Mulch -1 0 0 0 1 1 0 0 0 -1/E;
contrast "Interaction-Lime Quadratic*Mulch" Ca_Trt*Mulch 1 0 0 -2 1 -1 0 0 2 -1/E;
contrast "Interaction-Lime vs Gypsum*Mulch" Ca_Trt*Mulch 0 1 1 -1 -1 0 -1 -1 1 1/E;
```

pH

Contrasts				
Label	Num DF	Den DF	F Value	Pr > F
Main Effect-Gypsum Linear	1	18	0.09	0.7615
Main Effect-Gypsum Quadratic	1	18	0.96	0.3407
Main Effect-Lime Linear	1	18	11.20	0.0036
Main Effect-Lime Quadratic	1	18	1.34	0.2625
Main Effect-Lime vs Gypsum	1	18	11.10	0.0037
Interaction-Gypsum Linear*Mulch	1	18	2.18	0.1575
Interaction-Gypsum Quadratic* Mulch	1	18	1.14	0.2998
Interaction-Lime Linear*Mulch	1	18	0.70	0.4138
Interaction-Lime Quadratic*Mulch	1	18	0.30	0.5914
Interaction-Lime vs Gypsum*Mulch	1	18	0.10	0.7591

Ca

Contrasts				
Label	Num DF	Den DF	F Value	Pr > F
Main Effect-Gypsum Linear	1	18	2.22	0.1536
Main Effect-Gypsum Quadratic	1	18	0.61	0.4464
Main Effect-Lime Linear	1	18	9.49	0.0065
Main Effect-Lime Quadratic	1	18	1.38	0.2560
Main Effect-Lime vs Gypsum	1	18	3.72	0.0697
Interaction-Gypsum Linear*Mulch	1	18	0.45	0.5131
Interaction-Gypsum Quadratic* Mulch	1	18	1.00	0.3304
Interaction-Lime Linear*Mulch	1	18	0.09	0.7697
Interaction-Lime Quadratic*Mulch	1	18	0.00	0.9844
Interaction-Lime vs Gypsum*Mulch	1	18	0.04	0.8376

which is a large value in comparison with the mean. This indicates that the precision may be considered inadequate for the research objective.

The results of the planned and multiple comparisons answer some questions but also raise others, such as: What difference in soil Ca is of practical importance? Were the number of replications or cores sampled too few or error control inadequate for meaningful differences in soil Ca also be deemed statistically different? Would higher rates of gypsum result in significant increases in soil Ca? Is there some unknown reason why the gypsum treatments did not increase the soil Ca concentration? So far, these are interesting results to be interpreted as a piece of an unfolding puzzle. To learn more about the analysis and interpretation of factorial experiments, readers are encouraged to read Chapter 7 by Vargas et al. (2018).

ANOVA Combined Over Fixed Locations

The linear model for ANOVA combined over locations (Table 8) adds terms for the location main effect and interactions, allowing tests of significant differences between and across locations. Location was considered a fixed effect because too few locations were studied to be considered as a representative random sample of all locations in the State of Reality (Piepho et al., 2003). Thus, inferences about location were narrow and limited in scope to the three locations studied.

SAS Code– PROC MIXED- pH Combined Over Locations

The revised PROC MIXED statements for the combined ANOVA are given in Table 12. Because the model included numerous factors and interactions, the SAS bar operator (|) was used in the Model statement. The shortcut notation Loc|Mulch|Ca_Trt substituted for Loc, Mulch, and Ca_Trt main effects and all their possible interactions, which is not only convenient but also safeguards against inadvertently overlooking interactions.

Checking for Heterogeneity of Variance

We have already assessed the residual plots for each location and deemed that the classical ANOVA assumptions were appropriate. Before using the combined ANOVA, we also need to assess the residual plots generated from fitting the combined linear model. One common concern is that the error variances may not be the same at all locations, violating the assumption for homogeneity of variance and indicating that the pooled error term (Residual MS) may not be appropriate for tests of significance. The residual error MS's obtained from the ANOVA at each location (not shown) were 0.068, 0.040, and 0.016. Using the Hartley's Fmax, the ratio of the largest and smallest error MS, as a rough measure of heterogeneity of variance, we note a four-fold difference between the highest and lowest error variances (Hartley's Fmax = 4.38), which indicates a need to further investigate the nature and extent of possible error heterogeneity.

Good news, the residual panels from the combined analysis did not reveal any serious data problems (Fig. 8). Although the residuals near the predicted pH of 6.0 are more dispersed than the low and high predicted pH values, this pattern is slight. Evidence of either heterogeneity or non-normality of the residuals which can affect the *p*-values of *F*-tests was not seen in the residual panels.

ANOVA Results– Combined over Locations

The PROC MIXED results of the ANOVA combined over locations for soil pH include estimates of the random variance components, tests of significance for fixed effects, and an ANOVA table with the EMS used to construct the Type III tests of significance (Table 13). The combined ANOVA has two random error variance components. The

TABLE 12. SAS Code – PROC MIXED for pH combined over locations.

```
Title 'Mixed pH ANOVA combined locations';
proc mixed data=anova.statbean plots=residualpanel method=type3;
class Loc Blk Mulch Ca_Trt;
model pH=Loc|Mulch|Ca_Trt;
random Blk(Loc);
lsmeans Loc|Mulch|Ca_Trt;
run;
```

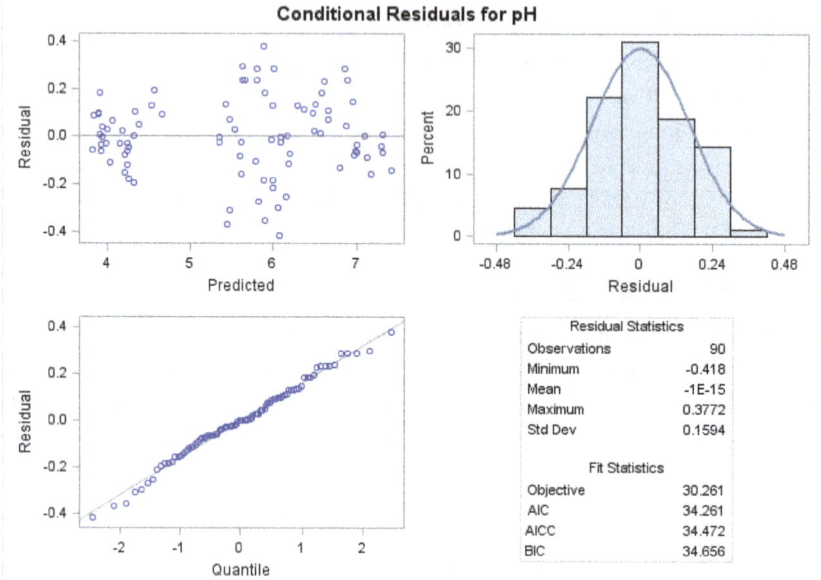

FIG. 8. Residual panels for pH combined over locations.

TABLE 13. PROC MIXED output for pH combined over locations using Type 3 estimation method.

Model Information	
Data Set	ANOVA.STATBEAN
Dependent Variable	pH
Covariance Structure	Variance Components
Estimation Method	Type 3
Residual Variance Method	Factor
Fixed Effects SE Method	Model-Based
Degrees of Freedom Method	Containment

Covariance Parameter Estimates	
Cov Parm	Estimate
Blk(Loc)	0.07653
Residual	0.04161

Type 3 Tests of Fixed Effects				
Effect	Num DF	Den DF	F Value	Pr > F
Loc	2	6	65.08	<.0001
Mulch	1	54	1.88	0.1766
Loc*Mulch	2	54	0.30	0.7408
Ca_Trt	4	54	16.23	<.0001
Loc*Ca_Trt	8	54	0.47	0.8708
Mulch*Ca_Trt	4	54	1.42	0.2389
Loc*Mulch*Ca_Trt	8	54	0.49	0.8612

Type 3 Analysis of Variance								
Source	DF	Sum of Squares	Mean Square	Expected Mean Square	Error Term	Error DF	F Value	Pr > F
Loc	2	105.036909	52.518454	Var(Residual) + 10 Var(Blk(Loc)) + Q(Loc,Loc*Mulch,Loc*Ca_Trt,Loc*Mulch*Ca_Trt)	MS(Blk(Loc))	6	65.08	<.0001
Mulch	1	0.078028	0.078028	Var(Residual) + Q(Mulch,Loc*Mulch,Mulch*Ca_Trt,Loc*Mulch*Ca_Trt)	MS(Residual)	54	1.88	0.1766
Loc*Mulch	2	0.025109	0.012554	Var(Residual) + Q(Loc*Mulch,Loc*Mulch*Ca_Trt)	MS(Residual)	54	0.30	0.7408
Ca_Trt	4	2.701038	0.675259	Var(Residual) + Q(Ca_Trt,Loc*Ca_Trt,Mulch*Ca_Trt,Loc*Mulch*Ca_Trt)	MS(Residual)	54	16.23	<.0001
Loc*Ca_Trt	8	0.157002	0.019625	Var(Residual) + Q(Loc*Ca_Trt,Loc*Mulch*Ca_Trt)	MS(Residual)	54	0.47	0.8708
Mulch*Ca_Trt	4	0.236856	0.059214	Var(Residual) + Q(Mulch*Ca_Trt,Loc*Mulch*Ca_Trt)	MS(Residual)	54	1.42	0.2389
Loc*Mulch*Ca_Trt	8	0.161624	0.020203	Var(Residual) + Q(Loc*Mulch*Ca_Trt)	MS(Residual)	54	0.49	0.8612
Blk(Loc)	6	4.841627	0.806938	Var(Residual) + 10 Var(Blk(Loc))	MS(Residual)	54	19.39	<.0001
Residual	54	2.247173	0.041614	Var(Residual)				

Blk(Loc) is the average of the Blk effects and the Residual is the errors pooled over locations. The Blk(Loc) variance component estimate is 0.076, almost twice the Residual variance estimate of 0.042. It can be seen in the ANOVA table that the Blk(Loc) MS is the error term for the Loc effect and the Residual MS is the error term for the other fixed effects. Because the Blk(Loc) MS is based on fewer df (6 df vs. 54 df) and was significantly larger (0.81 vs 0.04) than the Residual MS, the test of significance for Loc has less power and a higher probability of a Type 2 error than the tests of the other fixed effects. Regardless, the main effects of Loc and Ca_Trt were highly significant ($p < 0.001$), while the main effect of Mulch and all interactions were not significant ($\alpha = 0.05$).

Planned and Multiple Pairwise Comparison

Planned comparisons similar to those used for the ANOVA at the Central location can also be used for the combined ANOVA. For the combined ANOVA, the contrasts in Table 11 can be recoded to test the lime and gypsum effects both averaged over locations and/or within locations using the pooled residual error term. Contrasts testing specific Loc x Ca_Trt interactions can also be conducted. Details regarding contrasts for the combined ANOVA are beyond the scope of this chapter but can be found in Chapter 7 (Vargas, et al., 2018).

As previously mentioned, the soil pH and soil Ca concentration means provide information important for statbean production. If statbean is to become a new crop, growers will want to know if the pH and Ca concentration of their soil are suitable for statbean and if it will "pay" to add lime. And they probably want recommendations that are based on scientifically and statistically sound research.

Means should usually be given in tables or graphs to highlight the significant effects, especially significant interactions. For both soil pH and soil Ca concentration, the Loc and Ca_Trt effects were significant but not the Loc × Ca_Trt interaction. Therefore, bar graphs are shown for the Ca_Trt means at each location (Fig. 9). These graphs illustrate the large and significant differences between locations, the smaller yet significant differences between Ca_Trt means within each location, and that the differences among Ca_Trt means are similar at each location. As in Fig. 7, LSD(0.05) bars centered around the means give a confidence interval for the mean and show which means are significantly different from other means. The graphs can also be used to interpret the results of the planned comparisons.

Presentation of ANOVA Results

For research to be published as a graduate thesis or in a quality scientific journal, it must meet rigorous scientific standards, which include demonstrating that the statistical design and the analysis are appropriate and support the research conclusions (Table 1). This entails

TABLE 14. Statistical information that can be determined from ANOVA components in a SIMPLE ANOVA.

ANOVA Component	What the reader can directly or indirectly determine
Source of Variation	effects in the linear model explanatory and design effects construction of the effects (additive, nested, or crossed) treatment factors and interactions
Effect Type	random and fixed model effects scope of inference for an effect (narrow or broad) model type (random, mixed, fixed) experimental design and error terms
Degrees of Freedom	number of treatments or levels of each factor number of replications, samples/replication adequacy of df for estimates of variances of random effects adequacy of sample size for tests of significance
F-value - fixed effects	relative magnitude of the effect size
Mean Square - random effects	test significance for additional mean comparisons test for significant differences between error terms standard deviations and standard errors
$P > F$ or *, **, ****	probability used for the test of significance

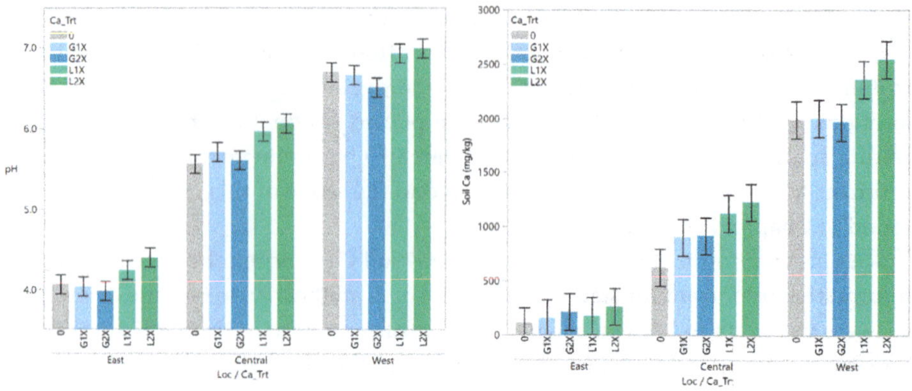

FIG. 9. Bar graph of Ca_Trt means and LSD(0.05) for soil pH and soil Ca concentrations at each the eastern, western, and central locations. The LSD(0.05) is centered around each mean so means are significantly different if the LSD bars do not overlap. The LSD(0.05) for pH = 0.24; LSD(0.05) for Ca concentration= 341 mg kg^{-1}.

TABLE 15. SIMPLE ANOVA for soil pH at three locations.

			Loc=Central		Loc=Western		Loc=Eastern	
Source	Effect	DF	*F* Value	P > *F*	*F* Value	P > *F*	*F* Value	P > *F*
Blk	R	2	-	-	-	-	-	-
Ca_Trt	F	4	4.3	**	5.9	**	11.2	**
Mulch	F	1	< 1	ns	1.4	ns	< 1	ns
Mulch*Trt	F	4	< 1	ns	< 1	ns	1.2	ns
Error	RE	18	(0.07)†		(0.04)		(0.02)	

† MS of random error in parentheses

TABLE 16. SIMPLE ANOVA for pH, Ca, and Ptotal.

			pH		Ca		Ptotal	
Source	Effect†	DF	*F* (MS)‡	*P* > *F*	*F* (MS)	*P* > *F*	*F* (MS)	*P* > *F*
Loc	F	2	65.1	0.01	50.2	0.01	3.9	0.08
Blk(Loc)	RE	6	(0.076)		(603646)		(1093)	
Mulch	F	1	1.9	0.18	0.4	0.50	3.8	0.06
LocxMulch	F	2	0.3	0.74	0.6	0.52	12.1	0.001
Ca_Trt	F	4	16.2	0.01	6.5	0.01	0.9	0.49
LocxCa_Trt	F	8	0.5	0.87	1.1	0.34	1.3	0.24
MulchxCa_Trt	F	4	1.4	0.24	0.0	1.0	0.3	0.88
LocxMulch×Ca_Trt	F	8	0.5	0.86	0.5	0.85	0.4	0.93
Residual	RE	54	(0.042)		(86858)		(65)	

† Effect Type: fixed (F) or random error (RE).

‡ *F*-values for fixed effect, MS values in parentheses for random effects

describing the research objectives, experimental design, dependent and explanatory variables, and statistical tests in sufficient detail for reviewers and readers to independently judge their computational, statistical, and especially biological soundness. The description should also enable other researchers to repeat the experiment. An ANOVA table can be constructed to encapsulate most of this statistical information. Analysis of variance tables are especially effective for research with complex designs, many dependent variables, multiple factors, multiple error terms, and/or mixed models. In addition, ANOVA tables provide readers a convenient place to find, verify, and interpret ANOVA results as they read through a paper. An ANOVA table has been proposed that contains ANOVA components chosen to concisely quantify, summarize, and organize the key details of the design, analysis, and results of an experiment (McIntosh, 2015). This ANOVA table format, termed SIMPLE (Simple, Informative, Meaningful, Powerful, Logical, Effective), includes selected ANOVA components for all SOV in the linear model to provide the statistical information listed in Table 14. Just as ANOVA is not "one-size-fits all", neither is a SIMPLE ANOVA table. Instead, it is a framework for authors to consider for presentation of ANOVA results, which can be compared to other options.

Examples of SIMPLE ANOVA tables for soil pH analyzed separately at each location and for the three dependent variables (soil pH, soil Ca, and Ptotal) combined over locations are shown in Tables 15 and 16. Statistics were rounded to show only the informative digits and not exaggerate their precision. Probability values only report two decimal places or significance levels no less than 0.01 (**). The SIMPLE ANOVA tables differ from more common tables of significance tests of fixed effects because they also give the SOV and df of random effects, including error terms. Thus, an ANOVA table containing SOV for each term in the linear model can facilitate a holistic understanding of the ANOVA that includes attributes about the experimental design and error variation. The ANOVA process, which included planned comparisons and LSD values to rank and compare means provided the statistical basis for interpreting the effects of the factors and their interactions on the means of the dependent variables. It has also provided estimates of random error variance components that can be useful in designing future experiments with desired level of precision.

Conclusions

One goal of this chapter was to reinforce and enhance your practical knowledge of ANOVA. Maybe you also have increased your appreciation for ANOVA as a powerful research tool for enhancing scientific inquiry and defeating the spread of junk (irreproducible) science. An underlying goal of this chapter is to raise awareness of practices for using ANOVA that will encourage rigorous statistical standards for publishing research and require rationale for and description of statistical methods with sufficient detail to both understand and justify their use. The concepts and suggestions in this chapter are meant to support your efforts to ensure that your research meets these standards but do not involve the analysis of complex models and non-normal data. These require a deeper understanding of statistical concepts beyond those covered here. Hopefully you will find helpful, practical discussions of more complex topics in other chapters in this book. Also, always keep in mind that it is wise to involve a skilled statistician from the start to finish of an experiment. Finally, if you want to try statbean as a herbal supplement to enhance your statistical skill, it will be available as soon we can convince the producers that our new statbean cultivar is significantly better than the old cultivar. In our first trial we chose a significance level of 0.05 and the new cultivar was not significantly better than the old cultivar ($P > F = 0.0541$). Thus, thus the new cultivar is no better than the old. Do you agree?

Key Learning Points

- Basic concepts and terms of ANOVA.
- ANOVA process to test hypotheses.
- Use and relevance of the separate components of ANOVA.
- How to conduct a mixed model ANOVA using SAS or R.
- How to construct an effective ANOVA table for scientific publication.
- The role of ANOVA to maintain standards and advance science.

Review Questions (T/F)

1. ANOVA is a statistical approach used to discover the factors that contribute to the variation of the dependent variable.
2. A linear model was the mathematical basis of the traditional ANOVA but is not used for the contemporary mixed model.
3. The ratio of Var(Treatment)/Var(Residual) is used to test if a treatment effect is significant. If Var(Treatment) is larger than Var(Residual), then the *F*-value will be greater than 1 and the treatment effect will be considered significant.
4. The probability of a Type 1 error (incorrectly concluding that treatment means are significantly different) can be decreased by increasing the number of replications.
5. To ensure scientific standards, research papers are subjected to peer review prior to publication. These papers need to include sufficient detail for peer reviewers and readers to judge the validity and merit of the experimental design and statistical analyses.

Exercises

1. Find an interesting article in a journal in your field of study to "peer review" and answer the following questions.
   a. Do the descriptions of the experimental design and statistical analyses provide adequate detail and clarity? Explain your answer using Table 1 as a guide.
   b. Are the design, analysis, and variables measured appropriate for the experimental objectives? Justify your answer.
   c. Are the results and conclusions substantiated by the statistics provided? Justify your answer.
2. Researcher A published a paper that found that the treatment effect was significant at the 0.05 level. Researcher B had conducted an experiment with the same

treatments but concluded that the treatment effect was not significant at the 0.05 level. Not surprisingly, Researcher B did not publish these non-significant results.

a. Give at least two likely reasons for the different results.
b. Based on one or more of your reasons in 2a, describe a scenario that justifies publishing Researcher A's findings but not Researcher B's findings. Explain your justification, including the expected long-term consequences.
c. Based on one or more of your reasons in 2a, describe a scenario that justifies publishing both Researcher A's and Researcher B's findings. Explain your justification, including the expected long-term consequences.

3. *P*-values are quantitative statistics that are often reduced to two categories (significant and non-significant) to make inferences about the data. What are the advantages and disadvantages of this common practice?

References

Acutis, M., B. Scaglia, and R. Confalonieri. 2012. Perfunctory analysis of variance in agronomy, and its consequences in experimental results interpretation. Eur. J. Agron. 43:129–135. doi:10.1016/j.eja.2012.06.006

Amrhein, V., F. Korner-Nievergelt, and T. Roth. 2017. The earth is flat (p > 0.05): Significance thresholds and the crisis of unreplicable research. PeerJ 5:e3544.

Boos, D.D., and L.A. Stefanski. 2011. *P*-value precision and reproducibility. Am. Stat. 65:213–221. doi:10.1198/tas.2011.10129

Box, G.E. 1976. Science and statistics. J. Am. Stat. Assoc. 71:791–799. doi:10.1080/01621459.1976.10480949

Casler, M.D. 2018a. Blocking principles for biological experiments. In: B. Glaz and K.M. Yeater, Applied statistics in the agricultural,biological, and environmental sciences. ASA, CSSA, SSSA, Madison, WI.

Casler, M.D. 2018b. Power and replication-Designing powerful experiments. In: B. Glaz and K.M. Yeater, Applied statistics in the agricultural,biological, and environmental sciences. ASA, CSSA, SSSA, Madison, WI.

Dixon, P.M., K.J. Moore, and E. van Santen. 2018. The analysis of combined experiments. In: B. Glaz and K.M. Yeater, Applied statistics in the agricultural,biological, and environmental sciences. ASA, CSSA, SSSA, Madison, WI.

Fisher, R.A. 1921. Studies in crop variation. I. An examination of the yield of dressed grain from Broadbalk. J. Agric. Sci. 11:107–135. doi:10.1017/S0021859600003750

Fisher, R.A. 1925. Statistical methods for research workers. Oliver and Boyd, Edinburgh, UK.

Fisher, R.A. 1926. Arrangement of field experiments. J. Minist. Agric. (G. B.) 33:503–513.

Fisher, R.A., and W.A. Mackenzie. 1923. Studies in crop variation. II. The manurial response of different potato varieties. J. Agric. Sci. 13:311–320. doi:10.1017/S0021859600003592

Franco, A., N. Malhotra, and G. Simonovits. 2014. Publication bias in the social sciences: Unlocking the file drawer. Science 345:1502–1505. doi:10.1126/science.1255484

Garland-Campbell, K. 2018. Errors in statistical decision-making. In: B. Glaz and K.M. Yeater, Applied statistics in the agricultural, biological, and environmental sciences. ASA, CSSA, SSSA, Madison, WI.

Gbur, E.E., W.W. Stroup, K.S. McCarter, S. Durham, L.J. Young, M. Christman, M. West, and M. Kramer. 2012. Analysis of generalized linear mixed models in the agricultural and natural resources sciences. ASA, CSSA, SSSA, Madison, WI.

Gelman, A., and H. Stern. 2006. The difference between "significant" and "not significant" is not itself statistically significant. Am. Stat. 60:328–331. doi:10.1198/000313006X152649

Kozak, M. and H.P. Piepho. 2017. What's normal anyway? Residual plots are more telling than significance tests when checking ANOVA assumptions. Journal of Agronomy and Crop Science 2017: 1-13. doi:10.1111/jac.12220.

Lehmann, E.L. 1993. The Fisher, Neyman-Pearson theories of testing hypotheses: One theory or two? J. Am. Stat. Assoc. 88:1242–1249.

Littell, R.C., G.A. Milliken, W.W. Stroup, R.D. Wolfinger, and O. Schabenberger. 2006. SAS for Mixed Models. Second edition ed. SAS Institute, Cary, NC.

Loy, A., L. Follett, and H. Hofmann. 2016. Variations of Q–Q Plots: The power of our eyes! Am. Stat. 70:202–214. doi:10.1080/00031305.2015.1077728

Majumder, M., H. Hofmann, and D. Cook. 2013. Validation of visual statistical inference, applied to linear models. J. Am. Stat. Assoc. 108:942–956. doi:10.1080/01621459.2013.808157

McIntosh, M.S. 2015. Can analysis of variance be more significant? Agron. J. 107:706-717. doi:10.2134/agronj14.0177

Mervis, J. 2014. Why null results rarely see the light of day. Science 345:992. doi:10.1126/science.345.6200.992

Moore, K.J., and P.M. Dixon. 2015. Analysis of combined experiments revisited. Agron. J. 107:763–771. doi:10.2134/agronj13.0485

Murdoch, D.J., Y.-L. Tsai, and J. Adcock. 2008. *P*-values are random variables. Am. Stat. 62:242–245. doi:10.1198/000313008X332421

Neyman, J., and B. Tokarska. 1936. Errors of the second kind in testing "Student's" hypothesis. J. Am. Stat. Assoc. 31:318–326. doi:10.2307/2278560

Nuzzo, R. 2014. Scientific method: Statistical errors. Nature 506:150–152. doi:10.1038/506150a

Piepho, H.P., A. Buchse, and K. Emrich. 2003. A hitchhiker's guide to mixed models for randomized experiments. J. Agron. Crop Sci. 189:310–322.

Saville, D.J. 2015. Multiple comparison procedures—Cutting the Gordian knot. Agron. J. 107:730–735. doi:10.2134/agronj2012.0394

Saville, D.J. 2018. Multiple comparison procedures: The ins and outs. In: B. Glaz and K.M. Yeater, Applied statistics in the agricultural,biological, and environmental sciences. ASA, CSSA, SSSA, Madison, WI.

Spilke, J., H.P. Piepho, and X. Hu. 2005. Analysis of unbalanced data by mixed linear models using the MIXED procedure of the SAS system. J. Agron. Crop Sci. 191:47–54. doi:10.1111/j.1439-037X.2004.00120.x

Stroup, W. 2018. Analysis of non-Gaussian data. In: B. Glaz and K.M. Yeater, Applied statistics in the agricultural, biological, and environmental sciences. ASA, CSSA, SSSA, Madison, WI.

Vargas, M., B. Glaz, J. Crossa, and A. Morgounov. 2018. Analysis and interpretation of interactions of fixed and random effects. In: B. Glaz and K.M. Yeater, Applied statistics in the agricultural,biological, and environmental sciences. ASA, CSSA, SSSA, Madison, WI.

Wasserstein, R.L. and N.A. Lazar. 2016. The ASA's statement on *p*-values: Context, process, and purpose. The American Statistician 70(2): 129–133. doi:10.1080/00031305.2016.1154108.

Yang, R.C. 2010. Towards understanding and use of mixed-model analysis of agricultural experiments. Can. J. Plant Sci. 90:605–627. doi:10.4141/CJPS10049

Yeater, K.M. and M.B. Villamil. 2018. Multivariate methods for agricultural research. In: B. Glaz and K.M. Yeater, Applied statistics in the agricultural, biological, and environmental sciences. ASA, CSSA, SSSA, Madison, WI.

CHAPTER 3: BLOCKING PRINCIPLES FOR BIOLOGICAL EXPERIMENTS

Michael D. Casler

Blocking designs represent one of the fundamental tools available to all biological researchers. Blocking designs can be used when experimental units can be organized into blocks, which can be either complete or incomplete, that is, containing all or a portion of the treatments in the experiment. Blocks are intended to organize experimental units into groups that are more uniform or homogeneous compared to the full sample of experimental units that comprise the experiment. In so doing, blocks can be used to account for significant amounts of spatial or temporal variability, or both, among experimental units, thereby reducing residual variances and increasing precision of an experiment. This chapter reviews a wide range of blocking designs, philosophies, and methodologies, providing three examples that can assist researchers in making informed decisions during the design, execution, analysis, and interpretation of biological experiments.

All biological experiments are subject to variability—some more than others. It is that variability that concerns us when we design experiments in the field, glasshouse, or laboratory. Our general goal is to maximize treatment effects and minimize residual variation. Most of our research is comparative in nature; that is, we are comparing or contrasting two or more practices or systems (treatments) that we hypothesize to have some impact on a particular measurement or response variable. Specifically, we try to create a circumstance in which the variability among treatments is sufficiently large, such that we have strong evidence to claim that one treatment is better than another given the background noise within a system. Tradition demands significance (or α) levels of 0.05 or 0.01, that is, low probabilities of falsely rejecting a null hypothesis of no treatment effect. As such, there are very specific requirements or thresholds for among-treatment variability as compared to within-treatment variability.

Variability among treatments is under the control of the researcher, who decides both the size and scope of the experimental conditions. Once the size and scope are determined, expectations about differences between specific treatment means can

Abbreviations: AIC, Akaike's information criteria; BIC, Bayesian information criteria; CRD, completely randomized design; df, degrees of freedom; LSCOD, Latin square change-over design; p-rep, partially replicated; RCBD, randomized complete block design; SBCBD, spatially balanced complete block designs.

Michael D. Casler, USDA-ARS, US Dairy Forage Research Center, 1925 Linden Dr., Madison, WI 53706-1108 (mdcasler@wisc.edu, michael.casler@ars.usda.gov).

doi:10.2134/appliedstatistics.2015.0074

Applied Statistics in Agricultural, Biological, and Environmental Sciences
Barry Glaz and Kathleen M. Yeater, editors

be generated. These expectations can be used to estimate the amount and type of replication required to meet experimental objectives (see Chapter 4, Casler, 2018). Conversely, variability within treatments is only partially under the control of the researcher. There are two general types of within-treatment variability: (i) predictable, estimable, or descriptive variation and (ii) inherent variation. The latter is that variation which exists within the population of experimental units over which we have absolutely no control. Walk into a maize *Zea mays* (L.) or soybean [*Glycine max* (L.) Merr.] field, and all the plants look identical to each other until you begin to take detailed measurements on variables such as seed size, number of seeds per ear or pod, plant height, etc. We can control this variability only up to a point. For example, a field planted to an open-pollinated variety of maize would possess significantly more inherent variability than a field planted to an inbred line or a hybrid because every seed is genetically different in an open-pollinated population, but almost genetically identical within an inbred line. Likewise strains of laboratory rats were inherently more variable in the early 20th century before they were inbred to create highly uniform strains. Once the researcher chooses the subject matter, the experimental material, and the research sites, a certain amount of inherent variation is "locked in" and will always be an important component of the statistical data analysis. This chapter deals with numerous mechanisms whereby researchers can manipulate or control the other portion of within-treatment variability, the portion that is predictable, estimable, and/or descriptive.

Randomization

Randomization provides two key services in comparative research: (i) unbiased estimation of treatment means, treatment differences or contrasts, and experimental error and (ii) precaution against a disturbance that may or may not arise (Greenberg, 1951; Kempthorne, 1977). The fundamental principle of randomization in biological experiments has been well understood for nearly a century (Fisher, 1925). Nevertheless, systematic experiments, in which one or more replicates of the treatments are applied in a systematic or ordered manner, are still common (Piepho et al., 2013). Randomization is an insurance policy. Like insurance, it carries a cost in time spent conducting randomizations and putting up with some inconvenience during the conduct of the experiment, having to use a field map or diagram to identify randomized treatments. The benefits always outweigh the costs because disturbances such as spatial trends or unexpected intrusions (e.g., emergency helicopter landings, wrong turns by armored personnel carriers, herbicide application mistakes, irrigation pipe breaks during holiday weekends, motorcycle joyrides, and mischievous college students, each of which has occurred on my research plots) cannot be anticipated or entirely avoided. If there were no field gradients, no hidden spatial variation, no unexpected disturbances, and independent observations of our treatments provided uniform results, we would not need to randomize treatments, a situation that might actually occur if we were chemists or physicists.

Failure to properly randomize biological experiments results in biased estimates of treatment means and therefore treatment differences, biased estimates of experimental error, and therefore incorrect p values (Piepho et al., 2013). Indeed,

Piepho et al. (2013) argued that test statistics developed by statisticians for randomized experiments have no meaning or validity in systematic designs, that they are unable to generate accurate probability statements. While randomization is a strict mathematical process in which a random number generator is used to order the treatments, randomized designs can fail the "eye test" when they give the appearance of "clumping" or grouping of certain treatments. Such undesirable patterns may reduce confidence on the part of some researchers who may question the design's ability to overcome spatial trends. Interspersion of treatments is a mechanism to avoid the possibility of biases that may arise from confounding or correlation between treatment effects and spatial trends (Hurlbert, 1984; Martin, 1986). Interspersion is accomplished by using spatially balanced designs that ensure that all pairs of treatments are (more-or-less) equally spaced from each other.

Blocking

Concepts of Blocking

Blocking serves two purposes in biological experimentation. First, it serves as a mechanism of replication. In its simplest form, a block contains one experimental unit per treatment. An experimental unit is the smallest unit to which a treatment is independently applied. Multiple blocks are used as the form of replication, so that an experiment with four complete blocks has, by definition, four replicates of each treatment. Multiple and independent experimental units for each treatment is the fundamental requirement for proper replication at the proper scale (Casler, 2015). Second, blocking is used as a mechanism of error control, largely to remove spatial variation from the residual variance. Most of the remainder of this chapter expands on that topic.

A block is nothing more than a group of experimental units. In general, the goal of the researcher is to define blocks so that the experimental units within blocks are more homogeneous than between blocks. In field or glasshouse experiments, members of a block should be contiguous and blocks should be arranged into regular shapes, such as squares or rectangles, not irregular shapes such as esses or ells. Irregular shapes defeat the entire purpose of blocking because they greatly increase the likelihood of spatial variation within the block. Furthermore, square blocks are preferred to rectangular blocks because they minimize the mean Euclidean distance between two random points within the block. Biological researchers use blocking designs for one or both of two reasons: (i) precision, to create groups of experimental units that are more homogeneous than would occur with random sampling of the entire population of experimental units or (ii) convenience, to allow different sizes of experimental units when larger plots or larger experimental areas are required for application of one factor compared to other factors.

Experimental units can be blocked in a variety of ways to create blocking designs. Spatial blocking is the classical approach described above and in most statistical textbooks. The classical approach to spatial blocking is for the researcher to identify sources of variation that can be used as an effective blocking criterion, e.g., a fertility, moisture, or soil gradient in the field; a temperature or light gradient in the glasshouse or growth chamber; or livestock traits (body mass or condition, sex, or stage of lactation). There are times when this is not possible. Many researchers are assigned fields that, by all appearances, are highly uniform with no information on spatial variability. Situations like these, in which nothing is known about spatial variability, will frequently result in blocking designs that are ineffective, i.e., no significant treatment differences, especially when you guess one direction and the gradient is in another direction (Casler, 1992). When in doubt use (i) square blocks, (ii) small blocks (few treatments), and (iii) bi-directional blocking to the greatest extent possible (e.g., Jones et al. 2015).

Temporal blocking is a legitimate practice in many situations. Growth chamber experiments are frequently not properly replicated, with researchers frequently applying each treatment to one chamber and using multiple plants within a chamber as an inadequate form of replication (Casler, 2015). Repetition of the experiment in time, with re-randomization of treatments to chambers, is a solution to this problem, with each repetition of the experiment defined as a block. Field or glasshouse experiments that are too large or unwieldy to be properly replicated on a spatial plane can be repeated across multiple years in which each year is treated as a block. This is a common practice for many pathology and entomology studies designed to evaluate plant treatments or genotypes for resistance to a pest, especially when the disease inoculum or insect larvae are available only in limited supply.

Multiphase experiments present a different challenge (Piepho et al., 2003). Some variables may be measured directly on the experimental units, while other variables are measured in the laboratory after numerous sampling and processing steps have had the opportunity to introduce additional sources of variability to tissue samples. For example, many field studies and most livestock trials have a laboratory component in which grain, forage, or feed quality traits are measured to support treatment analysis. Time of day for tissue sampling, humidity of storage conditions, sharpness of grinder or mill blades, and ambient humidity surrounding some laboratory equipment can all introduce new and uncontrolled sources of variability to laboratory samples and thus, to treatment variability. There are two generally accepted solutions to this problem: (i) use the field design to confound these new sources of variation with field design factors, to the extent that this is acceptable, or (ii) create a multiphase sampling design in which field spatial variability is orthogonal with temporal variability associated with tissue sampling and processing. The first solution is accomplished by using the field design at all phases of the experiment, so that samples are harvested, ground, and analyzed in randomized order, but these activities occur block-by-block, in blocks defined by the field, glasshouse, or livestock design (e.g., Casler, 1992). The second solution is created by using two design phases in which the first-phase design is used to physically arrange treatments in the field or glasshouse and the second-phase design is used to independently arrange samples for laboratory analysis (e.g., Brien et al., 2011; Smith et al., 2006). As pointed out by Brien et al. (2011, their Principles 8 and 10), a combination of the two approaches may be optimal in many situations.

The simplest design for biological experiments is the completely randomized design (CRD), in which there is one block that contains all replicates of all treatments (Table 1, Fig. 1A). In the CRD, spatial analysis is the only mechanism to deal with nonuniformity of experimental units (see Chapter 12, Burgueño, 2018). This can be a significant limitation to the inferences that can be drawn from a CRD experiment because spatial analyses are intended to supplement blocking designs, not to replace them (Piepho et al., 2013). The major advantage of the CRD is that it maximizes error degrees of freedom (df), compared with any blocking design (Table 2). Thus, if the researcher is reasonably certain that the complete experimental area is homogeneous, the CRD design can be advantageous. However, for many biological experiments, experimental areas are rarely homogeneous across the entire spatial range, suggesting that blocking designs are more advantageous.

Complete Block Designs

The simplest blocking design is the randomized complete block design (RCBD), in which each block contains one experimental unit per treatment, and each treatment

TABLE 1. Experimental design families organized according to type and complexity of blocking arrangements.

Number of potential blocking levels	Treatment design (from Table 2)	Experimental design options	Defining characteristics	References
One	Any structure	Randomized complete block	Block size (k) = number of treatments (t).	Steel et al. (1996)
Bidirectional and structured	Any structure	Latin square, Graeco-latin square, lattice square, incomplete Latin square	Bi-directional blocking in perpendicular directions. Number of replicates and treatments are highly restricted in some designs.	Cochran and Cox (1957), Petersen (1985), Steel et al. (1996)
Multiple and flexible	Full factorial	Split-plot and variations, split-block (Strip-plot)	Design contains multiple sizes of experimental units, one for each level (or "split"). Larger experimental units may be required for convenience, but will be associated with increased error if Error(a) > Error(b).	Cochran and Cox (1957), Cox (1958), Petersen (1985), Steel et al. (1996)
Multiple and flexible	Nested structure	Blocks in reps (sets in reps) reps in blocks (reps in sets)	Treatments are randomly divided into sets. Block size = number of treatments per set. Good for inferences on random effects.	Schutz and Cockerham (1966), Casler (1998)
Multiple and flexible	Any structure	Balanced or partially balanced incomplete blocks	Potentially large reduction in block size with flexibility in both structure and field layout. Block size (k) may be $t^{1/2}$ or $t^{1/3}$.	Cochran and Cox (1957), Cox (1958), Petersen (1985)
Multiple and flexible, bidirectional	Any structure	Alpha, row-column	Treatments arranged in rows and columns. Extremely flexible with regard to number of treatments, number of replicates, and block size.	John and Eccleston (1986), Patterson and Robinson (1989), Williams (1986)
Multiple and flexible	Any structure with variable replication across treatments	Control-plot, Augmented, Modified augmented, p-rep	Designed to accommodate extremely unbalanced treatment structures and unequal replication across treatments. No restrictions on numbers of treatments and replicates.	Chandra (1994), Lin and Poushinsky (1983, 1985), Wolfinger et al. (1997), Williams et al. (2011)

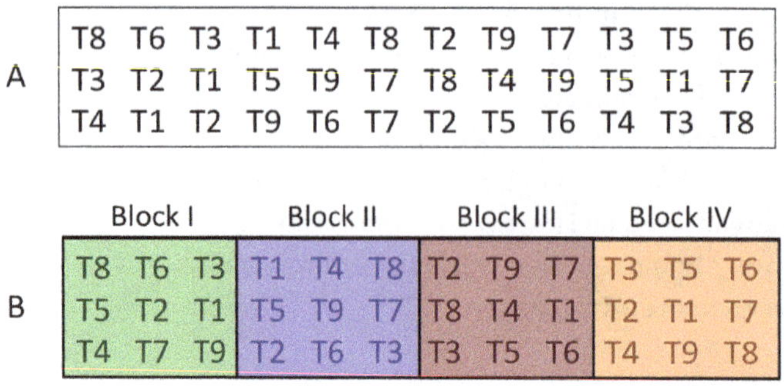

FIG. 1. Four replicates ($r = 4$) of nine treatments ($t = 9$) (36 experimental units) randomized in two different experimental designs. (A) Completely randomized design. (B) Randomized complete block design with a block size equal to the number of treatments and number of blocks equal to the number of replicates ($r = k$). Each complete block is shown by a different color.

TABLE 2. Full-model sources of variation and degrees of freedom for eight experimental designs illustrated in Fig. 1 through 4, each with four replicates of nine treatments for a total of 36 experimental units in each case.

Design/source of variation	df	Design/source of variation	df	Design/source of variation	df
Completely randomized (Fig. 1A)		**Lattice square (Fig. 2D)**		**Split-plot in RCBD† (Fig. 4G)**	
Treatments	8	Complete replicates (*R*)	3	Complete replicates (*R*)	3
Residual	27	Rows/*R*	8	Factor A	2
		Columns/*R*	8	Whole-plot error (*a*)	6
		Treatments (unadjusted)	8	Factor B	2
		Intrablock error	8	A × B Interaction	4
				subplot error (*b*)	18
RCBD† (Fig. 1B)		**Sets within reps (Fig. 3E)**			
Blocks (Complete replicates, *R*)	3	Complete replicates (*R*)	3		
Treatments	8	Sets	2		
Residual	24	Sets × *R*	6		
		Treatments/sets	6	**Split-block in RCBD† (Fig. 4H)**	
		Residual	20	Complete replicates (*R*)	3
				Factor A	2
Lattice (Fig. 2C)		**Reps within sets (Fig. 3F)**		Strip-plot error (*a*)	6
Complete replicates (*R*)	3	Sets	2	Factor B	2
Blocks/*R*	8	Replicates/sets	9	Strip-plot error (*b*)	6
Treatments (unadjusted)	8	Treatments/sets	6	A × B Interaction	4
Intrablock error	16	Residual	20	Subplot error (*c*)	12

† RCBD, randomized complete block design.

occurs once (Table 1; Fig. 1B). This design can be used with any number of treatments or any treatment structure (Table 3). The RCBD is ideal for relatively small experiments or other situations in which it can be assumed that groups of uniform experimental units can be created to accommodate all treatments. The cost is minimal, generally just a few df and a minor effort to accomplish the randomization. Many researchers will set up one block in a systematic manner and the remaining blocks with randomized treatments. Piepho et al. (2013) strongly discouraged this practice because the lack of randomization can lead to biased estimates of treatment differences. Lastly, the RCBD can be organized into single or bidirectional blocking, depending on the number of blocks. The RCBD can be organized into any shape of block, ranging from square blocks to long and narrow rectangular blocks. If there is a known gradient or known pattern of spatial variation along a directional vector, then long and narrow rectangular blocks are the best choice, oriented perpendicular to the gradient. If nothing is known about gradients or spatial variation, then square blocks are the best choice.

Many researchers create plot numbers in a RCBD using three digit numbers where the first digit is the block number and the last two digits are an incremental code for plots arranged in an ordered, often serpentine manner, starting at 01 and ending at the number of treatments (e.g., 101 to 124, 201 to 224, etc. for 24 treatments). An alternative numbering system is to code each plot by a simple row and column number for the entire experiment, which will greatly facilitate the use of spatial analyses to supplement the RCBD analysis (Chapter 12, Burgueño, 2018), especially if the blocking arrangement fails to account for spatial variability. In both cases, each

TABLE 3. List of some common treatment structures utilized in designing comparative or manipulative experiments.

Design name or definition	Number of factors	Defining characteristics
Unstructured	One	There is no structure or organization to the treatments.
Nested structure	Two or more	Factors have levels that are not repeated or have the same meaning at all levels of the other factors (Schutz and Cockerham, 1966).
Full factorial design	Two or more	Each factor has a specific number of levels that are repeated (have the same definition and meaning) over all levels of the other factors. The number of treatment combinations is the product of the number of levels of each factor (Cochran and Cox, 1957).
Confounding design	Two or more	A full factorial in which a higher-order interaction term is sacrificed as a blocking factor to achieve a reduction in block size (Cochran and Cox, 1957; Cox, 1958).
Composite design	Two or more	A subset of a factorial designed to severely reduce the number of treatments required to evaluate the main effects and first-order interactions using regression-based modeling (Draper and John, 1988; Draper and Lin, 1990; Lucas, 1976).
Fractional factorial	Two or more	A partial factorial arrangement, in which only a subset of the factorial treatments is included, usually based on choosing a higher-order interaction term as the defining contrast (Cochran and Cox, 1957; Cox, 1958).
Repeated measures	Two or more	One or more of the treatment factors is observed over multiple time points without re-randomization of treatments to experimental units (Milliken and Johnson, 2009).

plot has a unique numerical identifier, and blocks can be indicated by color coding, for example.

Bidirectional blocking is readily facilitated within the Latin square family of designs (Table 1). The Latin square design includes two sources of blocking that are orthogonal with each other and with treatments. The Latin square design consists of three orthogonal dimensions: an equal number of rows, columns, and treatments. Classically, they are arranged as complete rows and complete columns, perpendicular to each other in a spatial arrangement, with each treatment represented once in each row and once in each column to create orthogonality. Latin squares also have great utility in livestock feeding studies in which animals correspond to columns and time periods correspond to rows (Jones and Kenward, 2003). The Graeco-Latin (or Euler) square design builds on the same principle, but with a fourth orthogonal dimension. The fourth dimension could be used as a second factor to allow the use of a balanced factorial set of treatments (e.g., 3×3, 4×4, 5×5, etc.) or it could be used to create another blocking factor. Latin squares are valuable for experiments with a relatively small number of treatments, but error degrees of freedom are generally small. This problem can be solved at the cost of repeating the design in multiple squares, a common practice in livestock feeding trials.

The restriction requiring equal numbers of rows, columns, and treatments in the Latin and Graeco-Latin squares severely limits their utility for many biological research applications. Incomplete Latin square (Youden square) designs eliminate this restriction by using only a subset of rows from a Latin square design (Youden, 1940). Each row remaining in the design forms a complete block, but the columns are treated as incomplete blocks in which each pair of treatments occurs together once within one of the columns. Trojan squares represent an alternative approach in which several mutually orthogonal Latin squares are combined together into a single design. For example, three 4×4 Latin squares are each assigned four treatments at random from an experiment with 12 treatments. The three squares are combined row-by-row and randomized to create four rows of 12 treatments. Trojan squares are a special form of semi-Latin squares that have utility in agricultural research (Edmondson, 1998).

Spatially balanced complete block designs (SBCBD) represent a special case in which randomization is restricted to meet two criteria: (i) spatial balance among treatment contrasts and (ii) disallowing treatments to be placed in the same position in multiple blocks (van Es et al., 2007). The purpose of the SBCBD is to improve the accuracy of treatment comparisons or contrasts by removing biases that may be due to trends or gradients, temporal variation in sampling, or spatial autocorrelation. For example, if the randomization process results in two treatments that have a tendency to be very far apart from each other, compared to two random treatments, the difference between their means may not reflect the true treatment difference. van Es et al. (2007) published SBCBD for up to 15 treatments in up to eight blocks. Following the random assignment of treatments, each SBCBD represents one of many possible RCBD randomization results (van Es et al., 2007).

Finally, on-farm research presents a unique set of problems for developing proper and efficient field designs. On-farm research typically demands considerable patience and compromise for design factors, for example, with little regard for true replication, randomization, blocking structure, and statistical data analysis. For situations in which the research team is required to produce publishable results, two relatively simple solutions are frequently acceptable to growers: (i) use of control-plot designs in which each treatment is paired with a control treatment that is repeated throughout the experiment, or (ii) repetition of the study on multiple farms, using farms as blocks or replicates. These two design options can also be used in combination with each other. Riley and Alexander (1997) summarized a number of studies that provide design and statistical guidelines for conducting on-farm research.

Incomplete Block Designs

Balanced and partially balanced incomplete block designs represent a large family of designs that can handle any treatment structure, any number of treatments, and a very large reduction in block size relative to the number of treatments (Table 1). This family includes all of the various lattice and lattice square designs, which have some restrictions on treatment numbers, as well as a wide range of generic designs that are more flexible with respect to number of treatments and block size (Cochran and Cox, 1957; Dey, 2010). Some of these designs will allow for bidirectional blocking to remove both row and column effects. Lattice designs have historically been extremely effective for controlling spatial variation in field sites that are routinely used for cultivar testing (Casler, 1999; Patterson and Hunter, 1983). Lattice designs require a block size equal to the square root of the number of treatments, severely restricting the number of treatments in the experiment. Lattice designs classically involve one-way blocking for error control (Fig. 2C), but blocks can be arranged in any direction desired. Lattice square designs (Fig. 2D) allow for two-way blocking and error control, using the basic concept of the Latin square design, but for significantly greater numbers of treatments. Classically, lattice designs are employed for experiments with hundreds or thousands of treatments and relatively few replicates (two to four). Balanced designs are those in which all treatment pairs occur an equal number of

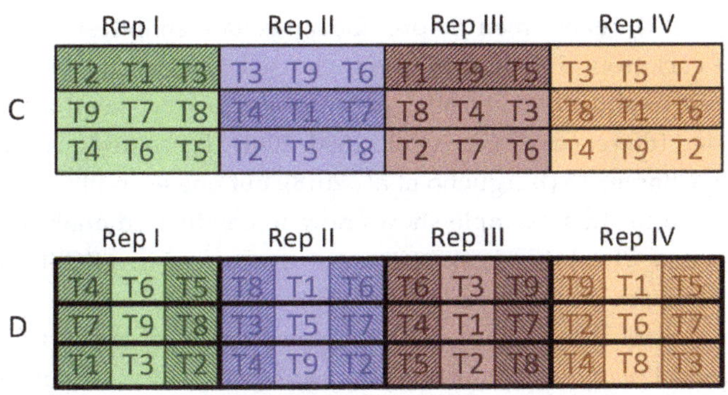

FIG. 2. Four replicates ($r = 4$) of nine treatments ($t = 9$) (36 experimental units) randomized in two incomplete block designs. (C) Balanced lattice design with an incomplete block size $k = 3$ and all treatment pairs occurring once within an incomplete block; each complete block (Rep) is shown by a different color; cross-hatching is used to illustrate the four incomplete blocks ($k = 3$) that contain treatment number 1 (T1). (D) Lattice square with block size $k = 3$ and all treatment pairs occurring once with a column block and once within a row block; cross-hatching (left, none, or right) is used to delineate column blocks ($k = 3$) and bold lines are used to delineate row blocks ($k = 3$) within each of the four reps.

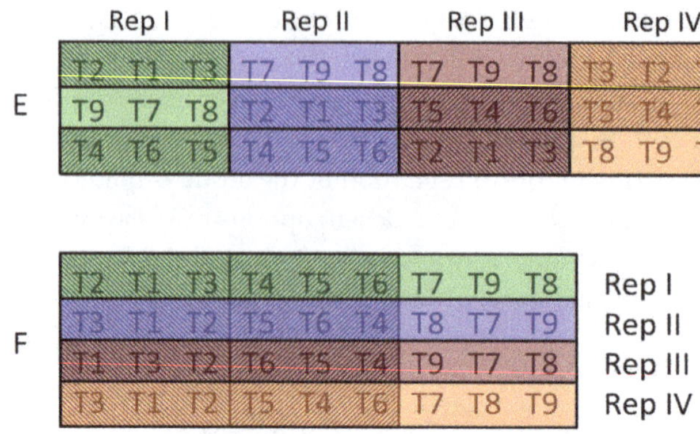

FIG. 3. Four replicates ($r = 4$) of nine treatments ($t = 9$) (36 experimental units) randomized in two incomplete block designs. (E) Blocks-in-reps (sets-in-reps) design with three sets of three treatments each. (F) Reps-in-blocks (reps-in-sets) design with three sets of three treatments each. For both designs, each complete block (Rep) is shown by a different color; cross-hatching is used to illustrate the sets or groups of treatments that remain constant across reps.

times within incomplete blocks. As such, there are many more options for variation in number of treatments and replicates within the subfamily of unbalanced designs.

More recently, α designs were created to provide a mechanism to construct resolvable incomplete block designs for any number of treatments, *t*, and any block size (number of treatments) that is an integer divisor of *t* (Patterson and Williams, 1976). Lattice designs and row-column designs are special cases of α designs, the latter having the advantage of greater flexibility in number of treatments and block size, as well as having bidirectional control of spatial variability associated with both rows and columns (John and Eccleston, 1986; Patterson and Robinson, 1989).

Lastly, augmented, control-plot, and partially replicated (p-rep) designs were created to handle an extremely large number of treatments, generally composed of genotypes or families in plant breeding programs. The basic principle of these designs is that some treatments are replicated and some are not replicated. Replicated treatments are generally control treatments that are repeated numerous times throughout the experiment to provide a mechanism for estimation of spatial variation and background error, which is then used to adjust the observed data on all unreplicated treatments. Typically these designs are repeated at multiple locations to broaden inferences (Casler et al., 2000, 2001). These designs are covered in more detail in Chapter 13 (Burgueño et al., 2018), but one example is provided here (Box 1). The code for this example shows how to conduct an analysis of a simple augmented design where the check cultivars are arranged in a RCBD and each test cultivar is present as a single experimental unit.

Blocks-in-reps or reps-in-blocks are variations on the split-plot theme that will be described in detail in the next section. They are specifically useful in breeding and genetic studies with very large numbers of treatments (genotypes or families) that can be arranged into sets in which genotypes or families are a nested factor (Schutz and Cockerham, 1966; Casler, 1998). In both designs, treatments are randomly assigned to groups or sets and random effects are estimated within each set, and then pooled across sets. Block size is $k = t/s$ where *s* is the number of sets and *t* is the number of treatments. The blocks-in-reps design arranges each set within a complete block of all treatments (Fig. 3E), while the reps-in-blocks design arranges repeated experiments, each of which contains all blocks of one set of treatments (Fig. 3F).

Table 2 illustrates a general principle of blocking: "you never get something for nothing" or "everything has a price." For the hypothetical experiment with nine treatments and four replicates, the RCBD allocates only 3 df to blocking, a number that rises to a maximum of 19 for the lattice square, with all other designs falling in between (Table 2). Each df that is allocated to a blocking factor is removed from the residual variance, affecting power of the hypothesis tests (Chapter 4, Casler, 2018), so that residual variances range in df from 6 to 24 in the various blocking designs shown in Table 2. The challenge is to choose a research design that accounts for spatial variation on the scale and dimensions that occur within the entire experimental area. Degrees of freedom for blocking factors should be "put to work" accounting for significant amounts of variability, but also leaving sufficient df in the residual variance to provide adequate hypothesis tests.

The power of a hypothesis test for a hypothetical experiment can be predicted with prior estimates of variances and some knowledge of the treatment differences to be detected (Chapter 1, Garland-Campbell, 2018; Chapter 4, Casler, 2018). Block designs and estimates of block effects can be incorporated into power analyses using the procedures described in Chapters 1 and 4, so that the direct impact of allocating df to blocking factors (Table 2) can be readily evaluated in terms of predicted power of hypothesis tests. These types of power analyses are the best way to compute the required number of blocks to achieve the goals of any experiment.

Split-Plot Blocking Patterns

The factorial treatment arrangement (Table 3) is the principal driver for the broad family of split-plot blocking patterns (Table 1). Split-plot patterns or arrangements represent an unlimited family of randomization restrictions on a parent design, such as CRD, RCBD, or Latin square design. A common misconception of the split-plot randomization is that it represents an experimental design per se, for example, a split plot with Factor A as the whole-plot factor and Factor B as the subplot factor. In fact, this represents an incomplete description of the experimental design because it is missing the basic design for Factor A, the whole-plot factor. The RCBD is the most common, but many other designs can be used to organize the whole-plot factor into randomized experimental units, such as Latin square (Casler et al., 2000, 2001).

Split-plot randomizations are frequently used strictly for convenience, when the whole-plot factor requires larger experimental units than the subplot factor(s). For example, in Fig. 4G, the whole plot is three times the size of each subplot, with each whole plot containing three subplots. Common examples of whole-plot treatments include tillage treatments, irrigation treatments, and planting dates. The versatility of split plots is illustrated by the fact that multiple splits can be easily incorporated into the arrangement for logistical or convenience purposes; that is, subplots in Fig. 4G can be further split as desired and as logistics allow. A further variation is the split block or strip plot in which there are two whole-plot factors that are stripped across each other (Fig. 4H). Care must be taken to ensure that both whole-plot factors are randomized independently and differently in each replicate of the experiment, rather than strip one factor across the entire experiment without rerandomization.

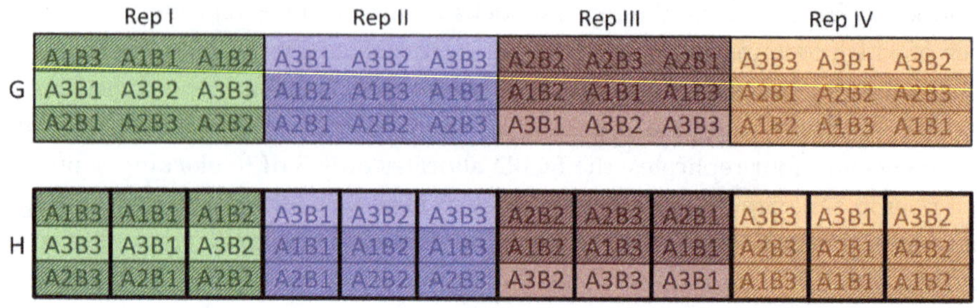

FIG. 4. Four replicates (r = 4) of nine treatments (t = 9) (36 experimental units) randomized in two split-plot arrangements, both based on a 3 ´ 3 factorial treatment structure. (G) Split-plot with Factor A as the whole-plot factor and Factor B as the subplot factor with Factor A arranged in a randomized complete block design; each complete block (Rep) is shown by a different color; cross-hatching is used to illustrate the whole plots. H. Split-block (strip-plot) with both Factor A and Factor B as whole plot factors stripped across each other and re-randomized within each complete block; each complete block (Rep) is shown by a different color; cross-hatching is used to illustrate Factor A strip plots, and thick lines define borders between Factor B strip plots.

Combined use of both strip-plot and traditional split-plot "splits" in one experiment (e.g., Riesterer et al., 2000) illustrate both the complexity and versatility available in these randomization restrictions. For example, each subplot in Fig. 4H could be assigned multiple levels of a factor *C*, creating a strip-split-plot randomization within the RCBD.

Split-plot randomizations are also employed strictly for statistical reasons. In the simplest split-plot, with two factors, there are two error terms: Error(a) is the whole plot error (Ea) and Error(b) is the subplot error (Eb). Their expected values are: Ea = $\sigma^2(1 + \rho)$ and Eb = $\sigma^2(1 - \rho)$, where σ^2 is the unit variance and ρ is the autocorrelation coefficient. The statistical success of the design relies on the empirical relationship Ea > Eb, which results when $\rho > 0$. Split plots employed for statistical reasons are meant to take advantage of this relationship, increasing precision for the subplot factor and the interaction at the expense of precision on the whole-plot factor and whole plot–based simple effects. Note that the split-block design sacrifices precision on both main effects, both of which are whole-plot factors, and increases precision only on the interaction. Researchers who design experiments using the split-plot concept strictly for logistics or convenience will often hope for $\rho \approx 0$, or Ea $\approx$ Eb, especially when they prefer not to have a difference in precision between the whole-plot and subplot factors.

Change-Over (Crossover) Designs

Change-over designs represent a family of designs in which individual subjects receive multiple treatments over the course of the experiment, with each subject receiving each treatment (usually) an equal number of times (Damon and Harvey, 1987; Petersen, 1985). In livestock feeding trials, human nutrition trials, food preference trials, and drug trials, subjects are frequently expensive and highly variable. As such, these types of trials generally have relatively few subjects, and the variability among subjects is sufficiently large that simple blocking designs are not adequate for error control. Change-over designs employ blocking in two directions: (i) spatial blocking is accomplished by grouping subjects into groups to promote within-group homogeneity to the extent possible and (ii) temporal blocking achieved by repeating the experiment across multiple time periods. Temporal blocking is achieved by assigning a series of treatments to each subject over time.

The most common design for livestock feeding trials is the Latin square change-over design (LSCOD), in which the number of subjects, treatments, and time periods are all equal. Each subject receives each treatment in a different order, and each treatment is applied to one subject within a time period. Due to limited df for the residual variance, the LSCOD is typically repeated across multiple squares or groups of livestock. Because each subject receives all treatments in some linear order, residual effects of one treatment on a subsequent treatment are a potential concern in the data analysis phase. These potential residual or carry-over effects are resolved in one of two ways: (i) employing a rest period or uniform-treatment period of sufficient length to reduce or eliminate carry-over effects or (ii) conducting a residual analysis in which the treatment effects and residual effects of each treatment are both estimated and analyzed. The latter is far more complicated than the former and is generally not used in most published livestock trials. Lack of orthogonality between treatment effects and residual effects is a common problem with residual analyses, often solved by employing an extra time period to include each treatment followed by itself, which would otherwise never occur within the LSCOD.

The Special Needs of Glasshouse and Growth Chamber Experimentation

Glasshouse and growth chamber experimentation deserves special mention due to one particular factor: mobility of experimental units (usually pots). One of the most common designs for this type of research is to use the CRD in combination with frequent and systematic rotation of experimental units (Brien et al., 2013; Hardy and Blumenthal, 2008; Kempthorne, 1957). This is a bad practice for a number of reasons. First, rearrangement of experimental units increases the workload and creates more opportunities for mistakes in randomization or damage to plants (Kempthorne, 1957). Second, successful application of this approach requires that experimental units spend similar amounts of time within each microclimate (Brien et al., 2013). Third, moving experimental units results in averaging different residual values over time, leading to underestimation of background variation. Fourth, there is no evidence that this approach will lead to greater precision than the use of formally randomized designs because the rearrangement either eliminates or creates significant challenges in accounting for spatial or microclimate variation (Brien et al., 2013). Finally, there are numerous alternative experimental design options ranging in both complexity and expected effectiveness, sufficient to satisfy any need or situation (Table 1). Microclimate effects within glasshouse rooms or growth chambers are common and predictable over time (Brien et al., 2013; Guertal and Elkins, 1996; Wallihan and Garber, 1971). Thus, randomized blocking designs, especially incomplete block designs tailored to match the scale of variation in the room or chamber, are more effective than pot-rearrangement approaches (Brien et al., 2013). The best approach to glasshouse and growth chamber design is to use designs with small blocks, combined with spatial terms that account for additional sources of variation that may arise during the course of the experiment (see Chapter 12 for more details, Burgueño, 2018).

Fixed vs. Random Effects

Blocking effects should generally be considered to be random effects, except in very rare and special circumstances, as noted in the next paragraph. First, blocks are nearly always intended to represent random samples from a population of circumstances under which the treatments are evaluated. For most experiments, they simply represent a random choice of experimental units, essentially random positions within a field, glasshouse, or other physical situation. Second, if blocks are chosen to have fixed effects, and treatments are fixed, as is nearly always the case, then the block × treatment interaction must also be a fixed effect. Of course, this is completely illogical because a fixed effect cannot serve as an error term. The expected value of the block and treatment mean squares do not contain any terms that allow the use of a fixed block × treatment interaction to be used in the denominator of an F test (Table 4); that is, there is no valid F test for treatments. Treating blocks as a random effect allows the researcher to then assume block × treatment interactions are random effects, effectively allowing them to be used as error terms for treatment F tests. This is particularly critical in the split-plot and split-block design restrictions where the block × treatment interaction is split into multiple error terms (Table 2). It is also critical for recovery of intra- and interblock information in the analysis of many incomplete block designs.

Exceptions can and should be made when there is an obvious and meaningful blocking "factor," something that defines the blocks and that likely has an impact on the measurement variable. Field gradients, glasshouse benches, growth chamber shelves, and livestock traits are all factors that can strongly influence the measurement variable. In cases where these factors are used to define blocks and the researcher is likely to conduct a statistical test of the "blocking factor," followed by an inferential discussion, the "blocking" factor loses its value as replication and essentially becomes another treatment factor. In these cases, researchers should anticipate this problem during the design of the experiment and add an additional blocking factor, such as repeating the experiment at multiple sites or across multiple years. The multiple sites or years would then be treated as a random effect, and its interactions with treatments would then serve as error terms.

To Pool or Not to Pool

The concept of dropping nonsignificant terms from the model has been around for a long time, having been treated very thoroughly by Carmer et al. (1969). This has become a common practice with modern mixed linear models approaches in which many researchers use Akaike's or Bayesian information criteria (AIC or BIC) to determine which terms to include in ANOVA models. For example, many researchers will run a split-plot analysis and may observe that the whole-plot error term is too small to be significant. Rerunning the ANOVA without the whole-plot error term may result in an improved AIC or BIC, giving the researcher justification to leave that term out of the model. The main effect of blocks is often handled in the same manner—

dropped from the ANOVA if that practice leads to a better "fit." Researchers should be aware of the implications of this approach. Dropping terms from the model based on AIC or BIC results in pooling those terms into error terms (Carmer et al., 1969).

There are three distinct philosophies about pooling. Consider the ANOVA on the left side of Table 4, and assume that there is evidence that the blocking effect is either nonsignificant or meaningless (either a nonsignificant *F* test or AIC or BIC were improved by dropping it from the ANOVA model). First, you can take Carmer's "never pool" philosophy that blocks are a fundamental part of the design and must be part of the fundamental ANOVA model, whether significant or not. This tends to be a fairly conservative philosophy because it results in some df that are not "put to work" or attached to meaningful variation.

Second, you can take Carmer's "always pool" philosophy that *any* evidence for a nonsignificant blocking effect should be taken as license to drop it from the model, which amounts to pooling blocks with the error term. This is a fairly liberal or non-conservative approach that avoids lazy degrees of freedom and results in a possible increase in power of treatment *F* tests in small experiments. One cautionary note: it is not a good idea to drop fixed effects from the model in this manner because that will automatically result in pooling nonsignificant fixed effects with random error terms. Pooling fixed and random effects may result in violation of fundamental assumptions regarding normality and independence of residual effects.

Third, you can choose a moderate philosophy in which you make a conscious design "to pool or not to pool" on a case-by-case basis (Carmer et al., 1969). Before the use of mixed models approaches, this was done via *F* test of the block effect in our example from Table 4. The Carmer et al. (1969) middle-of-the-road approach can be implemented using likelihood ratio tests within modern mixed linear models software packages. Start by running the full model with all random effects. Eliminate one random effect at a time, rerunning the model and computing the likelihood ratio test for the eliminated term. Use $\alpha = 0.25$ or 0.50 as a cutoff value to protect against Type 2 errors in pooling random effects that are nonhomogeneous; that is, if $p < 0.25$ or 0.50, the tested term should remain in the model. Box 2 illustrates the use of the likelihood ratio test to make a pooling decision, using a split-plot experiment as an illustration.

The Value of Retrospective Analyses

Research programs that have been in place for many years may benefit substantially from retrospective analyses of historical experiments under the conditions typically

TABLE 4. Expected values of mean squares and *F* tests for treatments in a randomized complete block design in which blocks are considered as a random effect or a fixed effect.

Blocks are random and treatments are fixed				Blocks, treatments, and their interaction are all fixed effects			
Source of variation	df	Expected values of mean squares	*F* test	Source of variation	df	Expected values of mean squares	*F* test
Blocks (B)	$b-1$	$\sigma^2 + \sigma^2_B$		Blocks (B)	$b-1$	$\Sigma\,\beta_j^2/(b-1)$	
Treatments (T)	$t-1$	$\sigma^2 + \Sigma\,\tau_i^2/(t-1)$	MST/MSe	Treatments (T)	$t-1$	$\Sigma\,\tau_i^2/(t-1)$	none
B×T (error)	$(b-1)(t-1)$	σ^2		B×T	$(b-1)(t-1)$	$\Sigma\,\tau\beta_{ij}^2/(t-1)(b-1)$	

used by the program. For example, retrospective analyses have been used to make significant design changes to field evaluations of forage grass cultivars (Casler, 2013, using 28 yr of historical data), field evaluations of cereal-crop cultivars (Patterson and Hunter, 1983, using 7 yr of historical data) and a perennial ryegrass (*Lolium perenne* L.) breeding program (Conaghan et al. (2008), using 14 yr of historical data). The greatest value of retrospective analyses derives from historical data generated on individual experiments conducted over a long period of time on individual fields, within glasshouse rooms, or within growth chambers. Historical experiments provide estimates of error variances and treatment effects that can be modeled to develop inferences for optimizing plot size, block size, block shape, and block orientation (Casler, 2013; Lin and Binns, 1984, 1986; Patterson and Hunter, 1983). Sometimes, such a retrospective analysis can lead to surprising conclusions, such as the advantage of a reduced plot size in combination with sophisticated incomplete block designs that capture spatial variation at the proper scale (Casler, 2013).

Some of our fundamental principles of designing experiments were established during the early and middle of the 20th century as a result of numerous uniformity trials conducted on many species under many experimental conditions (Smith, 1938). While uniformity trials are no longer in vogue and, more realistically, considerably wasteful of scarce resources (Edmondson, 1989), comparative research experiments can be coaxed into providing similar information of use in designing more efficient experiments in the future. Linear mixed models can be used to account for fixed and random effects in an experiment, leaving all unexplained spatial variation in the residual effects. These residuals, which are typically used to conduct diagnostic analyses of the data (Koenker, 2013; Schützenmeister et al., 2012), can be empirically modeled using spatial or trend analyses (see Chapter 12, Burgueño, 2018) or dummy designs (Table 1) with different sizes and shapes of blocks to identify optimal designs. This type of retrospective analysis can be particularly powerful if it is conducted using many years of experimental data and results in reasonably consistent predicted results (e.g., Casler, 2013; Conaghan et al., 2008; Patterson and Hunter, 1983).

Lastly, retrospective analyses can also be conducted on any individual blocking experiment that has been completed. Box 3 provides an illustration of a RCBD experiment that was completed, analyzed, and published. Once the data were used for their intended purpose, the ANOVA residuals were generated and used to conduct an analysis of spatial variation on the experimental site, mainly to determine the relative importance of rows and columns and the scale on which blocking should be conducted for future experiments on this particular field site.

Conclusions

Proper design of biological experiments involves significant advance thought, attention, and planning. Blocking designs are available for a wide range of experimental circumstances that include various treatment structures, a wide range in the numbers of treatments, different experimental goals, and a multitude of blocking patterns. Informed decisions on optimal or desirable blocking designs require some prior knowledge of the experimental units. What is the experimental unit? Can the experimental units be organized into blocks? Is there spatial and/or temporal variability among experimental units? If so, what is the scale and nature of that variability? What degree of homogeneity is required to detect treatment differences? What block size (number of experimental units) is required to achieve that level of homogeneity? Retrospective analyses of prior experiments, either a limited number of recent experiments or a long-term series of experiments conducted over many years, can go a long ways toward helping researchers answer these questions and design efficient and effective blocking experiments.

Key Learning Points

- A block design should be employed in any circumstance in which the researcher expects some level of spatial or temporal variation among observations.
- The most informed choice of a proper block design includes an analysis of the impact of blocking on the *F* tests for treatments, including impacts on error degrees of freedom. Blocking robs degrees of freedom from the error or residual term—it carries a cost, and the researcher should weigh expected benefits vs. costs.
- Blocking can be employed for statistical reasons, logistical reasons, or both. Sometimes the statistical and logistical reasons are at odds with each other, forcing the researcher to make difficult decisions.
- In certain situations, complex block designs can be collapsed into simpler designs if blocks fail to account for significant variability. Fundamental principles of pooling mean squares and model fitting in a restricted maximum likelihood format should be used to make informed decisions.
- Spatial analyses can be used to supplement block designs, but complex block designs can be just as effective, or more so, than spatial analyses.
- Retrospective analyses that examine the long-term empirical impacts of blocking for a particular crop–location combination can be highly useful in planning efficient and effective changes to design protocols.

Exercises

1. Conduct a linear mixed model ANOVA from an augmented design.

PROBLEM. Augmented designs are unbalanced, specifically with reference to test treatments that are typically unreplicated.

2. Make a logical and objective decision regarding whether or not random design effects should be retained in the final model for publication purposes.

PROBLEM. Modern mixed models analysis is often taught in a manner that encourages researchers to used reduced models, containing only those terms that are important. This practice results in pooling random design effects with residual effects. There are three philosophies that can be employed in pooling when it is clear that a design component is small or non-significant: always pool, never pool, or pool using an objective decision tool that seeks to avoid Type 2 errors.

3. Arranging blocks and blocking patterns for future experiments.

PROBLEM. Decisions on exactly how to arrange blocks in many field experiments are very difficult, because there is often little information available to determine if there are gradients and the patterns of any gradients. This is especially true on agricultural experiment stations, which are often located on sites that have a uniform visual appearance.

References

Brien, C.J., B. Berger, H. Rabie, and M. Tester. 2013. Accounting for variation in designing greenhouse experiments with special reference to greenhouses containing plants on conveyor systems. Plant Methods 9:5 doi:10.1186/1746-4811-9-5

Brien, C.J., B.D. Harch, R.L. Correll, and R.A. Bailey. 2011. Multiphase experiments with at least one later laboratory phase. I. Orthogonal designs. J. Agric. Biol. Environ. Stat. 16:422–450. doi:10.1007/s13253-011-0060-z

Burgueño, J. 2018. Spatial analysis of field experiments. In: B. Glaz and K.M. Yeater, editors, Applied statistics in agricultural, biological, and environmental sciences. ASA, CSSA, SSSA, Madison, WI.

Burgueño, J., J. Crossa, F. Rodríguez, and K.M. Yeater. 2018. Augmented design—experimental design with treatments replicated once. In: B. Glaz and K.M. Yeater, Applied statistics in agricultural, biological, and environmental sciences. ASA, CSSA, SSSA, Madison, WI.

Carmer, S.G., W.M. Walker, and R.D. Seif. 1969. Practical suggestions on pooling variances for F-tests of treatment effects. Agron. J. 61:334–336. doi:10.2134/agronj1969.00021962006100020051x

Casler, M.D. 1992. Usefulness of the grid system in phenotypic selection for smooth bromegrass fiber concentration. Euphytica 63:239–243.

Casler, M.D. 1998. Genetic variation within eight populations of perennial forage grasses. Plant Breed. 117:243–249. doi:10.1111/j.1439-0523.1998.tb01933.x

Casler, M.D. 1999. Spatial variation affects precision of perennial cool-season forage grass trials. Agron. J. 91:75–81. doi:10.2134/agronj1999.00021962009100010012x

Casler, M.D. 2013. Finding hidden treasure: A 28-year case study for optimizing experimental designs. Commun. Biometry Crop Sci. 8:23–28.

Casler, M.D. 2015. Fundamentals of experimental design: Guidelines for designing successful experiments. Agron. J. 107:692–705. doi:10.2134/agronj2013.0114

Casler, M.D. 2018. Power and replication—designing powerful experiments. In: B. Glaz and K.M. Yeater, editors, Applied statistics in agricultural, biological, and environmental scineces. ASA, CSSA, SSSA, Madison, WI.

Casler, M.D., S.L. Fales, A.R. McElroy, M.H. Hall, L.D. Hoffman, and K.T. Leath. 2000. Genetic progress from 40 years of orchardgrass breeding in North America measured under hay management. Crop Sci. 40:1019–1025. doi:10.2135/cropsci2000.4041019x

Casler, M.D., S.L. Fales, D.J. Undersander, and A.R. McElroy. 2001. Genetic progress from 40 Years of orchardgrass breeding in North America measured under management intensive rotational grazing. Can. J. Plant Sci. 81:713–721. doi:10.4141/P01-032

Chandra, S. 1994. Efficiency of check-plot designs in unreplicated field trials. Theor. Appl. Genet. 88:618–620. doi:10.1007/BF01240927

Cochran, W.G., and G.M. Cox. 1957. Experimental designs. John Wiley & Sons, New York.

Conaghan, P., M.D. Casler, P. O'Keily, and L.J. Dowley. 2008. Efficiency of indirect selection for dry matter yield based on fresh matter yield in perennial ryegrass sward plots. Crop Sci. 48:127–133. doi:10.2135/cropsci2007.05.0274

Cox, D.R. 1958. Planning of experiments. John Wiley & Sons, New York.

Damon, R.A., and W.R. Harvey. 1987. Experimental design, ANOVA, and regression. Harper & Row, New York.

Dey, A. 2010. Incomplete block designs. World Sci. Publ. Co., Hackensack, NJ.

Draper, N.R., and J.A. John. 1988. Response-surface designs for quantitative and qualitative variables. Technometrics 30:423–428. doi:10.1080/00401706.1988.10488437

Draper, N.R., and D.K.J. Lin. 1990. Small response-surface designs. Technometrics 32:187–194. doi:10.1080/00401706.1990.10484634

Edmondson, R.N. 1989. Glasshouse design for repeatedly harvested crops. Biometrics 45:301–307. doi:10.2307/2532054

Edmondson, R.N. 1998. Trojan square and incomplete Trojan square designs for crop research. J. Agric. Sci. 131:135–142. doi:10.1017/S002185969800567X

Fisher, R.A. 1925. Statistical methods for research workers. Oliver and Boyd, Edinburgh.

Garland-Campbell, K. 2018. Errors in statistical decision making. In: B. Glaz and K.M. Yeater, editors, Applied statistics in agricultural, biological, and environmental sciences. ASA, CSSA, SSSA, Madison, WI.

Guertal, E.A., and C.B. Elkins. 1996. Spatial variability of photosynthetically active radiation in a greenhouse. J. Am. Soc. Hortic. Sci. 121:321–325.

Greenberg, B.G. 1951. Why randomize? Biometrics 7:309–322. doi:10.2307/3001653

Hardy, E.M., and D.M. Blumenthal. 2008. An efficient and inexpensive system for greenhouse pot rotation. HortScience 43:965–966.

Hurlbert, S.H. 1984. Pseudoreplication and the design of ecological field experiments. Ecol. Monogr. 54:187–211. doi:10.2307/1942661

John, J.A., and J.A. Eccleston. 1986. Row-column α-designs. Biometrika 73:301–306.

Jones, B., and M.G. Kenward. 2003. Design and analysis of cross-over trials. Chapman & Hall, CRC Press, NY.

Jones, M., R. Woodward, and J. Stoller. 2015. Increasing precision in agronomic field trials using Latin square designs. Agron. J. 107:20–24. doi:10.2134/agronj14.0232

Kempthorne, O. 1957. 126. Query: Arrangements of pots in greenhouse experiments. Biometrics 13:235–237. doi:10.2307/2527805

Kempthorne, O. 1977. Why randomize? J. Stat. Plan. Inference 1:1–25. doi:10.1016/0378-3758(77)90002-7

Koenker, R. 2013. Quantile regression. Encyclopedia of environmetrics. John Wiley & Sons, New York. doi:10.1002/9780470057339.vnn091

Lin, C.S., and M.R. Binns. 1984. Working rules for determining the plot size and numbers of plots per block in field experiments. J. Agric. Sci. 103:11–15. doi:10.1017/S0021859600043276

Lin, C.S., and M.R. Binns. 1986. Relative efficiency of two randomized block designs having different plot sizes and numbers of replications and plots per block. Agron. J. 78:531–534. doi:10.2134/agronj1986.00021962007800030029x

Lin, C.S., and G. Poushinsky. 1983. A modified augmented design for an early stage of plant selection involving a large number of test lines without replication. Biometrics 39:553–561. doi:10.2307/2531083

Lin, C.S., and G. Poushinsky. 1985. A modified augmented design (type 2) for rectangular plots. Can. J. Plant Sci. 65:743–749. doi:10.4141/cjps85-094

Lucas, J.M. 1976. Which response surface design is best. Technometrics 18:411–417. doi:10.1080/00401706.1976.10489472

Martin, R.J. 1986. On the design of experiments under spatial correlation. Biometrika 73:247–277. doi:10.1093/biomet/73.2.247

Milliken, G.A., and D.E. Johnson. 2009. Analysis of messy data. Vol. 1. Designed experiments. 2nd ed. CRC Press, Boca Raton, FL.

Patterson, H.D., and E.A. Hunter. 1983. The efficiency of incomplete block designs in National List and Recommended List cereal variety trials. J. Agric. Sci. 101:427–433. doi:10.1017/S002185960003776X

Patterson, H.D., and D.L. Robinson. 1989. Row-and-column designs with two replicates. J. Agric. Sci. Cambridge 112:73–77.

Patterson, H.D., and E.R. Williams. 1976. A new class of resolvable incomplete block designs. Biometrika 63:83–92. doi:10.1093/biomet/63.1.83

Petersen, R.G. 1985. Design and analysis of experiments. Marcel-Dekker, New York.

Piepho, H.-P., A. Büchse, and K. Emrich. 2003. A hitchhiker's guide to mixed models for randomized experiments. J. Agron. Crop Sci. 189:310–322. doi:10.1046/j.1439-037X.2003.00049.x

Piepho, H.-P., J. Möhring, and E.R. Williams. 2013. Why randomize agricultural experiments? J. Agron. Crop Sci. 199:374–383. doi:10.1111/jac.12026

Riesterer, J.L., M.D. Casler, D.J. Undersander, and D.K. Combs. 2000. Seasonal yield distribution of cool-season grasses following winter defoliation. Agron. J. 92:974–980. doi:10.2134/agronj2000.925974x

Riley, J., and C.J. Alexander. 1997. Statistical literature for participatory on-farm research. Exp. Agric. 33:73–82. doi:10.1017/S0014479797000185

Schutz, W.M., and C.C. Cockerham. 1966. The effect of field blocking on gain from selection. Biometrics 22:843–863. doi:10.2307/2528078

Schützenmeister, A., U. Jensen, and H.-P. Piepho. 2012. Checking normality and homoscedasticity in the general linear model using diagnostic plots. Comm. Stat. Simul. Comput. 41:141–154. doi:10.1080/03610918.2011.582560

Smith, A.B., P. Lim, and B.R. Cullis. 2006. The design and analysis of multi-phase plant breeding experiments. J. Agric. Sci. 144:393–409. doi:10.1017/S0021859606006319

Smith, H.F. 1938. An empirical law describing heterogeneity in the yields of agricultural crops. J. Agric. Sci. 28:1–23. doi:10.1017/S0021859600050516

Steel, R.G.D., J.H. Torrie, and D.A. Dickey. 1996. Principles and procedures in statistics. 3rd ed. McGraw-Hill, New York.

van Es, H.M., C.P. Gomes, M. Sellman, and C.L. van Es. 2007. Spatially-balanced complete block designs for field experiments. Geoderma 140:346–352. doi:10.1016/j.geoderma.2007.04.017

Wallihan, E.F., and M.J. Garber. 1971. Efficiency of glasshouse pot experiments on rotating versus stationary benches. Plant Physiol. 48:789–791. doi:10.1104/pp.48.6.789

Williams, E.R. 1986. Row and column designs with contiguous replicates. Aust. J. Stat. 28:154–163. doi:10.1111/j.1467-842X.1986.tb00594.x

Williams, E.R., H.-P. Piepho, and D. Whitaker. 2011. Augmented p-rep designs. Biometrical J. 53:19–27. doi:10.1002/bimj.201000102

Wolfinger, R.D., W.T. Federer, and O. Cordero-Brana. 1997. Recovering information in augmented designs, using SAS PROC GLM and PROC MIXED. Agron. J. 89:856–859. doi:10.2134/agronj1997.00021962008900060002x

Youden, W.J. 1940. Experimental designs to increase the accuracy of green-house studies. Contrib. Boyce Thompson Inst. 11:219–228.

CHAPTER 4: POWER AND REPLICATION—DESIGNING POWERFUL EXPERIMENTS

Michael D. Casler

Power is the probability of correctly rejecting a null hypothesis that two or more treatment means are equal to each other when in fact they are different. Designing experiments with high power is critical for detecting small, but biologically meaningful, treatment mean differences and for situations in which the researcher expects the null hypothesis of no treatment differences to represent the true state of nature. Biological researchers should be able to define the experimental unit in every biological research scenario and should be able to replicate treatments at the level of the experimental unit. Power analyses can be extremely effective to provide researchers with an objective mechanism to chose the number of replicates to balance statistical and logistical concerns. Power analyses can also be used to efficiently allocate resources among various types or forms of replication, including locations, years, and sampling units, among others.

"Everything is different from everything else," so it was always said by Prof. Frank N. Martin of the University of Minnesota to the students in his introductory statistics courses. By far, most of the comparative experiments conducted in biological research are designed to detect differences between treatments or systems. As such, researchers create treatment designs in which the individual treatments are viewed as "different" from each other and likely to result in rejection of the null hypothesis for measurement variables of interest to the researcher. Power is simply the probability of correctly rejecting a null hypothesis that two or more treatment means are equal to each other when in fact they are different—in nontechnical terms, the likelihood of "getting it right" with a high degree of confidence, for example, $\alpha = 0.05$ or 95% confidence. The trick, or secret, according to Prof. Martin, is to design experiments that are unlikely to fail in this regard.

Conversely, we occasionally find ourselves in the situation in which our null hypothesis that there are no treatment differences is exactly what we expect to happen. For example, in a breeding and selection program, we often test whether promising new candidate cultivars have disease resistance similar to that of a commercial reference cultivar. In these situations it is imperative that powerful

M.D. Casler, USDA-ARS, US Dairy Forage Research Center, 1925 Linden Dr., Madison, WI 53706-1108 (mdcasler@wisc.edu, michael.casler@ars.usda.gov).

doi:10.2134/appliedstatistics.2015.0075

Applied Statistics in Agricultural, Biological, and Environmental Sciences
Barry Glaz and Kathleen M. Yeater, editors

experiments be designed so that researchers can conclude with reasonable certainty that the true treatment means actually do not differ from each other. Consider a simple experimental situation in which the experiment is completed, the analysis is finished, and the final result is a failure to reject the null hypothesis of no differences among treatment means. What can you conclude? At this point, there are several possible explanations:

1. The treatments are truly not different from each other for the variable of interest.
2. The experimental design failed to account for spatial or temporal variation among experimental units.
3. There were inconsistencies, incongruities, or disturbances introduced during the course of the experiment that compromised the results.
4. There was insufficient replication or the type of replication was poorly chosen.

Explanation 1 relates to the choice of treatments, while the latter three relate to the design and conduct of the experiment, that is, its power. If there is any doubt about the quality, integrity, or power of the design, then it is virtually impossible to develop a definitive conclusion regarding the true differences among treatment means, so one cannot choose Explanation 1. The goal in this situation is to separate Explanation 1 from Explanations 2 through 4. Power, which is 1 – the probability of a Type 2 error (β) (see Chapter 1, Garland–Campbell, 2018), is decreased when any or all of Explanations 2 through 4 occur. Thus, when an experiment fails to detect treatment differences at a desired Type 1 error rate *and* power is low, the reasons for lack of significant differences cannot be clarified. Was the lack of significant differences due to low power, resulting from a design deficiency, a disruption or disturbance, or a lack of differences being the true state of nature?

One goal in designing biological experiments should be to reduce this form of doubt in a reasonable and economical manner, in other words, to create powerful experiments with low probability of a Type 2 error (see Chapter 1, Garland-Campbell, 2018). Blocking, randomization, and spatial analysis are important mechanisms for designing powerful experiments (see Chapters 3, Casler, 2018, and Chapter 12, Burgueño, 2018). Adherence to proper experimental methods, maintaining careful control over experimental conditions and protocols, and taking good notes are all mechanisms to help avoid problems related to Explanation 4 above. Replication is one of the most important factors that can be manipulated by the researcher to enhance power of biological experiments. Most of this chapter will deal with number and type of replicates as they influence power.

Concepts of Replication

What is an Experimental Unit?

First and foremost, proper replication demands that treatments be replicated at the scale of the experimental unit (Casler, 2015). The experimental unit is the smallest unit to which a treatment is applied. Experimental error can only be estimated when treatments are applied to multiple and independent experimental units and data are collected on each individual experimental unit. Replicate observations must occur at a spatial and temporal scale to match the application of treatments (Quinn and Keough, 2002). Replication at the appropriate scale is essential because of the inherent variability that exists within biological systems and to avoid confounding treatment differences with other factors that may vary among experimental units.

A common mistake in designing biological experiments occurs due to pseudoreplication, in which treatments are not truly replicated. Figure 1A represents this design, with each column representing a single treatment or, alternatively, each treatment represents a "block" of samples. In this case, each treatment is applied to only one experimental unit. Multiple samples or data points may be generated within each column, but no matter how many samples of each treatment are collected, columns are completely confounded with treatments. Common examples include: (i) large agricultural or biological engineering experiments designed to compare two or more harvesting systems; (ii) harvesting, storage, or logistical experiments in which each system is assigned one physical space or site; (iii) ecological experiments designed to compare different in situ ecosystems or habitats; (iv) glasshouse or growth chamber experiments designed to compare different environmental conditions; or (v) laboratory experiments in which duplicates or triplicates of each treatment are submitted for analysis as a group. In the latter case, any temporal variation that exists in the laboratory analysis will be confounded with treatments. The lack of randomization and replication are the missing key elements that could be used to avoid this problem. Many laboratory experiments, for example, those dealing with conversion of biomass to bioenergy, simply state that samples were run in duplicate or triplicate, providing no additional information on the randomization or arrangement of samples in time (Casler et al., 2015).

In each of the five examples above (Fig. 1A), the only form of replication is to collect multiple samples within each experimental unit (or column in Fig. 1A), for

A

T1	T2	T3	T4	T5
T1	T2	T3	T4	T5
T1	T2	T3	T4	T5
T1	T2	T3	T4	T5

B

T1	T2	T3	T4	T5
T1	T2	T3	T4	T5
T1	T2	T3	T4	T5
T1	T2	T3	T4	T5

C

T1	T4	T4	T3	T2
T4	T3	T2	T1	T3
T1	T5	T4	T2	T5
T2	T4	T1	T3	T5

D

T1	T4	T2	T3	T5
T4	T1	T3	T5	T2
T5	T2	T4	T3	T1
T1	T3	T5	T4	T2

FIG. 1. Four possible designs of an experiment with five treatments (T1-T5) and four observations per treatment. (A) The experimental unit is a column with one experimental unit per treatment and four observations made on each experimental unit. (B) There are four experimental units per treatment, but they are not randomized. This is a common design in benchtop experiments. (C) There are four independent experimental units per treatment, all randomized together in a completely randomized design. (D) There are four independent experimental units per treatment, grouped into four blocks, randomized within each block in a randomized complete block design.

example, multiple hay bales within a field, multiple pots within a room or chamber, or multiple flasks in the laboratory. For many of these experiments, researchers have made the a priori decision that replication at the experimental unit scale is too expensive or time consuming. The consequences of this decision are twofold: (i) treatments are confounded with experimental units, such that all physical and biological effects that vary among experimental units are intermixed with treatment effects, and (ii) error variances or residual terms are underestimated (Casler, 2013; Casler et al., 2015). The direct consequence of confounding is to mix treatment effects with environmental effects, such as field plots, glasshouse benches, animals, etc. The direct consequence of underestimating the residual variance is that the true Type 1 error rate will be $> \alpha$ by an unknown amount. This is a very undesirable result that is quite common in some research areas (Casler et al., 2015).

Consider hay storage experiments as an example. A common treatment effect is the storage system, which typically includes variations on the following themes: indoor vs. outdoor; grass, gravel, or concrete pad base; or different types of coverings. Numerous hay bales are placed within each of the storage conditions for a certain period of time, without replication of the storage conditions. Each storage "site" is an experimental unit, and each hay bale is a sampling unit as shown in Fig. 1A, where columns correspond to storage sites and treatment and each alphanumeric character corresponds to one hay bale. All differences between storage sites are confounded with treatments. Researchers implicitly assume that the treatment effect itself accounts for all variation among experimental units. Proper replication at the experimental unit scale would involve: (i) two or more independent storage sites for each defined treatment (i.e., two concrete pads, two gravel pads, two indoor sheds, etc.) or (ii) complete repetition of the experiment at another location, during another harvest period, or in another year. A close colleague of mine, a chemist, often jokes, "Why ruin a good experiment by trying to replicate it?" This philosophy seems to be prevalent in quite a few research areas that deal with biological substrates, which are subject to sources of variation that are frequently ignored, misunderstood, or underestimated by non-biologists. As an example, this is a common problem in benchtop digestion experiments of biomass conversion to bioenergy (Casler et al., 2015).

Figure 1B represents one step in the right direction: avoiding the confounding associated with grouping together all units of one treatment. Treatments are "blocked" in either space or time so that the variation associated with the four row blocks can be removed from the residual variance. However, the lack of randomization confounds treatment variation with either spatial or temporal variation within each row block, leading to improper estimates of treatment means and probable underestimation of the residual variance. The design in Fig. 1B is frequently employed for benchtop or laboratory experiments by researchers who are not well versed in the importance and nuances of randomization and replication. For example, in forage, grain, or biomass quality analyses, all tissue samples must be processed through a grinder or mill with blades that become progressively dull over time. As blades lose their sharpness, particle-size distribution gradually changes, resulting in predictable changes in the ability of laboratory procedures to extract or accurately assay chemical compounds. The systematic nature in which treatments are processed through the grinding mill results in confounding of treatments with blade sharpness and particle size, leading to inaccurate estimates of treatment differences.

Proper replication and randomization, as shown in Fig. 1C and 1D, are critical steps toward eliminating this problem for laboratory assays (Casler, 1992). These two figures represent two different and relatively simple designs, the completely randomized design (1C) and the randomized complete block design (1D), which are described in more detail in Chapter 3 (Casler, 2018). The design in 1C should be employed in situations for which the researcher expects limited or no potential for spatial or temporal variation. This has usually been established by the researcher from previous experiments. Conversely, design 1D should be employed if previous experiments have demonstrated any potential for spatial or temporal variation, allowing the row blocks to remove these sources of variation.

Replication on Multiple Scales

Experimental replication can occur in many forms, which can be classified into four basic levels within the experiment: (i) the experimental unit, as discussed above; (ii) replication of the entire experiment, as with multiple locations and/or years; (iii) sampling at one or more levels within experimental units; or (iv) repeated measures. The experimental unit is the most critical and should be given first and primary attention. Once the basic experiment has been designed with multiple and independent experimental units per treatment, then consideration can be given to replication at additional scales or levels. Many experiments are repeated at multiple locations or years, largely to broaden inferences by evaluating treatment differences across a range of environmental conditions, but also allowing estimation of treatment × environment interactions. In some cases, data cannot be collected on the entire experimental unit, requiring multiple samples to be collected within each experimental unit, each sampling unit representing a subset of the experimental unit. Lastly, some variables are measured repeatedly over time during the course of the experiment. If analyzed together in a single ANOVA, these factors are treated as repeated measures (see Chapter 10, Gezan and Carvalho, 2018). Repeated measures can be construed to represent true replication of treatments only when the autocorrelation between repeated measures is zero, which is probably quite rare.

Mixed models analysis provides estimates of random effects that impact a variable of interest. For most mixed model experiments, those in which fixed treatment effects are of primary interest, estimates of random effects are not of specific interest. However, estimates of random effects from previous experiments provide a mechanism for researchers to conduct retrospective exercises in resource allocation. For example, consider an experimental situation in which field studies repeated at multiple locations are an annual activity, such as uniform cultivar evaluations (e.g., Conaghan et al., 2008; Yaklich et al., 2002). Decisions regarding the relative numbers of locations, years, and replicates are critical for designing efficient future experiments, especially if there is a perceived deficiency or inefficiency in the existing design, or some desire arises to improve on the existing design. An ANOVA or mixed models analysis provides information that can be used to develop these inferences. Estimates of random effects from previous experiments are a frequently untapped resource, especially if they have sufficient degrees of freedom to be reliable estimates of true values. Methodology described in the next section can be adapted to predict the impact of changes in resource allocation and design factors on the power of hypothesis tests for future biological experiments.

Replication and Power

The number of replicates required for any given experiment is a function of four variables: (i) the desired Type 1 error rate, (ii) the desired Type 2 error rate, (iii) the residual variance, and (iv) the desired detection level or the difference between treatment means to be detected. The relationships among these four factors can be evaluated in a relatively simple formula that applies to a two-sample t test. The minimum number of replicates required to detect a specified treatment mean difference (n) is

$$n = (t_{\alpha/2} + t_{\beta})^2\sigma^2_D/\delta^2$$

where the two t values are Student's t for the desired Type 1 and Type 2 error rates (α and β, respectively), σ^2_D = the variance of differences or $2\sigma^2$, and δ = the desired detection level between treatment means (Steel et al., 1996, p. 123). Accurate estimates of n require a small number of iterations, because t is dependent on degrees of freedom, which are, in turn, dependent on n. Alternatively, if the number of treatments is sufficiently large and error degrees of freedom can be assumed to be > 30, $Z_{\alpha/2}$ and Z_{β} can be used in place of the two t values.

The formula above illustrates the impact of four factors on the number of replicates for a given experiment. Reductions in either or both Type 1 and Type 2 error rates require an increase in the number of replicates. Thus, increasing the number of replicates results in increased power of hypothesis tests. Larger experimental error or residual variance leads to greater requirements for replication, all other things being equal. Finally, the smaller the desired detection level (δ), the more replicates are required, all other things being equal.

Gbur et al. (2012, Chapter 7) developed a more efficient and generalized manner of solving this problem, illustrating this approach for numerous scenarios. Their approach takes into account all four variables in the above formula, but it can be generalized to include multiple sources of variation and different designs that may optimize allocation of resources across different forms of replication. Five steps are required to implement the probability distribution method to estimate power of a future hypothetical experiment:

1. Obtain an estimate of the experimental error variance and variances of any other pertinent random factors from previous experiments conducted in the appropriate field or laboratory setting or from the literature.
2. Identify the appropriate distribution of the variable of interest, for example, normal, binomial, binary, etc.
3. Determine the p value that will be used to set detection limits [α = probability of a Type 1 error under the null hypothesis] and the minimum difference between treatments to be detected (δ).
4. Choose an experimental design structure that includes all desired blocking arrangements and one or more levels of replication [see Chapter 3 (Casler, 2018) and the previous section of this chapter for more details].

5. Create a representative data set that matches the desired experimental design structure and contains two "dummy" treatments with constant values across all experimental and observational units of the data set. To create such a representative data set, one must choose a single value for all the replication factors in the hypothetical experiment.

Exercise 1 shows SAS code to accomplish a single power analysis in three coding steps: (i) create a representative data set for a completely randomized design with two treatments, four experimental units per treatment (replicates), and two observational units per experimental unit; (ii) use generalized linear mixed models (Proc GLIMMIX) to compute the noncentrality parameter of the F distribution with the appropriate degrees of freedom under the alternative hypothesis that the two treatments have difference δ (100 − 95 = 5 in this example); and (iii) compute power from the noncentrality parameter. The SAS code in Exercise 1 computes power for a test with the following parameters and assumptions: treatment means = 95 and 100, variance components = 5 and 10 (experimental error and sampling error, respectively), $r = 4$ replicates, $s = 2$ observational units per experimental unit, and the assumption of normally distributed errors. All of these values can be changed within the SAS code, including the Type 1 error rate ($\alpha = 0.05$ in this example). In so doing, it is relatively easy to study the direct impact of design decisions on the predicted power of a future experiment. The predicted power for a future experiment with this design is 0.47.

Exercise 2 shows a modified version of this SAS code that contains a macro, automating the code to allow estimation of power for numerous experimental design parameterizations. This code estimates power for this design with number of replicates ranging from 4 to 6 and number of observations per experimental unit ranging from 2 to 3, with a wider range of values displayed in Fig. 2. Such a graphical display easily allows any researcher to choose one of multiple design arrangements that meet the expected target for power, which is often set at 0.8 (80%). For example, the results of this exercise easily illustrate a general principle of resource allocation, that the largest sized replication factors (experimental units in this case) have the greatest direct impact on improving power of future experiments. In the case of Fig. 2, using five different values for number of observational units per experimental unit, any one of five different scenarios can be chosen for use in future experiments, each with an expected power ≈ 0.8.

Using this example and those provided by Gbur et al. (2012), power can be estimated for any imaginable design and sampling scenario, as illustrated in Fig. 2. Furthermore, once the decision is made and the experiment is completed, feedback can be obtained by a retrospective analysis of the experiment and its empirical detection limits. For example, Casler (1998) completed an extensive and expensive series of experiments that were designed to provide means and variance estimates to support a cultivar development program. The critical step involved direct comparison of numerous progeny populations created by various selection methods for which very small differences between means were expected. A power analysis was essential planning before designing the second experiment. Exercise 3 illustrates this power analysis and the process used to select the relatively unusual design employed by Casler (2008). While statistical theory, common sense, and the power

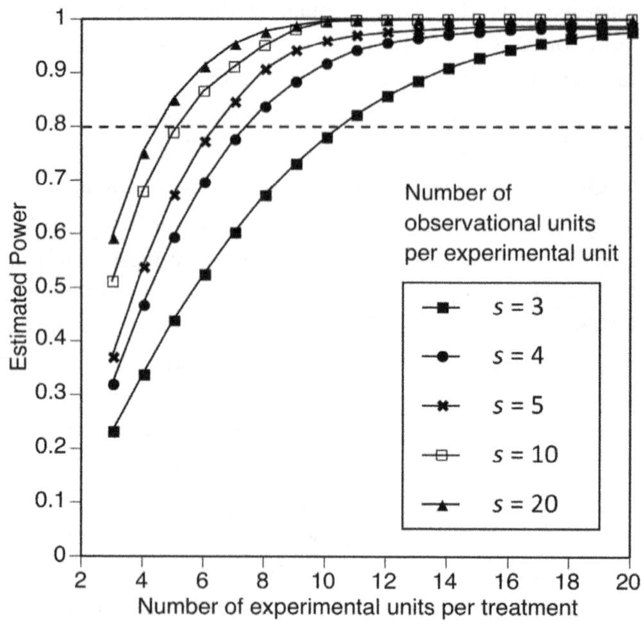

FIG. 2. Estimated power of a hypothesis test designed to detect a treatment difference of 5% of the mean with a Type 1 error rate of 0.05, variance component estimates of 5 and 10 (experimental and sampling errors, respectively), and varying numbers of experimental units and observational units (*s* = 3 to 20). The dashed line represents Power = 0.8 and illustrates that different replication and sampling scenarios can be created to provide similar expected results.

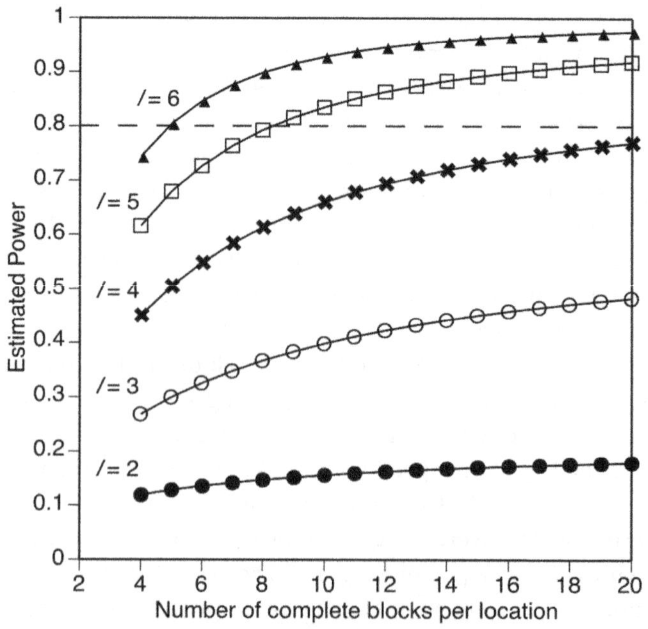

FIG. 3. Estimated power of a hypothesis test designed to detect a treatment difference of 5% of the mean with a Type 1 error rate of 0.05, variance component estimates of 0.02 and 0.2 (treatment × location interaction and experimental error, respectively) and varying numbers of locations (*l* = 2 to 6) and blocks. The dashed line represents Power = 0.8 and illustrates that different combinations of number of locations and replicates can be created to provide similar expected results.

analysis all suggested the need for four to six locations (Fig. 3), practical and logistical considerations led to the choice of three locations and 16 replicates per location (yes, 16 replicates per location—that's not a misprint), as a compromise, arranged in a randomized complete block design (Casler, 2008). The net result of this power exercise was a series of experiments with least significant differences (LSD) ranging from 2 to 3% of the mean and a high frequency of P values < 0.01 (Casler, 2008), that is, an extremely successful and satisfying result with LSD values about one-half of the target (5% of the mean; Exercise 3). This is an excellent practical example of using an a priori power analysis, combined with prior data and expectations, to alter normal replication strategies for a successful experimental outcome.

Conclusions

The potential economic, ecological, and sociological impacts of agricultural research are enormous compared with the low cost of research. However, public and private researchers who conduct experimental research are generally under severe budget constraints. Thus, the economic and emotional costs are high to researchers. As such, biological researchers should always follow the mantra, "failure is not an option." A failed experiment is one with generally high p values combined with low power, leaving the researcher with uncertain or equivocal conclusions: Are the treatments really not different from each other? Was my experimental design faulty due to poor planning and decision making? Was there some unknown and unseen disturbance that occurred to the experiment, causing errors to be inflated? Rarely can these questions be answered when p values are high and power of hypothesis tests is low, typically resulting in difficult-to-publish results and wasted time and money. To borrow an experimental design term, these causal explanations are confounded with each other when treatment effects are nonsignificant and power is low. It is generally impossible to assign cause to one or the other explanation. Designing experiments with predicted power levels as high as economically and logistically possible is critical for detecting expected differences among treatment means and for those experiments in which the null hypothesis of no treatment differences is the expected result.

Key Learning Points

Designing experiments with sufficient power to detect real treatment differences minimizes the chances that a researcher will be faced with inconclusive or equivocal results. Consider the following ideas and guidelines in planning future experiments.

- Be able to understand and define an experimental unit in any experimental situation in which you are involved.
- Strive to always replicate experiments at the level of the experimental unit and include additional levels or types of replication as necessary to improve precision or to broaden inferences.
- Always know the costs associated with various forms of replication so that you can easily conduct cost–benefit analyses of various replication strategies and make conscious decisions about which strategy gives the highest probability of success.

- Have a realistic idea of the treatment differences that you expect from any given experiment so that you can conduct a meaningful power analysis before making final decisions on the form and amount of replication.
- Do not become complacent with regard to replication and power. Conduct retrospective analyses of experiments, comparing your ability to detect differences among treatment means with your a priori power analyses, adjusting both experimental designs and replication strategies as needed to improve power of future experiments.

Exercises

1. Predict the power of a future hypothetical experiment using the probability distribution method.

PROBLEM. Guessing the appropriate number of blocks, replicates, or samples for any biological experiment is not a scientific method and is likely to result in experiments with low power of detecting treatment mean differences.

2. Simultaneously predict the power of several future hypothetical experiments using the probability distribution method and a SAS macro that allows for numerous power computations in a single run.

PROBLEM. There are potentially many satisfactory designs with sufficient predicted power to satisfy the goals of most researchers. A direct comparison of many possible designs is a desirable step toward choosing a statistically and logistically efficient design.

3. Design a new experiment to evaluate treatments that are expected to have very small mean differences, but for which it is critically important to be able to detect differences.

PROBLEM. The experiment to be designed is a follow-up experiment to that described by Casler (1998). Typical designs involve two or three locations and rarely more than four replicates per location. Least significant difference (LSD) values from these typical experiments were perceived to be too large, so a power analysis was conducted to identify the number of replicates and locations necessary to achieve a desired result.

References

Burgueño, J. 2018. Spatial analysis of field experiments. In: B. Glaz and K.M. Yeater, editors, Applied statistics in the agricultural, biological, and environmental sciences. ASA, CSSA, SSSA, Madison, WI.

Casler, M.D. 1992. Usefulness of the grid system in phenotypic selection for smooth bromegrass fiber concentration. Euphytica 63:239–243.

Casler, M.D. 1998. Genetic variation within eight populations of perennial forage grasses. Plant Breed. 117:243–249. doi:10.1111/j.1439-0523.1998.tb01933.x

Casler, M.D. 2008. Among-and-within-family selection in eight forage grass populations. Crop Sci. 48:434–442. doi:10.2135/cropsci2007.05.0267

Casler, M.D. 2013. Finding hidden treasure: A 28-year case study for optimizing experimental designs. Commun. Biometry Crop Sci. 8:23–38.

Casler, M.D. 2015. Fundamentals of experimental design: Guidelines for designing successful experiments. Agron. J. 107:692–705. doi:10.2134/agronj2013.0114

Casler, M.D., W. Vermerris, and R.A. Dixon. 2015. Scale, form, and degree of replication for bioenergy research experiments. BioEnergy Res. 8:1–16. doi:10.1007/s12155-015-9580-7

Casler, M.D. 2018. Blocking principles for biological experiments. In: B. Glaz and K.M. Yeater, editors, Applied statistics in the agricultural, biological, and environmental sciences. ASA, CSSA, SSSA, Madison, WI.

Conaghan, P., M.D. Casler, P. O'Keily, and L.J. Dowley. 2008. Efficiency of indirect selection for dry matter yield based on fresh matter yield in perennial ryegrass sward plots. Crop Sci. 48:127–133. doi:10.2135/cropsci2007.05.0274

Garland-Campbell, K. 2018. Errors in statistical decision making. In: B. Glaz and K.M. Yeater, editors, Applied statistics in the agricultural, biological, and environmental sciences. ASA, CSSA, SSSA, Madison, WI.

Gbur, E.E., W.W. Stroup, K.S. McCarter, S. Durham, L.J. Young, M. Christman, M. West, and M. Kramer. 2012. Analysis of generalized linear mixed models in the agricultural and natural resources sciences. ASA, CSSA, and SSSA, Madison, WI. doi:10.2134/2012.generalized-linear-mixed-models

Gezan, S.A., and M. Carvalho. 2018. Analysis of repeated measures for the biological and agricultural sciences. In: B. Glaz and K.M. Yeater, editors, Applied statistics in the agricultural, biological, and environmental sciences. ASA, CSSA, SSSA, Madison, WI.

Quinn, G.P., and M.J. Keough. 2002. Experimental design and data analysis for biologists. Cambridge Univ. Press, Cambridge.

Steel, R.G.D., J.H. Torrie, and D.A. Dickey. 1996. Principles and procedures in statistics. 3rd ed. McGraw-Hill, New York.

Yaklich, R.W., B. Vinyard, M. Camp, and S. Douglass. 2002. Analysis of seed protein and oil from soybean northern and southern region uniform tests. Crop Sci. 42:1504–1515. doi:10.2135/cropsci2002.1504

CHAPTER 5: MULTIPLE COMPARISON PROCEDURES: THE INS AND OUTS

David J. Saville*

Multiple comparison procedures (MCPs), or mean separation tests, have been the subject of great controversy since the 1950s. In part, this is because any discussion of MCPs raises many related issues that must be discussed simultaneously. These include the whole idea of hypothesis formulation and testing, how to specify statistical contrasts of interest from biological ideas of interest, the definitions of Type 1 and 2 error rates, and the ideas of comparison-wise and experiment-wise error rates. These concepts are at the very heart of the scientific method. This chapter covers the above topics and presents two examples of how *not* to statistically analyze a well-designed experiment—as well as *how to*. Moving on to the main topic, MCPs are an attempt at simultaneously formulating and testing pair-wise comparison hypotheses using data from a single experiment. An unacceptable operating characteristic of most MCPs is their "inconsistency," an idea that is illustrated by way of an example involving Goldilocks and the Four Bears. This leads to the development of a "practical solution" to the MCP problem, which is to simply abandon any attempt at simultaneous formulation and testing. Instead, I recommend using the simplest multiple comparison procedure, the unrestricted least significant difference (LSD) procedure, to: (i) formulate new hypotheses at a known "false discovery rate" (in the null case) such as 5% and (ii) independently test interesting new hypotheses in a second experiment.

The study of multiple comparison procedures (MCPs) is a many-stranded topic that has attracted a spaghetti-like plethora of opinions, both rational and emotive. If a researcher asks for advice as to which procedure is best, wise statisticians stroke their chins and reply, "Well, it all depends." Then, if pressed as to what it depends on, answers often become vague, and time runs out for further discussion. Being young and less wise (back in the early 1980s), I somehow got embroiled in the MCP controversy in my role as a consulting agricultural research biometrician working for the New Zealand Ministry of Agriculture and Fisheries. This came about because I noticed that certain anomalies arose when the restricted LSD procedure was employed for routine data analysis. This led me to make a recommendation as to which MCP was

Abbreviations: HSD, honest significant difference; LSD, least significant difference; MCP, multiple comparison procedure.

Saville Statistical Consulting Limited, 147 Bells Road, West Melton, R.D. 1, Christchurch 7671, New Zealand. *Corresponding author (savillestat@gmail.com).

doi:10.2134/appliedstatistics.2015.0085

Applied Statistics in Agricultural, Biological, and Environmental Sciences
Barry Glaz and Kathleen M. Yeater, editors

best, and in what manner it should be used by applied researchers (Saville, 1985, 1990, 2003, 2015; Saville and Rowarth, 2008). Here I discuss the ins and outs of this whole tangled issue.

In practice, MCPs are commonly used to test for "significant" differences between treatment means in experiments, even in cases when the set of treatments has clear structure and has been derived with obvious questions in mind. In these cases the use of MCPs is inappropriate, as has been pointed out a multitude of times (e.g., Swallow, 1984; Little, 1978). To quote Swallow (1984), MCPs "were developed for cases where the treatment set lacked structure, that is, where the treatments were just a collection of varieties or perhaps chemicals with no particular interrelationships. Most treatment designs are not of this type. Usually, the treatment set has a structure, and the statistical analysis should recognize that structure." This can be achieved by specifying appropriate contrasts between the treatments, with each contrast addressing a particular question of interest to the researcher. In many cases, these contrasts can be chosen to be "orthogonal" (mutually independent), but this is not essential. To explain this more fully, two examples are given of experiments, one real-life and one pseudo real-life, along with the relevant orthogonal contrasts, the statistical analysis, how this leads to a tidy and elegant presentation of the results, and the consequences of ignoring this structure in the set of treatments.

Journals often encourage researchers to thus tailor their statistical analysis to the objectives of the research, but the specification of appropriate contrasts is not a skill easily acquired by researchers, and help from a biometrician is not always available, so this encouragement is often to no avail. As an example of such advice, in the instructions to authors of *Agronomy Journal* (https://dl.sciencesocieties.org/publications/aj/instructions-to-authors, accessed 14 Apr. 2016), the statistical methods section warns of the limitations of MCPs, with a closing statement: "When treatments have a logical structure, orthogonal contrasts among treatments should be used."

Thus, the long-running debate on the relative merits of the many different MCPs is relevant only to the minority of studies in which such a procedure is appropriate.

I first introduce some necessary statistical ideas and terminology, give two examples of correct usage of orthogonal contrasts, and discuss the general topic of MCPs in relation to various types of error rate and in relation to the levels of conservatism of some of the better known MCPs. The idea of inconsistency is then introduced and discussed, with particular attention to Fisher's restricted LSD procedure, the MCP most commonly used in *Agronomy Journal* (Saville and Rowarth, 2008). Finally, a practical solution to the problem of best choice of MCP is described. This consists of using the simplest of procedures, the unrestricted LSD procedure, with the proviso that it be regarded as a hypothesis formulation tool, with any interesting pair-wise hypotheses thus formulated requiring testing in a second, independent experiment.

Types of Statistical Error

In drawing statistical conclusions from an experiment, the hope is that all of your decisions will be correct. For example, if there is truly no difference (e.g., between two treatment means), then the correct decision is to decide that there is no difference (Table 1). Similarly, if there is truly a difference, then the correct decision is to decide that there *is* a difference (Table 1).

In real life, not all decisions will be correct, and statisticians refer to two types of error that one can make. Loosely speaking, one error is to find things that are not there and the second is to not find things that *are* there. These errors are sometimes referred to as false positives and false negatives, respectively.

Formally, the first type of error (Type 1) is to erroneously declare a null effect to be real, or non-zero (Table 1). If your statistical test has a "5% level of significance", this means it has been constructed so that the probability of making a Type 1 error is $\alpha = 0.05$ (or 5%). Or with a 1% level test, the Type 1 error rate is $\alpha = 0.01$ (or 1%).

Conversely, a Type 2 error is to fail to find a real (non-zero) effect (Table 1). When designing your experiment, you may carry out a "power analysis" (see Chapter 4 of this volume, Casler, 2018) to determine the sample size required to reduce the Type 2 error rate to a specified low level. Here the "power" of the test is the probability of deciding there is a non-zero difference when the difference truly is non-zero; that is, the power is the inverse of the Type 2 error rate (so the two add up to 1.00).

In an ideal experiment you are able to minimize the chance of both types of error. In real life, resources are limiting, so if you specify a low chance of one type of error, you automatically increase the chance of the other type of error. For example, if you decide to reduce the chance of a Type 1 error by insisting on a 1% significant difference rather than a 5% significant difference (before you report the difference as non-zero), then you increase the chance of a Type 2 error ("not finding" a truly non-zero difference) if all other features of the experiment remain unchanged. The only way to reduce both chances is to increase the number of replications or adopt a more sophisticated experimental design, perhaps involving a more sophisticated block layout to reduce the residual error term.

When studying error rates associated with different MCPs, statisticians have found it relatively easy to study experiments in which there are truly *no* effects, but have found it less convenient to study experiments in which there are real effects (which can occur in an infinite variety of ways, a few of which were studied by Carmer and Swanson, 1971). As a result, statisticians in general have spent more time studying Type 1 errors than Type 2 errors. Researchers, on the other hand, are often more worried about not detecting a real effect, so are more concerned about Type 2 errors than Type 1 errors. Historically, there has been a difference in mind set—statisticians have tended to be more preoccupied with Type 1 errors, while

TABLE 1. The four cases that can occur in relation to a statistically based decision concerning whether a particular difference is zero. In two cases, the correct decision is made, and in two cases an error occurs (reproduced from Saville, 2015).

		The truth	
		No difference	Difference
Decision	No difference	Correct decision	Type 2 error
	Difference	Type 1 error	Correct decision
		Probability of Type 1 error = $\alpha = 0.05$ (for 5% level significance test)	Probability of correct decision (above) = "Power of test"

researchers have tended to be more preoccupied with Type 2 errors. (See also Chapter 1 of this volume, Garland-Campbell, 2018, for a more detailed discussion of Type 1 and 2 errors.)

Traditional Hypothesis Testing Scenario

The traditional hypothesis testing scenario goes something like the following:

1. A researcher has an idea (i.e., hypothesis). For example, "legumes out-yield non-legumes."
2. The researcher plans an experiment to test this idea (hypothesis). For example, they may decide to include six treatments, consisting of three legumes and three non-legumes.
3. The biometrician advises that the "contrast" (i.e., comparison) corresponding to the idea (i.e., hypothesis) of interest is the average of the treatment mean yields for legumes minus that for non-legumes. This is a *pre-specified contrast*. For information on how to match ideas (hypotheses) to contrasts/comparisons, refer to Cochran and Cox (1957) or Saville and Wood (1991).
4. After the data have been collected, an analysis of variance (ANOVA) is performed. An F test of the hypothesis "true contrast value = 0" is performed as part of the ANOVA; this is either significant or not.
5. Conclusion: The idea (hypothesis) that "legumes out-yield non-legumes" is either confirmed or not by the experiment.

How To Get The Statistical Analysis Wrong

Garlic Example

To illustrate the above ideas, I present an example of a well-designed experiment that examined the effects of plant spacing within and between rows on the yield of garlic (*Allium sativum* L.) (Table 2) (Saville and Wood, 1991).

If naively statistically analyzed using ANOVA with four blocks of six treatments, results are as given in Table 2, with lettering assigned to treatment means on the basis of the unrestricted LSD procedure ($\alpha = 0.05$). But where does this get us? It tells us that the "5-cm spacing, 5 rows" treatment is the highest yielding, although not significantly higher than the "5-cm spacing, 4 rows" treatment, but it does not highlight the relative effects of spacing within and between rows.

A much more enlightening ANOVA is one that recognizes the treatment structure, which is a two (spacings within rows) by three (number of rows per bed) factorial design, with linear and quadratic polynomial components specified for the second factor, as in Saville and Wood (1991). The resulting ANOVA table (Table 3) tells us that the "main effect" of spacing within rows is statistically significant ($p < 0.001$), as is the linear component of the "main effect" of number of rows per bed ($p < 0.01$). The quadratic (curvature) component of the "main effect" of number of rows per bed is not significant, and the linear and quadratic components of the interaction of the two treatment factors are not significant (using a 5% level test). (For an explanation of the statistical term *interaction*, see Chapter 7 of this volume, Vargas et al., 2018.)

TABLE 2. Weight of garlic (kg ha^{-1}) for a randomized complete block design experiment carried out in Marlborough, New Zealand. Plant spacings within each row were 5 and 10 cm, and there were 3, 4, and 5 rows per one meter wide bed. Treatment means are also given, with lettering assigned on the basis of an unrestricted LSD procedure (two means do not differ significantly at $p = 0.05$ if they have a letter in common).†

Treatment	Weight of garlic					
	Block 1	2	3	4	Mean	
	kg ha^{-1}					
5-cm spacing, 3 rows	5110	4040	8405	7045	**6150**	**bc**
5-cm spacing, 4 rows	6675	7000	8150	8145	**7493**	**ab**
5-cm spacing, 5 rows	7265	7230	6660	9259	**7604**	**a**
10-cm spacing, 3 rows	3710	3785	4637	4635	**4192**	**d**
10-cm spacing, 4 rows	4095	3730	4985	6540	**4838**	**cd**
10-cm spacing, 5 rows	6415	5355	5070	5875	**5679**	**c**
LSD(5%)					**1392**	

† Reproduced from Table 12.8 in Chapter 12 of Saville and Wood, 1991, with kind permission from Springer Science+Business Media.

TABLE 3. Analysis of variance table for the weight of garlic (kg ha^{-1}) data given in Table 2. The statistical model includes a block term and the five single degree of freedom contrasts for a 2 × 3 factorial design. The first contrast compares 5-cm with 10-cm spacing within rows. The next two contrasts (indented) are the linear and quadratic curvature contrasts for the factor "number of rows per bed" (3, 4, and 5). The next two contrasts (indented) are the interactions of the first contrast with the last two contrasts (based on Table 12.10 in Saville and Wood, 1991).

Source of variation	Degrees of freedom	Sum of squares	Mean square	*F* ratio	*p* value
Blocks	3	10,823,281			
Main effect of spacing (5 vs. 10 cm)	1	28,496,963	28,496,963	33.40	< 0.001
Main effect of no. of rows per bed	2	9,004,306			
Linear trend	1	8,646,540	8,646,540	10.14	0.006
Quadratic curvature	1	357,765	357,765	0.42	0.527
Spacing × (no. rows/bed) interaction	2	679,899			
Spacing × rows/bed (linear)	1	1,122	1,122	0.00	0.972
Spacing × rows/bed (quadratic)	1	678,776	678,776	0.80	0.386
Residual (or error)	15	12,796,691	853,113		
Total	23	61,801,139			

This information inspires us to present our results in graphical form (Fig. 1). The 5- and 10-cm spacings differ significantly, so they are shown separately. Neither of the curvature terms is significant, so straight lines are presented rather than quadratic curves. The average slope of the lines is significant, so the lines are shown with non-zero slope. The slopes of the two lines do not differ significantly, so parallel lines are plotted. Lastly, a vertical bar is plotted to display the

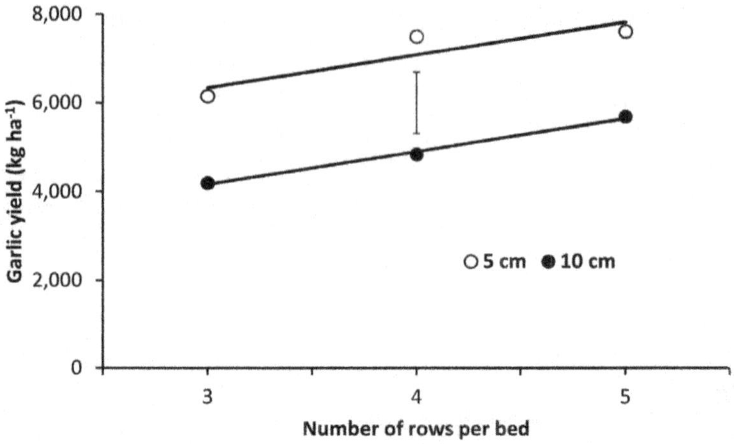

FIG. 1. Mean garlic yields (kg ha^{-1}) for the 5-cm (○) and 10-cm (●) treatments plotted against number of rows per bed (3, 4, and 5). The lines are parallel regression lines. The vertical bar gives the length of the LSD (5%) (based on Fig. 12.6 in Saville and Wood, 1991).

LSD(0.05) in case the reader needs to statistically compare two treatment means for an unforeseen special purpose.

In Fig. 1, the eye is drawn to the main features of the results. The 5-cm spacing is seen to give higher yields than the 10-cm spacing, the difference in yield between the 5- and 10-cm spacings is seen to be roughly constant across the three numbers of rows per bed (3, 4, and 5), and the yield is seen to increase with increasing number of rows per bed.

In summary, to analyze such data using an MCP is not only wrong, it is also less enlightening than a more correct analysis that recognizes the ideas that the researcher had at the design stage.

Alfalfa Example

As a further example, Table 4 presents data from an experiment involving three rates of K fertilizer applied to three varieties of alfalfa (*Medicago sativa* L.). These data came from a university statistics course, and I do not know the original source of the data and have altered the treatment means to make it a more interesting example (by adding or subtracting a constant to or from all of the yields for each treatment). That is, these data are "pseudo real-life."

If statistically analyzed naively using ANOVA as four blocks of nine treatments, the result is the ANOVA table given in Table 5a, which tells us that *overall,* there is evidence of differences between the nine treatment means ($p < 0.01$). If we then proceed to assign letters to the means on the basis of the unrestricted LSD procedure for $\alpha = 0.05$, we get the lettering given in the last column of Table 4. But where does this get us? It tells us that the three highest yielding treatments, those with the letter "a" assigned, are the three treatments where the rate of K is at the highest rate, 30 kg K ha^{-1}. It also tells us that the US cultivar at the lowest rate of 10 kg K ha^{-1}, assigned the letter "c," was significantly lower yielding than the above three treatments (the three cultivars at the highest rate of K) and the NZ_1 cultivar with 20 kg K ha^{-1}.

A somewhat more enlightening ANOVA is one that recognizes the treatment structure, which is a three (cultivars) by three (rates of K fertilizer) factorial design

TABLE 4. Yield of alfalfa (kg plot^{-1}) for a randomized complete block design experiment (fictitious data, in that the original treatment means have been altered). Two New Zealand (NZ) cultivars are compared with a US cultivar, at three rates of K fertilizer. Treatment means are also given, with lettering assigned on the basis of an unrestricted LSD procedure (two means do not differ significantly at $p = 0.05$ if they have a letter in common).

Treatment	Yield					
	Block 1	2	3	4	Mean	
	kg plot^{-1}					
NZ_1, 10 kg K ha^{-1}	2.17	1.88	1.62	2.34	**2.002**	**bc**
NZ_1, 20 kg K ha^{-1}	2.98	1.66	1.62	1.99	**2.062**	**b**
NZ_1, 30 kg K ha^{-1}	2.69	2.00	2.07	2.31	**2.267**	**ab**
NZ_2, 10 kg K ha^{-1}	2.33	2.01	1.70	1.78	**1.955**	**bc**
NZ_2, 20 kg K ha^{-1}	2.26	1.70	2.25	1.49	**1.925**	**bc**
NZ_2, 30 kg K ha^{-1}	2.36	2.10	2.21	1.94	**2.152**	**ab**
US, 10 kg K ha^{-1}	1.69	1.65	1.83	1.48	**1.662**	**c**
US, 20 kg K ha^{-1}	2.18	1.87	2.20	1.77	**2.005**	**bc**
US, 30 kg K ha^{-1}	2.75	2.41	2.62	2.36	**2.535**	**a**
LSD(5%)					**0.383**	

(Table 5b). Here the "main effect" of cultivar is not statistically significant ($p = 0.650$), the "main effect" of fertilizer is very significant ($p < 0.01$), and the interaction between cultivar and fertilizer is not statistically significant ($p = 0.123$). This suggests, after an examination of the means, that the highest rate of K fertilizer gives rise to the highest alfalfa yield, regardless of cultivar.

The most enlightening ANOVA, however, comes from zeroing in on the comparisons of most interest. For the cultivar factor, the questions of interest were: "Did the NZ cultivars differ in yield from the US cultivar?" and "Did one NZ cultivar differ in yield to the other NZ cultivar?" These two questions give rise to contrasts of the *class comparisons* type, where the class of NZ cultivars consists of two cultivars, and the class of US cultivars consists rather trivially of just one cultivar. For the fertilizer factor, the questions of interest were: "Was there a trend for alfalfa yields to increase or decrease as the rate of K increased?" and "Was there any curvature about such a trend line, when yield was plotted against K rate?" These two questions give rise to contrasts of the *polynomial* type, in particular, the linear (trend) contrast and the quadratic (curvature about the line) contrast. (In the case of three rates of K, a maximum of $2 = 3 - 1$ orthogonal polynomial contrasts can be specified; nothing is lost by specifying the quadratic contrast as well as the linear contrast, and new information may perhaps come to light.)

When specified in the context of a factorial treatment structure, these contrasts generate "main effect" contrasts and "interaction" contrasts, which are contrasts of the *factorial* type.

The bolded rows in Table 5c present the results of significance tests for each of the eight orthogonal contrasts, of which the first four are main effect contrasts, and the last four are interaction contrasts. The cultivar main effect contrasts revealed no

TABLE 5. Analysis of variance tables for the yield of alfalfa (kg plot^{-1}) data given in Table 4. (a) The statistical model includes block and treatment terms. (b) The treatment term is broken up to recognize the 3 × 3 factorial structure. (c) Each of the treatment factors has two appropriate orthogonal contrasts specified. For the cultivar factor, these are class comparisons comparing (i) NZ with US cultivars and (ii) one NZ cultivar with the other NZ cultivar. For the rates of K factor, these are linear and quadratic orthogonal polynomial contrasts. Rows of special interest are in bold italic type.

Source of variation	Degrees of freedom	Sum of squares	Mean square	*F* ratio	*p* value
(a) Terms block and treatment					
block	3	1.24047	0.41349	6.00	0.003
treatment	***8***	***1.88299***	***0.23537***	***3.42***	***0.009***
Residual	24	1.65390	0.06891		
Total	35	4.77736			
(b) Terms block, cultivar, fertilizer and their interaction					
block	3	1.24047	0.41349	6.00	0.003
cultivar	***2***	***0.06036***	***0.03018***	***0.44***	***0.650***
fertilizer	***2***	***1.26551***	***0.63275***	***9.18***	***0.001***
cultivar × fertilizer	***4***	***0.55713***	***0.13928***	***2.02***	***0.123***
Residual	24	1.65390	0.06891		
Total	35	4.77736			
(c) Terms block and contrasts for both cultivar and fertilizer and their interactions					
block	3	1.24047	0.41349	6.00	0.003
cultivar	2	0.06036	0.03018	0.44	0.650
(NZ vs. US)	***1***	***0.00036***	***0.00036***	***0.01***	***0.943***
(NZ_1 vs. NZ_2)	***1***	***0.06000***	***0.06000***	***0.87***	***0.360***
fertilizer	2	1.26551	0.63275	9.18	0.001
K(Linear trend)	***1***	***1.18815***	***1.18815***	***17.24***	***< 0.001***
K(Quadratic curvature)	***1***	***0.07736***	***0.07736***	***1.12***	***0.300***
cultivar × fertilizer	4	0.55713	0.13928	2.02	0.123
(NZ vs. US) × K(Lin)	***1***	***0.54827***	***0.54827***	***7.96***	***0.009***
(NZ_1 vs. NZ_2) × K(Lin)	***1***	***0.00456***	***0.00456***	***0.07***	***0.799***
(NZ vs. US) × K(Quad)	***1***	***0.00008***	***0.00008***	***0.00***	***0.972***
(NZ_1 vs. NZ_2) × K(Quad)	***1***	***0.00422***	***0.00422***	***0.06***	***0.807***
Residual	24	1.65390	0.06891		
Total	35	4.77736			

significant difference between the US and the NZ cultivars, nor between the two NZ cultivars. The K fertilizer main effect contrasts revealed a highly significant positive trend of increasing yield (averaged across the three cultivars) with increasing K rate ($p < 0.001$), along with no evidence of curvature about the linear trend ($p = 0.300$).

That is, the four main effect contrasts reveal little that was not shown by the 3 × 3 factorial analysis shown in Table 5b.

The four interaction contrasts reveal new information, however. In particular, the interaction between the "NZ vs. U.S." contrast and the linear contrast for K rate is statistically significant ($p < 0.01$). This suggests that the U.S. and the NZ cultivars are responding differently to increasing rate of K. This new information has come to light because of the much improved power associated with single degree of freedom contrasts, which zero in on questions of special interest, thereby avoiding the dilution effects associated with the four degrees of freedom test of interaction given in Table 5b.

This new information inspires us to present our results in graphical form (Fig. 2) in a similar manner to our preceding example (Fig. 1). Even though the two NZ cultivars have not been proven different in any way, I shall present their mean values and fitted lines separately. None of the curvature terms is significant, so for all cultivars, straight lines are presented rather than quadratic curves. Lastly, a vertical bar is plotted to display the LSD(5%) in case the reader needs to statistically compare two treatment means for an unforeseen special purpose.

In Fig. 2, the eye is drawn to the main features of the results. The two NZ cultivars are seen to be relatively nonresponsive to increasing the rate of K fertilizer, while the U.S. cultivar is seen to be responsive.

In summary, to analyze such data using an MCP is not nearly as helpful as to carry out an analysis that incorporates as much detail as possible concerning meaningful comparisons that are of interest to the researcher. In this example, the use of a factorial structure without the specification of tailor-made contrasts was also not as helpful as the analysis that employed a full set of eight orthogonal contrasts (here eight is one less than the number of treatments, nine).

Ideas, Hypotheses, and Contrasts of Various Types

In the last example we saw three types of contrast used. The fourth type is somewhat trivial in nature; it is the pairwise comparison. That is, there are four main types of contrast:

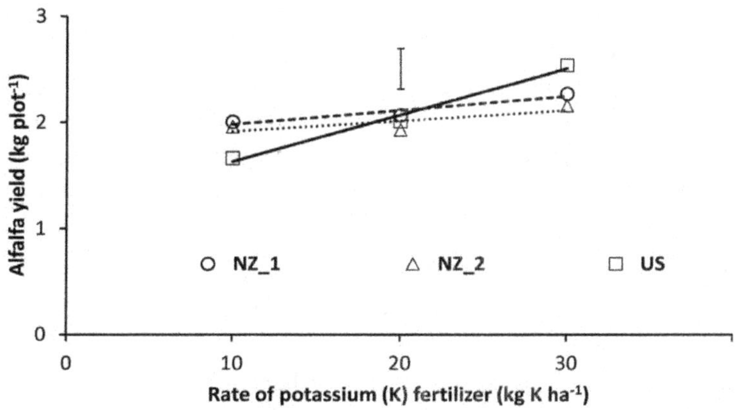

FIG. 2. Mean alfalfa yields (kg plot^{-1}) for three cultivars (NZ_1, NZ_2, and US) plotted against rate of K fertilizer (10, 20, and 30 kg K ha^{-1}). The lines are independently fitted regression lines. The vertical bar gives the length of the LSD (5%).

- Class comparisons.
- Factorial contrasts.
- Polynomial contrasts.
- Pairwise comparisons.

In any real-life experiment, the specification of an appropriate set of contrasts may involve several of these four types of contrast, as it did in our last example. In a tidy mathematical world, these contrasts are "orthogonal" (i.e., statistically independent), but in a real-life experiment they are not always orthogonal and need not be. The matching of ideas to contrasts and hypothesis tests is an art that post-graduate students in agriculture, and agricultural researchers in general, find difficult to master, as I discovered while teaching these ideas in the Department of Agronomy and Range Science at the University of California at Davis in 1985. For information on how to match contrasts to data sets and the related geometric ideas see Saville and Wood (1991, p. 133–295).

In summary, the above examples and related discussion are aimed at encouraging researchers to tailor their statistical analyses so that they are the best possible fit to the ideas underlying their research objectives. If necessary, this may involve obtaining assistance from a biometrician about how to specify the appropriate contrasts or how best to specify and fit some other appropriate model (like an asymptotic regression). Only when such possibilities are exhausted should a multiple comparison procedure (MCP) be considered. In practice, this means that MCPs are appropriate in only a minority of studies. In the rest of this chapter, I shall discuss, in some detail, how to handle this minority case.

What If There Are No Prior Ideas?

If there are no prior ideas (hypotheses), there are two cases that occur. The first, rather trivial case is that the experiment involves *only two treatments*, in which case there *is* presumably a prior plan, which is to compare the two treatments. In this case, there is good news—statisticians agree on how to statistically compare the two treatment means: use an ANOVA which, by default, includes an F test of the hypothesis "true difference between the two treatment means = 0."

The second case is that the experiment involves *more than two treatments*, but there are no prespecified contrasts (or ideas/hypotheses). Here the researcher often *did* have reasons for doing the experiment, but the underlying ideas were not articulated. In this case, there is bad news—statisticians do not agree on how to statistically compare the treatment means (two at a time) and have suggested scores of different MCPs, which have been the subject of great controversy since the 1950s.

Multiple Comparison Procedures

Essentially, MCPs are an attempt at simultaneously formulating and testing pairwise comparison hypotheses using data from a single experiment. Statisticians view this as similar to data-dredging. They think "What if all treatments are truly equal?" and worry about the number of false 5% level significances, or Type 1 errors, that can occur in this case. For example, with 20 truly identical treatments, there are ${}_{20}C_2 = 20$

$\times$ 19/2 = 190 null comparisons, so with a Type 1 error rate of 0.05, there would be, on average, 190 $\times$ 0.05 = 9.5 Type 1 errors.

Such thinking about the probability of committing Type 1 errors in the null case generates a desire to build conservatism into an MCP procedure. The question is: "How much conservatism should be built into an MCP?" Unfortunately, no one agrees on the answer. To date, scores of MCPs have been proposed, and new MCPs are currently being proposed, all with differing amounts of conservatism.

Ordering of Multiple Comparison Procedures by Level of Conservatism

Of those MCPs that are commonly used by research workers, the three *most conservative* MCPs, in terms of the ability of the researcher to declare significant differences between means, are Bonferroni, Tukey's Honest Significant Difference (HSD), and Student–Newman–Keuls. The first two of these MCPs are based on thinking about the *experiment-wise* Type 1 error rate, which is the probability that at least one Type 1 error occurs in an experiment that truly includes no treatment effects.

On the other hand, the *least conservative* MCPs are Duncan's multiple range test and the unrestricted LSD procedure. The latter is based on thinking about the *comparison-wise* Type 1 error rate, which is the probability of Type 1 error per null comparison.

Fisher's restricted LSD procedure is somewhere in between the above extremes in terms of level of conservatism. If the overall *F* test is statistically significant, then it reduces to the unrestricted LSD procedure. Overall, it is *variable* in its level of conservatism.

What Is the Natural Unit?

A related question is: "What is the natural unit for the statistical analysis?"

If the answer is the *comparison* (between any two treatment means), then the MCP to use is the unrestricted LSD procedure (with, e.g., $\alpha = 0.05$), which will falsely declare 5% of null differences to be significant. In this case, the comparison-wise Type 1 error rate is 5%.

If the answer is the *experiment,* then the MCP to use is the more conservative Tukey's HSD procedure, which will falsely declare a null difference to be significant in only 5% of null experiments. In this case, the experiment-wise Type 1 error rate (as defined in the last sub-section) is 5%. This MCP has a much reduced comparison-wise Type 1 error rate, but pays the price of a much increased Type 2 error rate.

But why stop there? If the answer is the *project,* consisting of several experiments, then the MCP to use is an even more conservative, yet-to-be-invented procedure that will falsely declare a null difference to be significant in only 5% of null projects. This procedure would have an even lower comparison-wise Type 1 error rate, but would pay an even higher price in terms of an increased Type 2 error rate.

The logical next step in this sequence of possible natural units is the *research program,* consisting of several projects. Here the MCP to use would be even more conservative. (Or, if your statistician is particularly keen on not making errors of the Type 1 variety, he or she may want the natural unit to be the *lifetime of your statistician.* This would have particularly dire consequences in terms of your hope of achieving any statistically significant effects!)

In all seriousness, I would argue that for the researcher, the *individual comparison* is the natural unit. Once you depart from it, there is no natural stopping point (experiment, project, research program, lifetime, etc.). Consequently, in situations in which usage of an MCP is appropriate, the unrestricted LSD should be used with the full understanding that the false discovery rate is whatever the researcher chooses his/her Type 1 error rate (α) to be.

Inconsistency of Multiple Comparison Procedures

In general, the more conservative an MCP, the more inconsistent it is. The term *inconsistency* is now defined, and its undesirability is explained.

By definition, an MCP is called inconsistent if for any two treatment means, the probability of judging them to be "significantly" different depends on either the number of treatments included in the analysis or the values of the other treatment means (Saville, 1990).

Goldilocks and the Four Bears

To illustrate this idea, I reproduce Table 1 and accompanying text from Saville and Rowarth (2008):

In this fictitious example, I borrow the terminology of Carmer and Walker (1982), Saville (1985 and 2015) and Saville and Rowarth (2008), and consider the case of a statistician, Goldilocks, who has analyzed data for four clients, Baby Bear, Mama Bear, Papa Bear, and Grandpa Bear (the latter has recently come to live with the family). The Bears are all keen porridge eaters and each had performed an experiment with eight oat (*Avena sativa* L.) cultivars in an attempt to increase their oat production. Each of the four experiments was laid out in a randomized complete block design with four replications. By coincidence, each experiment included six common cultivars, and these cultivars happened to have identical oat yield data in all four experiments (Table 6). The other two cultivars differed between the experiments, and their names are not specified; their oat yield data differed in their means but not in their standard deviations between experiments. Goldilocks' statistical analysis using the Fisher's restricted LSD procedure for each of the four experiments is summarized in Table 6 in terms of mean oat yield in kg ha^{-1} for each cultivar, the pooled standard error of the mean (SEM) (which turned out to be 200 in all four experiments), the overall F value and its significance, and the LSD for 5%, 1%, and 0.1% level tests (α = 0.05, 0.01, and 0.001 respectively). To illustrate the notion of "inconsistency," I also present the significance of the difference between two of the cultivars, MiteyOat and TrustyOat, as determined by Goldilocks using the Fisher's restricted LSD procedure.

Astonishingly, the significance of the difference between MiteyOat and TrustyOat varied widely between the four experiments (from not significant to significant at the 0.1% level), in spite of the fact that their mean yields were identical, the pooled SEMs were identical, and the residual degrees of freedom (21) were identical between experiments. The reason for this variation can be traced back to the decision on which two "various" cultivars were included. When the two "various" cultivars had yields that were similar to the experimental average, a low overall F value was calculated (Table 6). When the two "various" cultivars had extreme yields

TABLE 6. Mean oat grain yields from four fictitious experiments, one per bear, with data analysis carried out by Goldilocks using the Fisher's restricted LSD procedure. The pooled SEM, overall *F* value, and its significance are also presented, and the last row gives the "significance" of the difference between MiteyOat and TrustyOat for each experiment.†

Oat cultivar	Mean oat grain yield for experiment by:			
	Baby Bear	Mama Bear	Papa Bear	Grandpa Bear
	kg ha^{-1}			
WonderOat	5030	5030	5030	5030
"Various #1"	5160	5260	5460	5760
MegaOat	4910	4910	4910	4910
MiteyOat	4450	4450	4450	4450
TrustyOat	5550	5550	5550	5550
FlakeyOat	4870	4870	4870	4870
"Various #2"	5120	5120	4520	4320
TrendyOat	4910	4910	4910	4910
SEM	200	200	200	200
Overall *F* value	2.43	2.57	3.82	5.98
Significance of overall *F*	NS‡	*	**	***
LSD(5%)	NS	588	588	588
LSD(1%)	NS	NS	801	801
LSD(0.1%)	NS	NS	NS	1080
Significance of difference between MiteyOat & TrustyOat:				
	NS	*	**	***

* Significant at the 0.05 probability level.

** Significant at the 0.01 probability level.

*** Significant at the 0.001 probability level.

† Reproduced from Saville and Rowarth (2008).

‡ NS, not significant.

(one low, one high), a high overall *F* value was calculated (Table 6). That is, the mean yields for the two "various" cultivars determined the statistical significance of the overall *F* value, and hence the significance of the difference between MiteyOat and TrustyOat. However, such statistical subtleties aside, one can imagine what Grandpa Bear would say about the way in which Goldilocks had handed out significant differences: "Why would Goldilocks be so kind to an old bear, but so unfair to a young fledgling-experimenter bear?"

This sort of inconsistency in the results is something that no practicing biometrician would ever want to have to defend (like: "would you like to share your office block with four angry bears?"). In all four experiments, the t value for the comparison of MiteyOat with TrustyOat is given by $t = 1100/(200 \times \sqrt{2}) = 3.889$, which is significant at the 0.1% level (given 21 df). It defies statistical common sense to override this simple test in Baby Bear's experiment by arguing that the other cultivar means are not sufficiently spread out (hence a low F value), so this comparison should be declared "not significant." This nonsignificant result also contradicts what a statistically literate journal reader would decide after inspecting a bar graph of the results such as that shown in Fig. 3; the usual rule of thumb is that means that differ by more than about 3 SEs are significantly different ($p < 0.05$), yet in this case the two means differ by 5.5 SEs and are declared "not significantly different" by Fisher's LSD procedure. The logical response of researchers such as the Bears would be to ensure good results by including an *old, low-yielding* oat cultivar in their trials (to increase the overall F value). Thus, the insistence of Goldilocks that they use Fisher's restricted LSD procedure would lead to a nonsensical waste of resources by the researchers she was trying to help.

If the data from the four experiments were to be re-analyzed using the unrestricted LSD procedure, then the significance of the difference between MiteyOat and TrustyOat would be significant at the 0.1% level ($\alpha = 0.001$) in all four experiments. That is, the unrestricted LSD procedure is "consistent"; in fact, it is the only consistent procedure.

The Inconsistency of Other Multiple Comparison Procedures

The four data sets given in Table 6 were constructed to illustrate the inconsistency of Fisher's restricted LSD procedure since it is the MCP most commonly used in *Agronomy Journal* (Saville and Rowarth, 2008). Other well-known MCPs behave in a fairly consistent manner between these four data sets, and different examples are required to illustrate their inconsistency. Such examples are given in Saville (1990) for Tukey's HSD procedure.

Interestingly, if six well-known MCPs are used to analyze Baby Bear's data in Table 6, the significance of the difference between MiteyOat and TrustyOat varies markedly, from not significant to significant at the 0.1% level ($\alpha = 0.001$) (Table 7).

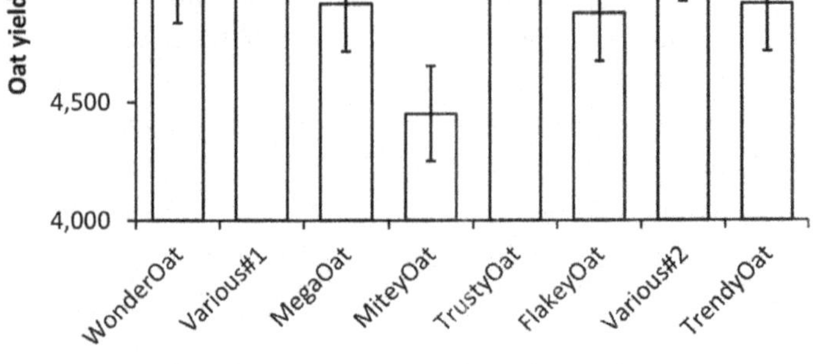

FIG. 3. Cultivar mean oat yields (kg ha^{-1}) in Baby Bear's experiment. Vertical bars are SE bars (using pooled SEM of 200). Note the mean difference between MiteyOat and TrustyOat is 1100 kg ha^{-1}, which is 5.5 times the SE, yet Fisher's restricted least significant difference procedure declares this difference to be not statistically significant (reproduced from Saville, 2015).

Fisher's restricted LSD procedure is at one extreme, yielding a nonsignificant result. With the Bonferroni, Tukey's HSD, or Student–Newman–Keuls procedures, the difference between MiteyOat and TrustyOat is significant at the 5% level ($\alpha = 0.05$) (Table 7). Using Duncan's multiple range test, the difference is significant at the 1% level ($\alpha = 0.01$), while with the unrestricted LSD procedure, the difference between MiteyOat and TrustyOat is significant at the 0.1% level ($\alpha = 0.001$). The above ordering of the significance levels of the results from the various MCPs reflects the level of conservatism and inconsistency of these MCPs. However, Fisher's restricted LSD is variable, and was unfairly disadvantaged by this example. Typically, Fisher's restricted LSD would be placed between Student–Newman–Keuls and Duncan's multiple range test in the ordering.

In general, the most conservative MCPs, such as Tukey's HSD, Bonferroni procedures, and the Student–Newman–Keuls test are also the most "inconsistent" procedures. Duncan's multiple range test is the most consistent of the alternatives to the unrestricted LSD procedure. More recently, in genomics applications, the false discovery rate was introduced by Benjamini and Hochberg (1995) and developed further by Storey and Tibshirani (2003) and other authors; such procedures, of necessity, also suffer from the problem of "inconsistency."

The Significance Level and Power of Some Well-Known Multiple Comparison Procedures

In 1971 a key paper was published in *Agronomy Journal* by Carmer and Swanson (1971). Here, computer simulations were used to compare the Type 1 and 2 (and 3) error rates of five MCPs for a range of hypothetical experiments with various hypothetical patterns of treatment means. The MCPs compared were the unrestricted LSD, Fisher's restricted LSD, Tukey's HSD, Duncan's multiple range test, and the Waller and Duncan Bayesian k ratio LSD procedure.

Their results are interesting. In their Table 2, they presented comparison-wise Type 1 error rates for experiments in which all treatment means were equal. For the unrestricted LSD, their estimates were 5.0% (rounded to one decimal place), while

TABLE 7. The significance of the difference between MiteyOat and TrustyOat in Baby Bear's experiment for multiple comparison procedures of varying levels of conservatism. †

Multiple comparison procedure	Significance
Fisher's restricted LSD	NS‡
Bonferroni	*
Tukey's HSD	*
Student Newman Keuls	*
Duncan's multiple range test	**
Unrestricted LSD	***

* Significant at the 0.05 probability level.
** Significant at the 0.01 probability level.
*** Significant at the 0.001 probability level.
† Reproduced from Saville (2015).
‡ NS, not significant.

for the other procedures, their estimates were less than 2.6%, with the estimates for Tukey's HSD being less than 0.2%. That is, in terms of comparison-wise Type 1 error rate, the unrestricted LSD was the only procedure that was "true to label," meaning that its actual comparison-wise Type 1 error rate (5%) was equal to the rate set by the researcher (note that the vast majority of researchers would assume a Type 1 error rate to be a *comparison-wise* error rate rather than an *experiment-wise* error rate). This explains why some procedures refer to "nominal 5%" error rates (meaning "in name only"). The fact that an MCP has an actual comparison-wise Type 1 error rate less than the nominal rate is dangerous for the researcher because the effect of this reduction is to increase the Type 2 error rate, which is the probability of not detecting real effects. If a researcher needs to use an MCP and it is important to protect against Type 1 errors, then the researcher is advised to use the unrestricted LSD and set the significance level (α) to a lower level, such as 0.01 or 0.001.

In most of the hypothetical experiments which were simulated, the hypothetical treatment means were not equal, and varied from "close to equal" to "very unequal." Results given by Carmer and Swanson (1971, their Tables 5 and 6) give the power of the test (correct decision rate) for each hypothetical experiment. Tukey's HSD consistently had a markedly lower power than the other four procedures. For large treatment differences, the other four procedures were quite similar. For medium-sized treatment differences, the Waller and Duncan Bayesian *k* ratio LSD procedure was slightly more powerful than the unrestricted LSD and Fisher's restricted LSD, with Duncan's multiple range test being less powerful than the other three MCPs. For small treatment differences, the unrestricted LSD was consistently more powerful than all other MCPs.

My overall interpretation of these results by Carmer and Swanson (1971) is that the unrestricted LSD procedure is the MCP of choice, having a significance level that is identical to the stated level (e.g., 5%) and having maximum, or close to maximum power. This differs from the conclusions reached by Carmer and Swanson (1971), who stated: "Ruling out use of the ordinary LSD, without the preliminary *F* test, is justifiable, since its sensitivity is not appreciably greater than that of the Fisher's restricted LSD and Waller and Duncan Bayesian *k* ratio LSD," and went on to say they favored the Fisher's restricted LSD over the Waller and Duncan Bayesian *k* ratio LSD. This reflects the general bias against the unrestricted LSD that was prevalent at that time.

Power Analysis

A consequence of the large variation in power between MCPs is that if a researcher intends to use a particular MCP to analyze their data (assuming no better methods are suitable), then the required sample size needs to be calculated with the particular MCP in mind. For example, if Tukey's HSD is to be used, then the sample size required for a specified power (e.g., 90% power) may be double that required if the unrestricted LSD is to be used; see Saville (2015) for a more detailed discussion of this point.

Note that in most statistics textbooks, the traditional power analysis implicitly assumes that if an MCP is used, it is the unrestricted LSD procedure.

The Practical Solution

The practical solution is as follows:

1. *Abandon* the idea of *simultaneously formulating and testing* all possible pair-wise comparison hypotheses using data from a single experiment.
2. Instead, use the *simplest* multiple comparison procedure (the unrestricted LSD procedure) to *formulate new hypotheses* at a known "false discovery rate" (e.g., 5%, 1%, or 0.1% of null comparisons), then *independently test* these new hypotheses in a second experiment.
3. This is *normal scientific practice*, so this solution fits well with the way in which reputable scientists operate.

That is, on the basis of the discussion above and as in Saville (1985, 1990, 2003, 2015) and Saville and Rowarth (2008), the author recommends to researchers that if use of an MCP is appropriate, the unrestricted LSD procedure is the best choice, given the proviso that it should be treated solely as a method to generate a hypothesis, not as a method for simultaneous formulation and testing of hypotheses.

This is a very similar conclusion to those reached independently by authors such as Rothman (1990) and Perneger (1998), who have written papers pointing out the undesirability of making adjustments for multiplicity.

Advantages of Using the Unrestricted LSD Procedure

Advantages of using the unrestricted LSD procedure are as follows:

1. It's the only *consistent* multiple comparison procedure.
2. It's the *simplest* procedure.
3. Its calculation is the *easiest to check*, so arithmetic errors are minimized.
4. It's the *most flexible* MCP, catering to *unequal sample sizes* and, if necessary, to *unequal variances*.
5. It has a *constant comparison-wise Type 1 error rate* (e.g., 5%), with all other MCPs having variable, nominal ("in name only") comparison-wise Type 1 error rates. That is, if all means are truly equal, the unrestricted LSD is the only MCP that will result in a comparison-wise Type 1 error rate equal to the rate set by the researcher.
6. It has *maximum power*, so it has the greatest chance of generating an interesting new pair-wise comparison hypothesis.
7. The *required sample size* is easily calculated, and the formula is given in standard statistics textbooks (for other MCPs, different formulas are required, as discussed in Saville, 2015).

General Contrasts and Report Writing

The above "practical solution" applies also to *general contrasts* (such as "legumes versus non-legumes"), not just pair-wise contrasts.

Within a single experiment, there may be a mix of prespecified contrasts (including some pair-wise comparisons perhaps), and contrasts or comparisons which have become interesting as a result of the experiment. For all contrasts (/ideas/hypotheses), the *key thing* when writing a report on an experiment is to be completely honest and to clearly describe which ideas you had *before the experiment*, and which ideas *were formulated as a result* of the experiment. *For pre-planned* pair-wise comparisons and general contrasts, your experiment *confirms or denies* each hypothesis. However, for *post-planned* pair-wise comparisons and general contrasts, your experiment has led you to *formulate* each hypothesis which needs to be confirmed or denied in a second experiment.

Example of Usage of the Practical Solution in Research

An example of the usage of this approach by a team of my clients arose recently. In brief, many microbial isolates were screened for efficacy against take-all [*Gaeumannomyces graminis* var. *tritici* (*Ggt*)], and promising isolates were followed up in a subsequent study. To quote, "This study tested 133 microbial isolates (mostly *Trichoderma* spp.) for their ability to suppress infection of wheat roots by *Ggt*. In spite of a targeted approach to the selection of isolates tested, and initially statistically significant results, no isolate demonstrated consistent disease suppression..." (Bienkowski et al., 2015).

In this research, five screening assays were performed to determine the most promising isolates (with each isolate included in at least two of the five assays), and five isolates were then taken forward to further assays. These were four isolates that were significantly different to the positive controls (on the basis of a combined statistical analysis of the five assays) plus the next highest ranked isolate. In this retesting, none of the five isolates demonstrated any suppression of take-all (*Ggt*) root infection. For the oral presentation of these results at the 2015 New Zealand Plant Protection Society Conference, the authors and I prepared the following explanation:

> When using the unrestricted LSD at $p = 0.05$ (following an ANOVA) to compare treatment means to a control mean, even if no treatment produces an actual effect, on average one treatment in 20 will pass the threshold of being significantly different to the control. Therefore, for every 100 ineffective treatments you test, you should expect, on average, five false positive results (inextricably mixed with any true positive results). However, for those that are significantly different from the control by chance alone, it is likely (with 95% certainty) that in each future assay they would not be significantly different from the control.

In my opinion, this said it all. In this particular case, there were no "true positive results" confirmed in the follow-up study, suggesting the five candidate isolates corresponded to "false positives" (assuming adequate power of the follow-up study). In other cases, true positive results may have been confirmed as such. (This explains why most reputable journals ask that experiments be repeated in time or space before manuscripts will be accepted.)

In the above, 133 microbial isolates were screened using an unrestricted LSD procedure with $\alpha = 0.05$, with five candidates carried through for further testing. In genomics research, however, the number of genes for screening may be in the thousands or tens of thousands, and an unrestricted LSD procedure with $\alpha = 0.05$ may generate a candidate list that is too large for the researcher to handle. In this situation, an unrestricted LSD procedure with $\alpha = 0.01$ or 0.001 (or even lower) can be used to generate a candidate list of a more appropriate size.

Conclusions

In summary, I suggest using the same statistical procedure in all cases:

- *Pre*-planned *pair-wise* comparisons.
- *Pre*-planned *general* contrasts.
- *Post*-planned *pair-wise* comparisons (the MCP case).
- *Post*-planned *general* contrasts.

The statistical procedure is the traditional F test of a contrast, which is mathematically equivalent to carrying out a t test of a contrast, which is also equivalent to performing an LSD test (or, more precisely, a *least significant contrast* test). This suggestion, of statistically analyzing both pre-planned and post-planned contrasts in an identical manner, is statistically defensible only if the author of the report on the experiment *distinguishes clearly* between *testing* and *forming* ideas and hypotheses.

Key Learning Points

- When planning an experiment, write down your *ideas of interest*. Make sure your experimental design allows adequate power for testing the idea(s) of *most* interest.
- When statistically analyzing data from your experiment, specify the *contrasts* that correspond to your ideas of interest, and incorporate these contrasts into your analysis of variance. If necessary, ask a statistician to help.
- On the rare occasions that you have no special ideas of interest (e.g., when comparing assorted chemicals or cultivars), and you decide to use an MCP, use the simplest MCP, the unrestricted LSD procedure, but think of it as a procedure for generating new ideas or hypotheses rather than for testing pre-existing (prior) ideas or hypotheses. Then, any interesting ideas thus generated will require confirmation in a second, independent experiment.
- The advantages of using the unrestricted LSD procedure are as follows:
    - It is the only *consistent* multiple comparison procedure.
    - It is the *simplest* procedure.
    - Its calculation is the *easiest to check*, so arithmetic errors are minimized.
    - It is the *most flexible* MCP, catering to *unequal sample sizes* and, if necessary, to *unequal variances*.
    - It has a *constant comparison-wise Type 1 error rate* (e.g., 5%), with all other MCPs having variable, nominal ("in name only") comparison-wise Type 1 error rates.
    - It has *maximum power*, so it has the greatest chance of generating an interesting new pair-wise comparison hypothesis.
    - The *required sample size* is easily calculated, and the formula is given in standard statistics textbooks (for other MCPs, different formulas are required).
- In the case of contrasts other than pair-wise comparisons, the same advice applies. If you are using the contrast to test a *prior* hypothesis, then the F test (or equivalent t test) provides either confirmation or contradiction of

your prior hypothesis. Alternatively, if you are using the contrast to test an hypothesis that you formed by looking at the results from the current experiment (a *posterior* hypothesis), then you are using the *F*-test (or equivalent *t* test) to generate a new hypothesis that requires independent confirmation in a second experiment (if it is interesting to you).

- In summary, use the same *F* test (or equivalent *t* test or equivalent LSD test) for all cases, both general and pair-wise contrasts, both pre-planned and unplanned, but in your report state clearly which contrasts/ideas were pre-planned and which were unplanned.

Exercises

1. The main effect means for the "within row spacing" factor are 7082 kg ha^{-1} of garlic (5-cm spacing) and 4903 kg ha^{-1} (10 cm spacing). Work out the effective sample size for each of these means (n). Then calculate the LSD(5%) which is appropriate for comparing these two means using the formula

LSD(5%) = t value × √[2 × (Residual mean square)/n]

2.

(a) Copy the six treatment means and the LSD(5%) from Table 2 (garlic data), hide from sight the letters given alongside the treatment means, and work out for yourself what letters should be assigned. (Hint: First, sort the means into descending order.)

(b) For each within-row spacing (5 and 10 cm), is there a 5% significant difference between the "3 rows per bed" treatment mean and the "5 rows per bed" treatment mean? Similarly, for each within-row spacing (5 cm and 10 cm), is there a 5% significant difference between the "4 rows per bed" treatment mean and either the "3 rows per bed" or "5 rows per bed" treatment means?

(c) For each between-row spacing (3, 4, and 5 rows per bed), is there a 5% significant difference between the "5 cm" treatment mean and the "10 cm" treatment mean?

Acknowledgments

Thanks to Barry Glaz for his helpful suggestions on what to include in this chapter and to three anonymous referees for their helpful comments.

References

Benjamini, Y., and Y. Hochberg. 1995. Controlling the false discovery rate: A practical and powerful approach to multiple testing. J. R. Stat. Soc., B 85:289–300.

Bienkowski, D., E.E. Hicks, and M. Braithwaite. 2015. Wheat take-all: Lessons learned during a search for effective biological control. N.Z. Plant Protect. 68:166–172.

Garland-Campbell, K. 2018. Errors in statistical decision making. In: B. Glaz and K.M. Yeater, editors, Applied statistics in agricultural, biological, and environmental sciences. ASA, CSSA, and SSSA, Madison, WI. doi:10.2134/appliedstatistics.2016.0007

Carmer, S.G., and M.R. Swanson. 1971. Detection of differences between means: A Monte Carlo study of five pairwise multiple comparison procedures. Agron. J. 63:940–945. doi:10.2134/agronj1971.00021962006300060036x

Carmer, S.G., and W.M. Walker. 1982. Baby Bear's dilemma: A statistical tale. Agron. J. 74:122–124. doi:10.2134/agronj1982.00021962007400010031x

Casler, M.D. 2018. Power and replication—Designing powerful experiments. In: B. Glaz and K.M. Yeater, editors, Applied statistics in agricultural, biological, and environmental sciences. ASA, CSSA, and SSSA, Madison, WI. doi:10.2134/appliedstatistics.2015.0075

Cochran, W.G., and G.M. Cox. 1957. Experimental designs. 2nd ed. John Wiley & Sons, New York.

Little, T.M. 1978. If Galileo published in *HortScience*. HortScience 13:504–506.

Perneger, T.V. 1998. What's wrong with Bonferroni adjustments. BMJ 316:1236–1238. doi:10.1136/bmj.316.7139.1236

Rothman, K.J. 1990. No adjustments are needed for multiple comparisons. Epidemiology 1:43–46. doi:10.1097/00001648-199001000-00010

Saville, D.J. 1985. Multiple comparison procedures. Proc. Agron. Soc. N.Z. 15:111–114.

Saville, D.J. 1990. Multiple comparison procedures: The practical solution. Am. Stat. 44:174–180.

Saville, D.J. 2003. Basic statistics and the inconsistency of multiple comparison procedures. Can. J. Exp. Psychol. 57:167–175. doi:10.1037/h0087423

Saville, D.J. 2015. Multiple comparison procedures: Cutting the Gordian knot. Agron. J. 107:730–735. doi:10.2134/agronj2012.0394

Saville, D.J., and J.S. Rowarth. 2008. Statistical measures, hypotheses, and tests in applied research. J. Nat. Resour. Life Sci. Educ. 37:74–82.

Saville, D.J., and G.R. Wood. 1991. Statistical methods: The geometric approach. Springer-Verlag, New York.

Storey, J.D., and R. Tibshirani. 2003. Statistical significance for genome-wide studies. Proc. Natl. Acad. Sci. USA 100:9440–9445. doi:10.1073/pnas.1530509100

Swallow, W.H. 1984. Those overworked and oft-misused mean separation procedures—Duncan's, LSD, etc. Plant Dis. 68:919–921.

Vargas, M., B. Glaz, J. Crossa, and A. Morgounov. 2018. Analysis and interpretation of interactions of fixed and random effects. In: B. Glaz and K.M. Yeater, editors, Applied statistics in agricultural, biological, and environmental sciences. ASA, CSSA, and SSSA, Madison, WI. doi:10.2134/appliedstatistics.2015.0084

CHAPTER 6: LINEAR REGRESSION TECHNIQUES

Christel Richter* and Hans-Peter Piepho

Using eight examples, each posing some specific challenges, we show the basics and possible extensions of the simple and multiple linear regression models. We discuss similarities and differences if both the regressor and the regressand are random variables and if only the regressand is a random variable. Several methods to evaluate model assumptions are illustrated. Transformations are applied where needed to achieve linearity between variables or to resolve issues of variance heterogeneity. Further possibilities to allow for variance heterogeneity are shown in the context of the mixed linear model. The integration of fixed or random classification variables (treatment factor or disturbance factor) into regression models are discussed. In two examples, we show that linear regression is sometimes only a first approach, and for such cases, we provide guidance on completing the analyses with nonlinear models or models with covariances between the residuals.

Most researchers are familiar with regression problems, either from the published literature or from their own experiments. In this chapter, we focus primarily on problems of linear regression. To set the stage, a first example will be given.

Example 1: The relation between the weight (g) and length (cm) of female eels at the age of 6 yr will be analyzed. $N = 34$ eels were caught from several lakes with similar habitats. In this chapter we symbolize the total sample size (not the population size) by N, and we use n for the sample size of a treatment or a given x-value. The boxplots on the margins of Fig. 1 show the empirical distributions of the weight and the length. Because the boxplots consider each variable separately, these distributions are also called empirical marginal distributions. The scatterplot

Abbreviations: AIC, Akaike criterion; BIC, Bayesian criterion; BLUE, best linear unbiased estimator; BQUE, best quadratic unbiased estimator; CRD, completely randomized design; ewLS, empirical weighted least-squares estimators; LL, logarithm of the likelihood; ML, maximum likelihood; RCBD, randomized complete block design; REML, restricted maximum likelihood; ResLL, restricted logarithm of the likelihood function; SBC, Schwarz criterion.

C. Richter, Humboldt-Universität zu Berlin, Faculty of Life Sciences, Albrecht Daniel Thaer-Institute of Agricultural and Horticultural Sciences, Germany; H.-P. Piepho, University of Hohenheim, Faculty of Agricultural Sciences, Institute of Crop Science, 70599 Stuttgart, Germany (Hans-Peter.Piepho@uni-hohenheim.de). *Corresponding author (ch.richter@agrar.hu-berlin.de).

doi:10.2134/appliedstatistics.2015.0080

Applied Statistics in Agricultural, Biological, and Environmental Sciences
Barry Glaz and Kathleen M. Yeater, editors

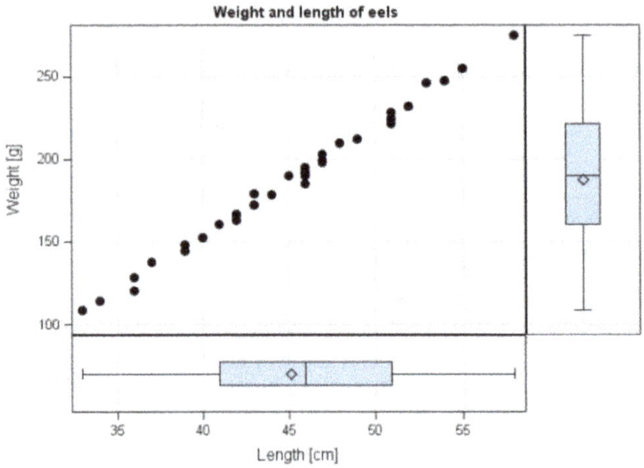

FIG. 1. Example 1. Scatterplot and boxplots for weight (g) and length (cm) of eels.

gives an impression of the empirical joint distribution of length and weight. We see that the assumption of linear dependency seems to be reasonable.

In the following, we will quantify the relation of eel length and weight by a linear function and test hypotheses about the parameters of the function assuming that the 34 samples are a representative sample for the eels under the given habitat conditions. Furthermore, we will discuss the suitable measures to describe the goodness of model fit. Later on, we will take up this example for illustration of various methods. First, however, we discuss some basic ideas and concepts of regression analysis.

Historical Background

The term *regression* comes from the Latin root word *regressio* and in common linguistic usage has the meaning of regress, decline, degeneration, relapse, and reduction. In addition to agronomy, it is used in several scientific disciplines, such as geology, psychoanalysis, epidemiology, software development, and psychology, and it focuses on different meanings depending on the subject of the discipline. In statistics, the usage originates with Francis Galton (1822–1911). In 1885, he reported on experiments with parent and offspring seeds of sweet peas and observed "that the offspring did not tend to resemble their parent seeds in size, but to be always more mediocre than they—to be smaller than the parents, if the parents were large; to be larger than the parents, if the parents were small. [...] The experiments showed further that the mean filial regression towards mediocrity was directly proportional to the parental deviation from it" (Galton, 1885). In his first reports about the pea experiment, Galton still used the term *reversion* instead of *regression* (Galton, 1877). Around the same time, he conducted a study in a human population (known as Galton's family records) and detected similar results. Galton named this effect the "regression toward mediocrity in hereditary stature" (Galton, 1886).

Pearson (1896) discussed Galton's results and fitted linear functions to his data by the least-squares method. He referred to the slope of a linear function as the regression coefficient. The slopes of the functions of Galton's data were smaller than one, and here we still see the original sense of the term *regression*. The heritability (h^2), an important parameter in animal and plant breeding, can be estimated by exactly this slope of a parent–offspring regression using the parental mean as the explanatory variable. Pearson, however, expanded and used the term *regression coefficient* independently of the concrete estimated value and independently of the meaning in genetics. Since that time, regression analysis has come to mean that an observed quantitative response variable will be related to one or more influencing (explanatory) quantitative variables by a function. This function is termed the *regression function*. In this context, the response variable is often termed the *regressand* and the explanatory variables the *regressors*. *Responses* and *predictors* are also terms commonly used in other texts.

For special scientific problems, a similar effect as that noticed by Galton can be observed. The term *regression to the mean* draws on his concept of regression. It can "be described as the tendency of observations that are extreme by chance to move closer to the mean when repeated" (Beach and Baron, 2005).

The Basic Idea

Basically, the term *regression* means any quantitative description of the relation between response and explanatory variables. Many regression methods are known where each uses any inherent specific properties of the data. The concepts linear, nonlinear, robust, non-parametric, ridge, spline regression or regression on the basis of the generalized model (e.g., with the logistic or Poisson regression as special cases) reflect on such specifics. Linear regression, which is our main focus, aims at the calculation of a regression function that is linear in its parameters and has to be determined on the basis of an optimization criterion. Several criteria are common in this context; some of them use distributional assumptions on the variables, while others have weaker assumptions. If the regression function is only needed to provide a best fitting curve (best fit in regard to the criterion) and a measure that describes the goodness of fit, the analysis would be complete. If, however, the observed pairs of variates are considered a sample of all possible results of a randomly influenced process, then statistical inferences, like confidence interval estimates or tests, are intended. These require distributional assumptions for the variables in any case. Specifically, if one or more variables are random variables, then this must be clarified. Further each random variable must be characterized by a corresponding distribution function. The regressand is always assumed to be a quantitative random variable. The regressor must also be quantitative, but it may be a fixed or random variable. In the following, we give examples of quantitative variables, where due to their specifics, not all are suited for a linear regression analysis.

Quantitative traits are random variables that are assessed on a metric scale mostly by measuring in physical units or counting of items per unit (e.g., per defined area, time interval, plant, litter, or others). Typical examples are: weight, size, and content characteristics of plants, animals, or their organs; grain yield components (yield per area, ears per area, kernel number per ear, kernel weight), number of fruits, flowers, pests; chemical, physical, and biological soil traits (nitrate N, acidity, moisture, bulk density, content of sand, silt, and clay, organic biomass); or weather traits (air and soil temperature, precipitation, relative humidity, number of days with extreme meteorological events). If the realizations of a quantitative random variable can be any real number in an interval, as is the case for measured traits, then it is a *continuous random variable*. Which values are actually measured depends only on the accuracy of the measuring instrument. Each measuring is connected with a discretization of a continuous variable.

Traits that are measured by counting using integers ≥ 0 are also referred to as *discrete quantitative variables*. Traits that are calculated from other quantitative traits (e.g., the percentage of pest or disease infested leaf area is the ratio of infested leaf area to total leaf area) are also quantitative.

Sometimes, a known upper limit exists for the observed variable. For instance, the observed amount of organic material in a 100-g soil sample cannot be larger than 100 g, a percentage is maximally 100%, the number of male piglets in litters of 8 piglets cannot be larger than 8, and the number of germinated seeds is bounded by the number of the tested seeds.

The differentiation between continuous and discrete quantitative variables and the possibly existing known upper limit of the observed data associated with their distributions are important aspects for the selection of an appropriate regression method that will be discussed in more detail later and in Chapter 16 (Stroup, 2018).

Nonrandom quantitative variables that are regressors may correspond to levels of quantitative treatment factors of a planned experiment, for example, several amounts or concentrations of a fertilizer, fungicide, plant growth regulator, food quantities in animal trials, or points in time. The experimenter plans exactly the levels of the treatments; they are fixed values, not randomly selected and not influenced by a random process. In cases of observational studies, like in monitoring, these variables, however, may be random quantitative variables too. In this case, their observed values are a random sample from a defined population.

A further and more complicated situation may arise when the observations of the regressor and/or the regressand are tainted with measurement errors that must be considered in the analysis (Fuller, 1987). This is especially relevant to technical sciences. Fuller (1987) explained his concerns by using an agronomic example: The study used a survey to analyze the relationship between yield and nitrogen content in the soil. Several fields were chosen randomly, and samples were drawn. The N content was determined in a laboratory. The analytical laboratory results were influenced both by laboratory errors, as well as by sampling errors in the field. The regression approach appropriate in this situation is known as a *regression model with errors-in-variables* or *measurement error models* (Fuller, 1987; Webster, 1997). In planned experiments, it is common to regard the treatment levels of a fixed factor as uninfluenced by errors, although an extension to allow for errors in the treatment levels would be possible. If it is planned, for instance, to apply a certain amount of fertilizer, then the actual amount can slightly differ due to technical inaccuracies of the fertilizer spreader. In classical regression analysis, however, the model is simplified in such a way that levels of a fixed factor are considered as uncontaminated with random errors.

If the regressors are fixed variables, then the regression problem is sometimes referred to as *Model I* of regression analysis. If the regressors are random variables, then the regression is referred to as *Model II* (Sokal and Rohlf, 1995). We will use this terminology. If both fixed and random regressors are considered, then a mixed situation arises. For estimation of the regression parameters and their tests, the differentiation between these models is not necessary. The different theoretical background, however, has consequences for the design of an experiment. Whereas in Model II, the statistical design refers to the determination of the sample size, it refers additionally to the allocation of the x values in Model I. Furthermore, an additional analysis of the correlation is only justified for Model II and never for Model I.

Linear vs. Nonlinear Regression

Here we will only deal with linear regression. Sometimes, this term is misinterpreted and is also not coherently used. If the regressand y is a linear function of one regressor x or several regressors $x_1,\ldots,x_p$, this is definitely a linear regression problem. Sometimes, a linear function can be achieved by transformation of the regressors or/and the regressand. For example, the function $y(x)=\beta_0+\beta_1\sqrt{x}$ (for simplicity we dropped the error term) will be handled as a linear function $y(z) = \beta_0 + \beta_1 z$ by transformation of the regressor $z=\sqrt{x}$ (see Example 4 given below); in contrast, the function $y(x)=\beta_0 e^{\beta_1 x}$ can be linearized by taking the natural logarithm of both sides (log transformation of the regressand) or can be analyzed in the original nonlinear form. Some authors include under the term *linear regression* only such additional cases where the linear function is achieved by transformation of the regressors and discriminate between linear and nonlinear models by the second derivatives of $y(x)$ with respect to the parameters. If they are equal to zero, then the function is linear in its parameters, otherwise nonlinear (Archontoulis and Miguez, 2015). In this sense, the function $y(x)=\beta_0 e^{\beta_1 x}$ is nonlinear. If, however, the log transformation of $y(x)$ is allowed, then $\log[y(x)]$ is linear in its parameters. If the regressand has been transformed, then, indeed, some specifics need to be considered. Some authors denote the linearized forms as intrinsically linear. Consensus exists that all nonlinear functions that cannot be linearized have to be considered in the class of nonlinear regression problems; for example, $y(x)=\beta_0+\beta_1 e^{\beta_2 x}$. We will not discuss one or the other terminology in general but rather refer to the specifics by the examples.

The Simple Linear Regression

Here we will describe simple linear regression analysis where we have only one regressor and will establish simultaneously the basis for the multiple linear regression model with several regressors. As we have seen in the preceding section, the simple linear model can result from the fact that the relation between the regressand and the regressor is linear for the original data or only after transformation. The following derivations apply to both cases. All problems where linearity can be achieved by transformation can be handled as special cases of the general linear model.

For both Models I and II, we will discuss four examples with one regressor where each example has its specifics. In the following, random variables will be written in boldface, while fixed variables and model parameters will not be boldfaced. Thus, $\boldsymbol{x}$ (boldfaced) will be a random variable, whereas x (not boldfaced) will be a fixed quantity.

Example 1: See the eel example presented above. Both the length (*x*) and the weight (*y*) are continuous random variables. It is an example for Model II.

Example 2: In chemical laboratories, the regression analysis is often used to calibrate the measuring instruments. In the calibration phase for given concentrations or amounts of a substance (x), a measuring signal will be observed (y) and, as appropriate, a linear regression $y = f(x)$ will be conducted (Model I). Later on, when sample material has to be evaluated, the measured value is a y value, and it is used to predict the corresponding x value. This problem is known as inverse regression. In this example, we want to predict the phosphorus content in the soil (x) that is accessible to plants based on colorimetric working measuring instrument. Five calibration concentrations are used.

Example 3: In a pasture survey, the dependency of the crude fiber content (g kg^{-1} biomass) (y) on the cutting date (x) needs to be analyzed. Both variables are quantitative; y is a continuous random variable, and the time points are fixed (Model I). Five dates were fixed at intervals of 5 d. The first date was set to zero. Four samples were randomly drawn from the field at each date so that for each x value $n = 4$ observations exist, and these $N = 4 \times 5 = 20$ observations can be assumed to be independent.

Example 4: We analyze the dependency of plot yield (g $plot^{-1}$) (y) on weed infestation (number of panicle-bearing wind grass [*Apera spica-venti* (L.) P. Beauv.] culms $plot^{-1}$) (x). Both are random variables (Model II), where y is a continuous variable and x is discrete because it has been counted by monitoring.

The data of the examples can be found in the Appendix in the supplemental material online. The first step in the data analysis with only two variables should be the examination of a scatterplot of the observed data. It gives a hint as to whether the assumption of a linear dependency is justified or a preceding transformation is necessary.

The scatterplots and the boxplots for the random variables are given in Fig. 1 and 2. It should be noted that the presented boxplots of the y values have different meanings in Models I and II. In Model II, the boxplot describes the empirical marginal distribution of the corresponding random variable. In Model I, its shape depends on the chosen x values and therefore on the experimental design. The boxplot provides information only about the location and dispersion of the y values in the whole experiment that we want to explain by the regression; its symmetry or asymmetry is not relevant for our further analysis. A better alternative would be to present one boxplot per fixed value of x where the symmetry should be given for each x value in case of the simple linear regression. This however, is not possible or makes no sense in Examples 2 and 3 due to the small number of y values per x value.

In the case where both variables are random (Examples 1 and 4), the scatterplots represent the observed joint distributions. From the boxplots of Example 1 it can be seen that both random variables are almost symmetrically distributed. In Example 4, the boxplot of the number of weed plants shows a strong right skewness that is diminished for the main body of points by a square root transformation of the

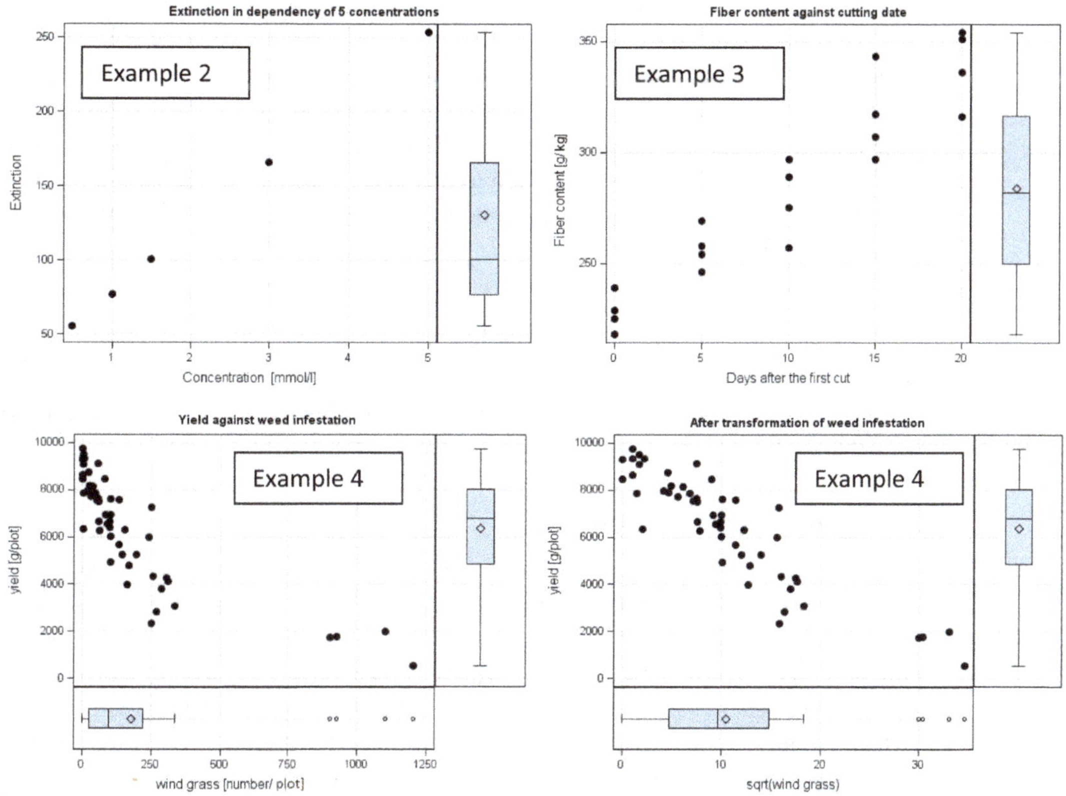

FIG. 2. Examples 2 to 4. Scatterplots and boxplots for the random variables.

regressor (Fig. 2 bottom). But, four unusual data points still remain. Whereas a linear regression function seems inappropriate for the untransformed data, it is conceivable after transformation. The skewness would totally vanish with a cubic root transformation, however, at the cost of the linearity.

The aim of the regression analysis with both models is to find a function describing the dependency of one variable on the other. Depending on the scientific background of the two variables, it may only make sense to consider $\mathbf{y}$ as a function of x or $\mathbf{x}$ and not vice versa. This applies to Examples 2, 3, and 4 where we can consider $\mathbf{y} = f(x)$ or $\mathbf{y} = f(\mathbf{x})$ but not $\mathbf{x}$ as a function of $\mathbf{y}$. A fixed variable x as a function of a random $\mathbf{y}$ is not possible in any case (Example 2). In Example 1, the relation is not unique because both $\mathbf{y} = f(\mathbf{x})$ and $\mathbf{x} = f(\mathbf{y})$ are conceivable.

The assessment of the mutual dependency of the two variables is the aim of correlation analysis. It assumes that both variables are random; that is, Model II must be assumed (Example 1 and 4).

The Model

In the following, the model for both situations, Models I and II, will be introduced. Recall that the variables x and/or y may or may not be transformed to get the simple linear model.

In the first step, we formulate the model equation. For Model I ($\mathbf{y}$ random and boldfaced, x fixed and not boldfaced), we can write:

$$y_i = \beta_0 + \beta_1 x_i + \boldsymbol{e}_i \qquad i = 1 \ldots N \qquad [1]$$

where N is the sample size, β_0 is the regression constant or intercept, β_1 the regression coefficient or slope, and $\boldsymbol{e}_i$ are the random errors, also called "residual effects" or "residuals". The residual effects are the random deviations of the observed y values from the y value on the linear function for a given x value.

The corresponding equation for Model II is ($\boldsymbol{y}$ and $\boldsymbol{x}$ random and boldfaced):

$$\boldsymbol{y}_i = \beta_0 + \beta_1 \boldsymbol{x}_i + \boldsymbol{e}_i \qquad i = 1 \ldots N \qquad [2]$$

It should be noticed that these models always assume an additive effect of the residuals. Later, we will come back to this assumption.

In the second step, we make assumptions about the distribution of each random variable in the model equation.

Assumptions about the Residuals

In the context of simple linear regression, it is generally assumed that the residuals follow the normal distribution with an expected value of zero (for the average of all imaginable samples, positive and negative deviations from the regression line of the observed y values add to zero), with the same variance σ^2, and that all samples are independent of each other. This can be written as: $\boldsymbol{e}_i \sim \mathrm{NI}(0, \sigma^2)$ for each $i = 1 \ldots N$, which means that all residual effects are normally and independently distributed (NI) with expected value 0 and variance σ^2.

Assumptions about the Regressor and Regressand

1. If the regressor is a fixed variable (Model I, Examples 2 and 3), we do not need to make any assumption about the distribution of x. This implies that the $\boldsymbol{y}_i$ follow the normal distribution with $\boldsymbol{y}_i \sim \mathrm{NI}(\beta_0 + \beta_1 x_i, \sigma^2)$.
2. If the regressor is a random variable (Model II, Examples 1 and 4), then two approaches are possible:

   a. We are interested in the correlation analysis between the random variables $\boldsymbol{x}$ and $\boldsymbol{y}$. For this, the two-dimensional random variable $(\boldsymbol{x}, \boldsymbol{y})$ is characterized by its joint distribution. The characterization needs five parameters: the expected values μ_x and μ_y of $\boldsymbol{x}$ and $\boldsymbol{y}$, respectively, their variances $\sigma_x^{\ 2}$ and $\sigma_y^{\ 2}$ and their covariance σ_{xy}. Especially, if $(\boldsymbol{x}, \boldsymbol{y})$ follows a two-dimensional normal distribution, we can write $(\boldsymbol{x}, \boldsymbol{y}) \sim \mathrm{N}([\mu_x, \mu_y], \boldsymbol{\Sigma})$ where $\boldsymbol{\Sigma}$ is the variance–covariance matrix with

   $$\boldsymbol{\Sigma} = \begin{bmatrix} \sigma_x^2 & \sigma_{xy} \\ \sigma_{xy} & \sigma_y^2 \end{bmatrix}$$

   Pearson's correlation coefficient ρ is the standardized covariance $\rho = \sigma_{xy} \Big/ \sqrt{\sigma_x^2 \sigma_y^2}$, which lies between −1 and +1 with each extreme value indicating a perfect deterministic dependency. In general, ρ only measures the linear part of a dependency between $\boldsymbol{x}$ and $\boldsymbol{y}$. Its estimation does not require the two-dimensional normal distribution, whereas confidence intervals and tests require this assumption. An interesting relation to the regression analysis is that in the case of the two-dimensional (or with more regressors: of higher-dimensional) normal distribution, the regression function is always a linear function. Furthermore, only if normality holds, does $\rho = 0$ mean stochastical independence. If a nonlinear relation between $\boldsymbol{x}$ and $\boldsymbol{y}$ exists, the calculation of Pearson's ρ makes no sense. Especially, if $\boldsymbol{x}$ is a nonlinear transformed variable of the observed variable, as in Example 4, then one is interested in the correlation of $\boldsymbol{y}$ and the observed $\boldsymbol{x}$ variable, not of $\boldsymbol{y}$ and the transformed $\boldsymbol{x}$ variable. Due to the nonlinear relation between them, instead of Pearson's, other correlation coefficients should be used, for example, rank-based measures such as Spearman's correlation.

b. We want to predict $\boldsymbol{y}$ for given values of $\boldsymbol{x} = x_i$ and assume a linear relation. This is the situation of regression analysis. For this analysis, we consider the variable $\boldsymbol{y} \mid \boldsymbol{x}$ (read: $\boldsymbol{y}$ conditional on given values of $\boldsymbol{x}$). Then, the Eq. [2] changes to [3]

$$(\boldsymbol{y}_i \mid \boldsymbol{x}_i = x_i) = \beta_0 + \beta_1 x_i + \boldsymbol{e}_i \qquad i = 1 \ldots N \qquad [3]$$

The corresponding distribution function is called the conditional distribution, not the joint distribution as in Assumption 2a, and we write: $\boldsymbol{y} \mid \boldsymbol{x} = x \sim \mathrm{NI}[\beta_0 + \beta_1 x, \sigma_Y^2(1 - \rho^2)]$. We see that this is nearly the same situation as in Model I, meaning that all following calculation steps concerning regression analysis are the same for both models. Therefore, we will refrain from the notation with the condition on the x values in the following. Any model-specific assumptions and peculiarities will be indicated.

For a better understanding of generalizations discussed later, we will show that the simple linear regression model is a special case of the general linear model (Schabenberger and Pierce, 2002) and write the model (Assumption 1 above) in a matrix notation. Capital letters in normal Roman text stand for vectors and matrices with nonrandom components; capital letters in bold Roman text stand for vectors and matrices with random components.

$$\mathbf{Y} = \mathrm{X}\beta + \mathbf{E} \qquad [4]$$

where

$$\mathbf{Y} = \begin{bmatrix} y_1 \\ y_2 \\ \vdots \\ y_N \end{bmatrix}, \mathrm{X} = \begin{bmatrix} 1 & x_1 \\ 1 & x_2 \\ \vdots & \vdots \\ 1 & x_N \end{bmatrix}, \beta = \begin{bmatrix} \beta_0 \\ \beta_1 \end{bmatrix}, \text{ and } \mathbf{E} = \begin{bmatrix} e_1 \\ e_2 \\ \vdots \\ e_N \end{bmatrix}$$

and

$$\mathbf{E} \sim \mathrm{N}_N([0], \mathrm{R}), \text{ with } \mathrm{R} = \sigma^2 \mathrm{I} \qquad [5]$$

In Eq. [5], "$\sim \mathrm{N}_N$" means the N-dimensional normal distribution. [0] is an N-dimensional vector of zeros and is the vector of expected values of $\mathbf{E}$. $\mathrm{R} = \sigma^2\mathrm{I}$ is the variance–covariance matrix of **E** with I the identity matrix. Hence, the variance σ^2 stands on the major diagonal of R and the covariances stand on the minor diagonals which are all equal to zero in this case. The components in β are the two fixed parameters of the regression function to be estimated.

Equation [4] with [5] is a special case of the general linear model.

The Estimation of the Fixed Parameters β_0 and β_1, of the Variance σ^2, the Coefficient of Determination, and in the Case of Model II, of Pearson's Correlation Coefficient

The best known method for the estimation of the two fixed parameters β_0 and β_1 is the ordinary least-squares method. By this method, β_0 and β_1 are determined in such a way that for the observed values x_i and y_i the sum (S) of squared residuals will be minimal. For this, it is required that the expression

$$S = \sum_{i=1}^{N} e_i^2 = \sum_{i=1}^{N} (y_i - \beta_0 - \beta_1 x_i)^2 = (\mathrm{Y} - \mathrm{X}\beta)^T (\mathrm{Y} - \mathrm{X}\beta) \qquad [6]$$

is derived with respect to β_0 and β_1 (respectively to β in vector notation) and the derivative equated to zero. In contrast to Eq. [4], the vector Y in Eq. [6] is written in normal Roman text because it consists of the observed (nonrandom) y values. This gives

$$(\mathrm{X}^T\mathrm{X})\beta = \mathrm{X}^T\mathrm{Y}$$

and the estimator of β, denoted by $\hat{\beta}$, is

$$\hat{\beta} = (\mathrm{X}^T\mathrm{X})^{-1}\mathrm{X}^T\mathrm{Y} \qquad [7]$$

where $(\mathrm{X}^T\mathrm{X})^{-1}$ is the inverse of $\mathrm{X}^T\mathrm{X}$. For the simple linear regression, the inverse exists if at least two unequal x values are chosen (Model I) or are observed (Model II). This, however, should be a given; otherwise, a regression analysis would be pointless.

Inserting in Eq. [7] the above given matrix X, one gets the more well-known formulas

$$\widehat{\beta_0} = b_0 = \bar{y} - b_1\bar{x} \text{ and } \widehat{\beta_1} = b_1 = \frac{\mathrm{SPXY}}{\mathrm{SSX}} \qquad [8]$$

where

$$\mathrm{SPXY} = \sum_{i=1}^{N} (x_i - \bar{x})(y_i - \bar{y}) \text{ and } \mathrm{SSX} = \sum_{i=1}^{N} (x_i - \bar{x})^2$$

b_0 and b_1 are the point estimators of β_0 and β_1. SSX is the sum of squares for x (more precisely: sum of squared deviations of x values from the mean) and SPXY is the sum of products of x and y (more precisely: sum of products of the deviations of x and y values from the corresponding means).

The estimated regression function is: $\hat{y} = b_0 + b_1 x$.

For each x_i we can write:

$$\hat{y}_i = b_0 + b_1 x_i \text{ or } \hat{Y} = X\hat{\beta} \tag{9}$$

The fitted residuals are:

$$\hat{e}_i = y_i - \hat{y}_i \text{ or } \hat{E} = Y - \hat{Y} \tag{10}$$

Inserting the estimates b_0 and b_1 in the estimated function gives:

$$\hat{y} = b_0 + b_1 x = (\overline{y} - b_1\overline{x}) + b_1 x = \overline{y} + b_1(x - \overline{x})$$

so that the regression function goes through the point $(\overline{x}, \overline{y})$.

By the regression, the sum of squares of y (the total sum of squares) SS_{total} is partitioned into two parts: the sum of squares, which is explained by the regression ($SS_{explained}$), and the residual sum of squares (RSS):

$$SS_{total} = SSY = \sum(y_i - \overline{y})^2 = \underbrace{\sum(\hat{y}_i - \overline{y})^2}_{SS_{explained}} + \underbrace{\sum(y_i - \hat{y}_i)^2}_{RSS} \tag{11}$$

The residual sum of squares is the achieved minimum of S (see Eq. [6]).

The part explained by the regression can be more simply calculated than in Eq. [11] by

$$SS_{explained} = b_1 SPXY \tag{12}$$

because

$$\sum(\hat{y}_i - \overline{y})^2 = \sum(b_0 + b_1 x_i - \overline{y})^2 = \sum([\overline{y} - b_1\overline{x}] + b_1 x_i - \overline{y})^2$$
$$= b_1^2 SSX = b_1 SPXY$$

Analogously to SS_{total}, the corresponding degrees of freedom df can be partitioned. The df of the explained part is equal to the number of fixed parameters p [= 2] in the regression function minus 1, so that for the linear regression:

df of $SS_{total} = N - 1 =$ df of $SS_{explained}$ + df of RSS = 1 + (N − 2)

An estimate of σ^2 can be obtained by the analysis of variance method (ANOVA method):

$$\widehat{\sigma^2} = s^2 = \frac{RSS}{\text{df of RSS}} = \frac{RSS}{N-2} \tag{13}$$

This ratio is also referred to as the mean squared error (MSE).

The result of this partition is often given in an ANOVA table (Table 1A, Columns 1–4). To recognize the similarities between the simple and the multiple linear regressions with p parameters, the corresponding partitioning of SSY and df is presented in Table 1B and will be discussed later.

TABLE 1. Analysis of variance tables of the linear regression.

Source	Sum of squares	df	Mean squares	*F* value
A. Simple linear regression with 2 parameters (b_0 regression constant, b_1 regression coefficient)				
Model (explained part)	$\sum(\hat{y}_i-\bar{y})^2 = \text{SS}_{\text{explained}} = b_1\text{SPXY}$	1	$\text{SS}_{\text{explained}}$	$\frac{\text{SS}_{\text{explained}}}{s^2}$
Error (residual part)	$\sum(y_i-\hat{y}_i)^2 = \text{RSS} = \text{SSY}-b_1\text{SPXY}$	$N-2$	$\frac{\text{RSS}}{N-2}=s^2$	
Total†	$\sum(y_i-\bar{y})^2 = \text{SSY}$	$N-1$	$\frac{\text{SSY}}{N-1}$	
B. Multiple linear regression with *p* parameters (b_0 regression constant, $b_1\ldots b_{p-1}$ regression coefficients)				
Model (explained part)	$\sum(\hat{y}_i-\bar{y})^2 = \text{SS}_{\text{explained}} = \sum_{j=1}^{p-1} b_j\text{SPX}_j\text{Y}$	$p-1$	$\frac{\text{SS}_{\text{explained}}}{p-1}$	$\frac{\text{SS}_{\text{explained}}}{(p-1)s^2}$
Error (residual part)	$\sum(y_i-\hat{y}_i)^2 = \text{RSS} = \text{SSY}-\sum_{j=1}^{p-1} b_j\text{SPX}_j\text{Y}$	$N-p$	$\frac{\text{RSS}}{N-p}=s^2$	
Total†	$\sum(y_i-\bar{y})^2 = \text{SSY}$	$N-1$	$\frac{\text{SSY}}{N-1}$	

† Total is sometimes denoted as corrected total. The term *corrected* is used to point out explicitly that from all observed values the mean has been subtracted. In contrast, the *uncorrected total* sum of squares is Σy_i^2, which we will use later.

To quantify the dispersion of the observed values around the function (the goodness of fit), we can use s^2 = MSE as the dispersion of the errors around the function, the corresponding standard deviation s (also called the root mean square error, RMSE) or the coefficient of variation $\text{CV}\% = 100 \times s/\bar{y}$ (also denoted by s%). Mostly, however, the coefficient of determination, symbolized by R^2, is presented due to its better comparability between different regression problems. It is estimated by

$$R^2 = 1-\frac{\text{RSS}}{\text{SSY}} = \frac{\text{SS}_{\text{explained}}}{\text{SSY}} \quad [14]$$

or in its adjusted form (adjusted for the corresponding df) by

$$\text{adj. } R^2 = 1-\frac{s^2}{s_y^2} = 1-\frac{\text{RSS}}{\text{SSY}}\frac{N-1}{\text{df of residuals}} = 1-\frac{\text{RSS}}{\text{SSY}}\frac{N-1}{N-2} \quad [15]$$

for df of residuals > 0 (i.e., $N > 2$ in the simple current case). Equation [14] can also be written for the regression with one regressor as

$$R^2 = \frac{SS_{explained}}{SSY} = \frac{b_1 SPXY}{SSY} = \frac{(SPXY)^2}{SSX\,SSY} \qquad [16]$$

These measures allow to compare the fit of regression functions because of their normalized form. Values of R^2 are between 0 and 1 and the adj. R^2 lies between –1/(df of residuals) and 1. If $R^2 = 0$ then nothing of the sum of squares of y can be explained by the regression function ($SS_{explained} = 0$ and SSY = RSS). In this case, the adj. R^2 is smaller than or equal to zero. $R^2 = 1$ as well as adj. $R^2 = 1$ means that RSS = 0, i.e., all observed points lie on the regression function. Both measures can be interpreted as the percentage of the variability of the y values which can be explained by the regression function.

It can be seen that unless RSS = 0, $R^2 >$ adj. R^2. If RSS = 0, then both R^2 and adj. $R^2 = 1$. The adj. R^2 should be used if for a given dataset several regression functions with different numbers of parameters are fitted and the goodness of fit needs to be compared. Because we desire not only a function with a good fit, but also a function with as few parameters as possible, regression models with more parameters should be penalized. The more parameters the function has, the smaller is the df of the residuals. This reduction in df must be more than offset by the associated reduction in RSS for the adjusted R^2 to be enlarged for more complex models. If $N = 2$, then R^2 is equal to one (we can always find a straight line passing through two points) and the RSS would be zero), the adj. R^2 is not defined.

If for Model II the correlation analysis is desired, Pearson's correlation coefficient ρ can be estimated by r with

$$r = \hat{\rho} = \frac{\hat{\sigma}_{xy}}{\sqrt{\hat{\sigma}_x^2\,\hat{\sigma}_y^2}} = \frac{SPXY}{\sqrt{SSX\,SSY}} \qquad [17]$$

Comparing Eq. [16] and [17], we see the relation to the coefficient of determination: it is $r^2 = R^2$. Sometimes, for the estimated correlation coefficient, the symbol R is used and why this is done is clearly based on the relation between the estimates of the two parameters. We use the two different symbols because their calculation and interpretation are based on different models (see Assumptions 2a and 2b above). The correlation coefficient should only be used with Model II and never with Model I. In Model I, while it is mathematically correct that $r = \sqrt{R^2}$, this r, however, does not estimate ρ, because neither σ_x^2 nor σ_{xy} exist.

The described estimators of β and σ^2 have some desirable characteristics under the assumptions that the residuals have the same variance, are uncorrelated, and their expected values are zero: the estimators of β are unbiased and have the smallest variance under all linear estimators (best linear unbiased estimator, BLUE), and the estimator of σ^2 is also unbiased and has the smallest variance among all quadratic estimators (best quadratic unbiased estimator, BQUE). If the assumptions are not fulfilled, the estimators lose these features. As we have seen, up to this point we did not need any assumptions about the distribution function and if no confidence intervals or tests are desired, the analysis would be complete.

Aside from the chosen estimation methods (ordinary least squares for the fixed parameters and ANOVA method for the variance), other methods with other optimization criteria exist. The maximum likelihood (ML) method and the restricted maximum likelihood (REML) method use the knowledge of the distribution. By the ML method, the fixed effects and the variance–covariance parameters are determined by maximization of the logarithm of the likelihood function (LL). The REML method maximizes the restricted logarithm of the likelihood function (ResLL) and yields estimators of the variance–covariance parameters. These estimators are subsequently used to estimate the fixed parameters of the model and yield the empirical weighted least-squares (ewLS) estimators (Schabenberger and Pierce, 2002). In the following, we refer to the REML method as the combination of the estimation procedures of both parameter types. Sometimes in subsequently discussed generalizations, we will use the REML or the ML method. In simple linear regression, if a normal distribution can be assumed, we would obtain the same results with the least-squares method used above and with the REML method.

Statistical Inferences

For the following, the distribution assumption in Eq. [5] is used.

1. To construct confidence intervals of β_0 and β_1 and to test them, it is necessary to estimate the variances of their estimators.

   Based on Eq. [7], it can be shown that the estimated variance–covariance matrix of $\hat{\beta}$ is

   $$s_{\hat{\beta}}^2 = (X^T X)^{-1} s^2 \qquad [18]$$

   The variances for both parameter estimators are found on the major diagonal, their covariance on the minor diagonal. In the simple linear regression, we can write the estimated variances in an explicit form:

   $$s_{b_0}^2 = s^2 \left[\frac{1}{N} + \frac{\bar{x}^2}{\mathrm{SSX}} \right] \text{ and } s_{b_1}^2 = \frac{s^2}{\mathrm{SSX}} \qquad [19]$$

   The estimated covariance of the two parameters $s_{b_0 b_1}$ is always negative if $\bar{x} > 0$:

   $$s_{b_0 b_1} = \frac{-\bar{x}}{\mathrm{SSX}} s^2 \qquad [20]$$

   The two-sided $1 - \alpha$ confidence intervals are

   $$\left[b_0 \pm s_{b_0} t_{1-\alpha/2}(N-2) \right] \text{ and } \left[b_1 \pm s_{b_1} t_{1-\alpha/2}(N-2) \right]$$

   To compare β_0 or β_1 with any constants, such as const_0 and const_1, the two-sided hypotheses H_0: $\beta_0 = \mathrm{const}_0$ and H_A: $\beta_0 \neq \mathrm{const}_0$, respectively, H_0: $\beta_1 = \mathrm{const}_1$ and H_A: $\beta_1 \neq \mathrm{const}_1$ can be tested by t statistics:

   $$|t| = \frac{|b_0 - \mathrm{const}_0|}{s_{b_0}}, \text{ resp. } |t| = \frac{|b_1 - \mathrm{const}_1|}{s_{b_1}}$$

If the calculated t value $|t|$ is larger than the $(1 - \alpha/2)$ t quantile with df = $(N - 2)$ [= df of residuals] the alternative hypothesis will be accepted with a Type 1 error rate of α.

In particular, the test of β_1 against zero is often of interest. If the null hypothesis H_0 is not rejected, then we may conclude that y does not linearly depend on x. We cannot conclude that y does not depend on x. It could be a quadratic or any other relation between the variables. To decide this question, graphical representations of the observed data and the residuals are useful. We will discuss these options later in this chapter.

Another manner in which to test H_0: $\beta_1 = 0$ against H_A: $\beta_1 \neq 0$ is to use the F test based on the ANOVA table (Table 1A.). If the calculated F value $F = b_1\text{SPXY}/s^2$ is larger than the $(1 - \alpha)$ quantile of the F distribution with df of the numerator = 1 and df of the denominator = $N - 2$, the alternative hypothesis will be accepted with a Type 1 error rate of α. The t test and the F test of β_1 give exactly the same result because (t value)^2 = F value and $[t(1 - \alpha/2; \text{df})]^2 = F(1 - \alpha; 1,\text{df})$. The F test can also test whether the theoretical coefficient of determination is equal to or larger than zero.

The determinant of the variance–covariance matrix in Eq. [18] is the generalized variance and in the simple linear regression case is

$$\det\left[(X^T X)^{-1} s^2\right] = \frac{s^4}{N\text{SSX}} \qquad [21]$$

The generalized variance provides a summary measure for the variance of all parameters considering their covariance. The determinant plays an important role in the construction of optimal designs under Model I as well in the evaluation of individual data points regarding its influence on the estimated function using the COVRATIO (see below).

2. The $(1 - \alpha)$ confidence interval for the expected value of the regressand for a given value of $x = x'$ can be obtained by

$$\left[\hat{y}(x') \pm s_{\hat{y}(x')} t_{1-\alpha/2}(N-2)\right]$$

where

$$s^2_{\hat{y}(x')} = s^2 \text{L}(\text{X}^T\text{X})^{-1}\text{L}^T \text{ and } \text{L} = (1,\ x') \qquad [22]$$

In the simple case, the explicit formula for $s^2_{\hat{y}(x')}$ can be given

$$s^2_{\hat{y}(x')} = s^2\left[\frac{1}{N} + \frac{(x' - \bar{x})^2}{\text{SSX}}\right]$$

It can be seen that for $x' = 0$ $s^2_{\hat{y}(x')} = s^2_{b_0}$. If for each measured value x_i (i = 1…N), the variance $s^2_{\hat{y}(x_i)}$ needs to be determined, then Eq. [22] can be used. We replace the L vectors in Eq. [22] by the matrix X because it contains the L vectors for all observed values. The result is the matrix $s^2\text{X}(\text{X}^T\text{X})^{-1}\text{X}^T$ with the corresponding variances on its major diagonal and the covariances on the minor diagonals. The matrix H

$$\mathrm{H} = \mathrm{X}(\mathrm{X}^T\mathrm{X})^{-1}\mathrm{X}^T \quad [23]$$

is known as the *Hat matrix* (the term will be explained later).

So finally one can say that

$$s^2_{\hat{y}(x_i)} = s^2 h_{ii} \quad [24]$$

where h_{ii} is the ith element on the major diagonal of H. The h_{ii} are also referred to as *leverages* and are used in the diagnostic methods of the results. The farther an x_i value is away from $\overline{x}$, the larger is the value of h_{ii}. The value of h_{ii} lies between $1/N$ and 1 and is exactly 1 if only two different x_i values exist and $N = 2$. However, if this is the case, then one has only two points that can be connected by a straight line, and this is not really a problem of regression analysis.

3. The $(1 - \alpha)$ prediction interval for the regressand for future observations at the point $x = x'$ (also known as the interval of individual prediction) is

$$\left[\hat{y}(x') \pm \sqrt{s^2_{\hat{y}(x')} + s^2}\, t_{1-\alpha/2}(N-2)\right] = \left[\hat{y}(x') \pm s\sqrt{1 + \frac{1}{N} + \frac{(x' - \overline{x})^2}{SSX}}\, t_{1-\alpha/2}(N-2)\right] \quad [25]$$

Using the formulation with the Hat matrix at the measured x_i values ($i = 1 \ldots N$) gives

$$s^2_{\hat{y}(x_i)} + s^2 = s^2(1 + h_{ii}) \quad [26]$$

It can be seen that the width of the prediction interval is larger than the width of the confidence interval and both depend on x'. The width is minimal where $x' = \overline{x}$.

4. In Model II, some additional considerations are meaningful, assuming that (x,y) follows a two-dimensional normal distribution. Thus, based on the joint distribution a confidence ellipse for the mean vector (μ_x, μ_y) and a prediction ellipse for further observations (x, y) can be determined (formula will not be given here, see Sokal and Rohlf, 1995), a confidence interval for Pearson's correlation coefficient can be determined or the hypotheses H_0: $\rho = 0$ can be tested against H_A: $\rho \neq 0$ by a t test. If the calculated test statistic $|t| = |r|\sqrt{(N-2)/(1-r^2)}$ is larger than the $(1 - \alpha/2)$ t quantile with df $= (N - 2)$, the alternative hypothesis will be accepted with a Type 1 error rate of α.

Results of Examples 1 to 4

For the analysis of linear regression problems, PROC GLM (general linear model) and PROC MIXED (linear mixed model), as well as the regression procedure PROC REG, of SAS can be used. The REML method is implemented as the default method in the procedure PROC MIXED. The PROC REG procedure uses the least-squares method. PROC REG has several options that were developed especially for regression problems; PROC GLM and PROC MIXED allow for the additional consideration of qualitative factors in the model; PROC MIXED has additional options suited for some generalizations concerning the variance–covariance matrix R of the residuals from Eq. [5] (see Examples 6, 7, 8, and the modified Example 3). For Examples 1 to 4, we obtain the same results with all three procedures, however, PROC MIXED does not provide the prediction intervals. With the procedure PROC CORR, several correlation coefficients can be estimated, including Pearson's correlation coefficient ρ. The procedure also provides confidence intervals and significance tests for the correlation. The confidence and prediction ellipses based on the joint distribution can also be calculated.

Graphical Representation

If only one regressor exists, we can represent the scatterplot of observed points, the fitted function, the confidence and the prediction intervals in the two-dimensional coordinate system. Doing these things provides an impression as to whether the function is suited for the problem, of the goodness of fit, and whether conspicuous observations (outliers) exist that are far from the main body of points (Fig. 3). Some numerical results are shown in the plots in Fig. 3, and some points are marked, which we will discuss later.

Numerical Results

Numerical results for all examples are partially provided in Fig. 3. We tested the regression coefficients and regression constants in all examples. The alternative hypotheses that they are different from zero were always accepted at $\alpha = 0.05$. In the following, we will only focus on the specifics of the four examples. In Examples 1 and 2, the dependency between x and y is strong. That is why the distinction between the estimated function and the confidence interval is hardly detectable in Fig. 3.

Example 1 (SAS Code in Appendix)

The estimated regression function for the dependency of weight from length is

$$\widehat{\text{weight}} = -114.20043 + 6.6808 \text{ length, with } R^2 = 0.9943 \text{ and adj.} R^2 = 0.9941$$

It should be considered that the interpretation of the parameters is only meaningful for the observed domain of the length. We can say that within this domain, for each increase in length of 1 cm, we predict an average expected increase by 6.6808 g in weight. The regression constant cannot be interpreted because an eel with 0 cm in length will not have a weight of –114.2 g. Obviously, the linear relation applies only in the observed region. There is no scientific basis for making predictions of weight if length is < 33 cm or > 58 cm; doing so is known as extrapolating.

If one is interested in a function which describes the dependency of length from weight, the estimated regression function weight = f(length) cannot only be solved for length. To solve an equation for any variable is a common approach in

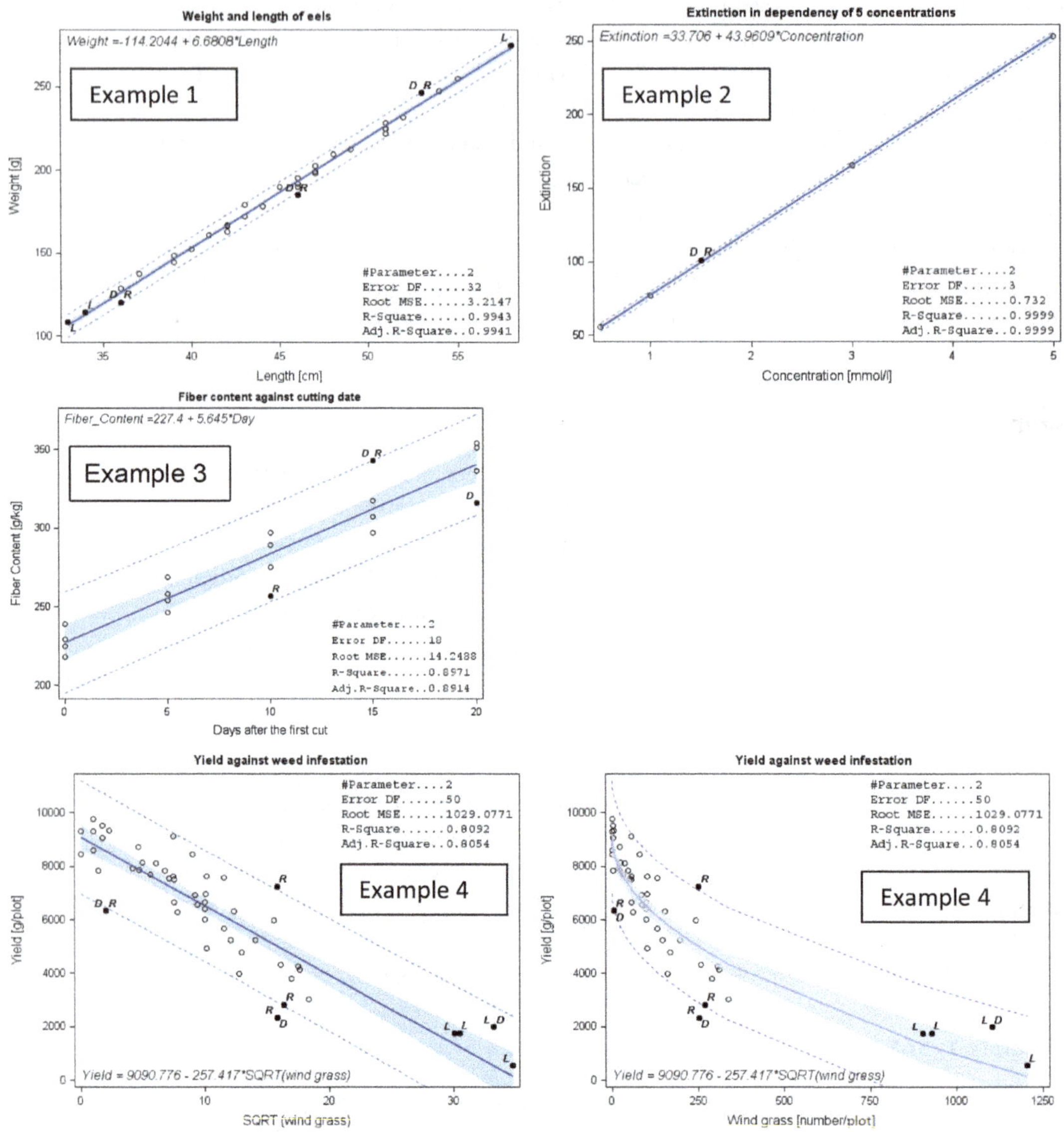

FIG. 3. Examples 1 to 4. Observed data, regression function, confidence interval (blue shaded), and prediction interval (dotted lines) with $1 - \alpha = 0.95$ with conspicuous points: *R* = Externally Studentized residual (RSTUDENT), *D* = DFITTS, *L* = leverage (see Table 4 and explanations in text).

mathematics, where strong deterministic and no stochastic relations between variables are analyzed. In this example, it would result in

$$\text{length} = \frac{\text{weight} + 114.20043}{6.6808} = 17.093826 + 0.14968\ \text{weight}$$

In contrast, the estimated regression function with the model length = *f*(weight) is

$\widehat{\text{length}} = 17.25476 + 0.14883\text{weight}$, with the same R^2 and adj. R^2 as above

The differences between the parameters intercept and slope of both procedures are small in this case, but are not due to rounding errors. Both procedures would only yield the same result if R^2 were equal to 1; that is, a strong deterministic dependency exists between the variables. The weaker the correlation between the two variables, the larger will be the differences between the parameters of the two procedures. In the example, a close relation between both variables exists; therefore, the differences are small. The R^2 and adj. R^2 are the same for both $\boldsymbol{y} = f(\boldsymbol{x})$ and $\boldsymbol{x} = f(\boldsymbol{y})$. To explain the relation between the estimated regression coefficients in the two models, we write them as

$$\hat{y} = b_{0(y.x)} + b_{1(y.x)}x \text{ and } \hat{x} = b_{0(x.y)} + b_{1(x.y)}y$$

where the subscripts $(y.x)$ and $(x.y)$ denote which variable is regressed on which.

From Eq. [8] we know that $b_{1(y.x)}$ = SPXY/SSX, and analogously we have $b_{1(x.y)}$ = SPXY/SSY, so that with Eq. [16] $b_{1(y.x)}b_{1(x.y)} = (\text{SPXY})^2/(\text{SSY SSX}) = R^2$. Only if $R^2 = 1$ (strong deterministic dependency), $b_{1(x.y)}$ is the reciprocal of $b_{1(y.x)}$.

In this example, the regressor is a random variable so it belongs to Model II. From Fig. 1, it does not seem that anything speaks against the two-dimensional normal distribution so that a correlation analysis corresponding to the remarks in (2a) is also acceptable. The estimation of ρ gives $r = 0.9971$ and the test against zero accepts H_A (p value < 0.0001). The confidence ellipse for the mean vector (μ_x, μ_y) (Fig. 4, left) and the prediction ellipse for further observations (x, y) (Fig. 4, right) can be determined.

The comparison of the prediction intervals in Fig. 3 and 4 illustrates the differences between the conclusions based on the conditional and the joint distribution.

Example 2 (SAS Code in Appendix)

As the result of the calibration, we quantified the dependency of the extinction E from the given concentrations C: $\hat{E} = 33.706 + 43.9609C$, with an $R^2 = 0.9999$. Now, we want to analyze an actual new sample. For this new sample, we measure the extinction $E = 230$ and want to infer the concentration. The R^2 is large, but not equal to 1. As in Example 1, it is not proper to solve the regression function for the concentration and insert the value $E = 230$. Instead, based on the prediction interval of the extinction, one may ask at which concentrations an extinction of 230 would be expected. As the

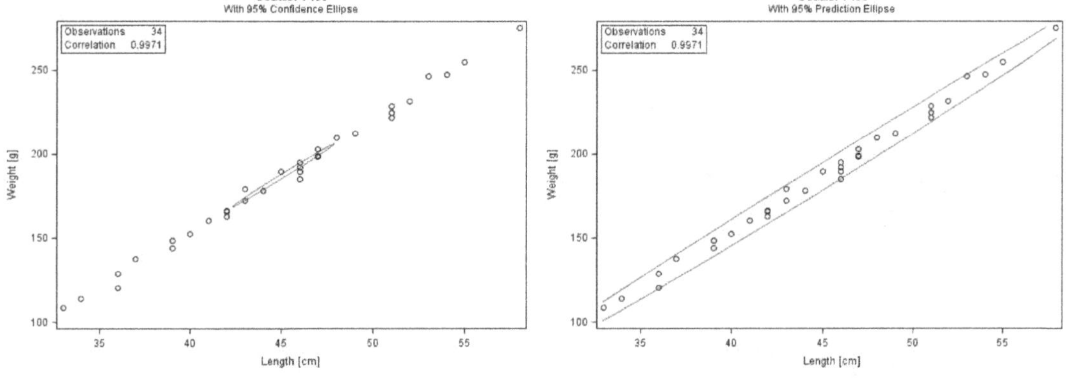

FIG. 4. Example 1. Observed data, confidence ellipse for the mean vector (μ_x, μ_y) (left panel) and prediction ellipse based on the joint distribution (right panel).

result, one gets an interval of the concentration with a lower and upper limit $[C_L; C_U]$ so that it will contain the true concentration with the probability of $1 - \alpha$. This means that using Eq. [25] we need to solve the following equations for C_L and C_U:

$$\hat{E}(C_L) + s\sqrt{1+\frac{1}{N}+\frac{(C_L-\bar{C})^2}{SSC}}\, t_{1-\alpha/2}(N-2) = 230$$

and

$$\hat{E}(C_U) - s\sqrt{1+\frac{1}{N}+\frac{(C_U-\bar{C})^2}{SSC}}\, t_{1-\alpha/2}(N-2) = 230$$

Here, we use a numerical iterative solution method for example by using the procedure PROC MODEL of SAS. Alternatively, taking squares on both sides, we obtain quadratic equations that are readily solved analytically (Seber, 1977). This method is known as Fieller's method. In Fig. 5, the interesting region of Fig. 3 is enlarged and the solution is represented with $C_L = 4.399$ and $C_U = 4.532$.

Example 3 (SAS Code in Appendix)

The results of parameter estimation as well the variance tables are given in Table 2. In contrast to examples with only one observation for each x value, here, there are four. Having more than one observation per x value is a typical situation in designed experiments with a fixed treatment factor. As a result the RSS from Eq. [11] can be split into two parts: the SS for lack of fit and the SS for pure error. For this, we write all observed y_i values with two indices y_{ij} ($i = 1,..,a$; $j = 1,\ldots, n_i$) where a is the number of different values of x, and n_i is the number of replications of x_i. In our example, it is $a = 5$ and $n_i = n = 4$ replicates per day. The partition of RSS results in

$$\text{RSS} = \sum_{i=1}^{a}\sum_{j=1}^{n_i}(y_{ij}-\hat{y}_i)^2 = \underbrace{\sum_{i=1}^{a}\sum_{j=1}^{n_i}(y_{ij}-\bar{y}_i)^2}_{\text{SS pure error}} + \underbrace{\sum_{i=1}^{a}n_i(\bar{y}_i-\hat{y}_i)^2}_{\text{SS lack of fit}}$$

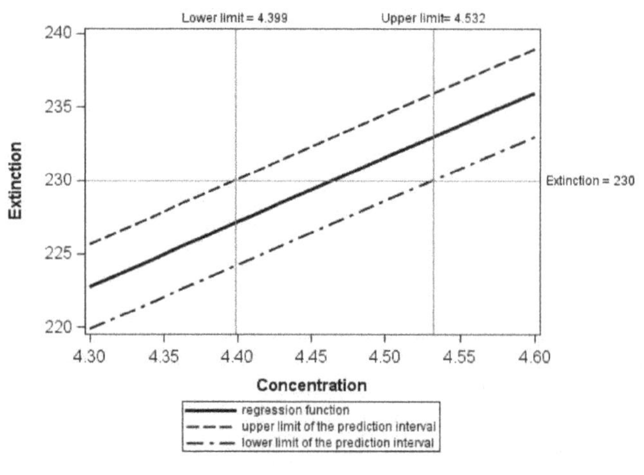

FIG. 5. Example 2. Detail from Fig. 3 with inference of the concentration from the extinction = 230, including Fieller's confidence interval.

with the corresponding partition of the degrees of freedom:

df residuals = $N - 2$ = df pure error + df lack of fit = $(N - 1 - a) + (a - 1)$

The SS pure error describes the deviation of the observed values that pertain to a given x value from the corresponding mean. In general, we have RSS $\geq$ SS pure error. The two SS are equal only when the SS for lack of fit is zero. In other words, the two SS would be equal if the means of y_i values lie exactly on the regression function for each x_i. Although, polynomial regression will be discussed later, one remark is given now: The lack of fit would be zero if instead of fitting a linear function of Day a polynomial of the fourth (= $a - 1$) order of Day is fitted. In this case, the sum of SS for the quadratic, cubic, and quartic term is equal to SS lack of fit (see Appendix in the supplemental material for explanation of the lack of fit). The SS for pure error (Table 2B) is exactly the SS error of an analysis of variance in which day is handled like a qualitative treatment factor in a completely randomized design (Table 2C). The results in Table 2C can be obtained by the SAS procedure PROC GLM or PROC MIXED. Dividing the SS error by the corresponding df in Table 2A and 2C, we get the MSE of the regression analysis (s^2 = 203.03) and the MSE of the analysis of variance with day as a qualitative treatment factor (233.82). The lack of fit can be tested

TABLE 2. Example 3. Analyses of variance (ANOVA), estimation and tests of fixed parameters.

Source	df	Sum of squares	Mean square	*F* value	*p* > *F*
A. ANOVA table for the regression analysis without lack of fit test					
Model	1	31866	31866	156.95	< 0.0001
Error	18	3654.53	203.03		
Corrected total	19	35521			
B. ANOVA table for the regression analysis with lack of fit test					
Model	1	31866	31866	156.95	< 0.0001
Error	18	3654.53	203.03		
Lack of fit	3	147.28	49.09	0.21	0.8879
Pure error	15	3507.25	233.82		
Corrected total	19	35521			
C. ANOVA table if Day is handled as a qualitative treatment factor (Class variable in SAS)					
Treatment (Day)	4	32013.3	8003.33	34.23	< 0.0001
Error	15	3507.25	233.82		
Corrected total	19	35521			

D. Estimations and tests of the fixed parameters

Variable	df	Parameter estimate	SE	*t* value	*p* > |*t*|	95% confidence limits	
Intercept	1	227.4	5.52	41.21	< 0.0001	215.81	238.99
Day	1	5.65	0.45	12.53	< 0.0001	4.70	6.59

by the ratio MS lack of fit/MS pure error with the corresponding degrees of freedom. Here it was not significant, which indicates that the chosen regression model with the estimated function $\hat{y} = 227.4 + 5.645x$ (Table 2D) is appropriate. The tests of β_0 and β_1 against zero are based on the error in Table 2A and result in the acceptance of the alternative hypotheses at $\alpha = 0.05$. The intercept $b_0 = 227.4$ (g kg^{-1}) describes the mean fiber content at the beginning of the experiment (day = 0) and the slope $b_1 =$ 5.645 (g kg^{-1} per day) the mean daily increase of the fiber content. The result of these tests can also be derived from the confidence intervals of the parameters: If they do not contain zero, the alternative hypotheses will be accepted. The *F* test of the model gives exactly the same result as the *t* test of β_1.

In publications, one can often find that the regression analysis of experiments with a quantitative treatment factor and n_i replications of each of its *a* levels ($i = 1, \ldots, a$) has been conducted with the means per treatment level. We discuss the corresponding results for this example conducting the analysis with the means of the four replications per day (Table 3).

Because the number of replications n_i is equal for all days in this case, the estimates b_0 and b_1 are the same in Table 2D and Table 3B. With unequal numbers of replications the estimations would differ. Furthermore, it is due to the special parameter constellation ($n_i = n = 4$ and $n = a - 1$) that some more relations exist between the ANOVA tables of Table 2B, 2C, and Table 3A.

The main reason for the presentation of the results on the basis of mean values is the seemingly better precision. It sounds better to have $s = 3.503$, CV% = 1.234, $R^2 = 0.9954$ and adj. $R^2 = 0.9939$ instead of the corresponding values on the basis of the replicate values ($s = 15.766$, CV% = 5.568, $R^2 = 0.8776$ and adj. $R^2 = 0.8708$). The MSE in Table 3A is the MS lack of fit/n from Table 2B. The R^2 would be the same for the mean-based and the single value-based analysis only if the SS pure error is equal to zero (which means that all values to a given treatment level are identical). Depending on the df, however, in most cases, the criteria *s*, CV%, R^2, and adj. R^2 appear to be more favorable with the mean values. The MSE and R^2 in the mean-based approach, however, only inform about the dispersion of the mean values around the function (the lack of fit of the function). The *F* test of the model in Table 3A is the same as MS Model/MS lack of fit in Table 2A. The tests of the regression parameters (Table 3B), the confidence, and prediction intervals of the function are also only determined by the MS for lack of fit and have no meaningful interpretation when a lack of fit cannot be ruled out.

Example 4 (SAS Code in Appendix)

As explained previously, in the example, we use regression to determine the relationship between weed infestation (*x*) and yield (*y*). Because Model II applies here (the regressor is a random variable), calculation of an association measure between regressor and regressand may be useful. In this example, the linearity was only approximately achieved by a nonlinear (square root) transformation of the regressor values. Therefore, Pearson's correlation coefficient would be meaningful only if computed between *y* and the square root of *x*. To describe the association between yield (*y*) and the number of wind grass plants, Spearman's correlation coefficient (also Spearman's rank correlation coefficient) ρ_{Spearman} can be used. Its

TABLE 3. Example 3 based on the mean values per day. Analyses of variance (ANOVA), estimation and tests of the fixed parameters.

A. ANOVA table for the regression analysis

Source	df	Sum of squares	Mean square	*F* value	$p > F$
Model	1	7966.51	7966.51	649.11	0.0001
Error	3	36.82	12.27		
Corrected total	4	8003.33			

B. Estimations and tests of the fixed parameters

Variable	df	Parameter estimate	SE	*t* value	$p > \|t\|$	95% confidence limits	
Intercept	1	227.4	2.71	83.80	< 0.0001	218.76	236.04
Day	1	5.65	0.22	25.48	0.0001	4.94	6.35

calculation and interpretation, however, is only reasonable if a monotonically increasing or decreasing relation between the two variables exists. From Fig. 2 and 3, it can be seen that there is a monotonically decreasing relation. The estimation of ρ_{Spearman} is based on the assignment of ranks to both variables: the smallest value gets the rank 1, the next gets the rank 2, and so on, up to the largest value which gets the rank *N*. If several values are identical, the mid-ranks will be used. In this way, one gets the ranks *Ry* of the *y* values and the ranks *Rx* of the *x* values. Using Eq. [17], but replacing *x* by *Rx* and *y* by *Ry*, we get Spearman's correlation coefficient

$$r_{\text{Spearman}} = \hat{\rho}_{\text{Spearman}} = \frac{\text{SPRxRy}}{\sqrt{\text{SSRx}\cdot\text{SSRy}}} = \frac{\sum(Rx_i - \overline{Rx})(Ry_i - \overline{Ry})}{\sqrt{\sum(Rx_i - \overline{Rx})^2 \sum(Ry_i - \overline{Ry})^2}}$$

which lies between −1 and 1 like Pearson's ρ. In the example, we have $r_{\text{Spearman}} = -0.90403$. The test can be performed as for Pearson's ρ and here the conclusion is that $\rho_{\text{Spearman}} \neq 0$ (*p* value < 0.0001). Spearman's correlation coefficient is a nonparametric measure to describe the dependency between the two variables.

Diagnostics of the Residuals

We assumed that the residuals e_i follow a normal distribution, have the same variance, and are independent of each other. The fitted residuals (also: raw residuals) can be calculated by Eq. [10]. For the evaluation whether the assumptions are fulfilled, the raw residuals, however, are not suited because they are not independent (because of $\sum_{i=1}^{N} \hat{e}_i = 0$) and because they have nonconstant variances.

As $\hat{E} = Y - \hat{Y}$ and the variance–covariance matrix of $\hat{Y}$ for the given x_i values is $H\sigma^2$, the variance–covariance matrix of the fitted residuals $\hat{E}$ is equal to $(I - H)\,\sigma^2$, where *H* is the *H* at matrix from Eq. [23]. Points near $\bar{x}$ have a smaller h_{ii} and therefore the raw residuals near to $\bar{x}$ have larger values of $1 - h_{ii}$ and larger variances. Hence, raw residuals do not have the same variance even if the errors have the same variance. For this reason, we standardize the raw residuals by dividing them by their standard deviation. These standardized residuals have the expected value zero and the variance 1. Replacing σ^2 by its estimate s^2 (= mean squared error), we get the Studentized residuals $\hat{e}_i^*$ (also called: internally Studentized residuals, and in SAS this is called the STUDENT residuals)

$$\hat{e}_i^* = \frac{\hat{e}_i}{\sqrt{(1-h_{ii})s^2}} \qquad [27]$$

These residuals should be used with caution (the influence of specific variances is excluded, but their dependency still exists) to evaluate the assumptions and whether the chosen model equation is appropriate.

1. Assumption of Normal Distribution

The histogram, the boxplot, the Q–Q plot (quantile–quantile plot), or the P–P plot of the $\hat{e}_i^*$ can indicate serious violations of the normality assumption. To check normality, there are several tests, e.g., the Kolmogorov–Smirnov test and the Shapiro–Wilk test. The tests, however, also assume independence of residuals. Only $N - 2$ residuals are linearly independent. There exist other suggestions to overcome this problem, such as to work with recursive residuals (see Schabenberger and Pierce, p. 122 ff.). However, these calculations bring up other difficulties, so that we do not consider them here. For Examples 1, 3, and 4, we represent the results of the two tests and the Q–Q plots in Fig. 6. The result for Example 2 is not shown because this analysis is meaningless since there are only five points. For the plot, the $\hat{e}_i^*$ are sorted according to size such that $\hat{e}_{(1)}^* \leq \ldots \leq \hat{e}_{(N)}^*$, where $\hat{e}_{(i)}^*$ is the ith element in the ordered sequence. If the N residuals follow the standard normal distribution, the expected quantile of the ith value would be $\Phi^{-1}[(i - 0.375)/(N + 0.25)]$, where Φ is the standard normal cumulative distribution function. The constants added and subtracted in the argument are continuity corrections. The Q–Q plot plots the $\hat{e}_{(i)}^*$ against the corresponding quantiles. If the normal distribution holds, the plotted points should lie approximately on a 1:1 diagonal. None of the tests rejects normality, which agrees with the visual impression, but it should be pointed out again that these diagnostics can give only hints to serious violations when samples are small.

2. Assumption of Variance Homogeneity and Assessment of Model Adequacy

The simple regression analysis assumes that for each x value the variance of the y values is the same. To get a visual impression about the validity of this assumption, we map the Studentized residuals $\hat{e}_i^*$ against the x values (Fig. 7). The dispersion of the Studentized residuals around the zero line should not be functionally related to the regressor. The probability that its values exceed the boundaries −2 or +2 is approximately 0.05 for sufficiently large sample sizes. If any $\hat{e}_i^*$ exceeds these boundaries, the corresponding observation could be an outlier. In Examples 1, 3, and 4, we find some data that meet these criteria. These boundaries, however, only provide some orientation for closer inspection of these data. They should not be interpreted like a test decision to exclude these data. Exclusions should be primarily motivated by scientific reasons. The same holds true for all other benchmarks for criteria given below.

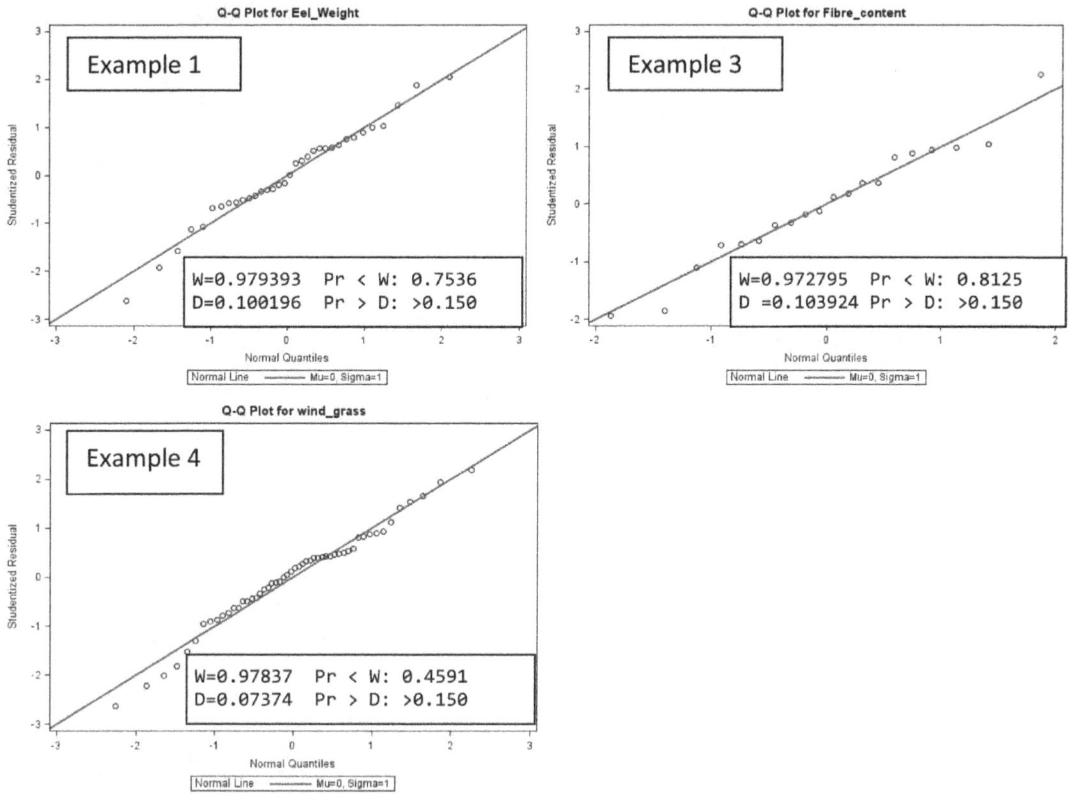

FIG. 6. Examples 1, 3, and 4. Q–Q plot of the Studentized residuals $\hat{e}_i^*$. *W* and *D* are the test statistics of the Shapiro–Wilk test and Kolmogorov–Smirnov test, respectively.

Furthermore, if the model equation is adequate, no systematic pattern should be visible for the $\hat{e}_i^*$ in relation to the function values $\hat{y}_i$. Compared with the plots in Fig. 7, we did not find any new insights and so do not show additional plots.

The plot of Example 4 is conspicuous. It seems that for intermediate x values, the $\hat{e}_i^*$ have a larger variance and tend to be negative, whereas for the four largest x values, the $\hat{e}_i^*$ are all positive. This indicates that variance heterogeneity and/or a model inadequacy exist.

3. Evaluation of Each Data Point in Regard to Its Leverage and Its Influence on the Estimated Function

In the calculation of the variances in Eq. [24, 25, and 26] as well as for the influence diagnostics, the Hat matrix $H = X(X^TX)^{-1}X^T$ from Eq. [23] plays a prominent role. The Hat matrix describes the dependence of the predicted $\hat{y}$ values (the "*y* hat" values) on the observed y values, and this relation has earned it the name. The dependency arises inserting Eq. [7] in Eq. [9]: $\hat{Y} = X\hat{\beta} = X(X^TX)^{-1}X^TY = HY$ or for the *i*th predicted y value $\hat{y}_i = \sum_{j=1}^{N} h_{ij} y_j$ for each $i = 1 \ldots N$.

The h_{ij} constitute weighting factors of the observed data for the calculation of the predicted values. The values of h_{ii} lie between $1/N$ and 1, and the sum over all h_{ii} (the elements on the major diagonal of H) is equal to the number of the fixed parameters p in the model [= 2 in the simple model]. With N observations, its mean is p/N. If h_{ii}

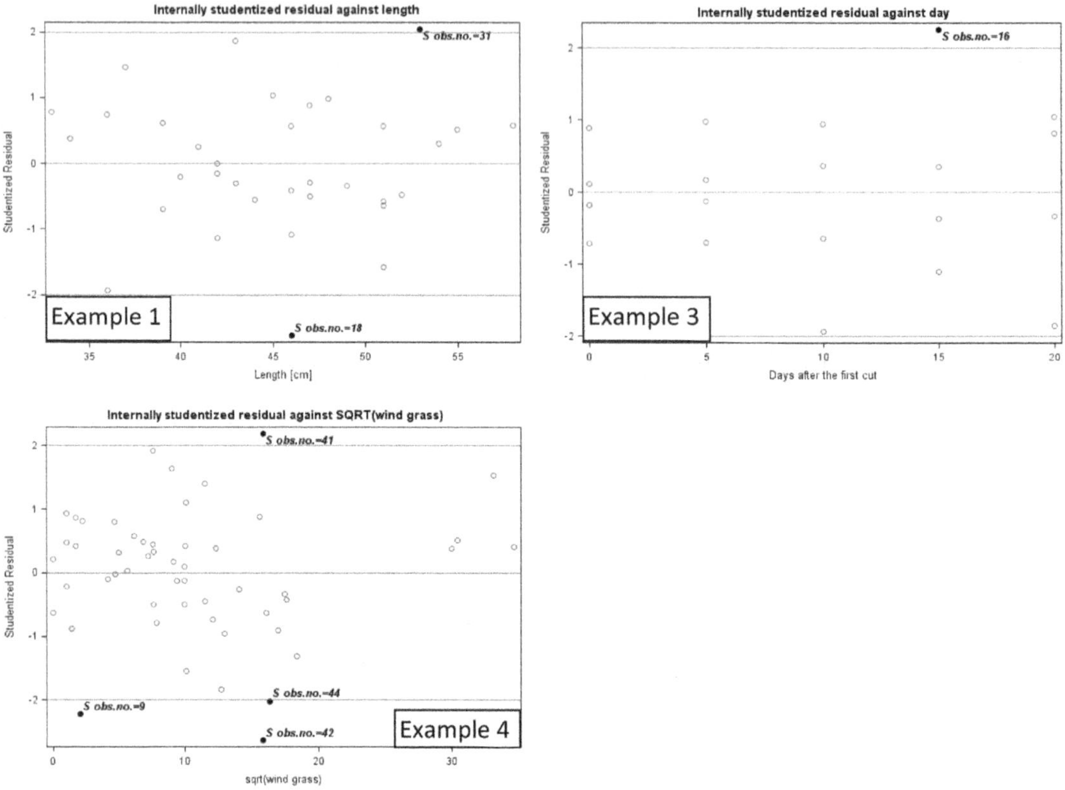

FIG. 7. Examples 1, 3, and 4. Studentized residuals $\hat{e}_i^*$ plotted against regressor values. S = observations with $|\hat{e}_i^*| > 2$ and the corresponding observation number in the data file (see Table 4 and Appendices 1, 3, and 4).

is larger (the difference of the x value to the mean of x is larger, see above), then the corresponding observation has potentially a large influence on the estimation of the regression function. The leverages are often classified as remarkable if they are two times larger than the mean leverage; that is, if $h_{ii} \geq 2p/N$ [$= 4/N$ in the simple linear model]. In Model I, the leverages of the points are determined by the experimental design and, therefore, in contrast to Model II, are not randomly influenced.

In Fig. 3 and in Table 4 all points with leverages larger than the benchmark are marked. They are relevant in Examples 1 and 4 and occur only for relatively small or large x values. This conforms to the above statement that the larger the difference to $\overline{x}$, the larger the leverage.

One can inspect whether the influence of a data point is really large, by excluding it from the data set and recalculating the regression analysis without this value. Whether its influence is small or large can be assessed by several criteria:

(i) The externally Studentized residual (also called RSTUDENT) $\hat{e}_i^{**}$ differs from the internally Studentized residual in Eq. [27] by using the mean square error when estimating the model without the ith point, denoted as $s_{(i)}^2$, instead of its estimate s^2 from the full data set

$$\hat{e}_i^{**} = \frac{\hat{e}_i}{\sqrt{(1-h_{ii})s_{(i)}^2}}$$

In Table 4 for all examples, the points with $|\hat{e}_i^*| > 2$ or $|\hat{e}_i^{**}| > 2$ are given. The differences are mostly small. Only Example 2 shows a remarkable $\hat{e}_i^{**}$ in the middle of the x values. This reveals how sensitive the criteria partially react. From Fig. 2 or 3 one would not expect any conspicuous data point. In Fig. 3, points with $|\hat{e}_i^{**}| > 2$ are marked with R, in Fig. 7 points with $|\hat{e}_i^*| > 2$ with S.

(ii) The influence of the ith data point is large if the difference between the original estimated $\hat{y}_i$ and the estimated $\hat{y}_{(i)}$ after deleting the ith point is remarkable. The measure DFFITS (difference of fits, scaled) is the scaled difference between $\hat{y}_i$ and $\hat{y}_{(i)}$

$$\mathrm{DFFITS}_i = \frac{\hat{y}_i - \hat{y}_{(i)}}{\sqrt{s_{(i)}^2 h_{ii}}}$$

As a threshold for declaring a large influence of a data point under consideration of the sample size, it is recommended to compare $|\mathrm{DFFITS}_i|$ with $2\sqrt{p/N}$ $\left[=2\sqrt{2/N}\right]$ (Belsley et al., 2004). In Fig. 3 the corresponding points are marked by D.

(iii) DFBETAS (difference of β, scaled) measures the influence of a deleted point i on the jth parameter of the function (in the simple linear model these are $j = 0$ and 1 for the intercept and the slope, respectively) with

$$\mathrm{DFBETAS}_{j(i)} = \frac{b_j - b_{j(i)}}{\sqrt{s_{(i)}^2 \left[(\mathrm{X}^T\mathrm{X})^{-1}\right]_{jj}}}$$

where $[(\mathrm{X}^T\mathrm{X})^{-1}]_{jj}$ is the jth element on the major diagonal of $(\mathrm{X}^T\mathrm{X})^{-1}$. Here, Belsley et al. (2004) recommended for $|\mathrm{DFBETAS}_j|$ the benchmark $2/\sqrt{N}$. From Table 4, it can be seen, that if a deleted point would lead to an increase of the intercept, then this increase is primarily due to a decrease of the slope and vice versa. This is the result of its negative covariance (see Eq. [20]).

(iv) Cook's D has similarities with the DFFITS measure. It is

$$D_i = \frac{\left(\hat{y}_i - \hat{y}_{(i)}\right)^2}{ps^2} \quad (p = 2 \text{ in the simple linear regression}).$$

Due to the relation to DFFITS, Rawlings et al. (1998) suggest the threshold $4/N$.

(v) The COVRATIO of point i expresses its effect on the precision of the parameter estimates. From Eq. [21], we know the generalized variance as a summary measure for the variance of the parameter estimates. The reciprocal of the variance can be used as a measure for the precision. The COVRATIO is the ratio of each precision estimate, that is, the estimate calculated for the full data set and the data set reduced by the ith point.

$$\mathrm{COVRATIO}_i = \frac{\text{Precision (full data set)}}{\text{Precision (reduced data set)}} = \frac{\det\left[s_{(i)}^2\left(\mathrm{X}_{(i)}{}^T\mathrm{X}_{(i)}\right)^{-1}\right]}{\det\left[s^2\left(\mathrm{X}^T\mathrm{X}\right)^{-1}\right]}$$

TABLE 4. Examples 1 to 4. Summary of all points conspicuous by the criterion (i) to (iv) or with $|\hat{e}_i^*| \geq 2$. Bold: value exceeds the corresponding benchmark. (In Fig. 3, the corresponding points are marked by R, L, and D and in Fig. 7 by S.)

Example	Obs. No.	Raw residual $\hat{e}_i$	Internally Studentized residual (*S*)	Externally Studentized residual (*R*)	Leverage (*L*)	Press	Cook's	DFFITS (*D*)	DFBETAS intercept	DFBETAS slope	Covratio
Benchmark for $p = 2$:			$\lvert\hat{e}_i^*\rvert \geq 2$	$\lvert\hat{e}_i^{**}\rvert \geq 2$	$h_{ii} \geq 4/N$		$D_i \geq 4/N$	$\lvert\text{DFFITS}_i\rvert \geq 2\sqrt{2/N}$	$\lvert\text{DFBETAS}_i\rvert \geq 2/\sqrt{N}$		$\lvert\text{Covratio}_i - 1\rvert \geq 6/N$
$N = 34$	Benchmark for given sample size:				0.1176		0.1176	0.4851	0.3430	0.3430	0.1765
1	1	2.34	0.79	0.78	**0.14**	2.73	0.05	0.32	0.30	−0.29	**1.20**
1	2	1.16	0.38	0.38	**0.13**	1.32	0.01	0.14	0.13	−0.13	**1.21**
1	3	−5.91	−1.93	**−2.02**	0.09	−6.52	**0.19**	**−0.65**	**−0.58**	**0.54**	0.92
1	5	4.51	1.46	1.49	0.08	4.91	0.09	0.44	**0.39**	**−0.35**	1.01
1	18	−8.31	**−2.63**	**−2.92**	0.03	−8.57	0.11	**−0.51**	0.004	−0.07	**0.68**
1	31	6.32	**2.05**	**2.16**	0.08	6.85	**0.18**	**0.63**	**−0.44**	**0.49**	0.87
1	34	1.72	0.58	0.58	**0.16**	2.04	0.03	0.25	−0.21	0.26	**1.24**
$N = 5$	Benchmark for given sample size:				0.8		0.8	1.2649	0.8944	0.8944	1.2
2	3	−0.02	−1.65	**−4.33**	0.23	−0.03	0.41	**−2.39**	−0.28	**−1.87**	0.03
$N = 20$	Benchmark for given sample size:				0.2		0.2	0.6325	0.4472	0.4472	0.3
3	9	−26.85	−1.93	**−2.11**	0.05	−28.26	0.10	−0.48	−0.28	0.00	0.74
3	16	30.93	**2.26**	**2.59**	0.08	33.43	**0.21**	**0.74**	0.00	0.43	**0.62**
3	17	−24.30	−1.85	−1.998	0.15	−28.59	**0.30**	**−0.84**	0.28	**−0.69**	0.87
$N = 52$	Benchmark for given sample size:				0.0769		0.0769	0.3922	0.2774	0.2774	0.1154
4	9	−2236	**−2.22**	**−2.31**	0.04	−2330.68	**0.10**	**−0.48**	**−0.47**	**0.35**	**0.88**
4	41	2219	**2.19**	**2.28**	0.03	2282.09	0.07	0.38	0.03	0.21	**0.88**
4	42	−2673	**−2.63**	**−2.81**	0.03	−2748.36	**0.10**	**−0.47**	−0.03	−0.26	**0.79**
4	44	−2047	**−2.02**	**−2.09**	0.03	−2108.53	0.06	-0.36	−0.01	−0.21	0.91
4	49	376.02	0.39	0.39	**0.13**	432.91	0.01	0.15	−0.08	0.14	**1.19**
4	50	496.72	0.52	0.52	**0.14**	575.18	0.02	0.20	−0.10	0.19	**1.19**
4	51	1435	1.53	1.55	**0.17**	1730.41	**0.24**	**0.71**	**−0.38**	**0.66**	**1.14**
4	52	381.26	0.41	0.41	**0.19**	471.71	0.02	0.20	−0.11	0.19	**1.28**

A $\text{COVRATIO}_i > 1$ means that *i*th point improves the precision; a $\text{COVRATIO}_i < 1$ means that precision deteriorates. As a benchmark, Belsley et al. (2004) recommended $|\text{COVRATIO}_i - 1| \geq 3p/N$ [$= 6/N$].

They showed that a large COVRATIO tends to be connected with a large leverage and a small COVRATIO with a large $\hat{e}_i^{**}$. In Table 4, not all points which exceed the benchmark are listed. For the remaining points, the tendency can be confirmed.

(vi) The PRESS measure for the ith point is

$$\text{PRESS}_i = \frac{\hat{e}_i}{1-h_{ii}}$$

the predicted residual after deleting the ith point. More important is the sum of the squared individual PRESS measures, the PRESS (predicted residual SS), which is equal to

$$\text{PRESS} = \sum_i \text{PRESS}_i^2 = \sum_i \left(\frac{\hat{e}_i}{1-h_{ii}}\right)^2 = \sum_i (y_i - \hat{y}_{(i)})^2$$

where $\hat{y}_{(i)}$ is the predicted value of y after deleting the ith point from the estimation. The PRESS value should be similar to the sum of squared residuals (RSS) for the full data set. In this case, deleting of single values has no remarkable influence on the results and the prediction value of the model is high (Table 5).

A summary of all points which are conspicuous by any of the criteria (i) to (iv) or by $|\hat{e}_i^*| > 2$ are given in Table 4. In Fig. 3, all points with $|\hat{e}_i^{**}| > 2$, with a high leverage and with a high DFITTS are marked.

A high leverage is not automatically associated with a high influence of the point. If it is close to the estimated regression function, it has no influence (Fig. 3, Example 1 and 4). We see that many points which we identified by the influence criteria are also well identifiable in the plots.

If a point has a large residual $|\hat{e}_i^{**}| > 2$ and/or DFITTS is large, then its exclusion does not automatically mean that the regression parameters would change. Here too, the distance to the center of the observations is crucial: if the distance is large, it is more probable that the parameter changes. We reviewed all conspicuous points for adherence to the model and decided not to exclude any point from the analysis.

The Problem of Regressand Transformation—Reconsideration of Example 4

Due to the conspicuous residuals, we reanalyzed this example with other transformations of the regressor, but we did not find a better fit. Therefore, we tried a transformation of the regressand. Above, we said that in case of a regressand transformation, some specifics need to be considered. These issues will now be illustrated.

For the example, we used a log-transformation resulting in the model $\log(y_i) = \beta_0 + \beta_1 x_i + e_i$. If this model is retransformed, it results in the model $y_i = \exp(\beta_0 + \beta_1 x_i + e_i) = \beta_0^* \exp(\beta_1 x_i)\, e_i^*$ with $\beta_0^* = \exp(\beta_0)$ and $e_i^* = \exp(e_i)$. Alternatively, one could fit the model $y_i = \beta_0 \exp(\beta_1 x_i) + e_i$, which cannot be linearized—it is a nonlinear model. The difference is that the retransformed model assumes multiplicative and the second model additive residuals. The fitted residuals of the first model are $\hat{e}_i^* = y_i / \hat{y}_i$ and the residuals of the second model are $\hat{e}_i = y_i - \hat{y}_i$ (see also Eq. [10]). The assumption of multiplicative errors may be of advantage if variance heterogeneity exists. A similar situation arises with the linearized model (transformation of the regressor and the regressand) $\log(y_i) = \beta_0 + \beta_1 \log(x_i) + e_i$, which after retransformation is $y_i = \beta_0^* x_i^{\beta_1} e_i^*$.

TABLE 5. Examples 1 to 4. PRESS and sum of squared residuals (RSS).

Example	PRESS	RSS
1	371.729	330.696
2	0.00168	0.00083
3	4456.56	3654.525
4	57044697	52949986

With other transformations, implications for the model on the original scale are more complex. If, for example, using the square root of the regressand, one gets after retransformation $\boldsymbol{y}_i = (\beta_0 + \beta_1 x + \boldsymbol{e}_i)^2$ where after solving the right-hand side of the equation it can be seen that both additive and multiplicative error effects occur. Also, the log-transformation as well as the square-root transformation of the regressand are special cases of the class of Box–Cox transformation (Box and Cox, 1964) and will be discussed in more detail later.

We tried the log-transformation resulting in $\widehat{\log(y)} = 8.98065 - 0.00186x$ with $R^2 = 0.8267$, adj. $R^2 = 0.8232$ and RMSE = 0.2323. The retransformation yields $\hat{y} = 7947.8\exp(-0.00186x)$. Concerning the coefficient of determination, these results are a little better than the former analysis (Fig. 3); the two RMSE values are not comparable. The corresponding nonlinear approach with additive residuals is again an improved fit. We found $\hat{y} = 8673.6\exp(-0.0026x)$ and with $R^2 = 0.8239$, adj. $R^2 = 0.8204$, and RMSE = 988.658 (this value can be compared with Fig. 3). In both cases, however, some conspicuous residuals do not vanish.

Finally, we ascertained that the best result has been achieved by a three-parametric nonlinear model $\boldsymbol{y}_i = \beta_0 + \beta_1\exp(\beta_2 x_i) + \boldsymbol{e}_i$, which also cannot be linearized but assumes constant variance on the observed scale. The estimation is $\hat{y} = 1252.77 + 7551.15\exp(-0.003476x)$. Results describing the fit indicate that $R^2 = 0.836$, adj. $R^2 = 0.829$, and RMSE = 963.3. Also based on the pattern of residuals, the results appear to be acceptable (Fig. 8). Possibly this (nonlinear) model with additional consideration of variance heterogeneity is better suited; here, however, we are focused on linear models and will discuss variance heterogeneity only in this context.

Some Aspects of Planning Experiments for Linear Regression

The construction of optimal designs and the planning of sample sizes are not in the main focus of our chapter. However, we provide some basic ideas.

From Eq. [19], we have seen that the variance of β_1 as well the generalized variance Eq. [21] are inversely proportional to SSX. In Model I, SSX depends on the chosen levels of x (the allocation of the x values) and hence on the experimental design. The theory of optimal designs aims at the allocation of the x levels and only applies to Model I (regressors are fixed variables). If we want to minimize these variances then SSX should be as large as possible. The SSX is maximal if the linear function holds in the interval $[x_L, x_U]$ and measurements are only conducted at both boundaries of the interval. For a sample size of N, where N is an even number, $N/2$ observations should be taken for $x = x_L$ and as well for $x = x_U$. Such a design is called *D-optimal* because it minimizes the determinant in Eq. [21] as well as the variance of β_1. In this case, SSX is $N(x_U - x_L)^2/4$. A design that minimizes the maximal expected width of the confidence interval in Eq. [22] is called *G-optimal*. If N is even, the D-optimal and the G-optimal designs are identical. If N is an odd number, then for D-optimality $(N-1)/2$ observations should be chosen at $x = x_L$ and $(N+1)/2$ at $x = x_u$ or vice versa. The G-optimal design consists of $(N-1)/2$ observations at the boundaries of the interval and one point in the middle. How large the N should be chosen is determined separately. It should be stressed, however, that the optimal features strongly depend on the linearity assumption in the interval. If no information exists about the shape of the function before conducting the experiment, often equidistant x levels will be chosen (as in Example 3). Let us assume for Example 3 that we are confident before conducting the experiment that a linear function applies in the interval [0, 20] and we have the capacity for 20 observations. Then a design with 10 observations at the beginning of the experiment (day = 0) and 10 observations at the end (day = 20) would be optimal with SSX = 2000. Note that SSX was 1000 for the original design. For the original design, $N = 20$ was necessary to obtain the confidence interval for β_1 of [4.69835; 6.59165], which has a half width of $d = 0.94665$ (Table 2D). Using the optimal design and assuming the same residual variance, $N = 12$ samples (6 for $x = 0$ and 6 for $x = 20$) would be sufficient to result in an interval width which is no larger than 0.9467 or with $N = 20$ the half width would be $d = 0.6694$. Whether $N = 20$ is sufficient to answer the experimental question with given precision requirements is a second (and extremely important) aspect of planning the experiment. It should be added here that often there is no certainty that the response will be linear for the x domain of interest. In such cases one will usually use one or two intermediate levels of x in addition so as to be able to test the lack of fit of the linear model and extend the model to allow for a nonlinear response if needed.

In Model II, the theory of optimal designs is not relevant because the x values are random. The second design aspect, however, is also relevant to this model. Assume that it is necessary to determine the sample size N for the estimation of the confidence interval of β_1. The precision requirements are given by the targeted half width of the interval d and the confidence level $1 - \alpha$. Depending on the model, prior information on several quantities is necessary.

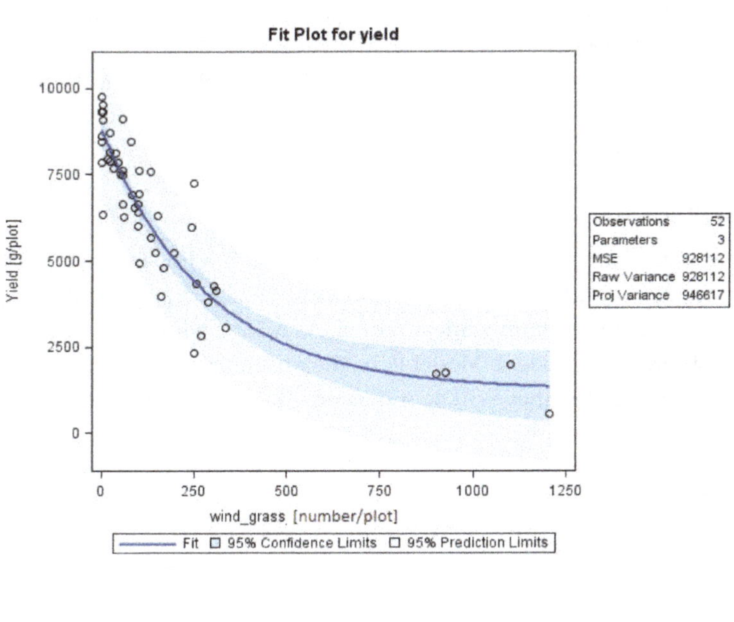

FIG. 8. Example 4. Analysis as a nonlinear model which cannot be linearized resulting in $\hat{y} = 1252.77 + 7551.1 \exp(-0.003476x)$.

For Model I, assuming a D-optimal plan has been chosen, N can be calculated by

$$N=\left\lceil \frac{4\sigma^2}{d^2(x_U-x_L)^2}t^2\left(N-2;1-\alpha/2\right)\right\rceil \quad [28]$$

where $\lceil z \rceil$ means the smallest integer which is larger or equal to z. Use of Eq. [28] requires information about σ^2. If the allocation did not follow the optimality criterion and a different x_i values have been chosen denoted by x_i^*, the SSX can be calculated by

$$\text{SSX}=\frac{N}{a}\sum_{i=1}^{a}(x_i^*-\overline{x})^2$$

and the required sample size by

$$N=\left\lceil \frac{a\sigma^2}{d^2\sum_{i=1}^{a}(x_i^*-\overline{x})^2}t^2\left(N-2;1-\alpha/2\right)\right\rceil \quad [29]$$

For Model II, N can be calculated by Eq. [30] which requires information about σ_y^2, σ_x^2, and β_1:

$$N=\left\lceil \frac{t^2(N-1;1-\alpha/2)}{d^2}\left(\frac{\sigma_y^2}{\sigma_x^2}-\beta_1^2\right)\right\rceil+1 \quad [30]$$

(Rasch et al., 1998; procedure 4/32/3011). In fortunate circumstances, the necessary information can be obtained from literature or by preliminary experiments. In practice, the difficulties in obtaining this information, however, are often the reason why the used sample size is not chosen based on statistical considerations. In Eq. [28, 29, and 30], N appears on both sides of the equations so that a solution has to be determined iteratively, for example, by PROC MODEL of SAS.

The construction of designs using D-optimality is also possible for simple linear and multiple linear functions as well as for functions with polynomial terms with PROC OPTEX in SAS (Atkinson et al., 2009).

Multiple Linear Regression

The multiple linear regression model arises if the regressand is a linear function of p fixed parameters $\beta_0, \beta_1, \ldots \beta_{p-1}$ that are linked multiplicatively with $p-1$ regressors. The case $p = 2$ corresponds to the simple linear model. As before, while we can often refrain from the distinction of Model I, Model II, or the mixed situation, if there are model specifics in our examples, they will be indicated. The model equation can be written as

$$y_i=\beta_0+\beta_1x_{1,i}+\beta_2x_{2,i}+\ldots+\beta_{p-1}x_{p-1,i}+e_i \qquad i=1\ldots N \quad [31]$$

where the $p - 1$ regressors may be $p - 1$ different variables or may arise by any transformation of one or of several variables. For example, in the model $y = \beta_0 + \beta_1 x_1 + \beta_2 x_2 + \beta_3 x_3$ the regressors x_1, x_2, and x_3 may be three different traits, or the original model was $y = \beta_0 + \beta_1 x + \beta_2 x^2 + \beta_3 x^3$, and we identified x with x_1, x^2 with x_2, and x^3 with x_3 or the original model was $y = \beta_0 + \beta_1 x_1 + \beta_2 x_2 + \beta_3 x_1 x_2$ and we set $x_1 x_2 = x_3$. Thus, all polynomial functions can be handled as functions of multiple linear regression functions, meaning that a quadratic, cubic, quartic, etc. function is a multiple linear regression.

In matrix notation the model can be written as

$$\mathbf{Y} = \mathrm{X}\beta + \mathbf{E}$$

where

$$\mathbf{Y} = \begin{bmatrix} y_1 \\ y_2 \\ \vdots \\ y_N \end{bmatrix}, \mathrm{X} = \begin{bmatrix} 1 & x_{1,1} \cdots & x_{p-1,1} \\ 1 & x_{1,2} \cdots & x_{p-1,2} \\ \vdots & \vdots \;\; \vdots & \\ 1 & x_{1,N} \cdots & x_{p-1,N} \end{bmatrix}, \beta = \begin{bmatrix} \beta_0 \\ \beta_1 \\ \vdots \\ \beta_{p-1} \end{bmatrix}, \text{ and } \mathbf{E} = \begin{bmatrix} e_1 \\ e_2 \\ \vdots \\ e_N \end{bmatrix} \quad [32]$$

and

$$\mathbf{E} \sim \mathrm{N}_N([0], \mathrm{R}).$$

As in the simple case we assume $\mathrm{R} = \sigma^2 \mathrm{I}$ (see Eq. [5]). The estimators and their properties derived for the simple linear regression with variance homogeneity and no existing covariances can immediately be transferred because we used the matrix notation in the relevant cases. The estimations of the fixed parameters correspond to Eq. [7], the estimation of σ^2 to Eq. [13] with df of RSS = $N - p$, respectively, to Table 1B, and the variance of $\hat{\beta}$ to Eq. [18]. Referring to Eq. [14 and 15] and Table 1B, the model fit can be described by the multiple coefficient of determination or by its adjusted form

$$R^2_{\mathrm{mult}} = \frac{\mathrm{SS}_{\mathrm{explained}}}{\mathrm{SSY}} = \frac{\sum_{j=1}^{p-1} b_j \mathrm{SPX}_j Y}{\mathrm{SSY}} \text{ and adj. } R^2_{\mathrm{mult}} = 1 - \frac{\mathrm{RSS}}{\mathrm{SSY}} \frac{N-1}{N-p} \quad [33]$$

Specifics of the Multiple Linear Regression Model Compared to the Simple Linear Regression Model

In contrast to the simple linear regression, some specifics exist which will now be described.

Sequential and Partial Evaluation of Regressors

The role of the several regressors can be jointly evaluated by two different strategies: the partial and the sequential evaluation.

The *sequential evaluation* assumes that a hierarchy exists between the regressors, and for a given regressor, it is sequentially included in the model or excluded from the model corresponding to its position in the hierarchy. The hierarchy can be given by the analyst independently of the observed data (e.g., some variables are theoretically especially important to describe the regressand or in a polynomial regression function, one prefers a model with terms of low order) or it can be dependent on the observed data combined with a test-based approach. The test-based approach is used in several model selection techniques; it generates the hierarchy by a sequence of t or F tests, where the significance of each regressor is assessed in each step of the sequence. For each regressor x_i, a $SS_{sequential}$ value can be calculated that describes its contribution to $SS_{explained}$ (Table 1B), depending on its position in the sequence. The sum of all $SS_{sequential}$ values is equal to $SS_{explained}$. The F test of the whole model (Table 1B) is based on $SS_{explained}$.

The *partial evaluation* of a regressor x_i assumes that all regressors except x_i are already in the model and assesses its significance if it is included in addition to all other regressors. Usually in software packages, the final tests of the regression parameters (t or F test) are based on the partial approach, although final tests based on the sequential approach are also possible. In the partial approach, the test statistic of the regression parameter β_i corresponding to the regressor x_i is

$$t(\beta_i)_{\text{partial}} = \frac{b_i}{s_{b_{i,\text{partial}}}} = \left[\frac{SS_{\text{partial}}(x_i)}{s^2_{\text{full model}}}\right]^{1/2}$$

$$\text{or } F(\beta_i)_{\text{partial}} = [t(\beta_i)_{\text{partial}}]^2 = \frac{SS_{\text{partial}}(x_i)}{s^2_{\text{full model}}} \quad [34]$$

If these tests are based on the sequential approach, then one must replace each partial value in Eq. [34] by its corresponding sequential value. In this case, the test decision on β_i depends on the position of x_i in the sequence.

Understanding these different approaches is necessary for properly interpreting the F test for the whole model and the tests of regression parameters (partial or sequential approach). To explain the difference between both approaches and their effects on proper interpretation of results, we will use the syntax $SS(A|B)$. $SS(A|B)$ describes the increase of $SS_{explained}$ by including the variable A when the variable B (or set of variables B) is already included (Searle, 1971).

The described partial and sequential evaluations are also possible for qualitative variables and will be used in Example 5.

Besides the multiple coefficient of determination, partial coefficients of determination can be calculated in multiple regression problems. The multiple-partial coefficient will not be discussed here (Neter et al., 1989). Regarding the partial coefficient, we will consider it only in the context of the partial approach explained above. Assuming the model Eq. [31], the partial coefficient of determination of a regressor x_i, $R^2_{y|x_i}$, describes which part of uncertainty in the linear model $y = f(x_1 \ldots x_{i-1} x_{i+1} \ldots x_{p-1})$ can be explained by including x_i additionally.

$$R^2_{y|x_i} = \frac{R^2_{\text{mult}} - R^2_{y \bullet x_1 \ldots x_{i-1} x_{i+1} \ldots x_{p-1}}}{1 - R^2_{y \bullet x_1 \ldots x_{i-1} x_{i+1} \ldots x_{p-1}}} \quad [35]$$

where $R^2_{y \bullet x_1 \ldots x_{i-1} x_{i+1} \ldots x_{p-1}}$ is the multiple R^2 for the dependency of y on all regressors except x_i. It can be calculated by

$$R^2_{y|x_i} = \frac{\text{SS}_{\text{partial}}(x_i)}{\text{SS}_{\text{partial}}(x_i) + \text{RSS}_{\text{full model}}}$$

Techniques for Model Selection

If doubts exist about which of the given regressors should be included, they can be evaluated regarding their influence on the model fit. The influence of a regressor can be measured by a test statistic (or its corresponding p value) or by its contribution to the improvement of suitable fit criteria.

Test-based techniques are referred to as the *forward, backward,* and *stepwise methods,* which work sequentially. The forward method builds up the model, the backward method starts with the full model (the model with all regressors) and reduces it sequentially, and the stepwise method is a combination of both. All three methods are based on an F statistic or equivalently on a t statistic of the regressors in each step of the model construction. A critical point of these strategies is that thresholds for the p values of the tests that decide on the inclusion or exclusion of the regressors need to be preset. Regressors are sequentially included as long as their p value is smaller than the threshold (forward selection) or excluded in stepwise fashion as long as their p value is larger than the threshold (backward selection). Depending on the thresholds, the resulting models may differ widely. Furthermore, the methods are statistically questionable because the successive significance tests are not independent and cause uncontrolled risks of test decisions. The only justification of the forward and backward selection is that the inclusion or exclusion of a variable with the smallest or largest p value results in the largest increase or smallest decrease of R^2. To avoid these difficulties, fit criteria are an alternative for model selection. Because we seek not only a good model fit but also a model with as few parameters as possible, the criteria should consider a penalty term depending on the number of parameters p. The information criteria are often the only possibility to compare models discussed in the chapter *extensions of the models.* The Akaike (AIC), Schwarz (SBC), and the Bayesian (BIC) criterion are the most well known. These information criteria are functions of RSS, N, and p and are used in SAS PROC REG in the following forms:

$$\text{AIC} = N \log\left(\frac{\text{RSS}}{N}\right) + 2p$$

$$\text{SBC} = N \log\left(\frac{\text{RSS}}{N}\right) + p \log(N)$$

$$\text{BIC} = N \log\left(\frac{\text{RSS}}{N}\right) + 2(p+2)q - 2q^2 \quad [36]$$

where $q = N\widehat{\sigma^2}/\text{RSS}$ and $\widehat{\sigma^2}$ is a given estimate of the error variance (independently of the observed data). In SAS, the error variance estimate is used for the full model. The smaller the values of the criteria, the better will be the model fit. Another strategy aimed at the prediction ability of a model uses the PRESS statistic discussed above in relation to RSS.

Collinearity Problems

Numerical problems may arise if the regressors are nearly linearly dependent of each other; in this case, the regressors are described as collinear or multicollinear. If regressors exist that are exactly linearly dependent of each other, the matrix X is not of full rank. As rank(X) = rank (X^TX), the inverse of (X^TX) cannot be calculated in this case. The inverse, however, is necessary, e.g., in Eq. [7 and 18]. If two regressors are not exactly but nearly linearly dependent, then perhaps an inverse can be calculated. The solution, however, may be numerically instable, meaning that it strongly depends on the numerical algorithm used. From the practical point of view, collinearity between variables means that at least one variable can be excluded because most of the information about this one variable is included in the other(s). Whether multicollinearity occurs can be evaluated, for example, by the tolerance of each regressor TOL(x_i), its variance inflation factor VIF(x_i) = 1/TOL(x_i), by the eigenvalues of the matrix X^TX, or the condition number of this matrix (condition = maximum eigenvalue/minimum eigenvalue). TOL(x_i) is 1 minus the multiple coefficient $R^2_{x_i \bullet x_1 \ldots x_{i-1} x_{i+1} \ldots x_{p-1}}$. This multiple coefficient describes the dependency of one regressor x_i on the others corresponding to the linear model $x_i = f(x_1 \ldots x_{i-1} x_{i+1} \ldots x_{p-1})$. If the regressors are all fixed values (Model I), its calculation is meaningless because its value is determined by the experimental design. TOL(x_i) ≈ 0 means large dependency and TOL(x_i) ≈ 1 no dependency of x_i on the other regressors. Likewise, an eigenvalue near 0 indicates large dependency and induces a high condition number with possible problems of inverting the matrix. This problem may especially arise in regression functions with polynomial or interaction terms.

Example 5 (SAS Code in Appendix)

Mead (1988) described a randomized complete block experiment with rice (*Oryza sativa* L.), in which 10 different spacing treatments were compared. The 10 spacings were all combinations of pairs of interseedling distances (15, 20, 24, and 30 cm) in row and column directions. The number of blocks was four and the response variable was yield (kg plot^{-1}). Initially, Mead analyzed the data regarding the treatment factor as qualitative (Table 6A). To get these and the following results, SAS PROC GLM or PROC MIXED are suitable because the underlying models require qualitative variables (block and treatment) in addition to quantitative variables. To understand the differences found between the treatment means and using the quantitative dimension of treatment levels, Mead quantified their effects by regression on the stand area per seedling (Table 6B). Now, qualitative and quantitative variables are simultaneously included in the model. For this analysis, we used the sequential approach to build up the model hierarchically, which results in the need to use Type I sum of squares for the analyzed sources of variation. The theoretical background of the inclusion of quantitative and qualitative variables in the model is explained in the section "Extensions of the Linear Regression Models." The SS lack of fit in Table 6B corresponds to that portion of SS treatment in Table 6A, which is not explained by the variable area. It can be seen that the treatment differences are determined primarily by the stand area; the area is significant at the 5% level, whereas the lack of fit is not significant. The estimated regression function is $\widehat{\text{Yield}} = 9.11 - 0.00336$ area. Mead (1988) remarked, however, that there seems to be a further influence on the treatment means by the shape of the stand area—better conditions are given by quadratic areas than by extreme rectangular shapes. To formally test this hypothesis, it is useful to compute an index reflecting the shape of the spacing. We tested two indices. As shape index 1 we used the ratio of length x_1 to width x_2 of the area and shape index 2 was the ratio of the circumference of the standing area at hand to the minimal circumference that could be obtained from a square arrangement with the same area. The circumference of a rectangle is $2(x_1 + x_2)$, while the area is x_1x_2. If the rectangle were a square with area x_1x_2, then the circumference would be $4\sqrt{(x_1x_2)}$. Thus, the shape index 2 was $(x_1 + x_2)/[2\sqrt{(x_1x_2)}]$. For a square, both indices are equal to unity. The more elongated the rectangular area, the larger becomes the shape index. Adding these regressors yields the results in Tables 6C and 6D; the SS lack of fit corresponds to that portion of the SS treatment that cannot be explained by area and the index. Both shape indices are significant and the lack of fit (the nonexplainable part of treatment effects) is nonsignificant in both cases. Compared with Table 6B, in both cases an improvement (a reduction of the lack of fit) occurred; the shape index 2 is better suited than the other index. For the final evaluation of the variables area and shape index 2, we omitted the treatment (lack of fit) from the analysis, resulting in MSE = 0.4787, which is smaller than the MSE = 0.53628 (Table 6B), but larger than the MSE = 0.4598 (Table 6A). Compared with the original analysis, the advantage of the regression analysis is the quantification of the effect of the treatments and the option to use the fitted model for interpolating to other stand areas and shapes. The improvement of the analysis with the shape index used in addition to area can also be seen from the residuals in Table 7.

Finally, the fitted multiple regression model is

$$\widehat{\text{yield}} = 23.365 - 0.00357\,\text{area} - 13.8208\,\text{shape} - \text{index }2$$

Both regression coefficients differ significantly from zero.

Example 6 (SAS Code in Appendix)

An experiment analyzed the dependency of the weight of tubers **y** from their size **x** for a given potato (*Solanum tuberosum* L.) variety. Size was measured by sorting tubers in 5-mm intervals of the side length using squares of increasing size through which tubers were passed. The sample size was $N = 524$. Both the x and y variables are continuous random variables where the x variable is discretized in 5-mm intervals and its values are the midpoints of the intervals.

The scatterplot and the univariate analysis imply that the assumption of a linear dependency may not be appropriate and that the variance of the weight increases with increasing size (Fig. 9). Thus, a strategy is needed to account for variance

TABLE 6. Example 5. ANOVA tables.

Source	df	Sum of squares	Mean square	*F* value	$p > F$
A. Treatments considered as qualitative levels of the factor spacing					
Block	3	5.88	1.96	4.26	0.0138
Treatment	9	23.14	2.57	5.59	0.0002
Error	27	12.41	0.46		
Corrected total	39	41.44			
B. Type I sum of squares, *F* tests of block, area and lack of fit (treatment)					
Block	3	5.88	1.96	4.26	0.0138
Area	1	16.79	16.79	36.51	< 0.0001
Lack of fit	8	6.36	0.79	1.73	0.1372
Error	27	12.41	0.46		
Corrected total	39	41.44			
C. Type I sum of squares, *F* tests of block, area, shape index 1, and lack of fit (treatment)					
Block	3	5.88	1.96	4.26	0.0138
Area	1	16.79	16.79	36.51	< 0.0001
Shape index 1	1	1.98	1.98	4.30	0.0477
Lack of fit	7	4.38	0.63	1.36	0.2619
Error	27	12.41	0.46		
Corrected total	39	41.44			
D. Type I sum of squares, F tests of block, area, shape index 2, and lack of fit (treatment)					
Block	3	5.88	1.96	4.26	0.0138
Area	1	16.79	16.79	36.51	< 0.0001
Shape index 2	1	2.49	2.49	5.42	0.0276
Lack of fit	7	3.86	0.55	1.20	0.3363
Error	27	12.41	0.46		
Corrected total	39	41.44			

heterogeneity. Here, we try to find a model equation with relevant regressors in the first step. Then, using the information from the first step, in a second step, we describe specifics due to the variance-covariance structure. Because the results of the first step may influence the results of the second step, it may be necessary to revise the results of the first step once we have the results of the second step.

First Step—Searching for a Suitable Model Equation (Relevant Regressors)

The basic idea for the derivation of the model was that the weight was determined by its volume, and the volume can be approximated by a cubic function of a size measure. Therefore, a polynomial function with a cubic term including all terms

TABLE 7. Example 5. Mean residuals corresponding to the analyses in Table 6 based on the sum of lack of fit and error.

Spacing (x1 × x2)	Observed mean	Residuals pursuant to Table 6B	Residuals pursuant to Table 6C	Residuals pursuant to Table 6D
30 × 30 cm	6.02	−0.07072	−0.19832	−0.16444
30 × 24 cm	6.48	−0.22506	−0.20920	−0.27131
30 × 20 cm	7.19	0.08455	0.25572	0.21142
30 × 15 cm	6.92	−0.68156	−0.18194	−0.03354
24 × 24 cm	7.85	0.66647	0.47477	0.50320
24 × 20 cm	7.78	0.26916	0.20171	0.14274
24 × 15 cm	7.50	−0.40873	−0.21342	−0.23499
20 × 20 cm	7.71	−0.06194	−0.28846	−0.26299
20 × 15 cm	8.70	0.58982	0.58226	0.51052
15 × 15 cm	8.20	−0.16199	−0.42313	−0.40060

with a lower order was fit: $y = \beta_0 + \beta_1 x + \beta_2 x^2 + \beta_3 x^3$ with the transformation $x_1 = x$, $x_2 = x^2$ and $x_3 = x^3$. In addition to the question of whether this model is suited, we must also determine whether all polynomial terms should be included.

The results for the full model are given in Table 8, where $x_1 = x$ (= size1), $x_2 = x^2$ (= size2) and $x_3 = x^3$ (= size3). For each variable, the sequential and the partial sums of squares are provided. For the sequential approach, two different sequences have been chosen: ($x_1 \rightarrow x_2 \rightarrow x_3$) means that at first the linear, then the quadratic, and finally the cubic term will be included, and ($x_3 \rightarrow x_1 \rightarrow x_2$) corresponds to the original idea, that we expect a cubic function which is added by the other terms. If no concrete idea of the underlying function exists, as is often the case in polynomial regression, then the sequence ($x_1 \rightarrow x_2 \rightarrow x_3$) would be appropriate. It may be argued that this is the only meaningful sequence because other sequences would involve models violating the principle of functional marginality (Nelder, 2000). This principle implies that a polynomial regression function of a given order should include all terms of lower order. Quadratic polynomials can describe responses with decreasing rates of return, and they can have a maximum anywhere on the observed x range, provided the linear term is present as well. Cubic polynomials can describe S-shaped response patterns because they have one turning point, and this may be anywhere on the x axis provided the quadratic terms are present as well. When a polynomial of third degree shows a lack of fit, this usually means that a different kind of nonlinear model is needed, for example, a model that approaches an upper yield level. It is therefore not helpful to consider polynomials of order higher than three.

We see that $SS_{explained}$ can be partitioned to the sequential SS of the regressors for the different sequences (Table 8A and 8B):

$$\begin{aligned} SS_{explained} &= SS(x_1 \mid \beta_0) + SS(x_2 \mid \beta_0, x_1) + SS(x_3 \mid \beta_0, x_1, x_2) \\ &= SS(x_3 \mid \beta_0) + SS(x_1 \mid \beta_0, x_3) + SS(x_2 \mid \beta_0, x_2, x_3). \end{aligned}$$

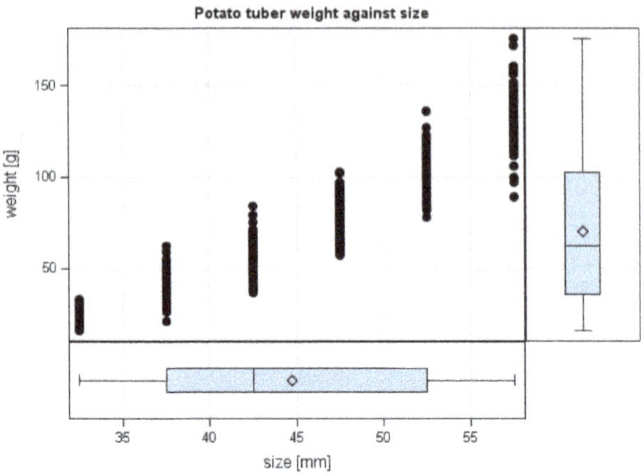

Size	n	Mean	Variance
32.5	92	22.93	13.468
37.5	89	38.26	56.239
42.5	97	55.06	89.434
47.5	83	75.18	123.906
52.5	68	103.16	166.764
57.5	95	134.23	265.648

FIG. 9. Example 6. Scatterplot, boxplots and some univariate measures.

The SS value for a regressor and its F value depend on its position in the sequence. For the partial SS, such a relation to $SS_{explained}$ does not exist and the F or t value (Table 8B and 8C) of the coefficients do not depend on any sequence. It is conspicuous that the F test for the whole model shows high significance (Table 8A), whereas all of the t tests of the single parameters in the partial approach accept the null hypothesis that they are not different from zero (Table 8B and 8C). This apparent contradiction can be explained as follows: The F test for the whole model tests the null hypothesis H_0: $\beta_1 = \beta_2 = \beta_{p-1} = 0$ against H_A: at least one β_j is unequal to zero. This hypothesis is equivalent to the hypothesis whether the theoretical multiple coefficient of determination is equal to or larger than zero. Conversely, the F tests in Table 8B (last columns) and the t tests in Table 8C assume the partial view point: Can the model fit be improved by adding a regressor if all other regressors are already in the model? These are two different approaches. In contrast, it can be seen from the sequential approach that there are regressors that have a significant parameter (Table 8B, first columns). If, as in the example, the partial tests accept the null hypotheses while the F test of the whole model rejects the null hypothesis, then it is a hint that the model is overparameterized. This is also evident from $R^2_{y|x_i}$ and the small TOL values.

In this case, where only three regressors have been considered, all submodels of the full model can be analyzed without any problems (Table 9A). Based on the adj. R^2_{mult} and s^2 values, the differentiation between the first four models and the last model is small and the model with the linear and the cubic term would be selected as the best; the coefficient of the linear term, however, does not differ significantly from zero. The model fit evaluated by AIC shows the same ranking as of BIC and SBC (not shown). Considering additionally the difference |RSS – PRESS|, it can be concluded that the model with only the cubic term has not only the best model fit (based on the information criteria) but also the best prediction quality. For the (temporary) final analysis, we use this model (Fig. 10), which is also supported by our original idea with the sequence ($x_3 \rightarrow x_1 \rightarrow x_2$) (Table 8B). Table 9A shows that the ranks of the model fit criteria, of s^2, and of the adj. R^2_{mult} give nearly the same ranking of the models; the ranks of |RSS – PRESS|, however, differ from them. This makes obvious that, depending on the aim of the regression analysis (best model fit

TABLE 8. Example 6. Analysis of variance table, parameter estimates, and tests of the fixed parameters.

A. Analysis of variance for the regression analysis of the full model

Source	df	Sum of squares	Mean square	*F* value	$p > F$
Model	3	783828	261276	2218.95	< 0.0001
Error	520	61229	117.75		
Corrected total	523	845057			

B. Sequential and partial approach. SS for all parameters, *F* and *p* values for the coefficients

Sequence ($x_1 \rightarrow x_2 \rightarrow x_3$)		Sequence ($x_3 \rightarrow x_1 \rightarrow x_2$)		Partial approach	
Sequential SS	*F* value (*p* value)	Sequential SS	*F* value (*p* value)	Partial SS	*F* value (*p* value)
$SS(\beta_0,\cdot) = 2593284$		$SS(\beta_0,\cdot) = 2593284$		$SS(\beta_0 \mid x_1,x_2,x_3) = 118$	
$SS(x_1 \mid \beta_0) = 767104$	6514.80 (< 0.0001)	$SS(x_3 \mid \beta_0) = 783609$	6654.98 (< 0.0001)	$SS(x_1 \mid \beta_0,x_2,x_3) = 107$	0.91 (0.3416)
$SS(x_2 \mid \beta_0,x_1) = 16475$	139.91 (< 0.0001)	$SS(x_1 \mid \beta_0,x_3) = 123$	1.04 (0.3074)	$SS(x_2 \mid \beta_0,x_1,x_3) = 96$	0.81 (0.3672)
$SS(x_3 \mid \beta_0,x_1,x_2) = 249$	2.12 (0.1460)	$SS(x_2 \mid \beta_0,x_1,x_3) = 96$	0.81 (0.3672)	$SS(x_3 \mid \beta_0,x_1,x_2) = 249$	2.12 (0.1460)

C. Estimates of the fixed parameters for the full model. *t* tests based on the partial approach

Variable	df	Parameter estimate	SE	*t* value	$p > \lvert t \rvert$	$R^2_{y \mid x_i}$	TOL
Intercept	1	−101.33	101.19	−1.00	0.3171		
Size1	1	6.68	7.02	0.95	0.3416	0.00174	0.00006108
Size2	1	−0.14	0.16	−0.90	0.3672	0.00156	0.00001452
Size3	1	0.00172	0.00118	1.46	0.1460	0.00406	0.00005435

to the observed data or best prediction), the selected models may differ. Therefore in publications, the selection criterion used should be specified.

From the prediction interval and the scatter of points (the interval width is independent of the scattering of points) it can be seen that the variance heterogeneity has not been considered in the analysis (Fig. 10, left). Therefore, we need a second step as described before. But before that step, we will have a closer look at some issues arising if a model without an intercept is assumed.

The No-Intercept Problem in Simple and Multiple Linear Regression

Up to this point, we assumed in Examples 1 through 6 that each regression model comprised an intercept β_0. Sometimes, however, one feels strongly from a scientific point of view that the intercept should be equal to zero; that is, the regression should pass through the origin ($x = 0$, $y = 0$). For Example 1, it is obvious that for length = 0, the weight should be zero. Nonetheless the estimated intercept was significantly different from zero, which was the rationale to conclude that the linear relation holds only in the observed length region. In Example 6, the intercept does not differ significantly from zero in the final selected model and theoretically it also seems reasonable. If we drop the intercept from a model, several peculiarities occur. In the no-intercept model, the first column of X in Eq. [4] or in Eq. [32] is removed and the estimate (Eq. [8]) for β_1 in the model with only one regressor changes to

$$\widehat{\beta_1} = b_1 = \frac{\text{SP}_{\text{uncorr}}\text{XY}}{\text{SS}_{\text{uncorr}}\text{X}} \quad [37]$$

where

$$\text{SP}_{\text{uncorr}}\text{XY} = \sum_{i=1}^{N} x_i y_i \text{ and } \text{SS}_{\text{uncorr}}\text{X} = \sum_{i=1}^{N} x_i{}^2$$

The suffix uncorr means "uncorrected" and indicates that the mean has not been subtracted from the observed values. In contrast to the model that includes an intercept, this function does not go through the point $(\bar{x}, \bar{y})$. The partition of SS_{total} given in Eq. [11] is also no longer valid; it can be shown, however, that

$$\text{SS}_{\text{uncorr}}\text{Y} = \sum y_i{}^2 = \underbrace{\sum \hat{y}_i^2}_{\text{SS}_{\text{explained (uncorr)}}} + \underbrace{\sum (y_i - \hat{y}_i)^2}_{\text{RSS}}$$

and this requires modifications in the ANOVA table compared with Table 1 (Table 10). If there is only one regressor, then $p = 1$ (Table 10).

Referring to the multiple coefficient of determination in Eq. [33] and considering the modifications in Table 10, several suggestions have been made as to how to describe the model fit for the no-intercept model:

$$R^2_{\text{mult-noint}} = 1 - \frac{\text{RSS}}{\text{SS}_{\text{uncorr}}\text{Y}} = \frac{\text{SS}_{\text{expl.(uncorr)}}}{\text{SS}_{\text{uncorr}}\text{Y}} = \frac{\sum_{j=1}^{p} b_j \text{SP}_{\text{uncorr}}\text{X}_j\text{Y}}{\text{SS}_{\text{uncorr}}\text{Y}} \quad [38]$$

and

$$\text{adj. } R^2_{\text{mult-no int}} = 1 - \frac{\text{RSS}}{\text{SS}_{\text{uncorr}}\text{Y}} \frac{N}{N-p}$$

These coefficients lie between 0 and 1, respectively, between −1|(df of residuals) and 1 as R^2 and adj. R^2 or as its multiple forms in Eq. [33]. Their disadvantage is that they are only comparable with the coefficients of the with-intercept models and are interpretable as those coefficients if the mean of the y values is zero, a practically uninteresting case. For models with intercept, the coefficient of determination describes the percentage of the variability of the y values (SSY), which can be explained by the regression. In Eq. [38], it is the percentage of $\text{SS}_{\text{uncorr}}\text{Y}$ that can be explained and $\text{SS}_{\text{uncorr}}\text{Y}$ has no contextual interpretation. Therefore, apart from the estimated residual variance s^2 as measure of model fit, Schabenberger and Pierce (2002) suggest to replace $\text{SS}_{\text{uncorr}}\text{Y}$ by SSY in Eq. [38] and term it Pseudo-R^2. The Pseudo-R^2, however, is also not really interpretable and might have negative values if RSS of the no-intercept model

TABLE 9. Example 6. Parameter estimations and selected model fit criteria of the full and all sub-models with and without regression constants.

A. Models with intercept

Model	Intercept	Size1	Size2	Size3	adj. R^2_{mult} (Rank)	s^2 (Rank)	PRESS	|RSS-PRESS| (Rank)	AIC (Rank)
	Parameter estimates								
Size1, 2, and 3	-101.330	6.681	-0.144	0.001716	0.9271 (4)	117.75 (4)	62218.605	989.7 (7)	2502.7 (4)
Size1 and 2	44.422*	-3.490*	0.088*		0.9270 (5)	118.00 (5)	62236.918	758.5 (4)	2502.8 (5)
Size1 and 3	-10.465	0.355		0.000653*	0.9272 (1)	117.71 (1)	62092.654	767.9 (5)	2501.5 (2)
Size2 and 3	-5.130		0.008	0.000599*	0.9271 (3)	117.73 (3)	62114.700	779.1 (6)	2501.6 (3)
Size1	-127.587*	4.428*			0.9076 (7)	149.34 (7)	78603.229	650.1 (3)	2625.2 (7)
Size2	-31.935*		0.049*		0.9233 (6)	123.87 (6)	65220.094	561.9 (2)	2527.3 (6)
Size3	-0.125			0.000709*	0.9271 (2)	117.72 (2)	62008.124	560.4 (1)	2500.6 (1)

B. Models without intercept

Model	Size1	Size2	Size3	adj. R^2_{mult}-no int (Rank)	adj. Pseudo-R^2_{mult}-no int (Rank)	s^2 (Rank)	PRESS	|RSS-PRESS| (Rank)	AIC (Rank)
	Parameter estimates								
Size1, 2, and 3	-0.338	0.015	0.000548*	0.9821 (4)	0.9270 (4)	117.75 (4)	62137.37	790.4 (6)	2501.7(4)
Size1 and 2	-1.477*	0.066*		0.9817 (5)	0.9257 (5)	119.82 (5)	63115.33	569.6 (2)	2509.9(5)
Size1 and 3	-0.001		0.000708*	0.9821 (3)	0.9270 (3)	117.72 (3)	62034.42	584.9 (3)	2500.6(3)
Size2 and 3		0.0001	0.000706*	0.9821 (2)	0.9270 (2)	117.72 (2)	62058.12	608.9 (4)	2500.6(2)
Size1	1.676*			0.8878 (7)	0.5433 (7)	736.52 (7)	386814.48	1616.6 (7)	3460.4(7)
Size2		0.036*		0.9618 (6)	0.8444 (6)	250.96 (6)	131882.92	630.3 (6)	2896.3(6)
Size3			0.000708*	0.9821 (1)	0.9271 (1)	117.49 (1)	61844.40	394.8 (1)	2498.6(1)

* Significantly different from zero at $\alpha = 0.05$ (partial approach).

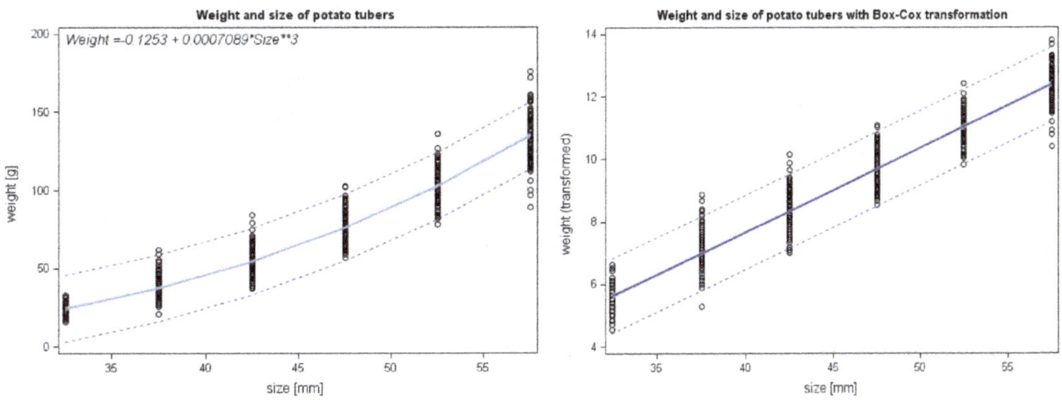

FIG. 10. Example 6. Observed data, regression function, confidence and prediction interval with $1 - \alpha = 0.95$. (Left) Untransformed; (right) Box–Cox transformed with $\lambda = 1/3$.

is larger than SSY. The problem will be demonstrated by a small example in Fig. 11. It can be seen that the function without intercept is not reasonable; nonetheless $R^2_{\text{no int}} > R^2$. The negative sign of Pseudo-R^2 emphasizes that the no-intercept model should not be selected (SSY = 34.83 < RSS = 655.71), its concrete value, however, cannot be interpreted. Apparently, the problem is connected with the significant intercept in the with-intercept model. In any case, these R^2 statistics are not suited for the comparison of without- and with-intercept models; only for several models within the two model classes (with or without intercept) is it justified and interpretable.

The residual variances and the information criteria show clearly the distinction between the model fits (Table 11). The problems are not always so obvious. For Example 6, the full model and all submodels were also analyzed without intercepts (Table 9B). It can be seen that all AIC values are larger in Table 9B than in Table 9A if the intercept was significant. This is a reasonable result, but it does not fully hold for s^2. Both the adj. $R^2_{\text{mult-no int}}$ and the adj. Pseudo-$R^2_{\text{mult-no int}}$ should not be compared with adj. R^2_{mult}. The difference |RSS – PRESS| depends strongly on the number of parameters, and except for the cubic approach, it shows a different rank order. Finally, as the result of the first step of model selection, we can use the no-intercept model with the cubic term; its graphical representation looks quite similar to that of Fig. 10, left (therefore not shown).

Second Step—Box–Cox Transformations of the Regressand as a Way to Consider Variance Heterogeneity

To handle the variance heterogeneity, we take up the idea of regressand and/or regressor transformation. Such transformations of the regressand are special cases from the Box–Cox family of transformations (Box and Cox, 1964). This class contains the log-transformation and the nontransformation as special cases and is given by

$$y(\lambda)=\begin{cases}\dfrac{y^{\lambda}-1}{\lambda} & \text{for } \lambda\neq 0\\ \log(y) & \text{for } \lambda=0\end{cases}$$

where λ is a transformation parameter. The back-transformation of $\boldsymbol{y}_i(\lambda) = \beta_0 + \beta_1 x_i + \boldsymbol{e}_i$ is $\boldsymbol{y}_i = [1 + \lambda(\beta_0 + \beta_1 x_i + \boldsymbol{e}_i)]^{1/\lambda}$ (for $\lambda \neq 0$) or $\boldsymbol{y}_i = \exp(\beta_0 + \beta_1 x_i + \boldsymbol{e}_i)$ (for $\lambda = 0$). The aim is to find a λ that fits the data best. To compare the fit of different transformations, we use the ML method and use a fit criterion based on a corrected form of the achieved maximum of the log-likelihood (LL) function of the transformed data $\text{LL}_{\text{uncorr}}$. The correction term $\log(J)$ is the logarithm of the Jacobian (J) of the transformation so that

$$\text{LL}=\text{LL}_{\text{uncorr}}+\log(J)=\text{LL}_{\text{uncorr}}+(\lambda-1)\sum_{i=1}^{N}\log(y_i) \quad [39]$$

and it accounts for the change of variable in the likelihood function (Atkinson, 1987). To make the fit comparable with results given later, we use –2LL. The fit improves (better fit) as the value (not the absolute value) of –2 LL becomes smaller. The SAS procedure PROC TRANSREG can be used to search for the optimal λ, its confidence

TABLE 10. Analysis of variance table of a multiple linear regression without intercept and $b_1 \ldots b_p$ regression coefficients.

Source	Sum of squares	df	Mean squares	*F* value
Model (explained part)	$\sum \hat{y}_i^2 = \mathrm{SS}_{\mathrm{expl.(uncorr)}} = \sum_{j=1}^{p} b_j \mathrm{SP}_{\mathrm{uncorr}} \mathrm{X}_j\mathrm{Y}$	p	$\frac{\mathrm{SS}_{\mathrm{expl.(uncorr)}}}{p}$	$\frac{\mathrm{SS}_{\mathrm{expl.(uncorr)}}}{ps^2}$
Error (residual part)	$\sum (y_i - \hat{y}_i)^2 = \mathrm{RSS} = \mathrm{SS}_{\mathrm{uncorr}}\mathrm{Y} - \sum_{j=1}^{p} b_j \mathrm{SP}_{\mathrm{uncorr}} \mathrm{X}_j\mathrm{Y}$	$N-p$	$\frac{\mathrm{RSS}}{N-p} = s^2$	
Uncorrected total	$\sum y_i^2 = \mathrm{SS}_{\mathrm{uncorr}}\mathrm{Y}$	N	$\frac{\mathrm{SS}_{\mathrm{uncorr}}\mathrm{Y}}{N}$	

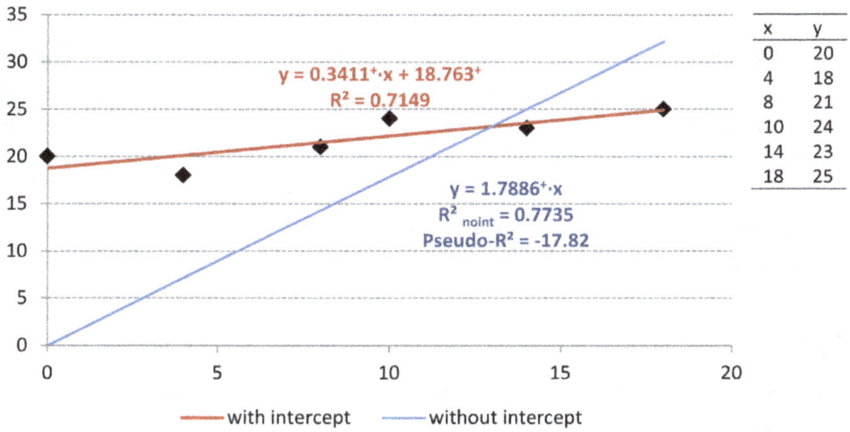

FIG. 11. Observed data, regression functions, and R^2–like fit criteria with and without intercept.

TABLE 11. Fit criteria for the example in Fig. 11.

Model	s^2	|RSS-PRESS|	AIC	BIC	SBC
With intercept	2.48	12.1	7.02	1.52	6.61
Without intercept	131.14	160.0	30.16	32.48	29.96

interval, and the estimation of the model parameters. The value of LL estimated by PROC TRANSREG needs to be corrected by a constant part of the maximum likelihood function which equals $N/2\log(2\pi)$ + df residuals/2. Table 12 gives the corrected –2 LL.

The comparison with the approach without considering the variance heterogeneity is given in the last row of Table 12. The corresponding value of –2LL shows the worst fit. With log(size) as regressor and no transformation of the regressand, the 95% confidence interval of β_1 is [3.00367; 3.14136] and does not contain "3", the exponent we assumed above. By setting $\beta_1 = 3$, the value of –2LL is 3821.92, and the loss is considerable. With the same regressor, the optimal value for the transformation of the regressand is $\hat{\lambda} = 0.14$, and its 95% confidence interval does not contain the value 0 which would correspond to the preceding situation. This is also the approach with the smallest value of −2LL among the considered ones. Taylor (2006) suggests that to simplify interpretation and for justification to non-statisticians it is sometimes preferable to limit λ to a finite set (−2, −1, −1/2, 0, 1/4, 1/3, 1/2, 2/3, 1, 2). Thus, the back-transformed function with $\hat{\lambda} = 0.14$ would surely be considered as formalistic (Table 12), especially because in this example a clear scientifically motivated idea of the underlined model exists. The back-transformed function describes the median of y for given x values. That is also the reason why Fitzmaurice et al. (2007) described the Box–Cox transformation as an alternative to the so-called median regression. The back-transformation to the expected value of y for given x is more laborious to compute (Freeman and Modarres, 2006). With size as regressor, the optimal value is $\hat{\lambda} = 0.339$. The 95% confidence interval of λ contains the value 1/3, so for simplification we estimated the function with this parameter value. This approach shows the second-best fit and results in a polynomial of third order including all terms of lower order. If a linear regression is raised to the power of three for back-transformation, we obtain a model that involves multiplicative error terms of the form xe, xe^2, and x^2e, as well as an additive term of the form e^3. Figure 10, right, shows that the variance heterogeneity has been adequately considered. In the next section further possibilities regarding the variance heterogeneity are described and will be compared with the approach discussed here (x as regressor and $\hat{\lambda} = 1/3$).

Extensions of the Linear Regression Models

The extensions considered in this chapter concern three aspects:

Integration of Qualitative Fixed or Random Variables in the Model

The additional consideration of qualitative variables in connection with regression often arises in planned experiments or in monitoring when the relation between quantitative variables are analyzed under different qualitative conditions. The qualitative variables may be fixed or random and are treatment factors or disturbance factors, as for example, the factor block in randomized block designs. The simultaneous inclusion of qualitative and quantitative variables in a model is originally a concern of analysis of covariance. In Example 5 such a situation was already considered. Here, we will examine two typical examples often arising in conjunction with regression in more detail (Example 7 and modified Example 3).

Consideration of Variance Heterogeneity of the Residual Effects

In Example 6, we noticed that the variances of the regressand increased with increasing regressor values. As demonstrated above, transformations may be an option. Another option is to model variances by a monotonically increasing (in other examples perhaps decreasing) function of the continuous regressor variable or by a function of the expected value of the regressand (i.e., the linear predictor). If for each regressor value several observations exist, then it is also possible to estimate an individual variance for each regressor value without assuming any functional relation between variance and mean. The latter option will often be used if a qualitative factor occurs in the model and a level-specific variance may be appropriate.

TABLE 12. Example 6. Comparison of the results of several transformations to describe variance heterogeneity with the Box–Cox transformation.

Regressor	λ	Estimated function	Back-transformed function	−2 LL
$\log(x)$	0	$\widehat{\log(y)} = -7.54405 + 3.07252\log(x)$	$\hat{y} = 0.000529x^{3.07252}$	3817.64
$\log(x)$	0.14 (optimum)	$\widehat{\frac{y^{0.14}-1}{0.14}} = -14.89424 + 5.40408\log(x)$	$\hat{y} = [-1.0852 + 0.75657\log(x)]^{7.1429}$	3799.21
x	1/3 (optimum = 0.339)	$\widehat{3(\sqrt[3]{y}-1)} = -3.1849 + 0.2706x$	$\hat{y} = [-0.061642 + 0.090199x]^3$	3803.78
x^3	no transformation		$\hat{y} = 0.000708x^3$	

Consideration of Covariances of the Residual Effects

Covariances need to be considered when observations are measured on the same experimental units several times so that serial correlations between the data of observations on the same unit at several time points arise. This is the situation of repeated measurements and may arise in time series. Correlations may also arise if a spatial correlation between the experimental units is modeled, for example, between plots in a field, or in breeding programs due to kinship relations (Example 8).

Whereas the first extension can be handled within the framework of the general linear model, the others require the mixed linear model that generalizes the general linear model. For a consistent description, we will consider all extensions in the context of the linear mixed model. In SAS, the general linear model is implemented in the procedure PROC GLM. However, some useful options are not available in PROC GLM if random factors are included in the model. The linear mixed model is implemented in PROC MIXED with the REML method as the default, whereas the ML method can be selected as an option.

The Linear Mixed Model

In matrix notation, the linear mixed model can be written as

$$\mathbf{Y} = \mathrm{X}\beta + \mathrm{Z}\mathbf{u} + \mathbf{E} \qquad [40]$$

where $\mathbf{Y}$ is the N-dimensional vector of the random observations, the $(p \times 1)$ vector β contains the unknown fixed-effect parameters (the regression parameters and possible effects of fixed classification factors), and the $(N \times p)$ matrix X consists of the values of the regressor variables as before as well as dummy variables (0–1 variables) indicating existing fixed factor levels. We use $\mathbf{u}$ to denote a $(q \times 1)$ vector of unknown random effects and Z is a $(N \times q)$ matrix, which may consist of values of continuous and/or dummy variables referring to the random effects. For the random variables, it is assumed that the residuals $\mathbf{E}$ and the elements in $\mathbf{u}$ follow multivariate normal distributions with $\mathbf{E} \sim \mathrm{N}_N([0], \mathrm{R})$ and $\mathbf{u} \sim \mathrm{N}_q([0], \mathrm{G})$, respectively.

From this it follows for the vector $\mathbf{Y}$:

$$\mathbf{Y} \sim \mathrm{N}_N(X\beta; \mathrm{V}) \text{ and } \mathrm{V} = \mathrm{ZGZ}^{\mathrm{T}} + \mathrm{R} \qquad [41]$$

In all examples discussed so far, we assumed $R = \sigma^2 I_{N\times N}$, and Z and **u** did not exist so that $V = \sigma^2 I_{N\times N}$.

We will distinguish two cases of Eq. [40] and [41] for considering the extensions formulated above:

1. Z and **u** do not exist and we know R except for a constant factor, let us say σ_f^2 so that $R = \sigma_f^2 R^*$ and R* is a known matrix. In particular, if no covariances between the residuals exist, then R* and R are diagonal matrices. In this case, we can use weighted regression. The basic idea of a weighted regression is that points with a larger variance are assigned a smaller weight than points with a smaller variance. This means that the diagonal elements of R^{*-1} can be used as weights. The implementation of this special case is possible with PROC REG and PROC GLM as well with PROC MIXED of SAS. The estimators of the fixed parameters are weighted least square estimators (wLS).
2. Z and **u** may or may not exist and at least two variance–covariance components are to be estimated. Several components may arise, for example, from variance heterogeneity and/or existing covariances of the residuals and/or consideration of one or several variance components for **u**. Here, we usually use the REML method. As noted above, the estimators of the fixed parameters are empirical weighted least square estimators (ewLS). To compare the fit of these models with models using transformations of *y* or with different numbers of fixed parameters, the ML method may need to be used.

In Table 13, the basic ideas of the solution of these problems are contrasted to the ordinary least squares method and selected consequences are summarized. In all cases where a matrix is not of full rank and its inverse does not exist (see the following examples), the inverse has to be replaced by the generalized inverse.

For the construction of confidence intervals or tests of the fixed parameters, consider that in the case of general V (Table 13, last column), additional challenges arise that result from the necessity to estimate V. First, $\left(X^T\hat{V}^{-1}X\right)^{-1}$ underestimates the variance of $\hat{\beta}$ because the variance of the estimator of V is not taken into account, and second, the corresponding *F* and *t* test statistics of the fixed parameters follow an exact *F* distribution and an exact *t* distribution only in exceptional cases. In the majority of cases, the denominator degrees of freedom need to be suitably approximated. There exist several approximation methods for this purpose. We recommend the first-order Kenward–Roger method (Kenward and Roger, 2009, referred to as the Prasad–Rao estimator in Harville and Jeske, 1992), which uses a correction of $\left(X^T\hat{V}^{-1}X\right)^{-1}$ and then the corrected estimator to approximate the denominator degrees of freedom (Richter et al., 2015).

Using the REML or ML method with general V and two or more variance components, the model fit cannot be described any longer by R^2 or MSE. Instead, information criteria can be used. They are based on the achieved minimum of −2ResLL (REML method) or of −2LL (ML method) and a penalty term. Similar to Eq. [36], several criteria exist. We will concentrate on Akaike's criterion, AIC, which is given by AIC = −2ResLL + 2*d* (REML method with *d* = number of variance–covariance parameters) and AIC = −2LL + 2*d* (ML method with *d* = number of variance–covariance parameters and fixed effect parameters). Because the penalty term of the REML method only considers the number of variance–covariance parameters and because the restricted likelihood is free of the fixed effects, the ML method should be chosen to compare approaches with different numbers and types of fixed effects. Conversely, the REML-based information criteria can be used only to compare models having the exact same fixed-effects structure. The seven models from Table 9A can also be analyzed by PROC MIXED with the REML method. The parameter estimations and the test results are the same as with PROC REG. The comparison between the models, however, should not be based on the AIC values of the REML method because these values do not consider the different numbers of regressors. Using PROC MIXED with the ML method yields the same estimations of the fixed parameters; the standard errors and the test decisions are slightly different from those of PROC REG. The AIC values of the ML method, however, penalize the number of regressors and may serve as fit criteria for the comparison of the seven models.

Besides the information criteria, a test-based comparison of a tested model against a basic model is possible. This assumes that the basic model is a special case of the tested model. The difference [–2ResLL(basic)] – [–2ResLL(tested)] or [–2LL(basic)] – [–2LL(tested)] is χ^2–distributed. The corresponding df is the difference of the number of variance–covariance parameters. If the alternative is accepted, the tested model shows a better fit. This test is known as likelihood-ratio (LR) test.

Example 6 Continued.

Second Step—Other Ways to Consider Variance Heterogeneity

In this example, Z and **u** do not exist, and X and β are the same as before. We try different approaches to describe the variance heterogeneity:

(i) We assume that $V = R = \sigma_f^2 R^*$, and R* is a known positive definite diagonal matrix. Our first naïve idea is that the variances are proportional to the x values (sizes) so that the sizes stand on the main diagonal of R* and the variances are $\sigma_j^2 = \sigma_f^2\,\text{size}_j$, with j = 1 to 6. The variance σ_f^2 needs to be estimated. Of course, other specifications for the diagonal elements in R* would be possible, for example size^2.

(ii) We assume that the variances are a nonlinear function of the size so that $\sigma_j^2 = \sigma_f^2 \exp(\theta\,\text{size}_j)$. σ_f^2 and θ need to be estimated. This is the power-of-x model.

(iii) We assume that the variances are a nonlinear function of the expected value of y so that $\sigma_j^2 = \sigma_f^2 \left|\hat{y}(\text{size}_j)\right|^{\theta}$. σ_f^2 and θ need to be estimated. This is the power-of-the-mean model.

(iv) Because multiple observations for each value of the regressor variable exist, an individual variance per size can be estimated that does not assume any functional dependency on the regressor variable. This approach is not typically used for regression problems because often the number of observations per x value does not allow for robust estimations of the variances, and it is more natural to assume a variance depending on the regressor or regressand values. For the approaches (i) to (iii), multiple observations are not required.

TABLE 13. Comparison of the basic idea of the ordinary least squares, the weighted least squares, and the REML method and resulting consequences. If the inverse of a matrix does not exist, it has to be replaced by a generalized inverse.

$V = R = \sigma^2 I$	$V = R = \sigma_f^2 R^*$ and R^* known matrix	General V
Optimization criterion of the different estimation methods		
Ordinary least squares method $\underset{\beta}{\text{Min}}\, S = \underset{\beta}{\text{Min}}\,(Y - X\beta)^T (Y - X\beta)$ (See Eq. [6])	Weighted least squares method $\underset{\beta}{\text{Min}}\, S = \underset{\beta}{\text{Min}}\,(Y - X\beta)^T R^{-1} (Y - X\beta)$	REML method Minimum of −2 Restricted log-likelihood function (−2 Res LL)
Estimation of β		
$\hat{\beta} = \left(X^T X\right)^{-1} X^T Y$ (See Eq. [7])	$\hat{\beta} = \left(X^T \hat{R}^{-1} X\right)^{-1} X^T \hat{R}^{-1} Y$ $= \left(X^T \hat{R}^{*-1} X\right)^{-1} X^T \hat{R}^{*-1} Y$	$\hat{\beta} = \left(X^T \hat{V}^{-1} X\right)^{-1} X^T \hat{V}^{-1} Y$ using the estimation $\hat{V}$ of V from the REML method
Estimation of σ^2	Estimation of σ_f^2 and R	Estimation of V
by ANOVA method $\hat{\sigma}^2 = s^2 = \frac{\text{RSS}}{\text{df of RSS}} = \frac{\text{RSS}}{N - p}$ (See Eq. [13])	by ANOVA method $\widehat{\sigma_f^2} = s_f^2 = \frac{\text{RSS(weighted)}}{\text{df of RSS}} = \frac{\text{RSS(weighted)}}{N - p}$ $\hat{R} = s_f^2 R^*$	by REML method Explicit formula for most cases not possible
Variance of $\hat{\beta}$		
$s_{\hat{\beta}}^2 = \left(X^T X\right)^{-1} s^2$ (See Eq. [18])	$s_{\hat{\beta}}^2 = \left(X^T R^{*-1} X\right)^{-1} s_f^2$	$s_{\hat{\beta}}^2 = \left(X^T \hat{V}^{-1} X\right)^{-1}$

Approach (i) is a special case of weighted regression with weights specified before the analysis. Approaches (ii) to (iv) also cause a weighting of the data whereby the weights are estimated as part of the analysis.

As a starting point for the inclusion of the variance heterogeneity by Approaches (i) to (iv), we choose the cubic function without intercept from above. These approaches have been implemented by the REML method. For comparing the fit with the results of the Box–Cox transformation, −2LL and the AIC value of the ML method are also given in Table 14.

The "unweighted regression" corresponds to the regression model from Table 9B (last row). The estimated variances are compared with those observed in Fig. 12.

Using the power-of-x model, convergence problems of the iterative REML algorithm arose, which we assume were due to the fact that the default starting values for the covariance parameters were poor. To overcome this problem, we specified alternative initial values for the covariance parameters. In such cases, the reported null-model LR test tests the fit of the final estimated model against the fit of the model with the initial values. Because the result of the LR test depends on the initial values and initial values were chosen just to help convergence but do not correspond to a meaningful null model, the test should not be interpreted. The other results of this model given in Table 14 can be interpreted.

From AIC (REML and ML) and the LR test (only shown for REML) we see that the approach with an individual variance per size shows the best fit even though it needs the largest number of df and the largest number of estimated variance components. The estimated variances nearly reproduce the observed variances (Fig. 12). The comparison of the Studentized residuals of the unweighted regression and Approach (iv) shows that their original pattern disappeared (Fig. 13).

A comparison of the considered models based on an R^2 statistic or MSE is no longer possible. Caution is required when using the weighted regression in PROC REG and PROC GLM where a R^2 statistic is reported that cannot be compared with the R^2 of the unweighted regression. In the Approaches (i) to (iii), the estimated factor s_f^2 is designated as "residual variance." This factor is also not comparable with the residual variance of the unweighted regression. The differences in the estimated β_1 are small. If the first-order Kenward–Roger method were not used, in the last three approaches the standard error would be smaller, the df would be 523, and the t value would be larger. The p value would be < 0.0001 in all cases.

Third Step—Final Decision about the Model

To finally decide on the best fitting model, we compare the results of the Box–Cox transformation (Table 12; x as regressor, λ = 1/3) and of the power-of-x, power-of-mean, and individual variance per size where the last three used the cubic function without intercept (Table 14). The −2LL value of the Box–Cox transformation was 3803.78. To make it comparable with the other AIC values, we set d = 4 (two fixed effects, one residual variance and λ), so that AIC = 3811.78. Thus, the model $\hat{y} = [-0.061642 + 0.090199x]^3$ gives the best fit among all considered models.

Using this strategy—searching for relevant regressors at first and then considering variance heterogeneity by the methods given in Table 14—there is no guarantee that the best of all possible combinations will be found. This approach, however, is often the only practicable way to minimize the effort. In our case, for demonstration, we fitted all models of Table 9A and 9B with the Approaches (ii) to (iv) to check the information loss by this strategy. With the ML method, the best fit of the unweighted regression was found with $y_i = \beta_3 x_i^3 + e_i$ (AIC-ML = 3987.6), and for the Approaches (ii) to (iv) the best fit was found with $y_i = \beta_0 + \beta_1 x_i + \beta_2 x_i^2 + \beta_3 x_i^3 + e_i$ with AIC-ML = 3830.9 for Approach (ii), AIC-ML = 3819.0 for Approach (iii) and AIC-ML = 3814.2 for Approach (iv). In all cases, however, β_0, β_1, and β_2 do not differ significantly from zero. Again, the Box–Cox transformation shows the better fit and requires much less effort to conduct.

With the REML method, the best-fitting model was $y_i = \beta_0 + \beta_1 x_i + \beta_2 x_i^2 + e_i$ for the unweighted regression as well as for Approaches (ii), (iii), and (iv). For the unweighted regression, we found AIC-REML = 3997.5, AIC-REML = 3840.4 for Approach (ii), AIC-REML = 3831.2 for Approach (iii), and AIC-REML = 3823.4 for Approach (iv). In this case, all fixed regression parameters are significantly different from zero. These values cannot be compared with the result of the Box–Cox transformation because of the differences in scale.

It is a drawback of the Box–Cox transformation that with the estimated λ the retransformed function is sometimes difficult to interpret and the implicit error structure on the original scale is not immediately obvious if the approach to analysis

TABLE 14. Example 6. Comparison of unweighted regression with several approaches to describe the variance heterogeneity for the approach $y_i = \beta_1 x_i^3 + e_i$ using the REML method and the Kenward–Roger approximation. Values of −2LL and AIC-ML of the ML method are added.

Approaches without and with variance heterogeneity	−2ResLL (−2LL)	AIC-REML (AIC-ML)	LR statistic df (p value)	Estimated variance components	Estimation of β_1 (SE)	t test of β_1 t value df (p value)
Unweighted regression	4006.6 (3983.6)	4008.6 (3987.6)		$s^2 = 117.49$	0.000708 (4.176E-6)	169.53 523 (< 0.0001)
(i) Weighted regression with weight =1/size	3943.8 (3920.9)	3945.8 (3924.9)		$s_f^2 = 2.3765$	0.000708 (4.271E-6)	165.77 523 (< 0.0001)
(ii) Power-of-x model	3850.3 (3827.6)	3854.3 (3833.6)	see text	$s_f^2 = 1.0822$ $\hat{\theta} = 0.09819$	0.000707 (4.728E-6)	149.63 472 (< 0.0001)
(iii) Power-of-mean model	3843.1 (3820.4)	3847.1 (3826.4)	163.5 1 (< 0.0001)	$s_f^2 = 0.1937$ $\hat{\theta} = 1.4905$	0.000707 (4.717E-6)	149.83 476 (< 0.0001)
(iv) Individual variance per size	3829.6 (3806.9)	3841.6 (3820.9)	176.9 5 (< 0.0001)	14.87 56.95 89.67 122.73 166.42 263.77	0.000704 (4.682E-6)	150.35 467 (< 0.0001)

is not explained. For example, the fitted model $\hat{y} = [-0.061642 + 0.090199x]^3$ (Table 12) could also be the result of a polynomial regression with additive errors. Furthermore, the transformation causes the usual tests of β_0 and β_1 in the transformation approach to be liberal, and the ML method provides biased variance estimators.

This example reveals how important a predetermined strategy for the model selection is, how sensitive the results are, and that for reproducibility and proper interpretation, the reader should be informed about the details of the strategy for analysis.

Example 7 (SAS Code in Appendix)

This example considers the regression model with integration of a qualitative fixed treatment factor and the option of possible variance heterogeneity for the levels of the factor.

In studies with perennial crops, as in fruit production, it is desirable to predict long-term performances on the basis of short-term observations. In this example, the relation between the cumulative yield of apple trees after 10 yr and the yield after 4 yr is analyzed. Two varieties (Variety A and B) have been selected, and both are combined with 30 different root stocks. The goal is to predict the yield after 10 yr by using the yield after 4 yr. If the prediction has sufficient precision it will be acceptable to discontinue testing of under-performing root stocks at an early stage in future experiments. Both the cumulative yield of the first 4 yr (x) and the cumulative yield of 10 yr (y) are continuous random variables (Model II). It is important to consider that the relation between x and y and the precision of prediction may depend

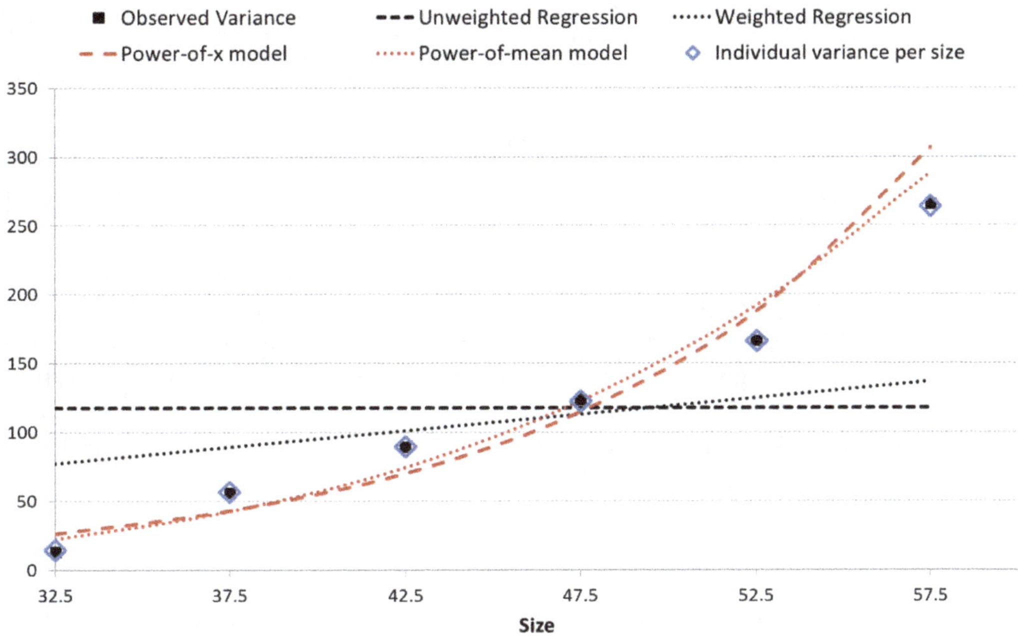

FIG. 12. Example 6. Estimated variances for the approaches in Table 14.

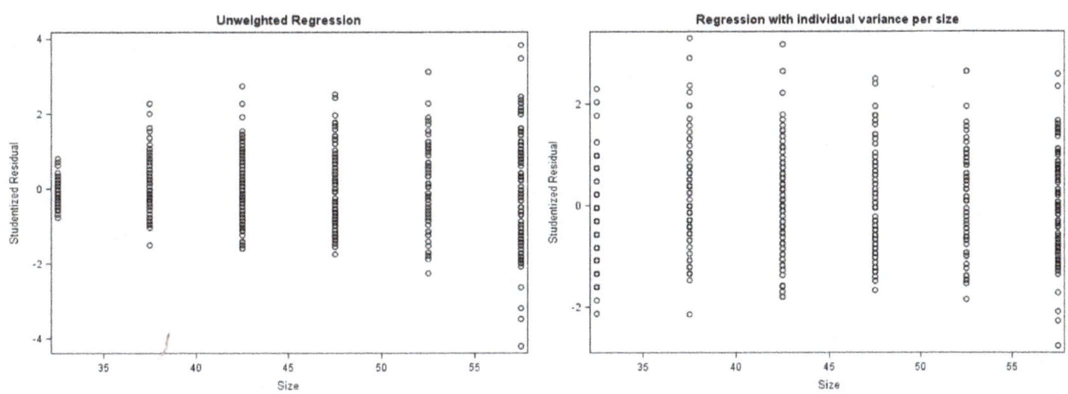

FIG. 13. Example 6. Studentized residuals for unweighted regression and weighted approach (iv).

on the variety. In a first step, each variety was analyzed separately (Table 15 and Fig. 14). In a second step, we wanted to know whether or not the regression relation is the same for both varieties.

For the second step, we insert the fixed classification factor variety in the model. The y and x values now need to have two subscripts: i for the variety ($i = 1,2$) and j for the replication of the ith variety ($j = 1 \ldots n_i$, here: $j = 1 \ldots 30$).

(a) Assuming that the intercepts as well as the slopes are identical, the model can be written as

$$y_{ij} = \beta_0 + \beta_1 x_{ij} + e_{ij}$$

If any of its parameters is variety-specific, then the model needs to be extended.

(b) If the regression constants (intercepts) and the regression coefficients (slopes) are different, the model changes to

$$y_{ij} = \beta_{0i} + \beta_{1i} x_{ij} + e_{ij} \qquad [42]$$

This model can also be written as

$$y_{ij} = \underbrace{\beta_0 + \beta_{0i}^{+}}_{\beta_{0i}} + \underbrace{\beta_1 x_{ij} + \beta_{1i}^{+} x_{ij}}_{\beta_{1i} x_{ij}} + e_{ij} \qquad [43]$$

where β_0 and β_1 are the common regression parameters and β_{0i}^{+} and β_{1i}^{+} the variety-specific deviations of the parameters from the common parameters. By the decomposition of β_{0i} and β_{1i} in Eq. [43] we are able to test whether β_{0i}^{+} and β_{1i}^{+} are significantly different from zero for both varieties. If they are not, then the corresponding variety-specific parameter is unnecessary.

(c) If only the regression coefficients are different, the model is (with both parametrizations)

$$y_{ij} = \beta_0 + \beta_{1i} x_{ij} + e_{ij} = \beta_0 + \beta_1 x_{ij} + \beta_{1i}^{+} x_{ij} + e_{ij}$$

(d) If only the intercepts are different, the model is (with both parametrizations)

$$y_{ij} = \beta_{0i} + \beta_1 x_{ij} + e_{ij} = \beta_0 + \beta_{0i}^{+} + \beta_1 x_{ij} + e_{ij}$$

For Case (b) and parametrization (Eq. [43]), we give the corresponding matrix notation; matrix expressions for the other models can be obtained by reduction. The vectors and the X matrix are sorted by varieties.

$$\mathbf{Y} = \begin{bmatrix} y_{11} \\ y_{12} \\ \vdots \\ y_{1;30} \\ y_{21} \\ y_{22} \\ \vdots \\ y_{2;30} \end{bmatrix}, \; \mathrm{X} = \begin{bmatrix} 1 & 1 & 0 & x_{11} & x_{11} & 0 \\ 1 & 1 & 0 & x_{12} & x_{12} & 0 \\ \vdots & \vdots & \vdots & \vdots & \vdots & \vdots \\ 1 & 1 & 0 & x_{1;30} & x_{1;30} & 0 \\ 1 & 0 & 1 & x_{21} & 0 & x_{21} \\ 1 & 0 & 1 & x_{22} & 0 & x_{22} \\ \vdots & \vdots & \vdots & \vdots & \vdots & \vdots \\ 1 & 0 & 1 & x_{2;30} & 0 & x_{2;30} \end{bmatrix}, \; \beta = \begin{bmatrix} \beta_0 \\ \beta_{01}^{+} \\ \beta_{02}^{+} \\ \beta_1 \\ \beta_{11}^{+} \\ \beta_{12}^{+} \end{bmatrix}, \text{ and } \mathbf{E} = \begin{bmatrix} e_{11} \\ e_{12} \\ \vdots \\ e_{1;30} \\ e_{21} \\ e_{22} \\ \vdots \\ e_{2;30} \end{bmatrix} \qquad [44]$$

A fundamental distinction compared with previous situations is that the matrix X is not of full rank; that is, the rank(X) is smaller than the number of fixed parameters. The number of parameters is 6 and the rank(X) is only 4, meaning that only 4 columns are linearly independent (the first column is equal to the sum of the second and third columns and the fourth column is equal to the sum of the fifth and sixth columns). Because the rank(X) is equal to the rank(X^TX), the inverse of X^TX does not exist. Therefore, the estimation equation for the parameters (Eq. [7]), their variance estimator (Eq. [18]), and the Hat matrix (Eq. [23]) need to be modified. Thus, we replace the inverse $(X^TX)^{-1}$ by a generalized inverse $(X^TX)^{-}$. A generalized inverse of a matrix A is a matrix A^{-}, which satisfies the condition $AA^{-}A = A$ (Searle, 1971). An inverse of a matrix is unique, whereas a generalized inverse is not and therefore the results for the parameter estimates are also not unique. Hence, we

TABLE 15. Example 7. Parameter estimates, tests, and confidence intervals of the fixed regression parameters ($1 - \alpha = 0.95$) for the separate analysis of the two varieties.

Effect	df	Estimate	SE	*t* value	*p* > |*t*|	Lower	Upper
A. Variety A							
Intercept	1	51.43	6.92	7.44	< 0.0001	37.27	65.60
year1_4	1	2.43	0.18	13.82	< 0.0001	2.07	2.79
B. Variety B							
Intercept	1	46.68	2.81	16.59	< 0.0001	40.92	52.45
year1_4	1	1.10	0.07	15.91	< 0.0001	0.96	1.24

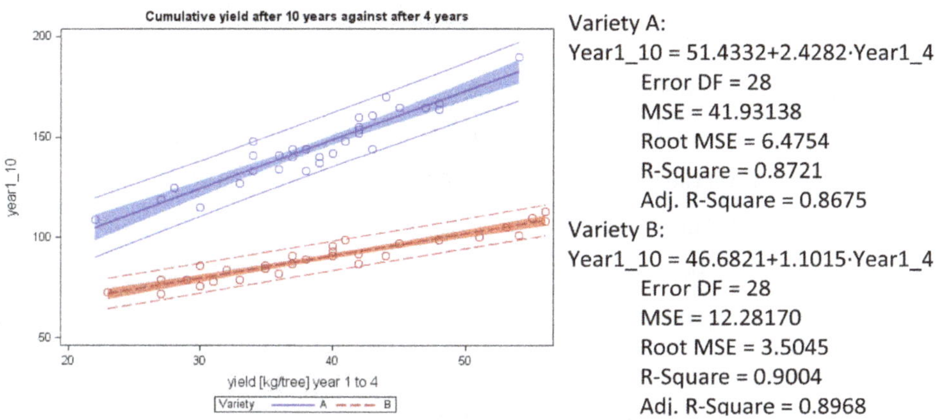

FIG. 14. Example 7. Observed data, regression functions with confidence and prediction intervals with $1 - \alpha = 0.95$ for variety A and B.

speak of *solutions* and not of *estimates* for the parameters, and some care is needed in interpreting the results. Only estimable functions of the parameters should be interpreted. In Approach (b), for example, $\beta_0 + \beta_{0i}^+$ and $\beta_{1i} + \beta_{1i}^+$ are estimable functions, corresponding to the variety-specific intercepts and slopes, whereas the individual parameters are not estimable. Furthermore, it should be noticed that the df of residuals is N-rank(X^TX). In Examples 1 to 5, X was of full rank and rank(X^TX) was equal to the number of parameters p. If we had used the parametrization in Eq. [42] combined with the no-intercept option, in the matrix X from Eq. [44] the first and fourth columns would be dropped, the resulting X matrix would be of full rank, and these complications would not arise. With Eq. [42], however, the tests of equality of the two intercepts or the two slopes would be considered separately and furthermore, Eq. [44] simplifies the discussion for the following modified Example 3.

The results of Approaches (a), (b), and (c) are given in Table 16, and the fitted functions for Approaches (a) and (c) in Fig. 15. Approach (a) describes the mean behavior of the varieties. For Approach (b), the effect of the generalized inverse we used can be seen in the lower part of Table 16B. The solutions for b_{02}^+ and b_{12}^+ are set to zero. In Table 16B, it is also demonstrated how the variety-specific functions can be deduced. These functions correspond exactly to the functions for the separate analyses in Table 15; due to the same sample size, the estimate of the residual variance in the joint analysis is the mean of the two MSE in the separate analysis.

TABLE 16. Example 7. Analysis of variance for Approaches (a) and (b). Estimates [Approach (a)] and solutions [Approach (b) and (c)], tests, and 95% confidence intervals of the fixed regression parameters.

A. Approach (a). Variance table and estimates (MSE = 878.08; Adj. $R^2 = 0.1144$).

Source	df	Sum of squares	Mean square	F value	$p > F$
Model	1	7571.73	7571.73	8.62	0.0048
Error	58	50928.85	878.08		
Corrected total	59	58500.58			

$\hat{y} = 63.5403 + 1.3883x$

Effect	Estimate	SE	df	t value	$p > \|t\|$	Lower	Upper
Intercept	$b_0 = 63.5403$	18.9084	58	3.36	0.0014	25.69	101.39
Year1_4	$b_1 = 1.3883$	0.4728	58	2.94	0.0048	0.44	2.33

B. Approach (b). Variance table (partial approach) and solutions (MSE = 27.11; Adj. $R^2 = 0.9727$)

Source	df	Sum of squares	Mean square	F value	$p > F$
Model	3	56982.62	18994.21	700.72	< 0.0001
Error	56	1517.97	27.11		
Corrected total	59	58500.58			
Year1_4	1	11053.81	11053.81	407.79	< 0.0001
Variety	1	12.65	12.65	0.47	0.4974
Year1_4*Variety	1	1561.50	1561.50	57.61	< 0.0001

Variety A: $\hat{y} = (46.6821 + 4.7511) + (1.1015 + 1.3267)x = 51.4332 + 2.4282x$

Variety B: $\hat{y} = (46.6821 + 0) + (1.1015 + 0)x = 46.6821 + 1.1015x$

Effect	Estimate	SE	df	t value	$p > \|t\|$	Lower	Upper
Intercept	$b_0 = 46.6821$	4.1799	56	11.17	< 0.0001	38.31	55.06
Year1_4	$b_1 = 1.1015$	0.1029	56	10.71	< 0.0001	0.90	1.31
Variety A	$b_{01}^{+} = 4.7511$	6.9561	56	0.68	0.4974	−9.18	18.69
Variety B	$b_{02}^{+} = 0$	–	–	–	–	–	–
Year1_4*Variety A	$b_{11}^{+} = 1.3267$	0.1748	56	7.59	< 0.0001	0.98	1.68
Year1_4*Variety B	$b_{12}^{+} = 0$	–	–	–	–	–	–

C. Approach (c) with MSE = 26.85. Adj. $R^2 = 0.9729$

Variety A: $\hat{y} = 48.3976 + (1.0604 + 1.4438)x = 48.3976 + 2.5042x$

Variety B: $\hat{y} = 48.3976 + 1.0604x$

Effect	Estimate	SE	df	t value	$p > \|t\|$	Lower	Upper
Intercept	$b_0 = 48.3976$	3.3254	57	14.55	< 0.0001	41.74	55.06
Year1_4	$b_1 = 1.0604$	0.08303	57	12.77	< 0.0001	0.89	1.23
Year1_4*Variety A	$b_{11}^{+} = 1.4438$	0.03366	57	42.89	< 0.0001	1.38	1.51
Year1_4*Variety B	$b_{12}^{+} = 0$	–	–	–	–	–	–

To test whether the regression functions of both varieties can be assumed as having identical intercepts and slopes [Approach (a)], one can use the following F test (coincidence test), which uses the results of Table 16A and 16B:

$$F = \frac{\text{SS Model}_{(\text{Approach b})} - \text{SS Model}_{(\text{Approach a})}}{\left[\text{df Model}_{(\text{Approach b})} - \text{df Model}_{(\text{Approach a})}\right] \text{MSE}_{(\text{Approach b})}} =$$

$$\frac{56982.62 - 7571.73}{2 \times 27.11} = 911.42$$

In our case, the calculated F value is larger than the $(1 - \alpha)$ quantile of the F distribution with df of the numerator = 2 and df of the denominator = df error [Approach (b)] = 56. Therefore, we reject the coincidence with a Type 1 error rate of $\alpha = 0.05$ (p value < 0.001). It can be seen from the detailed results in Table 16B whether the rejection results from different intercepts and/or different slopes.

The t test of β_{01}^{+} against zero result in the nonrejection of this null hypothesis. Therefore, we can assume that the regression constants of both varieties are equal. The test of β_{11}^{+} accepts the alternative hypothesis, so that we have variety-specific regression coefficients. In the Appendix in the supplement material online, it is also shown how β_{01} and β_{02} as well as β_{11} and β_{12} can be compared by t tests (see parametrization in Eq. [42]). The F tests of Variety and of year1_4*Variety in the ANOVA table show the

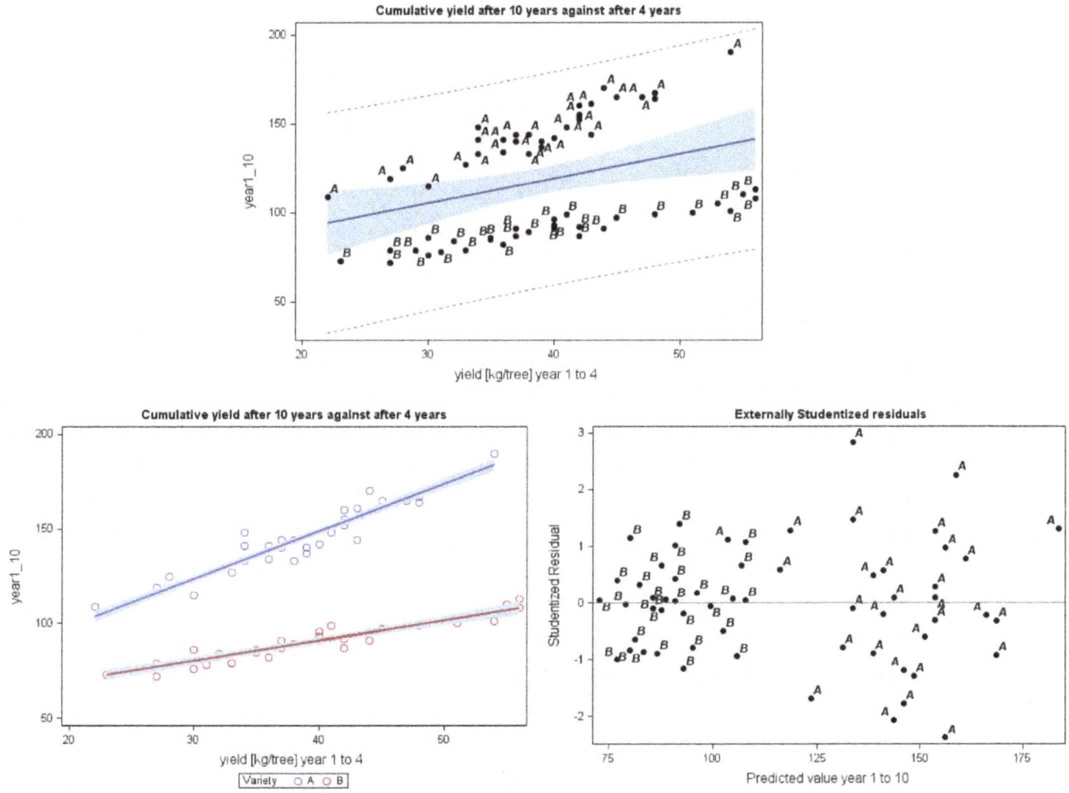

FIG. 15. Example 7. (Upper) Regression function, confidence, and prediction interval for Approach (a). (Lower) Regression function and confidence interval for Approach (c) with variance homogeneity and its externally Studentized residuals. A = Variety A, B = Variety B.

TABLE 17. Example 7. Comparison of the approaches without and with variance heterogeneity assuming Approach (c) as well as the parameter estimations.

A. Estimations of the variance components, fit statistics and likelihood ratio test

Approaches without and with variance heterogeneity	−2ResLL	AIC	LR statistic df (p value)	Estimated variance components	
Variance homogeneity	371.8	373.8		26.8528	
Variance heterogeneity	361.9	365.9	9.87 1 (0.0017)	Variety A Variety B	41.1900 12.2447

B. Estimation and tests of the fixed parameters for Approach (c) with variance heterogeneity.

Variety A: $\hat{y} = 47.3655 + (1.0852 + 1.4449)x = 47.3655 + 2.5301x$

Variety B: $\hat{y} = 47.3655 + 1.0852x$

Effect	Estimate	SE	df	t value	$p > \|t\|$
Intercept	$b_0 = 47.3655$	2.6439	34.7	17.91	< 0.0001
Year1_4	$b_1 = 1.0852$	0.06429	36.8	16.62	< 0.0001
Year1_4*Variety A	$b_{11}^{+} = 1.4449$	0.0338	44	42.75	< 0.0001
Year1_4*Variety B	$b_{12}^{+} = 0$	–	–	–	–

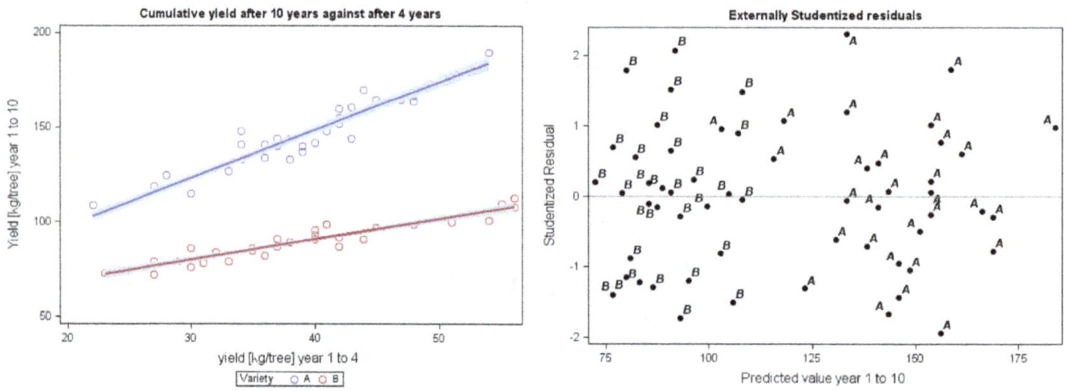

FIG. 16. Example 7. Regression function and confidence interval for Approach (c) with variance heterogeneity and its externally Studentized residuals.

same results because we have only two levels of the treatment factor (variety). If we would have had more levels of the treatment factor, e.g., 3 varieties, the F test of variety would show whether the three intercepts differ significantly, and the F test of year1_4*Variety would show whether the three slopes differ significantly. These tests are useful before pairwise comparisons of intercepts or slopes.

In Table 16C and the lower part of Fig. 15, the analysis by Approach (c) is given. The lower right part of Fig. 15 emphasizes that the model equation is adequate but not the description of the residual variance.

Now we consider the larger variance of Variety A compared with Variety B by estimating their individual variances on the basis of Approach (c) (Table 17, Fig. 16). Based on the AIC and the representation of the Studentized residuals, the model with the individual variances fits much better.

We can conclude that by using the yields of the first 4 yr, we can predict the yield of each variety after 10 yr. However, the predictions are variety-specific, and the prediction accuracy is smaller for Variety A.

Example 3 (Modified) (SAS Code in Appendix)

We now take up Example 3 again. Up to this point, it has been assumed that the data came from a completely randomized design (CRD). Now, we modify the example imagining that the design was a randomized complete block design (RCBD) with four blocks each with five plots for the five cutting dates. Block is a qualitative factor with four levels. We illustrate the analysis for the case that the block effects are considered as fixed or random.

As in the first discussion of Example 3, the response has two subscripts: i (i = 1–5) for the days and j = 1 to 4 for the blocks (in the CRD j was the index for the replications).

First, we assume that the block effects are fixed. In contrast to Example 7, we have four instead of two levels of a classification factor so that basically, the Approaches (a) to (d) could also be used, noting that j now is the index of the classification factor. Approach (a) corresponds to the analysis in the CRD given above. In Example 7, Approaches (b) to (d) took into consideration that different regression functions per variety were possible. If we apply the same approaches to this example, we are able to estimate block-specific functions. This is the aim only for special cases, but not generally for an experiment laid out as a RCBD. A RCBD will be chosen instead of a CRD if a disturbance factor is present that could have an effect on the observations. By capturing and eliminating these effects, the hope is to get more precise estimates. If the model is extended by fixed block effects ρ_j, it can be written as

TABLE 18. Example 3 (modified) with fixed block effects. Analysis of variance, estimates, and tests of the fixed parameters.

A. ANOVA table with decomposition of SS model

Source	df	Sum of squares	Mean square	F value	$p > F$
Model	4	34770.98	8692.74	173.95	< 0.0001
Day	1	31866.03	31866.03	637.68	< 0.0001
Block	3	2904.95	968.32	19.38	< 0.0001
Error	15	749.58	49.97		
Corrected total	19	35520.55			

B. Parameter estimations for Approach (d) with MSE = 49.97. Adj. R^2 = 0.9733

Block 1: $\hat{y} = 210.35 + 5.645x$. Block 2: $\hat{y} = 223.75 + 5.645x$.

Block 3: $\hat{y} = 232.15 + 5.645x$. Block 4: $\hat{y} = 243.35 + 5.645x$.

Effect	Estimate	SE	df	t value	$p > \|t\|$	Lower	Upper
Intercept	$b_0 = 243.35$	3.8719	15	62.85	< 0.0001	235.10	251.60
Day	$b_1 = 5.645$	0.2235	15	25.25	< 0.0001	5.17	6.12
Block 1	$\widehat{\rho_1} = -33.00$	4.4709	15	−7.38	< 0.0001	−42.53	−23.47
Block 2	$\widehat{\rho_2} = -19.60$	4.4709	15	−4.38	0.0005	−29.13	−10.07
Block 3	$\widehat{\rho_3} = -11.20$	4.4709	15	−2.51	0.0243	−20.73	−1.67
Block 4	$\widehat{\rho_4} = 0$	–	–	–	–	–	–

$$y_{ij} = \underbrace{\beta_0 + \rho_j}_{\beta_{0j}} + \beta_1 x_{ij} + e_{ij}$$

Summarizing the intercept with the block effect one obtains automatically Approach (d) from above with the results in Table 18.

The F test in Table 18A shows significant block effects. The consideration of this source of variation entails a considerable reduction in the RMSE compared to the CRD (from 14.249 to 7.069). In Table 18B, the regression coefficient is the same as in the CRD, and the regression constant of the CRD is equal to the constant in Table 18B plus the mean of the solutions for the block effects (227.4 = 243.35 + [−33 − 19.6 −11.2 + 0]/4). Due to the block-specific intercept, the confidence intervals of the regression functions would also be block specific. The estimation of a confidence interval for the mean of all blocks is possible (which corresponds to the exclusion of the disturbance factor) by estimating a special contrast for each day. Let us assume that the matrix X has in the first column only ones, in the second column the days (0–20), and in the third to sixth columns the dummy variables for the four blocks. Then with the contrast L^T = (1 day 1/4 1/4 1/4 1/4) where day must to be replaced by 0, 5, 10, 15, and 20, $L^T\hat{\beta}$ gives the expected mean per day (the value on the regression function). The standard errors $\sqrt{L^T\left(X^T\hat{V}^{-1}X\right)^{-}L}$ are the basis of the confidence interval in the mean of the blocks (Table 20).

Whether block effects should be considered as random or fixed is frequently controversial. We now demonstrate the consequences if the blocks are considered as levels of a random factor. In this case, the model changes to

$$y_{ij} = \beta_0 + \beta_1 x_{ij} + \boldsymbol{\rho}_j + \boldsymbol{e}_{ij}$$

We assume that $\boldsymbol{\rho}_j \sim NI(0; \sigma^2_{Block})$, $\boldsymbol{e}_{ij} \sim NI(0; \sigma^2)$, and that the $\boldsymbol{\rho}_j$ and $\boldsymbol{e}_{ij}$ are independent of each other. In this case, we have a special case of Eq. [40] with Eq. [41]. The vector $\mathbf{u}$ contains the random block effects $\mathbf{u}^T = [\rho_1, \rho_2, \rho_3, \rho_4]$ with $q = 4$, and the matrix Z consists of four columns with 0–1 variables indicating the random block effects for the N values. The variance–covariance matrices are $R = \sigma^2 I_{20\times20}$ and $G = \sigma^2_{Block} I_{5\times5}$. If we sort the values in **Y**, X, Z, and **E** by blocks, the variance–covariance matrix V of **Y** has the block-diagonal form with four blocks on the main diagonal and off the block-diagonal all elements are zero.

$$V = \begin{bmatrix} \square & 0 & \cdots & 0 \\ 0 & \square & 0 & \vdots \\ \vdots & 0 & \square & 0 \\ 0 & \cdots & 0 & \square \end{bmatrix}$$

Each block in the V matrix is a 5 × 5 matrix with $\sigma^2 + \sigma^2_{Block}$ on the diagonal and σ^2_{Block} off the diagonal.

One block on the diagonal of the V matrix is given by:

$$\begin{bmatrix} \sigma^2 + \sigma^2_{Block} & \sigma^2_{Block} & \cdots & \sigma^2_{Block} \\ \sigma^2_{Block} & \ddots & & \vdots \\ \vdots & & \ddots & \sigma^2_{Block} \\ \sigma^2_{Block} & \cdots & \sigma^2_{Block} & \sigma^2 + \sigma^2_{Block} \end{bmatrix}$$

This means, that all y_{ij} have the variance $\sigma^2 + \sigma^2_{\text{Block}}$ and all y_{ij} from the same block (with the same j) have the covariance σ^2_{Block}. The covariances of the components in Y from different blocks are equal to zero.

The REML method gives the variance component estimates: $\hat{\sigma}^2_{\text{Block}} = 183.67$ and $s^2 = 49.9717$. The regression coefficient is the same as in the CRD and in the above RCBD analysis with fixed block effects; the intercept is the same as in the CRD (Table 2D). Elements in the **u** vector, the random block effects, are estimated by a method called best linear unbiased prediction (BLUP; Searle et al., 1992) (Table 19B). The confidence interval for the mean over the predicted random block effects for the same contrasts as above is exactly the same as in the mean of the fixed block effects. We say that this corresponds to the narrow inference space. If we do not exclude the random block effects, we get intervals that correspond to confidence intervals for the broad inference space (Schabenberger and Pierce, 2002).

Summarizing Example 3, we have demonstrated the analysis for the CRD and the RCBD with fixed and random block effects with single values as well as with means per day. We do not recommend the analysis with the means because the estimated MSE is only determined by the MS for lack of fit of the model. The analysis with fixed block effects results initially in an analysis per block, which is often not the aim. Considering the average of the regression function over the block effects, the estimated regression function was the same as in all other cases. Differences occurred in the standard errors of its parameter estimates and, correspondingly, in the confidence intervals of the function. In Table 20 the intervals are given for all analyzed models. If the layout is a RCBD, the blocks are considered random and there is no averaging over the predicted block effects, the confidence interval is essentially like a prediction interval. If the layout is a RCBD with fixed or random blocks, the calculations for the mean of the corresponding block effects consider the blocks as a disturbance factor that has to be excluded. For a RCBD, we normally recommend the narrow inference space. If the blocks (or observed levels of another factor) are a random sample of a defined population, then we would use the broad inference space.

Example 8 (SAS Code in Appendix)

At an experimental station, a trend analysis of annual mean air temperature (°C) is performed. The temperature (y) is a continuous random variable and year (x) is a fixed variable (Model I). The data are available for the years 1960 to 2013.

TABLE 19. Example 3 (modified) with random block effects. Estimates, predictions, and corresponding tests.

A. Estimates of fixed regression parameters.							
Effect	Estimate	SE	df	*t* value	$p > \|t\|$	Lower	Upper
Intercept	227.40	7.3084	3.64	31.11	< 0.0001	206.30	248.5
Day	5.6450	0.2235	15	25.25	< 0.0001	5.1685	6.1215
B. Predictions of random effects.							
Effect	Estimate	SE pred	df	*t* value	$p > \|t\|$	Lower	Upper
Block 1	−16.17	7.3243	3.54	−2.21	0.1007	−37.60	5.26
Block 2	−3.46	7.3243	3.54	−0.47	0.6641	−24.89	17.97
Block 3	4.50	7.3243	3.54	0.62	0.5759	−16.93	25.94
Block 4	15.13	7.3243	3.54	2.07	0.1169	−6.30	36.56

TABLE 20. Example 3. Comparison of the 95% intervals for four models from Example 3 and Example 3 (modified).

		CRD				RCBD fixed blocks		RCBD random blocks	
		s^2 = 203.03‡				s^2 = 49.97‡		s^2 = 49.97‡ s^2_{block} = 183.67‡	
		Prediction interval		Confidence interval		Confidence interval†		Confidence interval for broad inference space	
Day	Expected mean	df	Lower...upper limit	df	Lower...upper limit	df	Lower...upper limit	df	Lower...upper limit
0	227.4	18	195.3 … 259.5	18	215.8 … 239.0	15	221.56 … 233.24	3.64	206.3 … 248.5
5	255.6	18	224.6 … 286.7	18	247.4 … 263.8	15	251.50 … 259.75	3.16	233.8 … 277.4
10	283.9	18	253.1 … 314.5	18	277.2 … 290.5	15	280.48 … 287.22	3	261.7 … 306.0
15	312.1	18	281.0 … 343.1	18	303.9 … 320.3	15	307.95 … 316.20	3.16	290.3 … 333.8
20	340.3	18	308.2 … 372.4	18	328.7 … 351.9	15	334.46 … 346.14	3.64	319.2 … 361.4

† These confidence limits are calculated for the mean of all fixed blocks. They are equal to the limits for the model with random blocks averaged over the block predictions (narrow inference space).

‡Estimated variance components.

TABLE 21. Example 8. Analysis of variance, estimates, and tests of the fixed parameters.

A. ANOVA table for the regression analysis

Source	df	Sum of squares	Mean square	*F* value	$p > F$
Model	1	8.09	8.09	16.21	0.0002
Error	52	25.94	0.50		
Corrected total	53	34.02			

B. Estimates and tests of the fixed regression parameters

Variable	df	Parameter estimate	SE	*t* value	$p > \|t\|$	95% confidence limits	
Intercept	1	−40.35	12.25	−3.29	0.0018	−64.93	−15.77
Year	1	0.025	0.0062	4.03	0.0002	0.012	0.037

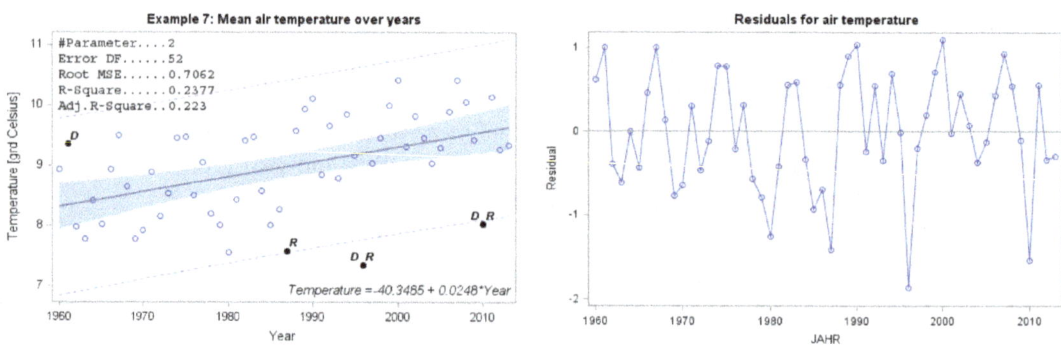

FIG. 17. Example 8. (Left) Observed data, fitted regression function, confidence, and prediction intervals with $1 - \alpha = 0.95$ with conspicuous points highlighted as follows: *R* = externally Studentized residual (RSTUDENT), *D* = DFITTS. (Right) Studentized residuals $\hat{e}_i^*$ against regressor values.

TABLE 22. Example 8. Estimates (REML) and tests of the fixed parameters assuming a first-order autocorrelation.

Variable	df	Parameter estimate	SE	*t* value	$p > \|t\|$	95% confidence limits	
Intercept	11.3	−38.88	16.32	−2.38	0.0357	−74.67	−3.10
Year	11.3	0.0241	0.0082	2.93	0.0132	0.0061	0.0421

In this special case, the data constitute a time series. For time series, one needs to take into account correlations between consecutive observations. In such cases, one speaks of autocorrelation or serial correlation. Beyond doubt, autocorrelations exist between diurnal temperatures. For the year data, our first approach assumed, however, that they are independent as in simple regression analysis. The results are shown in Table 21 and Fig. 17.

The Durbin–Watson test was used to test whether autocorrelations between the residual effects exist (Draper and Smith, 1998). In PROC REG, this test can only be performed with a lag of 1 yr. The Durbin–Watson test uses the test statistic DW, which lies between 0 (autocorrelation = 1) and 4 (autocorrelation = −1) and DW = 2 if the autocorrelation of the errors is zero.

$$\mathrm{DW} = \sum_{i=2}^{N} (\hat{e}_i - \hat{e}_{i-1})^2 \Big/ \sum_{i=1}^{N} \hat{e}_i^{\,2}$$

The autocorrelation r for lag = 1 can be estimated by

$$r(\mathrm{lag}=1) = \sum_{i=2}^{N} \hat{e}_i \hat{e}_{i-1} \Big/ \sum_{i=1}^{N} \hat{e}_i^{\,2}$$

Here, we find $r(\mathrm{lag} = 1) = 0.234$, implying a weak autocorrelation of the residuals which is significantly larger than zero according to the Durbin–Watson test at $\alpha = 0.05$ (DW = 1.513 with a p value of 0.0242). Based on this result, we analyzed the data considering the correlation of the residuals in the variance–covariance matrix R from Eq. [41] by a first-order autoregressive model (AR(1)). This model assumes that the covariance between two residuals e_i and e_j, which are $|i-j|$ apart in the time sequence is $\sigma^2\rho^{|i-j|}$, where ρ is the correlation of adjacent years. The REML procedure of PROC MIXED yields $\hat{\rho} = 0.2827$, $\hat{\sigma}^2 = 0.5127$, and the fixed regression parameter estimates in Table 22.

The comparison of the models with and without autocorrelation is possible by AIC and the likelihood ratio (LR) test. The AIC value of the original analysis is 126.9 and the AIC value of the analysis with autocorrelation is 124.8; the LR test statistic is 4.09 (df = 1) with a p value of 0.0432. Obviously, the model with autocorrelation fits better than a model with independent errors.

To give an outlook on following chapters, we will briefly indicate further approaches to the analysis for this example. With PROC AUTOREG in SAS it can be examined whether a trend combined with an autoregressive model of higher order possibly fits even better. We found periodic fluctuations around the trend and by using a backward algorithm the best fit was achieved with a model of fourth order with $r(\mathrm{lag} = 4) = -0.2799$. Fig. 18 (upper) shows the results of the trend model combined with the AR(1) model (left) and with the AR(4) model (right). Alternatively, the fluctuations can be considered as a deterministic part of the model by using a sine or cosine function that yielded a period of 8 yr, which conforms to the fourth order of the AR model. This approach results in $\widehat{\mathrm{Temp}} = -41 - 0.5451 \sin[(\mathrm{year} - 1968.9)2\pi/8.0138] + 0.0256\,\mathrm{year}$ (Fig. 18 lower). We guess that there is a relation to the sun-spot activity. A detailed consideration of repeated measurements (Chapter 10; Gezan and Carvalho, 2018) and nonlinear regression models (Chapter 15; Miguez et al., 2018) will be given in the corresponding chapters.

If one is only interested in the trend, considering Fig. 17 one could have the impression that a breakpoint exists so that up to and after this time point different models could describe the temperature development in time. This is called piecewise

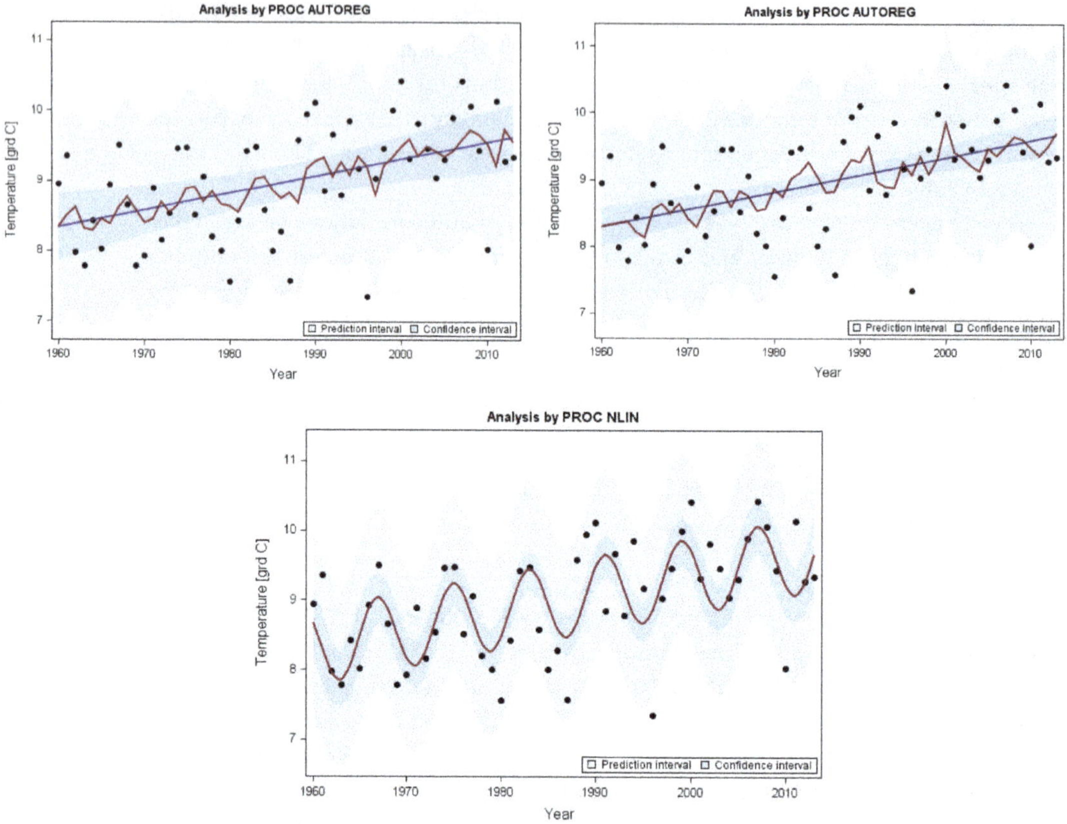

FIG. 18. Example 8. (Upper) Analysis as AR(1) with MSE = 0.48005 (left) and AR(4) with MSE = 0.46078 (right). (Lower) Analysis with fluctuations as a deterministic part with MSE = 0.3693.

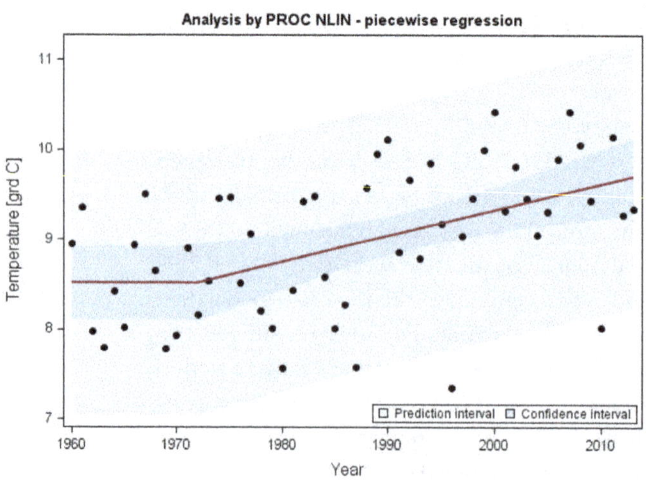

FIG. 19. Example 8. Piecewise regression.

regression. We assumed here that up to the breakpoint, the linear function was constant, and after that, the temperature followed another linear function. With PROC NLIN the breakpoint was found to be in year 1972 (MSE = 0.5002) and the function is $\widehat{\text{Temp}} = 0.029\text{year} - 48.6868$ for year $\geq$1972 and $\widehat{\text{Temp}} = 8.5012$ otherwise (Fig. 19).

Concluding Remarks

1. Whenever possible, model selection should be driven by scientific motivation and not by statistical criteria alone. A formal procedure based on purely statistical considerations can be helpful, but the result should be scientifically challenged.
2. A regression function describes dependencies between variables that are not necessarily in a cause–effect relationship. Dependencies found may be spurious correlations and not an expression of causation of y by x. The simultaneous dependency of x on a hidden variable z and of y on z may give the false impression of a causal relationship between x and y. Here, z is a confounding variable that should be identified by the scientist. For complex relations with several variables, path analysis may be helpful (Loehlin, 2004).
3. Besides the Box–Cox transformation, other transformations may achieve normality, variance homogeneity, and/or additive effects. Because these assumptions must be met simultaneously, a single best suited transformation is sometimes difficult to identify. A flexible alternative is to use generalized linear (mixed) models, which can assume distributions other than the normal and which imply specific forms of heterogeneity of variance. The search for a suitable transformation, called the link function in this context, can then be entirely focused on the assumption of additivity/linearity on the transformed scale (Bolker et al., 2009; Stroup, 2015). Some of the quantitative variables given in the introduction are a priori better suited for an analysis as a generalized linear mixed model. For example, the Poisson regression uses a logarithmic link function to model the expected value of a Poisson-distributed regressand. Similarly, logistic regression can be used to model a regressand with Binomial or multinomial distribution (see Chapter 16; Stroup, 2018).

Key Learning Points

- Regression deals with the quantitative description of the dependency of a regressand y from one or more regressors. The regressand always must be a random quantitative variable, and the regressors may be random (Model II) or fixed quantitative variables (Model I), for example, levels of a quantitative treatment factor. Mixed situations are also possible.
- Despite their different theoretical backgrounds, it is not necessary to differentiate between Models I and II to estimate and test regression parameters. However, correlation analyses are only possible with Model II. The distinction between Models I and II is also important for the planning of an experiment. Whereas sample size needs to be planned for both models, in Model I the levels of the regressors must also be allocated.

- "Linear" regression does not only mean that the regressand y must be a linear function of the regressor x or of the regressors $x_1 \ldots x_{p-1}$. The adjective "linear" means that the function is linear in its parameters. A function is linear in its parameters if the second derivatives of the function with respect to its parameters are equal to zero.
- Sometimes, linearity can be achieved by nonlinear transformations of the regressor(s) or of the regressand.
- If the regressor x has been transformed, the resulting model can often be handled as a simple linear model, or, for example, if y is a polynomial function of x, the resulting model can be considered as a multiple linear regression model. When there are several regressors (transformed or not), the result may be a multiple linear regression model. Both simple linear regression and multiple linear regression models assume an additive error structure that can be preserved despite the transformation of regressors. In contrast, if a nonlinear transformation has been applied to the regressand, then the retransformed model will not have an additive error structure. For example, for the logarithmic transformation, the retransformed model will have a multiplicative error structure. This entails consideration of variance heterogeneity of the regressand. To avoid this issue, consider not transforming the regressand and refer to Chapter 15, which covers Non-Linear Regression (Miguez et al., 2018).
- The Box–Cox transformation and some approaches as special cases of the mixed linear model are useful for evaluating variance heterogeneity of the regressand.
- We discussed differences between the sequential and partial approaches for evaluating several regressors. A sequential approach is especially useful if a scientifically motivated hierarchy of the regressors exists or in polynomial regression, where terms with low order need to be fitted before terms of higher order.

Review Questions

You described the dependency of a random variable y on one regressor x by a simple linear regression function and obtained $R^2 = 0.8$ and $R^2_{adj} = 0.77$.
Decide which answers are correct:

Answer	Correct	Not Correct
y depends on x at 80% (or 77%).		x
80% (or 77%) of the observed points lie on the regression function.		x
80% (or 77%) of the variability of y can be explained by the variability of x.		x
By the regression function, 80% of SSY can be explained.	x	
77% of the variance of y can be explained by the regression function.	x	

You described the dependency of a random variable y on a regressor x by a simple linear regression function with the result $\hat{y} = 0.68 + 2.34x$.
Decide which answers are correct:

Cases	Decision	Correct	Not Correct
The intercept is not significantly different from zero, the slope differs significantly from zero ($\alpha = 0.05$).	Because the intercept is not significantly different from zero you inform about the result in the form $\hat{y} = 2.34x$		x
Both parameters are significantly different from zero ($\alpha = 0.05$).	The scientific interpretation of $\hat{y} = 0.68$ for x = 0 is acceptable		
	• if $x = 0$ is in the observed region of x-values.	x	
	• if $x = 0$ is not in the observed region of x-values		x†
The intercept is significantly different from zero, the slope does not differ significantly from zero ($\alpha = 0.05$).	If x increased by one unit, y increases by 2.34 units.		x
You observe an individual with $x = x'$.	To predict the y-value of this individual you use		
	• the (1-α) - prediction interval for the response at $x = x'$	x	
	• the (1-α) - confidence interval for the expected response at $x = x'$		x
You observe individuals with $x = x'$.	To predict the mean y-value of these individuals you use		
	• the (1-α) - prediction interval for the response at $x = x'$		x
	• the (1-α) - confidence interval for for the response at $x = x'$	x	

† In most cases it is not acceptable. Only, if it seems scientifically justified to extrapolate the function to $x = 0$ the interpretation is acceptable (it is acceptable, e.g., in Example 6, but not in Example 1; see the discussion of the no-intercept problem in this chapter).

You described the dependency of a random variable y on x_1, x_2, and x_3 by a multiple linear regression model. The F-test of the whole model gives an F-value = 105 with the p-value = 0.001. The p-values of the partial t-tests of β_1, β_2, and β_3 are 0.12, 0.07, and 0.23. What do you conclude?

Answer	Correct	Not Correct
y does not depend linearly on x_1, x_2, and x_3.		x
Not all three regressors should be included in the model.	x	
You decide for a sequential forward variable selection. Which regressor should be on the top of hierarchy if you have no scientific motivation of the hierarchy?		
x_1		x
x_2	x	
x_3		x

Exercises

In a field experiment with winter wheat, the dependency of the yield [dt ha^{-1}] on N [kg ha^{-1}] and organic fertilization were analyzed. There were three levels of organic fertilization (factor Org; levels 0 = no organic fertilization, 1 and 2 = two different variants of organic fertilization) and four levels of N fertilization (factor N, levels

Org	N	Block	Yield
0	0	1	32.36
0	0	2	40.58
0	0	3	36.40
0	100	1	42.21
0	100	2	48.99
0	100	3	46.82
0	150	1	51.75
0	150	2	53.74
0	150	3	49.80
1	0	1	36.53
1	0	2	33.74
1	0	3	32.50
1	50	1	59.93
1	50	2	54.31
1	50	3	56.95
1	100	1	60.74
1	100	2	66.10
1	100	3	57.55
1	150	1	54.42
1	150	2	57.97
1	150	3	53.02
2	0	1	37.51
2	0	2	39.10
2	0	3	33.06
2	50	1	50.79
2	50	2	51.62
2	50	3	46.27
2	100	1	55.61
2	100	2	55.52
2	100	3	52.78
2	150	1	46.76
2	150	2	49.66
2	150	3	47.36

0, 50, 100, 150 kg ha^{-1}). All levels of Org were combined with all levels of N with the exception of the combination (Org = 0, N = 50 kg ha^{-1}). The design was a randomized complete block design with three blocks. The blocks are considered as fixed.
The data are given in the table.

Org is a qualitative factor, N is a quantitative factor. The researcher wants to estimate regression functions describing the dependency of yield on the N levels. It needs to be decided whether the function and the function type depend on the levels of the organic fertilization.

Compared with the examples of this chapter, in this example we have several problems to be solved simultaneously:

- we need to decide on the function type (see Example 6)
- there is one qualitative treatment factor (see Example 7)
- the field layout was a randomized complete block design with block as a qualitative disturbance factor (see Example 3 modified)

QUESTIONS

1. Which function type could be appropriate to describe the response to increased levels of N?
2. Do the regression functions of the three levels of Org coincide? Consider especially the result for the variant without organic fertilization.
3. Estimate the regression functions for the three levels of Org corresponding to your conclusions in 2.

ACKNOWLEDGMENTS

We thank our colleagues Bärbel Kroschewski, Kirsten Weiß, and Michael Baumecker from the Humboldt-Universität zu Berlin, Albrecht Daniel Thaer-Institute, for providing of sample data to illustrate several problems connected with linear regression problems.

References

Archontoulis, S.V., and F.E. Miguez. 2015. Nonlinear regression models and applications in agricultural research. Agron. J. 107:786–798. doi:10.2134/agronj2012.0506

Atkinson, A.C. 1987. Plots, transformations, and regression: An introduction to graphical methods of diagnostic regression analysis. Oxford Statistical Science Series. Oxford, Oxford, UK.

Atkinson, A.C., A.N. Donev, and R.D. Tobias. 2009. Optimum experimental designs with SAS. Oxford Univ. Press, Oxford, UK.

Beach, M.L., and J. Baron. 2005. Regression to the mean In: P. Armitage and T. Colton, editors, Encyclopedia of biostatistics. Vol. 7. 2nd ed. Wiley, New York.

Belsley, D.A., E. Kuh, and R.E. Welsch. 2004. Regression diagnostics. Identifying influential data and sources of collinearity. Wiley, New York.

Bolker, B.M., M.E. Brooks, C.J. Clark, S.W. Geange, J.R. Poulsen, M.H.H. Stevens, and J.-S.S. White. 2009. Generalized linear mixed models: A practical guide for ecology and evolution. Trends Ecol. Evol. 24:127–135. doi:10.1016/j.tree.2008.10.008

Box, G.E.P., and C.R. Cox. 1964. An analysis of transformations (with discussion). J. R. Stat. Soc., B 26:211–252.

Draper, N.R., and H. Smith. 1998. Applied regression analysis. 3rd ed. John Wiley and Sons Inc., New York.

Fitzmaurice, G.M., S.R. Lipsitz, and M. Parzen. 2007. Approximate median regression via the Box–Cox transformation. Am. Stat. 61:233–238. doi:10.1198/000313007X220534

Freeman, J., and R. Modarres. 2006. Inverse Box–Cox: The power-normal distribution. Stat. Probab. Lett. 76:764–772. doi:10.1016/j.spl.2005.10.036

Fuller, W.A. 1987. Measurement error models. John Wiley & Sons, New York.

Galton, F. 1877. Typical laws of heredity. Nature 15:492–495, 512–514, 532–533. doi:10.1038/015492a0

Galton, F. 1885. Opening address Section H. Anthropology. Nature 32:507–510.

Galton, F. 1886. Regression towards mediocrity in hereditary stature. J. Anthropol. Inst. Great Britain Ireland 15:246–263. doi:10.2307/2841583

Gezan, S. and M. Carvalho. 2018. Analyzing repeated measures for the biological and agricultural sciences. In: B. Glaz and K.M. Yeater, editors, Applied statistics in the agricultural, biological, and environmental sciences. ASA, CSSA, SSSA, Madison, WI.

Harville, D.A., and D.R. Jeske. 1992. Mean squared error of estimation or prediction under a general linear model. J. Am. Stat. Assoc. 87:724–731. doi:10.1080/01621459.1992.10475274

Kenward, M.G., and J.H. Roger. 2009. An improved approximation to the precision of fixed effects from Restricted Maximum Likelihood. Comput. Stat. Data Anal. 53:2583–2595. doi:10.1016/j.csda.2008.12.013

Loehlin, J.C. 2004. Latent variable models: An introduction to factor, path, and structural analysis. 4th ed. Lawrence Erlbaum Associates, Hillsdale, NJ.

Mead, R. 1988. The design of experiments. Cambridge Univ. Press, Cambridge.

Miguez, F., S. Archontoulis, and H. Dokoohaki. 2018. Non-linear regression models and applications. In: B. Glaz and K.M. Yeater, editors, Applied Statistics in Agricultural, Biological, and Environmental Sciences. ASA, CSSA, SSSA, Madison, WI.

Nelder, J.A. 2000. Functional marginality and response-surface fitting. J. Appl. Stat. 27:109–112. doi:10.1080/02664760021862

Neter, J., W. Wasserman, and M.H. Kutner. 1989. Applied linear regression models. 2nd ed.IRWIN, Inc., Boston.

Pearson, K. 1896. Mathematical contributions to the theory of evaluation. III. Regression, heredity, and panmixia. Philos. Trans. R. Soc. Lond. A 187:253–318. doi:10.1098/rsta.1896.0007

Rasch, D., G. Herrendörfer, J. Bock, N. Victor, and V. Guiard. 1998. Verfahrensbibliothek. Versuchsplanung und–auswertung. Band II. R. Oldenburg München Wien.

Rawlings, J.O., S.G. Pantula, and D.A. Dickey. 1998. Applied regression analysis. A research tool. 2nd ed. Springer, New York.

Richter, C., B. Kroschewski, H.-P. Piepho, and J. Spilke. 2015. Treatment comparisons in agricultural field trials accounting for spatial correlation. J. Agric. Sci. 153:1187–1207. doi:10.1017/S0021859614000823

Schabenberger, O., and F.J. Pierce. 2002. Contemporary statistical models for the plant and soil sciences. CRC Press, Boca Raton.

Searle, S.R. 1971. Linear models. John Wiley & Sons, New York.

Searle, S.R., G. Casella, and C.E. McCulloch. 1992. Variance components. John Wiley and Sons, New York.

Seber, G.A.F. 1977. Linear regression analysis. New York, Wiley.

Sokal, R.R., and F.J. Rohlf. 1995. Biometry. 3rd ed. W.H. Freeman and Company, New York.

Stroup, W.W. 2015. Rethinking the analysis of non-normal data in plant and soil science. Agron. J. 107:811–827. doi:10.2134/agronj2013.0342

Stroup, W.W. 2018. Analysis of non-Gaussian data. In: B. Glaz and K.M. Yeater, editors, Applied Statistics in Agricultural, Biological, and Environmental Sciences. ASA, CSSA, SSSA, Madison, WI.

Taylor, J.M.G. 2006. Transformations-II. In: S. Kotz et al., editors, Encyclopedia of statistical science. Vol. 14. John Wiley and Sons, New York.

Webster, R. 1997. Regression and functional relations. Eur. J. Soil Sci. 48:557–566.

CHAPTER 7: ANALYSIS AND INTERPRETATION OF INTERACTIONS OF FIXED AND RANDOM EFFECTS

Mateo Vargas, Barry Glaz, Jose Crossa,* and Alex Morgounov

Good agronomic research is often characterized by experiments designed with more than one factor identified as a fixed effect. However, many researchers do not recognize and report the maximum impact of their research because they do not appropriately analyze and interpret the well-designed interactions between factors. One purpose of this chapter is to provide guidance on properly analyzing and interpreting interactions between fixed effects. We demonstrate why it is crucial to consider all interactions of all factors and optimize your model to reach proper conclusions and make correct recommendations that maximize the impact of your research. We provide examples of the analysis and interpretation of interactions, including single degree of freedom contrasts of linear, quadratic, and cubic responses when a quantitative fixed effect is one component of an interaction. A second purpose of this chapter is to discuss considerations that will help researchers determine if effects should be analyzed as random or fixed and to clarify how this affects inferences. The discussion on random effects is based on examples using 2 and 10 yr of data. Key concepts learned by the reader in this chapter and its appendices include how to use the GLIMMIX Procedure in SAS to calculate a LSD, how to develop contrast statements involving regression responses, how to interpret interactions involving regression responses, and precautions necessary in interpreting traditionally random effects such as years or locations if it is prudent to analyze them as fixed effects.

Agricultural production systems are complex. Crops are affected by weather conditions, such as rainfall and temperature, and these weather conditions change from year to year. Water and nutrient availability are influenced by soil properties and agronomic inputs made by farmers. Farmers also make decisions on such things as what crop species to grow, what cultivars of a particular species to use, when to plant, when to apply inputs, and when to harvest. Clearly, there are many decisions to make when growing a crop.

Abbreviations: DDF, Denominator degrees of freedom; GY, Grain yield; NCOI, Noncrossover interaction; NDF, Numerator degrees of freedom.

M. Vargas, Universidad Autónoma Chapingo, Chapingo, Mexico; B. Glaz (retired) USDA-ARS Sugarcane Field Station, 12990 U.S. Highway 441, Canal Point, FL 33438, USA; J. Crossa, Biometrics and Statistics Unit, International Maize and Wheat Improvement Center (CIMMYT), Apdo. Postal 6-641, 06600 México DF, Mexico; A. Morgounov, CIMMYT, P.O. Box 39, Emek, 06511 Ankara, Turkey. *Corresponding author (j.crossa@cgiar.com).

doi:10.2134/appliedstatistics.2015.0084

Applied Statistics in Agricultural, Biological, and Environmental Sciences
Barry Glaz and Kathleen M. Yeater, editors

The ultimate goal of most agronomic research is to help farmers. This help can be in the form of increased production or profits, practices that facilitate crop management, or recommendations to conserve resources or ameliorate ecological challenges. With these goals in mind, and being cognizant of the large number of inputs and decisions that farmers must make, agronomic experiments often study the responses to more than one factor. By combining factors, researchers can provide more realistic information to farmers. Sometimes recommendations are complex because in farmers' fields, as in researchers' plots, two or more farmer inputs (typically factors in an experiment) may affect each other.

For example, let us consider that geneticists in a breeding program test new genotypes and that they only test those genotypes using one standard set of agronomic inputs (the same fertilizer, pesticides, and irrigation) and one standard set of decisions (only plant on one soil type on the same date and harvest only on one date). Also, they only conduct this experiment in 1 yr. Eventually they will probably find a new high yielding genotype (perhaps with 30% higher yields) and release this genotype as a new cultivar. Should these researchers communicate to farmers that they can expect a 30% increase in production? Most researchers know that based on the research described, a prediction like this could result in many disappointed farmers. Farmers who have a different soil type, use different fertilizers and irrigation practices, or who use different planting and harvest dates may not obtain a 30% increase in yields. Perhaps some farmers will even observe a yield *decrease* with the new cultivar compared with the other cultivars they have used. On the other hand, some farmers may have yield increases that are more than 30%. Also, farmers may find shifts in all of these results from year to year. It is easy to see that optimally, geneticists in a breeding program should test their promising genotypes for more than 1 yr, on more than one soil type, using several planting and harvest dates, and they should use a broad range of fertilizers, irrigation, and other practices. These "factors" in their experiments should match options that farmers in the target region use. When researchers consider more than one input or decision, they create experiments with more than one factor, and this gives them the opportunity to identify important interactions.

Perhaps the breeding program will identify a new genotype that produces high yields at all locations (or soil types) tested in all years, with all standard fertilizer inputs, with and without irrigation, and irrespective of planting and harvest dates. However, it is much more likely that the researchers in the breeding program will identify new genotypes that do well only at some locations and only when certain agronomic practices are used. This type of information can only be discovered by researchers who conduct experiments with more than one factor and then analyze the interactions between these factors.

Designing research that tests more than one factor is only one important step in developing useful recommendations based on sound research. It is also crucial to properly analyze and interpret factor interactions. Using the example of the breeding program, the researchers would not be able to recommend certain cultivars for certain locations, soil types, and agronomic inputs unless they conducted research with factors that accounted for these conditions and then properly analyzed and interpreted the interactions between these factors.

Looking for the Best Model

Experimental Data

We use one data set in this section to help explain proper methodologies for analyzing interactions of factors as fixed effects. The data set is from research conducted on spring wheat (*Triticum aestivum* L.) in a high-altitude region of northern Kazakhstan. Data from this experiment were also used by Vargas et al. (2015), who assessed the interactions of fixed effects factors. The purpose of this experiment was to learn about the effects of N and P fertilizers on two soil types in this region. The experiment was conducted at two locations in Akmola region of Kazakhstan, one with a Chestnut soil and a second location with a Black soil (chernozem). On each soil, two replications (two complete blocks) of N and P fertilizer treatments were arranged in a randomized complete block design. Nitrogen rates were 0 and 30 kg ha^{-1}, and the four unevenly spaced P rates were 0, 50, 150, and 250 kg ha^{-1}. The researchers were interested in the effects of these treatments (N and P fertilizer rates on two soil types) on the grain yield (GY) (Mg ha^{-1}) of one wheat genotype. Our major goal in this section is to provide guidance on properly analyzing and interpreting interactions between fixed effects. We also demonstrate why it is crucial to consider all interactions between all factors to optimize your model and ultimately reach proper conclusions and make correct recommendations that maximize the impact of your research. These data are provided in the supplemental material as the Data Wheat file online.

The experiment was conducted in two separate years. Year is generally considered as a random effect and is often analyzed as such. However, later in this chapter, we discuss why it can be counterproductive to analyze an effect with a low number of levels as random. Therefore, consistent with that explanation, we analyze Year as a fixed effect here, and we understand that we need to take care in our interpretations to convey that our results do not apply to years generally, but only to the two specific years for which we have data. The proper analysis of fixed and random effects was discussed by McIntosh (1983), and her concepts were later updated by Moore and Dixon (2015) for analysis of mixed models with modern software. We further recommend readers to Chapter 8 (Dixon et al., 2018) for excellent discussions on analysis of fixed and random effects. Finally, in the next section, we analyze a data set with several models to discuss some key issues involved in determining if a factor should be analyzed as a fixed or random effect.

The Complete Model: Analyzing Main Effects and Their Two-Way, Three-Way, and Four-Way Interactions

We will use different linear mixed models to describe our research, and we will explain the analysis pertaining to each model. For people familiar with general linear model theory based on fixed models and least squares estimation procedures, instead of the maximum likelihood estimation procedures used in mixed model theory, it is important to point out that least squares *F* tests and likelihood based *F* tests do not always give the same results.

The first step recommended in a factorial experiment is to adjust the complete model including all main effects and all possible interactions. In our case, we have four factors: year, soil, N, and P. The model including all these terms and their interactions is:

$$y_{ijklm} = \mu + g_i + s_j + r(gs)_{ijk} + n_l + p_m + (gs)_{ij} + (gn)_{il} + (gp)_{im} + (sn)_{jl} + (sp)_{jm} \\ + (np)_{lm} + (gsn)_{ijl} + (gsp)_{ijm} + (gnp)_{ilm} + (snp)_{jlm} + (gsnp)_{ijlm} + \varepsilon_{ijklm} \quad [1]$$

where:

r_k is a random effect for replications (k = 1, 2,..., R, where R is the number of replications) nested within years and soils, while all other terms, main effects, two-way, three-way, and four-way interactions, are considered as fixed effects;
g_i is the main effect for year (i = 1, 2, ..., G, where G is the number of years);
s_j is the main effect of soil (j = 1, 2,..., S, where S is the number of soils);
n_1 is the main effect for nitrogen (l = 1, 2,..., N, where N is the number of rates);
p_m is the main effect for phosphorus (m = 1, 2,..., P, where P is the number of rates);
$(gs)_{ij}$ is the two-way interaction effect of the ith year with the jth soil;
$(gn)_{il}$ is the two-way interaction effect of the ith year with the lth N rate;
$(gp)_{im}$ is the two-way interaction effect of the ith year with the mth P rate;
$(sn)_{jl}$ is the two-way interaction effect of the jth soil with the lth N rate;
$(sp)_{jm}$ is the two-way interaction effect of the jth soil with the mth P rate;
$(np)_{lm}$ is the two-way interaction effect of the lth N rate with the mth P rate;
$(gsn)_{ijl}$ is the three-way interaction effect of the ith year, j^{th} soil, and the lth N rate;
$(gsp)_{ijm}$ is the three-way interaction effect of the ith year, j^{th} soil, and the mth P rate;
$(gnp)_{ilp}$ is the three-way interaction effect of the ith year, l^{th} N rate, and the mth P rate;
$(snp)_{jlm}$ is the three-way interaction effect of the jth soil, l^{th} N rate, and the mth P rate,
$(gsnp)_{ijlm}$ is the four-way interaction effect of the ith year, j^{th} soil, l^{th} N rate, and the mth P rate;
ε_{ijmkl} is the error residual associated with each effect that was previously defined.

In this mixed model, both $r(gs)_{ijk}$ and ε_{ijklm} are assumed to be identically, independently, and normally distributed (NII) random variables with $N(0, I_{ijk}\sigma^2_r)$ and $N(0, I_{ijklm}\sigma^2_e)$, respectively where the matrices $\mathbf{I}_{ijk}$ and $\mathbf{I}_{ijklm}$ are the identity matrices of order GSR × GSR and GSRNP × GSRNP, respectively. Also, it is assumed that correlations among replications and residuals are equal to zero.

The basic SAS code for running the above mixed model using the GLIMMIX procedure and the bar notation, without including the lsmeans or contrast statements, has the following form:

```
ODS select CovParms Tests3;
Proc GLIMMIX data = Wheat;
     Class Year Soil N P Rep;
     Model Yield = Year | Soil | N | P/Dist = Normal Link = Identity;
     Random Rep(Year Soil);
Run;
```

Although the Normal or Gaussian distribution is used by default in Proc GLIMMIX we have included here the DIST = Normal and LINK = Identity options in the MODEL statement, to show how to use this option because not necessarily the response variable is always normally distributed. For non-normally distributed random variables Proc GLIMMIX is a useful procedure.

The results (Table 1) show that all four main effects (year, soil, N, and P), five of six two-way interactions, and three of four three-way interactions were significant. The only effects that were not significant were the two-way interaction N × P, the three-way interaction Y × N × P, and the four-way interaction Y × Soil × N × P.

Before interpreting the significant effects and calculating the adjusted means or orthogonal polynomial contrasts for the four rates of P, we recommend testing a reduced model that removes the nonsignificant terms found in the complete model.

Therefore, the second step is to run the following code in SAS. Note that now we cannot use the bar notation because the model does not include all the terms. Therefore, we must specify each specific term in the reduced model.

```
ODS select CovParms Tests3;
Proc GLIMMIX data = Wheat;
     Class Year Soil N P Rep;
     Model Yield = Year Soil N P
          Year*Soil Year*N Year*P Soil*N Soil*P
          Year*Soil*N Year*Soil*P Soil*N*P;
     Random Rep(Year Soil);
Run;
```

The results for the reduced model are shown in Table 2, where we can see now that all the remaining terms in the model were significant. Therefore, this is the final optimized model. Our task is to interpret the main effects and interactions of this reduced model. It is well known that researchers should not interpret the results of any significant main effect that is part of a significant interaction. Therefore, for our

TABLE 1. *p* value for grain yield (Mg ha^{-1}) using the complete model (main effects, two-way, three-way, and four-way interactions) in an experiment that tested the effects of N and P fertilizer rates (0 and 30 kg N ha^{-1} and 0, 50, 150, and 250 kg P ha^{-1}) for 2 yr on Black and Chestnut soils.

Effect	NDF†	DDF‡	*p* values
Year (Y)	1	4	< 0.0001
Soil	1	4	< 0.0001
N	1	4	0.0016
P	3	28	< 0.0001
Y × Soil	1	28	< 0.0004
Y × N	1	28	0.0036
Y × P	3	28	0.0013
Soil × N	1	28	< 0.0001
Soil × P	3	28	< 0.0001
N × P	3	28	0.2400
Y × Soil × N	1	28	< 0.0001
Y × Soil × P	3	28	0.0004
Y × N × P	3	28	0.5892
Soil × N × P	3	28	0.0047
Y × Soil × N × P	3	28	0.1183
Covariance parameter	Variance component	SE	
Rep(Year Soil)	0.000292	0.000955	
Residual	0.008036	0.002148	

† NDF, numerator degrees of freedom in the Type III test of fixed effects.

‡ DDF, denominator degrees of freedom in the Type III test of fixed effects.

results, we will interpret only the effects of significant interactions because each main effect is a factor in a significant interaction.

When there are more than two quantitative levels or rates of a factor, Swallow (1984), Little (1978), Saville (2015), and Vargas et al. (2015) recommended analyzing those factors by calculating polynomial orthogonal contrasts. In Appendices 1 and 2 of the supplemental material, we provide SAS code for calculating those contrasts. When there are no missing data and levels or rates are equally spaced, these coefficients can be determined as explained by Steel and Torrie (1980). Kuehl (1999) also explained how to calculate these regression coefficients. We recommend that readers review these materials, so they use this procedure properly. However, as shown in Appendix 1 of the supplemental material, Proc IML of SAS is an excellent tool for generating precise coefficients even if rates are not equally spaced, as is our case for the P rates. Table 2 contains the analysis of variance results of all main effects and interactions in our final, reduced model, as well as the results from the orthogonal polynomial contrasts for the main effect of P and for all the two- and three-way interactions including P that were part of the final model.

In Table 2, we see that in all three of the possible preplanned single degree of freedom contrasts (linear, quadratic, and cubic) at least one was significant for each interaction effect. For the main effect P and for the two-way interaction Y × P, both the linear and quadratic effects were significant, while for the two-way interaction Soil × P and the three-way interaction Y × Soil × P, only the linear effect of each

TABLE 2. *p* value for grain yield (Mg ha^{-1}) for main effects, two-way, and three-way interactions in an experiment that tested N and P fertilizer (0 and 30 kg N ha^{-1} and 0, 50, 150, and 250 kg P ha^{-1}) in 2 yr on Black and Chestnut soils. Rates of P are partitioned into linear, quadratic, and cubic polynomial contrasts.

Source of variation	NDF†	DDF‡	*p* values			
			Main effect or interaction	Partitioned contrasts §		
				Linear	Quadratic	Cubic
Year (Y)	1	4	< 0.0001			
Soil	1	4	< 0.0001			
N	1	34	0.0019			
P	3	34	< 0.0001	< 0.0001	0.0016	0.2103
Y × Soil	1	4	0.0004			
Y × N	1	34	0.0041			
Y × P	3	34	0.0015	0.0006	0.0361	0.6391
Soil × N	1	34	< 0.0001			
Soil × P	3	34	< 0.0001	< 0.0001	0.6734	0.9807
Y × Soil × N	1	34	< 0.0001			
Y × Soil × P	3	34	0.0004	< 0.0001	0.8825	0.4969
Soil × N × P	6	34	0.0131	0.1032	0.7931	0.0013
Covariance parameter	Variance component	SE				
Rep(Year Soil)	0.000222	0.000953				
Residual	0.008593	0.002084				

† NDF, numerator degrees of freedom in the Type III test of fixed effects.

‡ DDF, denominator degrees of freedom in the Type III test of fixed effects.

§ Numerator degrees of freedom = 1 for each linear, quadratic, and cubic contrast.

was significant. Finally, in the three-way interaction Soil × N × P, only the cubic contrast was significant among the three preplanned single degree of freedom contrasts partitioned.

When a significant main effect or interaction is also part of a higher order significant interaction, and in our case, when at least one of the polynomial contrasts of a higher order interactions is significant, it is advisable to begin the results by discussing the meaning of that higher order interaction. The focus is always on the highest order interactions that are significant. Therefore, in this example, we will focus on the three three-way interactions Y × Soil × N, Y × Soil × P, and Soil × N × P because any significant main effect or lower order interaction is also part of one of these significant three-way interactions, with at least one significant polynomial contrast.

Thus, by discussing these significant interactions, we take into account the importance of all other lower order significant interactions and main effects because in this study all significant main effects and two-way interactions comprised part of a significant three-way interaction. Therefore, rather than discussing each of these lower order interaction effects separately, we discuss each from the much more useful perspective of how three effects combined to interact significantly. If the researcher feels that it is scientifically important to point out the importance of a lower order interaction or main effect, then this is perfectly acceptable as long as the importance of the results of the highest order interaction that is significant is ultimately fully explained, and it is clear that the lower order interaction is explained within the context of the significant highest order interaction. Another way of saying this is that the lower order interaction can be discussed if doing so helps the scientist explain the more complex higher order interaction.

For example, to clarify the point that the researcher should focus on the higher order interaction, let us consider the significant Soil × N interaction (Table 2). While it is important for us to know how N affected GY differently in each soil, just having this information would not be complete because the Year × Soil × N interaction and the cubic response to P partitioned within the overall Soil × N × P interaction were also significant. This means that when we explain the effects that N had on soil type, we also need to include the effects of either Year or P in those interpretations. If we discuss only the results of the Soil × N interaction in the report that communicates our research, then we could possibly miss an opportunity for our research to have important impact and misinform our readers because Year and P each affected the Soil × N interaction.

Discussing the three-way interactions is facilitated by graphs with four response functions like those shown in Fig. 1, 2, and 3. We recommend using graphs to interpret and communicate the importance of three-way interactions such as these.

Interpreting the Year × Soil × N interaction

In Fig. 1, it is immediately seen that this interaction was a noncrossover interaction (NCOI), which means that the four response functions never cross each other in the exploration space used in this study. A careful look at the figure shows that the reason for the significant Y × Soil × N interaction was because GY increased significantly on the Black soil due to the N_{30} treatment only in 2007. There was no GY response to N_{30} in 2008 on Black soil or in either year on Chestnut soil (Fig. 1).

We can see that adding N to the Black soil was beneficial in only one of the 2 yr in which the study was conducted. We can speculate that on a Black soil, adding N will be useful some years and other years not useful. Because we only analyzed 2 yr and Year was a fixed effect, we would need to make it clear that N_0 and N_{30} need to be tested for more years on the Black soil. It would also probably be prudent to suggest that one or two higher rates of N also be tested in subsequent years.

While it is important for us to know how N affected GY differently in each soil between years, just having this information would not be complete because the cubic response to P partitioned within the overall Soil × N × P interaction was also significant (Table 2). This means that when we explain the effects that N had on soil

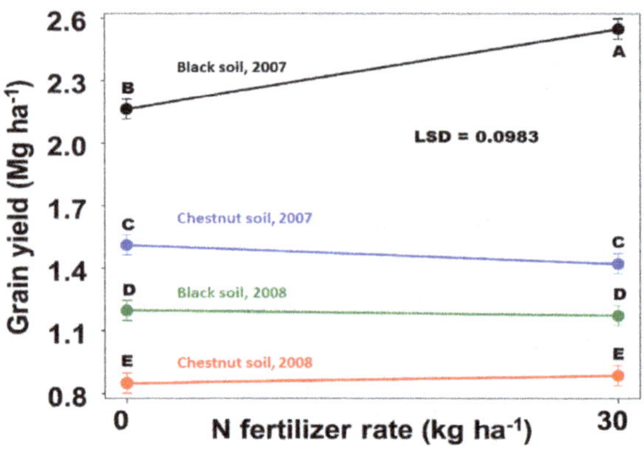

FIG. 1. Responses of grain yield to N fertilizer on Black and Chestnut soils during 2007 and 2008. Means with the same letter are not significantly different at $p = 0.05$. The vertical bars represent the LSD interval for the means that are located at the center of each interval.

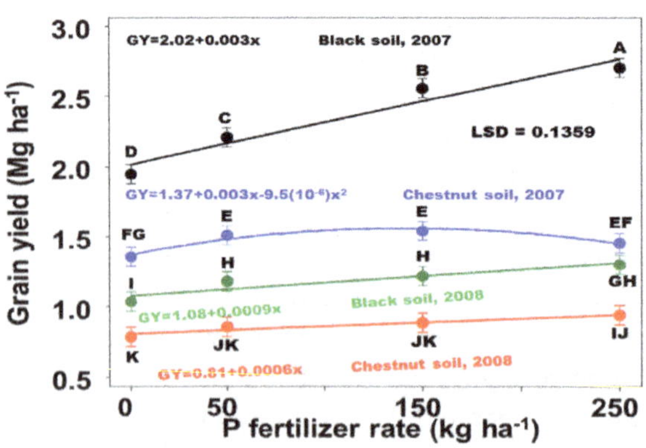

FIG. 2. Responses of grain yield to phosphorus fertilizer on Black and Chestnut soils during 2007 and 2008. Means with the same letter are not significantly different at $p = 0.05$. The vertical bars represent the LSD interval for the means that are located at the center of each interval

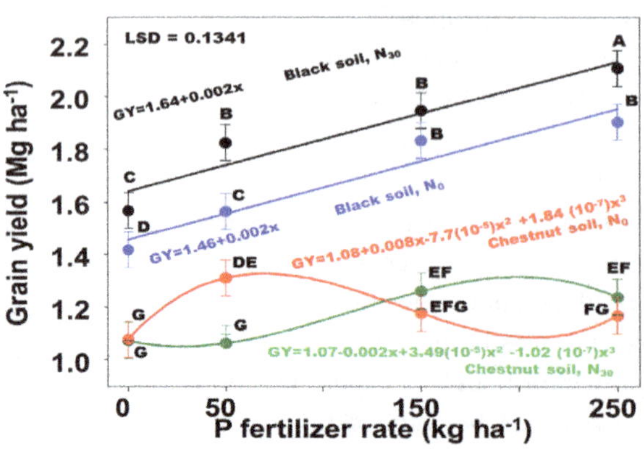

FIG. 3. Responses of grain yield to phosphorus fertilizer on Black and Chestnut soils at 0 (N_0) and 30 (N_{30}) kg N ha^{-1}. Means with the same letter are not significantly different at the 0.05 significance level. The vertical bars represent the LSD interval for the means that are located at the center of each interval.

type, in addition to considering the effects of year, we also need to include the effects of P in those interpretations. If we discuss only the results of the Soil × N interaction in the report that communicates our research, then we could possibly miss an opportunity for our research to have important impact and misinform our readers because Year and P each affected the Soil × N interaction.

Interpreting Year × Soil × P Interaction

In addition to N, GY also reacted differently to P across years and soils because the linear response of the Y × Soil × P interaction was significant (Table 2). The significant differences in the linear responses of GY to P in the 2 yr and soils, tested by the Year × Soil × P interaction, were a NCOI (Fig. 2).

There were several reasons for this significant NCOI. First, GY on the Black soil had a stronger linear response to increasing P in 2007 than in 2008. On the Chestnut soil, there was a significant quadratic response of GY to increasing P rates in 2007, indicating that GY was maximized by adding around 142 kg ha^{-1} of P fertilizer. In 2008, the response to P was linear on the Chestnut soil. Yields on Chestnut soil at all P rates were higher in 2007 than in 2008, but the response to P fertilizer was similarly moderate in each year. Thus, on the Black soil, the response to P was greater in 2007 than 2008, and the response on the Chestnut soil was quadratic in 2007 and linear in 2008, but of a similar order of magnitude in each year.

The significant GY responses to added P in 2007 and 2008 on the Chestnut soil were small, suggesting that an economic analysis would indicate not to use P fertilizer on the Chestnut soil in these years. Our interaction analyses have identified that N and P fertilizer management will differ between these two soils and will probably—we are speculating—differ from year to year based on climatic variables.

Interpreting Soil × N × P interaction

The significant response of GY to increasing rates of P fertilizer on each soil at each rate of N identified in the predetermined cubic contrast partitioned from the Soil × N × P interaction indicates that GY response to P on each soil was affected differently by N (Table 2 and Fig. 3).

The rate of linear increase of GY to P on the Black soil was 0.002 Mg ha^{-1} with each incremental increase in P at each N rate. The GY on Black soil for N_{30} was significantly greater than the GY for N_0 at each corresponding rate of P except at 150 kg P ha^{-1} (Fig. 3). In addition, based on the LSD, the overall GY at N_{30} across all P rates was significantly higher than the overall GY without N across all P rates. Grain yield had a distinct cubic response to P fertilizer at each N rate on the Chestnut soil. In practical agronomic terms, the key conclusions of these different cubic responses on the Chestnut soil are that the optimum P rate was about 196 kg ha^{-1} when N was applied at 30 kg ha^{-1}, and the optimum P rate was about 70 kg ha^{-1} when N was not applied. Thus, our analysis of this three-way interaction has determined that GY can be optimized on the Black soil with the highest rates of N (30 kg ha^{-1}) and P (250 kg ha^{-1}), while on the Chestnut soil, it is preferable not to apply N and to maintain P at only 70 kg ha^{-1} to maximize GY. We can determine that for the Chestnut soil, N and P = 0 and 70 kg ha^{-1}, respectively, are preferable fertilizer rates to N and P = 30 and 196 kg ha^{-1}, respectively, by common sense and a quick look at Fig. 3. The cubic response of

P with no N on the Chestnut soil had a slightly higher peak than the cubic response of P when N = 30 kg ha^{-1} on the Chestnut soil. It is not important whether these peak GYs for each curve differ significantly. With no other information available, by common sense, we choose the one with the lower fertilizer rates as the preferable option.

Thus, this three-way interaction, with the significant polynomial contrast proved to be extremely useful in identifying the sensitive and narrow range of responses of GY to fertilizer on the Chestnut soil and how the interactions of N and P rates on this soil need to be monitored and managed more closely than on the Black soil. Our research has identified important agronomic and environmental impacts for farmers who aim to maximize yields and minimize environmental impacts of N and P enrichment.

Interpreting the Adjusted Means

A multiple comparison procedure is often appropriate when levels of a factor are discrete (Carmer and Swanson, 1971, 1973; Carmer, 1976; Saville, 2015; Chapter 5, Saville, 2018). In the experimental data we present in this chapter, there were only two levels for both Year and Soil and also only two rates of N fertilizer. Therefore, for these factors, the *F* test was sufficient because there was only one comparison. We did not need to test the two rates of N with linear regression because the *F* test already indicated that these two rates differed significantly. Therefore, these LSD analyses are not needed, but because this is the information we had, we will use it as an example of how to obtain least squares means and their LSD groupings using the GLIMMIX procedure.

We only need to include the LSMeans statement in the SAS code used previously to calculate the adjusted means for all the terms that were found to be significant in the final reduced model shown in Table 2. Below we have added this LSMeans statement:

```
ODS select CovParms Tests3 LSMLines;
Proc GLIMMIX data = Wheat;
Class Year Soil N P Rep;
Model Yield = Year Soil N P
              Year*Soil Year*N Year*P Soil*N Soil*P
              Year*Soil*N Year*Soil*P Soil*N*P;
Random Rep(Year Soil);
LSMeans Year Soil N P
        Year*Soil Year*N Year*P Soil*N Soil*P
        Year*Soil*N Year*Soil*P Soil*N*P/Lines;
ODS output lsmeans = ADMEANS diffs = DIFFS tests3 = DOF;
Run;
```

The results are shown in Tables 3 and 4. Although we have shown previously the grouping of the least squares means in the figures for interpreting the interactions, we show these here again in Tables 3 and 4 so we can report their precise values. For the effects that include P, we emphasize again that using regression and presenting the results with figures was a far more useful way to report these results than using the LSD values. However, we show these here for researchers who might have had a factor with four qualitative rather than quantitative levels as a possible approach to presenting their data in table format.

As an example, we will look at the LSD results for the Y × Soil × P interaction (Table 4). We see that in each year, GY on the Black soil was higher than on the Chestnut soil. There is also a clear positive response of GY to increasing rates of P on the Black soil in 2007. However, more than anything, trying to interpret the importance of these results with the LSD shows clearly why statisticians recommend using polynomial contrasts rather than the LSD when factors are quantitative. It is much easier to identify and explain the important differences that are part of significant interactions with the polynomial contrasts.

Caution must be taken when interpreting the grouping obtained using the LINES option in the LSMeans statement, because there are situations where the output can have a pair of means followed by the same letter but really they can be significantly different, especially when the interactions structure is much too complex. The theoretical basis here is beyond the scope of the basic level of this book.

Mixed Models

Practical Recommendations Related to the Fixed vs. Random

TABLE 3. Least significant difference (LSD0.05) mean comparison for the main effects and two- way interactions of Experiment 1, which tested N and P fertilizer rates (0 and 30 kg N ha−1 and 0, 50, 150, and 250 kg P ha−1) on Black and Chestnut soils in Years 2007 and 2008. Only for the terms that were found to be significant in the reduced model.

Effect		Least squares means and LSD grouping			
Year		2007	2008		
		1.913 A†	1.029 B		
Soil		Black	Chestnut		
		1.771 A	1.171 B		
N		N_0	N_{30}		
		1.432 B	1.510 A		
P		P_0	P_{50}	P_{150}	P_{250}
		1.284 C	1.441 B	1.555 A	1.604 A
Year × Soil		Black	Chestnut		
	2007	2.356 A	1.470 B		
	2008	1.187 C	0.871 D		
Year × N		N_0	N_{30}		
	2007	1.838 B	1.987 A		
	2008	1.026 C	1.032 C		
Soil × N		N_0	N_{30}		
	Black	1.680 B	1.862 A		
	Chestnut	1.183 C	1.158 C		
Year × P		P_0	P_{50}	P_{150}	P_{250}
	2007	1.654 C	1.860 B	2.054 A	2.084 A
	2008	0.914 F	1.021 E	1.056 DE	1.125 D
Soil × P		P_0	P_{50}	P_{150}	P_{250}
	Black	1.492 D	1.695 C	1.891 B	2.006 A
	Chestnut	1.075 F	1.186 E	1.219 E	1.202 E

†All means in the cells corresponding to the same interaction, followed by a different letter are significantly different based on the Fisher's test LSD(0.05).

TABLE 4. Least significant difference ($LSD_{0.05}$) mean comparison for the three-way interactions of Experiment 1, which tested N and P fertilizer rates (0 and 30 kg N ha^{-1} and 0, 50, 150, and 250 kg P ha^{-1}) on Black and Chestnut soils in Years 2007 and 2008, only for the terms found to be significant in the reduced model.

Effect			Least squares means and LSD grouping			
			Black		Chestnut	
Year × Soil × N			N_0	N_{30}	N_0	N_{30}
		2007	2.162 B†	2.549 A	1.514 C	1.426 C
		2008	1.199 D	1.175 D	0.852 E	0.890 E
			P_0	P_{50}	P_{150}	P_{250}
Year × Soil × P	2007	Black	1.947 D	2.207 C	2.560 B	2.707 A
		Chestnut	1.360 FG	1.512 E	1.547 E	1.460 EF
	2008	Black	1.037 I	1.182 H	1.222 H	1.305 H
		Chestnut	0.790 K	0.860 JK	0.890 JK	0.945 IJ
			P_0	P_{50}	P_{150}	P_{250}
Soil × N × P	Black	N_0	1.417 D	1.565 C	1.835 B	1.905 B
		N_{30}	1.567 C	1.825 B	1.947 B	2.107 A
	Chestnut	N_0	1.077 G	1.310 DE	1.177 EF	1.167 FG
		N_{30}	1.072 G	1.062 G	1.260 EF	1.237 EF

† All means in the cells corresponding to the same interaction followed by a different letter are significantly different based on the Fisher's test LSD(0.05).

Effects Debate

One of several practical recommendations when using mixed models is that there should be enough information in the data to estimate variance and covariance parameters of random effects with sufficient precision; this is crucial when data have a multifactorial structure. Stroup and Mulitze (1991) argued that a factor should have a large number of levels before it is considered a random effect. This is sound advice when considered from a practical standpoint. For example, if a researcher uses only two locations, it would probably be difficult to make valid inferences about a diverse production region of 100,000 ha. With only two locations, the best approach may be to consider location as a fixed effect and make inferences only on those two locations. However, as the number of research locations increases and better represents the diversity of the 100,000 ha production region, the researcher should consider analyzing and interpreting location as a random effect, so that inferences can be made on the full production region. Using examples, we illustrate some issues for the researcher to consider when analyzing a multifactorial experiment with a mixed model.

We use a data set from a of 10-yr (1988–1997) study on agronomic practices conducted in Ciudad Obregon, Mexico and analyzed in Vargas et al. (1999). This experiment was a complete factorial with four main effects. Each annual experiment was arranged in a randomized complete block design with three replications. The four main effects were two levels of tillage (with and without deep knife), two levels of summer crop (sesbania (*Sesbania* spp.) and soybean [*Glycine max* (L.) Merr.]), two levels of manure (with and without chicken manure), and three rates of N fertilization (0, 100, and 200 kg N ha^{-1}). Thus, the experiment was a 2 × 2 × 2 × 3 factorial with 24 treatments per replication. The three rates of inorganic N fertilizer represented a baseline (0 kg ha^{-1}), a moderate rate (100 kg ha^{-1}), and a relatively high rate (200 kg ha^{-1}). The raw data are provided the supplemental material file online, Data FixRnd2.csv.

Example 1

We start by using only the first 2 yr (1988 and 1989) of our data to fit six mixed models including different numbers of random and/or fixed terms. In this example, we use the MIXED procedure instead of the GLIMMIX procedure, however the same results will be obtained using any of these two procedures, with only slight changes in the code.

Model 1: Year and Rep(Year) are random effects, and all four agronomic factors are fixed effects that do not interact with Year.

The partial code used in Proc Mixed of SAS was:

```
Title1 "Evaluate Fixed and Random Effects";
Title2 "Model 1: only year and rep(year) as random";
Proc Mixed Data = FixRnd2 covtest noinfo noitprint;
     Class Year Rep Till SumCrop Man N;
     Model Yield = Till | SumCrop | Man | N;
     Random Year Rep(Year);
     ODS listing exclude dimensions Nobs FitStatistics;
Run;
```

The results show that the variance components for the two random terms are different from zero, indicating that it was possible to estimate each variance component (Table 5, Column 2). Even though we used only 2 yr of data, because we partitioned only one additional random term related with Year (and this was Rep(Year)), it was possible to estimate its variance component. For the fixed effects, the four main effects are significant as well as the interactions Summer crop × N and Manure × N, while all other interactions are not significant.

Model 2: Each main effect interacts with Year and Rep(Year) and each Year × main effect interaction is a random effect, while the four main agronomic effects and their interactions are fixed effects.

The SAS code for Model 2 is:

```
Title2 "Model 2: year, rep(year), and year × main effects as random";
Proc Mixed Data = FixRnd2 covtest noinfo noitprint;
     Class Year Rep Till SumCrop Man N;
     Model Yield = Till | SumCrop | Man | N;
     Random Year Rep(Year)
            Year*Till Year*SumCrop Year*Man Year*N;
     ODS listing exclude dimensions Nobs FitStatistics;
Run;
```

Each Year × main effect interaction has a variance component different from zero (Table 5, Column 3). Also, Rep(Year) and the estimate of the residual are different from zero. However, the main random effect of Year has a variance component equal to zero. Because we used only 2 yr (one degree of freedom), and Rep(Year) has four degrees of freedom, there was not sufficient information (degrees of freedom) to simultaneously model the main effect of Year, Rep(Year), and the four Year × main effect interactions as random effects, in addition to the residual. The four fixed main effects are no longer significant, but the two-factor interactions that were significant in Model 1 (Summer crop × N and Manure × N) remained significant. Why were the four fixed main effects no longer significant? Because we included the Year × main effect interactions as random effects there were substantial changes in denominator degrees of freedom (DDF) for each of the four fixed main effects. In Model 1, all fixed terms were tested using 115 DDF. However, in this second model, the DDF = 1 for tillage, summer crop, and manure (with 2 levels each), and DDF = 2 for N (which had 3 levels). All other fixed terms in Model 2 had 110 DDF.

Model 3: Each interaction of Year × main effect is random and each interaction of Year × two-factor interaction is a random effect. The four agronomic main effects and their interactions are fixed effects.

The resulting SAS code follows:

```
Title2 "Model 3: year, rep(year), year × main effects, and year × two-factor interactions as random";
Proc Mixed Data = FixRnd2 covtest noinfo noitprint;
     Class Year Rep Till SumCrop Man N;
     Model Yield = Till | SumCrop | Man | N;
     Random Year Rep(Year)
                    Year*Till Year*SumCrop Year*Man Year*N
                    Year*Till*SumCrop Year*Till*Man Year*Till*N Year*SumCrop*Man
                    Year*SumCrop*N Year*Man*N;
     ODS listing exclude dimensions Nobs FitStatistics;
Run;
```

The variance components for the interactions of Year with the four main effects of agronomic factors are still different from zero, as are the covariance parameter estimates of Rep(Year) and residual as in the previous model (Table 5, Column 4). However, of the six Year × two-factor interactions, only one was different from zero, Year × Manure × N. When analyzing these interactions as random effects, those that included N were more likely to be different from zero because N had three rates (levels) while each of the other fixed main effects had only two levels. However, from the three interactions of Year with two agronomic factors, when one of those factors was N, (Year × Tillage × N, Year × Summer crop × N, and Year × Manure × N), only Year × Manure × N had a variance component different from zero, and this indicates that this specific interaction was the most informative. This result occurred independently of the order in which the interactions, including N, were listed in the model.

Only two of the two-factor interactions were significant in the first two models, Summer crop × N and Manure × N. The most important results for the fixed terms in the third model were that these interactions were no longer significant. Thus, all the fixed terms in the third model were not significant, including main effects and all of their interactions. The reason for the reduced number of significant fixed effects was that the random effect of Year extracted variability from each fixed term. These extractions resulted in terms with reduced DDF. In Model 2, the two-way, three-way, and four-way interactions had 110 DDF. In Model 3, the two-way interactions were tested using DDF = 1 or DDF = 2, which contributed to their non-significance. In Model 3, the three-way and four-way interactions were tested using 101 DDF.

Model 4: The interactions of Year × main effects, Year × two-way interactions of main effects, and Year × three-way interactions of main effects are random effects, and the remaining four agronomic factors and their interactions are fixed effects.

For this model the SAS code is:

```
Title2 "Model 4: year, rep(year),year × main effects, year × two way,";
Title3 "and year × three way interactions as random";
Proc Mixed Data = FixRnd2 covtest noinfo noitprint;
     Class Year Rep Till SumCrop Man N;
     Model Yield = Till | SumCrop | Man | N;
     Random Year Rep(Year)
            Year*Till Year*SumCrop Year*Man Year*N
            Year*Till*SumCrop Year*Till*Man Year*Till*N Year*SumCrop*Man
            Year*SumCrop*N Year*Man*N
            Year*Till*SumCrop*Man Year*Till*SumCrop*N Year*Till*Man*N
            Year*SumCrop*Man*N;
     ODS listing exclude dimensions Nobs FitStatistics;
Run;
```

The results for this model for the variance components of the random terms, including the interactions of Year with the main effects and with two factors are similar to those obtained from the previous model (Table 5, Column 5). The three factor interactions with Year, Year × Tillage × Summer crop × Manure, Year × Tillage × Summer crop × N, and Year × Tillage × Manure × N had variance components of zero. However, the Year × Summer crop × Manure × N interaction had a variance component different from zero but equal to zero for all practical purposes, considering that it had a variance component of 68, which compared with other variance parameters between 40,000 and 275,000. For all the fixed main effects and their two-factor interactions, the *p* values were similar to those obtained in the previous model. However, the four three-factor interactions now had larger *p* values compared with the previous model because the random Year × three-factor interactions extracted some of the variability associated with the three-factor interactions of fixed effects. Also, in Model 3, the three-way and four-way interactions were tested using 101 DDF, while in Model 4, the three-way interactions were tested using only 1 or 2 DDF and the four-way interaction was tested using 94 DDF. The *p* values of the four-factor interaction Tillage × Summer crop × Manure × N in this and previous models were similar.

Model 5: This is the Saturated Model that Includes All Possible Random and Fixed Terms.

The code describing the final model was:

```
Title2 "Model 5: Saturated model including all possible fixed and random terms";
Proc Mixed Data = FixRnd2 covtest noinfo noitprint;
     Class Year Rep Till SumCrop Man N;
     Model Yield = Till | SumCrop | Man | N;
     Random Year Rep(Year)
            Year*Till Year*SumCrop Year*Man Year*N
            Year*Till*SumCrop Year*Till*Man Year*Till*N Year*SumCrop*Man
            Year*SumCrop*N Year*Man*N
            Year*Till*SumCrop*Man Year*Till*SumCrop*N Year*Till*Man*N
            Year*SumCrop*Man*N
            Year*Till*SumCrop*Man*N;
```

TABLE 5. Results from six mixed models, using only 2 yr (1988 and 1989) in dataset "FixRand2." An empty cell indicates that the term was not included in the corresponding model. Effects were Year (Y), Tillage (T), Summer crop (S), Manure (M), and Nitrogen (N)

Source of variation	Model 1	Model 2	Model 3	Model 4	Model 5	Model 6
	Covariance parameter estimate					*p* values
Rep (Y)	56,421	68,607	71,180	71,181	71,181	(70,512)†
Y	237,357	0	0	0	0	0.0380
Y × T		200,503	206,075	206,077	206,076	<0.0001
Y × S		38,679	40,547	40,524	40,535	0.0275
Y × M		155,883	68,629	68,630	68,629	< 0.0001
Y × N		389,588	259,942	259,944	259,943	< 0.0001
Y × T × S			0	0	0	0.6099
Y × T × M			0	0	0	0.6014
Y × T × N			0	0	0	0.8644
Y × S × M			0	0	0	0.8003
Y × S × N			0	0	0	0.9921
Y × M × N			274,128	274,095	274,095	0.0001
Y × T × S × M				0	0	0.1804
Y × T × S × N				0	0	0.7505
Y × T × M × N				0	0	0.7589
Y × S × M × N				68	68	0.1061
Y × T × S × M × N					0	0.7709
Residual	701,195	398,434	339,449	339,431	339,430	(363,020)
			Type III tests of fixed effects			
Source of variation				*p* values		
T	< 0.0001	0.4118	0.4147	0.4147	0.4147	< 0.0001
S	0.0056	0.3280	0.3286	0.3286	0.3286	0.0002
M	< 0.0001	0.2541	0.2557	0.2557	0.2557	< 0.0001
N	< 0.0001	0.1612	0.1629	0.1629	0.1629	< 0.0001
T × S	0.9384	0.9184	0.9294	0.9294	0.9294	0.9145
T × M	0.3357	0.2024	0.3972	0.3971	0.3971	0.1824
T × N	0.8515	0.7538	0.7504	0.7504	0.7504	0.7335
S × M	0.2802	0.1529	0.3630	0.3632	0.3632	0.1350
S × N	0.0488	0.0055	0.1351	0.1352	0.1352	0.0036
M × N	< 0.0001	<0.0001	0.1997	0.1997	0.1997	< 0.0001
T × S × M	0.6386	0.5334	0.5001	0.6212	0.6212	0.5144
T × S × N	0.7385	0.5872	0.5358	0.6143	0.6143	0.5580
T × M × N	0.3399	0.1520	0.1107	0.3077	0.3077	0.1278
S × M × N	0.4037	0.2048	0.1566	0.3465	0.3465	0.1768
T × S × M × N	0.3655	0.1723	0.1281	0.1285	0.3229	0.1465

† Values in parentheses in Model 6 are covariance parameter estimates.

```
     ODS listing exclude dimensions Nobs FitStatistics;
Run;
```

For the saturated model, the additional random effect was the Year × Tillage × Summer crop × Manure × N interaction and this had a zero variance component (Table 5, Column 6).The significance for the interaction of the fixed four-factor interaction of Tillage × Summer crop × Manure × N diminished compared with Models 2 through 4.

Model 6: Fixed Effects Model.

In this final saturated model, all five main effects (including Year) were fixed factors, only Rep(Year) was random. The SAS code is:

```
Title2 "Model 6: Saturated model, all terms as fixed";
Proc Mixed Data = FixRnd2 covtest noinfo noitprint;
     Class Year Rep Till SumCrop Man N;
     Model Yield = Year | Till | SumCrop | Man | N;
     Random Rep(Year);
     ODS listing exclude dimensions Nobs FitStatistics;
Run;
```

The main effect of Year, the four interactions of Year × each main effect, and the Year × Manure × N interaction were significant (Table 5, Column 7). All of these terms had variance components different from zero in previous models in which they were included. The interaction of Year × Summer crop × Manure × N was not significant in this saturated model. This confirms what we mentioned earlier regarding Models 4 and 5 where this interaction had a variance component of 68 which is different from zero, but for all practical purposes a value of 68 was similar to 0 because it compared with values of other variance parameters between 40,000 and 275,000. The significance results for all significant main effects and interactions were similar for the saturated model and the first model. In the first model, only the Year and Rep(Year) terms were considered as random effects and no Year × fixed effect factor interaction was considered. In mixed models some parameters can represent both variances and covariance and it is possible for covariances (and hence, correlations) to be negative. In this example we preferred to constrain variances to be bounded to a zero value. In the Mixed procedure, the variance components have a default lower boundary constraint of 0, and the NOBOUND option in the MODEL statement requests the removal of boundary constraints on covariance parameters, therefore allowing variance estimates to be negative.

Example 2

Example 2 illustrates results of two models using all 10 yr of data from the experiment analyzed in Vargas et al. (1999) (Table 6).

First Model: Saturated Model, including All Possible Random and Fixed Terms.

In this example, we use the saturated model that includes Year, Rep(Year) and all interactions of Year with the four agronomic factors as random terms, while all four agronomic treatments and their interactions are fixed terms. The SAS code was the same as in the Model 5 from Example 1. The difference between Model 5 from

Example 1 and this model is that we are now analyzing all 10 yr of data compared with only 2 yr of data analyzed in Example 1.

```
Title1 "Evaluate Fixed and Random Effects, Example 2";
Title2 "Saturated model: Including all possible fixed and random terms";
Proc Mixed Data = FixRnd10 covtest noinfo noitprint;
     Class Year Rep Till SumCrop Man N;
     Model Yield = Till | SumCrop | Man | N;
     Random Year Rep(Year)
            Year*Till Year*SumCrop Year*Man Year*N
            Year*Till*SumCrop Year*Till*Man Year*Till*N Year*SumCrop*Man
            Year*SumCrop*N Year*Man*N
            Year*Till*SumCrop*Man Year*Till*SumCrop*N Year*Till*Man*N
            Year*SumCrop*Man*N
            Year*Till*SumCrop*Man*N;
     ODS listing exclude dimensions Nobs FitStatistics;
Run;
```

Of the 16 terms representing random effects, including Year or interactions of Year with fixed main effects, 10 had a variance component different from zero (Table 6, Column 2). This result was not surprising because Year has ten levels. The results of significance for the fixed terms were similar to those obtained with Model 1 of Example 1, which included only 2 yr of data; that is, the four main effects and the two-factor interactions Summer crop × N and Manure × N were significant. Also, the Tillage × Manure × N interaction was significant, while the interactions Tillage × Manure and Tillage × Summer crop × N were almost significant at $p = 0.05$ ($p = 0.0589$ and 0.0501, respectively). The four-factor interaction (Tillage × Summer crop × Manure × N) was also almost significant ($p = 0.0778$).

For situations like these, the researcher could use different strategies to identify the most useful model. For example, one strategy is to consider a model that does not include the random terms found with variance components equal to zero and then to sequentially delete the fixed terms that were not significant in a previous analysis. However, this approach has the potential disadvantage of identifying as significant interactions of higher order that have become confounded (aliased) with interactions of lower order. This can occur because when we delete a lower order interaction (such as a two-factor interaction), the degrees of freedom assigned to a remaining higher order interaction (perhaps a three-way or four-way interaction) may be higher than what actually corresponds to the interaction. In other words, the higher order interaction would include degrees of freedom from at least one lower order interaction that was removed from the model. Thus, by removing a lower order interaction, we have created a confounded or aliased structure between the factors. If this occurs, we recommend interpreting the highest order interactions that are not confounded.

Second Model: In Searching for the Most Useful Model, We Will Include Only the First 10 Lowest Order Random Terms and All Possible Fixed Terms.

Instead of the strategy discussed above, we will review a second strategy, which is to use all main effects and their interactions in our revised model but only the 10 random terms that are of lowest order. As stated above, the reason we chose to use 10 random terms is because Year has 10 levels. Under this strategy, we include Year, the four interactions of Year with the four agronomic factors that are main effects, and then select the remaining five random terms from the six possible Year × two-factor

TABLE 6. Results from two mixed models using all 10 yr of data from Vargas et al. (1999). An empty cell indicates that the term was not included in the model. Fixed effects were Tillage (Till), Summer crop (SumCrop), Manure (Man), and Nitrogen (N).

Covariance parameter	Covariance parameter estimate	
	Model 1	Model 2
Rep(Year)	13,123	12,826
Year	411,641	411,436
Year × Till	56,323	55,659
Year × SumCrop	0	0
Year × Man	67,505	67,415
Year × N	192,038	189,402
Year × Till × SumCrop	737	3487
Year × Till × N	0	2224
Year × Till × Man	0	
Year × SumCrop × Man	0	0
Year × SumCrop × N	54,490	61,204
Year × Man × N	120,237	124,527
Year × Till × SumCrop × Man	0	
Year × Till × SumCrop × N	7394	
Year × Till × Man × N	0	
Year × SumCrop × Man × N	7251	
Year × Till × SumCrop × Man × N	5057	
Residual	241,025	248,107
Type III tests of fixed effects		
Effect	$P > F$	
Till	0.0275	0.0275
SumCrop	0.0020	0.0022
Man	0.0009	0.0009
N	< 0.0001	< 0.0001
Till × SumCrop	0.4377	0.4356
Till × Man	0.0589	0.0284
Till × N	0.2273	0.2001
SumCrop × Man	0.2532	0.2125
SumCrop × N	0.0020	0.0019
Man × N	< 0.0001	< 0.0001
Till × SumCrop × Man	0.4544	0.4272
Till × SumCrop × N	0.0501	0.0140
Till × Man × N	0.0488	0.0252
SumCrop × Man × N	0.7368	0.6871
Till × SumCrop × Man × N	0.0778	0.0482

interactions. Because in the first model, Tillage or Manure was included in each of the three terms that had a variance component equal to zero, we took out the Year × Tillage × Manure interaction that included each of these effects. The SAS code for this model is:

```
Title2 "Second model: Including only year and the nine lowest random
interaction terms and all possible fixed effects";
Proc Mixed Data = FixRnd10 covtest noinfo noitprint;
     Class Year Rep Till SumCrop Man N;
     Model Yield = Till | SumCrop | Man | N;
     Random Year Rep(Year)
            Year*Till Year*SumCrop Year*Man Year*N
            Year*Till*SumCrop Year*Till*N Year*SumCrop*Man
            Year*SumCrop*N Year*Man*N;
     ODS listing exclude dimensions Nobs FitStatistics;
Run;
```

In this output, we found that Year, Rep(Year), and seven other variance components differed from zero (Table 6, Column 3). The results for the fixed effects in this and the previous (saturated) model were similar except that now we have also determined that the two-factor interaction Tillage × Manure and the three-factor interaction Tillage × Summer crop × N and the four-factor interaction Tillage × Summer crop × Manure × N were significant. Based on these enhanced significance values of the fixed effects, this model is more useful than the previous model.

In a next step, readers could delete the fixed terms that were not significant in this model. Another useful exercise for readers would be to generate two orthogonal polynomial contrasts (linear and quadratic) for the three rates of N and then use graphs for interpreting the behavior of the significant interactions of the linear and quadratic responses of N with other factors. The major focus would be on the significant four-factor interaction if using the strategy we demonstrated (Table 6, Column 3) because this was the highest order interaction that was significant and it included all fixed main effects.

Conclusions

It is essential to analyze and interpret all possible interactions between factors in experiments with more than one fixed effect (treatment). In a table, the researcher should report on the significance of all the interactions in the study and explain the nature of the significant interactions in the Results and Discussion section. Then, the researcher should strategize on how best to communicate the results of the study. There are no rules of statistics that mandate how results must be described based on which interactions were significant. The researcher's final explanation may focus on two-way interactions when higher order interactions were significant, or it may focus on main effects when two-way interactions were significant. However, it will always be important to discuss main effects or lower order interactions based on their relationship to the highest order significant interaction. Our most important message to the researcher is to be sure to analyze all meaningful research by interpreting all possible interactions, to account for all meaningful interactions that are statistically significant (this can be done by only focusing on the highest order significant interaction or reporting lower order effects in the context of the highest order significant interaction), and to report meaningful results properly and fully while seeking to do so in a simplified manner. Final reports of results and conclusions should be based on biological knowledge and interpretation of statistical procedures that were properly conducted and interpreted.

When using mixed models for multifactorial data, one should exercise caution when using random effects. A good rule of thumb is to not consider as random more factors than the levels of the main effect being considered as random. Factors considered as random effects should have a large number of levels. When using multifactorial structured data, be careful with the number and order of random × fixed term interactions used. The number of interactions of random terms depends on the number of levels of the random main effect. We recommend using the interactions of the lowest possible order among the random interactions when it is not prudent to use all of the interactions. The above concerns become less important as there are fewer possible interactions and a structure of years, locations, treatments, etc., each with an increasing number of levels.

Key Learning Points

- Understand the importance of analyzing interactions of fixed effects.
- Learn to interpret significant interactions.
- Learn when it is appropriate to interpret mean differences of main effects or when mean differences among interactions should be analyzed instead.
- Learn how to use the Contrast statement in GLIMMIX procedure of SAS to analyze contrasts involving linear, quadratic, cubic, and higher order regression responses to a treatment variable.
- Learn (included in Appendices in the supplemental material) how to calculate the LSD value using the GLIMMIX procedure in SAS, and learn how to use results from the Pdiff to calculate the LSD option. Identify a strategy for when results of the Pdiff option differ from the results of the calculated LSD.
- Learn how to interpret interactions involving orthogonal polynomial trend contrast responses.
- For an effect that can also be analyzed as a random effect (e.g., an effect such as Year), learn how to analyze and interpret the results as a fixed effect and the care that is needed in explaining the results.
- Learn about the importance of numbers of levels of random effects and considerations in deciding which random interactions to include in a model.

Review Questions

Answers are provided in Appendix 3 of the supplemental material online.

True or False

1. When the researcher can identify predetermined contrasts, then these and not the LSD should be used for mean separation.

2. It is not important whether you include all of the interactions in your model, your F test result for each effect will be identical in the model with and without the interactions. For example, if a study includes genotypes and compost rates, we can ignore the interaction of Genotype × Compost and each *p* value for Genotype and Compost will be identical to each *p* value for Genotype and Compost in the analysis that includes the Genotype × Compost interaction.
3. In a study with three factors, the initial analysis must include main effects, two-way interactions, and the three-way interaction. If the three-way interaction is significant, then the researcher cannot properly discuss any main effect or two-way interaction.
4. It is never appropriate to speculate about what causes a significant effect.
5. A researcher conducts an experiment in each of 2 yr. When analyzing such an experiment, it is always necessary that Year be considered as a random effect.
6. The more levels a researcher has for an effect that is normally considered as a random effect (such as Year or Location), the more likely it would be useful to analyze it as a random effect.
7. It is not important when making inferences whether an effect is fixed or random.

Exercises

Code and output for exercises are provided in Appendix 3 of the supplemental material online.

1. Using the data (Wheat) from Section 1 of this chapter, construct a SAS program that calculates the *F* values for the main effects, two-way, three-way, and the four-way interactions as well as the single degree of freedom contrasts for the linear, quadratic, and cubic responses of GY to P for all effects involving P. Rather than using the method that we provided in Appendices 1 and 2 of the supplement material that automatically calculates the coefficients, in this exercise we ask that you insert the actual coefficients into each line of code. Also, although we showed earlier that it is advantageous to use a reduced model that includes only the interactions that were significant in the complete model, we ask that you provide code for calculating all contrasts, including those of the four-way interaction, as a means of practicing how to develop these Contrast statements.

2. Using the data (Wheat) from this chapter, calculate the Mean Squared Error (MSE) for Rep(Year Soil) for the complete model using two approaches. Then, using the residual variance (which is another name for MSE) or the MSE of Rep(Year Soil) if appropriate, calculate the LSD for all effects and interactions.

Acknowledgments

We thank the researchers that performed all the field trials analyzed in this article.

References

Carmer, S.G. 1976. Optimal significance levels for application of the least significant difference in crop performance trials. Crop Sci. 16:95–99. doi:10.2135/cropsci1976.0011183X001600010024x

Carmer, S.G., and M.R. Swanson. 1971. Detection of differences between means: A Monte Carlo study of five pairwise comparison procedures. Agron. J. 63:940–945. doi:10.2134/agronj1971.00021962006300060036x

Carmer, S.G., and M.R. Swanson. 1973. An evaluation of ten pairwise multiple comparison procedures by Monte Carlo methods. J. Am. Stat. Assoc. 68:66–74. doi:10.1080/01621459.1973.10481335

Dixon, P., K. Moore, and E. van Santen. 2018. The analysis of combined experiments. In: B. Glaz and K.M. Yeater, editors, Applied statistics in agricultural, biological, and environmental sciences. ASA, CSSA, SSSA, Madison, WI.

Kuehl, R.O. 1999. Design of experiments: Statistical principles of research design and analysis. 2nd ed. Duxbury Press/Brooks-Cole, Belmont, CA.

Little, T.M. 1978. If Galileo published in HortScience. HortScience 13:504–506.

McIntosh, M. 1983. Analysis of combined experiments. Agron. J. 75:153–155. doi:10.2134/agronj1983.00021962007500010041x

Moore, K.J., and P.M. Dixon. 2015. Analysis of combined experiments revisited. Agron. J. 107:763–771. doi:10.2134/agronj13.0485

Saville, D. 2015. Multiple comparison procedures—Cutting the Gordian knot. Agron. J. 107:730–735. doi:10.2134/agronj2012.0394

Saville, D. 2018. Multiple comparison procedures: The ins and outs. In: B. Glaz and K.M. Yeater, editors, Applied statistics in agricultural, biological, and environmental sciences. ASA, CSSA, SSSA, Madison, WI. doi:10.2134/appliedstatistics.2015.0085

Steel, R.G.D., and J.H. Torrie. 1980. Principles and procedures of statistics. A biometrical approach. 2nd ed. McGraw-Hill, New York.

Stroup, W.W., and D.K. Mulitze. 1991. Nearest neighbor adjusted best linear unbiased prediction. Am. Stat. 45:194–200.

Swallow, W.H. 1984. Those overworked and oft-misused mean separation procedures—Duncan's, LSD, etc. Plant Dis. 68:919–921.

Vargas, M., J. Crossa, F.A. van Eeuwijk, M.E. Ramirez, and K. Sayre. 1999. Using partial least squares regression, factorial regression, and AMMI models for interpreting genotype × environment interaction. Crop Sci. 39:955–967. doi:10.2135/cropsci1999.0011183X003900040002x

Vargas, M., B. Glaz, G. Alvarado, J. Pietragalla, A. Morgounov, Y. Zelenskiy, and J. Crossa. 2015. Analyses and Interpretation of Interactions in Agricultural Research. Agron. J. 107:748–762. doi:10.2134/agronj13.0405

CHAPTER 8: THE ANALYSIS OF COMBINED EXPERIMENTS

Philip M. Dixon,* Kenneth J. Moore, and Edzard van Santen

Agronomic experiments are often replicated over time (e.g., seasons or years), space (e.g., locations), or both to evaluate treatments across multiple environments; the subsequent statistical analysis is commonly called the analysis of combined experiments. The details of the analysis depend on the desired scope of inference and the model for treatment effects. Narrow-sense inferences are conclusions about the years, locations, and treatments included in the experiment. In a narrow-sense analysis, the environment × treatment interaction is modeled as a fixed effect. Broad sense inferences make conclusions about treatment effects in larger population of years and locations. In this analysis, the treatment × environment interaction is modeled as a random effect. Treatment effects can be modeled as a factor; this provides estimates of treatment means or linear contrasts among those means. Or, treatments can be considered as levels of a continuous variable; this estimates a response curve. In this chapter we provide guidance on how to choose among the various models and provide details for both narrow- and broad-sense inference. We use a meat triglyceride study and an oat (*Avena sativa* L.) cultivar trial to illustrate the analysis of group means and a fertilizer response trial to illustrate the analysis of a response curve. Finally, we discuss heterogeneity and pooling of variances.

Agronomic experiments are often repeated in multiple growing seasons and/or multiple geographic locations to expand the scope of the conclusions. Collectively, times, locations, or their combinations will be called environments. Experiments repeated in multiple environments come in two slightly different forms with slightly different goals. One form, which we will call an agronomic practice experiment, has the goal of understanding treatment effects across a range of environments. This provides information about both the treatment effects averaged over a broad range of environments and the consistency of effects over such environments (Blouin et al., 2011). Analysis of this form of study is often called the combined analysis of multiple experiments. The second form of multienvironment study, commonly done by plant breeders, evaluates the performance of many genotypes across multiple environments. This type of study often has the goal of understanding details of the interaction between geno-

Abbreviations: AIC, Akaike Information Criterion; EMS, Expected Mean Square; LR, Likelihood ratio; REML, Restricted maximum likelihood; RCBD, Randomized complete block design.

P.M. Dixon, Department of Statistics, Iowa State University, IA 50011; K.J. Moore, Department of Agronomy, Iowa State University, IA 50011; E. van Santen, IFAS Statistical Consulting Unit and Department of Agronomy, University of Florida, FL 32611. *Corresponding author (pdixon@iastate.edu)

doi:10.2134/appliedstatistics.2016.0004

Applied Statistics in Agricultural, Biological, and Environmental Sciences
Barry Glaz and Kathleen M. Yeater, editors

types and environments, for which specialized models, such as the AMMI model (see e.g., Gauch 2006), are frequently used. In this chapter, we only discuss agronomic practice experiments and the combined analysis of multiple experiments. We only consider analyses of the plot-level data from all environments and do not consider two-stage analyses (Piepho et al., 2012) or meta-analysis (Koricheva et al., 2013).

A combined analysis of multiple experiments extracts more information from a given body of collected data than can be gained from a series of experiment-specific analyses. For example, Thompson et al. (1993), describe the additional information provided by a combined analysis of 12 grazing studies conducted at nine locations over a period of 13 yr: "The mixed models procedure permitted estimation of the fixed effects of treatments over a broad inference space of future years and different tall fescue pastures over a wide geographic range, detected relationships that had not been apparent in the individual studies, such as the interactions between clover presence and E+ infestation levels, and provided a more coherent body of information than did the results obtained from each discrete study."

One of the first instances of analysis of a combined experiment, actually a multi-environment plant breeding trial, is described in an exchange of letters between the USDA geneticist R.F. Immer and R.A. Fisher about the analysis of a six-location nine-year barley yield trial (Wright, 2013). Fisher (1935, p. 211–215) discusses the concepts of a combined experiment. He specifically contrasts separate analyses for each location and the combined analysis of all locations. Cochran and Cox (1957) in Chapter 14 of their classic textbook on experimental design describe the analysis of combined experiments, mentioning three rationales for conducting an analysis of a series of experiments, viz. planned recommendations for an "extensive population in space and/or time", investigating the effect of "external conditions on some measurement", and testing interlaboratory agreement of an assay. Laird and Cady (1969) provided one of the first examples for the analysis of a large-scale combined experiment that involved 76 locations. Laird and Cady (1969) discussed the problem of arriving at the proper error term when conducting a combined analysis, foreshadowing comments by Binns et al. (1983) regarding the nature of random error components in combined experiments in their response to the seminal paper by McIntosh (1983).

McIntosh (1983), to our knowledge, was the first to pull together in an easily accessible form the information a researcher needs to conduct an analysis of experiments across environments. She describes the analyses of combined experiments based on the constraint that the effects of random interactions, for example treatment by year interactions, sum to zero over each level of a random factor, such as year. This assumption reflects a tradition in the analysis of fixed effects that, in most cases, is not appropriate for a random effect (Hocking 1985, p. 332–334). Moore and Dixon (2015) describe the analysis of combined experiments under the assumption that all random effects are mutually independent, which many, such as Hocking (1985, p 332) and McLean et al. (1991), consider more reasonable.

In 1983, virtually all researchers in agronomy and crop science, except for quantitative geneticists, were operating in a fixed-effects-only environment, that is, the only random effect term was the residual; all other effects in the model, even blocks, were fixed. This is not to imply that 'modern approaches' didn't exist, just that researchers were not utilizing them. Applied researchers in those fields were not aware of the contributions of Yates (1937) and Nelder (1965a, b) and if they had been, the computational resources needed would not have been available to them. The state of the art data analysis in the 1980s was described as "Making ordinary least squares the 'usual case'- essentially treating the mixed model as something of a curiosity of occasional importance– creates a number of difficulties in data analysis, some unrecognized or not fully appreciated, others ignored" (Stroup, 1989). At the heart of the matter, as Stroup (1989) pointed out, is that real-world error structures are more complex than the simple models assumed under fixed effects models. There were notable exceptions to the widespread use of fixed effects models, for example by researchers supported by the Scottish Agricultural Statistics Service after the implementation of restricted maximum likelihood (REML) in the 1970s (Robinson et al., 1982), but most agronomic researchers did not have this tool available to them.

Developments in statistical computing over the last 50 yr have enabled researchers to use mixed models procedures on a regular basis. Two key developments were the development of the Wilkinson algorithm for balanced multistratum problems (Wilkinson, 1970), implemented in the early 1970s in the GenStat package, and the REML algorithm (Patterson and Thompson, 1971), which enabled the analysis of unbalanced data. The credit for initiating a generally applicable software application in the United States goes to the Southern Regional Project S-189 entitled "Statistical computing methodology for research planning and analysis", a project of the University Statisticians of Southern Experiment Stations, commonly known as USSES, a group that since its inception in 1962 has had a successful career in providing information for appropriate statistical analyses of agronomic data. SAS PROC MIXED has its roots in the cooperation between USSES and the SAS Institute. Giesbrecht (1989) pointed out "The initial goal in the development of MIXMOD [the progenitor of PROC MIXED] was to create a relatively general tool that researchers could use to estimate variance components in unbalanced datasets. The emphasis has always been to preserve generality while achieving reasonable computational efficiency." The Statistics Department at Rothamsted Experiment Station in England played a similar role in the late 1960s with the development of the GenStat package. Subsequently, contributions such as (Piepho et al., 2003; Piepho et al., 2004; Piepho et al., 2006) have provided practical road maps for the use of mixed model methodology in the analysis of agronomic data.

It is understood that the preceding discussion should not be construed as a rigorous and exhaustive historical treatise of worldwide efforts in the development of practical approaches to mixed modeling. It only functions in setting the stage for the discussion following.

In this chapter, we describe two models commonly used for the analysis of combined experiments and elaborate on some practical details. These include deciding whether interactions should be fixed or random, choosing whether to subdivide the treatment by environment interaction, and using the data to decide whether or not to assume equal residual variances for all environments. The most important SAS code is provided in boxes in the paper; additional SAS code is in the Supplemental Material. R code is provided in the Supplemental Material.

A Linear Model for Qualitative Treatments in a Combined Experiment

We consider an experiment with qualitative treatments, for example different forms of tillage or weed management strategies, conducted in a randomized complete block design with each treatment occurring once per block. If this experiment is conducted in one environment, a model commonly used to evaluate differences among treatments is:

$$Y_{jk} = \mu + \beta_j + \tau_k + \varepsilon_{jk} \qquad [1]$$

where:

Y_{jk} is the response measured on the plot with treatment k in block j,

β_j is the effect of block j,

τ_k is the effect of treatment k, and

ε_{jk} is the experimental error for plot jk.

We use the common model for a randomized complete block design (RCBD), because RCBDs are very commonly used in agronomic investigations. The model in Eq. [1] can be altered in standard ways when some other field design (e.g., completely randomized, Latin square, lattice) is used.

Commonly, the goal of the study is to estimate and describe differences between treatment means or linear contrasts among treatment means. To simplify descriptions in the rest of the chapter, we refer to treatment differences, but everything said about treatment differences applies more generally and equally well to linear contrasts among treatment means. When treatment differences are the focus of a study, the overall mean and the treatment effects are modeled as fixed effects and the experimental errors are modeled as random effects. Blocks are sometimes modeled as random effects and sometimes modeled as fixed effects. The consequences of this choice are described by Dixon (2016) and summarized in the next section.

When this experiment is repeated in multiple environments, for example multiple years, multiple locations, or both, Eq. [1] is extended to account for variability among environments. A commonly used model for such an experiment is:

$$Y_{ijk} = \mu + E_i + \beta_{j(i)} + \tau_k + E\tau_{ik} + \varepsilon_{ijk} \qquad [2]$$

where:

Y_{ijk} = response measured on the plot with treatment k in block j in environment i,

μ = overall mean,

E_i = effect of environment i,

β_{ij} = effect of block j in environment i,

τ_k = effect of treatment k,

$E\tau_{ik}$ = interaction effect for treatment k in environment i, and

ε_{ijk} = experimental error for plot jk in environment i.

Again, the overall mean, μ, and the treatment effects, τ_k, are commonly modeled as fixed effects and the experimental error, ε_{ijk}, is modeled as a random effect. Each of the other terms in Eq. [2] may be modeled either as a fixed or random effect. The choice for environment and block effects can change some results, but often has no influence on the conclusions about treatment differences (see next section). However, the choice (fixed or random) for the environment by treatment interaction, $E\tau_{ik}$, is crucial, so that choice will be discussed separately.

The Choice of Fixed or Random Effect in General

Most of the terms in either Eq. [1], for a single-environment study, or Eq. [2], for a multi-environment study, may be modeled as fixed effects or random effects. Gelman (2005) summarizes five criteria that have been proposed to decide whether an effect should be modeled as fixed or as random. Of these, the most common is whether the effects in the model are a random sample from some probability distribution (e.g., Searle, 1971; p. 392; McCulloch, Searle and Neuhaus 2008, p. 16; Robinson 1991). Sometimes, the random sample part is dropped, for example, "do these effects come from a probability distribution"? (Robinson 1991). Sometimes, the criterion is elaborated, for example, "the context of the data, the manner in which they were gathered, and the environment from which they came" (Searle 1971, p. 382).

A clear answer to the question of fixed or random may not be available (McCulloch, Searle, and Neuhaus, 2008, p. 17). For example, consider a hypothetical agronomic field experiment. Because the study field is expected to be heterogeneous, and that heterogeneity is expected to be spatially organized, the field is divided into blocks of (usually) contiguous plots. To simplify the description, consider a study conducted in a 0.5 ha field divided into five blocks. Each block is 0.1 ha in size and contains four plots. Should block effects be modeled as a fixed effect or as a random effect? In the absence of a treatment, two plots in the same block are expected to be more similar to each other than are two plots in different blocks. This suggests a random effect model. In incomplete-block or row-column designs, block effects are almost always considered random to recover interblock information.

If the 0.1-ha blocks in our hypothetical study are considered as a random effect, it is not clear what population those random effects are sampled from. The blocks are either a complete enumeration of all five 0.1-ha parts of the 0.5-ha study field or they are a sample from some population of 0.1-ha size patches of land. If the latter, neither the population nor the mechanism used to draw the sample of blocks is known. The sampling issue is even more serious when blocks are chosen as internally homogeneous areas of a field, as suggested by Stroup (2013, p. 448). Similar issues apply to other sorts of blocks, such as groupings of animals based on initial weight (Dixon, 2016). Dixon (2016) argues that blocks should be considered a fixed effect, unless there are specific reasons to consider them random. Others (e.g., Casler, 2018 in Chapter 3) disagree.

For many study designs, the choice of random or fixed for block or environment effects changes statistical results about treatment means (Dixon, 2016) but has no consequence for statistical results about differences among treatment means (Giesbrecht and Gumpertz, 2004, p. 86). A very common study design with this property is the RCBD in which each treatment occurs once in each block. It is also true for generalizations where the columns of the design matrix for blocks are orthogonal to those for treatments. However, the interpretation of the statistical results is not the same. When block effects are considered fixed, the interpretation describes differences on these specific blocks. When blocks are considered random, the interpretation is about differences in the wider population. Although we refer frequently to differences among treatment means, all of our statements about those differences apply equally well to linear contrasts among treatment means.

The Choice of Fixed or Random Effect for the Environment by Treatment Interaction
The choice of fixed or random for the environment by treatment interaction effects, $E\tau_{ik}$, has major consequences for inferences on differences between treatment means (Newman et al., 1997). These consequences can be seen in two ways: by computing the variance of differences between two treatment means or by computing the Expected Mean Square (EMS) for the treatment effect (Table 1). The model used for these computations is Eq. [2]. To simplify the presentation, both experiments and blocks within experiments are considered fixed effects and we presume that the number of blocks, J, is the same in each of the I experiments. When blocks or environments and blocks are random, there are additional terms in the EMS or in the variance of a treatment difference, but the fundamental principles described here still hold. When the number of blocks is not the same in all experiments, the expressions for the EMSs and the variance of the treatment difference do not simplify as nicely as they do with equal numbers of blocks, but again, the fundamental principles still hold.

The quantities relevant to both the EMS and the variance of a treatment difference are the variance of the experimental errors, σ^2_{error}, and the variance of the interaction, σ^2_{ET}. If the environment by treatment interaction, $E\tau_{ik}$, is a random effect, the random values of that interaction effect are assumed to have the variance σ^2_{ET}, be mutually independent of each other, and be independent of the random experimental errors. That is, we do not impose sum-to-zero constraints, for example, for values of $E\tau_{ik}$, within a treatment.

When the environment by treatment interaction is considered fixed, the variance of a treatment difference depends only on the error variance and the total number of plots used for that treatment. Since there is one plot per treatment in each block of each environment, each treatment occurs on J plots in each experiment, so the total number of plots is IJ. For this study design, the only random effect in the Expected Mean Square for Treatments is the error variance, so the appropriate denominator for an F test of no differences among treatment means is the mean square that estimates σ^2_{error}, that is, MS error.

When the environment by treatment interaction is considered random, the variance of a treatment difference includes a term that depends on the size of the environment by treatment interaction. In Eq. [2], the environment by treatment effect is considered to have one value for each combination of environment and treatment. That value is shared by all J plots for one treatment in one environment. The treatment mean is an average over IJ plots but only I environment by treatment effects, so the contribution of the environment by treatment interaction to the variance of the treatment mean is σ^2_{ET}/I. From Table 1, we see that considering the environment by treatment interaction as a random effect increases the variance of the treatment difference. The magnitude of that increase depends on the size of the environment by treatment interaction, quantified by σ^2_{ET}, and the number of environments in the study.

The choice of model for the environment by treatment interaction corresponds to the choice of an inference space (McLean et al., 1991). McLean et al. (1991) distinguish between three inference spaces: narrow sense, intermediate sense, and broad sense. In a multienvironment study, narrow-sense conclusions describe average treatment effects in the environments used in the study. To obtain these, the environment by treatment interaction is modeled as a fixed effect. Broad-sense conclusions describe average treatment effects in new environments. These new environments are assumed to be a random sample from a population of potential environments. To obtain broad-sense inferences, environments, blocks, and the environment by treatment interaction are modeled as random effects. The environment by treatment interaction quantifies the consistency of treatment effects across multiple environments. The precision of broad-sense conclusions about average treatment effects depends on the magnitude of the environment by treatment interaction; the precision of narrow-sense conclusions does not.

McLean et al. (1991) use intermediate-sense inference to describe conclusions from a model where the main effect of environment is fixed but the environment by treatment interaction is random. In a balanced design, the difference between intermediate- and broad-sense does not influence the variance of the treatment difference. It only influences the variance of the treatment mean. Because the focus of an agronomic practice experiment is on the difference between treatments, we will use broad-sense inference as defined by McLean et al. (1991) in our examples. Modeling environments as a fixed effect, so using intermediate-sense inference, will lead to different results for treatment means but the same, or very similar, results for treatment differences.

Example: Triglyceride Measurements in Meat

These points are illustrated by a small repeated experiment on two methods to measure triglyceride concentrations in meat. The data are in the Supplemental Material. On day one of the study, four pieces of meat are randomly assigned to one of two methods for chemical extraction and processing. Each piece of meat is measured once resulting in two measurements of triglyceride concentration for each method. This is repeated for a total of four days, giving a total of 16 observations. We are considering days as independent repetitions of a basic experiment; one single day of the study is a valid experiment, with replicated randomization of treatments to pieces of meat. Although you could consider days to be blocks, we prefer to consider them as separate environments, because data for each day could be analyzed separately.

TABLE 1. Consequences of the choice of fixed effect or random effect for the environment by treatment interaction. The variance of the experimental errors is σ^2_{error}. $Q(T_k)$ indicates a quadratic function of the treatment effects. When the environment by treatment interaction is random, the random values of that interaction effect are assumed to have variance σ^2_{ET}. Sample sizes are *I*: number of environments and *J*: number of blocks per environment.

	Fixed	Random
Variance of a treatment difference, $\bar{Y}_{..k}-\bar{Y}_{..l'}$, averaged over I environments and J plots	$\frac{2\sigma^2_{error}}{IJ}$	$\frac{2\sigma^2_{ET}}{I}+\frac{2\sigma^2_{error}}{IJ}$
Expected Mean Square for Treatments	$Q(T_k)+\sigma^2_{\text{error}}$	$Q(T_k)+J\sigma^2_{\text{ET}}+\sigma^2_{\text{error}}$
Denominator MS to test H_0: $Q(T_k)=0$	MS Error	MS Expt × Trt

TABLE 2. ANOVA table for the triglyceride measurement data, with Expected Mean Squares (E MS) when day and the day*method interaction are considered a fixed effect and when both are considered a random effect. In both cases, the main effect of treatment is considered a fixed effect. The Q() notation indicates a quadratic function of the specified model coefficients.

Source	Degrees of freedom	Mean square	E MS: fixed	E MS: random
day	3	143.8	$\sigma_e^2+Q(d,dm)$	$\sigma_e^2+2\sigma_{dm}^2+4\sigma_d^2$
method	1	406.0	$\sigma_e^2+Q(m,dm)$	$\sigma_e^2+2\sigma_{dm}^2+Q(m)$
day*method	3	36.1	$\sigma_e^2+Q(dm)$	$\sigma_e^2+2\sigma_{dm}^2$
error	8	14.4	σ_e^2	σ_e^2

Within each day, there are no additional restrictions on randomization, so an appropriate model (Eq. [3]) is a simplification of Eq. [2] without block effects.

$$Y_{ijk}=\mu+d_i+m_j+dm_{ij}+\varepsilon_{ijk} \quad [3]$$

where d_i is the effect of day i, m_j is the effect of method j, and dm_{ij} is the interaction effect for day i and method j. The random error, ε_{jik}, is associated with a single measurement on an individual piece of meat.

A dot plot of the data (Fig. 1) shows the major patterns in the data:

1) the average measurement on Day 2 is lower than the average on other days,
2) the two methods have approximately similar measurements on Day 1,
3) on subsequent days, measurements for method A are substantially larger than those for method B, and
4) the average difference between the two methods is largest on Days 2 and 3.

The partial ANOVA table for these data, with degrees of freedom and Mean Squares, is given in Table 2. Table 2 also includes the Expected Mean Squares when the interaction is modeled as a fixed effect and when it is modeled as a random effect.

If we want to make conclusions about the average difference between methods A and B on these 4 d, we should use narrow-sense inference, so the day × method interaction is modeled as a fixed effect. The appropriate F test for the difference between the two methods is the ratio F = MS(method)/MS(error) = 28.16, which has a central F distribution with 1, 8 degrees of freedom under the null hypothesis of no difference between the two methods. This gives a p-value for the test of that null hypothesis of 0.0007. Alternatively, the standard error of the mean difference between the two methods is estimated as $\sqrt{2\left[\frac{14.4}{8}\right]}=\sqrt{3.60}=1.90$. A confidence interval for the mean difference will use a t statistic with the error degrees of freedom, here 8. The resulting 95% confidence interval for the mean difference is (5.70, 14.45). The SAS code to produce these results is in Box 1. R code to fit the same model is in the Supplemental Material.

We might instead believe that the true difference between the two treatments could differ from day to day, for many subject-specific reasons. Narrow-sense inference describes the treatment effects for the four days used in the study, but when there is day-to-day variation in the treatment difference, the results are unlikely to be repeatable on a new set of four days. Broad-sense inference makes conclusions about the average treatment difference on four new days, assumed to be randomly chosen from a population of potential study days. The appropriate F test for the difference between the two methods is the ratio F = MS(method)/MS(day*method), based on inspection of the expected mean squares in Table 2. Under the null hypothesis of no difference, this ratio has a central F distribution with 1, 3 degrees of freedom because the mean square for the interaction has three degrees of freedom. The F statistic is 11.25, which gives a p-value of 0.044. A confidence interval for the mean difference will use a t statistic with the degrees of freedom (here 3). The resulting 95% confidence interval for the mean difference is (0.52, 19.63).

SAS code to compute the broad-sense results is in Box 2. To use this code, replace the proc mixed step in Box 1 with the code in Box 2. The SAS code code in Box 2 uses proc mixed with REML estimates of the variance components (the default) and the

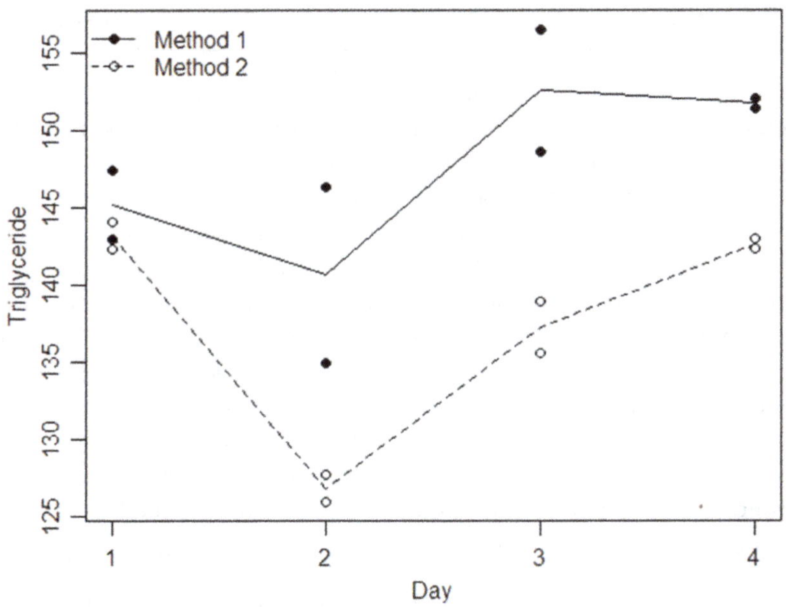

FIG. 1. Plot of the measured triglyceride concentration, as measured by two methods (A or B) on four days. Lines connect method-specific averages across the days.

BOX 1.

```
data trigly;
  infile 'triglybook.txt' firstobs=2;
  input method day y;
  run;

proc sgplot data=trigly;
  scatter x=day y=y /group=method;
  run;

proc mixed method=type3 data=trigly;
  class day method;
  model y = day method method*day;
  lsmeans method / diff cl;
  title 'Triglyceride example: Narrow sense inference';
  run;
```

BOX 2.

```
proc mixed data=trigly;
  class method day;
  model y = method /ddfm=satterth;
  random day method*day;
  lsmeans method /diff cl;
  title 'Triglyceride example: Broad sense inference';
  run;
```

TABLE 3. Comparison of broad- and narrow-sense inference for the mean difference between two measurement methods.

Inference about	Inference space	Number of days	Estimated difference	Standard error	Degrees of freedom	95% confidence interval
The 4 days in the study	narrow	4	10.08	1.90	8	(5.70, 14.45)
Four randomly chosen new days	broad	4	10.08	3.00	3	(0.52, 19.63)

generalized Satterthwaite approximation (Giesbrecht and Burns, 1985) for degrees of freedom. For this data set and design, other analysis choices, including method-of-moments estimates of the variance components (proc mixed method = type3), Kenward–Roger degrees of freedom adjustment, or use of proc glimmix, give identical results.

Table 3 compares the results from narrow- and broad-sense inference. The estimated differences are the same, but the broad-sense standard error of the difference is much larger. This reflects the additional uncertainty arising from apparent day-to-day variation in the treatment difference. The broad-sense 95% confidence interval is much wider because of both the larger standard error and the smaller degrees of freedom.

Should you use broad- or narrow-sense inference? The change from narrow- to broad-sense inference is the change from making a conclusion about the four days used in the study to making a conclusion about four new days randomly chosen from an estimated population of days. If we believe that the true difference between the two treatments on day *j*, that is, $(m_A+\mathrm{dm}_{\mathrm{A}j})-(m_B+\mathrm{dm}_{\mathrm{B}j})$, is the same for each day, then the choice of days on which measurements are made is irrelevant. Narrow-sense conclusions are sufficient to describe the treatment difference. In fact, repeating the experiment on multiple days is unnecessary, except perhaps for practical reasons. If you want eight replicates of each treatment, but you can only process four measurements in a day, or a field only has room for four plots, you need to use multiple days or multiple fields.

When you have replicates on the same day or from the same field, it is possible to test for a non-zero interaction. For the meat data, the *F* test of the day by method interaction gives an *F* statistic of 2.5 and a *p*-value of 0.13. However, we suggest this test not be used to decide on narrow- or broad-sense inference because the test is answering the wrong question. A nonsignificant interaction does not imply that there is no interaction (Cox, 1958, p. 103). Claiming that the true treatment difference is the same each day is trying to use a statistical test to demonstrate the null hypothesis. Interaction tests in a factorial ANOVA have much lower power than do tests of main effects (Pearce, 1988). The power to detect a non-zero interaction can be increased by using a relaxed α level, such as 0.20, as suggested by Sulc et al. (2001)

and Edwards (1985, p. 263). This doesn't change the fundamental logical problem: that you are trying to prove the null hypothesis.

We suggest that the choice of broad- or narrow-sense inference be made based on study goals, not on characteristics of the data. Are you interested in describing what happened on the specific days used in the study? If so, narrow-sense inference is appropriate. The uncertainty in your results describes what could have happened if treatments had been assigned differently to the various replicates. Or, are you interested in describing what might happen if the study is repeated on new days? Now, broad-sense inference is appropriate. The uncertainty in your results describes what could have happened if treatments had been assigned differently to the various replicates *and* the study had been performed on a new set of days. This approach makes intuitive sense, but it calls for thinking about the goal(s) of the study at hand, not blindly following other examples.

Example: Oat Cultivar Trial

A more involved example is the analysis of an oat cultivar trial conducted in two years at three locations in Iowa. This study compared harvest index (HI) of 10 oat cultivars. Since the study goal is to compare mean HI among these cultivars, cultivar will be considered as a fixed effect. In each year and location, the experiment was conducted in a randomized complete block design with three blocks. There were no missing observations. This study is repeated in both space (Location) and time (Year). The analysis described in this section will consider the six combinations of location and year as six separate environments. The discussion of pooling in the next section describes other models for locations and years. Figure 2 shows the average Harvest Index for three cultivars in each of the six environments.

The initial analyses use Model 2 (Eq. [2]), with additive effects for environment, block within environment, and treatment, an interaction between environment and treatment, and an additive residual error. Because each cultivar occurs only once in each block, the plot-to-plot variation and the interaction between blocks and cultivars are combined into the error variation. This is a classic example of confounding two effects; we know both are present but they cannot be separately estimated from these data.

When narrow-sense inference, that is, inference about cultivar differences averaged over the six environments, is desired, the analysis requires only one random component, the residual error. Blocks can be included as a random effect if desired, but statistical results about treatment differences are unchanged because the design is orthogonal. The F test for cultivars has the error MS as the denominator; confidence intervals for differences or linear combinations among cultivar means use the error variance. SAS code to read and plot the data and fit the model for narrow-sense inference is in Box 3. In this code, blocks are considered a random effect nested in environments (the Environment variable); method = type3 is used to obtain the Mean Squares for the ANOVA table. Using REML estimates of the variance components provides the same F tests and p-values so long as the nobound option is used to allow a negative estimate of the block variance. The three estimate statements illustrate estimates that are often of interest: a cultivar mean averaged over environments, a difference between two cultivar means averaged over environments, and a difference between two cultivar means in a single environment.

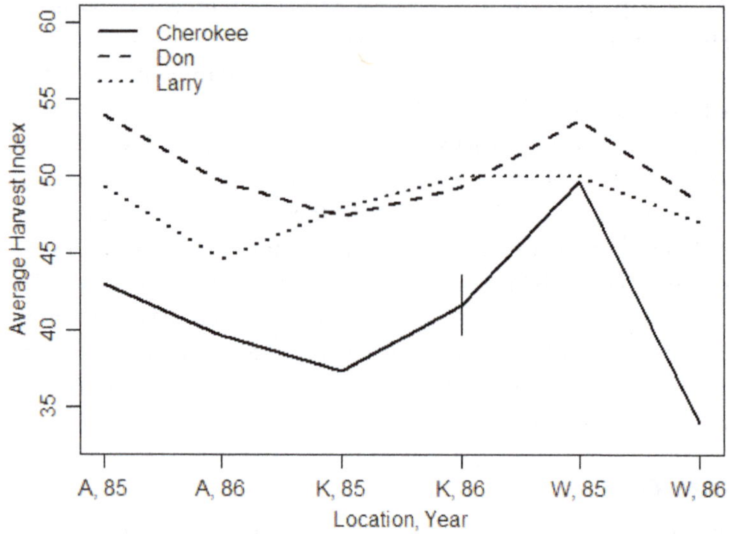

FIG. 2. Harvest index means for three varieties at each location and year, demonstrating the variety × environment interaction. Vertical bar shows + / - 1 se for the average harvest index for one variety and environment.

BOX 3.

```
proc import datafile='oats.csv' out=oats0 replace;
  guessingrows = 100;
  run;

data oats;
  set oats0;
  Environment = trim(Location) || ', ' || put(Year, 4.);
  run;

proc sort data=oats;
  by Environment;
  run;

proc sgplot data=oats;
  scatter x=Cultivar y=HI / group = Environment;
  title ;
  run;

proc mixed method=type3 data=oats;
  class Environment Block Cultivar;
  model HI = Environment Cultivar Environment*Cultivar /ddfm=kr;
  random Block(Environment);
  estimate 'Cultivar: Cherokee' intercept 1 Cultivar 1;
  estimate 'Cultivar: Cherokee - Don' Cultivar 1 -1;
  estimate 'Cult: Cherokee - Don in Ames 1985' Cultivar 1 -1
Environment*Cultivar 1 -1;
  lsmeans Environment*Cultivar /slice = Cultivar slice=Environment;
  title 'Oat HI: Narrow sense inference, random blocks';
  run;
```

Results of this analysis indicate a significant interaction between environment (year and location) and cultivar (Table 4). Two ways to understand this interaction are to plot the cultivar means in each environment and testing differences among cultivars in each environment. A plot of average Harvest Index for three cultivars across environments shows that some cultivars have similar means in all six environments, while other cultivars vary considerably across the environments (Fig. 2). This is confirmed by testing differences among environments within each cultivar. In SAS, this is accomplished by slicing the interaction by cultivar (Environment*Cultivar/slice = Cultivar). The table of cell means (i.e., means for each combination of environment and cultivar) is "sliced" into cultivar-specific components. The result is a collection of F tests, one for each cultivar, that evaluate differences among environments for that cultivar. Some cultivars (e.g., Larry) showed no evidence of difference in mean HI over the six environments, some (e.g., Don) showed weak evidence of difference, and others (e.g., Cherokee) performed differently in at least one environment (Table 5).

Testing cultivar-specific effects by slicing an interaction is similar, but not identical, to splitting the data into pieces, one for each cultivar, then testing differences among environments for each cultivar. It is not identical because the "slicing" F tests use pooled estimates of the error and block variances while the "splitting" tests use cultivar-specific estimates. The environment by cultivar interaction can be sliced either by cultivar or by environment. Slicing the interaction by environment, to

TABLE 4. ANOVA table for the narrow sense analysis of the oat variety trial example.

Source	Degrees of freedom	Mean Square	*F* statistic	*P*-value
Environment	5	294.27	33.65	< 0.001
Block(Environment)	12	8.74		
Variety	9	171.92	14.83	< 0.0001
Variety*Environment	45	26.57	2.29	0.0003
Error	108	11.59		

TABLE 5. *P*-values for slices of the cultivar by environment interaction. Each cultivar-specific test compares the average responses of a cultivar across six environments; each test has five numerator degrees of freedom. Each environment-specific test compares the average response of 10 cultivars in the specified environment; each test has nine numerator degrees of freedom.

Variety-specific tests		Environment-specific tests	
Variety	*P*-value	Location, Year	*P*-value
Cherokee	< 0.0001	Ames, 1985	0.0012
Don	0.074	Ames, 1986	0.0003
E77	0.0033	Kanawha, 1985	< 0.0001
Hytest	0.0064	Kanawha, 1986	0.0060
Lang	0.099	Washington, 1985	0.030
Larry	0.34	Washington, 1986	< 0.0001
Ogle	0.094		
Porter	0.0006		
Proat	< 0.0001		
Richland	< 0.0001		

BOX 4.

```
proc mixed nobound data=oats;
  class Environment Block Cultivar;
  model HI = Cultivar /ddfm=kr;
  random Environment Block(Environment) Cultivar*Environment;
  estimate 'Cultivar: Cherokee' intercept 1 Cultivar 1;
  estimate 'Cultivar: Cherokee - Don' Cultivar 1 -1;
  estimate 'Cultiver: Cherokee - Don in Ames 1985' Cultivar 1 -1 |
    Cultivar*Environment 1 -1;
  title 'Oat HI: Broad sense, 6 environments, REML variance
component
    ests';
  run;
```

produce *F* tests of cultivar differences within each environment, indicates that there was a significant cultivar response in each of the six environments (Table 5).

When broad-sense inference, that is, inferences about the differences among cultivars in new environments, is desired, the interaction between treatments and environments is considered a random effect. In addition, for broad-sense inference as defined by McLean et al. (1991), environment effects are considered random. Because a fixed effect cannot be nested inside a random effect, then block within environment must be a random effect. SAS code for broad-sense inference across 6 environments is in Box 4. Notice that the estimate statement to compute the difference between two cultivars in a specific environment requires a vertical bar (| symbol) to separate the fixed effects (cultivar) and the random effects (Cultivar × Environment).

As with the meat triglyceride example, the choice of narrow- or broad-sense inference has no effect on the estimated mean for one cultivar, averaged across environments, or the estimated mean difference between two cultivars, averaged across environments, but it does change the standard errors (Table 6). The standard errors are larger when using broad-sense inference because of variation between environments in both cultivar means and differences. However, broad-sense inference provides an estimate of the average performance of the cultivars over a population of environments.

The choice of narrow- or broad-sense inference does change the reported environment-specific difference between two cultivars (Table 6). This is because the narrow-sense value is an estimate of that difference, which involves two estimated interaction components. Under broad-sense inference, the interaction is a random effect, so components of that interaction are predicted using best linear unbiased predictions (BLUPs). In general, BLUPs are shrunk toward zero, so the narrow-sense estimates and broad-sense BLUPs are not the same (Robinson 1991). Hence, the broad-sense prediction is not the same as the narrow-sense estimate of an environment-specific difference.

Inferences using the broad-sense approach are to a larger population of environments than used in the actual study. Whether that population is hypothetical or real depends on how environments were chosen for the study. The population for broad-sense inference is clearly identifiable when the environments used in a study are a simple random sample from that population. For example, when six farms are randomly selected from a list of all farms in Iowa, the population for broad-sense inference is all farms in Iowa. If the environments are not chosen by a simple random sample, then broad-sense conclusions apply to a hypothetical population defined by the environments actually used in a study. Whether this hypothetical population reflects any real population depends on non-statistical information. In the meat example, the days used in the study are not a random sample from some population of days, but they could be imagined to represent a random sample from a hypothetical population of all the things that differ between days (e.g., lab conditions, properties of the piece of meat used that day).

The usual mixed models approach assumes infinite-size populations of random effects. Applied to a combined experiment, this means that the broad-sense variance calculations for treatment means or differences assume inference about an infinite population of environments. When the population of environments is not much larger than the number in the study, the calculated variances are too large because of the finite population correction factor (Thompson 1992, p. 15). Equations for variances of treatment differences and expected means squares for a finite population of environments are given in Bennett and Franklin (1954, Chapter 7). In the oat cultivar trial, the calculated variances are appropriate for inferences to a very large collection of locations and years. If the desired inferences are to five research farms in central Iowa, three of which were used in the study, the calculated variance is too large. In the extreme, if the entire population of environments were included in the study, after correctly adjusting for the finite population of environments, the variance of a treatment difference and the Expected Mean Square for treatments are the expressions in the Fixed column of Table 1. Again, this makes intuitive sense as you are in "possession" of all information regarding the target environments.

TABLE 6. Estimates and standard errors for the mean of one variety (Cherokee), a difference of two variety means (Cherokee – Don), and an environment-specific difference of two variety means (Cherokee – Don in Ames, 1995) for narrow- and broad-sense inference using six environments.

Quantity	Estimate for:		Standard Error for:	
	Narrow	Broad	Narrow	Broad
Mean for one variety	40.89	40.89	0.79	1.72
Difference of two varieties	-9.50	-9.50	1.13	1.72
Difference of two varieties in one environment	-11.00	-10.35	2.78	2.22

Choosing Whether to Subdivide the Treatment by Environment Interaction by Environment

The previous analysis of the oat cultivar data considered each combination of year and location as one of six unique environments. This assumes that the random effect associated with one year and location, for example, "Ames 1985", is independent of the random effects associated with any other year and location. Because all observations from a specific environment share one value of the

TABLE 7. Estimates, standard errors, and error degrees of freedom (df) for the mean of one variety (Cherokee), a difference of two variety means (Cherokee–Don), and an environment-specific difference of two variety means (Cherokee – Don in Ames, 1995) after subdividing environments into year, location and the year by location interaction. Degrees of freedom computed using the Kenward–Roger approximation.

Quantity	Estimate	Standard Error	Degrees of freedom
Mean for one variety	40.89	2.13	1
Difference of two varieties	-9.50	1.60	3.72
Difference of two varieties in one environment	-10.45	2.22	123

BOX 5.

```
proc mixed nobound data=oats;
  class Year Location Block Cultivar;
  model HI = Cultivar /ddfm=kr;
  random Year Location Year*Location Block(Year*Location)
    Cultivar*Year Cultivar*Location Cultivar*Year*Location;
  estimate 'Cultivar: Cherokee' intercept 1 Cultivar 1;
  estimate 'Cultivar: Cherokee - Don' Cultivar 1 -1;
  estimate 'Cult: Cherokee - Don in Ames 1985' Cultivar 1 -1 |
    Cultivar*Year 1 -1 Cultivar*Location 1 -1
    Cultivar*Year*Location 1 -1;
  title 'Broad sense: subdivided interaction';
  run;
```

environment random effect, a random effect for environment induces a correlation between observations from the same environment. Observations from different environments are independent, because they do not share any random variables. Six independent environments represent one of many possible ways to structure the random effects in the model. Other specifications of the random effects provide different models for the correlation among observations.

One alternative is to define three random effects: one for years (shared by all observations in the same year), one for locations (shared by all observations in the same location) and one for the interaction of year and location (shared only by observations in the same year and same location). This adds additional sources of correlation between observations. The model with random effects for year, location, and year × location allows observations from the same year to have one non-zero correlation, observations from the same location to have a different non-zero correlation, and observations from the same environment (year and location) to have a third non-zero correlation.

The same concept can be applied to subdividing a treatment by environment interaction into multiple components. For the oat cultivar data, the cultivar by environment interaction could be subdivided into three components: a cultivar by year interaction, a cultivar by location interaction and a cultivar by year by location interaction. Since the treatment by environment interaction represents the consistency of treatment differences across environments, subdividing the interaction allows one to assess consistency across years and a different consistency across locations.

SAS code to fit this model is in Box 5. Again, the nobound option is specified to allow variance components to be negative. Although a variance is strictly non-negative, when viewed as a parameter determining the correlation between pairs of observations, a negative variance component specifies a negative correlation, which is entirely legitimate.

Selected results are given in Table 7. The estimated mean for one cultivar and estimated difference of two cultivars are unchanged by splitting environments, but the standard errors are different from those given in Table 6. Some standard errors increase compared with their "six environment" values while others decrease. Because there is less information about each component of the interaction, the degrees of freedom for estimated means and estimated differences are reduced when environment is split into year, location, and the year by location components. The Kenward–Roger (Kenward and Roger, 1997) approximate degrees of freedom for a cultivar mean and difference are 1 and 3.72 (Table 7), compared with 15.3 and 45 when environment is not split.

Choosing Whether to Subdivide the Treatment by Environment Interaction by Treatment

The pooled treatment by environment interaction combines information across treatments as well as across environments. So, the interaction could be subdivided by treatment instead of by environment. If the 10 cultivars represent two breeding groups and five maturity groups, the set of 10 treatments could be divided into a 2 × 5 factorial combination of breeding group × maturity group. Then, you could consider three components of the environment by treatment interaction: breeding group by environment, maturity group by environment, and the three-way interaction. Each of these contributions represents the consistency of that set of treatment effects across environments. A variance component close to 0 for one component means that those treatment contrasts are relatively consistent across the environments. A large variance component means that treatment contrasts are quite different across the environments. For example, if the breeding group by environment interaction variance is large, the difference between the two breeding groups varies across environments. In the usual analysis of a combined experiment, each of these interaction components is a random effect and would serve as the error term for the associated treatment effect. For example, breeding group by environment would be the error term for testing the breeding group effect.

Further subdivision of interaction components is possible. The differences among five maturity groups are described by four orthogonal contrasts. The environment by maturity group interaction could be subdivided into contributions from each orthogonal contrast. For example, if the difference between Maturity Group 1 and the other four maturity groups varied considerably across locations, but differences among the other four maturity groups were relatively consistent then the single degree of freedom contrast between Maturity Group 1 and the rest should be separated from the other three df among maturity groups. Such an analysis would include two different interactions, one for each group of contrasts.

Even further subdivision of interaction components is possible. The differences among the ten cultivars are described by nine orthogonal contrasts. Each could be assumed to have a different interaction with environment. The resulting model would have nine variance components for the environment by treatment interaction, one for each orthogonal contrast. When each orthogonal contrast is analyzed separately, this approach is known as within subject contrasts analysis (see for example, Keppel and Wickens, 2004, p. 358–359) or derived variables analysis (Diggle et al. 2002, p. 17).

A classic example of treatment-specific interaction components is the experiment on early lentils in Syria described by Peterson (1994, p. 207–215). The treatment structure is a three-way complete factorial with two levels of fertilizer, two levels of weevil control, and two levels of weed control. This was repeated at eight locations. Peterson's model included eight location interactions, one for each component of the treatment structure. Each location interaction is estimated with seven degrees of freedom. The location by fertilizer interactions are all small relative to the location by weevil and location by weed interactions (Peterson, 1994, p. 213). Peterson speculates that the eight locations had different levels of infestation by weevils and weeds. The response to the weevil treatment or the weed control treatment would be large at locations with heavy infestation and small at locations with little infestation. In contrast, the data suggests that the response to fertilization was similar at all locations. An alternative to Peterson's model would have three components to the interaction: location × weevil, location × weed, and a pooled location × other term. The location × other term would have 35 degrees of freedom. If the assumption of equal contributions to that interaction is appropriate, the 35 df estimate is a more precise estimate than any of the seven degree of freedom estimates.

The decision to pool or not to pool should preferably be made in the planning stage of a study, using the experimenter's understanding of sources and magnitudes of variability. This will be especially important when each individual experiment has multiple sources of variation; for example, variation among main plots and variation among split plots. A narrow-sense analysis will pool the main-plot variability across environments and separately pool the split-plot variability. The appropriate pooling for a broad-sense analysis is less clear. Is the consistency across environments of the main-plot treatment different from the consistency of the split-plot treatment? If so, interaction components should not be pooled. If the two consistencies are expected to be similar, interaction components could be pooled (Carmer et al., 1969).

A second consideration in deciding whether or not to pool is the error df for important tests (Edwards, 1985, p. 263). Increasing the error df greatly reduces upper quantiles of t and F distributions when the error df is small. For example, the 0.975 quantiles of t distributions are approximately 4.3 for two degrees freedom, 2.6 for five degrees freedom, 2.2 for 10 df and 2.1 for 15 df. Increasing the error df has a much smaller effect when the error df exceeds 20 or 30. For example, the 0.975 quantiles of t distributions are approximately 2.08 for 20 df, 2.04 for 30 df, 2.01 for 50 df and 1.98 for 100 df. Because of this characteristic, Edwards (1985, p. 263) suggests never pooling effects that have more than 20 or 30 df. He also suggests designing a study so that it has sufficient replication to provide 30, or at least 20, error df.

Careful consideration of both sources of variability and appropriate amounts of pooling is especially important in studies with multifactor treatment structures repeated in different types of environments. When both treatments and environments are structured, an analysis might subdivide both and consider all combinations of contributions. One author (PMD) has seen a combined experiment with a 2 × 2 × 2 treatment design repeated in two locations and two years. The proposed analysis subdivided environment into location, year, and the location × year interaction and subdivided treatments into 7 single degree of freedom contrasts. The analysis had over 20 treatment by environment interaction terms, (e.g., year × A, or location × B), each with one degree of freedom. This is the ultimate in subdivision and leads to tests with very low power because denominator degrees of freedom will be close to one for each test. The practical analysis is to pool interaction terms, using subject matter knowledge to decide which terms should be similar.

Choosing Whether or Not to Assume Equal Residual Variances for All Environments

The analyses of both the meat and the oat cultivar data sets assumed that all observations had the same error variance. Because each repetition of a repeated study is a stand-alone experiment, an error variance can be estimated separately for each repetition. When each repetition can be assumed to have the same error variance, these repetition-specific estimates are pooled to estimate a single common variance. Is it reasonable to use a single pooled variance, or should each repetition be allowed to have a different error variance?

One way to answer that question relies on an experimenter's knowledge of the study sites. The error variance describes the variability between experimental plots treated alike. When the repetitions are at the same location, that is, repeated in different years, it might be more reasonable to use the pooled estimate. When a study is repeated at different locations, "common sense" suggests that the variability among plots differs among locations. Some locations may simply be more heterogeneous than others. In addition, plot size and experiment design may differ among locations. This understanding of the study sites argues for requiring different error variances for each location (Peterson, 1994).

Using environment-specific error variances complicates the analysis (Piepho, 2009). Environment-specific variances are estimated by restricted maximum likelihood (REML). The REML algorithm may fail because of infinite likelihood or failure to converge within the allowed number of iterations. The null distributions of F and t statistics have to be approximated by Satterthwaite or Kenward–Roger approximations. These may not maintain appropriate Type 1 error rates for hypothesis tests or appropriate coverage of confidence intervals (Richter et al., 2015).

When error variances differ among locations, the same common sense argument suggests that block variances should also be allowed to vary with location. If the small-scale heterogeneity among plots is not the same at two locations, then the medium-scale heterogeneity is also likely to differ among locations. Hence, the variance between blocks should be allowed to differ among locations. In our experience, this is rarely suggested and rarely done. SAS code to fit environment-specific block variances is in the supplemental material. Using environment-specific block variances in addition to environment-specific error variances further complicates the analysis.

The data can be used to assess equality of variances, but this has to be done carefully. Choosing between a pooled variance or separate variances for each environment is a specific example of the general problem of choosing a model for the variance-covariance matrix of the observations. This is commonly done using the Akaike Information Criterion, AIC, or related statistics based on the REML log-likelihood (Gbur et al. (2012), p. 80). The AIC statistic is computed for the model with a single pooled variance and compared with the AIC statistic for the model with separate variances for each environment. The model with the smaller AIC statistic is the more appropriate model. Models that have AIC statistics within 2 units of the smallest AIC are considered roughly comparable to the best model; models with AIC statistics that are more than 10 units larger than the best model are considered unlikely (Burnham and Anderson, 2002). The small-sample corrected AIC statistic (AICc), Schwartz Bayesian Criterion or other forms of Bayesian Information

Criterion (BIC) are sometimes used instead of AIC. Compared to AIC, BIC puts a larger penalty on the more complex model, so BIC more frequently selects simpler models, that is, the pooled variance model.

The operating characteristics of an AIC-based decision about pooling variances have not been well studied, but there is a strong argument that the use of AIC could be unduly sensitive to non-normality. AIC is similar to the likelihood ratio (LR) test and Bartlett's test of equal variances in that all three approaches depend on a log-likelihood that assumes normally distributed errors. Both the LR and Bartlett's tests are known to be very sensitive to violations of normality (Box 1953; Madansky 1988) so it seems likely that the AIC statistics will be also. An alternative is to use a test of equal variance that is less sensitive to normality, such as Levene's test (Madansky 1988, p. 67) or one of the many variations of Levene's test.

For the oat cultivar trial data, we consider four models for the error variance: one with a separate variance for each of the six environments (combinations of year and location), one with a separate variance for each year (but the same variance for all locations that year), one with a separate variance for each location (but the same variance in both years), and a pooled model with one error variance for all observations. The AIC statistics indicates that separate variances for each location (AIC = 698.4) should be used (Table 8). The AIC statistic for the pooled variance model (AIC = 701.1) is more than two units larger than the best model. The estimated location-specific variances are somewhat different (Ames: 17.6, Kawawha: 8.12, and Washington: 9.04), but they are not more than a factor of four different, which some use as a guide to assess when heterogeneity could be important (Fox 2008, p. 277). The Levene's test indicates no evidence of unequal variances between years, among locations, or in the year by location interaction (Table 8).

We suggest that in many cases, the use of a single pooled variance will be appropriate even when locations have somewhat different error variances unless the environment-specific sample sizes are very different. Our argument has three parts:

1. The overall treatment difference is an average of the environment-specific treatment differences. When variances are pooled, this is compared to the average error variance. This averaging leads to appropriate inferences, for the same reasons that an overall F statistic in a single-environment study is robust to variances that differ among treatments. Analogously, when sample sizes are not similar in the various environments, or if comparisons are to be made among specific pairs of environments, using a pooled error variance may be inappropriate.

TABLE 8. AIC statistics and Levene's test results to evaluate whether error variances can be pooled. The pooled model has one error variance for all observations. The location, year and year*location models fit separate error variances for each location, each year, or each combination of year and location.

Model	Number of variances	AIC	Levene's test results	
			Source	*P*-value
location	3	698.4	Year	0.79
pooled	1	701.1	Location	0.18
year*location	6	701.3	Year*Location	0.33
year	2	703.1		

2. In broad-sense inference, the variance that matters is the variance of an environment-specific treatment difference. This has two components: the error variance and the interaction variance. Either or both may differ between environments. When a single interaction variance is included in the model, the variances that matter will be more similar to each other than are the error variances.
3. Under narrow sense inference, treatment means and differences of treatment means are equally weighted averages of the environment-specific values, so the choice of pooled or not has no effect on the estimates. Under broad sense inference, with random interactions, the estimated means and their differences may or may not be equally weighted averages. When the variances are pooled, environment-specific estimates are equally weighted. When variances are separated, more weight is given to the estimates from the environments with smaller error variances. This situation is analogous to the distinction between Type II and Type III estimates and tests in fixed effects linear models. The choice of pooled or separate can change the estimates, but the desirable choice may not be clear.

To illustrate these points, we consider the three possible types of results in the oat cultivar trial: the overall mean harvest index for one cultivar, the overall mean difference between two cultivars, and the mean difference in one environment. We consider pooling error variances across all environments or using location-specific error variances with narrow-sense inference (fixed interaction, broad-sense inference (random environment by treatment interaction) and broad-sense inference with subdivided interactions (location by treatment, year by treatment, and location by year by treatment). SAS code to fit these models is in Box 6. The broad-sense model is fit without the nobound option because of convergence issues with the nobound model and default starting values.

Table 9 gives the estimates and standard errors when separate error variances are estimated for each location. The values in Table 9 can be compared to those using the pooled error variance (Table 6). When the focus is on the difference in the treatment means, the choice of pooled or separate error variances has almost no effect on the estimates or their standard errors (Table 9). One source of the small difference in standard errors is a different estimate of the treatment × environment variance component. When error variances are pooled, the interaction variance is estimated as 4.99; when error variances are estimated separately for each environment, the interaction variance is estimated as 5.47. When the focus is an environment-specific result, the choice of pooled or separate error variances affects both the estimates and their standard errors (Table 9), because of the much larger estimated error variance for the Ames location. If the sample sizes in each environment were substantially different, the choice of pooled or separate error variances would have a larger effect on the results.

Finally, we suggest that more attention needs to be given to treatment-specific error variances and treatment-specific components of the environment by treatment interaction. When the focus of a study is differences between treatments, the assumption that each treatment has the same error variance is much more important than the assumption that each environment has the same variance (Box, 1954).

BOX 6.

```
proc mixed nobound data=oats;
  class Environment Location Block Cultivar;
  model HI = Environment Cultivar Cultivar*Environment /ddfm=kr;
  random Block(Environment);
  repeated /group=Location;
  estimate 'Cultivar: Cherokee' intercept 1 Cultivar 1;
  estimate 'Cultivar: Cherokee - Don' Cultivar 1 -1;
  estimate 'Cult: Cherokee - Don in Ames 1985' Cultivar 1 -1
Environment*Cultivar 1 -1;
  title 'Narrow sense inference, 3 location specific error
variances';
  run;

proc mixed data=oats;
  class Environment Location Block Cultivar;
  model HI = Cultivar /ddfm=kr;
  random Environment Block(Environment) Cultivar*Environment;
  repeated /group=Location;
  estimate 'Cultivar: Cherokee' intercept 1 Cultivar 1;
  estimate 'Cultivar: Cherokee - Don' Cultivar 1 -1;
  estimate 'Cultivar: Cherokee - Don in Ames, 1985' Cultivar 1 -1
       | Cultivar*Environment 1 -1 ;
  title 'Broad sense inference, 3 location specific error
variances';
  run;
```

Some statisticians may take exception to our suggestions. We agree that ignorance of alternatives is a poor reason to pool. However, we encourage careful consideration of all sources of heteroscedasticity and return to Box's (1954) conclusion for treatment comparisons, which is that treatment-specific variances are a much larger concern than environment-specific variances.

Choosing the Model for a Quantitative Treatment

When a treatment is a quantitative independent variable, such as rate of fertilizer or Julian day of planting, it can be modeled in two different ways. One is to treat each level of the quantitative variable as a separate group and use the approaches discussed previously. The other is to use a regression model to estimate a response curve (e.g., Cox, 1958, p. 30). When a simple regression model is an adequate fit to the data, regression provides especially straight-forward conclusions. A regression model also allows asking new questions about the effects of the quantitative treatment. Using a regression model in the analysis of a repeated experiment introduces new issues, which we now discuss.

To illustrate the issues, we consider a study of corn yield response to fertilizer addition on Antigua Island in the Lesser Antilles. The data are given in Table 58.1 in Andrews and Herzberg (1985). The 12 treatments were combinations of amounts of N, P, and K fertilizer using a 2 × 2 × 2 factorial treatment design supplemented by four additional treatments (Springer, 1972). Treatments were randomized to plots in an incomplete block design. The response variables are the number of ears and the total grain weight, in kg per plot. The entire design was repeated at eight locations. We focus on the response of grain weight to nitrogen (N) addition.

We use a subset of the data to illustrate the combined analysis of a regression model. We deleted all data from all plots at two locations (NSAN and WLAN) and certain plots at two additional locations because of experimental irregularities, detailed in Andrews and Herzberg (1985). A thorough evaluation of fertilizer effects would model the response to all three fertilizers, N, P, and K. We focus on the effects of N addition and simplify the example by omitting consideration of P and K. We remove all data from the treatments labeled 131 and 113, which have a high amount of P and high amount of K, respectively. After removing these two treatments, it is appropriate to model the effect of N addition independently of the effects of P or K.

A plot of the grain weight and fitted quadratic response curves at each location is in Fig. 3. Two locations, DBAN and LFAN, visually show a large effect of increasing N amount, two locations, OVAN, and ORAN, show a smaller effect, and two locations, WEAN and TEAN, show a very small effect. Heterogeneity in error variances between locations or between N levels was assessed by fitting a model with locations, blocks within location, N level as a factor, and N × location interactions, calculating the residuals from that fitted model, then using Levene's test on those residuals. There was no evidence of unequal variance between locations ($p = 0.46$) or between N levels ($p = 0.77$).

In a single experiment with replicated levels of a quantitative treatment, e.g., of N, the effect of that quantitative treatment can be evaluated using (at least) three different approaches. One is to fit a model with a separate mean for each level of N, and follow that up with multiple comparisons among treatment means. This approach ignores the quantitative nature of the treatment. A second approach is to fit a means model, but follow that up by estimating linear, quadratic, or more complicated contrasts (e.g., Gbur et al., 2012, p. 131–132; Vargas et al., 2018; Chapter 7) to summarize responses to N level. A third approach is to fit a regression model with N level as a continuous covariate. This model may be simple, e.g., a linear response to N, more

TABLE 9. Estimates and standard errors for the mean of one variety (Cherokee), a difference of two variety means (Cherokee – Don), and an environment-specific difference of two variety means (Cherokee – Don in Ames, 1995) when error variances are pooled across environments and when separate error variances are estimated for each location (three estimated error variances).

Quantity	Model	Estimate when error variances:		Standard error when error variances:	
		Pooled	Separate	Pooled	Separate
Mean:	Narrow sense	40.89	40.89	0.792	0.792
	Broad sense	40.89	40.81	1.721	1.734
	Broad subdivided	40.89	40.81	2.130	1.734
Difference:	Narrow sense	-9.50	-9.50	1.135	1.128
	Broad sense	-9.50	-9.41	1.718	1.745
	Broad subdivided	-9.50	-9.41	1.604	1.745
Difference in one environment	Narrow sense	-11.00	-11.00	2.780	3.370
	Broad sense	-10.35	-10.21	2.227	2.576
	Broad, subdivided	-10.45	-10.21	2.225	2.576

BOX 7.

```
proc import datafile = 'antigua.csv' out=antigua replace;
run;

proc mixed data=antigua;
  where location='DBAN';
  class N block;
  model wt = block N;
  lsmeans N / diff;
  title "Antigua, DBAN location, analysis of means for each N
level";
  run;

proc mixed data=antigua;
  where location='DBAN';
  class N block;
  model wt = block N;
  estimate 'linear N slope' N -3 -1 1 3 /divisor=10;
  estimate 'quadratic N slope' N 1 -1 -1 1 /divisor=4;
  estimate 'cubic N slope' N -1 3 -3 1 / divisor=6;
  contrast 'lack of fit to linear' N 1 -1 -1 1, N -1 3 -3 1;
  title "Antigua, DBAN location, orthogonal polynomial trends";
  run;

proc mixed data=antigua;
  where location='DBAN';
  class block;
  model wt = block N N*N;
  estimate 'linear N slope' N 1;
  estimate 'quadratic N slope' N*N 1;
  title "Antigua, DBAN location, regression analysis";
  run;
```

complicated, e.g., a polynomial or nonlinear response to N, or nonlinear (discussed by Miguez et al., (2018) in Chapter 15). The regression model has the advantages of providing predictions of the response at N levels not used in the study and often being a more parsimonious description of the relationship between N level and the response. Box 7 shows SAS code to fit a means model, a means model with contrasts, and a polynomial regression to data from the DBAN location.

In a single experiment, these approaches differ in what is being described (means for each level or the overall pattern of response), how that pattern is estimated, and the definition of the error variance (pure error, i.e., only variability within replicate units at each level, or a combination of pure error and lack of fit to the proposed regression model). For example, when using linear or quadratic contrasts with a means model, the error variance is the pure error, but when using a regression model, the error variance is the combination of pure error and lack of fit. Contrasts and linear regression weight observations differently unless sample sizes are the same for each group. Linear regression weights each observation equally, so groups with large sample sizes get more weight. Contrasts weigh each mean equally, no matter what the sample size. Unless sample sizes are very unequal or the regression model has severe lack of fit, results from the means model followed by contrasts are similar to those from the regression model. We use the regression approach for the Antigua fertilizer response data.

For a repeated experiment, results from a means model and a regression model can be quite different because the two models describe experiment by treatment interactions differently, which leads to different variance–covariance matrices for the observations. For simplicity of exposition, we will focus on a linear response to N level and temporarily ignore all other features of the Antigua corn yield study. The means model for a repeated experiment is Eq. [3] with locations instead of days; the regression model for a repeated experiment is Eq. [5] in which location-specific linear regression models are jointly estimated.

$$Y_{ij} = (\beta_0 + b_{0i}) + (\beta_1 + b_{1i})X_{ij} + \varepsilon_{ij} \qquad [5]$$

where i indexes the location, j indexes the observation, and X_{ij} is the N level for observation j in experiment i. The b_{0i} terms quantify the between-location variability in the intercept of the regression. These are the equivalent of the day, that is, d_i, terms in Eq. [3]. The

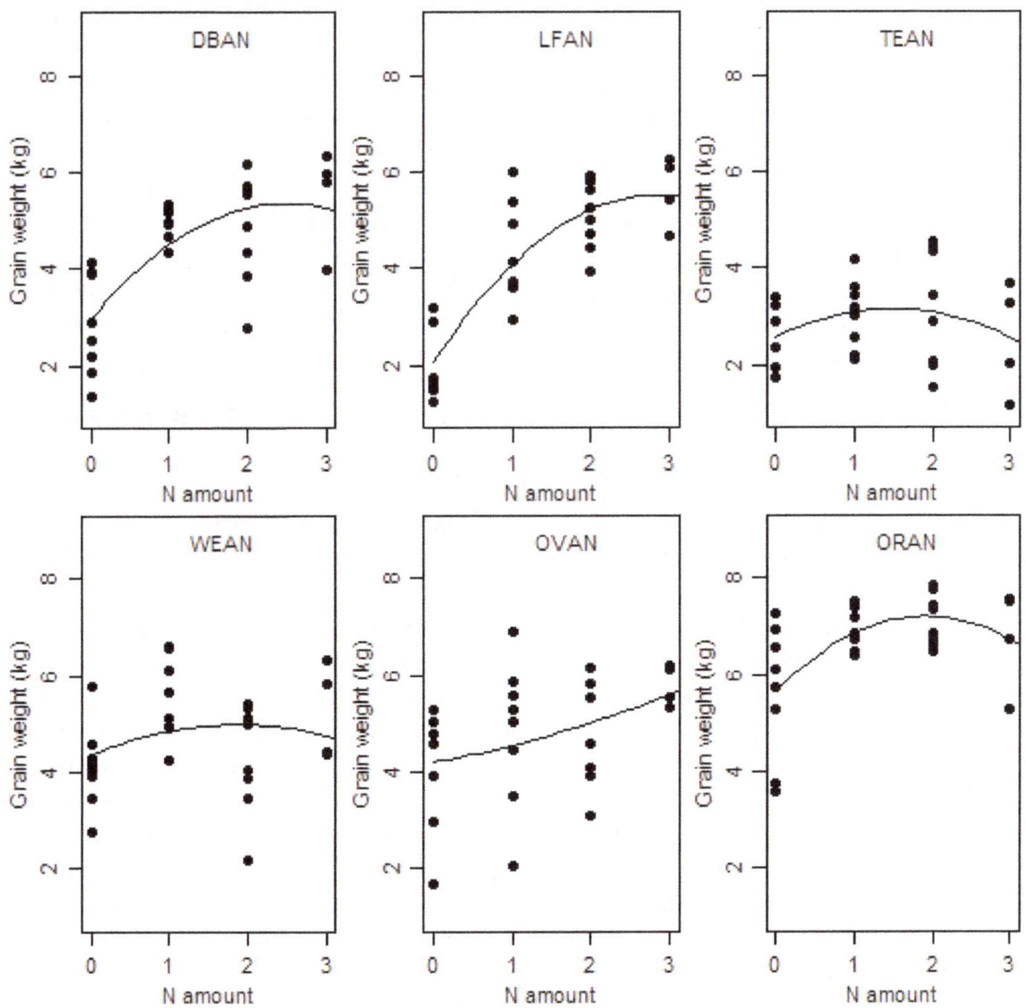

FIG. 3. Plot of grain weight vs N addition at the six Antigua locations used in the data analysis. Locations are indicated by their four letter abbreviations. The curve is the fitted quadratic from a separate analysis for each environment.

b_{1i} terms quantify the variability in the slopes of each regression; they are the equivalent of the day-by-treatment interaction terms in model (3). If every b_{1i} term is 0 in Eq. [5] the location-specific regression lines are parallel because they all have the same slope, β_1.

When broad-sense inference is used with Eq. [5], the location-specific intercepts and slopes become correlated random variables. These are commonly considered to be independent of the observation-specific errors. Hence, Eq. [5] becomes Eq. [6]:

$$Y_{ij}=(\beta_0+b_{0i})+(\beta_1+b_{1i})X_{ij}+\varepsilon_{ij} \quad [6]$$

$$\begin{bmatrix} b_{0i} \\ b_{1i} \end{bmatrix} \sim N\left(\begin{bmatrix} 0 \\ 0 \end{bmatrix}, \begin{bmatrix} \sigma_0^2 & \sigma_{01} \\ \sigma_{01} & \sigma_1^2 \end{bmatrix}\right)$$

where σ_0^2 is the between-location variance in the intercepts, σ_1^2 is the between-location variance in the slopes, and is the covariance between the intercept and slope. Fitting model (Eq. [6]) is often numerically more stable after centering the X values by subtracting the overall mean X from all values. Centering the X values changes the interpretation of the intercept but does not change the interpretation of the slope.

The means model (Eq. [3]) and the regression model (Eqs. [5-6]) both allow location-to-location variation in the difference between two treatments, but they do so in different ways. The location-by-treatment interaction coefficients in Eq. [3] allow very general location-specific variations in the differences between two treatments. A non-zero interaction allows any sort of location variation. Two of the possible patterns with a non-zero interaction in Eq. [3] are shown in panels a and b of Fig. 4. The location-specific slope in Eq. [5] allows treatment differences to vary between locations but only in a very specific way. Consider the interaction between Treatments A and B in Locations 1 and 2. Under Eq. [3], that interaction is $(dt_{1A}-dt_{1B})-(dt_{2A}-dt_{2B})$. According to Eq. [5], that difference is $(\beta_1+b_{11})(X_A-X_B)-[(\beta_1+b_{12})(X_A-X_B)]=(b_{11}-b_{12})(X_A-X_B)$. The interaction depends on the difference in the X values and the difference in the two location-specific slopes. Two of the possible patterns with a non-zero interaction in Eq. [5] are shown in panels c and d of Fig. 4.

How should you determine which is the more appropriate model for the data? The commonly-used data-based approach is to compare AIC statistics. Because the means model 3, (Eq. [3]) and the regression model 6 (Eq. [6]) have different fixed effects, the AIC statistics must be computed using ML, not the default REML. The other approach is based on your understanding of how the experiments might differ from each other. If the relationship between the response and an X variable is based on a well-understood mechanism and it is reasonable that different experiments will have different regression slopes, then the random coefficients regression, model 6 (Eq. [6]), is the more appropriate model. If the relationship between the response and an X variable is expected to be more complex in each experiment, then perhaps the general experiment-by-treatment interaction, model 3 (Eq. [3]) is more appropriate.

A regression model for repeated experiments requires both specifying the form of the regression function, for example, linear or quadratic, and choosing which coefficients vary across the experiments. The need for location-specific linear or quadratic slopes can be evaluated using either the narrow-sense or broad-sense model. We find it easier to use a narrow-sense model for model selection. The form of the N response was evaluated by the SAS code given in Box 8. The p-value for the quadratic effect of N is very small, but the p-value for the quadratic by location interaction is 0.48, so that interaction will be dropped. When the narrow-sense model is re-run without the quadratic by location interaction, the p-values for the quadratic N effect and the linear by location interaction are 0.0010 and 0.0001, respectively, so no further simplifications are made to the model. The location-specific plots of residuals against predicted values indicate no concerns.

Code to fit the broad-sense model is in Box 9. The estimated variance components are 1.62 for the locations, 0.07 for blocks within locations, 0.16 for the linear nitrogen slopes, and 0.96 for the pooled error variance. The estimated location-specific intercepts and

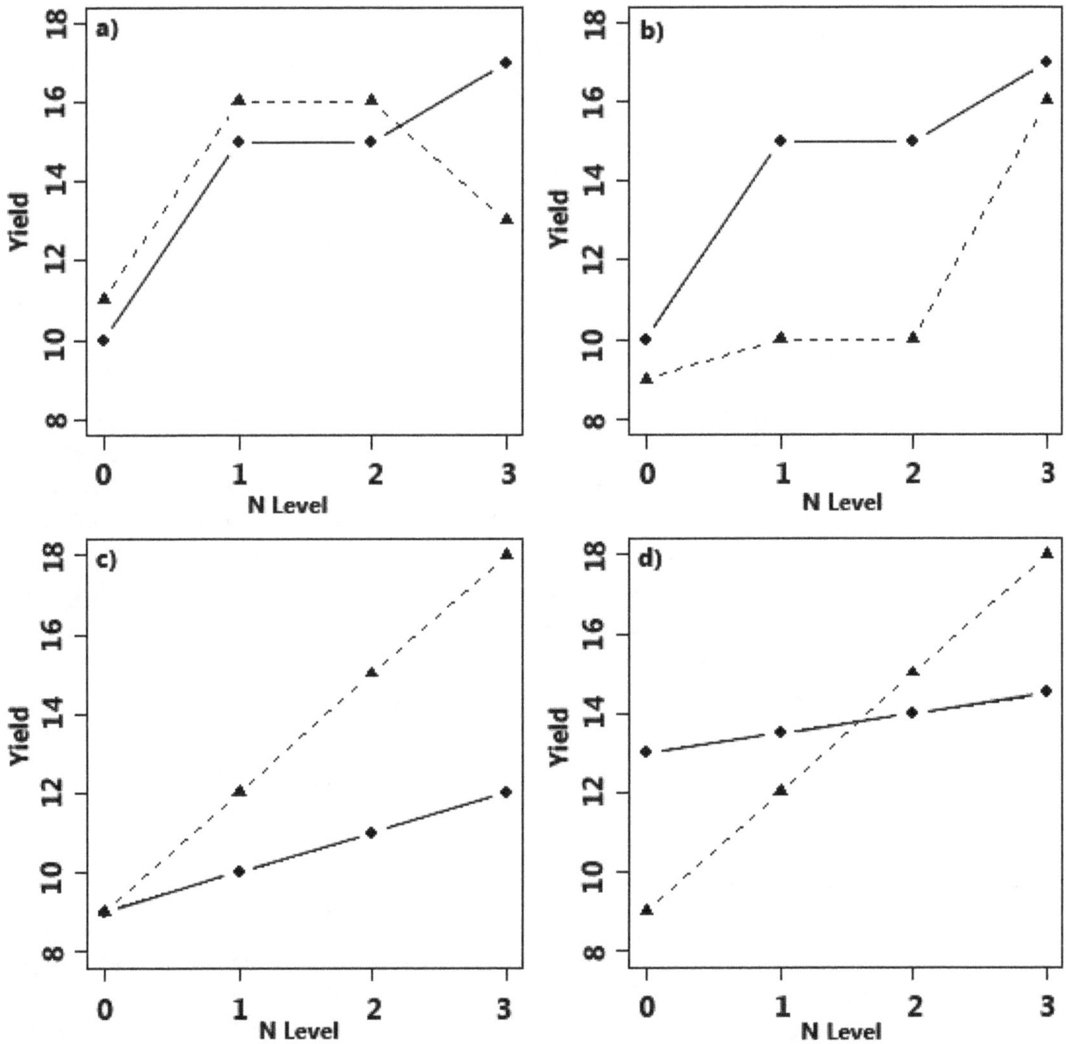

FIG. 4. Location by treatment Interaction patterns allowed by the means model (panels a and b) and the linear regression model (panels c and d). Treatment means for Location 1 are shown with circles and solid lines; treatment means for Location 2 are shown with triangles and dashed lines. Interaction in the means model (equation 3) allows any pattern of non-parallel responses to treatment. Interaction in the regression model (Eqs. 5 or 6) only allows patterns that correspond to different regression slopes.

BOX 8.

```
proc mixed data=antigua;
  class location block;
  model wt = N N*N N*location N*N*location / ddfm=kr;
  random location block(location);
  title 'Quantitative N, with quadratic interaction';
  run;

proc mixed data=antigua;
  class location block;
  model wt = N N*N N*location / ddfm=kr;
  random location block(location);
  title 'Quantitative N, only linear N*location';
  run;
```

BOX 9.

```
/* preceding code, not in box 9, creates prediction locations */
proc mixed data=full;
  class location block;
  model wt = N N*N / outpm=predmean outp=predloc ddfm=kr;
  random intercept N /subject=location type=un;
  random block(location);

  estimate 'intercept' intercept 1;
  estimate 'N linear' N 1;
  estimate 'N quadratic' N*N 1;

  title 'Quantitative treatments';
  run;
```

location-specific slopes are correlated. The estimated covariance is -0.18, so the estimated correlation is -0.36. Estimates and tests of the fixed effects describing the average response curve are shown in Table 10. More important are the predictions of the location-specific N response curves shown in Table 11. We estimate the parameters of the overall curve, because those are fixed effects. We predict the coefficients for each location because the location-specific coefficients are random variables. Because the model did not include a location by quadratic interaction, the quadratic coefficient is the same for all six locations. The predicted intercept and slope do vary between the locations.

The overall N response curve and the location-specific response curves are plotted in Fig. 5. The standard errors for the overall curve are larger than those for a location-specific curve because the uncertainty in the overall curve includes the variation among the six locations.

Summary

Combined analyses of repeated experiments provide estimates of both average treatment effects and their consistency across environments. Doing a combined analysis requires decisions about the scope of inference, pooling of error variances across environments, and pooling of contributions to treatment by environment interactions. The scope of inference (narrow- or broad-sense) has major consequences for the analysis. Narrow-sense conclusions apply to the set of environments in the study; broad-sense conclusions apply to a more general population of environments. Error variances are likely to differ across environments, especially when environments are different locations. Conclusions about environment-specific treatment effects depend on the choice of pooled or environment-specific error variances, but conclusions about average treatment effects are quite robust to that choice. The choice of pooling treatment by environment interactions is especially important when treatments differ in their consistency across environments. The analysis may also require deciding whether to model quantitative treatments using means models or regression models.

TABLE 10. Estimates and tests for fixed effects in the broad-sense model for the Antigua response to fertilization study. Denominator df are computed using the Kenward–Rogers approximation.

Parameter	Estimate (se)	Denominator df	*P*-value
Intercept	3.64 (0.54)	5.17	0.001
N	1.28 (0.29)	28.6	< 0.0001
N*N	-0.267 (0.079)	135	0.001

TABLE 11. Predictions of the location-specific coefficients for N response.

Location	Intercept	N	N*N
DBAN	3.14	1.54	-0.267
LFAN	2.39	1.86	-0.267
ORAN	5.70	1.21	-0.267
OVAN	3.90	1.20	-0.267
TEAN	2.53	0.87	-0.267
WEAN	4.19	0.98	-0.267

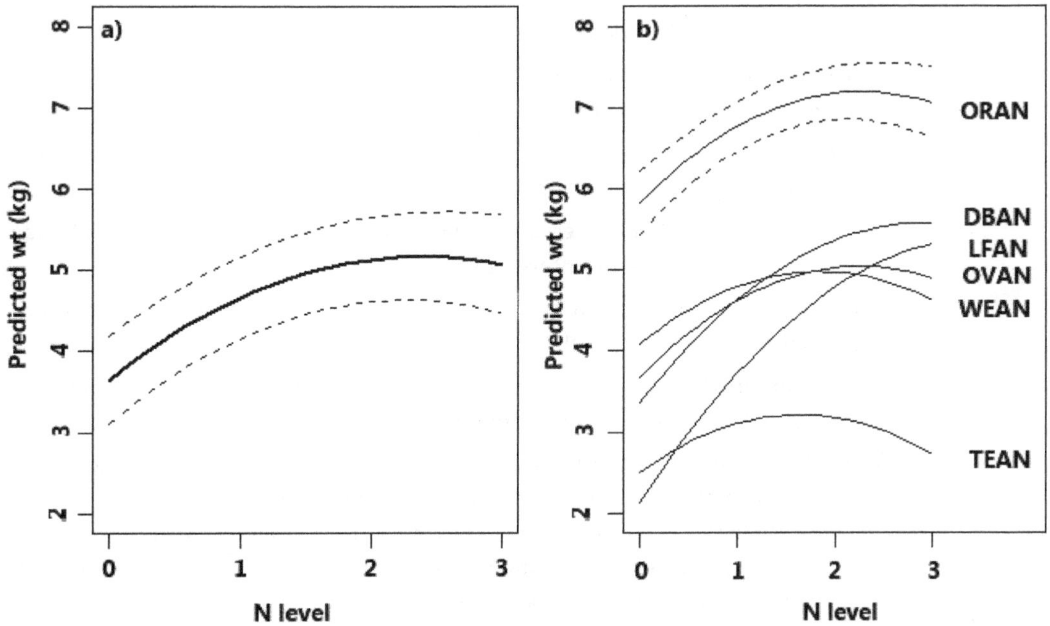

FIG. 5. Overall mean N response curve (panel a) and location-specific predicted N response curves (panel b). Dashed lines indicate +/- 1 standard error around the overall curve and the location-specific curve for the ORAN location.

Key Learning Points:

- What is the difference between broad- and narrow-sense inference about treatment effects?
- How do you specify a model to obtain narrow-sense inference? Broad-sense inference?
- What are consequences, benefits, and disadvantages of using environment-specific variances?
- How do you specify a model with environment-specific error variances?
- How can you decide whether to pool or subdivide treatment by environment interactions when doing broad-sense inference?
- How can a regression model be extended to multiple environments?

Review Questions (T/F)

For all questions, consider an experiment comparing three tillage treatments repeated in 18 environments. The study area at each location is divided into three blocks with three plots in each block. Tillage treatment is randomly assigned to plot within each block. Different plots are used in each environment.

1. In this study, modeling environment effects and block effects as fixed effects or as random effects leads to the same inference about the difference between the two types of tillage.
2. If conclusions about the effects of tillage in a new location are desired, you should use narrow sense inference.
3. Narrow sense conclusions about the effects of tillage usually have smaller standard errors than do broad sense conclusions.
4. To obtain narrow-sense conclusions, omit the treatment by environment interaction from the model.
5. To obtain broad-sense conclusions, define the treatment by environment interaction as a random effect.
6. Broad-sense confidence intervals for the difference between two types of tillage will be based on t distributions with 18 degrees of freedom.
7. The combined analysis across environments requires that plot-plot variation be pooled across environments.
8. The combined analysis across environments requires that variation between blocks be pooled across environments.
9. The 18 environments are actually six locations, each studied for three years. Tillage effects are expected to vary somewhat among locations because of different soil characteristics. Tillage effects are not expected to vary among years at the same location. In this case, subdividing the treatment by environment interaction will have minimal effect on the conclusions about the treatment effect.
10. Imagine that the three tillage treatments are three levels of some quantitative factor, for example, amount of soil disturbance. The data for each environment could be

analyzed using a regression model with a linear effect of soil disturbance. It is possible to construct a combined analysis of those regression models in all 18 environments.

Exercises

1. An experiment was conducted to assess the effect of a fungicide treatment on soybean yield (kg ha^{-1}). It was conducted as an on-farm strip-plot trial with six pairs of side-by-side strips of which one randomly received fungicide treatment. The experiment was repeated at eleven farms (environments). The data were extracted from a much larger dataset provided by the Iowa Soybean Association and are provided in the on-farm soybean dataset in the supplemental materials.

    a. Analyze the experiment separately for each environment.

    b. Evaluate the error variances to determine whether or not they may be considered homogeneous.

    c. You are interested in broad-sense inferences about the average difference between the fungicide and control treatments. Conduct a combined analysis assuming environment and replication to be random factors and treatment as fixed.

    d. Interpret the results of the experiment with respect to the efficacy of fungicide treatment in improving soybean yield.

2. Antonio Mallorino at Iowa State University has studied corn response to P fertilization since 2002. The Prate.csv file contains 13 yr of data from the SouthWest research farm. The design is a RCBD with three blocks of five plots each. Four P levels (0, 28, 56, and 112 lb ac^{-1}) were used; the 0 level was replicated twice in each block. Blocks and plots can be considered independent across years. The response variable is yield in bu $acre^{-1}$.

Consider years to be a fixed factor and Prate to be a continuous variable.

a. What sort of polynomial model is appropriate to describe yield response to Prate? Linear? Quadratic? With one coefficient for all years, or coefficients that differ among years?

Now consider years to be a random factor. Fit a quadratic model that allows the intercept and linear Prate coefficient to vary between years (but the quadratic coefficient is constant).

b. What is the equation that predicts yield as a function of Prate for a year not in the data set, (e.g., 2015)?

c. What is the year-to-year variability in the linear Prate slope? Use the standard deviation to describe that variability.

d. Examine the residuals. Is it appropriate to use yield as the response variable, or should yield be transformed?

e. Apply Levene's test to the residuals to assess whether the error variance differs among years.

f. Refit the model used in Parts b through e with year-specific error variances. Do the answers to Parts B and C change much?

References

Andrews, D.F., and A.M. Herzberg. 1985. Data: A collection of problems from many fields for the student and research worker. Springer, New York. doi:10.1007/978-1-4612-5098-2

Bennett, C.A., and N.L. Franklin. 1954. Statistical analysis in chemistry and the chemical industry. John Wiley & Sons, New York.

Binns, M.R., P.M. Morse, and B.K. Thompson. 1983. Re: "Analysis of combined experiments" by M.S. McIntosh. Agron. J. 75:1056. doi:10.2134/agronj1983.00021962007500060045x

Blouin, D.C., E.P. Webster, and J.A. Bond. 2011. On the analysis of combined experiments. Weed Technol. 25(01):165–169. doi:10.1614/WT-D-10-00047.1

Box, G.E.P. 1953. Non-normality and tests on variances. Biometrika 40:318–335. doi:10.1093/biomet/40.3-4.318

Box, G.E.P. 1954. Some theorems on quadratic forms applied to the study of analysis of variance problems. II. Effects of inequality of variance and of correlation between errors in the two-way classification. Ann. Math. Stat. 25:484–498. doi:10.1214/aoms/1177728717

Burnham, K.P., and D.R. Anderson. 2002. Model selection and multimodel inference, 2nd ed., Springer, New York.

Carmer, S.G., W.M. Walker, and R.D. Seif. 1969. Practical suggestions on pooling variances for F tests of treatment effects. Agron. J. 61(2):334–336. doi:10.2134/agronj1969.00021962006100020051x

Casler, M.D. 2018. Blocking principles for biological experiments. In: B. Glaz and K.M. Yeater, Applied statistics in agricultural, biological,and environmental sciences. ASA, CSSA, SSSA, Madison, WI.

Cochran, W.G., and G.M. Cox. 1957. Experimental designs, 2nd ed. John Wiley & Sons, New York.

Cox, D.R. 1958. Planning of experiments. John Wiley & Sons, New York.

Diggle, P.J., P. Heagerty, K.-Y. Liang, and S.L. Zeger. 2002. Analysis of longitudinal data, 2nd ed. Oxford University Press, Oxford, U.K.

Dixon, P. 2016. Should blocks be fixed or random? Annual Conference on Applied Statistics in Agriculture, Manhattan, KS. 1-3 May 2016. New Prairie Press, Manhattan, KS. http://newprairiepress.org/agstatconference/2016/proceedings/4 (verified 17 Nov. 2017).

Edwards, A.L. 1985. Experimental design in psychological research, 5th ed., Harper and Row, New York.

Fisher, R.A. 1935. The design of experiments. Oliver and Boyd, Edinburgh.

Fox, J. 2008. Applied regression analysis and generalized linear models, 2nd ed. Sage, Thousand Oaks, CA.

Gauch, H.G. 2006. Statistical analysis of yield trials by AMMI and GGE. Crop Sci. 46:1488–1500. doi:10.2135/cropsci2005.07-0193

Gbur, E.E., W.W. Stroup, K.S. McCarter, S. Durham, L.J. Young, M. Christman, M. West, and M. Kramer. 2012. Analysis of generalized linear mixed models in the agricultural and natural resources sciences. Agronomy Society of America, Madison, WI.

Gelman, A. 2005. Analysis of variance: Why it is more important than ever. Ann. Stat. 33(1):1–53. doi:10.1214/009053604000001048

Giesbrecht, F.G. 1989. A general structure for the class of mixed linear models. Application of mixed models in agriculture and related disciplines. Louisiana Agricultural Experiment Station, Baton Rouge, LA. p. 183-201.

Giesbrecht, F.G., and J.C. Burns. 1985. Two-stage analysis based on a mixed model: Large-sample asymptotic theory and small-sample simulation results. Biometrics 41:477–486. doi:10.2307/2530872

Giesbrecht, F.G., and M.L. Gumpertz. 2004. Planning, construction, and statistical analysis of comparative experiments. John Wiley & Sons, New York. doi:10.1002/0471476471

Hocking, R.R. 1985. The analysis of linear models. Brooks/Cole, Monterey, CA.

Kenward, M.G., and J.H. Roger. 1997. Small sample inference for fixed effects from restricted maximum likelihood. Biometrics 53:983–997. doi:10.2307/2533558

Keppel, G., and T.D. Wickens. 2004. Design and analysis: A researcher's handbook. Pearson, Prentice Hall, Upper Saddle River, NJ.

Koricheva, J., J. Gurevitch, and K. Mengersen, editors. 2013. Handbook of meta-analysis in ecology and evolution. Princeton Univ. Press, Princeton NJ. doi:10.1515/9781400846184

Laird, R.J., and F.B. Cady. 1969. Combined analysis of yield data from fertilizer experiments. Agron. J. 61:829–834. doi:10.2134/agronj1969.00021962006100060001x

Madansky, A. 1988. Prescriptions for working statisticians. Springer-Verlag, New York. doi:10.1007/978-1-4612-3794-5

McCulloch, C.E., S.R. Searle, and J.M. Neuhaus. 2008. Generalized, linear, and mixed models. 2nd ed. John Wiley & Sons, New York.

McIntosh, M.S. 1983. Analysis of combined experiments. Agron. J. 75:153–155. doi:10.2134/agronj1983.00021962007500010041x

McLean, R.A., W.L. Sanders, and W.W. Stroup. 1991. A unified approach to mixed linear models. Am. Stat. 45:54–64.

Miguez, F., S. Archontoulis, and H. Dokoohaki. 2018. Nonlinear regression models and applications. In: B. Glaz and K.M. Yeater, editors, Applied statistics in agricultural, biological, and environmental sciences. ASA, CSSA, SSSA, Madison, WI.

Moore, K.J., and P.M. Dixon. 2015. Analysis of combined experiments revisited. Agron. J. 107:763–771. doi:10.2134/agronj13.0485

Nelder, J.A. 1965a. The analysis of randomized experiments with orthogonal block structure. I Block structure and the null analysis of variance. Proc. R. Soc. London, Ser. A 283: 147-162.

Nelder, J.A. 1965b. The analysis of randomized experiments with orthogonal block structure. II Treatment structure and the general analysis of variance. Proc. R. Soc. London, Ser. A 283: 163-178.

Newman, J.A., J. Bergelson, and A. Grafen. 1997. Blocking factors and hypothesis tests in Ecology: Is your Statistics text wrong? Ecology 78:1312–1320. doi:10.1890/0012-9658(1997)078[1312:BFAHTI]2.0.CO;2

Patterson, H.D. and R. Thompson. 1971. Recovery of inter-block information when block sizes are unequal. Biometrika 58: 545-554.

Pearce, S.C. 1988. Analysis of data from agricultural experiments. Trop. Agric. (St Augustine) 65(1):2–5.

Peterson, R.G. 1994. Agricultural field experiments: Design and analysis. Marcel Dekker, New York.

Piepho, H.-P. 2009. Data transformation in statistical analysis of field trials with changing treatment variance. Agron. J. 101:865–869. doi:10.2134/agronj2008.0226x

Piepho, H.P., A. Büchse, and K. Emrich. 2003. A hitchhiker's guide to mixed models for randomized experiments. J. Agron. Crop Sci. 189:310–322. 10.1046/j.1439-037X.2003.00049.x

Piepho, H.P., A. Büchse, and C. Richter. 2004. A mixed modelling approach for randomized experiments with repeated measures. J. Agron. Crop Sci. 190:230–247. doi:10.1111/j.1439-037X.2004.00097.x

Piepho, H.-P., J. Möhring, T. Schulz-Streeck, and J.O. Ogutu. 2012. A stage-wise approach for the analysis of multi-environment trials. Biom. J. 54:844–860. doi:10.1002/bimj.201100219

Piepho, H.P., E.R. Williams, and M. Fleck. 2006. A note on the analysis of designed experiments with complex treatment structure. HortScience 41:446–452.

Richter, C., B. Kroschewski, H.-P. Piepho, and J. Spilke. 2015. Treatment comparison in agricultural field trials accounting for spatial correlation. J. Agric. Sci. 153:1187–1207. doi:10.1017/S0021859614000823

Robinson, G.K. 1991. That BLUP is a good thing– the estimate of random effects. Stat. Sci. 6:15–32. doi:10.1214/ss/1177011926

Robinson, D.L., Thompson, R. & Digby, P.G.N. 1982. REML – a program for the analysis of non-orthogonal data by restricted maximum likelihood. In: H. Caussinus and P. Ettinger, Compstat 1982 Proceedings in Computational Statistics, Part II (supplement), 231-232. Physica-Verlag, Vienna, Austria.

Searle, S.R. 1971. Linear models. John Wiley & Sons, New York.

Springer, B.G.F. 1972. Experimental design and analysis under limited resources. Proceedings of the Caribbean Food Crops Society, Tenth annual meeting, San Juan, Puerto Rico. 16 June 1972. Caribbean Food Crops Society, Mayaguez, Puerto Rico. p. 147-151.

Stroup, W.W. 1989. Why mixed models? Application of mixed models in agriculture and related disciplines. Louisiana Agricultural Experiment Station, Baton Rouge, LA. p. 1-8.

Stroup, W.W. 2013. Generalized linear mixed models: Modern concepts, methods and applications. CRC Press, Boca Raton, FL.

Sulc, R.M., E. van Santen, K.D. Johnson, C.C. Sheaffer, D.J. Undersander, L.W. Bledsoe, D.B. Hogg, and H.R. Willson. 2001. Glandular-haired cultivars reduce potato leafhopper damage in alfalfa. Agron. J. 93:1287–1296. doi:10.2134/agronj2001.1287

Thompson, R.W., H.A. Fribourg, J.C. Waller, W.L. Sanders, J.H. Reynolds, J.M. Phillips, S.P. Schmidt, R.J. Crawford, V.G. Allen, and D.B. Faulkner. 1993. Combined analysis of tall fescue steer grazing studies in the Eastern United States. J. Anim. Sci. 71:1940–1946. doi:10.2527/1993.7171940x

Thompson, S.K. 1992. Sampling. John Wiley & Sons, New York.

Vargas, M., B. Glaz, J. Crossa, and A. Morgounov. 2018. Analysis and interpretation of interactions of fixed and random effects. In: B. Glaz and K.M. Yeater, editors, Applied statistics in agricultural, biological, and environmental sciences. ASA, CSSA, SSSA, Madison, WI.

Wilkinson, G.N. 1970. A general recursive algorithm for analysis of variance. Biometrika 57: 19-46.

Wright, K. 2013. Revisiting Immer's barley data. Am. Stat. 67:129–133. doi:10.1080/00031305.2013.801783

Yates, F. 1937. The design and analysis of factorial experiments of the Commonwealth Bureau of Soil Science. Technical Communication No. 35. Commonwealth Agricultural Bureaux, Farnham Royal, UK.

CHAPTER 9: ANALYSIS OF COVARIANCE

Kevin S. McCarter*

Abstract

This chapter presents analysis of covariance from an applied perspective. A descriptive overview of the analysis of covariance model is provided, and its relationship to analysis of variance models and regression models is discussed. A series of examples illuminates certain problems that can occur when analysis of variance, based only on classification treatment effects, is used to analyze experimental data, and how analysis of covariance can be used to alleviate these problems and improve model performance and the quality of analyses by increasing power, improving precision, and reducing bias. From these examples, this chapter develops a general approach for performing analysis of covariance. Analysis of covariance is then used to analyze data from an actual field trial.

Introduction

This chapter presents the modeling methodology known as analysis of covariance. The chapter begins with an overview of the analysis of covariance model and a discussion of its relationship to analysis of variance models and regression models. The analysis of covariance is a member of the general class of linear models; the chapter identifies the features it shares with these models, and uses these features to provide a description of the model's general structure as well as to help motivate useful interpretations of it. Methods for model parameter estimation and procedures for hypothesis testing are identified. Then, a sequence of examples demonstrate how analysis of covariance can improve statistical inference and avoid certain problems that can occur when simpler analysis of variance models or regression models are used. From these examples, a general procedure for performing analysis of covariance is identified. In the final example, the analysis of covariance procedure is applied to the analysis of data from an actual experiment. The presentation in this chapter is aimed at researchers who are not statistical experts but who do possess knowledge of analysis of variance and regression. For background on analysis

161 Martin D. Woodin Hall, Department of Experimental Statistics, Louisiana State University, Baton Rouge, LA 70803. *Corresponding author (mccarter@lsu.edu)

doi:10.2134/appliedstatistics.2016.0006

Applied Statistics in Agricultural, Biological, and Environmental Sciences
Barry Glaz and Kathleen M. Yeater, editors

of variance, please refer to Chapter 2 (McIntosh, 2018); for background on linear regression, refer to Chapter 6 (Richter and Piepho, 2018); and nonlinear regression is covered in Chapter 15 (Miguez et al., 2018). This chapter expands on that information by focusing on model characteristics and analysis issues that help distinguish analysis of covariance as a distinct methodology. The methods and interpretations presented aim at the analysis of data from designed experiments rather than observational studies.

Description of the ANCOVA Model

The analysis of covariance (ANCOVA) model is related to analysis of variance (ANOVA) models and regression models, each of which has been presented in other chapters in this book. This chapter builds on the understanding that the reader has obtained from those expositions. Analysis of covariance models possess features of both ANOVA models and regression models. The features of these two types of models will be briefly reviewed to facilitate a discussion of the ANCOVA model.

Traditional ANOVA and regression models consist of a continuous response variable, the expected value of which is a function of one or more explanatory variables. In an ANOVA model, the explanatory variables are of the qualitative type, while in a regression model, the explanatory variables are quantitative. Analysis of variance models and regression models are special cases of the more general class of models known as linear models. A linear model is the sum of two modeling components: a deterministic component, which models the expected value, or mean, of the response, and a random component, which provides for the inevitable deviations of the response from its expected value (Graybill, 1976). In this so-called linear model, the relationship between the mean response and the explanatory variables can be quite complex. The explanatory variables may be linearly related to the mean response, but that is not always the case. It is common to have regression models with nonlinear polynomial effects and ANOVA models containing nonlinear, cross-product interaction terms. The name "linear model" derives from the fact that the deterministic part of the model is a linear function, not of the explanatory variables, but of the model parameters it contains (Graybill, 1976). For additional discussion of this class of models, see Chapter 16 (Stroup, 2018).

The ANCOVA model is also a type of linear model; hence, an understanding of the linear model, and of the ANOVA and regression models in particular, provides a useful conceptual framework with which to understand the ANCOVA model. Model parameter estimation and hypothesis testing for the ANCOVA model flow immediately from the general theory of linear models (Graybill, 1976; Milliken and Johnson, 1992). From a practical standpoint, this means that any general-purpose statistical software capable of performing analysis of variance and regression can be used to perform analysis of covariance.

Structurally, the deterministic component of the ANCOVA model combines the characteristics of both the ANOVA model and the regression model. The deterministic component of the ANCOVA model consists of at least one classification variable and at least one quantitative explanatory variable, or regression variable. In the context of ANCOVA, a regression variable is often called a *covariate*. The ANCOVA model can be complex. As in a pure regression model, the regression part of an ANCOVA model can have polynomial effects and interactions between covariates. The ANOVA part of the model can have more than one classification variable and can include their interactions. In addition, the model can include interactions between classification variables and regression variables. See Chapter 7, Vargas et al. (2018), for an in-depth discussion of interactions. This flexibility allows for a wide variety of possible models. As will be seen, how one goes about performing an analysis of covariance depends on the structure of the model.

A common goal in many experiments is to compare the effects of two or more treatments on the mean of a measured response variable. Treatments are randomly assigned to experimental units in some fashion. The method of assignment determines the design structure of the experiment, which must be properly accounted for in the statistical model to obtain valid inferences about the treatments. Ideally, experimental units receiving a given treatment would be as homogeneous as possible. In reality, experimental units often exhibit heterogeneity with respect to characteristics that could have an effect on the response variable of interest. If unaccounted for, this heterogeneity inflates the variation associated with error terms. This has the effect of diminishing the power of treatment comparisons and reducing the precision of estimates of treatment effects and their differences. In addition, heterogeneity across treatment groups that is not accounted for can result in biased estimates of treatment effects and their differences. If these underlying heterogeneous characteristics could be measured, the researcher could account for the concomitant variability in the response variable by inclusion of the underlying heterogeneous characteristics in the statistical model. Treatment comparisons would then be more powerful, and estimates of treatment effects and their differences could be made with greater precision and less bias.

As in ANOVA and regression settings, ANCOVA can be performed for experiments involving complex design structures. These design structures are typically accounted for by the inclusion of random effects or the specification of nonindependent covariance structures for the error terms (Milliken and Johnson, 1992; Littell et al., 2006). Because of the wide variety of possible design structures, and because the defining characteristic of the class of ANCOVA models is a deterministic component consisting of both classification variables and regression variables, the focus in this chapter will be on presenting ANCOVA in the context of relatively simple design structures. Extension of the ANCOVA model to the more complex design structures that one may encounter practice will be similar to the extension of basic ANOVA and regression models to these more complex situations. Analysis of covariance models extended by the inclusion of random effects or the specification of nonindependent error covariance structures are called *linear mixed ANCOVA models*. The same issues involved in other types of mixed models apply to linear mixed ANCOVA models. Excellent information regarding these extensions can be found in Milliken and Johnson (2002) and Littell et al. (2006).

The ANCOVA model combines features of both the ANOVA model and the regression model. Like an ANOVA model, it contains classification variables that account for the treatment structure of the experiment. This includes treatment main effects and can also include interactions among treatments. Like a regression model, it contains one or more quantitative explanatory variables, or covariates, and can contain polynomial covariate effects as well as interactions among the covariate terms. One interpretation of such a model is as an ANOVA-type model in which covariates account for heterogeneity among the experimental units. Another interpretation is as a regression model that fits the same relationship between the mean response and the covariates for all treatment groups, but which allows the regression models for different treatment groups to be offset from one another (Littell et al., 2002; Milliken and Johnson, 2002).

The deterministic part of an ANCOVA model can also include interactions between the two types of explanatory variables, the classification variables and the covariates. This greatly increases the flexibility of the model. From an ANOVA standpoint, such a model can be thought of as one in which treatment effects depend on the values of the covariates. From a regression standpoint, the model can be thought of as one in which the relationship between the mean response and the covariates is allowed to vary across treatment groups. This model fits nonparallel regression lines and/or surfaces, one for each treatment group (Milliken and Johnson, 2002).

The interpretation one chooses may depend on the nature of the experiment and the goals of the research, but quite often both interpretations can be useful for a given experiment. Regardless of the interpretation, inclusion of both classification effects and regression covariates increases the flexibility of the model, which improves the model's ability to account for variation in the response beyond that which would be possible by using a model containing either classification effects or regression covariates alone. The resulting reduction in unexplained variation in the response increases the power of treatment comparisons and the precision with which treatment effects and their differences are estimated.

In this chapter, the use of ANCOVA will be investigated via five examples. The first four examples use simulated data and were designed to illustrate several issues that can arise when performing an ANOVA and how ANCOVA can address these issues and improve the analyses in various ways. To help keep focus on the ANCOVA issues, the same simple experimental design is used across the first four examples. These examples will help readers to organize the various steps that have been utilized into a rough framework for performing ANCOVA in their own experiments. The last example applies the ANCOVA approach that has been developed to actual experimental data involving the comparison of the effect of soybean (*Glycine max* [L.] Merr.) variety on the growth of fall armyworm (*Spodoptera frugiperda* [J.E. Smith]) larvae. All statistical analyses will be performed using SAS software.

Analysis of Covariance Example 1

Consider an experiment to compare the effects of two treatments, denoted *A* and *B*, on the mean of a response variable *Y*. Twenty experimental units were available for the study. Ten experimental units were randomly assigned to Treatment A and the rest assigned to Treatment B. The experimental design consists of a one-way treatment structure and a completely randomized design structure. A single covariate *X* was measured on each experimental unit before its assigned treatment was applied. The response variable *Y* and the covariate *X* were expected to be positively correlated. The primary objective of the study was to determine whether a difference in the mean response existed between treatments, and if so, to quantify the difference via point and confidence interval estimates. The data for this example are presented in Table 1.

In looking at the data, it appears that the values of the response Y tend to be larger under Treatment B than under Treatment A, although the observed difference between the two samples is not large enough to draw a conclusion by visual inspection alone. It also appears that the covariate values tend to be larger for larger values of the response, which is consistent with the stated expectation of a positive correlation between response and covariate. Do the data provide sufficient evidence to allow a firm conclusion to be drawn about whether the treatment means differ? If so, what is the estimate of this difference and how precise is the estimated difference? Can the information provided by the covariate be used to help improve the conclusions? To answer these questions both ANOVA and ANCOVA will be performed and the results compared.

Prior to performing formal statistical analysis, it is a good idea to perform exploratory analysis of the data using graphical and numerical summaries to look for general trends and to check for violations of expected relationships and potential problems in the data. This is especially true when performing analysis of covariance. The SAS statements to enter the data into a SAS dataset are given in Fig. 1. Figure 2 gives the SAS statements to produce univariate summary statistics and boxplots of the response variable and the covariate for each treatment, as well as a scatter plot of the response versus the covariate for the two treatments, superimposed on the same plot.

Side-by-side boxplots of the response variable Y are shown in Fig. 3. From these boxplots it can be seen that the responses under Treatment B tend to be higher than under Treatment A. In particular, the sample mean response under Treatment B is larger than the sample mean response under Treatment A. The variation of the two samples appears to be similar, with Sample B showing only slightly less variation than sample A.

TABLE 1. Dataset for Example 1.

					Treatment A					
Y	71.3	63.4	55.0	54.0	54.6	47.7	49.1	88.1	59.4	70.5
X	12.7	13.3	8.6	7.3	8.2	6.4	7.0	14.2	8.6	10.7
					Treatment B					
Y	63.0	80.9	78.7	85.1	78.5	73.0	53.0	76.3	68.7	84.9
X	7.6	13.4	10.3	14.8	13.6	13.5	5.3	9.9	9.4	14.2

```
data example_1;
input grp $ y x @@;
datalines;
A 71.3 12.7 A 63.4 13.3 A 55.0  8.6 A 54.0  7.3 A 54.6  8.2
A 47.7  6.4 A 49.1  7.0 A 88.1 14.2 A 59.4  8.6 A 70.5 10.7
B 63.0  7.6 B 80.9 13.4 B 78.7 10.3 B 85.1 14.8 B 78.5 13.6
B 73.0 13.5 B 53.0  5.3 B 76.3  9.9 B 68.7  9.4 B 84.9 14.2
;
run;
```

FIG. 1. SAS statements to create the SAS dataset for Example 1.

```
title "Summary of the Dataset";
proc means data=example_1 mean std;
var y x;
by grp;
run;

title "Boxplots of Response Variable Y for Each Group";
proc sgplot data=example_1;
vbox y / group=grp extreme ;
run;

title "Boxplots of Covariate X for Each Group";
proc sgplot data=example_1;
hbox x / group=grp extreme ;
run;

title "Scatter Plot of Response Y vs Covariate X";
proc sgplot data=example_1;
scatter y=y x=x / group=grp;
run;
```

FIG. 2. SAS statements to compute summary statistics, boxplots, and scatter plot for Example 1.

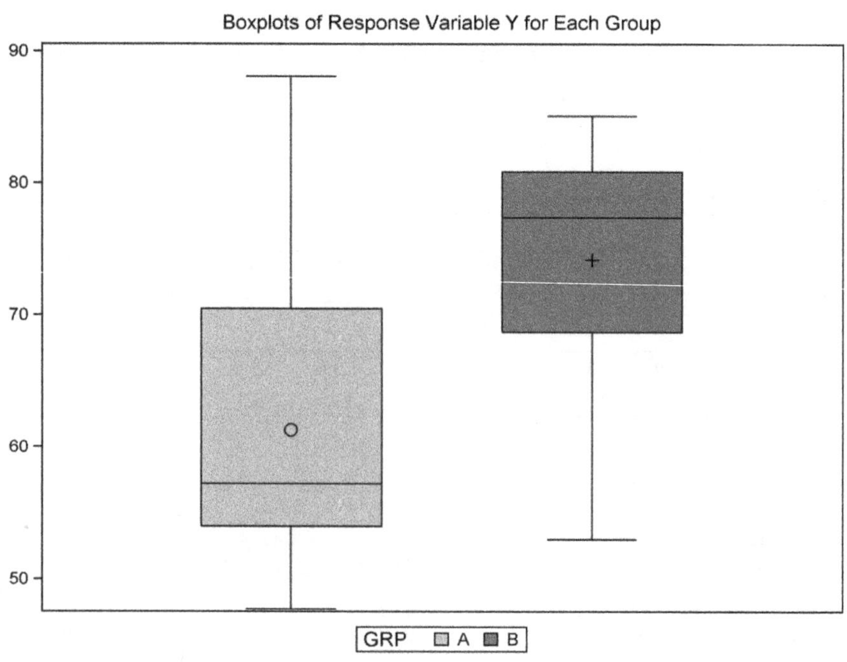

FIG. 3. Boxplots of the response *Y* for Example 1.

Boxplots for the covariate are given in Fig. 4. Based on these boxplots, the two samples appear to be similar with respect to the distribution of the covariate. The covariate mean of the Treatment B assignment group is slightly larger than that of the Treatment A assignment group. The level of variation of the covariate appears to be similar in the two assigned groups.

Summary statistics for both the response and the covariate are given in Table 2. The sample mean response under Treatment B is greater than the sample mean response under Treatment A by 12.9 units. As measured by the standard deviation, the variation in the response is similar under the two treatments. With a sample size of ten experimental units in each treatment group, the estimated standard error of the difference between sample means would be approximately five units, so the observed difference looks to be statistically significant, although perhaps not highly significant. The sample means and the sample standard deviations of the covariate are similar in the two samples. This suggests that the two treatment groups are similar with respect to characteristics which could affect the response variable, as measured by the covariate.

As stated previously, the primary objective is to determine whether a treatment difference exists, and if so, to quantify that difference. To do this, both an ANOVA as well as an ANCOVA of the response Y will be performed. The results from these analyses will then be used to compare the effectiveness of the two approaches. In addition, an ANOVA on the covariate X will be performed to formally test whether the covariate means differ across the two treatment groups. The SAS statements to perform these analyses are provided in Fig. 5.

The results of the ANOVA of Y are given in Table 3. The observed significance level of the test is $p = 0.0203$. The observed difference between treatment means is

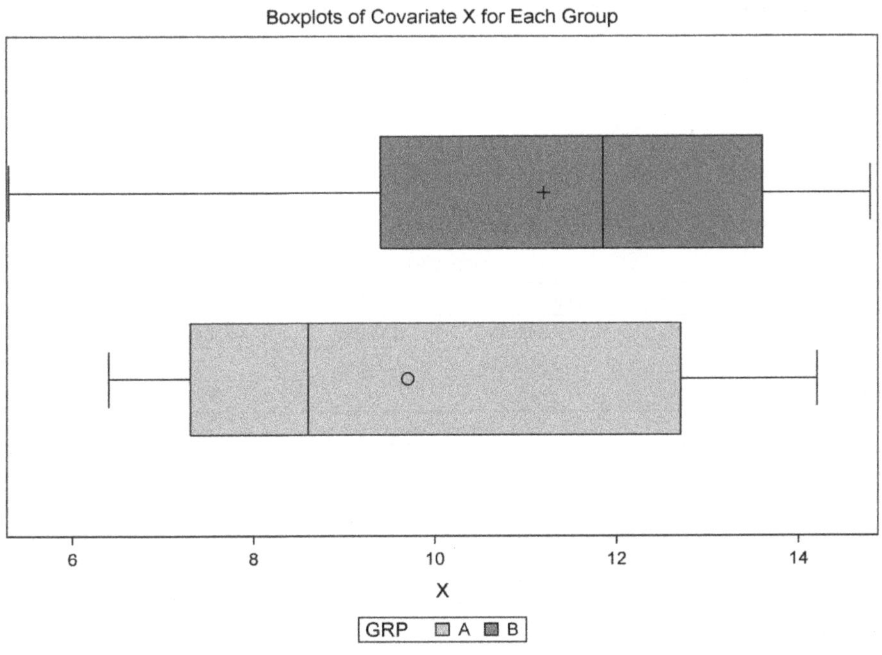

FIG. 4. Boxplots of the covariate X for Example 1.

TABLE 2. Summary statistics for treatments A and B for Example 1.

Treatment	Variable	Mean	Standard Deviation
A	Y	61.3	12.4
	X	9.7	2.8
B	Y	74.2	10.1
	X	11.2	3.2

```
title "ANOVA To Compare Mean of Y Across Groups";
proc mixed data=example_1;
class grp;
model y = grp ;
lsmeans grp / pdiff cl ;
run;

title "ANCOVA To Compare Mean of Y Across Groups, Adjusting for X";
proc mixed data=example_1;
class grp;
model y = x grp ;
lsmeans grp / pdiff cl ;
run;

title "ANOVA To Compare Mean of Covariate X Across Groups";
proc mixed data=example_1;
class grp;
model x = grp ;
run;
```

FIG. 5. SAS statements to perform ANOVA and ANCOVA for Example 1.

TABLE 3. ANOVA results for Example 1.

ANOVA Table					
Source	DF	Sum of Squares	Mean Square	*F* Value	*P* > *F*
Model	1	832.050	832.050	6.48	0.0203
Error	18	2310.478	128.360		
Corrected Total	19	3142.528			
Type III Tests					
Source	DF	Type III SS	Mean Square	*F* Value	*P* > *F*
GRP	1	832.050	832.050	6.48	0.0203

Least Squares Means for Effect GRP		
Point Estimate of TRT B – TRT A	95% Confidence Interval for TRT B – TRT A	
	Lower CL	Upper CL
12.90	2.26	23.54

significant, but not highly significant. Using a 5% significance level, the conclusion is that a treatment difference exists. The point estimate of the difference between the mean response under Treatment B and the mean response under Treatment A is 12.90 units. The 95% confidence interval estimate of the difference is [2.26, 23.54]. Because the interval does not contain zero, it could be used by itself to conclude that a difference exists between treatment means at the 5% level. Note that the width of this confidence interval is 21.28 units. Finally, note that from the ANOVA table, the point estimate of the variance of the error distribution is 128.360 units squared.

The magnitude of the error variance is important to consider, as it affects the overall significance of the model and the precision of confidence intervals. The error variance is a measure of the uncertainty with which the response is measured. If the model does not capture variation in the response, it contributes to the estimate of the error variance. To the extent that we can account for variation in a response variable with a model, we can increase the power of hypothesis tests and increase the precision of confidence interval estimates. Therefore, if we can add a variable to the model that accounts for significant additional variation in the response, then we should be able to increase the power and precision of our analysis. That is the reason for adding a covariate to the model and forming an ANCOVA model.

The results of the ANCOVA are given in Table 4. The only structural difference between this ANCOVA model and the ANOVA model described above is the inclusion of the covariate X in the model. Notice first what has happened to the error variance compared with that in the first model. The estimated error variance has been reduced from 128.360 to 29.258. Thus, the inclusion of the covariate X has resulted in a reduction in the estimated error variance of over 75%. Such a reduction has a very positive inferential impact on the analysis, increasing power as well as precision. The p-value for the overall model is now < 0.0001. The p-value of the covariate is < 0.0001, indicating that the linear relationship between the response and the covariate is highly significant. Having accounted for the variation in the response associated with the covariate, the p-value for the treatment effect has been reduced from 0.0203 to 0.0057. This is a 72% reduction in the treatment effect p-value compared with that obtained from the ANOVA model. The result of this improvement is that the treatment effect is much more significant now that the covariate has

TABLE 4. ANCOVA results for Example 1.

ANOVA Table					
Source	DF	Sum of Squares	Mean Square	*F* Value	*P* > *F*
Model	2	2645.139	1322.569	45.20	< 0.0001
Error	17	497.389	29.258		
Corrected Total	19	3142.528			

Type III Tests					
Source	DF	Type III SS	Mean Square	*F* Value	*P* > *F*
X	1	1813.089	1813.089	61.97	< 0.0001
GRP	1	291.926	291.926	9.98	0.0057

Least Squares Means for Effect GRP		
Point Estimate of TRT B – TRT A	95% Confidence Interval for TRT B – TRT A	
	Lower CL	Upper CL
7.90	2.62	13.18

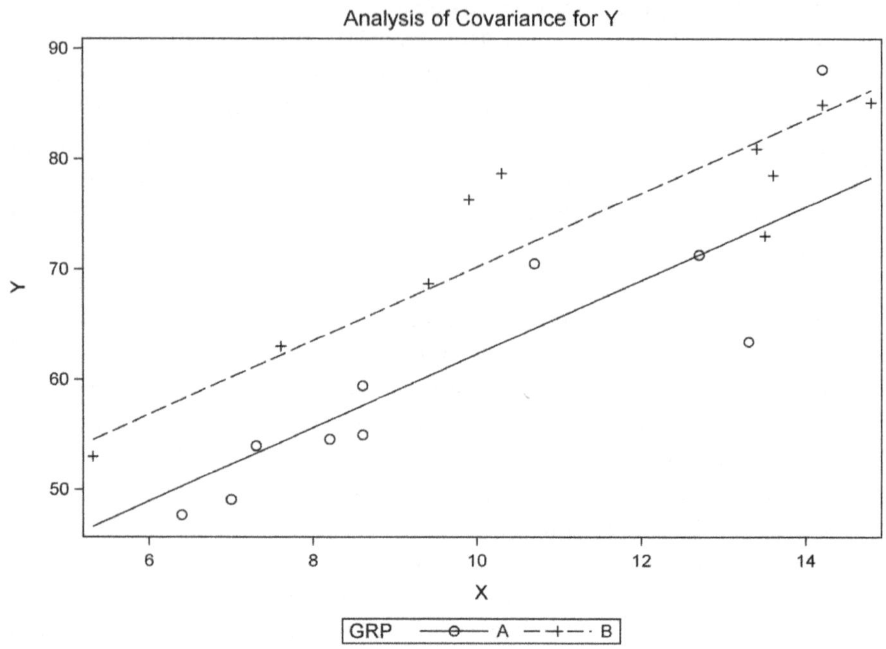

FIG. 6. Scatter plot of the response variable and covariate, with the fitted ANCOVA model for Example 1.

been added to the model. The estimate of the difference between the mean response under Treatment B and the mean response under Treatment A is 7.90 units. The 95% confidence interval estimate of the difference is [2.62, 13.18]. The important point to note here is that the width of this interval is 10.56 units, which is much shorter than the interval produced by the ANOVA model. Adding the covariate to the model has greatly improved precision by reducing the width of the confidence interval estimate of the treatment difference by 50%.

To gain insight into why this ANCOVA model provides greater power and better precision than the ANOVA model, it is helpful to compare the boxplots of the response variable in Fig. 3 to the plot in Fig. 6. The boxplots can be loosely thought of as a graphical representation of what the ANOVA model is fitting. In particular, the estimated difference in treatment means obtained from the ANOVA model corresponds to the difference in the location of the mean symbols in those boxplots. The explanatory variable in the ANOVA model is the classification variable identifying treatment group. The ANOVA model sees only the vertical axis. It is therefore capable of fitting only the means of the two treatment groups (or equivalently, the differences of the treatment means from the overall mean).

Figure 6 gives a scatter plot of the response Y versus the covariate X for both treatment groups, along with parallel regression lines running through the points from each treatment group. This plot is a graphical representation of the model that is fitted by the ANCOVA model. The explanatory variables of the ANCOVA model include the covariate as a continuous variable along with the treatment classification variable. The continuous covariate accounts for the variation in the response that is due to the linear relationship between the response variable and the covariate, while the treatment classification variable accounts for the overall difference in means between the two treatments. This graphic corresponds to the conceptualization of the ANCOVA model as the simultaneous fitting of separate parallel regression lines for each treatment group described previously. The estimated difference between treatments, estimated to be 7.90 units, corresponds to the vertical difference between the fitted parallel regression lines. Unaccounted-for variability corresponds to the squared deviations of the points within each treatment group from that treatment group's regression line, rather than from the treatment group's overall mean. The added perspective provided by the horizontal covariate axis makes it much easier to visually detect the difference in treatment means relative to the unaccounted-for variation than is possible with the boxplots, and the ANCOVA model sees the same sharper distinction. The result is that the ANCOVA model provides greater power to detect the treatment difference and greater precision in estimating that difference than does the ANOVA model.

Table 5 gives the results of an analysis of variance to compare the means of the covariate values across the two treatment assignment groups. The p-value of the test is 0.2799, and so there is insufficient evidence to conclude that a difference exists in the covariate mean across treatment groups. Hence, while the covariate values vary within each sample, overall the distribution of covariate values are similar in the two samples. We conclude that while there is heterogeneity within treatment assignment groups, experimental unit characteristics do not differ significantly across the groups.

Let's think conceptually about how the ANCOVA approach works. Variation in the response comes from two main sources: variation due to application of the treatments, and variation due to heterogeneity in characteristics of the experimental units that can affect the response. In the ANOVA model, variation due to the treatments is accounted for by the treatment classification variable. The rest of the variation is unaccounted for, and therefore ends up contributing to the variability of the error term. In a given situation, there may be a number of characteristics that vary from one experimental unit to another and that have an effect on the response. If all of these characteristics could be measured prior to treatment application, then they could be used to account for variation in the response that they induce. It is probably impossible to identify, let alone measure, all such characteristics. We do not need to do that, however. Suppose we can identify and measure a covariate that is highly correlated with the response variable. Then the variability in the response due to heterogeneity in the experimental units could be assessed indirectly by measuring the covariate before application of the treatment. Since the response and covariate are strongly correlated, the relationship between the two variables can be used to measure the underlying variation in the response prior to application of the treatment and then account for this variation in the model, preventing it from contributing to the estimate of the error variance. In this way, heterogeneity in the experimental material that causes variation in the response is extracted and accounted for, leading to more accurate and precise

TABLE 5. ANOVA comparing mean covariate values across treatments for Example 1.

ANOVA Table					
Source	DF	Sum of squares	Mean square	F value	$P > F$
Model	1	11.25	11.25	1.24	0.2799
Error	18	163.18	9.06		
Corrected total	19	174.43			

estimates of treatment effects and more powerful hypothesis tests. The resolving power of the model is therefore substantially enhanced by the inclusion of a quality covariate, one that is highly correlated with the response variable and is measured prior to application of the treatment.

Why must the covariate be measured prior to application of the treatment? The reason is that we want the covariate to extract only that variability in the response that is due to heterogeneity in the experimental units prior to application of the treatment. We do not want the covariate to extract variation that is the result of application of the treatment; treatment effects are included in the model for that purpose. If a covariate could be affected by the treatment, and if it were measured after the treatment had been applied, then the variation in the covariate would come from both the pre-treatment experimental unit heterogeneity and also the effect of the treatment. Its inclusion in the model would extract the treatment effect that we are trying to detect and isolate with the classification effects in the model. The measured treatment effect would be attenuated to the degree that the covariate is affected by the treatment, which would decrease power to detect differences and would also bias estimates of treatment effects. To prevent this, the covariate must be such that it cannot be affected by the treatment being applied, or it must be measured prior to application of the treatment. The safest approach for obtaining a covariate is to always measure the covariate before treatments are applied.

This need for independence of the covariate and the treatment is the reason why the ANCOVA procedures being presented in this chapter are better suited for data from designed experiments than for data from observational studies, as was indicated in the introduction to this chapter. In designed experiments, treatments are randomly assigned to experimental units, and so if covariates are measured before application of treatments, independence of treatment and covariate is guaranteed. On the other hand, in observational studies, rather than being randomly assigned, measured characteristics are typically inherent characteristics of the experimental units themselves and are often correlated. Hence, measured characteristics considered treatments and those utilized as covariates cannot be guaranteed to be independent. Results of ANCOVA from such data can be difficult to interpret correctly. The misuse of covariates in such cases have apparently led many journals in the behavioral sciences to not allow their use when treatments are not randomly assigned (Freund et al., 2010).

Lessons Learned from Example 1

The deterministic part of the ANOVA model includes terms for the treatment effects only, and therefore possesses no way to account for heterogeneity in the experimental material. As a result, any variability due to heterogeneity in the experimental units is left unaccounted for by the deterministic part of the model and ends up in the error terms. This has the effect of inflating the estimated error variance, which diminishes both power and precision. On the other hand, the ANCOVA model includes a covariate as an explanatory variable and therefore is able to account for variability in the response that is associated with heterogeneity in the experimental material. Unaccounted-for variation in the response is reduced, which in turn reduces the estimated error variance. The result is an increase in the power of tests

(Chapter 4, Casler, 2018) as well as an increase in the precision of estimates. This increase in power results in a reduction in the Type 2 error rate for any given significance level used (Chapter 1, Garland-Campbell, 2018).

In this example, based on the ANOVA, the treatment effect was significant, although not highly so. Using the ANCOVA model the treatment effect became highly significant, with the p-value for this effect being 72% smaller that the corresponding p-value from the ANOVA. In addition, the ANCOVA model provided a more precise estimate of the treatment difference, the 95% confidence interval being 50% shorter than the interval obtained from the ANOVA model. Note that because a 5% significance level was used, both models were significant. However, if a 1% significance level had been used, for example, then the treatment effects would still be significant in the ANCOVA model, but the treatment effect would not be significant in the ANOVA model. From the standpoint of both power and precision, including the covariate in the model had a positive inferential impact on the statistical analysis of the data.

Analysis of Covariance Example 2

This example uses the same experimental setup and goals as in Example 1. An experiment is conducted to compare the effects of two treatments, denoted A and B, on the mean of a response variable Y. Twenty experimental units were available for the study, ten randomly assigned to each of the two treatments. A single covariate X was measured on each experimental unit before treatments were applied. The response Y and the covariate X are expected to be correlated. The data for this example are presented in Table 6.

Perusing the data it is evident that the response Y tends to be larger under Treatment B than under Treatment A. In addition, the values of the covariate appear to be larger for larger values of the response. Also, the values of the covariate are larger under Treatment B than under Treatment A. These observations are consistent with the expectation of a correlation between the response and covariate.

The SAS statements to create a dataset containing these data, produce graphical and numerical summaries, and perform ANOVA and ANCOVA analyses are similar to those in Example 1, the only difference being the substitution of the data values for this example.

Side-by-side boxplots of the response variable Y for the two treatments are shown in Fig. 7. The boxplots show a large difference between the means of the two samples relative to the variation in the samples, with the mean response much higher under Treatment B than under Treatment A. Note that there is little overlap between the two samples; in fact, there is almost complete separation. A difference between response means is clear in these boxplots. The variation in the two samples appears to be similar.

TABLE 6. Dataset for Example 2.

					Treatment A					
Y	69.2	59.4	70.2	52.3	61.0	73.9	57.1	64.9	68.2	75.1
X	10.8	10.7	13.1	6.6	9.6	13.4	7.3	10.0	13.8	14.8
					Treatment B					
Y	89.9	101.3	73.2	96.4	86.4	74.8	81.2	97.3	99.4	79.3
X	17.4	21.2	13.4	20.1	17.6	14.8	17.2	20.2	21.9	13.9

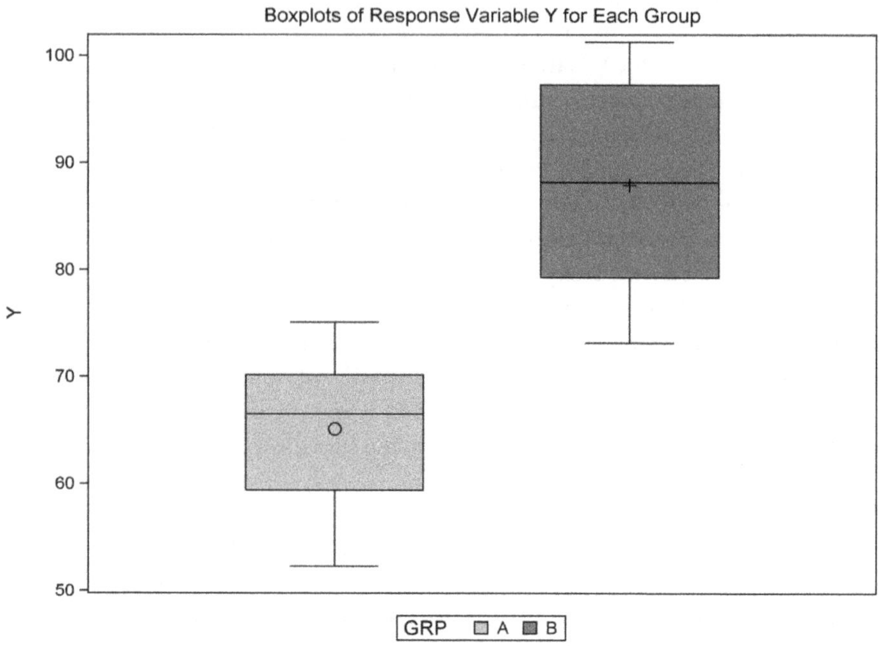

FIG. 7. Boxplots of the response Y for Example 2.

TABLE 7. ANOVA results for Example 2.

ANOVA Table					
Source	DF	Sum of Squares	Mean Square	*F* Value	*P* > *F*
Model	1	2596.921	2596.921	31.32	< 0.0001
Error	18	1492.257	82.903		
Corrected Total	19	4089.178			
Type III Tests					
Source	DF	Type III SS	Mean Square	*F* Value	*P* > *F*
GRP	1	2596.921	2596.921	31.32	< 0.0001
Least Squares Means for Effect GRP					
Point Estimate of TRT B – TRT A		95% Confidence Interval for TRT B – TRT A			
		Lower CL		Upper CL	
22.79		14.24		31.34	

Results of an ANOVA to compare the response means under the two treatments are given in Table 7. The p-value for comparison is < 0.0001, providing very strong evidence of a difference in the response means under the two treatments. The point estimate of the difference between the mean response of Treatment B and the mean response of Treatment A is 22.79 units. The 95% confidence interval estimate of the difference is [14.24, 31.34], the length of which is 17.10 units. Note also that the estimate of the error variance is 82.903 (units squared).

In Example 1, we saw that including the covariate in the model increased power and precision, and a significant treatment effect as assessed by an ANOVA became even more significant when the covariate was added to the model forming an ANCOVA model. Can we expect the same results in this example? Results of an ANCOVA for these data are given in Table 8. Note first that the error variance estimated by the ANCOVA model is 10.480. This is an 87% reduction in the estimated error variance of 82.903 from the ANOVA model. As in Example 1, inclusion of the covariate in the model has substantially reduced the amount of unexplained variation in the response. This will result in an increase in the power of statistical comparisons and also in an improvement in the precision of estimates.

The *p*-value in the ANCOVA table tests the overall null hypothesis that none of the explanatory variables in the model are significant. Specifically, it is testing that no linear relation exists between the response and the covariate, and that there is no treatment effect. The *p*-value of this test is < 0.0001. Hence, there is very strong evidence that this null hypothesis is false. We conclude that either a linear relationship exists between the response and the covariate, that a treatment effect exists, or both. This is expected based on a visual inspection of the raw data and the boxplots, which together suggested both a linear relationship between response and covariate and a pronounced difference in response means across treatment groups.

The section on Type III tests gives results for each of these two hypotheses. The *p*-value for the hypothesis of no linear relationship between response and covariate is < 0.0001, so there is very strong evidence that a linear relationship exists. On the other hand, the *p*-value for the test of no treatment effect is 0.2191. Based on this test, there is insufficient evidence to conclude that a treatment effect exists.

The point estimate of the difference between the response mean of Treatment B and the response mean of Treatment A is 2.92 units. The 95% confidence interval estimate of the difference is [-1.91, 7.75]. Since the interval includes zero, it could be used instead of the test above to conclude that the observed difference in mean response between treatment groups is not statistically significant at the 5% significance level. Note that the width of this interval is 9.66 units. Compared with the width of the interval based on the ANOVA model of 17.10 units, this is a 44% reduction in width. These results show that inclusion of the covariate in the model has once again improved precision as well as the power of tests.

TABLE 8. ANCOVA results for Example 2.

ANOVA Table					
Source	DF	Sum of Squares	Mean Square	*F* Value	*P* > *F*
Model	2	3911.053	1955.527	186.63	< 0.0001
Error	17	178.125	10.480		
Corrected Total	19	4089.178			
Type III Tests					
Source	DF	Type III SS	Mean Square	*F* Value	*P* > *F*
X	1	1314.132	1314.132	125.42	< 0.0001
GRP	1	17.061	17.061	1.63	0.2191

Least Squares Means for Effect GRP		
Point Estimate of TRT B – TRT A	95% Confidence Interval for TRT B – TRT A	
	Lower CL	Upper CL
2.92	-1.91	7.75

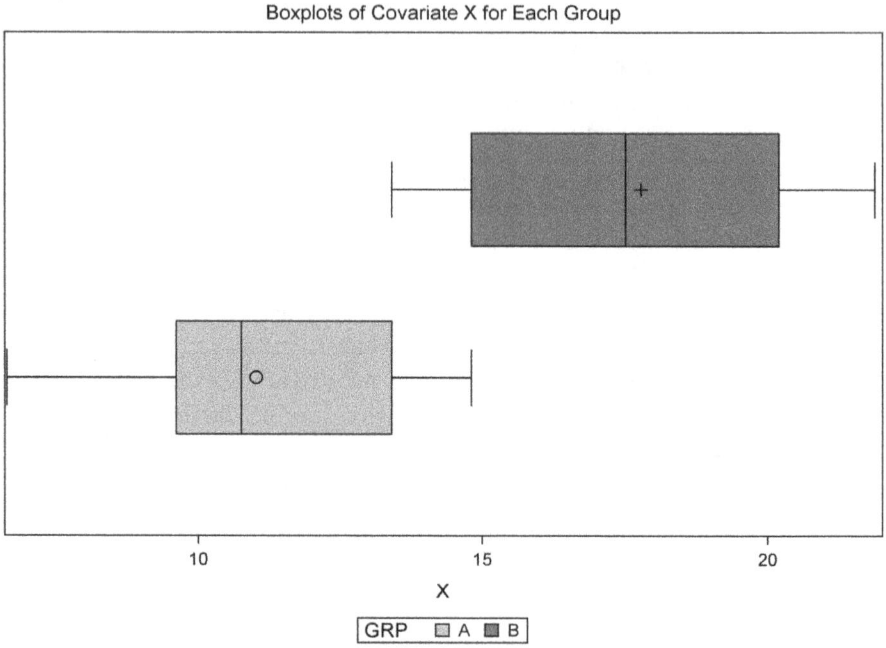

FIG. 8. Boxplots of the covariate *X* for Example 2.

The boxplots of the response and the ANOVA indicated strong evidence of a difference in response means across the two treatment groups; but according to this ANCOVA model, there is insufficient evidence to conclude that a treatment effect exists. Shouldn't the ANCOVA provide more power in making this comparison? A look at how the covariate X is distributed in each assigned treatment group will shed some light on what is going on. Figure 8 gives boxplots of the covariate for each treatment group. From the boxplots, it is clear that the covariate tends to be larger for experimental units assigned to Treatment B than for experimental units assigned to Treatment A.

An ANOVA comparison of the covariate means for the two samples is given in Table 9. Based on this analysis, there is strong evidence that the covariate means are different in the two groups. Note that because the covariate was measured prior to application of the treatment, any differences in the distribution of the covariate in the two samples would be the result of the randomization process.

Additional insight can be obtained by looking at scatter plots of the response variable and covariate for the two treatment groups in Fig. 9. The ANOVA models the mean of the response Y (along the vertical axis) as a function of the treatment group only, ignoring the covariate X (along the horizontal axis). Projecting the points onto the vertical axis provides the same perspective seen in the boxplots in Fig. 7. It is clear why the ANOVA sees a difference in the mean response for the two groups, for indeed the values of the response variable are larger under Treatment B than under Treatment A. Projecting the points onto the horizontal axis provides the same perspective seen in the covariate boxplots in Fig. 8. Again, it is clear from this scatter plot that the covariate values are greater in the sample assigned to Treatment B than in the sample assigned to Treatment A. Finally, from the scatter plot as a whole, it is clear that a strong linear relationship exists between the response and the covariate. This is why the linear effect was highly significant in the ANCOVA model. So why is the treatment effect not significant in the ANCOVA model?

Figure 10 shows the same scatter plot of the response variable and covariate superimposed with the model fit by the ANCOVA. In the ANOVA model, the estimated treatment difference corresponds to the difference between the means of the values projected onto the vertical axis, whereas in the ANCOVA model the estimated treatment difference corresponds to the vertical distance between the fitted regression lines, as indicated in the discussion of this type of plot in Example 1. Hence, it is clear why the treatment difference is not significant in the ANCOVA model. There is relatively little vertical separation between the regression lines fitted by the ANCOVA model. The significant difference in treatment means detected by the ANOVA model was driven mainly by heterogeneity in characteristics of the experimental units that affected the response, which registered as a difference in both the distribution of the response and the distribution of the covariate values across the two samples. This heterogeneity of experimental units across treatment groups

TABLE 9. ANOVA comparing mean covariate values across treatments for Example 2.

ANOVA Table					
Source	DF	Sum of Squares	Mean Square	*F* Value	*P* > *F*
Model	1	228.488	228.488	27.03	< 0.0001
Error	18	152.130	8.452		
Corrected Total	19	380.618			

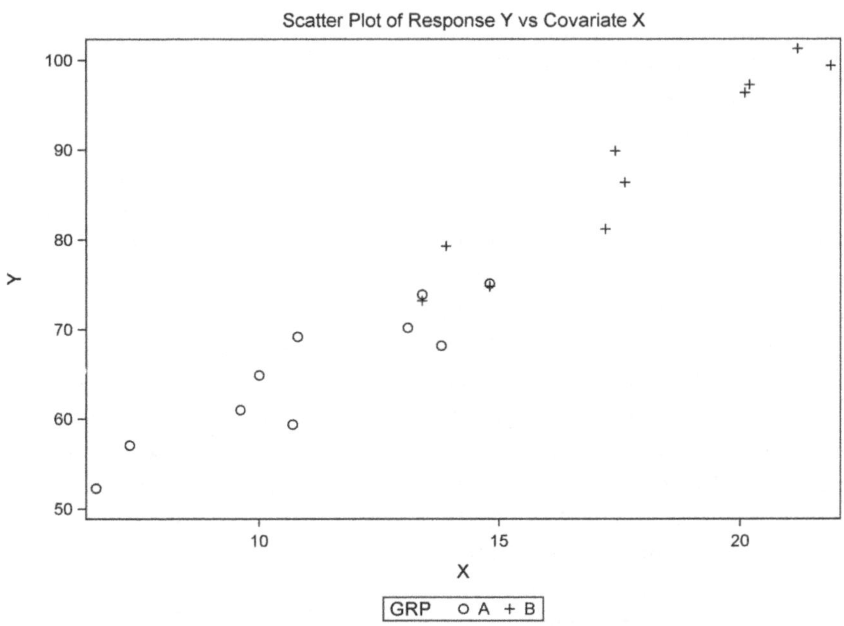

FIG. 9. Scatter plot of the response variable and covariate for Example 2.

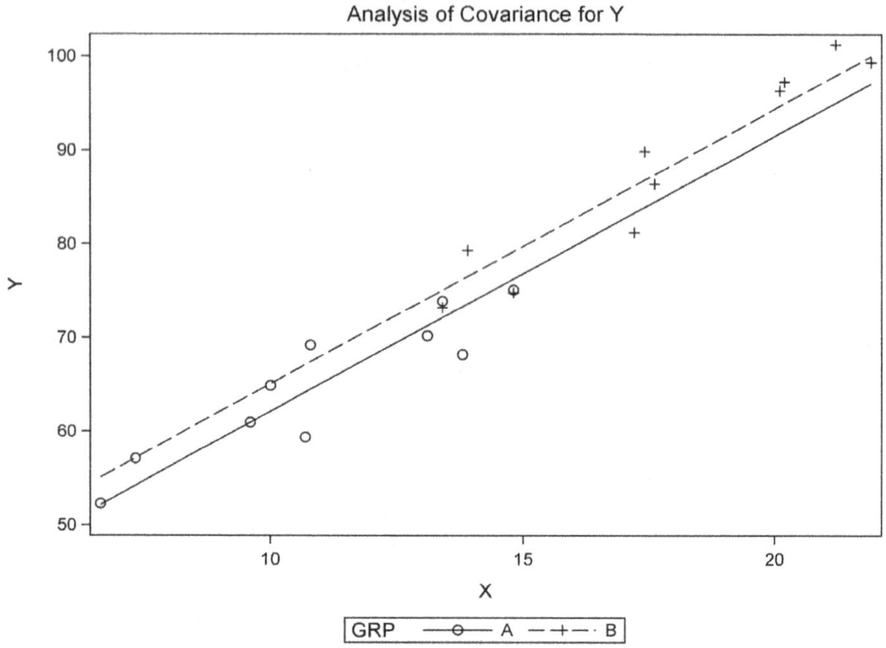

FIG. 10. Scatter plot of the response variable and covariate, with the fitted ANCOVA model for Example 2.

biases the estimate of the difference in treatment means by an amount that is proportional to the difference in covariate means. In addition to increasing power and precision, in this case addition of the covariate to the model has the added benefit of reducing, if not eliminating, this bias. Once this variation has been accounted for by the ANCOVA model, there is little variation remaining that is due to the treatments.

Lessons Learned from Example 2

In this example, the ANOVA model detected a highly significant difference between the treatment group means. However, this difference was not the result of the actual treatment effects, but instead was driven by differences between samples with respect to characteristics of the experimental units that affected the response variable. Because the covariate is correlated with the response, these differences were registered in the covariate, but because the ANOVA model does not take the covariate into account, it is not able to utilize this information. The estimate of the treatment difference was therefore biased, the amount of bias related to the difference in covariate means. On the other hand, because the ANCOVA model does incorporate the covariate, it is able to differentiate variation in the response due to heterogeneity in the underlying characteristics of the experimental units and variation due to the treatment. In other words, it removed the bias from the estimate of the treatment effects. The ANCOVA model therefore provides a more accurate assessment of the significance of the observed difference, as well as a more accurate estimate of the treatment difference. In addition to removing the bias, including the covariate also improved the power and precision of the analysis. This is evidenced by the 87% reduction in the estimated error variance and the 44% reduction in the width of the confidence interval estimate of the treatment difference compared with those obtained from the ANOVA model.

Analysis of Covariance Example 3

This example uses the same experimental setup and goals as in the first two examples. An experiment is conducted to compare the effects of two treatments, denoted A and B, on the mean of a response variable Y. Twenty experimental units were available for the study, ten randomly assigned to each of the two treatments. A single covariate X was measured on each experimental unit before treatments were applied. The response Y and the covariate X are expected to be correlated. The data for this example are presented in Table 10.

The statements to create a SAS dataset, produce graphical and numerical summaries, and perform ANOVA and ANCOVA are similar to those given in Example 1. As we inspect the data visually, it does not appear that the treatment groups differ substantially with respect to the distribution of the response variable Y. This is confirmed by the side-by-side boxplots in Fig. 11, where the means look to be almost identical and the levels of variation are similar.

Summary statistics for the response variable Y are provided in Table 11. From these values we see that the means and standard deviations are nearly identical in the two groups. Based on these summaries there appears to be no evidence of a difference in the two treatment groups.

To quantify this assessment with a formal test, Table 12 gives the results of an ANOVA to compare the mean responses across treatment groups. The p-value of 0.9802 confirms that based on the measured response variable Y alone, there is no evidence of a difference in means across treatments.

The point estimate of the difference between the mean responses under Treatment B and Treatment A is 0.12 units. The 95% confidence interval estimate of the difference is [-9.88, 10.12] units. Since the confidence interval contains zero, it can be used to draw the conclusion that no difference exists between treatment means. The width of this confidence interval is 20.00 units.

In Example 2, use of the ANOVA and ANCOVA models resulted in two qualitatively different conclusions. This difference in conclusions was driven by a difference in the distribution of the covariate across the two samples, which was unaccounted for by the ANOVA model but accounted for by the ANCOVA model. Before drawing final conclusions in this example, we will look at the distribution of the covariate data. We will then perform an ANCOVA and compare results with those of the ANOVA model.

TABLE 10. Dataset for Example 3.

					Treatment A					
Y	81.2	58.7	47.4	49.4	66.1	72.5	71.1	53.5	62.2	68.5
X	14.1	8.8	5.7	5.5	9.1	14.6	12.7	6.2	8.0	12.4
					Treatment B					
Y	56.6	57.5	75.6	68.5	58.0	57.7	62.6	73.9	77.0	44.4
X	15.9	17.9	21.6	19.9	14.1	15.1	16.8	20.6	20.9	13.0

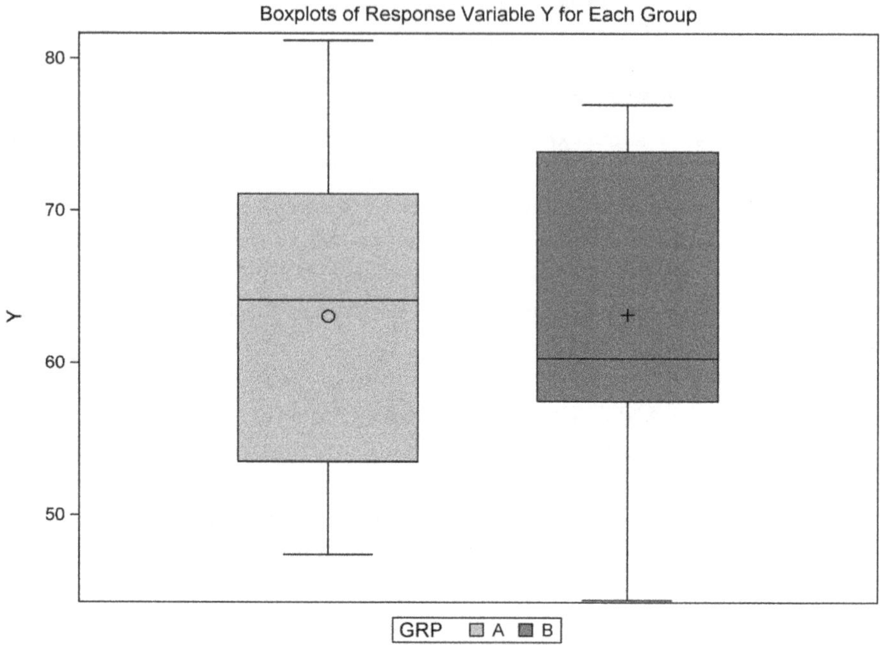

FIG. 11. Boxplots of the response *Y* for Example 3.

TABLE 11. Summary statistics for the response variable *Y* for Example 3.

Treatment	Mean	Standard deviation
A	63.06	10.88
B	63.18	10.40

TABLE 12. ANOVA results for Example 3.

ANOVA Table					
Source	DF	Sum of squares	Mean square	*F* value	*P* > *F*
Model	1	0.072	0.072	0.0006	0.9802
Error	18	2038.540	113.252		
Corrected Total	19	2038.612			
Type III Tests					
Source	DF	Type III SS	Mean square	*F* value	*P* > *F*
GRP	1	0.072	0.072	0.0006	0.9802

Least Squares Means for Effect GRP		
Point Estimate of TRT B – TRT A	95% Confidence Interval for TRT B – TRT A	
	Lower CL	Upper CL
0.12	-9.88	10.12

Figure 12 shows boxplots of the covariate in the two samples. The covariate sample means and medians for the two samples are very far apart relative to the variation in the two samples. In fact, the interquartile intervals do not overlap. The variation in the two samples appears to be about the same, as measured by both the range and interquartile range.

Table 13 gives summary statistics for the covariate for the two samples. The observed difference between sample means is 7.87 units. The standard deviations are close.

Table 14 shows the results of an ANOVA to compare the covariate means across assigned treatment groups. The p-value is < 0.0001, so the observed difference is highly significant. The covariate distributions clearly differ with respect to location in the two groups. Since the ANOVA does not take into account the variation in the response due to the covariate, the conclusion drawn from the ANOVA is highly suspect. As was seen in Example 2, an unaccounted-for difference in covariate distributions can result in biased estimates of treatment effects. Hence, we will perform an ANCOVA to account for the covariate.

Results of the ANCOVA are shown in Table 15. Note first that the estimate of the error variance is 14.486, which is 87% smaller than the estimate of 113.252 for the error variance from the ANOVA model. The p-value from the overall model is < 0.0001; the model is highly significant. The conclusion is that either the response and

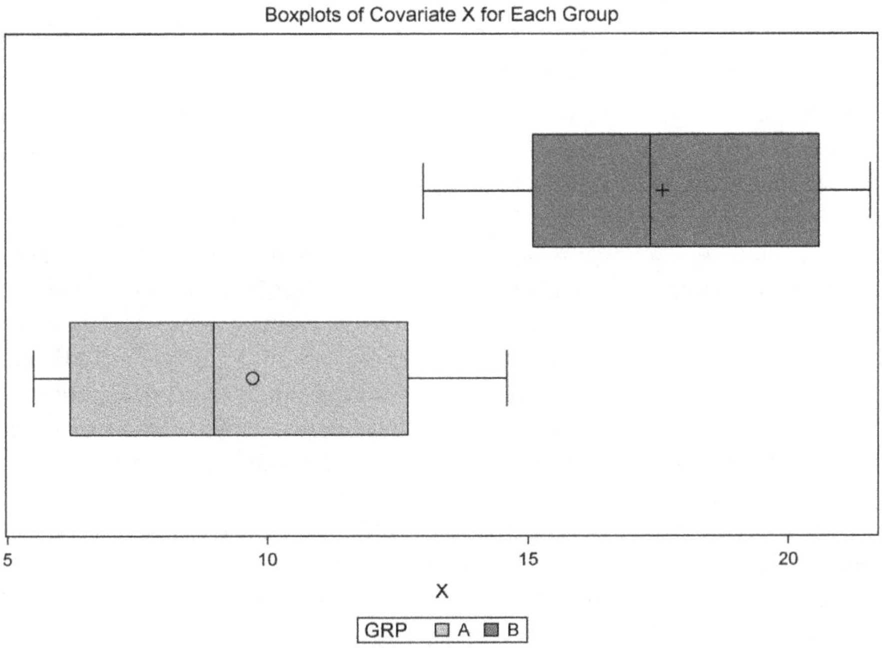

FIG. 12. Boxplots of the covariate X for Example 3.

TABLE 13. Summary statistics for the covariate X for Example 3.

Treatment	Mean	Standard Deviation
A	9.71	3.49
B	17.58	3.06

TABLE 14. ANOVA comparing mean covariate values across treatments for Example 3.

ANOVA Table					
Source	DF	Sum of Squares	Mean Square	*F* Value	*P* > *F*
Model	1	309.685	309.685	28.72	< 0.0001
Error	18	194.065	10.781		
Corrected Total	19	503.750			

TABLE 15. ANCOVA results for Example 3.

ANOVA Table					
Source	DF	Sum of squares	Mean square	*F* Value	*P* > *F*
Model	2	1792.342	896.171	61.86	< 0.0001
Error	17	246.270	14.486		
Corrected Total	19	2038.612			
Type III Tests					
Source	DF	Type III SS	Mean square	*F* Value	*P* > *F*
X	1	1792.270	1792.270	123.72	< 0.0001
GRP	1	1090.785	1090.785	75.30	< 0.0001

Least Squares Means for Effect GRP		
Point Estimate of TRT B – TRT A	95% Confidence Interval for TRT B – TRT A	
	Lower CL	Upper CL
-23.80	-29.58	-18.01

covariate are linearly related, or a treatment effect exists, or both. Results of the individual tests for these effects are given in the section on Type III tests. The test of the null hypothesis that no linear relationship exists between the response and covariate, given that the treatment effect is in the model, has a p-value of < 0.0001, so there is strong evidence of a linear relationship between the response and the covariate. The test of no treatment effect, given that the linear effect is in the model, also has a p-value < 0.0001, so there is strong evidence of a treatment effect as well. This is perhaps surprising, since the sample means were so close, differing by only 0.12 units, and since the ANOVA comparison of the treatment groups was so nonsignificant.

Note that the point estimate of the difference between the mean responses under Treatment B and Treatment A is -23.80 units, which is a much larger difference in magnitude than that estimated by the ANOVA model. The 95% confidence interval estimate of the difference is [-29.58, -18.01]. The width of this interval is 11.57 units, which is 42% smaller than the width estimated using results of the ANOVA model. It is clear that the ANCOVA model provides results that differ substantially from those provided by the ANOVA model.

Figure 13 gives a graphical representation of the fitted ANCOVA model superimposed over a scatter plot of the data. The vertical distance between the regression lines is equal to the estimated treatment difference. With this graph, it is easy to visually distinguish between the two regression lines; this is why the treatment difference is so significant in the ANCOVA model. On the other hand, the ANOVA ignores the covariate. Graphically, this is equivalent to ignoring the horizontal axis and projecting all of the points onto the vertical axis. When this is done, the observed treatment difference becomes very small and insignificant relative to the variability in the responses along the vertical axis. This illustrates clearly why the ANCOVA has so much more power than the ANOVA to detect the treatment difference in this type of situation.

Lessons Learned from Example 3

In this example, the treatment comparison was very nonsignificant based on the ANOVA model, but became highly significant when the ANCOVA model was used. This lack of significance of the ANOVA was the result of unaccounted-for heterogeneity in experimental units across samples with respect to characteristics that affect the response. This unaccounted-for heterogeneity induces extra variation in the response that was unaccounted for by the ANOVA model. This unaccounted-for variability in the response introduced bias in the estimated treatment effects and their difference. In this particular case, this bias substantially offset the actual treatment difference. The unaccounted-for variability in the response also resulted in an inflation of the estimated error variance, which decreased the power of the test and the precision of the confidence interval estimate.

Because the response variable and the covariate are correlated, the extra variation in the response variable that is induced by heterogeneity in the experimental units is also registered in the covariate. The ANCOVA model accounted for this heterogeneity in the experimental units indirectly by including the covariate in the model. Inclusion of the covariate reduced this heterogeneity-induced bias, reducing the estimated error variance by 87%, and reducing the length of the confidence interval estimate of the treatment difference by 42%. The result is a model that provides

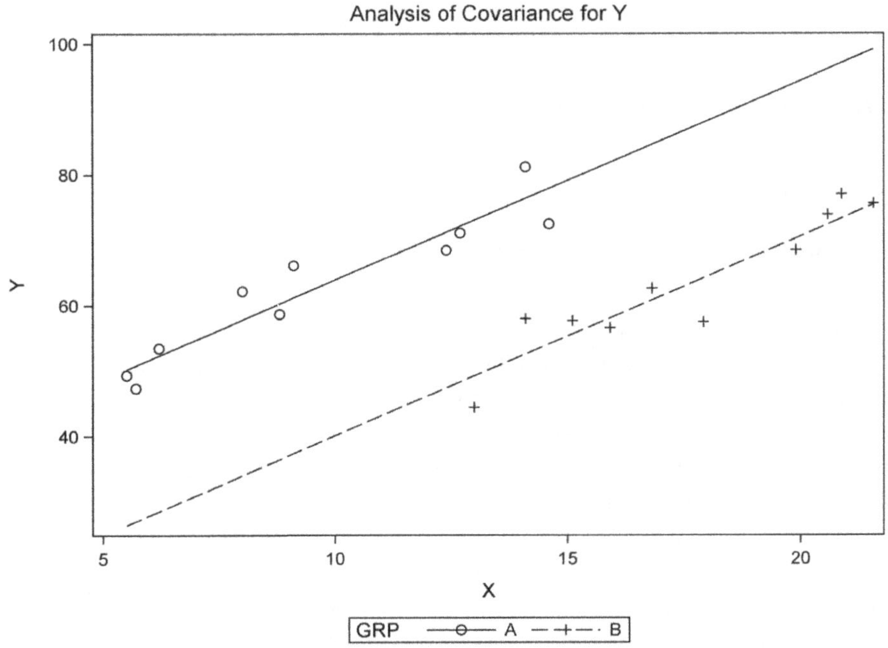

FIG. 13. Scatter plot of the response variable and covariate, with the fitted ANCOVA model for Example 3.

TABLE 16. Dataset for Example 4.

	Treatment A									
Y	55.1	67.1	73.6	64.6	76.4	45.5	47.3	57.4	78.5	61.9
X	8.2	11.0	14.8	9.3	13.8	5.3	6.4	6.6	12.9	8.4
	Treatment B									
Y	61.7	56.3	54.2	68.0	58.5	59.2	60.8	68.2	52.7	78.5
X	9.5	11.3	12.0	9.8	12.6	10.5	10.7	9.8	13.1	6.4

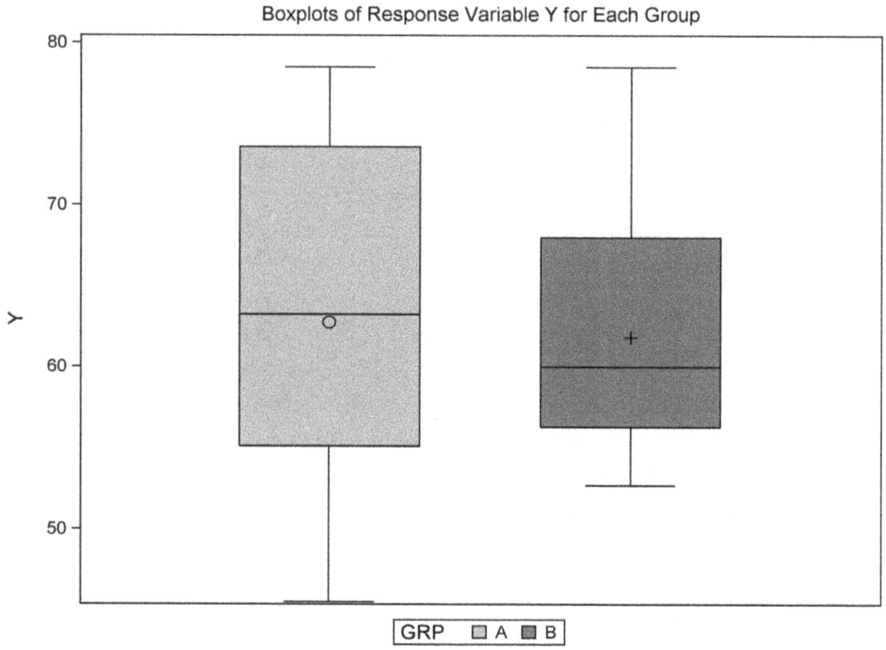

FIG. 14. Boxplots of the response Y for Example 4.

more power in detecting the treatment difference and a more accurate and precise estimate of that difference.

Analysis of Covariance Example 4

This example uses the same experimental setup and goals as the previous examples. An experiment is conducted to compare the effects of two treatments, denoted *A* and *B*, on the mean of a response variable *Y*. Twenty experimental units were available for the study, ten randomly assigned to each of the two treatments. A single covariate *X* was measured on each experimental unit before treatments were applied. The response *Y* and the covariate *X* are expected to be correlated. The data for this example are presented in Table 16.

Boxplots of the response variable under the two treatments are given in Fig. 14. From the boxplots we see that the observed response means are nearly equal. The range and interquartile range are both somewhat smaller under Treatment B than under Treatment A.

Summary statistics by treatment group for the response variable are given in Table 17. As can be seen, the sample means are close, and the standard deviation of the response is slightly smaller under Treatment B than under Treatment A. The p-value of the test of equal variances is $p = 0.2588$, so there is insufficient evidence to conclude that variation in the response differs under the two treatments. Note that the p-value for this comparison of variances is not produced as part of the ANOVA, but rather was obtained by hand using the F-test for comparing two population variances that is found in introductory statistical textbooks.

Results of an ANOVA to compare the response means across treatment groups are given in Table 18. The p-value of the comparison is 0.8352, so there is insufficient evidence to conclude that a difference exists between response means.

The point estimate of the difference between Treatment B and Treatment A is -0.93 units. The 95% confidence interval estimate of the difference is [-10.18, 8.32] units. The confidence interval contains zero, which also implies that there is insufficient evidence to conclude that a difference in response means exists at the 5% significance level. The width of the confidence interval is 18.50 units.

Examples 2 and 3 demonstrated that ANOVA can lead to wrong conclusions if differences exist in the covariate distribution across assigned treatment groups, and how ANCOVA can account for such differences and improve decision making. It is important to know if such differences exist, so we will now look at the distribution of the covariate. Boxplots of the covariate are given in Fig. 15. From the boxplots we see that the covariate means are similar. The range of covariate values is similar as well, although the interquartile range is quite a bit smaller in the group assigned to Treatment B.

Summary statistics for the covariate are given in Table 19. As can be seen, the covariate means and standard deviations are numerically similar across the two

TABLE 17. Summary statistics for the response variable Y for Example 4.

Treatment	Mean	Standard deviation
A	62.74	11.54
B	61.81	7.80

TABLE 18. ANOVA results for Example 4.

ANOVA Table					
Source	DF	Sum of Squares	Mean Square	F Value	$P > F$
Model	1	4.325	4.325	0.04	0.8352
Error	18	1746.363	97.020		
Corrected Total	19	1750.678			

Type III Tests					
Source	DF	Type III SS	Mean Square	F Value	$P > F$
GRP	1	4.325	4.325	0.04	0.8352

Least Squares Means for Effect GRP		
Point Estimate of TRT B – TRT A	95% Confidence Interval for TRT B – TRT A	
	Lower CL	Upper CL
-0.93	-10.18	8.32

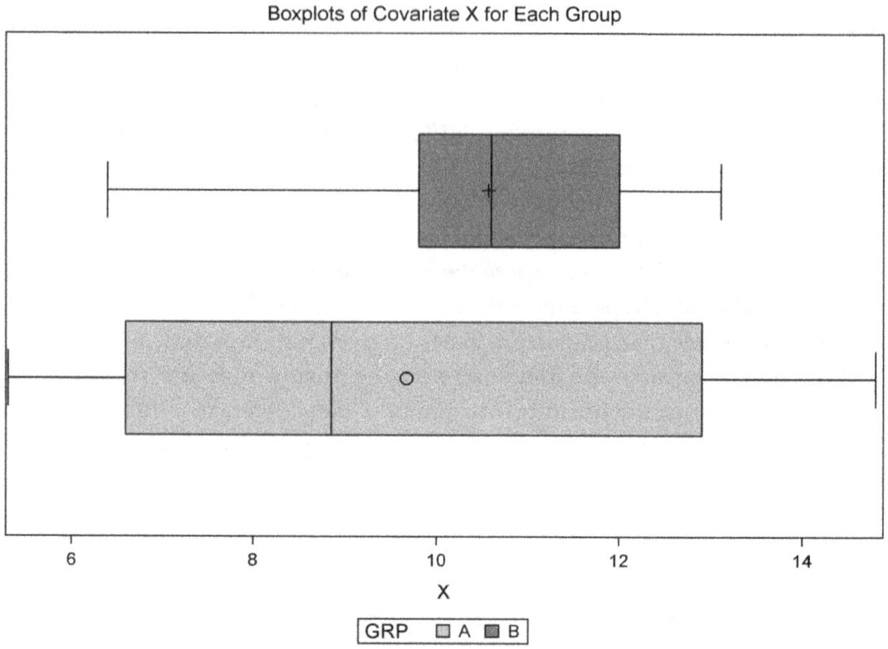

FIG. 15. Boxplots of the covariate X for Example 4.

TABLE 19. Summary statistics for the covariate *X* for Example 4.

Treatment	Mean	Standard deviation
A	9.67	3.31
B	10.57	1.91

assigned treatment groups. An *F*-test of the hypothesis of equal variances resulted in a *p*-value of 0.1170.

Results of an ANOVA to compare the covariate means across assigned treatment groups is given in Table 20. The *p*-value is 0.4662, and hence there is insufficient evidence to conclude that the assigned treatment groups are different with respect to the covariate means.

Even though the covariate means do not appear to differ, as we have seen in each of the previous examples including a quality covariate in the analysis can increase power and precision by accounting for variation in the response that is due to heterogeneity in the experimental units. Results of an ANCOVA to compare the response means across treatments are given in Table 21. The *p*-value for the overall ANCOVA model is 0.1965. Interestingly, the overall ANCOVA model is not significant.

This result is perhaps a bit surprising. While we may have expected the treatment effect to be nonsignificant based on the summary statistics and results of the ANOVA, given that the response and covariate were expected to be correlated, we would have expected the covariate to be significant, as it has been in previous examples. However, the covariate is not significant at the 5% significance level.

In previous examples, including the covariate has resulted in a significant decrease in the estimated error variance and in the width of the confidence interval estimate of the treatment difference. In this example, the estimated error variance in the ANOVA model is 97.020 and for the ANCOVA model it is 85.040, so they are about the same. Based on the ANOVA model, the width of the confidence interval estimate of the treatment difference is 18.50, while for the ANCOVA model it is 17.66, again about the same. In this example, including the covariate in the model has not resulted in much of a decrease in unexplained variation or in the precision of the confidence interval estimate. Why is this?

The answer can be obtained by careful inspection of Fig. 16, which gives the graphical representation of the fitted ANCOVA model superimposed over a scatter plot of the response and covariate values. Notice that in this case the two parallel regression lines do not fit the data well. As in previous examples, the assumption that the regression lines are parallel is implicit in this ANCOVA model. Looking back at the ANCOVA plots of previous examples, we can see that the assumption of parallel lines has been appropriate in those cases. In this example, however, it is clear from the ANCOVA plot that while the relationship between the response and the covariate does appear to be linear in each group, the linear relationships are not the same. In particular, in Treatment Group A there is an increasing relationship between the covariate and the response, whereas in Treatment Group B the relationship is decreasing. In this case, therefore, the ANCOVA model that imposes a parallel lines assumption is too restrictive and can therefore lead to erroneous conclusions. For these data, a more flexible ANCOVA model that allows the linear relationship between the response and the covariate to be different in the two treatment groups is needed.

TABLE 20. ANOVA comparing mean covariate values across treatments for Example 4.

ANOVA Table					
Source	DF	Sum of Squares	Mean Square	*F* Value	*P* > *F*
Model	1	4.050	4.050	0.55	0.4662
Error	18	131.542	7.308		
Corrected Total	19	135.592			

TABLE 21. ANCOVA results for Example 4.

ANOVA Table					
Source	DF	Sum of Squares	Mean Square	*F* Value	*P* > *F*
Model	2	304.999	152.499	1.79	0.1965
Error	17	1445.679	85.040		
Corrected Total	19	1750.678			
Type III Tests					
Source	DF	Type III SS	Mean Square	*F* Value	*P* > *F*
X	1	300.674	300.674	3.54	0.0773
GRP	1	25.453	25.453	0.30	0.5914

Least Squares Means for Effect GRP		
Point Estimate of TRT B – TRT A	95% Confidence Interval for TRT B – TRT A	
	Lower CL	Upper CL
-2.29	-11.12	6.54

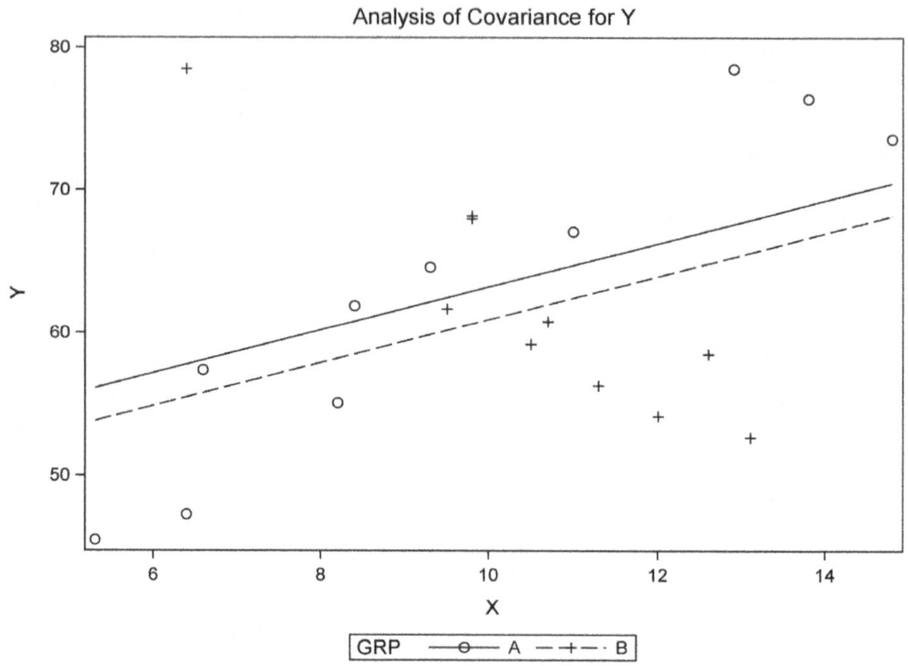

FIG. 16. Scatter plot of the response variable and covariate, with the fitted parallel-lines ANCOVA model for Example 4.

```
title1 "ANCOVA To Compare Mean of Y Across Groups, Adjusting for X";
title2 "Allowing for Separate Slopes";
proc mixed data=example_4;
class grp;
model y = x grp x*grp ;
lsmeans grp / pdiff cl ;
run;
```

FIG. 17. SAS statements to perform ANCOVA (nonparallel-lines model) for Example 4.

Fortunately, it is straightforward to extend the ANCOVA model and give it the flexibility to allow the slopes to vary across treatment groups. To do this, we include a term for the interaction between treatment group and the covariate. In such a model, the main effect for the covariate fits an overall average slope, and the interaction term fits a slope deviation from this average slope for each group. Fig. 17 shows the SAS statements for this extended model. The code also contains an LSMEANS statement with options for comparing treatment groups.

Table 22 shows the results of this analysis. Note that while the previous model was not significant, this model is highly significant. The *p*-value associated with the X×GRP interaction tests the null hypothesis that the slopes of the regression lines are the same in the two treatment groups. With the *p*-value < 0.0001, there is strong evidence that the slopes are not equal. The main effect for the covariate is not significant (*p*-value = 0.5173), whereas the main effect for treatment is significant (*p*-value < 0.0001). Because the interaction between treatment group and covariate is significant, however, as advised by Vargas et al. (2018) in Chapter 7, we will refrain from interpreting the main effects and their *p*-values and focus on the interaction when making comparisons.

Because the slopes can be different, this model provides a much better fit to the data than the previous ANCOVA model. This improvement in fit can be assessed visually in the ANCOVA plot in Fig. 18, which provides a graphical representation of this fitted ANCOVA model superimposed over a scatter plot of the response and covariate. From Table 22, we see that allowing the slopes to differ has resulted in an 83% reduction in the estimated error variance, which has decreased from 85.040 in the previous ANCOVA model to 14.060 in the current model, at the cost of a single error degree of freedom. This improvement in model fit results in more powerful tests and more precise estimates of treatment effects and their difference.

The p-value for the comparison of treatment means is 0.6858. The point estimate of the difference between the mean of Treatment B and the mean of Treatment A is -0.71 units, with the 95% confidence interval estimate being [-4.33, 2.92]. According to these results, there is insufficient evidence to conclude that a difference exists between treatment means. This may seem a bit surprising at first because the main effect for treatment was significant, but as stated previously, we need to be careful interpreting comparisons involving main effects when the model includes a significant interaction term. Another look at the ANCOVA plot will provide insight into what is going on.

Recall from discussions of the ANCOVA plot in previous examples that the vertical difference between the fitted regression lines is equal to the estimated treatment difference. From the ANCOVA plot in Fig. 18, it is clear that the vertical distance between fitted regression lines depends on the value of the covariate at which the vertical distance is computed. Hence, to compare treatment means through a hypothesis test or by estimating their difference, we need to specify the value of the covariate at which the comparison is to be made. Looking back at the LSMEANS statement in Fig. 17, note that no such specification was explicitly made, and yet SAS produced a test comparing the treatment means and estimated their difference. At what value of the covariate was this comparison made?

TABLE 22. ANCOVA (non-parallel slopes) results for Example 4.

ANOVA Table					
Source	DF	Sum of Squares	Mean Square	F Value	P > F
Model	3	1525.722	508.574	36.17	< 0.0001
Error	16	224.955	14.060		
Corrected Total	19	1750.678			

Type III Tests					
Source	DF	Type III SS	Mean Square	F Value	P > F
X	1	6.165	6.165	0.44	0.5173
GRP	1	1093.234	1093.234	77.76	< 0.0001
X*GRP	1	1220.724	1220.724	86.82	< 0.0001

Least Squares Means for Effect GRP					
			TRT B – TRT A		
TRT A	TRT B	P-value	Point Estimate	95% Confidence Interval	
				Lower CL	Upper CL
64.21	63.51	0.6858	-0.71	-4.33	2.92

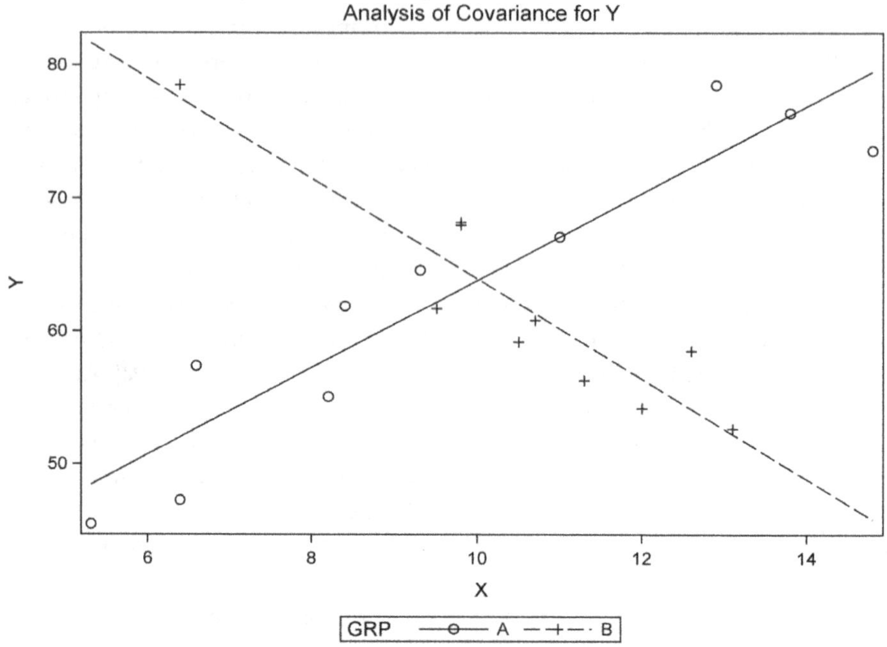

FIG. 18. Scatter plot of the response variable and covariate, with the fitted nonparallel-lines ANCOVA model for Example 4.

It turns out that the default behavior in SAS is to make such comparisons at the overall mean value of the covariate. To make a comparison at another value of the covariate, the AT option of the LSMEANS statement can be used. In this example, the overall mean of the covariate is 10.12. Figure 19 gives the code that explicitly instructs SAS to make the comparison of treatments at the covariate mean value of X = 10.12. This code will produce the same results obtained above using the code which did not specify the value at which to make the comparison. To make the comparison at a different value of the covariate, simply substitute that value in place of 10.12 in the AT statement.

Table 23 gives the results of comparing the two treatments at various values of the covariate. The results of the comparison depend on the particular value of the covariate at which the comparison is made. For example, using a 5% significance level, the treatment means would not be determined to be different at covariate values of 9.5, 10.0, or 10.5, but would be determined to be different at covariate values of 8.0, 8.5, 9.0, 11.0, and 11.5.

When slopes are different across treatment groups, the researcher should explicitly choose the covariate values at which to make treatment comparisons, rather than relying on the default behavior of the software. The covariate values at which to make comparisons must be determined by the researcher in each particular situation and will typically be driven by the research questions being investigated.

Lessons Learned from Example 4

When performing ANCOVA it is imperative that the covariate portion of the model be specified correctly (Milliken and Johnson, 2002). The relationship between the

```
title1 "ANCOVA To Compare Mean of Y Across Groups, Adjusting for X";
title2 'Allowing for Separate Slopes";
proc mixed data=example_4;
class grp;
model y = x grp x*grp ;
lsmeans grp / pdiff cl at x=10.12 ;
run;
```

FIG. 19. Performing comparisons at a particular covariate value using the AT keyword.

response and covariate must be determined and correctly specified to draw accurate, reliable conclusions from an ANCOVA. The first thing that must be determined is the nature of the relationship between the response and the covariate. Is the relationship linear, or does some higher order relationship exist? Scatter plots of the response and covariate can help determine the nature of the relationship. The next thing to determine is whether the relationship between the response and covariate is the same in each treatment group. Again, scatter plots can help answer this question. This can also be addressed formally via modeling by including a term for the interaction between the treatment classification variable and the covariate. This is an important step that should be routinely performed. The *p*-value for the interaction term tests the null hypothesis that the relationship is the same across all treatment groups. If the interaction term is significant, it can be retained, resulting in a model that fits nonparallel lines or surfaces. In this case, the researcher should explicitly specify the values of the covariate at which treatment comparisons are to be made. If the interaction term is not significant, then it can be removed and a parallel lines or surfaces model can reasonably be used. In this case, the software can be allowed to make treatment comparisons at the default value of the covariate, since the difference will be the same at every value of the covariate. In either case, it is only after the covariate part of the model has been correctly specified that treatment comparisons should be made.

Summary of Lessons Learned from Example 1 through Example 4

From the examples considered thus far, we have gleaned several things about ANCOVA.

A quality covariate can have a very positive inferential impact on an analysis to compare treatment effects. Specifically, a quality covariate can increase the power of

TABLE 23. ANCOVA comparison of treatments at specified values of the covariate for Example 4.

Value of Covariate	TRT A	TRT B	*p*-value	TRT B – TRT A		
				Point Estimate	95% Confidence Interval	
					Lower CL	Upper CL
8.0	57.28	71.50	< 0.0001	14.22	9.01	19.43
8.5	58.92	69.61	0.0002	10.70	6.03	15.36
9.0	60.55	67.73	0.0023	7.18	2.97	11.38
9.5	62.18	65.84	0.0613	3.66	-0.20	7.51
10.0	63.82	63.96	0.9364	0.14	-3.51	3.79
10.2	64.21	63.51	0.6858	-0.71	-4.33	2.92
10.5	65.45	62.07	0.0651	-3.38	-7.00	0.24
11.0	67.09	60.19	0.0013	-6.90	-10.66	-3.14
11.5	68.72	58.30	< 0.0001	-10.42	-14.47	-6.36

hypothesis tests and increase the precision of estimates by reducing unaccounted-for variation in the response variable that is due to heterogeneity in experimental units with respect to characteristics that affect the response. Including a covariate can also reduce or eliminate bias in estimates of treatment effects and their differences in situations where the assigned treatment groups vary with respect to such characteristics. As has been demonstrated, not including a covariate in such a case can lead to qualitatively incorrect conclusions regarding treatment effects, as well as biased estimates of treatment effects and their differences.

To be a quality covariate, the covariate must be correlated to the response variable. The stronger the correlation, the better. To avoid artificially diminishing treatment effects, the covariate should either be such that it cannot be affected by the treatment, or it must be measured before application of the treatment to the experimental units. The safest approach is always to measure the covariate before treatments are applied to the experimental units.

Before making treatment comparisons in the context of an ANCOVA, it is imperative that the relationship between the response and covariate be determined and correctly specified. As part of this, it must be determined whether the relationship between response and covariate is the same across all treatment groups. This can be formally tested by including appropriate interaction terms in the model and determining whether the interactions are statistically significant. Only when the covariate part of the model has been determined to be adequate should treatment comparisons be performed. If the relationship between response and covariate is determined to differ across treatment groups, the researcher should explicitly choose the covariate values at which to make treatment comparisons.

ANCOVA Example 5: Do Fall Armyworm Larvae Grow Better on Some Soybean Varieties than Others?

Researchers performed an experiment to determine whether fall armyworm (*Spodoptera frugiperda* J.E. Smith) larvae grow better on some soybean (*Glycine max* [L.] Merr.) varieties than others. Forty fall armyworm larvae were used in the experiment, with ten larvae randomly assigned to each of four soybean varieties. The experiment therefore consisted of a one-way treatment structure with four levels and a completely randomized design structure. On day zero of the experiment, each larva's initial weight was obtained for use as a covariate. Then for each armyworm larva, one leaflet was removed from that larva's assigned soy variety and used to feed the larva for two days. The primary response variable was final armyworm weight, which was measured at the end of the experiment on day two. It was believed that final weight may depend on initial weight. One way to account for variation in final armyworm weight due to variation in initial larval weight is to include initial weight as a covariate in an ANCOVA. To the extent that final weight and initial weight are correlated, this should improve the fit of the model and the performance of the resulting analyses. This was the approach used in this example.

Boxplots of the final larvae weights are given in Fig. 20. The levels of variability in the samples look similar. There is some variability in the sample means, with the biggest observed difference being between the weights of the larvae assigned to the Davis and Braxton varieties.

An ANOVA to compare the mean final weights across the four treatments is given in Table 24. The p-value for the comparison is 0.1284, so at the 5% significance level there is insufficient evidence to reject the null hypothesis that the mean weights are the same across all four soybean varieties. Because the overall ANOVA test is not significant, we will not perform pairwise comparisons. Note that for comparison with the subsequent ANCOVA analysis, the common width of the 95% confidence intervals for differences in mean larvae weights between soybean variety means is 12.30 units. Note also that the estimated error variance based on this ANOVA model is 49.78.

Recall from previous discussions that the degree to which a covariate will improve an analysis is dependent on the strength of its relationship with the response variable and on the degree to which treatment groups differ with respect to the distribution of the covariate. To get an idea of the improvement that we might expect from the covariate in this analysis, we will next look at both of these factors. Table 25 gives correlations between the initial and final armyworm weights, by

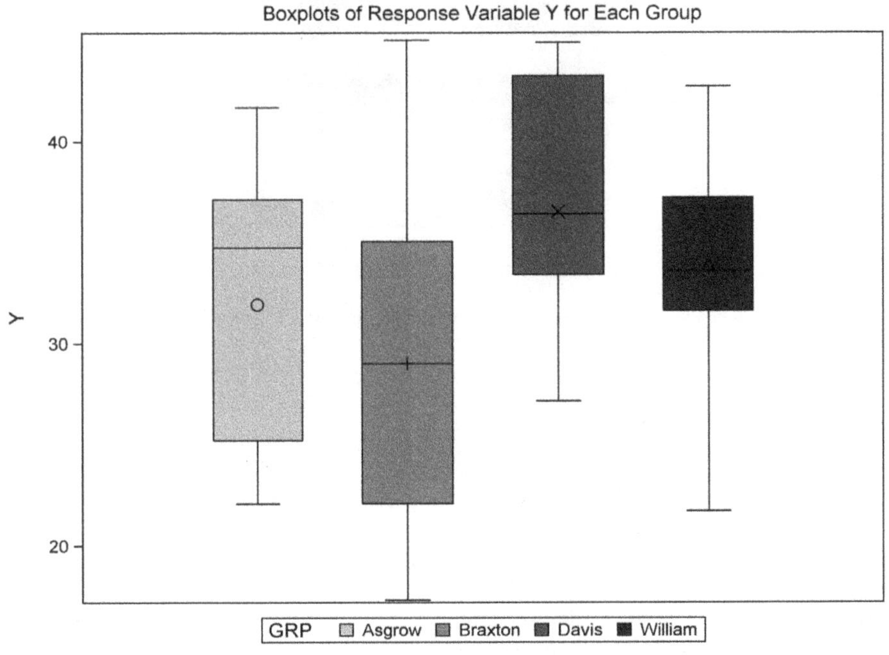

FIG. 20. Boxplots of armyworm final weights from each soy variety.

TABLE 24. ANOVA comparison of final armyworm weights across varieties.

ANOVA Table					
Source	DF	Sum of Squares	Mean Square	F Value	P > F
Model	3	301.77	100.59	2.02	0.1284
Error	36	1792.19	49.78		
Corrected Total	39	2093.96			
Type III Tests					
Source	DF	Type III SS	Mean Square	F Value	P > F
VARIETY	3	301.77	100.59	2.02	0.1284

TABLE 25. Correlation between final and initial armyworm weights.

Variety	Correlation Estimates		
	Point estimate	95% Confidence Interval	
		Lower CL	Upper CL
Asgrow	0.63	-0.03	0.90
Braxton	0.63	-0.03	0.90
Davis	0.78	0.24	0.94
William	0.60	-0.08	0.89
All Varieties Combined	0.61	0.36	0.77

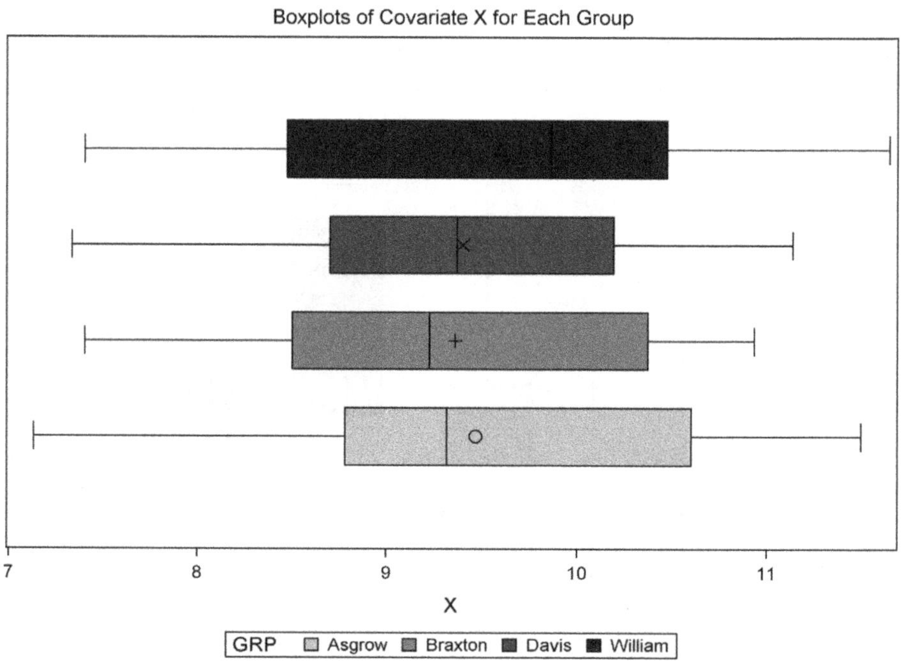

FIG. 21. Boxplots of initial larvae weights.

variety as well as over the entire sample. Within varieties, the correlation estimates range from 0.60 to 0.78. The 95% confidence interval estimates all have a broad range of overlap, and hence the assumption of a common correlation across all groups is reasonable at this point. The point estimate of the correlation obtained from the complete sample is 0.61. The 95% confidence interval estimate based on the complete sample is [0.36, 0.77]. Note that since this confidence interval does not contain zero, at the 5% significance level the conclusion is that the correlation between final and initial weights is nonzero, and in particular positive.

Figure 21 shows boxplots of the covariate. Based on the boxplots, the distributions of initial larval weights appear to be similar in the four assigned samples, both with respect to variability and mean values.

Results of an ANOVA to compare the initial weight means across the four varieties are given in Table 26. The p-value of 0.9762 reinforces the observation based on the boxplots that there is no evidence of a difference in covariate means.

The correlation between initial and final weights is significant, so we may expect that the inclusion of the covariate in the model may improve its power and precision. Because this correlation is smaller than the correlations in the previous examples, however, the degree of improvement may not be as great as seen in those models. In addition, since the distribution of the covariate does not appear to differ across samples, the covariate will not be correcting for any bias that would result from such differences.

Results of an ANCOVA that accounts for initial larvae weight in comparing the mean final larvae weights across the four soybean varieties are given in Table 27. Note first that the estimated error variance is 29.74, which is a 40.3% reduction from the estimate of 49.78 provided by the ANOVA model, at the cost of only a single degree of freedom. This should increase the power of hypothesis tests and improve precision. The overall model is highly significant, with a p-value < 0.0001. Because it is significant, we can evaluate the significance of each term in the model. The p-value associated with the covariate initial weight is < 0.0001, so the linear relationship between the response and covariate is highly significant. The p-value associated with the variety treatment effect on larva weights is 0.0357, and so we conclude that soybean variety significantly affected final weight means of larvae. Because of this, we will evaluate the pairwise comparisons. Based on the least-square means comparisons, we see that the mean final armyworm weights are different under the Braxton and Davis varieties (p-value = 0.0047). The observed difference between the Asgrow and Davis varieties is almost significant at the 5% significance level (p-value = 0.0549), and the observed difference between the Braxton and William varieties is almost significant at the 10% significance level (p-value = 0.1094). Note that the common width of the 95% confidence interval estimates of differences between means is 9.92 mg, which is a reduction of 22.6% from the width provided by the ANOVA model.

Figure 22 shows the fitted ANCOVA model with its four regression lines superimposed over a scatter plot of the final and initial weights. Each regression line represents expected mean final larvae weight as a function of initial larvae weight, one regression line for each of the four soybean varieties. Because this model does not include an interaction between soybean variety and initial larvae weight, it forces the fitted lines to be parallel. A formal test of the assumption that the parallel-lines model is adequate was performed by fitting an ANCOVA model containing the interaction between soybean variety and initial larvae weight. The p-value associated with the interaction term was 0.8369, so there was insufficient evidence to conclude that the lines were not parallel. The Braxton and Davis varieties, which are the only soybean varieties significantly different at the 5% level, are represented by the upper and lower regression lines, respectively, and the estimated difference

TABLE 26. ANOVA comparing mean initial armyworm weights across varieties.

ANOVA Table					
Source	DF	Sum of Squares	Mean Square	F Value	$P > F$
Model	3	0.332	0.111	0.07	0.9762
Error	36	57.935	1.609		
Corrected Total	39	58.267			

TABLE 27. ANCOVA comparison of final armyworm weights across varieties adjusted for initial weight.

ANOVA Table					
Source	DF	Sum of Squares	Mean Square	*F* Value	*P* > *F*
Model	4	1053.03	263.26	8.85	< 0.0001
Error	35	1040.93	29.74		
Corrected Total	39	2093.96			
Type III Tests					
Source	DF	Type III SS	Mean Square	*F* Value	*P* > *F*
INITIAL_WT	1	751.26	751.26	25.26	< 0.0001
VARIETY	3	284.15	94.71	3.18	0.0357

Least Squares Means for Effect VARIETY					
			VARIETY 1 – VARIETY 2		
VARIETY 1	VARIETY 2	*P*-value	Point estimate	95% Confidence interval	
				Lower CL	Upper CL
Asgrow	Braxton	0.3087	2.52	-2.43	7.47
Asgrow	Davis	0.0549	-4.85	-9.80	0.11
Asgrow	William	0.5442	-1.49	-6.45	3.46
Braxton	Davis	0.0047	-7.37	-12.32	-2.41
Braxton	William	0.1094	-4.02	-8.98	0.95
Davis	William	0.1790	3.35	-1.61	8.31

between the mean final larvae weights for these two varieties is equal to the vertical distance between these lines.

Summary of Example 5

An ANCOVA model, with its explanatory covariate, resulted in an analysis that improved power and precision compared with the ANOVA model, which included only the classification treatment effect as an explanatory variable. In the overall sample, the correlation between the response variable final weight and the covariate initial weight was 0.61, a moderate correlation. The ANOVA model was not significant at the 5% level. Inclusion of the covariate in the ANCOVA model accounted for additional variability in the response and decreased the estimated error variance by 40%. This reduction in unaccounted-for variation increased the power of the hypotheses tested by the model. The overall ANCOVA model was highly significant, as was the linear relationship between the response and the covariate. Having accounted for the variability in the response associated with the covariate, the variety effect on final larvae weight became significant, and larvae fed on two of the four varieties differed significantly in mean final weight. In addition, inclusion of the covariate increased the precision of the confidence interval estimates of differences between variety means by reducing their width by 22.6%. By including initial larvae weight as a covariate in an ANCOVA, we improved the overall analysis by increasing the power of treatment comparisons and improving the precision of estimates of treatment effects and their differences.

Summary

This chapter investigates analysis of covariance. It does so primarily through use of five examples designed to illustrate issues that can arise in the context of designed experiments where comparison of treatment effects is the purpose. One such issue is heterogeneity in the experimental units within treatment groups with respect to characteristics that affect the response. This heterogeneity induces variation in the response, which if not accounted for, decreases the power of significance tests and decreases the precision of estimates of treatment effects and their differences. Another potential issue is differences in such characteristics across assigned treatment assignment groups. Such differences bias estimates of treatment effects and their differences and can lead to qualitatively incorrect inferences. The examples in this chapter have demonstrated how using a quality covariate can reduce unaccounted-for variation in the response, reduce or eliminate bias when it exists, and as a result, provide more powerful hypothesis tests and more accurate and more precise estimates of treatment effects and their differences.

This chapter presents ANCOVA in the context of a simple design structure to focus on the essential features of this modeling approach that distinguish it from both ANOVA and regression. Like both ANOVA and regression, ANCOVA can be utilized to analyze data from a wide variety of experimental situations, from experiments with simple designs to those with complex designs. Random effects or nonindependent error structures are typically utilized in statistical models to accommodate more complex experimental designs. As long as the statistical software being used can correctly accommodate the design features of an experiment, that software can be used to perform ANCOVA utilizing that design structure.

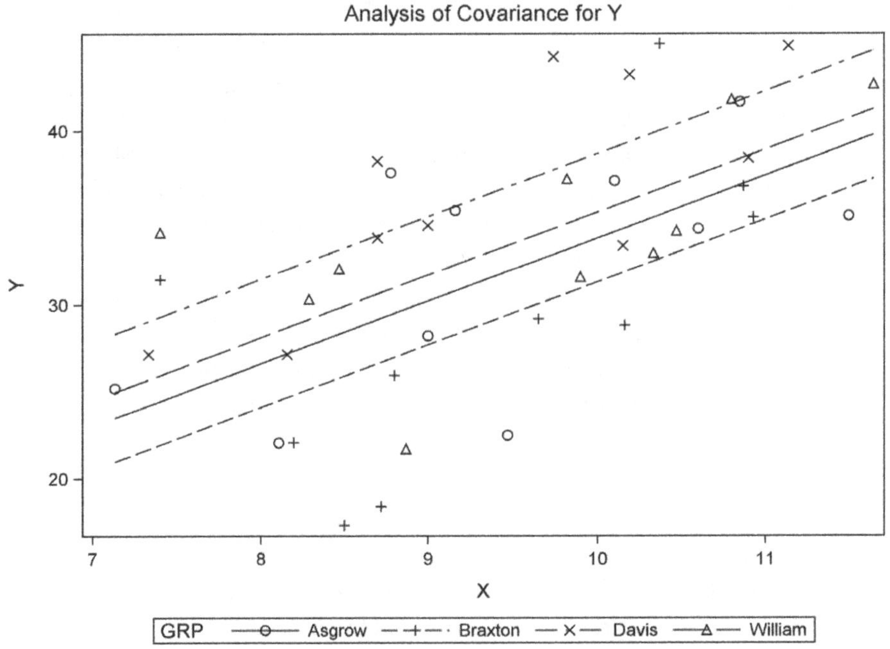

FIG. 22. Scatter plot of final weight versus initial weight, with the fitted ANCOVA model.

Since ANCOVA models are contained within the class of linear models, the issues involved in applying the ANCOVA model to more complex experimental designs are common to the issues involved in extending the class of linear models to the class of linear mixed models. In particular, they are similar to the issues faced when applying ANOVA models and regression models to data from complex experimental designs. Resources specializing in mixed model analysis and ANCOVA can be found in the reference section of this chapter.

Key Learning Points

- Unaccounted-for heterogeneity in experimental material that affects a measured response increases variation in the response. This reduces the power of hypothesis tests and decreases the precision of parameter estimates. Unaccounted-for heterogeneity can also bias estimates of treatment effects.
- ANCOVA models extend ANOVA models by accounting for heterogeneity in experimental material through the inclusion in the model of one or more covariates that have been measured on the experimental units.
- To be effective, covariates should be related to the response, either overall or within treatment groups.
- To avoid attenuating estimates of treatment effects, covariates should be measured before assigned treatments are applied, or be such that they cannot be affected by the treatment being applied. The safest approach is to measure covariates before treatments are applied whenever possible.
- The relationship between the response variable and the covariate must be determined and correctly specified in the ANCOVA model to draw valid inferences from the model.
- It is very important to determine whether the relationship between the response and the covariate is the same in all treatment groups, or whether the relationship varies from one group to another. This is accomplished in the model-development process by including a term for, and testing the significance of, the interaction between the treatment classification variable and the covariate.
- If the relationship between the response and the covariate differs across treatment groups, then treatment differences will depend on the value of the covariate. In order for treatment comparisons to be meaningful, the researcher must specify the values of the covariate at which treatment comparisons are made.

Review Questions

1. True or False: In an ANCOVA, the covariate should either be measured before application of the assigned treatment, or be such that it cannot be affected by the treatment if it is measured after the treatment is applied.

2. True or False: A covariate is effective only when a statistically significant difference

exists in the covariate means across the treatment assignment groups.

3. True or False: A covariate will be effective only when it has a statistically significant correlation with the response variable over the entire dataset.

4. True or False: When developing an ANCOVA model, the nature of the relationship between the response and the covariate must be determined.

5. True or False: When developing an ANCOVA model, one should check to see whether the relationship between the response and the covariate differs across treatment assignment groups, and to accommodate such a difference in the model if it exists.

6. In some of the examples considered in this chapter, the issue that ANCOVA has been used to address has been the result of an imbalance between treatment assignment groups with respect to the values of the covariate. In other words, one of the treatment assignment groups is comprised mostly of experimental units with large values of the covariate, while the other is comprised mostly of those with small values of the covariate. The good news is that large imbalances are unlikely when randomization is used to allocate experimental units to treatment groups. For example, suppose that twenty experimental units are available for an experiment to compare the effects of two treatments. Ten experimental units are to be randomly assigned to one treatment group, and the remaining ten to the other treatment group. Suppose that the covariate has already been measured prior to treatment assignment, and consider the subset of the experimental units with the ten largest values of the covariate and the subset with the ten smallest values of the covariate. The various randomization outcomes that can occur can be combined into the following six scenarios, from most balanced to most unbalanced with respect to these two subsets:

   A. Both treatment groups receive five of the experimental units with the ten largest covariate values;

   B. Either one of the treatment groups receives six of the ten experimental units with the largest covariate values, and the other treatment group receives four;

   C. Either one of the treatment groups receives seven of the ten experimental units with the largest covariate values, and the other treatment group receives three;

   D. Either one of the treatment groups receives eight of the ten experimental units with the largest covariate values, and the other treatment group receives two;

   E. Either one of the treatment groups receives nine of the ten experimental units with the largest covariate values, and the other treatment group receives one;

   F. Either one of the treatment groups receives all ten of the experimental units with the largest covariate values, and the other treatment group receives the experimental units with the ten smallest covariate values.

      a) Find the probability associated with each of the randomization scenarios,

A through F, above.

b) What is the probability that a randomization results in either of the two most balanced scenarios, A or B?

c) What is the probability that a randomization results in either of the two most unbalanced scenarios, E or F?

d) Even if a randomization does not result in an unbalanced scenario, can ANCOVA still provide a better analysis than ANOVA? If so, discuss the ways in which ANCOVA can provide improvement over ANOVA.

e) What do these results suggest about the importance of randomization in the treatment assignment process?

7. In this chapter interpretation of ANCOVA results has been based on the assumption that either the covariate has been measured prior to application of the treatment, or that it is impossible for the treatment to affect the value of the covariate. In some situations, however, neither of these conditions can be guaranteed to be satisfied. For example, in observational studies where ANCOVA-like models are often utilized, it is common for measured categorical factors to be used as "treatments" and other measured characteristics to be used as covariates. In such situations where application of the treatment is not under the control of an experimenter, the treatments and the covariates are likely to be related. Discuss how the results of an ANCOVA are to be interpreted in such situations, and how these interpretations differ from those in this chapter, where by virtue of how the experiment is performed, the value of the covariate is not affected by the treatment.

Data Analysis Exercises 1 Through 5

The five tables below give data for a series of data analysis exercises. In each case, the data come from hypothetical experiments to compare the effects of two treatments, denoted A and B, on the mean of a response variable, Y. Twenty experimental units were available for each study, ten being randomly assigned to treatment A and the remaining ten to treatment B. A single covariate, X, was measured on each experimental unit before its assigned treatment was applied. The response variable Y and the covariate X were expected to be related. Once all five data analysis exercises have been completed, write up a discussion of the various ways in which the use of ANCOVA has improved the analyses and led to better inferences in these exercises.

Questions To Be Considered For Each Data Analysis Exercise

1. Ignoring the covariate, *X*, perform an ANOVA to compare the mean of the response variable *Y* for the two treatments. As part of this analysis, also construct a 95% confidence interval estimate of the difference between the response variable means for the two treatments. What is the conclusion based on this analysis? Is there sufficient evidence to conclude that a difference exists between treatments with respect to the response variable means?

2. Are the distributions of the covariate the same for the two treatments, or are they different? Perform an ANOVA on the covariate X to compare the covariate means across the two treatment assignment groups.

3. Are the response variable Y and the covariate X related? Construct scatter plots of Y versus X to assess the nature and strength of their relationship graphically. Are they linearly related, and hence correlated? Is their correlation approximately the same in the two treatment groups? To answer these questions, perform a correlation analysis of Y and X to estimate their correlation and to test the null hypothesis that their overall correlation is zero, and construct a 95% confidence interval estimate of the correlation. Do this for the whole dataset, and then separately for each treatment group.

4. Perform an ANCOVA to compare the mean of the response variable Y under the two treatments, taking the covariate X into account. Is a parallel-lines model adequate, or is a nonparallel-lines model necessary? Perform the appropriate test in the context of the ANCOVA in making this determination. What are the conclusions based on this ANCOVA analysis? Is the covariate significant? Is there sufficient evidence to conclude that a difference exists between the treatments with respect to the response variable means? Construct a 95% confidence interval estimate of the difference between response variable means based on this analysis.

5. Discuss the impact of including the covariate on the estimate of the error variance. Is the mean squared error (MSE) smaller for the ANCOVA than for the ANOVA? If so, by what percentage does it change? Also, what happens to the width of the confidence interval for the difference of response variable means when the covariate is included in the model? Is the confidence interval width smaller for the ANCOVA model than for the ANOVA model? If so, by what percentage does it change? Finally, what happens to the point estimate of the difference in response variable means when the covariate is included in the model? Based on these considerations, what can be said about the impact of the inclusion of the covariate on the power, precision, and accuracy of the analysis?

6. Were the overall conclusions from the ANOVA and the ANCOVA the same, or were they different? If they were different, which analysis would you use, and why?

Data for Analysis Exercise 1										
Treatment A										
Y	49.4	72.2	61.6	63.4	71.0	49.6	56.6	61.1	66.1	71.9
X	5.2	13.2	9.2	11.3	12.3	7.2	8.3	6.7	9.7	12.0
Treatment B										
Y	95.3	96.7	75.8	102.6	78.5	101.4	100.8	88.2	79.1	108.0
X	20.2	20.5	12.5	21.9	15.3	20.8	21.3	17.2	14.4	21.2

Data for Analysis Exercise 2										
Treatment A										
Y	56.7	46.3	71.2	69.0	59.7	59.8	60.1	66.8	72.5	57.6
X	7.0	5.7	14.6	11.9	10.8	10.2	8.1	10.8	14.5	7.9
Treatment B										
Y	70.3	78.0	63.1	57.1	68.7	70.4	73.3	61.7	73.7	77.4
X	10.5	12.8	9.4	6.6	11.3	9.5	9.4	6.8	11.9	11.6

Data for Analysis Exercise 3										
Treatment A										
Y	54.8	54.6	52.5	68.2	59.6	76.8	56.3	71.2	62.0	68.0
X	8.9	8.3	10.0	12.6	9.4	13.2	7.9	15.0	9.2	13.8
Treatment B										
Y	54.4	55.8	69.4	51.7	67.2	43.9	71.6	66.7	42.4	55.7
X	14.7	19.1	20.3	15.3	21.3	12.2	21.6	20.5	13.4	15.9

Data for Analysis Exercise 4										
Treatment A										
Y	73.9	46.7	48.9	83.0	80.7	73.1	54.8	53.4	79.3	60.6
X	14.4	6.1	6.3	14.6	15.0	12.1	7.2	7.3	13.5	11.5
Treatment B										
Y	68.6	66.9	53.5	79.6	82.0	72.4	67.6	77.4	70.6	75.4
X	9.6	9.9	5.0	13.3	13.2	12.2	9.9	13.4	10.0	11.1

Data for Analysis Exercise 5										
Treatment A										
Y	70.9	71.8	61.0	71.4	56.9	55.8	62.1	80.8	54.8	50.8
X	12.6	13.5	10.9	13.9	6.4	9.1	7.4	13.2	8.8	5.6
Treatment B										
Y	52.4	57.9	68.9	62.7	62.4	78.5	55.0	79.2	82.6	51.6
X	12.3	10.7	11.1	11.6	10.2	6.5	14.8	6.8	5.5	13.7

Acknowledgments

Data for Example 5 kindly provided by Dr. Michael Stout, Department of Entomology, Louisiana State University, Baton Rouge, Louisiana.

References

Casler, M.D. 2018. Power and replication–Designing powerful experiments. In: B. Glaz and K.M. Yeater, Applied statistics in the agricultural, biological, and environmental sciences. ASA, CSSA, SSSA, Madison, WI.

Freund, R.J., W.J. Wilson, and D.L. Mohr. 2010. Statistical methods, Third ed. Academic Press/ Elsevier, Amsterdam, the Netherlands.

Garland-Campbell, K. 2018. Errors in statistical decision making. In: B. Glaz and K.M. Yeater, Applied statistics in the agricultural, biological, and environmental sciences. ASA, CSSA, SSSA, Madison, WI.

Graybill, F.A. 1976. Theory and application of the linear model. Wadsworth & Brooks/Cole, Pacific Grove, CA.

Littell, R.C., G.A. Milliken, W.W. Stroup, R.D. Wolfinger, and O. Schabenberger. 2006. SAS for mixed models, Second ed. SAS Institute Inc., Cary, NC.

Littell, R.C., W.W. Stroup, and R.J. Freund. 2002. SAS for linear models, Fourth ed. SAS Institute Inc., Cary, NC.

McIntosh, M. 2018. Analysis of variance. In: B. Glaz and K.M. Yeater, Applied statistics in the agricultural, biological, and environmental sciences. ASA, CSSA, SSSA, Madison, WI.

Miguez, F., S. Archontoulis, and H. Dokoohaki. 2018. Nonlinear regression models and applications. In: B. Glaz and K.M. Yeater, Applied statistics in the agricultural, biological, and environmental sciences. ASA, CSSA, SSSA, Madison, WI.

Milliken, G.A., and D.E. Johnson. 1992. Analysis of messy data, Volume I: Designed experiments. Chapman & Hall, London, UK.

Milliken, G.A., and D.E. Johnson. 2002. Analysis of messy data, Volume III: Analysis of covariance. Chapman & Hall/CRC, London, UK.

Richter, C., and H.-P. Piepho. 2018. Linear regression techniques. In: B. Glaz and K.M. Yeater, Applied statistics in the agricultural, biological, and environmental sciences. ASA, CSSA, SSSA, Madison, WI.

Stroup, W. 2018. Analysis of non-Gaussian data. In: B. Glaz and K.M. Yeater, Applied statistics in the agricultural, biological, and environmental sciences. ASA, CSSA, SSSA, Madison, WI.

Vargas, M., B. Glaz, J. Crossa, and A. Morgounov. 2018. Analysis and interpretation of interactions of fixed and random effects. In: B. Glaz and K.M. Yeater, Applied statistics in the agricultural, biological, and environmental sciences. ASA, CSSA, SSSA, Madison, WI.

CHAPTER 10: ANALYSIS OF REPEATED MEASURES FOR THE BIOLOGICAL AND AGRICULTURAL SCIENCES

Salvador A. Gezan* and Melissa Carvalho

Abstract

Biological and agricultural experiments often evaluate the same experimental unit over time. Here, a repeated measures analysis is required by combining all measurements into a single complex model that specifies the correlated structure of the experimental data. This is needed, as the assumption of independence between observations is no longer valid, and therefore, an appropriate linear mixed model must be fit. Repeated measures analysis is strongly recommended for analyzing repeated observations because it usually results in reduced standard error of the means, which then produce narrower confidence intervals and increased statistical power. In this chapter, an introduction to the topic of repeated measures analysis is presented in the context of biological studies and particularly those that deal with plant science with emphasis on the specification and evaluation of the variance–covariance matrix. A detailed example illustrates this topic and code is provided for SAS, R, and GenStat focusing on testing and comparison of alternative models.

Many biological and agricultural field and laboratory experiments evaluate one or more responses over a given period of time where repeated measurements are performed on the same experimental unit over the length of the study. Often, conclusions and analyses of these studies require statistical evaluations for each of the time points and the complete set of observations. Evaluating data collected at each of the time points is often straightforward and can be completed by fitting linear models, followed by

Abbreviations: AIC, Akaike information criterion; ANOVA, analysis of variance; AR(1), autoregressive of order 1; ARH(1), heterogeneous autoregressive of order 1; BIC, Bayesian information criterion; BLUE, best linear unbiased estimate; BLUP, best linear unbiased prediction; COR, homogeneous correlation; CORGH, heterogeneous general correlation; CS, compound symmetry; CSH, heterocedastic compound symmetry; DIAG, diagonal; EXP, exponential; FA, factor analytic; GLMM, generalized linear mixed model; ID, independent; LMM, linear mixed model; LRT, likelihood ratio test; ML, maximum likelihood; MVN, multivariate normal; NLMM, nonlinear mixed model; RCBD, randomized complete block; REML, restricted/residual maximum likelihood; ReslogL, residual maximum log-likelihood; SEM, standard error of the mean; TOEP, Toeplitz; TOEPH, heterogeneous Toeplitz; UN, unstructured; US, unstructured.

S.A. Gezan and M. Carvalho, School of Forest Resources and Conservation, University of Florida, 363 Newins-Ziegler Hall, P.O. Box 110410, Gainesville, FL, 32611. *Corresponding author (sgezan@ufl.edu)

doi:10.2134/appliedstatistics.2016.0008

Applied Statistics in Agricultural, Biological, and Environmental Sciences
Barry Glaz and Kathleen M. Yeater, editors

the use of Analysis of Variance (ANOVA) tables and prediction of treatment means or evaluation of contrasts of interests to make inferences and develop conclusions. Several statistics books (e.g., Kuehl, 2000; Welham et al., 2014), and other chapters in this book, deal with biological experiments and provide recommendations and guidelines for performing proper statistical analyses in single-point analyses. However, an analysis that combines all time-points is more complex. The main complication is that the assumption of independence of the data may no longer be valid as the same experimental unit is measured several times, and therefore, correlation between repeated measurements of the same experimental unit needs to be incorporated into the linear model. Hence, more sophisticated statistical tools and training are required. Other authors have reported agricultural analyses with repeated measures (e.g., Piepho et al., 2004). The objective of this chapter is to introduce repeated measures analysis by presenting a brief overview and by illustrating repeated measures analysis with a few examples. This chapter is intended as an introduction to this topic. Important recommendations and references are provided to facilitate further study.

Repeated measures occur in experiments in which the same experimental unit is observed several times over time as a result of repeated sampling. For example, when strawberry (*Fragaria* × *ananassa* Duchesne ex Rozier) yield (in kg) for a plot of a given variety comprised of six plants is measured every week over the season, then we have repeated measures over time. In this example, we have some form of temporal correlation between observations belonging to the same experimental unit. Alternatively, we could have spatial correlation; for example, in a given point, several records of soil carbon content are obtained at different depths (e.g., 2, 4, 8, and 20 cm). In this chapter, we focus primarily on temporal correlations, as are typically found in agricultural experiments, and we leave the details of spatial correlations to the literature (for example, Cressie, 1993 and Chapter 12 (Burgueño, 2018). Therefore, repeated measures analysis is the use of statistical tools that deal with correlations between observations. Several approaches exist to analyze this type of data including multivariate techniques (see Chapter 14, Yeater and Villamil, 2018). However, in this chapter, we will focus on extending the use of linear mixed models (LMM) with a single response variable by starting from the original experimental design and its structure. The focus here is based on the assumption that the residuals have an approximate normal distribution. An example of repeated measures with a non-normal response is presented in Chapter 16 by Stroup (2018).

Why perform repeated analyses? Often researchers need to test inferences over time or space on the same experimental unit. For these cases, use of repeated measures analysis using LMM has several important benefits including:

- More efficient analyses, because when data are correlated, repeated measures analysis provides more information.
- Greater statistical power (see also Chapter 4 by Casler, 2018) due to using a more efficient analysis and better control of the factors affecting the process.
- Influence of missing data on the analysis is reduced. This is a benefit of using LMM to model correlations.
- Further biological interpretation (and testing) can be performed with the variance component estimates, such as the temporal correlations.

The main challenge in repeated measures analysis is the definition and incorporation of the correlated structure of the data. This is done by fitting extended LMMs that modify the assumptions of independence of the experimental units by modeling complex error structures which consider correlations among units and heterogeneity of variances. This is the topic of the next section.

Linear Mixed Models

Linear mixed models extend the typical linear model by allowing for more complex and flexible specifications of errors and other random effects by incorporating correlation and heterogeneous variances between the observations or experimental units. An important distinction with LMMs is the definition of fixed and random effects. The former corresponds to factors whose levels are specifically selected (nonrandom) and include the complete population of levels of interest (see also Chapter 16 by Stroup, 2018). In contrast, a random effect corresponds to a factor whose levels are a random sample from a population of a large number of levels. The important distinction is that in the case of fixed effects, the statistical inferences are only made on the specific factor levels selected in the experiment, whereas for the random effects, the inferences are about the complete population of levels, not only those included in the study. For example, consider a LMM to describe a randomized complete block design (RCBD) where all plants from each plot were observed. This model has a fixed block and treatment factor and a random plot factor and can be written as:

$$y_{ijk} = \mu + a_i + t_j + p_{ij} + e_{ijk}$$

where y_{ijk} is the observation from the kth plant from the ith block and jth treatment; μ is the overall mean; a_i is a fixed effect of block i; t_j is a fixed effect of treatment j; p_{ij} is a random effect of plot ij, with $p_{ij} \sim N(0, \sigma_p^2)$; and e_{ijk} is a random error, with $e_{ijk} \sim N(0, \sigma_e^2)$.

As indicated above, an important assumption is that random effects follow a normal distribution with a given variance–covariance structure, as shown for the plot factor. In this particular case, the plot factor needs to be considered random as it is modeling the nature of the experiment where several measurement units (e.g., plants within a plot) with the same treatment are grouped together. Ignoring this will result in inflated degrees of freedom and leads to pseudoreplication (further discussion on this topic can be found in Welham et al., 2014). Also, note that the error term in any linear model is also a random effect that is assumed to be normally distributed with its corresponding variance–covariance structure, that is, $e_{ijk} \sim N(0, \sigma_e^2)$. These variance-covariance structures are the key to most LMMs, and are often described by a matrix of variance-covariance parameters that specifies the properties of these random effects. For example, in matrix notation we assume that $\boldsymbol{e} \sim \text{MVN}(\boldsymbol{0}, \boldsymbol{R})$, where the bold letters identify vectors or matrices, and here the vector of errors $\boldsymbol{e}$, of dimension $n \times 1$, is assumed to follow a multivariate normal distribution with a mean vector of zeros, $\boldsymbol{0}$, of dimension $n \times 1$, and a variance–covariance matrix $\boldsymbol{R}$, of dimension $n \times n$, where n is the total number of observations. In the example presented above, we have $\boldsymbol{R} = \sigma_e^2 \boldsymbol{I}_n$, where $\boldsymbol{I}_n$ is an identity matrix of dimension $n \times n$ with ones in the diagonal and zeros in the off-diagonal, therefore, this $\boldsymbol{R}$ matrix specifies that the errors are all independent and have the same variance σ_e^2 (i.e., homocedastic). Similarly, for the plot effects we have $\boldsymbol{p} \sim \text{MVN}(\mathbf{0}, \boldsymbol{W}_p)$, where $\boldsymbol{p}$ is a vector of plot effects of dimension $p{\times}1$, $\mathbf{0}$ is a vector of zeros of dimension $p{\times}1$ and $\mathbf{W}_p$ is a variance-covariance matrix of dimension $p{\times}p$, that in the above example is $\boldsymbol{W}_p = \sigma_p^{\,2}\boldsymbol{I}_p$. Both of these matrices ($\boldsymbol{R}$ and $\boldsymbol{W}_p$) can be defined in many forms or structures (see more below) to specify correlations and heterogeneity of variances between random effects.

The estimation of the variance components is done by a likelihood-based method, where the most common is the restricted and/or residual maximum likelihood (REML) (Patterson and Thompson 1971). These variance components are later used in the normal equations of the LMM to obtain estimates of the fixed effects (best linear unbiased estimates, BLUEs) and random effects (best linear unbiased predictions, BLUPs) as derived in detail by Henderson (1984). As with linear models, further hypothesis testing by ANOVA tables, predictions of means, and evaluation of contrasts is possible with approximated F- and t-tests.

In this chapter, we do not provide more details, but good textbooks are available with additional details about LMMs and their properties. We recommend Littell et al. (2006), Pinheiro and Bates (2006), and Chapter 16 (Stroup, 2018) of this book.

Several statistical packages can be used to fit LMMs. These include SAS (SAS Institute Inc. 2011), R (R Development Core Team, 2008), and GenStat (Payne et al., 2011). The SAS package has the procedure PROC MIXED for normal data and PROC GLIMMIX for non-normal data. The R package has a few libraries, such as lme4 (Bates et al., 2015) and nlme (Pinheiro et al., 2016) that can fit several types of linear models. All of these packages and their corresponding libraries have different implementations of LMM methodologies and they use different names for the same variance-covariance structures with an array of options and functions, and in some cases they provide different estimates of the variance components. For this reason, we recommend carefully checking their properties to determine if they can provide the required output and flexibility.

Variance–Covariance Structures

For repeated measures analysis, as indicated earlier, the key is the specification of the variance-covariance matrix of error ($\boldsymbol{R}$), as it is here where we model the correlated nature of the data. To facilitate an understanding of the $\boldsymbol{R}$ matrix, we will assume that we have a group of m individuals (or experimental units) each measured t times; hence, the total database has $n = m{\times}t$ observations. For now, we will consider that all measurement time points were done at regular intervals (for example, every two days) $\boldsymbol{R}_n = \boldsymbol{I}_m \otimes \boldsymbol{G}_t$. As before, here $\boldsymbol{I}_m$ is an identity matrix of dimension m, which indicates that each of the experimental units is independent from each other, and $\boldsymbol{G}_t$ is a complex variance–covariance of dimension $t{\times}t$ that will model the correlations between repeated measurements. Also, $\otimes$ is the Kronecker product between matrices, which is an operation between two matrices that produces a block matrix, and it is also known as direct product. Here the key element is the specification of the G_t matrix for which we have several structures. Wolfinger (1996) presents a complete description of the relevant structures for repeated measures, and below we will present some of the most common (see Fig. 1). Note here that for our example we will assume that we have $t = 4$ observations per individual, and we will define the structure in a generic way and not specifically for one or another statistical package.

The most basic reference structure is the independent matrix, or ID, that, as indicated earlier, assumes independence between observations. This can be extended to become heteroscedastic by assuming that each measurement will have a different error variance, and is often identified as DIAG. One of the simplest structures that allows for correlation between observations is compound symmetry, or CS. This structure and its heteroscedastic counterpart (CSH) have a variance component ρ that represents the correlation between any pair of observations. Hence, it assumes that the correlations between observations that are close or far apart in time will all have exactly the same correlation. For this reason, often this structure is not acceptable. However, if only a few observations are available (e.g., $t < 4$) CS or CSH could provide good approximations. The autoregressive error structure of order 1, AR(1), and its heterogeneous counterpart, ARH(1), are two of the most common structures used; they also have a single correlation variance component ρ, but in this case, there is an exponent that indicates the separation between repeated observations. For example, r^2, indicates that there are two time intervals (but one time point) between observations of the same experimental unit. One restriction of AR(1) and ARH(1) is that both assume that the intervals between observations are all identical. One equivalent structure that allows for modeling unequal intervals is the exponential structure (EXP), which depends on the variable d, that represents the distance (in time) between observations of the same experimental unit. If the intervals are all identical, then this model is a parametrization of the autoregressive. Another flexible model is the Toeplitz, or TOEP, also known as a diagonal constant matrix, which specifies a different correlation (or covariance depending on its parametrization) between observations. Hence, this has additional variance components but allows for greater flexibility. Finally, the unstructured (UN also identified as US), matrix is the most flexible. It specifies a different covariance for every pair and a different variance for each measurement point. This structure has the largest number of variance components to estimate, but as expected, offers the greatest flexibility. Sometimes, this matrix is expressed as an extension of correlation structures with homogeneous (COR) or heterogeneous variances (CORGH, also known as UNR in SAS). Note that different statistical packages (and R libraries) will have different names for the same structures, and they will also include a much larger number of alternative structures. Therefore, we recommend that the reader carefully review software manuals and literature. A good start on this topic is Wolfinger (1996).

New difficulties arise as the complexity of the structure increases (for example, from CS to UN). The first major issue is the increased computational complexity for estimating variance components. In all cases, REML is still used, but the minimization functions are more complex with the risk of nonconvergence of the model fit, or convergence to a local minimum instead of the global minimum. For this reason, it is recommended: i) to start with fitting a simpler model and to increase its complexity carefully; ii) if the software allows it, provide the routine with some starting values to aid with the convergence; and iii) always check that the variance component estimates are biologically reasonable.

ID: Identity

$$\sigma_0^2\begin{bmatrix}1&0&0&0\\0&1&0&0\\0&0&1&0\\0&0&0&1\end{bmatrix}$$

DIAG: Diagonal

$$\begin{bmatrix}\sigma_1^2&0&0&0\\0&\sigma_2^2&0&0\\0&0&\sigma_3^2&0\\0&0&0&\sigma_4^2\end{bmatrix}$$

CS: Compound Symmetry

$$\begin{bmatrix}\sigma_1^2+\sigma_0^2&\sigma_0^2&\sigma_0^2&\sigma_0^2\\\sigma_0^2&\sigma_1^2+\sigma_0^2&\sigma_0^2&\sigma_0^2\\\sigma_0^2&\sigma_0^2&\sigma_1^2+\sigma_0^2&\sigma_0^2\\\sigma_0^2&\sigma_0^2&\sigma_0^2&\sigma_1^2+\sigma_0^2\end{bmatrix}$$

CSH: Heterogeneous Compound Symmetry

$$\begin{bmatrix}\sigma_1^2+\sigma_0^2&\sigma_0^2&\sigma_0^2&\sigma_0^2\\\sigma_0^2&\sigma_2^2+\sigma_0^2&\sigma_0^2&\sigma_0^2\\\sigma_0^2&\sigma_0^2&\sigma_3^2+\sigma_0^2&\sigma_0^2\\\sigma_0^2&\sigma_0^2&\sigma_0^2&\sigma_4^2+\sigma_0^2\end{bmatrix}$$

AR(1): Autorregressive of order 1

$$\sigma_0^2\begin{bmatrix}1&\rho^1&\rho^2&\rho^3\\\rho^1&1&\rho^1&\rho^2\\\rho^2&\rho^1&1&\rho^1\\\rho^3&\rho^2&\rho^1&1\end{bmatrix}$$

ARH(1): Heterogeneous Autorreg. of order 1

$$\begin{bmatrix}\sigma_1^2&\rho^1\sigma_1\sigma_2&\rho^2\sigma_1\sigma_3&\rho^3\sigma_1\sigma_4\\\rho^1\sigma_1\sigma_2&\sigma_2^2&\rho^1\sigma_2\sigma_3&\rho^2\sigma_2\sigma_4\\\rho^2\sigma_1\sigma_3&\rho^1\sigma_2\sigma_3&\sigma_3^2&\rho^1\sigma_3\sigma_4\\\rho^3\sigma_1\sigma_4&\rho^2\sigma_2\sigma_4&\rho^1\sigma_3\sigma_4&\sigma_4^2\end{bmatrix}$$

TOEP: Toeplitz

$$\begin{bmatrix}\sigma^2&\sigma_1&\sigma_2&\sigma_3\\\sigma_1&\sigma^2&\sigma_1&\sigma_2\\\sigma_2&\sigma_1&\sigma^2&\sigma_1\\\sigma_3&\sigma_2&\sigma_1&\sigma^2\end{bmatrix}$$

TOEPH: Heterogeneous Toeplitz

$$\begin{bmatrix}\sigma_1^2&\rho_1\sigma_1\sigma_2&\rho_2\sigma_1\sigma_3&\rho_3\sigma_1\sigma_4\\\rho_1\sigma_1\sigma_2&\sigma_2^2&\rho_1\sigma_2\sigma_3&\rho_2\sigma_2\sigma_4\\\rho_2\sigma_1\sigma_3&\rho_1\sigma_2\sigma_3&\sigma_3^2&\rho_1\sigma_3\sigma_4\\\rho_3\sigma_1\sigma_4&\rho_2\sigma_2\sigma_4&\rho_1\sigma_3\sigma_4&\sigma_4^2\end{bmatrix}$$

CORGH: Heterogeneous General Correlation

$$\begin{bmatrix}\sigma_1^2&\rho_{12}\sigma_1\sigma_2&\rho_{13}\sigma_1\sigma_3&\rho_{14}\sigma_1\sigma_4\\\rho_{12}\sigma_1\sigma_2&\sigma_2^2&\rho_{23}\sigma_2\sigma_3&\rho_{24}\sigma_2\sigma_4\\\rho_{13}\sigma_1\sigma_3&\rho_{23}\sigma_2\sigma_3&\sigma_3^2&\rho_{34}\sigma_3\sigma_4\\\rho_{14}\sigma_1\sigma_4&\rho_{24}\sigma_2\sigma_4&\rho_{34}\sigma_3\sigma_4&\sigma_4^2\end{bmatrix}$$

UN: Unstructured

$$\begin{bmatrix}\sigma_1^2&\sigma_{12}&\sigma_{13}&\sigma_{14}\\\sigma_{12}&\sigma_2^2&\sigma_{23}&\sigma_{24}\\\sigma_{13}&\sigma_{23}&\sigma_3^2&\sigma_{34}\\\sigma_{14}&\sigma_{24}&\sigma_{34}&\sigma_4^2\end{bmatrix}$$

FIG. 1. Common variance–covariance structures used to fit linear mixed models for repeated measures analysis.

Since there are many alternative models to evaluate, it is helpful to have a procedure to select the most appropriate variance–covariance structure. Such a procedure should identify the most parsimonious structure that describes the response variable in a reasonable way. This can be done by using the likelihood ratio test (LRT), which is based on asymptotic derivations. It compares nested models by using a Chi-square test that compares the residual log-likelihood (ReslogL) values between a model with a complex structure (ReslogL_2) and a simpler counterpart (ReslogL_1). The statistic used is:

$$\chi^2_d = 2\ \text{ReslogL}_2 - 2\ \text{ReslogL}_1 \sim \chi^2_{k2\text{-}k1}$$

where ReslogL_1 and ReslogL_2 are the residual maximum log-likelihood values for the corresponding models 1 and 2, and k_2–k_1 is the degrees of freedom calculated as the difference between the number of variance components to estimate each of the models. It is important to note that when REML variance components are used, this test requires that the fixed effects between Models 1 and 2 are exactly the same, otherwise, the test is incorrect and a maximum likelihood procedure (ML) should be used (see also Chapter 11 by Payne, 2018).

In addition, and particularly for non-nested models, it is possible to calculate information criteria, such as the Akaike Information Criterion (AIC, Akaike 1974) and the Bayesian Information Criterion (BIC, Schwarz 1978). Both of these criteria are used to compare the fit between two or more alternative models. The AIC is more liberal and the BIC more conservative (i.e., it chooses models with less parameters), where the latter is not sensitive to prior distributions for large sample sizes. The BIC was developed using a Bayesian approach. Studies by Guerin and Stroup (2000) found that larger values of BIC were associated with larger Type 2 errors. The expressions for these criteria are:

$$\text{AIC} = -2\ \text{ReslogL} + 2k$$

$$\text{BIC} = -2\ \text{ReslogL} + k \log(n)$$

where ReslogL is the residual maximum log-likelihood value of the model, k is the number of effective variance parameters to estimate the model; and n is the sample size. Lower values for AIC and BIC indicate a better model fit. In the examples presented later, the selection of the variance structure will be illustrated in detail.

Fitting a Linear Mixed Model for Data with Repeated Measures

The process to fit a LMM for repeated measures data should be done with care. There are several steps to follow to select a reasonable model. Often, the first step is to fit every single measurement point individually. This step allows to clearly define the linear (mixed) model to use, but also facilitates the detection of departures from normality, and the detection of potential outliers. Also, evaluation of the ANOVA tables provides a preliminary idea of the factors that are significant for the response variable under study. It is also recommended to collect all variance components estimated from this step to be used later as starting values for a more complex model.

The second step consists of extending the original model to construct the repeated measurements LMM. This is done by adding time (as a factor or covariate) individually and with all its interactions (or in some cases as nested effects) with the original model terms. At this stage, it is important to define if the interest is to consider time as a factor, for which a different mean estimate is obtained for each level (or measurement point), or to assume it as a continuous variable (or covariate), for which the repeated measures model will now represent lines (or curves) if this is considered as an explanatory variable. These two approaches have different objectives, when time is a factor we are interested in comparing each of the time points and overall differences; however, when time is a continuous variable, we are interested in the describing the patterns over time of each of the treatments, and we might perform interpolations, as done with regression analyses.

For the construction of this model, the next step is the specification of the error structure. Here it is recommended to start with simpler structures, such as ID or DIAG and then evaluate some other more complex models. We recommend the use of AR(1) or AR1(H) as a baseline. It is often useful to fit the UN structure; however, convergence for the UN is usually difficult. Several error structures should be fitted for the current data, and evaluated using LRT (which is only valid when comparing nested models) or AIC and BIC goodness-of-fit statistics to select the error structure. Note here that we recommend to select the most parsimonious model as the main interest of this analysis is not focused on the interpretation of the error structure. For this reason, often simpler, and somehow incomplete error structures sometimes are selected. Finally, given that the error structure is clear, we can focus our attention on the ANOVA table, which we can follow with prediction of means and comparisons of predetermined contrasts of interest. It is possible at this stage, or it may be necessary to achieve convergence, to drop some factors from the model, or increase its complexity; however, changes in the model often require revisiting the selection of the error structure. Also, residuals need to be examined for potential departures from the basic assumptions. We recommend focusing on Studentized or standardized residuals, as these take into consideration the heterogeneity of variances.

Detailed Example

An experiment was established to assess the effects of three site-preparation treatments (*v*-plow, hand screef or removal of vegetation, and untreated control), two seedling species (Douglas-fir [*Pseudotsuga menziesii* {Mirb.} Franco] and lodgepole pine [*Pinus contorta* Douglas ex Loudon]), and two types of stock (bare root and plug) in a trial located in the Cariboo Forest Region (Nemec, 1996). The experiment used a RCBD with four blocks. Each block contained 12 plots, and a plot consisted of a single row of 25 seedlings. For this example, we will focus on the mean plot height (cm) of the plants. The plants in this experiment were observed annually for 6 yr (from 1984 to 1989), and an initial observation (before planting) is also available. The objective is to determine whether site-preparation treatment, species, stock type, or any of their interactions affect growth. The complete dataset is presented in Table 1.

For fitting this model, we will start by defining the complete factorial model for each of the observed measurement points as:

TABLE 1. Raw data used for example originating from a field trial located in the Cariboo Forest Region (Canada). Source: Nemec (1996). The response variable corresponds to the mean plant height (m) of a plot conformed by 25 seedlings. †

Spp	Stk	Prep	Trt	Blk	Initial	1984	1985	1986	1987	1988	1989
FD	B	S	FD-B-S	1	18.78	22.89	22.44	22.89	22.44	27.56	33.56
FD	B	S	FD-B-S	2	15.92	20.08	21.50	24.75	28.42	39.67	53.67
FD	B	S	FD-B-S	3	20.80	26.60	28.20	27.90	36.90	48.30	59.80
FD	B	S	FD-B-S	4	18.60	22.60	21.40	23.00	22.20	30.60	39.80
FD	B	U	FD-B-U	1	14.10	18.40	20.80	24.70	29.00	40.80	57.70
FD	B	U	FD-B-U	2	17.00	21.00	22.00	18.00	20.00	22.00	25.00
FD	B	U	FD-B-U	3	18.00	22.22	25.22	27.56	29.33	38.67	49.67
FD	B	U	FD-B-U	4	18.14	23.29	25.36	28.36	30.07	38.00	48.50
FD	B	V	FD-B-V	1	16.14	19.81	22.10	25.95	33.43	45.76	59.19
FD	B	V	FD-B-V	2	14.89	19.28	21.11	25.78	30.28	39.89	53.83
FD	B	V	FD-B-V	3	15.08	18.08	19.77	23.08	27.08	37.38	52.08
FD	B	V	FD-B-V	4	19.00	23.06	24.24	28.12	34.47	45.35	58.12
FD	P	S	FD-P-S	1	18.94	26.71	31.06	33.24	42.00	54.12	65.94
FD	P	S	FD-P-S	2	22.82	28.91	34.05	39.91	47.32	59.50	75.64
FD	P	S	FD-P-S	3	22.90	28.15	32.95	39.05	49.20	62.60	74.85
FD	P	S	FD-P-S	4	20.59	25.65	30.12	33.35	39.53	49.29	61.65
FD	P	U	FD-P-U	1	21.56	29.22	31.17	38.00	43.83	54.72	66.78
FD	P	U	FD-P-U	2	20.47	27.11	33.16	39.95	46.74	57.37	70.79
FD	P	U	FD-P-U	3	19.05	27.52	31.33	37.95	46.62	56.95	68.52
FD	P	U	FD-P-U	4	16.29	24.71	31.05	35.76	43.48	55.62	70.86
FD	P	V	FD-P-V	1	18.08	23.63	28.54	37.08	47.83	64.75	86.75
FD	P	V	FD-P-V	2	20.88	26.58	30.50	35.88	42.83	53.71	70.58
FD	P	V	FD-P-V	3	20.19	25.94	28.38	32.38	37.06	48.63	67.63
FD	P	V	FD-P-V	4	20.40	26.25	30.00	34.25	38.35	49.05	65.30
PL	B	S	PL-B-S	1	12.44	20.28	34.67	49.83	65.78	90.83	125.17
PL	B	S	PL-B-S	2	16.33	22.00	32.00	47.33	58.33	81.73	113.07
PL	B	S	PL-B-S	3	11.32	20.05	32.45	47.45	66.77	96.32	132.55
PL	B	S	PL-B-S	4	11.11	16.37	28.74	44.11	65.79	94.05	130.74
PL	B	U	PL-B-U	1	12.40	17.40	26.50	36.80	49.60	75.80	109.90
PL	B	U	PL-B-U	2	13.44	18.44	24.75	41.00	54.31	80.00	111.56
PL	B	U	PL-B-U	3	14.19	19.44	29.63	48.38	62.25	86.69	120.31
PL	B	U	PL-B-U	4	15.22	20.61	29.61	41.83	55.06	70.72	97.78
PL	B	V	PL-B-V	1	12.31	17.08	26.38	41.46	64.08	100.38	147.62
PL	B	V	PL-B-V	2	13.94	19.00	28.41	46.41	81.94	118.41	166.18
PL	B	V	PL-B-V	3	11.53	18.58	29.68	45.47	70.89	105.47	150.32
PL	B	V	PL-B-V	4	12.63	16.63	27.06	43.75	67.38	103.06	146.75
PL	P	S	PL-P-S	1	12.43	23.19	36.71	55.29	75.71	109.48	155.76
PL	P	S	PL-P-S	2	10.23	18.59	33.91	53.59	74.09	108.27	150.64
PL	P	S	PL-P-S	3	9.59	17.82	32.05	49.86	69.50	97.59	133.55
PL	P	S	PL-P-S	4	13.48	21.70	34.26	48.22	73.39	103.83	141.48
PL	P	U	PL-P-U	1	12.00	22.86	34.38	49.00	71.10	105.05	148.71
PL	P	U	PL-P-U	2	9.43	17.14	30.10	43.33	60.95	87.24	125.67
PL	P	U	PL-P-U	3	8.15	15.95	28.60	39.65	58.75	89.00	129.40
PL	P	U	PL-P-U	4	8.75	15.70	27.45	42.55	58.45	85.55	123.85
PL	P	V	PL-P-V	1	12.28	19.52	33.12	55.12	89.24	136.16	193.56
PL	P	V	PL-P-V	2	9.57	17.13	28.74	46.65	74.00	114.22	163.13
PL	P	V	PL-P-V	3	10.25	17.83	29.38	48.00	78.88	116.29	161.50
PL	P	V	PL-P-V	4	7.83	13.58	30.00	53.42	84.71	130.38	186.21

† Spp, seedling species (FD: Douglas-fir, PL: lodgepole pine); Stk, stock type (B: bare root, P: plug); Prep, site-preparation treatment (S, hand screw, U, untreated, V, v-plow); Trt, treatment combination of Spp, Stk and Prep; Blk, block; Initial, mean plot height at planting (1983).

TABLE 2. Summary of results from fitting single point measurement for Cariboo Forest Region data. P-values are presented for each of the model terms, together with residual variance (s_e^2), and average standard error of the mean (SEM) for the treatment combinations.

Effect	df	1984	1985	1986	1987	1988	1989
ht0	1	< 0.0001*	< 0.0001*	0.015*	0.089	0.408	0.821
Blk	3	0.082	0.441	0.814	0.751	0.498	0.402
Trt	11	< 0.0001*	< 0.0001*	< 0.0001*	< 0.0001*	< 0.0001*	< 0.0001*
s_e^2	–	1.020	2.845	9.724	26.450	55.483	105.980
SEM	–	0.615	1.026	1.898	3.130	4.533	6.265

* Significant at the 0.05 probability level.

$$y = \mu + \text{ht0} + \text{Blk} + \text{Spp} + \text{Stk} + \text{Prep} + \text{Spp}\times\text{Stk} + \text{Spp}\times\text{Prep} + \text{Stk}\times\text{Prep} + \text{Spp}\times\text{Stk}\times\text{Prep} + e$$

where all factors will be considered as fixed effects, with the exception of the error term e. Here, Blk represents the blocks, Ssp the species, Stk the stock, Prep the site-preparation treatment, and the other terms are the two- and three-way interactions. The term ht0 is a covariate representing the initial height (cm) of the plants at the beginning of the experiment (before the treatments were applied). For more details about analysis of covariance, see Chapter 9 by McCarter (2018).

For simplicity, at this stage we will combine all treatment factors (SSp, Stk, and Prep) into a single combined factor with 12 levels, identified as Trt (see also Table 1). Hence, the model is represented as:

$$y_{(t)} = \mu_{(t)} + \text{ht0}_{(t)} + \text{Blk}_{(t)} + \text{Trt}_{(t)} + e_{(t)}, \text{ for } t = 1, \ldots, 6$$

Here, the index (t) is used to identify the different measurement points, and the error terms for each are assumed to be $\boldsymbol{e}_{(t)} \sim \text{MVN}(\boldsymbol{0}, \sigma_e^2\boldsymbol{I}_n)$, with n the number of plots in the experiment. The fitting of the above model was done using SAS 9.3 (SAS Institute Inc. 2011) and the summary of each of the measurement points is shown in Table 2 (see Appendix 1 for SAS, R, and GenStat code). Note that the significance of some factors changes from measurement to measurement point (year to year), however, Trt is always significant. In addition, the estimated residual error, s_e^2, and the standard error of the mean (SEM) for Trt increase with time as would be expected as the plants become larger with time. The presence of heterogeneous variances among years, as with heterogeneous variances among time points or spaces in other experiments, is one of the major reasons that repeated measures analysis is useful.

The next step is to extend the above model to all six measurement points. To do this, we incorporate the factor Time that has a total of six levels. This factor is added alone and with all its potential interactions. Hence, we have:

$$y = \mu + \text{Time} + \text{ht0} + \text{Blk(Time)} + \text{Trt} + \text{Time}\times\text{Trt} + e$$

There are important elements in this model that need further clarification. First, y now represents a vector of dimension $n\times1$ that includes observations from all m experimental units and all t measurement times ($n = m\times6$). We will assume this is sorted by individual and then measurement point within individual; this is not critical here but for some statistical packages specific data sorting is often required. The constant μ now represents an overall mean across all observations, and therefore, often does not have a reasonable interpretation. The factor Time indicates that for each time point, there is a different expected value. Therefore, it is important to determine if there are trends for height over time in this study. The term ht0 constitutes the covariate that is correcting for initial plant height. The role of this covariate is to adjust the data due to different starting conditions of the plants at the beginning of the experiment. The factor Blk(Time) corresponds to blocks nested within time, and incorporates the different effects due to block at each time point. Having only the factor Blk instead of Blk(Time) would assume that the block effects are all the same for each time point, an assumption that might be incorrect. Also, note that if the Blk term from the single point model would have been assumed to be random, then there would have needed to be variance components associated with this factor. This implies that for the complete repeated measures model, the term Blk(Time) should have a different variance component for each measurement point (hence, six variance components), corresponding to a DIAG structure (see Fig. 1), hence, Blk(Time) ~ MVN($\mathbf{0}$, $\boldsymbol{D}_m \otimes \boldsymbol{I}_b$) with $\boldsymbol{D}_m$ a diagonal matrix of dimension 6 × 6. Note that the choice of random or fixed effects for blocks depends on the assumptions of the scientist, objectives of the study, and the characteristics of the experiment. For this reason we do not discuss this topic any further (see Littell et al., 2006 for an excellent explanation). The factor Trt represents the treatment effect across all time points. Thus, we can view Trt as the effect of a given treatment averaged over the t time points, which often is the main hypothesis of interest. The model factor Time×Trt

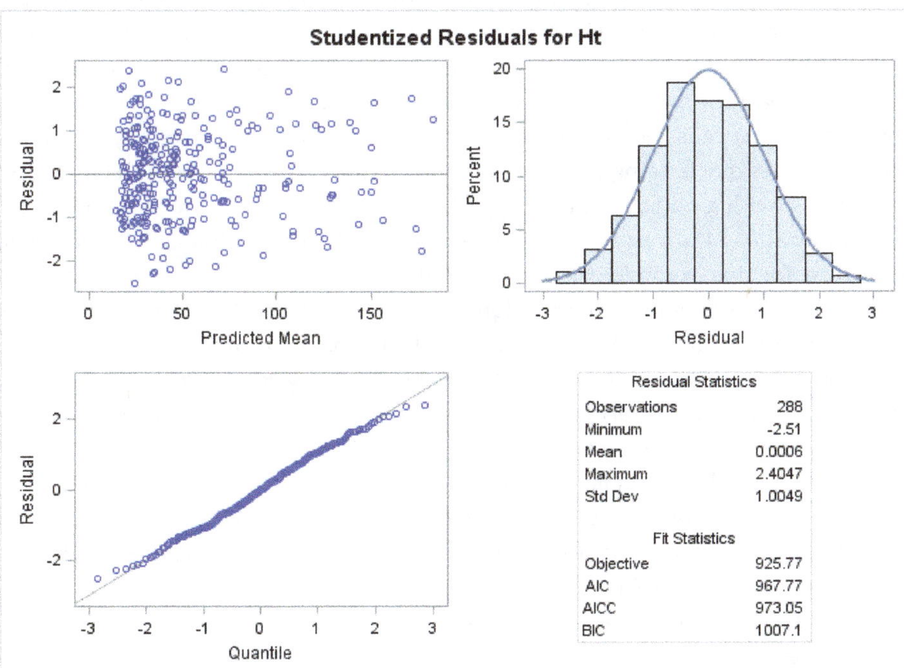

FIG. 2. Panel of residual plots for final model for repeated measures analysis with UN error structure produced with SAS v. 9.3.

TABLE 3. Goodness-of-fit statistics for different error structures evaluated in Cariboo Forest Region data. The statistics presented are: residual maximum log-likelihood (ResloglL), Akaike information criterion (AIC) and Bayesian information criterion (BIC). The parameter *k* indicates the number of variance components estimated by the corresponding error structure.

Structure	*k*	-2 ResloglL	AIC	BIC
ID	1	1390.5	1392.5	1394.3
CS	2	1320.3	1324.3	1328.1
AR(1)	2	1151.5	1155.5	1159.2
TOEP	6	1109.3	1121.3	1132.5
DIAG	6	1204.3	1216.3	1227.5
CSH	7	1092.8	1106.8	1119.9
ARH(1)	7	1008.1	1022.1	1035.2
TOEPH	11	1001.2	1023.2	1043.8
UN	21	925.8	967.8	1007.1

represents the interaction of treatment with time; this is probably one of the most important pieces of information from the repeated measures analysis. Finally, the error term is assumed to be $\boldsymbol{e} \sim \mathrm{MVN}(\mathbf{0}, \boldsymbol{I}_m \otimes \boldsymbol{G}_t)$, where, as indicated earlier, $\boldsymbol{G}_t$ is the matrix to represent repeated measures error structure.

For the above repeated measures model, we have fit several error structures. For all of these models, the number of variance components and the -2 ResloglL, AIC, and BIC are presented in Table 3. According to the AIC and BIC (smaller is better for both), the best error structure is UN, which has a total of 36 variance components. The ARH(1), a simpler model with only seven variance components is the second best model. For this dataset, the UN structure had no difficulties converging, but if this was not the case, ARH(1) would have been selected. To illustrate the use of the LRT, we can compare the models ARH(1) and DIAG. These two structures are nested, and the only difference is the presence of the temporal correlation ρ (see Fig. 1). Hence, the hypothesis that is being tested is H_0: $\rho = 0$ against H_1: $\rho \neq 0$, which is a two-sided hypothesis, with a critical value of χ^2_d = 1204.3– 1008.1 = 196.2, which is compared with a value of 3.84 corresponding to a critical χ^2 with 1 df for a 5% significance level. Therefore, for this example, we have more than enough evidence to conclude that this temporal correlation is highly significant. Comparisons between other nested structures are possible, and we leave this to the interested reader.

Now that we have selected the UN error structure for our model, and after checking for departures form normality and presence of outliers (see Fig. 2, which was part of our SAS output, an excerpt from a panel of Studentized residuals), we can proceed to check the Test of Fixed Effects (Table 4). It is clear form this table that there are several significant interactions, particularly relevant is the Spp×Stk×Prep×Time, interaction which means that we need to report the plant height mean for each combination of treatment and for each time point.

One difficulty that arises from fitting a complex LMM is that the *F*- and *t* tests are no longer valid (Littell et al., 2006). This occurs because these tests are derived assuming there is only one variance component in the linear model, σ_e^2. However, for most LMMs, we have several variance components that first are estimated and then used to construct ANOVA–like tables. For this reason, some packages report only asymptotic tests (such as χ^2 and z values by using Wald-type statistics), or others perform some corrections on the degrees of freedom (df) to allow for this additional level of approximation and uncertainty on the construction of these tests. One of the most recommended approaches is to use the Kenward–Rogers df correction (Kenward and Rogers, 1997). We have used this correction to generate the results in Table 4 where some denominator df have decimals.

Therefore, for this repeated measures analysis, it is possible to conclude that there are significant effects of most of the treatment factors and that the responses of the treatments (or treatment combinations) depend on the time point under study. We can proceed to predict means for some treatments with confidence intervals and generate different tables and graphs to report these results. In addition, it is also possible to perform some specific comparisons. For example, for this exploratory model we can evaluate, with the slice function in SAS, if there are significant differences between all 12 levels of treatments. These results are presented in Table 5 together with the SEMs for these treatments (these were also presented in Table 1 but for single time point analyses). Note that these SEMs increase with time. This results from having a different error variance for each year (see also Table 6). For the repeated measures analysis based on these data, there are no differences of conclusions compared with the previous model. This is because there were substantial differences among treatments. In addition, there are reductions of the SEM of ~10% in most years, with the exception of 1984. This is a result of a more efficient use of

TABLE 4. Test of fixed effects for final fit model based on an UN error structure for the Cariboo Forest Region data.

Effect	Numerator df	Denominator df	*F*-value	*P*-value
ht0	1	32	107.09	< 0.0001
Time	5	29	582.02	< 0.0001
Blk(Time)	18	49.8	0.84	0.6420
Trt	11	34	71.57	< 0.0001
Spp	1	46.9	653.97	< 0.0001
Stk	1	32.9	72.27	< 0.0001
Spp×Stk	1	35.4	0.38	0.542
Prep	2	33.1	15.90	< 0.0001
Spp×Prep	2	33	12.59	< 0.0001
Stk×Prep	2	33	0.40	0.6730
Spp×Stk×Prep	2	32.9	0.90	0.4170
Trt×Time	55	68.6	19.03	< 0.0001
Spp×Time	5	29	168.22	< 0.0001
Stk×Time	5	29	10.84	< 0.0001
Prep×Time	10	41.3	9.87	< 0.0001
Spp×Stk×Time	5	29	2.14	0.0890
Spp×Prep×Time	10	41.3	5.34	< 0.0001
Stk×Prep×Time	10	41.3	0.29	0.9792
Spp×Stk×Prep×Time	10	41.3	2.62	0.0146

TABLE 5. Summary of results from evaluating hypothesis of differences between treatment combinations (Trt) within a given measurement point from fitting full repeated measure model with UN error structure for the Cariboo Forest Region data. *P*-values are presented together with the average standard error of the mean (SEM) for the treatment combinations.

Effect	df	1984	1985	1986	1987	1988	1989
Trt	11	< 0.0001*	< 0.0001*	< 0.0001*	< 0.0001*	< 0.0001*	< 0.0001*
SEM	–	0.635	0.964	1.617	2.574	3.719	5.148

* Significant at the 0.05 probability level.

TABLE 6. Estimated variance-covariance and correlation matrices from fitting full repeated measures model with UN error structure for the Cariboo Forest Region data.

	Variance-covariance matrix						Correlation matrix					
	1984	1985	1986	1987	1988	1989	1984	1985	1986	1987	1988	1989
1984	1.018	1.077	0.834	1.412	1.639	1.984	1	0.606	0.264	0.275	0.220	0.192
1985	1.077	3.103	3.870	5.506	7.220	9.791	0.606	1	0.701	0.615	0.554	0.542
1986	0.834	3.870	9.830	12.668	18.027	25.013	0.264	0.701	1	0.794	0.778	0.777
1987	1.412	5.506	12.668	25.872	35.832	47.031	0.275	0.615	0.794	1	0.953	0.901
1988	1.639	7.220	18.027	35.832	54.682	74.469	0.220	0.554	0.778	0.953	1	0.981
1989	1.984	9.791	25.013	47.031	74.469	105.370	0.192	0.542	0.777	0.901	0.981	1

the information which translated into narrower confidence intervals and increased power to detect differences between treatment levels.

The inspection of the variance components for this analysis can provide some insight into the response variable of interest. For the current data, a matrix of variance-covariance and a matrix of correlations are presented in Table 6. These results, particularly the correlations, indicate that the pair-to-pair correlations increase with time, with high values for years 1988 and 1989. This indicates that the conclusions form these two years will be similar (data not shown). In addition, the first year (1984) has the lowest correlations, indicating that this measurement point is too early in the experiment to provide sufficient information to compare treatments across time.

Finally, it is possible to fit a slightly different, and simplified model, which is common in many repeated measures analyses. This model incorporates a random effect and has the following form:

$$y = \mu + \text{Time} + \text{ht0} + \text{Blk(Time)} + \text{Trt} + \text{Time} \times \text{Trt} + \text{Plot} + e$$

where all the terms were previously defined but here Plot is a random effect factor that identifies each experimental unit (i.e., plots, see Table 1), with Plot ~ MNV($\mathbf{0}$, $\sigma_p^2 \boldsymbol{I}_n$), and the residual errors are all assumed to be independent, that is, $\boldsymbol{e}$ ~ MNV($\mathbf{0}$, $\sigma_e^2 \boldsymbol{I}_n$). This model is equivalent to the previously fit model with CS for the error structure, with the difference that the modeling of the correlation structure is done through the model term Plot. Here, ReslogL values and variance component estimates are identical, and for this reason, this simplification is often preferred. However, it presents the same difficulties as the CS, mainly that they assume a common correlation between any pair of measurement points, which is only recommended whenever the number of time measurements is small ($t < 3$). Nevertheless, the issue of heterogeneity of variances for each of the measurement points still needs to be addressed properly, and this requires complex models, which are easily implemented by proper specification of the error structure.

Time as a Continuous Variable

In the example presented above we assumed that the time model term was a factor; however, if we were interested in trends, it is possible to change this factor into a continuous variable (or variate) as, for example:

$$y = \mu + \text{Timec} + \text{Timec}^2 + \text{ht0} + \text{Blk(Time)} + \text{Trt} + \text{Timec}\times\text{Trt} + \text{Timec}^2\times\text{Trt} + e$$

In this particular case, we are modeling a quadratic trend, as we are using the square of the continuous variate Timec, which is centered to avoid issues with multicollinearity between continuous variables. A centered variable is one where the mean of the data vector is subtracted to each observation (Welham et al., 2014). Note, that for the above model, we still need to specify the error structure and also, the term Blk(Time) is still included to model for a different block effect nested within each time point; therefore, for this model, Time is still considered as a factor rather than a variate.

In the example presented earlier, we focused on a normally distributed response variable. Often, we have discrete responses that follow a binomial distribution (such as survival with a yes/no or a 0/1 response) or a Poisson distribution (such as count of eggs). For these cases, we recommend the reader to review Chapter 16 by Stroup (2018), which describes in more detail a Generalized Linear Mixed Model (GLMM) approach. In short, this methodology allows the researcher to consider the specific probability distribution of the response, and allows for incorporating random effects

TABLE 7. Data from a soybean experiment that compares two varieties (P, introduction #416937, F, Forrest) measured weekly ten times, starting at 14 d after planting. Only data from year 1988 is considered. Source: Davidian and Giltinan (1995).

Plot	Variety	14	21	28	35	42	49	56	63	70	77
F1	F	0.106	0.261	0.666	2.110	3.560	6.230	8.710	13.350	16.342	17.751
F2	F	0.104	0.269	0.778	2.120	2.930	5.290	9.500	–	16.967	17.747
F3	F	0.108	0.291	0.667	2.050	3.810	6.130	10.280	18.080	20.183	21.811
F4	F	0.105	0.299	0.844	1.320	2.240	4.680	8.820	15.090	14.660	14.005
F5	F	0.101	0.273	0.848	1.950	4.770	6.010	9.910	–	19.262	–
F6	F	0.106	0.337	0.699	1.530	3.870	5.600	9.430	13.730	17.381	19.932
F7	F	0.102	0.275	0.767	1.450	3.950	4.940	9.640	–	17.880	17.724
F8	F	0.103	0.273	0.742	1.410	3.010	5.260	9.810	12.850	18.214	19.680
P1	P	0.131	0.338	0.701	1.660	4.250	9.240	12.150	16.780	15.925	17.272
P2	P	0.128	0.404	0.897	1.780	3.910	7.400	10.070	18.860	17.012	27.370
P3	P	0.131	0.379	1.126	2.440	3.890	6.910	12.490	15.670	23.763	21.491
P4	P	0.154	0.357	1.181	1.830	4.710	10.710	9.910	15.510	14.958	21.800
P5	P	0.139	0.328	0.932	1.990	3.460	7.020	11.790	15.830	15.921	17.442
P6	P	0.139	0.389	1.094	2.130	4.040	7.620	12.480	17.930	14.422	30.272
P7	P	0.145	0.366	0.799	1.610	3.510	6.790	9.950	14.540	19.280	22.573
P8	P	0.130	0.355	1.090	2.280	3.940	4.960	10.920	14.020	17.994	22.371

and specifying complex variance–covariance structures as done with LMMs. Most statistical packages have routines to fit these models. However, fitting a GLMM for repeated measures data, while possible, is often complex due to issues with convergence. We also recommend Gbur et al. (2012) for more information on general implementation of these models and Bolker et al. (2009) for further technical details.

So far, we have presented a few of the most common variance–covariance structures. However, it is possible to use other structures. One popular structure used in agriculture, and particularly in plant breeding, is the factor analytic (FA). This structure has the advantage of providing a good approximation of the UN structure while using a reduced number of variance components. Therefore, the FA structure tends to converge more easily than the UN. For further details and properties of the FA structure, we recommend Smith et al. (2015). Also, of particular relevance are the structures used for spatial statistics that are more flexible and consider irregular time or space measurement intervals, and even different measurement points per experimental unit. Good introductions of this topic are provided in the books from Cressie (1993) and Webster and Oliver (2007).

Finally, it is also possible to model repeated measures with the use of nonlinear models, more specifically, these will be nonlinear mixed models (NLMM). This is the typical case, for example, of a growth curve that models the development of a given experimental unit that is observed several times. The advantage is that the form of the nonlinear model often has good interpretability and better biological background than a linear model. For NLMM, as with LMM, we would need to model the error structure and follow a similar procedure as already described in this chapter. Nevertheless, use of these models is limited due to issues with convergence. Also, statistical tests and confidence intervals associated with NLMM's are based on asymptotic approximations. Additional details of these models, with some interesting examples, can be found in Davidian and Giltinan (1995).

Conclusions

In this chapter we have presented a brief introduction to the topic of repeated measures analysis in the context of biological studies, particularly those that deal with the plant sciences. These analyses rely strongly on the theory and practice of linear mixed models, which are powerful tools for many situations. We have presented only some general topics and illustrated these topics with an example. There are many more aspects that were not presented here. However, there are good references available for these topics. Nevertheless, the analysis of data with repeated measures requires a mixture of solid statistical modeling and practical experience, as each experiment and its corresponding datasets differ. Repeated measures analysis is strongly recommended for analyzing repeated observations of the same experimental unit because it usually results in reduced SEMs which then produce narrower confidence intervals and increased statistical power. However, there are usually concerns, difficulties, and diverse challenges with repeated measures analyses that need careful attention.

Key Learning Points

- If the same experimental unit is observed over time, then the assumption of independence of the data is no longer valid, and this correlation needs to be incorporated into the statistical analysis.

- Repeated measures analysis has several important benefits, particularly a reduction on the SEMs, which then produce narrower confidence intervals and increased statistical power.
- The specification of the variance-covariance matrix of errors is key for these analyses, and different structures need to be evaluated to properly describe the underlying biological process.
- The analysis of data with repeated measures requires a mixture of solid statistical modeling and practical experience, as each experiment and its corresponding datasets differ.

Review Questions

True or False

1. Spatial correlation is a type of correlation that is present between observations that belong to the same experimental unit.
2. If we have missing data, then repeated measures analysis can't be used.
3. Combining all data from several time points into a single analysis will provide greater statistical power than analyzing every time point separately.
4. For random effects, the statistical inferences are valid only for the levels that are considered in the corresponding factor.
5. The compound symmetry (CS) structure is the simplest structure that can model some form of correlation.
6. The AR(1) and ARH(1) structures do not need identical intervals between measurements.
7. Comparing two models by using the residual log-likelihood (ReslogL) requires that the fixed effects between models are the same.
8. The *F*- and *t*-tests from a repeated measures analysis are no longer valid tests because their degrees of freedom are incorrect.
9. Linear mixed models can only be used on normally distributed response variables.

Exercises

1. Consider the site-preparation experiment presented earlier that was fitted under a RCBD with a factorial model:

   a. Based on Table 3, perform a likelihood ratio test to compare the error structure of heterogeneous autoregressive of order 1 against the unstructured? Use a significance level of 5%. What do you conclude from this result?

b. Repeat the analysis of this data, but this time consider Time as a continuous variable (Timec) in its linear and quadratic form. Do you have similar conclusions to the ones obtained from Table 4? Do you need to consider the quadratic term? Use a significance level of 5%.

2. The data presented in Table 7 corresponds to a soybean experiment that was established to compare growth patterns of an experimental strain against a commercial variety (P: introduction #416937, F: Forrest). This experiment was repeated for several years but only the data from 1988 is considered here, and each plot was measured at weekly intervals eight to ten times starting at 14 d after planting. Average leaf weight from a sample of six plants was calculated. More details are presented in Davidian and Giltinan (1995).

   a. Fit a repeated measures analysis for this data with the fixed effects factors of Time, Variety, and their interaction considering a compound symmetry error structure. Consider a natural logarithm transformation of the data for your analysis. Do you have significant differences between the varieties evaluated? Is there a significant interaction? Use a significance level of 5%.

   b. Evaluate other error structures, such as DIAG and ARH(1), and use LRT, AIC, and BIC to compare your models. Which one do you recommend? Why?

   c. Based on your final selected model, can you indicate if there are significant differences between the varieties at the last measurement time? How about at the first week? Use a significance level of 5%.

References

Akaike, H. 1974. A new look at the statistical model identification. IEEE Trans. Automat. Contr. 19:716–723. doi:10.1109/TAC.1974.1100705

Bates, D., M. Maechler, B. Bolker, and S. Walker. 2015. Fitting linear mixed-effects models using lme4. J. Stat. Softw. 67(1):1–48. doi:10.18637/jss.v067.i01

Bolker, B.M., M.E. Brooks, C.J. Clark, S.W. Geange, J.R. Poulsen, M.H.H. Stevens, and J.S.S. White. 2009. Generalized linear mixed models: A practical guide for ecology and evolution. Trends Ecol. Evol. 24(3):127–135. doi:10.1016/j.tree.2008.10.008

Burgueño, J. 2018. Spatial analysis of field experiments. In: B. Glaz and K.M. Yeater, editors, Applied statistics in the agricultural, biological, and environmental sciences. ASA, CSSA, SSSA, Madison, WI.

Casler, M.D. 2018. Blocking principles for biological experiments. In: B. Glaz and K.M. Yeater, editors, Applied statistics in the agricultural, biological, and environmental sciences. ASA, CSSA, SSSA, Madison, WI.

Cressie, N. 1993. Statistics for spatial data: Wiley series in probability and statistics. Wiley & Sons, New York.

Davidian, M., and D.M. Giltinan. 1995. Nonliner models for repeated measurement data. Chapman & Hall/CRC Press, Boca Raton, FL.

Gbur, E.E., W.W. Stroup, K.S. McCarter, S. Durham, L.J. Young, M. Christman, M. West, and M. Kramer. 2012. Generalized linear models. In: E.E. Gbur, W.W. Stroup, K.S. McCarter, S. Durham, L.J. Young, M. Christman, M. West, and M. Kramer, editors, Analysis of generalized linear mixed models in the agricultural and natural resources sciences. ASA, CSSA, SSSA, Madison, WI.

Guerin, L., and W.W. Stroup. 2000. A simulation study to evaluate proc mixed analysis of repeated measures data. Proceedings of the 12th Annual Conference on Applied Statistics in Agriculture, Manhattan, KS. 30 Apr.–2 May 2000, Kansas State University, Manhattan, KS.

Kenward, M.G., and J.H. Rogers. 1997. Small sample inference for fixed effects from restricted maximum likelihood. Biometrics 53:983–997. doi:10.2307/2533558

Kuehl, R. 2000. Design of experiments: Statistical principles of research design and analysis. 2nd ed. Duxbury Press, Pacific Grove, CA.

Henderson, C.R. 1984. Applications of linear models in animal breeding. University of Guelph, Guelph, ON.

Littell, R.C., G.A. Milliken, W.W. Stroup, R.D. Wolfinger, and O. Schabenberger. 2006. SAS for mixed models. SAS Institute. Inc., Cary, NC.

McCarter, K.S. 2018. Analysis of covariance. In: B. Glaz and K.M. Yeater, editors, Applied statistics in agricultural, biological, and environmental sciences. ASA, CSSA, SSSA, Madison, WI.

Nemec, A.F.L. 1996. Analysis of repeated measures and time series: An introduction with forestry examples. Working Paper 15. Biometric information handbook No. 6. Province of British Columbia, Victoria, B.C.

Patterson, H.D., and R. Thompson. 1971. Recovery of inter-block information when block sizes are unequal. Biometrika 58(3):545–554. doi:10.1093/biomet/58.3.545

Payne, R.W. 2018. Long-term research. In: B. Glaz and K.M. Yeater, editors, Applied statistics in agricultural, biological, and environmental sciences. ASA, CSSA, SSSA, Madison, WI.

Payne, R.W., D.A. Murray, S.A. Harding, D.B. Baird, and D.M. Soutar. 2011. An introduction to GenStat for Windows. 14th edition. VSN International, Hemel Hempstead, UK.

Piepho, H.P., A. Büchse, and C. Richter. 2004. A mixed modelling approach for randomized experiments with repeated measures. J. Agron. Crop Sci. 190(4):230–247. doi:10.1111/j.1439-037X.2004.00097.x

Pinheiro, J., and D. Bates. 2006. Mixed-effects models in S and S-PLUS. Springer Science & Business Media, Berlin, Germany.

Pinheiro, J., D. Bates, S. DebRoy, D. Sarkar, S. Heisterkamp, and B. Van Willigen. 2016. nlme: Linear and nonlinear mixed effects Models. R package version 3.1-126, http://CRAN.R-project.org/package=nlme (verified 13 Dec. 2017).

R Development Core Team. 2008. R: A language and environment for statistical computing. R Foundation for Statistical Computing, Vienna, Austria. http://www.R-project.org (verified 13 Dec. 2017).

SAS Institute Inc. 2011. Base SAS 9.3 Procedures Guide. SAS Institute, Inc. Cary, NC.

Schwarz, G.E. 1978. Estimating the dimension of a model. Ann. Stat. 6(2):461–464. doi:10.1214/aos/1176344136

Smith, A.B., A. Ganesalingam, H. Kuchel, and B.R. Cullis. 2015. Factor analytic mixed models for the provision of grower information from national crop variety testing programs. Theor. Appl. Genet. 128:55–72. doi:10.1007/s00122-014-2412-x

Stroup, W. 2018. Non-Gaussian data. In: B. Glaz and K.M. Yeater, Applied statistics in agricultural, biological, and environmental sciences. ASA, CSSA, SSSA, Madison, WI.

Wolfinger, R.D. 1996. Heterogeneous variance-covariance structures for repeated measures. J. Agric. Biol. Environ. Stat. 1:205–230. doi:10.2307/1400366

Welham, S.J., S.A. Gezan, S.J. Clark, and A. Mead. 2014. Statistical methods in biology: Design and analysis of experiments and regression. Chapman and Hall, CRC Press, Boca Raton, FL.

Webster, R., and M.A. Oliver. 2007. Geostatistics for environmental scientists. John Wiley & Sons, Hoboken, NJ. doi:10.1002/9780470517277

Yeater, K.M., and M.B. Villamil. 2018. Multivariate methods for agricultural research. In: B. Glaz and K.M. Yeater, editors, Applied statistics in the agricultural, biological, and environmental sciences. ASA, CSSA, SSSA, Madison, WI.

CHAPTER 11: THE DESIGN AND ANALYSIS OF LONG-TERM ROTATION EXPERIMENTS

Roger William Payne*

Rotation experiments differ from ordinary field experiments in that they assess different sequences of crop (and possibly husbandry) combinations, rather than effects of treatments in a single year. Thus they aim to investigate longer-term effects of treatment strategies that may be more representative of use in practice. As the sequences take place over successive years, the conclusions will depend on the conditions during those years. It is therefore best to phase the first years of the experiment. Once a complete cycle has taken place, comparisons can then be made between the rotations in every subsequent year. If possible, it is best to have more than one replicate in every year. Interim analysis can then be done with the data from each year. Otherwise meaningful analyses will need several years' data and the assumption, for example, that higher order interactions can be ignored, or that responses over years can be modeled by low-order polynomials. Issues to consider in the analysis include the possibility that the within-year variances may be unequal, and that the correlation between observations on a plot may differ according to the distance in time between them. The old-fashioned method of analysis, feasible if the data are balanced, would be to do a repeated-measurements analysis of variance. A more satisfactory alternative is to do a mixed-model analysis using residual (or restricted) maximum likelihood, and investigate fitting models to the between-year correlation structure. These issues are illustrated using a long-term experiment on potato (*Solanum tuberosum* L.).

Rotation experiments play an important role in the study of alternative cropping systems, providing insights into the effects of proposed new strategies in more realistic situations than a single year's trial. For example, the advantages or disadvantages of a new strategy of pest control may take several years to become apparent. Likewise, the yields of a particular crop may be dependent on the previous cropping history of the field. Rotation experiments are also invaluable for the study of the long-term

---

Abbreviations: REML, Residual, or Restricted, Maximum Likelihood.

VSN International, Hemel Hempstead, Hertfordshire HP2 4TP, U.K., and Department of Computational and Systems Biology, Rothamsted Research, Harpenden, Hertfordshire AL5 2JQ, U.K.

*Corresponding author (roger.payne@vsni.co.uk)

doi:10.2134/appliedstatistics.2016.0001

Applied Statistics in Agricultural, Biological, and Environmental Sciences
Barry Glaz and Kathleen M. Yeater, editors

effects of cropping systems on aspects that are crucial for agricultural sustainability, such as soil organic matter.

The purpose of a rotation experiment is to compare different sequences of crop (and possibly husbandry) combinations. The separate crops in the sequence are usually called *courses*. Usually these occur at annual intervals, but the same principles apply with the shorter intervals that may be feasible, for example, in green houses. Sometimes, not all of the crops are of practical interest. For example a sequence may include a fallow year, where there is no crop to be measured or assessed, or it may include crops that form part of the treatment for a subsequent crop, but are not themselves of any interest. So *rotation experiment* here refers to experiments that aim to compare different rotations, following the example of Patterson (1964) who excluded the simpler *fixed rotation experiments* that study the effects of treatments on the crops of a single rotation. (These can involve similar problems to those discussed in the Analysis section below, but they are inherently much easier to handle.)

To illustrate the ideas, Table 1 shows one block of an experiment by Glynne and Slope (1959), which was designed to assess the effects of previous cropping by beans (*Vicia faba* L.) or potatoes (*Solanum tuberosum* L.) on the incidence of eyespot (*Oculimacula yallundae* and *Oculimacula acuformis*) in winter wheat (*Triticum aestivum* L.). The crops in Years 1 and 2 are *treatment crops* that establish the previous cropping regimes, ready for Year 3 when *test crops* are grown to enable the differences between the sequences to be assessed. The two winter wheat crops in Year 2 also act as partial test crops that can be used to assess the effects from the wheat and potato crops in Year 1. The bean and potato yields were not used for analysis.

This experiment is a short-term (or *fixed-cycle*) *rotation experiment*, where the sequences run through one simultaneous cycle to compare the sequences in the final year. These can be designed and analyzed in much the same way as ordinary single-year field experiments. More interesting design and analysis issues arise in the *long-term rotation experiments* that are intended to run through several cycles, and involve analyses of data from more than 1 yr. The aim of this chapter is to explore these issues, and explain how to design and analyze these experiments successfully.

Designing the Experiment

In the simplest long-term experiments, the rotations are all of the same length and have the test crops at the same points in the cycle. It is usual to phase the start of the experiment with new replicates of the rotations starting in successive years so that, once a complete cycle has taken place, comparisons can be made between the rotations in every subsequent year. The advantages of this scheme were pointed out by Yates (1949) where he noted that year-to-year variations mean that "To obtain

TABLE 1. One block from a short-term rotation experiment to study eyespot.

Year	Type of crop	Plot†			
		1	2	3	4
1	treatment	W	P	W	Be
2	treatment	W	W	P	P
3	test	W	W	W	W

† *Be*, beans; *P*, potato; *W*, winter wheat.

a proper measure of the effect of any treatment, therefore, it is necessary to repeat even 1-yr experiments in a number of years... The same holds for rotation experiments.... The first rule in the design of rotation experiments, therefore, is that such experiments should include all phases of the rotation." Within each year, comparisons will be made only between the plots with rotations that started at the same time. One straightforward and effective strategy is therefore to use a randomized-block design, with the rotations in each block all beginning at the same time. Year differences are then confounded with blocks, so no information is lost on the treatments. If sufficient resources are available to have more than one replicate block in each year, interim analyses can be performed with the data from individual years. Otherwise, meaningful analyses will need several years data and the assumption, for example, that there are some year-by-treatment interactions that can be ignored (these degrees of freedom can then provide the residual).

The situation becomes more complicated when rotations of different lengths are included, or the test years do not coincide. Furthermore, if there are many different rotations, the numbers of plots per block may become too large for the blocking to represent the fertility patterns in the field effectively. The rotations may then need to be partitioned into sets that can be placed into separate blocks. To enable this to be done effectively, Patterson (1964) introduced the concept of *comparable rotations.* Two rotations are defined to be comparable if they have at least 1 yr when they both grow the same test crop. So, ideally, only rotations that are not comparable should be placed in different blocks or, if that is not feasible, it should be those that are least comparable that are allocated to different blocks. Comparability is the key issue to consider if the rotations are of different lengths, or if the test years do not coincide. It may then be necessary to start the plots within a block in different years. The designer should construct a table showing the crop scheduled to be grown on each plot in each year, and check that there are years when sufficient plots are growing the same test crop in each block for meaningful analyses to be done.

Table 2 shows the phases of a rotation experiment performed from 1977 through 1988 on the experimental farm Westmaas, part of Wageningen UR, in the southwest of the Netherlands (51°47′ N, 4°27′ E). The farm is used to conduct research into arable farming, multifunctional agriculture and field production of vegetables. The aim of the experiment was to study the effect of short rotations, fumigation and N supply on the production of potato, sugar beet (*Beta vulgaris* L. subsp. *vulgaris*) and winter wheat. Before the start of the experiment in 1977, the experimental field was without any fumigation, in a four-course rotation of potato, winter wheat, sugar beet and winter wheat, ending in a winter wheat crop in 1976. In the experiment, three cropping plans were compared: a four-course crop rotation consisting of winter wheat,

TABLE 2. Rotation experiment at Westmaas experimental farm.

Crop	Rotation				
	IIf	III	IIIf	IV	IVf
Potato	x	x	x	x	x
Wheat after potato				x	x
Sugar beet	x	x	x	x	x
Wheat after sugar beet		x	x	x	x

sugar beet, winter wheat (with green manure) and potato; a three-course crop rotation consisting of sugar beet, winter wheat (with green manure) and potato; and a two-course rotation of sugar beet and potato. The leaves and heads of the sugar beets and the leaves and stems of the potatoes were left in the field after harvest. However, the wheat straw was removed from the field. Grass (*Lolium perenne* L.) was sown under the winter wheat that followed the sugar beet and preceded potato, to act as a green manure crop. Two versions of the three- and four-course rotations were performed, with and without soil fumigation to control nematodes. The fumigation consisted of applications of metam-sodium 300 L Monam ha^{-1} until 1982, and 1,3-dichloropropene, 160 L Telone II ha^{-1} from 1983 onward, following the potato crop. As suggested above, several instances of each rotation were included with different starting points, so that all phases of each rotation were present in every year. There were also two replicate blocks, each of which contained 16 plots (one for each of the rotation phases shown in Table 2). Table 3 shows the cropping sequence in the plots. Notice that we have two of each combination of crop and rotation in each year (one in each block). We would therefore have residual degrees of freedom available, if we wanted to analyze the data from a single year. We can also fit year × treatment interactions when we analyze the data from more than 1 yr. No data are available for potatoes from 1978. In that year, there were problems in planting the potatoes, and growth of the crop was irregular. Consequently, it was not harvested. The sugar beet crop was not harvested in 1984. The percentage of plants that emerged in that year was low (e.g., 22% for rotation II), and there was also damage caused by game animals.

Additional *auxiliary* treatment factors can be incorporated in a similar way as in designs for ordinary single-year experiments. These should be allocated at random to the subplots in the first year, and those allocations should be retained in subsequent years. In the Westmaas experiment, the plots were split into three subplots to study the effects of N fertilizer (levels N1, N2, and N3). The N2 level was chosen to represent the amount that was usual for each crop in the year concerned, N1 was 20% lower, and N3 was 20% higher.

Nonstatistical aspects must also be considered. For example, it is important to ensure that the plots are sufficiently large to avoid treatment effects spreading to adjacent plots, and cultivation techniques should aim to avoid movement of soil across plot boundaries. Also, if the experiment is to continue through several series of rotations, it may be beneficial to be able to split the plots later, to apply additional treatment factors. For a more detailed discussion of all these issues, see Dyke (1974, Chapter 7).

Analysis

The analysis of a long-term rotation experiment can use similar methods to those involved in ordinary field experiments, but as Payne (2015) noted, there are some special issues to consider.

1. Results will be recorded from several years, and these may show different amounts of random variation.

2. The same plot may be observed in several years and, unless these observations are well separated, the results may show a nonuniform correlation structure, where the

correlations between these observations decline with increasing distance in time.

3. The effect of a crop may depend on where it occurs within the rotation cycle.

4. There may be no replication, other than over years.

5. However, treatment effects may build up (or decline) over the period of the experiment. Basal treatments (fertilizers, cultivation practices, pesticides, etc.), or even the precise makeup of the rotations themselves, may have changed during the experiment to keep them relevant with current farming practices.

The first issue that occurs in many analyses when data are taken from several years, or from several sites, and analyzed together in a combined, or meta-analysis. (Note: this form of meta-analysis differs from the form generally used, for example,

TABLE 3. Cropping sequences in the plots of the Westmaas experiment

Block	Whole plot	Rotation	Year†											
			77	78	79	80	81	82	83	84	85	86	87	88
1	1	IVf	P	W	S	W	P	W	S	W	P	W	S	W
	2	IVf	W	S	W	P	W	S	W	P	W	S	W	P
	3	IVf	S	W	P	W	S	W	P	W	S	W	P	W
	4	IVf	W	P	W	S	W	P	W	S	W	P	W	S
	5	IIIf	P	S	W	P	S	W	P	S	W	P	S	W
	6	IIIf	S	W	P	S	W	P	S	W	P	S	W	P
	7	IIIf	W	P	S	W	P	S	W	P	S	W	P	S
	8	III	P	S	W	P	S	W	P	S	W	P	S	W
	9	III	S	W	P	S	W	P	S	W	P	S	W	P
	10	III	W	P	S	W	P	S	W	P	S	W	P	S
	11	IIf	P	S	P	S	P	S	P	S	P	S	P	S
	12	IIf	S	P	S	P	S	P	S	P	S	P	S	P
	13	IV	P	W	S	W	P	W	S	W	P	W	S	W
	14	IV	W	S	W	P	W	S	W	P	W	S	W	P
	15	IV	S	W	P	W	S	W	P	W	S	W	P	W
	16	IV	W	P	W	S	W	P	W	S	W	P	W	S
2	1	III	P	S	W	P	S	W	P	S	W	P	S	W
	2	III	S	W	P	S	W	P	S	W	P	S	W	P
	3	III	W	P	S	W	P	S	W	P	S	W	P	S
	4	IIf	P	S	P	S	P	S	P	S	P	S	P	S
	5	IIf	S	P	S	P	S	P	S	P	S	P	S	P
	6	IV	P	W	S	W	P	W	S	W	P	W	S	W
	7	IV	W	S	W	P	W	S	W	P	W	S	W	P
	8	IV	S	W	P	W	S	W	P	W	S	W	P	W
	9	IV	W	P	W	S	W	P	W	S	W	P	W	S
	10	IIIf	P	S	W	P	S	W	P	S	W	P	S	W
	11	IIIf	S	W	P	S	W	P	S	W	P	S	W	P
	12	IIIf	W	P	S	W	P	S	W	P	S	W	P	S
	13	IVf	P	W	S	W	P	W	S	W	P	W	S	W
	14	IVf	W	S	W	P	W	S	W	P	W	S	W	P
	15	IVf	S	W	P	W	S	W	P	W	S	W	P	W
	16	IVf	W	P	W	S	W	P	W	S	W	P	W	S

† *P*, potato; *S*, sugar beet; *W*, winter wheat.

in medical research where separate analyses are done on the individual data sets and the results are combined.) The traditional way to handle this, in ordinary analysis of variance, would be to analyze the years separately, test for homogeneity of variance (e.g., by using Bartlett's [1937]) test and then, if necessary, weight the data from each year by the reciprocal of that year's residual variance. If, as in many rotation experiments, there is no within-year replication, this will not be possible. Fortunately though, the more recent REML (*Residual,* or *Restricted, Maximum Likelihood*) method for the analysis of linear mixed models allows different residual variances for the years to be estimated during the combined analysis; see Patterson and Thompson (1971), Gilmour et al. (1995) and Chapter 2 of Payne (2014, p. 45—55). If there are additional random terms, for example whole-plots in a split-plot design, their variance components may also differ from year to year. This too can be handled in a REML analysis, as shown in the analysis of the Westmaas experiment below.

The second issue might traditionally be handled by "repeated-measures analysis of variance", which mitigates the effects of the non-uniform correlations by adjusting the numbers of degrees of freedom of the affected sums of squares (see Winer, 1962, 523, 594—599; or Payne, 2014, p. 110). This is feasible if the design is balanced that is, if the same plots have been measured in each of the years for which data are available, as shown, for example, by Christie et al. (2001), Liebman et al. (2008) and Barton et al. (2009). In long-term rotation experiments, however, different plots will usually have been measured in different sets of years, and the use of repeated-measures ANOVA becomes a difficult (if not impossible) task. Fortunately, here too, the REML methodology provides a solution, with the ability to fit models to the correlations (see e.g., Gilmour et al., 1997; Galwey, 2006; Section 9.7, Littell et al., 2006; Chapter 5, Payne, 2006; or Chapter 4, Payne, 2014, p. 38—49]. For examples, see Singh, Christiansen and Chakraborty, 1997; Singh and Jones, 2002; Richter and Kroschewski, 2006; Machado et al., 2008. In the analysis of the Westmaas experiment, below, correlation models are tried, but found to be unnecessary.

The third issue arises if the same test crop is grown several times in a particular rotation. It can be resolved by doing a separate analysis for each instance of the test crop in the rotation cycle, as would be necessary if they were actually different crops. (The only sensible way to perform an analysis combining data from several different test crops would be to assign some measure such as economic value to each one; however, these could be rather arbitrary and unlikely to remain constant through the whole experiment.) An alternative would be to include a factor for occurrence-within-cycle, so that the instances be included as separate "treatments", and compared in a combined analysis.

Issues four and five can be more difficult to resolve. If there are many auxiliary treatment factors, it may be acceptable to use some of the higher-order interactions among these and the rotation factor as the residual, that is, to use the traditional approach, for example, of treating second-order (and higher) interactions as the residual, and then feeling justified if the analysis detects no significant first-order interactions. (This reasoning tends to be rather circular, but can often be justified by experience from previous similar experiments.) A more easily justifiable variant of this approach was suggested by Patterson (1959) for an experiment that studied fertilizer response as one of the treatments. The fertilizer was applied at five different rates. So, on the assumption that the relationship between yield and fertilizer can be represented adequately by a second-order polynomial, the interactions between rotations and the cubic and the quartic polynomials could be used for the residual. (Under these circumstances, of course, it is arguable that it might have been safer to have had genuine replication and fewer levels of fertilizer, but this is the same issue that arises in any study of fertilizer response.) An alternative approach would be to model the year-by-treatment interaction. Again all the standard methods are available. For example, echoing Patterson's ideas for the fertilizer-response curves, one might fit interactions between the treatments and polynomial effects of year. The linear year effects and their interactions would assess whether the effects are increasing or decreasing in any uniform way with time (Issue 5), but again the success of this strategy is dependent on the assumption of a low-order polynomial response. If this is not feasible, an alternative would be to fit spline functions over years. For example, Verbyla et al. (1999) described how to fit random smoothing splines in REML. Polynomial models, though, may be easier to explain.

Payne (2015) showed an example, using the Woburn Ley–Arable Experiment, that had no within-year replication. The changes in fertility over the 20 yr of the experiment were modeled by polynomials. Initially, fourth-order polynomials were fitted. The higher order polynomials were assumed to be absent, and their degrees of freedom were used to estimate the variances between and within years. The analysis and conclusions thus depended on the appropriateness of this assumption. However, some justification for the approach was given by the fact that the analyses found no evidence for the inclusion of either the cubic or the quartic terms in the model, that is, a (relatively simple) quadratic relationship seemed to hold. So the cubic or the quartic terms were also omitted from the final analyses.

Changes in basal treatments (Issue 6) can be handled by including additional factors in the analysis, to indicate the underlying basal conditions applying on each year-plot observation. Ideally they should not have affected differences between rotations or any of the other treatments; provided the changes have not been too frequent, it should be possible to check this by fitting the relevant interactions. It may be more difficult, however, to accommodate changes in the makeup of the rotations themselves within a single analysis. For example, the test crops may not have remained the same, or the whole purpose of the experiment may have changed. In that case the analysis would need to start afresh, once a full cycle of the new rotations has been completed.

Example

The use of REML to analyze rotation experiments is illustrated by the analysis of the net yield of potato (t ha^{-1}) from the Westmaas experiment. (The net yield is the total yield after removing green tubers or those with disease damage.) This was done using the Genstat statistical system (VSN International, 2014) and in R using the ASReml R package (VSN International, 2009). The input scripts and output for the analyses are in Appendices 1 through 4, and the data are in the files wmpotato.xlsx and wmpotato.rda. The ASReml-R script was written by Chris Brien, and also makes uses of his asremlPlus package. A brief description of the Genstat commands is given below, but you can also use menus. If you are unfamiliar with Genstat, you may find Galwey (2006) and Payne (2014) helpful.

The treatment factors are Rotation and Nitrogen. Years are also treated as fixed, as their changes may be more appropriately regarded as systematic rather than as random. The fixed model thus consists of the main effects of these factors and their interactions.

The first step in the analysis is to establish the appropriate random model. This involves fitting various random models, all with the full fixed model, and examining the deviance of each one; this is defined as -2 times the log-likelihood for the model. If one random model is a generalization of another one (i.e., if it contains all the random parameters of that model, together with some additional ones), the difference between their deviances can be treated as a chi-square statistic with number of degrees of freedom equal to the number of additional parameters. If neither random model is a generalization of the other, Akaike or Schwarz Bayesian information criteria are generally used to assess which model to select [e.g., Payne (2014), p. 63—64]. The best model is the one with the smallest value of the chosen criterion. The choice of criterion is a matter of personal choice. The Schwarz Bayesian criterion tends to select models with fewer random parameters than the Akaike criterion, which can be an advantage if the aim is to avoid an overcomplicated random model. Conversely, the Akaike coefficient tends to include more parameters, and thus may provide a more detailed representation of the random variation.

Table 4 shows deviances, Akaike and Schwarz Bayesian information criteria, and numbers of degrees of freedom for all the random models that are investigated for the Westmaas experiment. The first analysis is based on the conventional split-plot analysis. In each year, there is a split-plot design with random-term blocks, and whole-plots within blocks. We treat these designs as being nested within the year factor, so that we now have random terms for blocks within years, and whole-plots within blocks (within years). At this stage we assume that the random variation is the same in each year, and that there is no correlation between the yields of each plot from year to year. These assumptions are investigated in subsequent models.

The model is analyzed in Genstat by

```
VCOMPONENTS [FIXED=Year*Rotation*Nitrogen] Year/Block/Wholeplot
REML [PRINT=components] Yield
VAIC [PRINT=deviance,aic,sic,dfrandom]
VRACCUMULATE [PRINT=*] 'Split-plot nested within years'
```

TABLE 4. Summary of the random models.

Model	Deviance	Akaike	Schwarz Bayesian	Random d.f
Split-plot nested within years	950.43	956.43	965.75	3
Nested split-plot meta analysis	891.60	917.60	957.98	13
Nested split-plot and power distance	950.38	958.38	970.80	4
Nested split-plot meta analysis and power distance	891.57	919.57	963.06	14
Meta analysis with different variance components	852.53	918.53	1021.02	33

The VCOMPONENTS command defines the fixed model to contain the main effects and interactions of Year, Rotation and Nitrogen. (The operator * defines a factorial relationship.) The random terms are defined to be Years, Block within Year, and Wholeplot within Blocks within Year. (The operator / defines a nested relationship.) However, as Year has also been defined as a fixed term, it is automatically removed from the random model. The REML command analyzes the variate Yield, and prints the variance components. The VAIC command prints the deviance, the information criteria and the number of degrees of freedom in the random model. The VRACCUMULATE command remembers this information, so that it can be displayed later for all the random models to provide Table 4. Setting option PRINT = * stops anything being printed at present.

In the equivalent ASReml-R commands, below, Year needs to be excluded explicitly from the random model.

```
model1.asr <- asreml(Yield ~ Year*Rotation*Nitrogen,
                     random = ~Year:Block/Wholeplot,
                     data = wmpotato)
summary(model1.asr)$varcomp
info.accumulate <- data.frame(Model = "Split-plot",
                     infoCriteria(model1.asr),
                     stringsAsFactors = FALSE)
```

In Genstat you could do this by specifying the random model as Year.Block/Wholeplot (the dot operator is the Genstat equivalent of the colon operator in R). Table 5 shows the estimated variance components.

The analysis is then extended to fit a different residual variance in each year, as required to satisfy Issue 1. The Genstat commands are:

```
VCOMPONENTS [FIXED=Year*Rotation*Nitrogen; EXPERIMENT=Year]\
  Year/Block/Wholeplot
REML [PRINT=components] Yield
VAIC [PRINT=deviance,aic,sic,dfrandom]
VRACCUMULATE [PRINT=*] 'Nested split-plot meta analysis'
```

The EXPERIMENT option in the VCOMPONENTS command specifies the factor within which different residual variances are to be estimated, and the character \ is the "continuation character" that continues the command onto the next line.

The ASReml-R commands are:

```
model2.asr <- asreml(Yield ~ Year*Rotation*Nitrogen,
                     random = ~Year:Block/Wholeplot,
                     residual = ~ idh(Year):Block:NWholeplot:Subplot,
                     data = wmpotato)
model2.asr <- update(model2.asr)
summary(model2.asr)$varcomp
info.accumulate <- rbind(info.accumulate,
                        data.frame(Model="Split plot meta analysis",
                                   infoCriteria(model2.asr),
                                   stringsAsFactors = FALSE))
```

As shown in Table 6, this "nested split-plot meta-analysis" generalizes the model by fitting 11 residual variances instead of the single residual variance contained in Table 5. The difference between the deviances of the two models, can be treated as a chi-square statistic on 10 degrees of freedom. Its value, 58.83, provides strong

TABLE 5. Variance parameters for the nested split-plot analysis.

Parameter	Estimate	SE
Block variance component	1.407	1.403
Whole-plot within block variance component	7.079	1.893
Residual variance	5.195	0.701

TABLE 6. Variance parameters for the nested split-plot meta analysis.

Parameter	Estimate	SE
Block variance component	2.117	1.168
Whole-plot within block variance component	0.912	0.517
1977 Residual variance	0.819	0.352
1979 Residual variance	2.684	1.149
1980 Residual variance	1.344	0.588
1981 Residual variance	5.231	2.558
1982 Residual variance	4.538	1.873
1983 Residual variance	1.988	0.887
1984 Residual variance	4.013	1.660
1985 Residual variance	5.415	2.185
1986 Residual variance	14.283	5.559
1987 Residual variance	58.211	22.425
1988 Residual variance	12.895	5.245

evidence that a different random variance is needed in each year, and this will usually be the situation. This conclusion is supported by both of the information criteria.

The next two analyses address Issue 2 by fitting a power-distance correlation structure over years. The assumption here is that correlation ρ between the residuals from two observations on a particular plot can be expressed as φ^d, where d is the distance in time between them, measured in years. The value of the parameter φ, which is usually assumed to be constant over the whole experiment, is estimated in the analysis. In the "nested split-plot and power distance" analysis the correlation structure is added to the nested split-plot analysis, while in the "nested split-plot meta-analysis and power distance" analysis it is added to the analysis where a different residual variance was fitted for each year.

The Genstat commands to fit the models are:

```
" Nested split plot: power-distance correlation structure over years"
VCOMPONENTS [FIXED=Year*Rotation*Nitrogen] Year/Block/Wholeplot
VARIATE Ycoord; VALUES=Year
VSTRUCTURE [TERM=Year.Block.Wholeplot; COORDINATES=Ycoord] power;\
  FACTOR=Year
REML [PRINT = *] Yield
VRACCUMULATE [PRINT=*] 'Nested split-plot and power distance'
" Nested split plot with different residual variance in each year
  and power-distance correlation structure over years "
VCOMPONENTS [FIXED = Year*Rotation*Nitrogen; EXPERIMENT = Year]\
  Year/Block/Wholeplot
VSTRUCTURE [TERM = Year.Block.Wholeplot; COORDINATES=Ycoord] power;\
  FACTOR=Year
REML [PRINT=*] Yield
VRACCUMULATE [PRINT=*] 'Nested split-plot meta analysis and power distance'
```

The fixed and random models are as before, and VSTRUCTURE command defines the power distance model on the Year.Block.Wholeplot random term. The information between each pair of double quotes is a comment to indicate which model is being fitted.

In ASReml-R the power-distance correlation model is known as the exp (or exponential) variance model, and is parameterized in exactly the same way as in Genstat. The commands to fit the models are:

```
model3.asr <- asreml(Yield ~ Year*Rotation*Nitrogen,
                     random = ~Year:Block/Wholeplot,
                     residual = ~ exp(Year):Block:NWholeplot:Subplot,
                     data = wmpotato)
summary(model3.asr)$varcomp
info.accumulate <- rbind(info.accumulate,
                       data.frame(Model = "Split-plot and EXP",
                                  infoCriteria(model3.asr),
                                  stringsAsFactors = FALSE))
model4.asr <- asreml(Yield ~ Year*Rotation*Nitrogen,
                     random = ~Year:Block/Wholeplot,
                     residual = ~ exph(Year):Block:NWholeplot:Subplot,
                     data = wmpotato)
model4.asr <- update(model4.asr)
summary(model4.asr)$varcomp
info.accumulate <- rbind(info.accumulate,
                    data.frame(Model = "Split-plot meta analysis and EXP",
                                  infoCriteria(model4.asr),
                                  stringsAsFactors = FALSE))
```

The deviances, information criteria and numbers of random degrees of freedom for these two models are in Table 4 in row 3 ("Nested split-plot and power distance") and row 4 ("Nested split-plot meta-analysis and power distance"). Both information criteria lead to the same conclusion, namely that there is no evidence to support the inclusion of a correlation model. (Note: the correlation should ideally be defined between the subplot residuals, rather than the plot residuals, but this leads to a very complicated random model that is difficult to fit. The use of this simpler model, however, is sufficient for the assessment of whether these correlations are necessary.)

Returning to Issue 1, we now consider whether the variance components of the block and whole-plot-within-block random terms vary from year to year. The final row of Table 4 ("Meta-analysis with different variance components") contains the deviance, information criteria and number of random degrees of freedom for a model where a different variance component is fitted for every random term, as well as a different residual variance, in every year.

In Genstat the commands are:

```
FORMULA [VALUE = Block/Wholeplot] differentvcterms
VRMETA [EXPERIMENTSFACTOR=Year; RANDOM=Random] 77,79,80...88;\
  LOCALTERMS=differentvcterms
VCOMPONENTS [FIXED=Year*Rotation*Nitrogen; EXPERIMENTS=Year] #Random
REML [PRINT=*; MVINCLUDE=explanatory] Yield
VRACCUMULATE 'Meta analysis with different variance components'
```

The FORMULA command defines a model–formula data structure containing terms for blocks and whole plots within blocks. The VRMETA command forms a random model to fit a different variance component for these terms within each of the years 1977, 1979, 1980, and so on to 1988. This model, Random, is then used in the subsequent VCOMPONENTS command. (The # operator substitutes the contents of Random into the command.)

The ASReml-R specification is:

```
model5.asr <- asreml(Yield ~ Year*Rotation*Nitrogen,
                     random = ~ idh(Year):Block/Wholeplot,
                     residual = ~ idh(Year):Block:NWholeplot:Subplot,
                     data = wmpotato)
summary(model5.asr)$varcomp
vcnames <- rownames(summary(model5.asr)$varcomp)[1:22]
model5.asr <- setvarianceterms(model5.asr$call, terms = vcnames,
                     bounds = "U", initial=0.01, ignore.suffices = FALSE)
model5.asr <- update(model5.asr)
summary(model5.asr)$varcomp
info.accumulate < - rbind(info.accumulate,
    data.frame(Model = "Meta analysis with different variance
components",
                        infoCriteria(model5.asr, bound.exclusions = "F"),
                        stringsAsFactors = FALSE))
```

This model has larger values for both information criteria than the "Nested split-plot meta-analysis" in row 2. The conclusion is that this further extension is unnecessary, and that the "split-plot meta-analysis" remains the best random model.

Once the appropriate random model has been decided, the treatment model can be assessed to see whether there are any unnecessary fixed terms. The standard way to do this is to examine their Wald statistics. These would have exact chi-square distributions if the variance parameters were known, but as those must be estimated, the statistics are only asymptotically distributed as chi-square. In practical terms, the chi-square values will be reliable if the number of residual degrees of freedom for the fixed term is large compared with its own degrees of freedom. Alternatively, statistical software systems such as ASReml, Genstat and SAS can use the method of Kenward and Roger (1997) to estimate the number of residual degrees of freedom relevant to each term. An *F* statistic can be used to provide the probability instead. The *F* statistic is initially calculated as the Wald statistic divided by the number of degrees of freedom of the term (i.e., the number of degrees of freedom for the numerator of the *F* test). However, the Kenward–Roger method makes minor adjustments to this initial calculation to remove biases and make the resulting probability more accurate. Simulations performed by Kenward and Roger (1997) showed that these *F* statistics are not subject to the biases of the chi-square statistics, and can thus be used with confidence.

Table 7 shows the *F* and Wald test statistics for the full fixed model in the split-plot meta-analysis. In the top half of the table, the fixed terms are added sequentially into the analysis, while in the lower half, the effect of dropping fixed terms from the full model is assessed. Note, a term should be dropped only if the model includes no higher-order terms that contain it: for example, you should not drop a main effect if the model includes an interaction term involving its factor. In a nonorthogonal design, the sizes of the *F* (or chi-square) statistics will depend on the order in which the terms are fitted. In the upper half of the table, each row shows the effect of the term concerned after eliminating the earlier terms, but ignoring the later ones. Alternatively, in the lower half of the table, their effects are assessed after eliminating all the other terms, and this provides a more definitive conclusion. In Table 7, the fixed terms are orthogonal, thus the lines in the lower half of the table contain the same numbers as in the upper half. However, it is still useful to present the lower

TABLE 7. Tests for terms in the full fixed model.

Fixed term	Wald statistic	n.d.f.†	*F* statistic	d.d.f.‡	*F* probability
Sequentially adding terms to the model					
Year	851.56	10	85.44	10.4	< 0.001
Rotation	106.91	4	26.73	27.0	< 0.001
Nitrogen	110.58	2	55.29	58.1	< 0.001
Year.Rotation	154.86	40	3.75	22.8	< 0.001
Year.Nitrogen	29.16	20	1.27	48.7	0.242
Rotation.Nitrogen	15.43	8	1.93	58.1	0.073
Year.Rotation.Nitrogen	108.45	80	1.19	40.6	0.271
Dropping feasible terms from the model					
Year.Rotation.Nitrogen	108.45	80	1.19	40.6	0.271

† n.d.f. contains the degrees of freedom for the Wald statistic, and the numerator degrees of freedom of the *F* statistic.

‡ d.d.f. contains the denominator degrees for freedom of the *F* statistic.

half, as this indicates the terms that are candidates to be dropped. Table 7 shows that there is no evidence for a three-way interaction of Year, N and Rotation. This interaction is therefore omitted in the next analysis, thereby making it feasible to assess the effect of dropping the two-factor interactions (see Table 8). The *F* (and Wald) tests now confirm that there is no evidence for interactions between N and Year or N and Rotation. Table 9 shows the tests after dropping these two terms.

The final fixed model thus contains main effects of Year, Rotation and N, and a Year × Rotation interaction. Predicted means for N (Table 10) show that yields increase (as expected) with the amount that was applied.

To fit these models in Genstat, remember that the original commands were:

```
VCOMPONENTS [FIXED = Year*Rotation*Nitrogen; EXPERIMENT=Year]\
  Year/Block/Wholeplot
REML [PRINT=components] Yield
```

To display the *F* and Wald statistics, set option PRINT = wald in the REML command. It is easiest to remove the terms from the fixed model by subtracting them

TABLE 8. Tests for terms in the fixed model excluding the three-factor interaction.

Fixed term	Wald statistic	n.d.f.†	*F* statistic	d.d.f.‡	*F* probability
Sequentially adding terms to the model					
Year	796.16	10	79.79	10.7	< 0.001
Rotation	91.94	4	22.98	32.8	< 0.001
Nitrogen	83.70	2	41.85	128.4	< 0.001
Year.Rotation	117.01	40	2.92	25.3	0.003
Year.Nitrogen	29.67	20	1.36	88.3	0.166
Rotation. Nitrogen	10.34	8	1.29	128.4	0.253
Dropping feasible terms from the model					
Year.Rotation	117.01	40	2.92	25.3	0.003
Year.Nitrogen	29.67	20	1.36	88.3	0.166
Rotation.Nitrogen	10.34	8	1.29	128.4	0.253

† n.d.f. contains the degrees of freedom for the Wald statistic, and the numerator degrees of freedom of the *F* statistic.

‡ d.d.f. contains the denominator degrees for freedom of the *F* statistic.

TABLE 9. Tests for terms in the fixed model after dropping non-significant terms.

Fixed term	Wald statistic	n.d.f.†	*F* statistic	d.d.f.‡	*F* probability
Sequentially adding terms to the model					
Year	780.43	10	78.21	10.8	< 0.001
Rotation	92.68	4	23.17	32.6	< 0.001
Nitrogen	83.21	2	41.60	142.0	< 0.001
Year.Rotation	117.47	40	2.93	27.0	0.002
Dropping feasible terms from the model					
Nitrogen	83.21	2	41.60	142.0	< 0.001
Year.Rotation	117.47	40	2.93	27.0	0.002

† n.d.f. contains the degrees of freedom for the Wald statistic, and the numerator degrees of freedom of the F statistic.

‡ d.d.f. contains the denominator degrees for freedom of the *F* statistic.

TABLE 10. Estimated mean yields of potato (t ha^{-1}) for N. †

N1	N2	N3
43.37	45.02	45.65

† Standard error for differences: 0.2585.

from the full factorial model, as shown below. The WORKSPACE option is set in the REML command to ensure that sufficient workspace is reserved for the Kenward–Roger calculations.

```
" drop unnecessary fixed terms "
VCOMPONENTS [FIXED=Year*Rotation*Nitrogen-Year.Rotation.Nitrogen;\
  EXPERIMENTS = Year] Year/Block/Wholeplot
REML [PRINT = wald; MVINCLUDE = explanatory; WORKSPACE = 100] Yield
VCOMPONENTS [FIXED = Year*Rotation*Nitrogen- Year.Rotation.Nitrogen\
  -  Rotation.Nitrogen; EXPERIMENTS = Year] Year/Block/Wholeplot
REML [PRINT = wald; MVINCLUDE = explanatory; WORKSPACE = 100] Yield
VCOMPONENTS [FIXED = Year*Rotation*Nitrogen- Year.Rotation.Nitrogen\
  -  Rotation.Nitrogen- Year.Nitrogen; EXPERIMENTS = Year]\
 Year.Block/Wholeplot
REML [PRINT=wald; MVINCLUDE=explanatory; WORKSPACE=100] Yield
```

The equivalent commands in ASReml-R are:

```
model2a.asr < - asreml(Yield ~ Year*Rotation*Nitrogen-Year:Rotation:Nitrogen,
                    random = ~ Year:Block/Wholeplot,
                    residual = ~ idh(Year):Block:NWholeplot:Subplot,
                    data = wmpotato)
wald(model2a.asr, denDF = "algebraic")
model2b.asr < - asreml(Yield ~ Year*Rotation*Nitrogen-Year:Rotation:Nitrogen
                            -  Rotation:Nitrogen,
                       random = ~ Year:Block/Wholeplot,
                       residual = ~ idh(Year):Block:NWholeplot:Subplot,
                       data = wmpotato)
wald(model2b.asr, denDF = "algebraic")
model2c.asr < - asreml(Yield ~ Year*Rotation*Nitrogen-Year:Rotation:Nitrogen
                         -  Rotation:Nitrogen - Year:Nitrogen,
                       random = ~ Year:Block/Wholeplot,
                       residual = ~ idh(Year):Block:NWholeplot:Subplot,
                       data = wmpotato)
wald(model2c.asr, denDF = "algebraic")
```

The one-way table of means for N, and the two-way table of means for Rotation and Year, are printed in Table 11 by the Genstat command

```
VDISPLAY [PRINT=means; PTERMS=Nitrogen+Year.Rotation]
```

TABLE 11. Estimated means of potato (t ha⁻¹) for year and rotation. † ‡ § ¶# †† ‡‡

	Rotation				
Year	IIf	III	IIIf	IV	IVf
1977	23.44	21.92	21.87	25.01	21.20
1979	23.44	21.92	21.87	25.01	21.20
1980	40.53	41.14	40.90	41.54	42.48
1981	30.37	38.04	39.45	39.43	38.14
1982	39.75	40.49	42.95	45.53	45.63
1983	31.14	36.55	39.53	37.96	37.61
1984	54.59	57.67	56.63	58.40	58.99
1985	51.48	47.80	49.79	52.54	49.75
1986	48.21	51.03	54.58	51.48	52.99
1987	42.99	48.76	41.56	45.67	51.85
1988	54.28	62.76	64.41	58.52	62.46

† Average standard error for rotation differences within year: 1.978.
‡ Maximum standard error for rotation differences within year: 3.686.
§ Minimum standard error for rotation differences within year: 1.413.
¶ Average standard error for year differences within rotation: 2.524.
# Maximum standard error for year differences within rotation: 3.527.
†† Minimum standard error for year differences within rotation: 2.074.
‡‡ Overall average standard error of differences: 2.483.

Genstat refers to these as *predicted means* to indicate that they are predictions estimated from the model, rather than ordinary averages. They may be called "least squares means" in output from other software systems, such as SAS. Table 11 gives a summary of the standard errors of differences between the means. To obtain the (large) full table, you can set the PSE option, like this.

```
VDISPLAY [PRINT=means; PTERMS=Nitrogen+Year.Rotation; PSE=alldifferences]
```

In ASReml-R the commands are

```
predict(model2c.asr, classify = "Nitrogen")$pvals
predYR <- predict(model2c.asr, classify = "Year:Rotation")$pvals
predYR
```

The means for Year and Rotation are plotted in Fig. 1, but it is difficult to pick out any consistent patterns. Rotation IIf is usually worse than the other rotations. However, there does not seem to be any consistent benefit from fumigation, nor from the choice between rotations III and IIIf versus IV and IVf.

Summary

The design and analysis of long-term rotation experiments present many interesting statistical challenges, which can be daunting in view of the amount of time and effort that will have been expended in collecting the data. However, an awareness of the special issues discussed above should avoid pitfalls and lead to clear and useful conclusions. On design, it is important to allow comparisons to be made between the rotations in several years. The key issue is to decide how to allocate the rotation treatment, and how to allow for several occurrences of the rotations, starting in

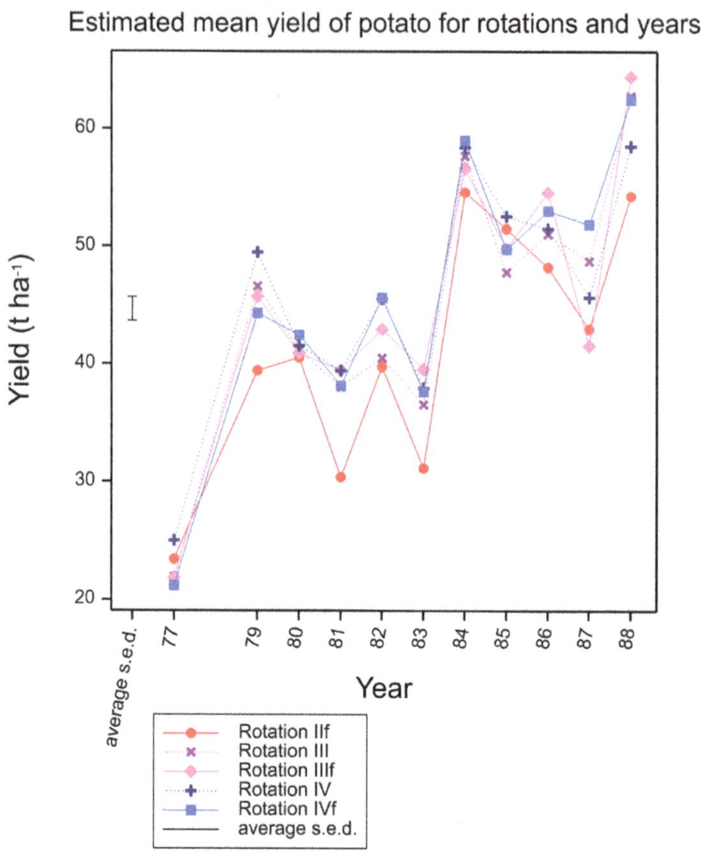

FIG. 1. Estimated mean yield of potato for rotations and years.

successive years. One effective method is to start occurrences of the rotations in successive years, each in a separate block, (i.e., to use a randomized-block design), with year-of-starting-point as the block factor, and rotation as the plots factor. As shown in the example in this chapter, other auxiliary treatments can be applied as in conventional single-year experiments, for example, by splitting the plots into subplots to give a split-plot design with the auxiliary treatment factor(s) as the split-plot factor(s).

This may lead to a rather large experiment, with insufficient resources to provide within-year treatment replication. Strategies are then needed in the analysis to provide degrees of freedom for the residual. These can again be drawn from the armory of conventional analysis techniques, for example, by allocating higher-order interactions to the residual, or by modeling responses by low-order polynomials or splines.

Other aspects to consider include the fact that variability is unlikely to be constant across years, and that there may be nonuniform correlation structures if there is an insufficient distance in time between the observations on each plot. As a result, conventional analysis of variance is unlikely to be suitable. However, appropriate analyses can be performed using the more recent REML methodology (Patterson and Thompson, 1971; Gilmour et al., 1995, 1997; Littell et al. 2006; Payne 2014). The Genstat and ASReml-R commands that performed the analyses above are listed in Appendices 1 and 3.

Acknowledgments
I thank Wim van den Berg of Wageningen UR for the use of the Westmaas data set and associated advice, and Chris Brien of the University of South Australia for the ASReml-R script. I am also grateful to VSN International and Rothamsted for their support.

Key Learning Points

- Rotation experiments take place over several years to allow sequences of crop and husbandry combinations to be compared, that is, they compare "treatments" that are applied over several, rather than a single year.
- It is best to phase the start of the experiment with new replicates of the rotations starting in successive years. The conclusions will then not be dependent on the conditions during one specific sequence of years.
- It is best to have more than one replicate in every year. If you cannot do this, you will be unable to do interim analyses on the date from a single year, and you will need to assume, for example, that higher-order interactions can be ignored or that responses over years can be modeled by low-order polynomials, to obtain a residual when you analyze data from all the years.
- In the analysis, you need to consider that the amount of random variation may differ from year to year.
- The correlation between observations on a plot may differ according to the distance in time between them.
- You can cater to these issues in the analysis by fitting a mixed model using REML (i.e., residual, or restricted, maximum likelihood), and modeling the between-year correlation structure.

Review Questions

1. Why would you do a long-term rotation experiment instead of several single-year experiments?
2. What is the difference between a short-term and a long-term rotation experiment?
3. Why might it be a problem if you ran the rotations over only one series of years?
4. How might you include auxiliary treatments, in addition to the rotation treatments?
5. Why might the analysis be more complicated than the analysis of a single-year experiment?
6. What are the advantages of REML compared with ordinary analysis of variance, and how would you exploit these in your analysis?
7. What statistics can you use to decide on the random model?
8. How would you assess the fixed terms?

Exercises

The files wmsugar.xlsx and wmsugar.rda contain the yields for the roots of sugar beet from the Westmaas experiment.

EXERCISE 1. Read the data. Are there any unexpected features? (Hint: Look at the Year factor.)

EXERCISE 2. Fit a model with a fixed model containing the factorial effects of Year, Nitrogen and Rotation, and a random model with split-plots nested within years.

EXERCISE 3. Extend the model to fit a different residual variance in each year. Calculate the difference of the deviances between this model and the first model. Is there evidence that the different residual variances are necessary?

EXERCISE 4. Fit models with a power–distance correlation structure over years. Is there any evidence that this is needed?

EXERCISE 5. Taking the best random model of those that you have fitted, see whether you can simplify the fixed model by omitting nonsignificant interactions.

EXERCISE 6. Plot the predicted means for the fixed terms that seem to be interesting in your final analysis. What aspects would you highlight in your report on the results?

References

Barton, L., G.G.Y. Wan, R.P. Buck, and T.D. Colmer. 2009. Nitrogen increases evapotranspiration and growth of a warm-season turfgrass. Agron. J. 101:17–24. doi:10.2134/agronj2008.0078

Bartlett, M.S. 1937. Some examples of statistical methods of research in agriculture and applied biology (with discussion). J.R. Stat. Soc. 4(Suppl.):137–183.

Christie, P., D.L. Easson, J.R. Picton, and S.C.P. Love. 2001. Agronomic value of alkaline-stabilized sewage biosolids for spring barley. Agron. J. 93:144–151. doi:10.2134/agronj2001.931144x

Dyke, G.V. 1974. Comparative experiments with field crops. Butterworths, London.

Galwey, N.W. 2006. Introduction to mixed modelling: Beyond regression and Analysis of Variance. Wiley, Chichester, UK. doi:10.1002/9780470035986

Gilmour, A.R., R. Thompson, and B.R. Cullis. 1995. Average information REML, an efficient algorithm for variance parameter estimation in linear mixed models. Biometrics 51:1440–1450. doi:10.2307/2533274

Gilmour, A.R., B.R. Cullis, and A.P. Verbyla. 1997. Accounting for natural extraneous variation in the analysis of field experiments. J. Agric. Biol. Environ. Stat. 2:269–273. doi:10.2307/1400446

Glynne, M.D., and D. Slope. 1959. Effects of previous wheat crops, seed-rate and nitrogen on eyespot, take-all, weeds and yields of two varieties of winter wheat: Field experiment 1954-56. Ann. Appl. Biol. 47:187–199. doi:10.1111/j.1744-7348.1959.tb02536.x

Kenward, M.G., and J.H. Roger. 1997. Small sample inference for fixed effects from restricted maximum likelihood. Biometrics 53:983–997. doi:10.2307/2533558

Liebman, M., L.R. Gibson, D.N. Sundberg, A.H. Heggenstaller, P.R. Westerman, C.A. Chase, R.G. Hartzler, F.D. Menalled, A.S. Davis, and P.M. Dixon. 2008. Agronomic and economic performance characteristics of conventional and low-external-input cropping systems in the Central Corn Belt. Agron. J. 100:600–610. doi:10.2134/agronj2007.0222

Littell, R.C., G.A. Milliken, W.W. Stroup, R.D. Wolfinger, and O. Schabenberger. 2006. SAS for Mixed Models. Second ed. SAS Institute Inc, Cary, NC.

Machado, S., S. Petrie, K. Rhinhart, and R.E. Ramig. 2008. Tillage effects on water use and grain yield of winter wheat and green pea in rotation. Agron. J. 100:154–162. doi:10.2134/agrojnl2006.0218

Patterson, H.D. 1959. The analysis of a non-replicated experiment involving a single four-course rotation of crops. Biometrics 15:30–59. doi:10.2307/2527600

Patterson, H.D. 1964. Theory of cyclic rotation experiments. J. R. Stat. Soc. B 26:1–45.

Patterson, H.D., and R. Thompson. 1971. Recovery of inter-block information when block sizes are unequal. Biometrika 58:545–554. doi:10.1093/biomet/58.3.545

Payne, R.W. 2006. New and traditional methods for the analysis of unreplicated experiments. Crop Sci. 46:2476–2481. doi:10.2135/cropsci2006.04.0273

Payne, R.W. 2014. The guide to the GenStat command language (Release 17), Part 2: Statistics. VSN International, Hemel Hempstead, U.K.

Payne, R.W. 2015. The design and analysis of long-term rotation experiments. Agron. J. 107:772–785. doi:10.2134/agronj2012.0411

Richter, C., and B. Kroschewski. 2006. Analysis of a long-term experiment with repeated measurement models. J. Agron. Crop Sci. 192:55–71. doi:10.1111/j.1439-037X.2006.00167.x

Singh, M., S. Christiansen, and B.K. Chakraborty. 1997. An assessment of the effect of covariances of plot errors over time on the precision of means of rotations with wheat. Exp. Agric. 33:469–475. doi:10.1017/S0014479797004018

Singh, M., and M.J. Jones. 2002. Modeling yield sustainability for different rotations in long-term barley trials. J. Agric. Biol. Environ. Stat. 7:525–535. doi:10.1198/108571102744

Verbyla, A.P., B.R. Cullis, M.G. Kenward, and S.J. Welham. 1999. The analysis of designed experiments and longitudinal data using smoothing splines (with discussion). Appl. Stat. 48:269–311.

VSN International. 2009. ASReml-R 3. VSN International, Hemel Hempstead, UK. Asreml.co.uk.

VSN International. 2014. GenStat for Windows. 17th ed. VSN International, Hemel Hempstead, UK. Vsni.co.uk

Winer, B.J. 1962. Statistical Principles in Experimental Design. McGraw-Hill, New York. doi:10.1037/11774-000

Yates, F. 1949. The Design of Rotation Experiments. Commonwealth Bureau of Soil Science Communication 46.

CHAPTER 12: SPATIAL ANALYSIS OF FIELD EXPERIMENTS

Juan Burgueño*

Well-planned experimental designs are able to deal with soil variability in field experiments; however, it is known that small-scale variation exists, and also variability introduced by experimental practices or environmental factors cannot be isolated adequately by the experimental designs. In those cases, spatial modeling of residual has been shown to be a good option to deal with this variability. Several spatial models have been studied to deal with this problem during the last hundred years, from Wood and Stratton in 1910 to the present day. In this chapter, we present one approach that is commonly used and has shown good results in many situations. We use the separable autoregressive model of order one (*AR1*) in row and column directions to model the experimental error. First, we discuss some practical issues about the spatial analysis and then we show the analysis of two real examples.

Field trials are under the effect of soil variability, which is the major uncontrolled factor that affects experimental results. Soil variability does not only include soil properties (soil fertility, soil water-holding capacity, soil physical properties) but also other factors related with soil characteristics like inoculum of a plague or disease, water distribution in both horizontal and vertical profile, among others. Other effects, like field management, previous history of cropping, irrigation system, plot trimming, and direction of cultivation or harvesting, are induced by humans.

Since the beginning of the experimental designs, researchers have been dealing with soil variability. Fisher, the pioneer of experimental design with statistical properties, described local control as a mechanism of dealing with field variability (Fisher, 1925). Local control is reached by clustering the experimental units in subsets of homogeneous units called blocks. It is assumed that experimental units in a block are similar enough to produce negligible differences in the results and that comparison of treatments in the same block is a fair comparison without any confounding effect. The randomized complete block design is probably the most frequently used experimental design. Some variants of completely randomized block design are useful in some specific situations like incomplete block design in variety

Abbreviations: AR1, autoregressive model of order one.

CIMMYT, Biometrics and Statistics Unit, Carretera Veracruz Km 45, El Batan, Texcoco, Edo. De Mexico, Mexico 56237. *Corresponding author (j.burgueno@cgiar.org).

doi:10.2134/appliedstatistics.2016.0011

Applied Statistics in Agricultural, Biological, and Environmental Sciences
Barry Glaz and Kathleen M. Yeater, editors

field trials when a large number of treatments must be compared or Latin-square design to control variation in two directions (see Chapter 3 by Casler, 2018). Wood and Stratton made one of the first allusions to spatial variability in 1910, 10 years before the pioneer works of Fisher.

However, block designs cannot control all the sources of variation present in an experimental field and must be recognized that experimental units in a block are not completely homogeneous; some intra-block variability exists. Gilmour et al. (1997) distinguished among global, natural, and extraneous variation. Global variability includes any major trends across the field ,while natural variation arises from uneven soil moisture or soil depth or other natural variations that occur in a minor scale. Extraneous variation includes effects introduced by experimental operations; these operations are usually aligned with rows or columns in the field. Natural and extraneous variation can be controlled by blocking; however,natural variation or small-scale variation are difficult to control except when very small blocks are used. Good experimental design can reduce the impact of some of these factors, but unless they are included appropriately in the statistical model, they will significantly reduce the precision of estimates of variety effects and variety contrasts. This is also true for a good statistical model, which must represent adequately the experimental design and other external factors that affect the results such as the small-scale spatial variability and some extraneous effects produced during the running of the experiment.

An appealing idea presented by Papadakis (1937) and developed by Wilkinson et al. (1983) is to adjust plot results for spatial variability using information from the neighbor plots. One useful measure for examining the heterogeneity patterns is the spatial autocorrelation of neighboring plots in the field in both, row and column directions, based on the correlation of residuals from pairs of plots at the same distance and this is made for several distances. If there is no spatial pattern, all the correlations will be low. If there is a pattern in the residuals, neighboring residuals will be more similar and so will have a higher correlation. Gleeson and Cullis (1987) proposed to sequentially fit a class of autoregressive-integrated-moving average models (ARIMA) to the plot errors in one direction (rows or columns); this was done in the context of randomized complete block experiments. They found that by differencing along the block and then fitting a moving average (MA) correlation structure to the residuals in that direction, a big gain in trial efficiency was obtained. Cullis and Gleeson (1991) extended the previous model to two directions (rows and columns) assuming that in the field rows and columns are regularly spaced. However, differencing in this two dimensional analysis was prone to discarding treatment information (as shown by Kempton et al., 1994). Grondona et al. (1996) analyzed 35 cereal yield trials using the two-dimensional spatial analysis proposed by Cullis and Gleeson (1991) and found that the autoregressive model in the direction of the rows and columns was most frequently the best model. An extended review of spatial models can be found in Martin (1996) and Hinkelmann and Kempthorne (2008).

Several others methods and models were proposed to model the spatial variability (Velazco et al., 2017; Hu and Spilke, 2009; Stringer et al., 2008; Kempton et al., 1994, Wilkinson et al.,1983), some of them are under development, such as splines models (Velazco et al., 2017); older models are not frequently used.

At least two different approaches are used to analyze experimental data. The first is based on the randomization process, which ensures independence of the residuals; then, a valid estimation of the experimental error variance and consequently valid statistical tests. Under this paradigm, a randomization-derived model is analyzed by analysis of variance. The analysis of variance has difficulties to analyze models with random effects and missing data. With the arrival of mixed models and maximum likelihood estimation, a new analytical paradigm based on considering that the observed data is a sample from a multivariate distribution, usually a normal distribution, was developed. For a detailed discussion see Hinkelmann and Kempthorne (2008, Chapter 2 and Section 6.3), and Cox (2009)

Spatial analysis has been closely related to variety evaluation trials since for modeling the spatial variability it is necessary a large number of rows and/or columns. It is usual in plant breeding to test hundreds or few thousands of genotypes. The main objective of variety trials is to obtain precise estimates of variety means and variety contrasts; using augmented designs along with spatial analysis allow the breeders to reach their objectives at a lower cost.

We use the term spatial analysis to refer to an analysis where we investigate the residual variance of the trial(s) and the use of an appropriate covariance structure to model it.

Practical Considerations

Experimental Design

The availability of computers and statistical programs has promoted extensively the use of statistical and mathematical methods. Nowadays it is easy to run several analyses on the same data set in a few minutes without considering how the data was collected and selecting the "best" ad hoc analysis. Spatial analysis is an *a posteriori* analytical method that is useful to improve the performance of a field experiment; however, the most important issues for the success of an experiment have to be decided *a priori,* before running the experiment. Only few works have been done in the direction to optimize the experimental design for the use of spatial analysis. The main problem is that we have to know the pattern and the magnitude of the spatial variability to optimize the experimental design (p-rep designs, Cullis et al., 2006), conditions that are hardly known. Unreplicated experiments, as the designs of Lin and Poushinsky (1983) and Rivera (1975) and the augmented designs of Federer (1956) were designed to model the spatial variability. In the last 10 years, some important references are Cullis et al. (2006) and Williams and Piepho (2013). Unfortunately there is not an experimental design whose use can be generalized. Residual error of any experimental design can be spatially modeled if the coordinates of the experimental units are known. This is easier for some experimental designs than for others because of the field layout. In experimental design, coordinates could be geographic coordinates, or in the case of a complete lattice structure, which is the most frequent in experimental designs, the number of rows and columns is enough. It is also possible to use the distances from an arbitrary selected zero-zero (row-column origin) point in the field to the center of each plot.

Spatial variability is the result of soil variability, distribution of treatments, interaction between treatments and plots, and the particular effect of the environment (agronomic management and climate) in which the experiment is conducted. No less importantly, spatial variability is trait-dependent. Some traits are known to have behavior that is predictable, like grain yield and plant height; however, for other traits like flowering date, some diseases, and incidence of plagues, it is very difficult to think that they will follow a spatial pattern. Spatial analysis is not adequate for all these situations. Especially for multi-trait experiments, it is difficult to assume that all of them will have the same spatial pattern of variation. The best we can do is to consider the most important trait, select the best design for it, taking into account that for other traits the experimental design may not be optimum. The knowledge of the experimental field and the conditions of the experiment as well as the kinds of treatments and traits of interest are very important at the time to decide the use of spatial analysis. Rather than obviate the need for a good experimental design, this approach increases it. Once a treatment effect is confounded with an environmental effect, the two cannot be disentangled. It is a good practice to collect the field position of the plots to have the possibility of modeling the spatial variability.

Due to the previous considerations, the use of a standard design (for instance complete block design, α lattice design or incomplete row-column design) could be a good option and for some traits, it might be possible to adjust a spatial model. For un-replicated experiments, the problem is different because spatial modeling is mandatory to adjust the observed value of the unreplicated treatments.

Spatial Model

In this chapter, we discuss the autoregressive models with separate components in row and column directions ($AR1_r \otimes AR1_c$; *r*, row direction; *c*, column direction) that have been mostly used in variety testing in CIMMYT, Australia, and the UK (Piepho et al., 2015). This model has shown that it performs well in most cases; it is the standard for variety trials in those places. One limitation of this model is that it is necessary to have a large number of plots in row and column directions to have a good estimation of the autocorrelation parameters. For time series, the recommendation is to use at least 50 points; however, in field experiments, it is common to estimate spatial autocorrelation with a smaller number of rows or columns.

The classical analysis of an experiment assumes that the experimental errors (ε) follow independent normal distributions with zero mean and common variance, $\varepsilon_i \sim NI(0,\sigma^2)$. In the case of an $AR1_r \otimes AR1_c$ model, the distributions continue being normal with zero mean, but now they are not independent. The matrix expression of the $AR1_r \otimes AR1_c$ model is

$$\begin{bmatrix} 1 & \rho_r & \rho_r^2 & \cdots & \rho_r^{R-1} \\ \rho_r & 1 & \rho_r & \cdots & \rho_r^{R-2} \\ \rho_r^2 & \rho_r & 1 & \cdots & \rho_r^{R-3} \\ \vdots & \vdots & \vdots & \ddots & \vdots \\ \rho_r^{R-1} & \rho_r^{R-2} & \rho_r^{R-3} & \cdots & 1 \end{bmatrix} \otimes \begin{bmatrix} 1 & \rho_c & \rho_c^2 & \cdots & \rho_c^{C-1} \\ \rho_c & 1 & \rho_c & \cdots & \rho_c^{C-2} \\ \rho_c^2 & \rho_c & 1 & \cdots & \rho_c^{C-3} \\ \vdots & \vdots & \vdots & \ddots & \vdots \\ \rho_c^{C-1} & \rho_c^{C-2} & \rho_c^{C-3} & \cdots & 1 \end{bmatrix}$$

where ⊗ is the Kronecker product, ρ_r and ρ_c are the autoregressive coefficients for row and column respectively, and R and C are the number of rows and columns in the field layout, respectively.

This correlation structure (covariance) is parsimonious, since there are only two parameters to estimate ρ_r and ρ_c. A scale parameter is required to convert the correlation matrix into a covariance matrix; three parameters have to be estimated to model the spatial variability. The structure is anisotropic; correlation in both directions can be different, and further correlation in a different angle than 0° (row) or 90° (column) can also be different. For instance, the correlation between two plots that are two rows apart and four columns away is $\rho_r^2\rho_c^4$, the exponents of the autocorrelation coefficients are the distances between the plots in each direction.

This structure supposes a smooth relationship between plots where adjacent plots are more correlated than plots far away and the correlations is a function of the distance and the angle only. Any pair of plots, with the same relative position to one another (same distance and same angle) will have the same correlation regardless of their position in the field. Some characteristics are advantages of the model and others are disadvantages but neither of them is absolute, at the end the adequacy of the model depends on the actual spatial relationships between plots. When the variability in the soil shows patches this smooth model is not good. As it was mentioned before, the goodness of the model will depend on the ability of the experimental design to discover the spatial structure, the trait, and other sources of variation.

Model Evaluation

Besides standard statistics and measurements to evaluate a model, in the case of the spatial model it is usual to evaluate it by the variogram (Fig. 1 to 4). The variogram is a figure of the semivariance at different distances (lags). In a time series or with only one spatial direction the variogram is simple but for an anisotropic model like $AR1_r \otimes AR1_c$ the variogram is multidirectional and different representations are possible.

Considering one direction the semivariance is the variance of the difference of the residuals at distance *h*, the calculus is done at different distances and the variogram can be constructed. If the model fits well, the variogram should show a flat pattern after the first distances. This is indicative of a stationary process, stationarity is a necessary condition to estimate the real semivariances (real variogram) and to analyze the spatial structure. When the stationary condition is not fulfilled, the semi variogram is considered a pseudovariogram; it is not an estimation of the real variogram. For a two-dimensional process, the semivariance can be calculated at several directions for an anisotropic model. In an isotropic model, the semivariance is the same in any direction. The number of directions that can be analyzed depends on the distribution and number of points in the space, the larger the number of points the more directions. A nonflat variogram is indicative of a nonstationary process. Some extra effects should be added to the model to have a stationary process. In a stationary process, the expected value of the variable is the same, unrespecting the specific position in the field.

For a set of data values $Z(s_i)$, in which s_i denotes the position of the point i, $i = 1,\ldots,n$ the semivariance at distance h is defined as

$$\gamma_Z(h) = \frac{1}{2}Var\left[Z(s_i + h) - Z(s_i)\right]$$

and it can be estimated by

$$\hat{\gamma}_Z(h) = \frac{1}{2|N_{(h)}|}\sum_{N_{(h)}}\left[Z(s_i) - Z(s_j)\right]^2$$

where the sets $N_{(h)}$ contain all the neighboring pairs at distance h.

$$N_{(h)} = \{i, j : s_i - s_j = h\},$$

is the number of such pairs (i, j) and s_i is the position of the data point.

For a multidirectional problem, the semivariance is estimated at several directions. One variogram can be constructed for each one, but different representations can be produced, including some in which the semivariance can be drawn in a three dimensional picture. Along with the examples, we will present two different approaches to analyze the spatial variability through the variogram. In the first one, a directional variogram is produced at different angles. The angles were selected to have a good number of observations in each one and at different distances. Because there are so many angles in a grid of points, and most of them with a lower number of distances at which the semi-variance can be calculated efficiently, it is usual to consider a tolerance parameter and aggregate the data at a small range of angles together. The second figure is produced by calculating the semivariogram for every row and every column; the semi-variance is plotted like a map with the row (column) on the x axis and the distance (lag) on the y axis. We found those representations to be useful for the purpose but of course other kinds of representations can be used.

Process of Analysis

The analysis of an experiment starts with the analysis of the data using the model that reflects the classification factors and their randomization. Under this analysis, residuals can be checked and outliers detected. Some other extraneous pattern in the residuals also can be detected like non-homoscedasticity and relationships between the mean and the variance.

The second step is to include in the model the spatial variability, possibly after deleting some outliers and making a transformation of the data. The third step is the fitting of a two-dimensional, separable, autoregressive spatial model of first order ($AR1_r \otimes AR1_c$) as the basic spatial model following the proposal of Gilmour et al. (1997). For this model, some kinds of variograms can be produced and analyzed. If the variograms show a good behavior, the model can be useful. Outliers should be checked again under this model. Otherwise, extra factors can be added to the model as suggested by the variograms until a reasonable result is obtained. This is an iterative process and it is necessary to discuss the models with the person who knows the details of trial management and physical site characteristics. The knowledge of these characteristics is useful to understand the pattern of variogram. Factors like agronomical and experimental practices and procedures such as irrigation flow, sowing and harvesting methods and direction, slope, proximity to roads, trees or rivers, machinery characteristics, and site history can affect the results of the experiments and the pattern of the residuals and the variograms. These factors can be modeled as spatial covariances, covariates, and functions of the spatial coordinates using polynomial functions or cubic splines. As mentioned before, some of those factors can be considered and should be considered before selecting the experimental site and cautions should be taken when the trials are running. Statistical models and analytical solutions are good; however, good data is the starting point for successful results.

The inclusion of extra factors should be considered based on statistical tests of their effects and not only in the visual inspection of the variogram; the most important considerations may come from the experimental field. Because those extra effects are not expected, they can be confounded with some experimental effects like replicate, block, and even treatment effects; therefore, it is preferable that those extra factors are included in the model as random effects. Otherwise, some effects of interest may become un-estimable. There is no best spatial model, but rather several reasonable spatial models. In our experience, the inclusion of extra factors in the model produces, in most cases, just a marginal gain.

This approach seems good, because it enhances experimental precision and researchers understand the problems related to the way data were generated. Nevertheless, as pointed out by Brownie et al. (1993), the presence of systematic variation within complete or incomplete blocks does not invalidate the use and the analysis of the classical complete block or incomplete block designs (such as lattices), but rather strengthens the need for the random allocation of varieties to plots within complete or incomplete blocks. Moreover, improved statistical tools never compensate for the precision loss due to poor design or inappropriate agronomic or experimental practices.

Example 1

This example is a variety trial established in a row–column design (Nguyen and Williams, 1993) with 64 varieties planted in two contiguous replicates laid out in eight rows and 16 columns with two replicates for each variety. This data set was also analyzed by Burgueño et al., 2000.

The appropriate model in a randomization-based approach should include blocking factors (row and column) and entry in the model (Hinkelmann and Kempthorne, 2008, p. 213–216).

$$y_{ijk} = \mu + \theta_i + \gamma_j + \alpha_k + \varepsilon_{ijk}$$

where

y_{ijk}, response variable for the jth variety in *i*th row and the *j*th column

μ, overall mean

θ_i, effect of the *i*th row

γ_j, effect of the *j*th column

α_k, effect of the *k*th variety

ε_{ijk}, experimental error for the jth variety in the *i*th row and the *j*th column, which received the variety *k*

Using the following code of SAS software V9. 4 (SAS Institute Inc., 2015) we obtain the analysis of variance for this data, which is presented in Table 1.

TABLE 1. Analysis of variance of Example 1, row-column design.

Source	DF	Sum of squares	Mean square	*F* value	*P* > *F*
Model	85	74.54	0.877	6.16	< 0.0001
row	7	24.93	3.562	25.01	< 0.0001
col	15	17.51	1.167	8.20	< 0.0001
entry	63	8.88	0.141	0.99	0.5215
Error	42	5.98	0.142		
Corrected Total	127	80.52			

```
PROC GLM data = tmp;
class row col entry;
model yld = row col entry/SS3;
lsmeans entry;
run;
```

Both row and column effects are significant. The estimation of the residual error variance is 0.142, with a coefficient of variation of 14.9%. Although the design controlled the field variability, no significant effect of the variety was found significant. With no significant effect of the variety and with a slightly high coefficient of variation, it is desirable to look for an analysis which improves the interpretation of the data. The first step is to analyze the residuals and the raw data. The residual analysis includes the testing of the assumptions of the model and the discovery of outliers. Second, we can inspect the use of covariates like number of stand plants or the number of harvested ears. The selection of the covariate depends on the trait we are analyzing and the effect that it may have on the trait. Another approach is the spatial analysis. We model the experimental error to deal with effects that were not controlled by the design. All extra analysis should be performed in light of observations in the field. Following that, we will include a spatial model for the residuals.

In this case, we are going to use the autoregressive model of order 1 (*AR1*) in row and column direction because it is a simple model and has been shown to be an adequate model in many situations. The expression of the statistical model is the same, but now we assume that the experimental error follows a normal distribution with a matrix of variance–covariance equals

$$\sum_{\varepsilon} = \sigma^2 \sum_{r} \otimes \sum_{c}$$

Where σ^2 is a scale parameter and $\sum_r \otimes \sum_c$ corresponds to the form shown above for the $AR1_r \otimes AR1_c$ model.

Variety effect can be considered fixed or random depending on the nature of the set of tested varieties; for this example, we will continue considering variety as a fixed effect. A factor with random effect is a factor whose levels in the experiment are a random sample of a population of levels (effects); these effects are considered as random variables, independent of the experimental error, all of them with zero mean and some matrix of variance–covariance. In this way, we need to estimate the matrix of variance–covariance of the effects and instead of estimating the effect of each level of the factor; we predict the random variable at each level of the sampled levels in the experiment (Robinson, 1991). This prediction is known as the best linear unbiased predictor (BLUP) under the frame of maximum likelihood method.

To include the separable spatial model ($AR1_r \otimes AR1_c$) in row and column directions for the residuals, we need to add a new line to the last code with the instruction repeated as shown in next SAS codes. First, we present the code using MIXED procedure and then using GLIMMIX procedure. GLIMMIX is an advanced procedure that allows the analysis of generalized linear mixed models. This procedure is covered in more detail within Chapter 16 by Stroup (2018). For the purposes of the current chapter, it allows us to analyze data that is not normally distributed. Count data, which follow a Poisson or binomial distribution, and also some nonsymmetrical distributions such as Log-Normal or Gamma can be analyzed with this procedure. Both MIXED and GLIMMIX procedures are similar in syntax with slight (but important) differences. For instance, GLIMMIX does not have a *repeated* statement, therefore the spatial part of the model must be coded with the *random* statement. There are also some differences in the produced results and in the way the output is presented.

```
PROC MIXED data = TMP covtest;
class row col entry;
model yld = row col entry/DDFM = SAT outp = TMP2;
lsmeans entry/pdiff;
repeated/subject = intercept type = SP(POWA)(row col);
run;

PROC GLIMMIX data = TMP;
class row col entry;
model yld = row col entry/DDFM = SAT;
lsmeans entry/pdiff;
random residual/type = SP(POWA)(row col);
output out = residual resid = resid;
run;
```

The option "outp = TMP2" must be added in the model line to obtain the residuals in a SAS format data file that will be used to construct the variogram, which is useful to evaluate the spatial model. In GLIMMIX, a complete output sentence must be included, *"output out = residual resid = resid;"*.

Table 2 shows the main results obtained with the code using the MIXED procedure provided. Results of the GLIMMIX procedure are identical, but with a slightly different presentation (not shown). In Table 2, covariance parameters of the spatial model are shown. Residual error is larger in the spatial model than in the nonspatial model (Table 1) but this is not really a good indicator to compare the models because the residual error of the spatial model is only a scale parameter. The spatial model includes two more parameters to model the experimental variability, that is, the autoregressive coefficients for row and column. Because both coefficients have large *p*-values, they are not significant.

Least square method seeks for a minimum in the residual sum of squares. When you have a defined model, it is the indicator of the adjustment of the model;

TABLE 2. Variance components for the spatial model in example 1. Row, column and entry are fixed effects.

Covariance parameters	Spatial model			
	Estimate	Standard error	*Z* value	*P* > *Z*
SP(POWA) row	-0.0546	0.2483	-0.22	0.8260
SP(POWA) col	0.1623	0.1973	0.82	0.4107
Residual	0.1436	0.0329	4.37	< 0.0001

however, when the model is not known and comparisons among different models have to be done, a smaller residual variance is not necessarily an indicative of a good model. For mixed models or models with an experimental error with residual structure different from that standard assumptions of normal distribution, independence, and homoscedasticity, maximum likelihood or residual maximum likelihood method must be used and other measurements of fitting are available. Table 3 shows those statistics for both models, nonspatial and spatial. To obtain those statistics in the nonspatial model, one must use the corresponding code of Proc Mixed rather than Proc GLM. All statistics are in favor of the no spatial model and in this case they made easy to select the "right" model, besides that spatial correlation coefficients were nonsignificant; however, this does not usually happen then the selection of the "right" model becomes an issue. The fitting statistics are all based in penalizing the -2 x Residual Log Likelihood regarding the number of parameters and number of observations. The relative weight and used scale of those parameters differentiate the fitting statistics as well as the properties to be considered when one of them will be used. However sometimes these statistics are not enough to evaluate a model, exploring the residuals and analyzing the spatial structure are good things to do.

Neither, row or column autoregressive correlations are significant (Table 2); consequently, adjusted least square means are very similar, with a correlation of 0.993 and the significance of entry effect is high (p-value = 0.5075, Table 4). Average standard error of an adjusted mean was 0.317 and 0.315 for the nonspatial and spatial model, respectively.

Exact distribution of the F or t tests is not known, due to the use of a residual spatial structure of covariance and for using its estimation. Classical test of hypothesis should be slightly modified to obtain accurate significance levels. In this example we used the option DDFM = SAT for SATTERTHWAITE method included in SAS and/or MIXED procedure which takes into account all the random effects that include the tested effect to estimate the degrees of freedom. The denominator degree of freedom available in SAS are BETWITHIN, CONTAIN, KENWARDROGERS, RESIDUAL and SATTERTHWAITE. Some methods perform well under certain conditions and others under other conditions. A good discussion of the topic can be found in Richter et al. (2015).

TABLE 3. Fitting statistics for non-spatial and spatial model of example 1.

Fit statistics	Non spatial model	Spatial model
-2 Res Log Likelihood	111.4	110.7
AIC (Smaller is Better)	113.4	116.7
AICC (Smaller is Better)	113.5	117.3
BIC (Smaller is Better)	115.1	121.9

TABLE 4. Type III tests of fixed effects for spatial model in example 1.

Effect	DF		Spatial analysis	
	Numerator	Denominator	F Value	$P > F$
Row	7	11.7	20.65	< 0.0001
Column	15	19.4	7.94	< 0.0001
Entry	63	31.4	1.58	0.5075

There is no unique measurement or statistic to evaluate or compare different models, and it is usually recommended to use more than one. It must be taken into account that the selected measurements must be in accordance with the objective of the model. Different models can be used for different purposes such as prediction, estimation of parameters or testing hypothesis.

Following the approach of Gilmour et al. (1997), we will examine the trends of the residuals as another way to evaluate the models; however, because we are using SAS instead ASReml, the kinds of figures will be different. The Gilmour approach can be reviewed also in Burgueño et al. (2000).

In addition to the different kinds of residuals that can be computed (residuals, Pearson residuals, standardized residuals) and figures produced with them in the case of spatial analysis, the most relevant figure is the semivariogram in which the semivariance is plotted against the lag distance. For an anisotropic two-dimensional model, the semivariance must be analyzed in several directions. However, just for a small number of directions, the semivariogram can be estimated accurately. Actually, in an ($AR1_r \otimes AR1_c$) model, the correlation in a direction different from zero or 90 degrees is computed by multiplying the row correlation by the column correlation power to the row and column distance values, respectively.

The procedure VARIOGRAM of SAS calculates the semivariogram at different lags and angles in the field layout. SAS automatically produces the variogram for each direction in which the semivariance of residuals is plotted. Stefanova et al. (2014) reported different ways to diagnose spatial models and discussed in detail the use of the semivariogram. A useful different approach is to calculate the variogram for each row in column direction, and vice versa, for each column in row direction. Figure 1 shows the variogram as produced by SAS for eight different directions with the following code.

```
PROC VARIOGRAM data = TMP2 plots = ALL;
 compute lagdist = 1 maxlag = 30 ndirections = 8;
 coord xc = col yc = row;
 var resid;
 ods output SemivariogramTable = TMP3;
run;
```

If the spatial model fits well, then the variogram must be flat after the first lags. In this case, the variogram looks flat at 0° (row direction) and 90° (column direction) with a slight linear trend for angles of 45°, 67.5° and 157.5°. These results indicate a lack of stationarity.

The following code produces the variogram for one direction, column or row, at several levels of the other direction row or column, respectively.

```
PROC SORT data = tmp2; by row;
PROC VARIOGRAM data = TMP2;
by row;
 compute lagdist = 1 maxlag = 15 ndirections = 1;
 coord xc = col yc = row;
 var resid;
 ods output SemivariogramTable = tmp4;
run;

PROC SORT data = tmp2; by col;
PROC VARIOGRAM data = TMP2;
by col;
```

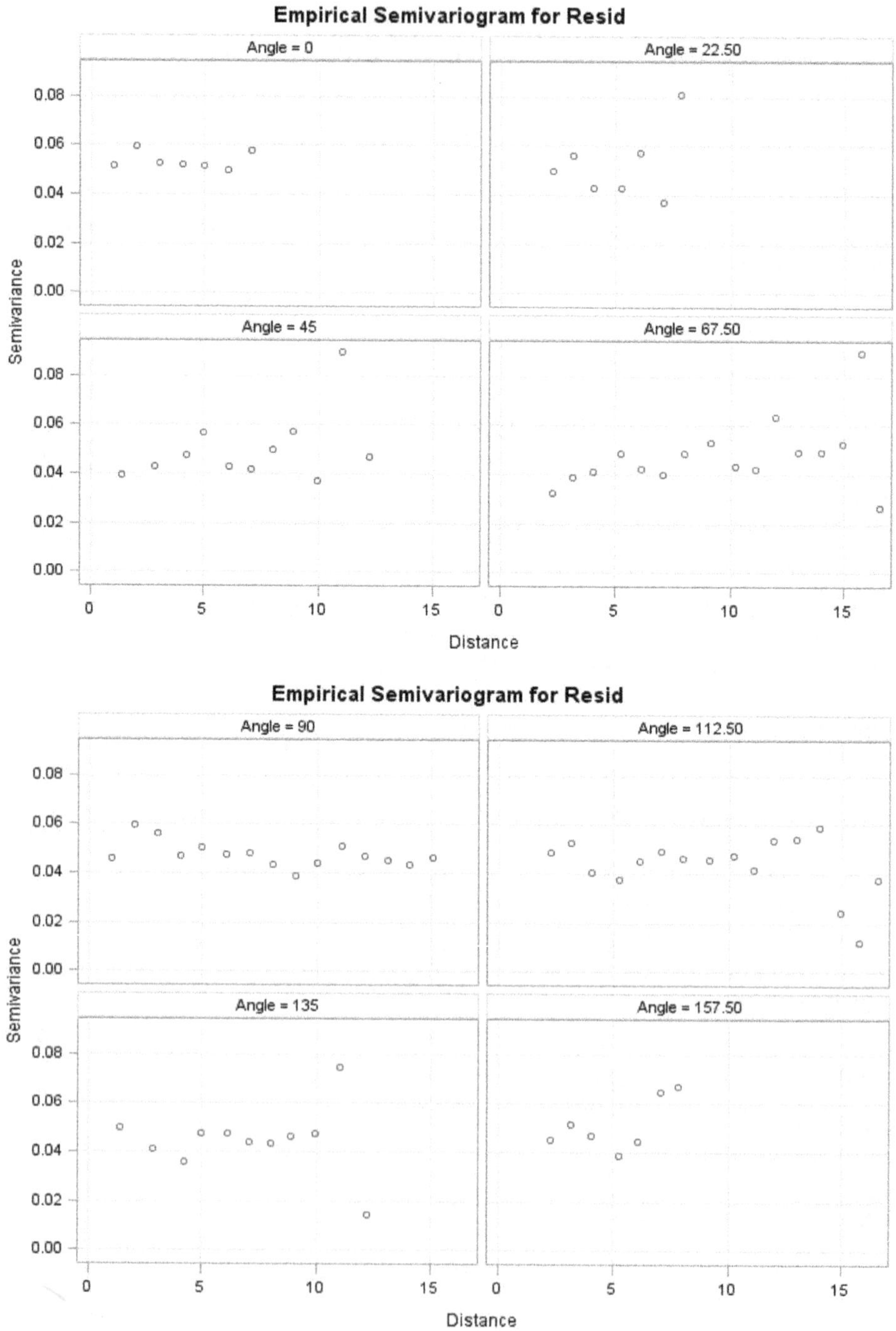

FIG. 1. Variogram of the spatial model at several directions, example 1. A 0° angel corresponds to row direction while a 90° angel corresponds to column direction.

```
 compute lagdist = 1 maxlag = 7 ndirections = 1;
 coord xc = col yc = row;
 var resid;
 ods output SemivariogramTable = TMP5;
run;

axis2 order = (1 to 8 by 1);
axis3 order = (1 to 16 by 1);
axis5 order = (1 to 7 by 1);
axis6 order = (1 to 15 by 1);
PROC GCONTOUR data = tmp4;
plot lag*row = semivariance/haxis = AXIS2 vaxis = AXIS6 pattern lev-
els = (0 to 0.4 by 0.05);
run;

PROC GCONTOUR data = tmp5;
plot lag*col = semivariance/haxis = AXIS3 vaxis = AXIS5 pattern lev-
els = (0 to 0.4 by 0.05);
run;
```

The VARIOGRAM procedure produces the variogram for each row or column by using the instruction *by row* or *by col*. *Axis* instructions are used to make up the figures created by the GCONTOUR procedure; with lag in one direction as vertical axis (column and row lag) and the levels of the other direction as the horizontal axis (row or column).

Figure 2 shows the graphs obtained with the last code. In column direction there is a trend to increase the semivariance at middle rows with a peak in the first row at large lags. For row direction semivariance decreases from the first column to the last. These results are related to what was previously seen in the multidirectional variogram in Fig. 1.

Because results using the spatial model were not improved, and considering the slight trends found in the variograms, other models can be explored. It is possible that the spatial model cannot capture any extra variability since the row and column design was efficient to do this. However we are probably not capturing a small-scale pattern of variability. To search for a better model, polynomial regression in row and/or column direction, as proposed by Federer et al. (1997), plus spatial model of the residuals could be a good option. After testing, both linear and quadratic regression coefficients of row and column were significant, therefore they were included in the final model that was analyzed with the following SAS code:

```
PROC GLIMMIX data = TMP;
class entry;
model yld = entry row row*row col col*col/DDFM = SAT;
lsmeans entry/pdiff;
random residual/subject = intercept type = SP(POWA)(row col);
output out = TMP20 resid = resid;
covtest/wald;
run;
```

In Table 5, main results for this analysis are shown. All the effects are almost significant, except entry effect and the correlation autoregressive coefficient in row direction. The average standard error of an adjusted mean (0.3179) is smaller than in the first two models. Due to the change in the fix part of the model, the fit statistics are not useful to compare these models because REML decomposes the

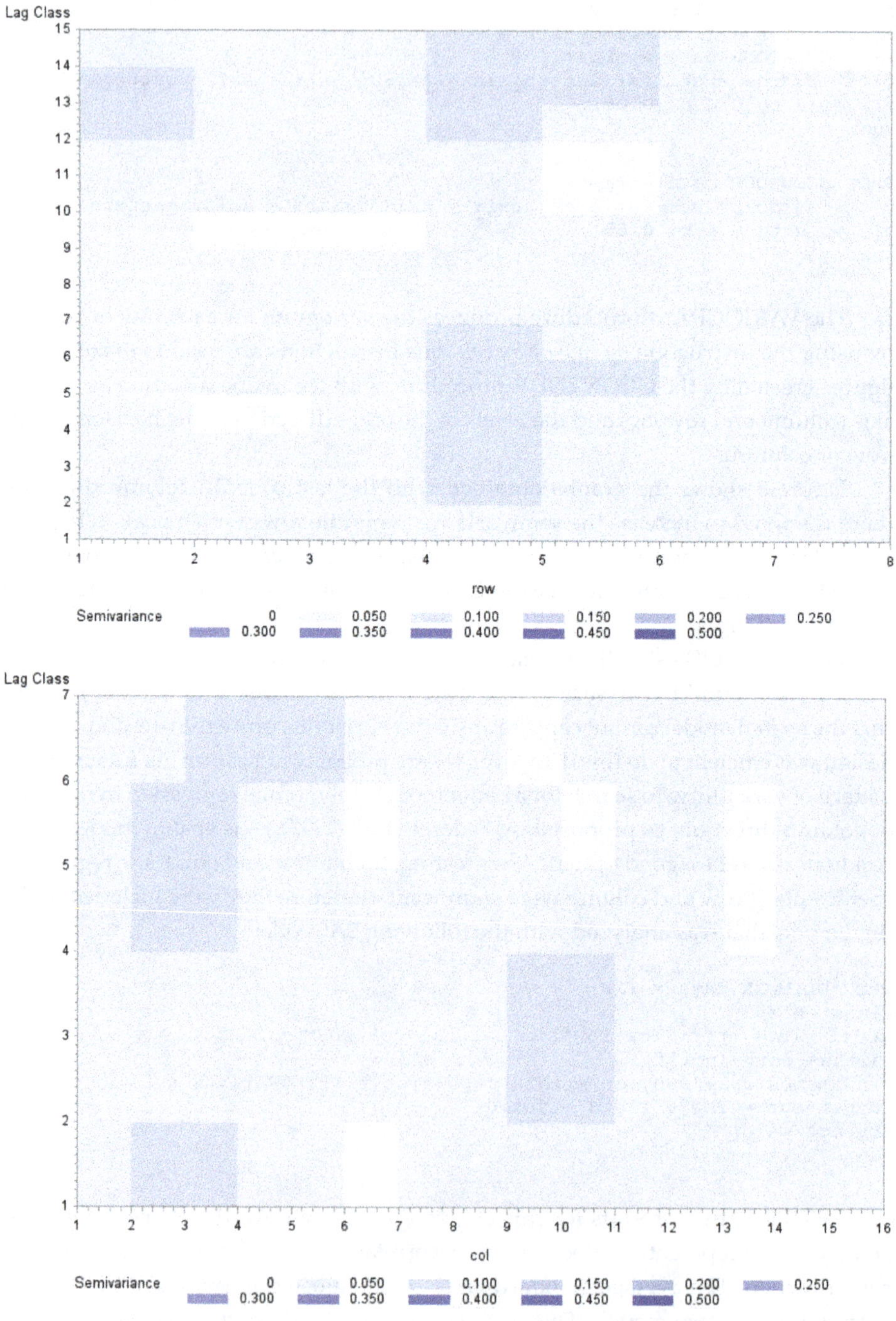

FIG. 2. One directional variogram in column direction for each row (above) and in row direction for each column (below) for spatial model in example 1

log-likelihood in two components, one related with zero expectation contrasts to estimate variance parameters and the other for the fixed effects. Actually, REML method is a two-stage method. Fitting statistics can be used in the case of using ML since this method does not split the log-likelihood. Model comparison is a complex topic out of the scope of this chapter and many criteria may be used. Criteria have to be chosen based on the objective of the analysis; for instance, the objective could be to select the best genotypes or to get BLUP to perform genome wide association analysis or to do prediction or to compare treatments against a control. Some references the reader can review are: Al-Marshadi, 2014; Thöle et al., 2013; Lin et al., 2013; Kehel et *al.*, 2010; Wang et al., 2010; Wang and Schaalje, 2009; Gurka, 2006.

With a similar code used to produce the last variograms we produce Fig. 3 and 4 by using the corresponding output data file (tmp2) and changing the levels in the plot instruction. No improvement can be seen, actually they look worse (see the scale).

Different models should be evaluated to select the most appropriate model to the objectives of the work. However, as Burgueño et al. (2000) noticed, for selection purposes the main change happens when the spatial structure is included in the model. Inclusion of extra effects helps to have a better estimation of the spatial structure and to model some extraneous effects that happened in the field but that are not related with the experimental design; extra effects have very small effect in the rank of genotypes. As Gilmour et al. (1997) asseverated, any effects that are included in the model must be justified on what happened in the field; however, it is not easy to find justification for most of the effects.

In this example, modeling spatial variability with an autoregressive model did not improve the results of the analysis, the row–column design was enough to capture spatial variability.

TABLE 5. Main results for the spatial model plus polynomial fixed regression in row and column in example 1.

Fit statistics				
-2 Res Log Likelihood			149.92	
AIC (smaller is better)			155.92	
AICC (smaller is better)			156.35	
BIC (smaller is better)			162.20	
Covariance Parameters				
Cov parm	Estimate	Standard error	*Z* value	*P* > Z
SP(POWA) row	0.1088	0.1897	0.57	0.5663
SP(POWA) col	0.4092	0.1817	2.25	0.0243
Residual	0.2341	0.05286	4.43	< 0.0001

Test Type 3 of fixed effects				
Effect	Num DF	Den DF	*F* value	*P* > F
entry	63	24.6	0.94	0.5866
row	1	10.15	19.83	0.0012
row*row	1	10.49	12.40	0.0051
col	1	10.13	11.23	0.0072
col*col	1	9.946	23.25	0.0007

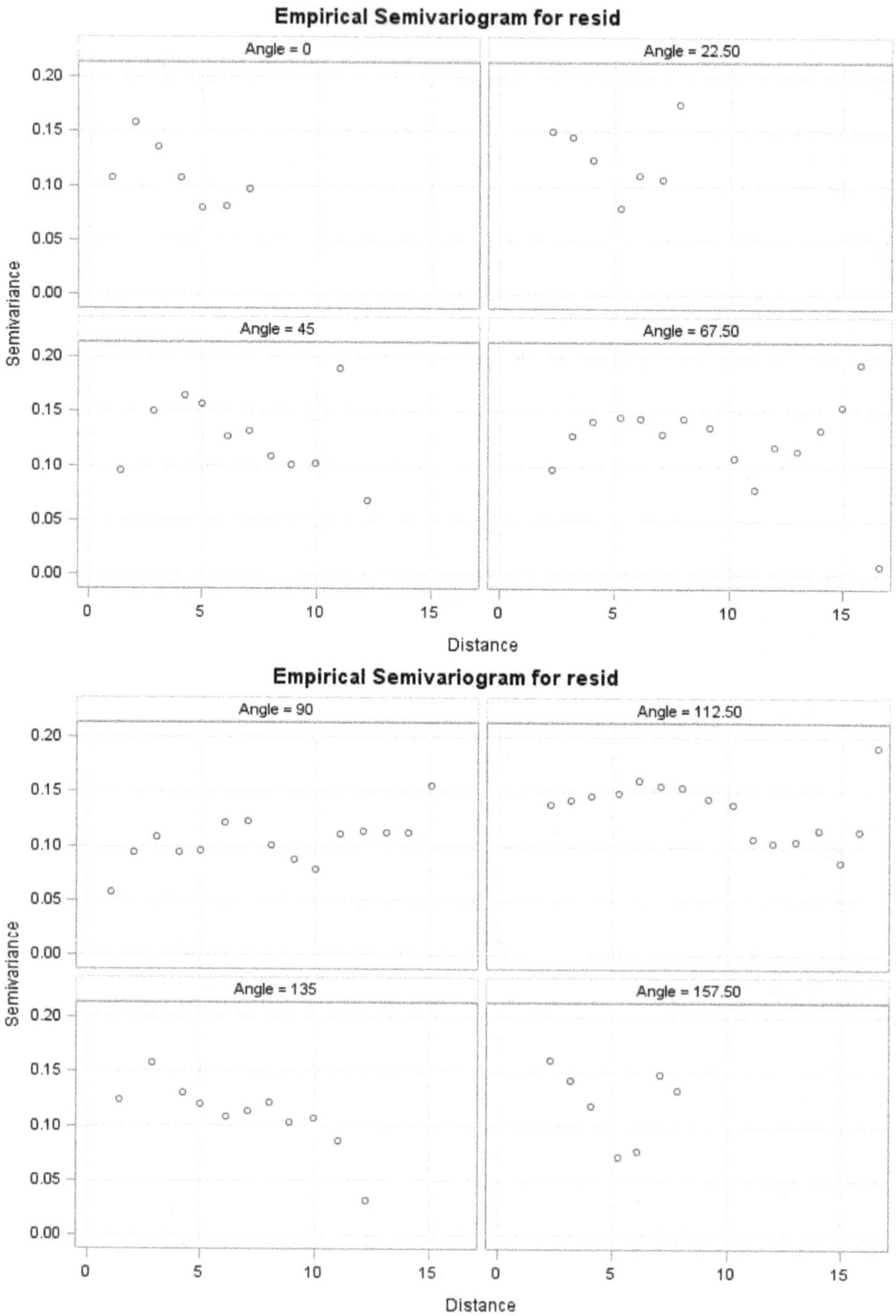

FIG. 3. Variogram of the polynomial row column effects plus spatial model at several directions, model 3, example 1. A 0° angle corresponds to row direction while a 90° angle corresponds to column direction.

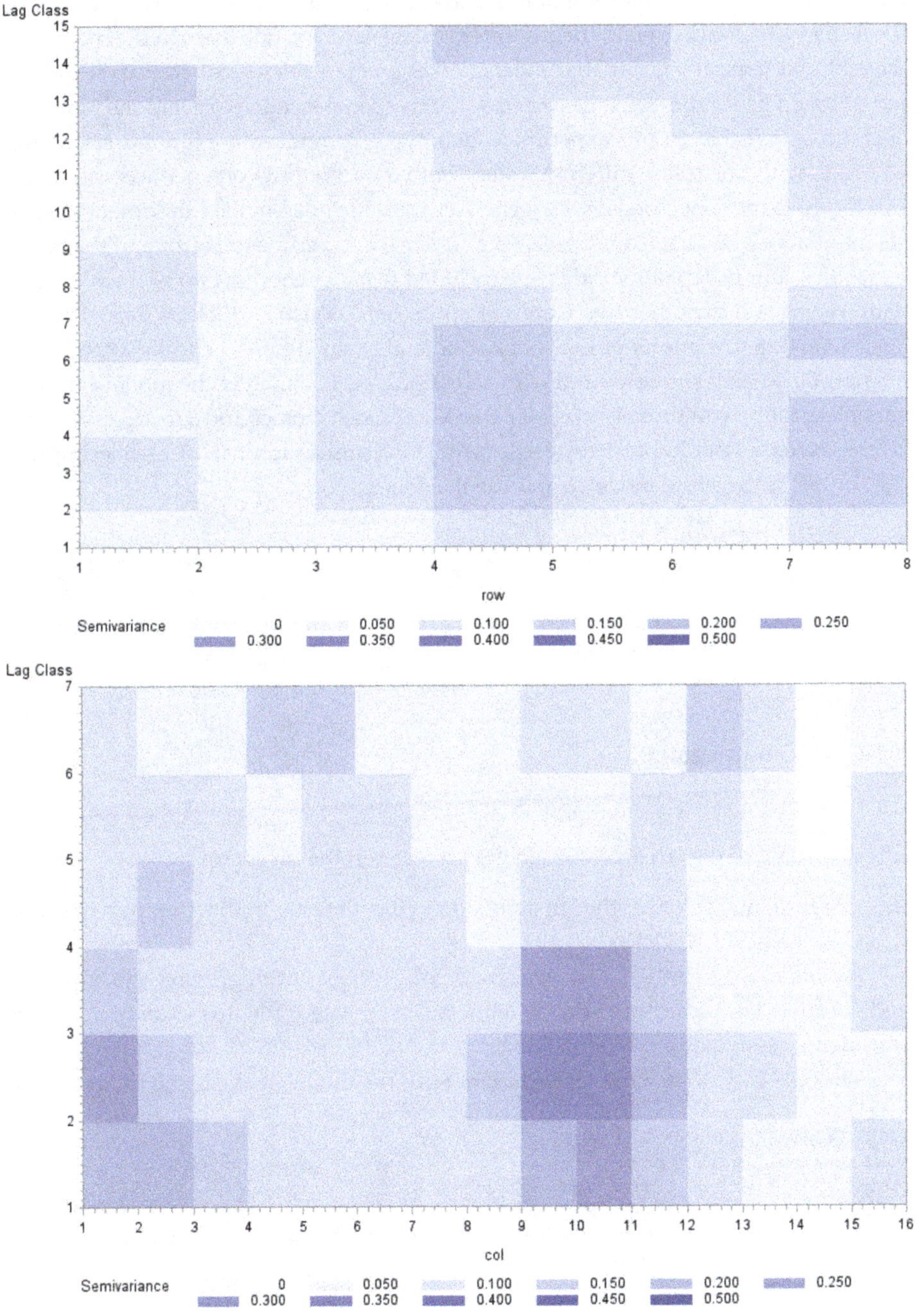

FIG. 4. One directional variogram in column direction for each row (above) and in row direction for each column (below) for polynomial in row and column direction effects plus spatial model, example 1.

Example 2

In this case, we will analyze the unreplicated design described in the previous chapter Augmented designs. In this experiment check plots were allocated systematically in the field. Four checks were allocated resembling a four by four Latin-square design, only to balance the checks in the rows and columns. There is not a randomization based-model with which to start the analysis; to build the statistical model, we just take in account the factors which can explain the response variable and include them in the model. For these kind of experiments, spatial analysis takes on more relevance because the data for unreplicated genotypes must be corrected for its spatial location to avoid the plot effect as much as possible. In this experiment, genotypes comprised the test crosses of each accession with one tester, different testers were used but only one per accession. Four control accessions not crossed with tester were also included in the experiment to evaluate the effect of the accessions per se. To estimate the experimental error and the spatial variability, four different hybrids were included to act as checks. Two of them are standard checks and the other two were susceptible and nonsusceptible to drought stress. To differentiate the effects of the checks (there are four different checks) with the differences between test-crosses and with accessions, we included in the model a dummy variable (group) with three levels (1 for checks, 2 for test crosses and 3 for accessions per se). See exercise 1 of Chapter 3 (by Casler, 2018) for a similar approach. Considering all of the above, the statistical model to analyze the data is:

$$y_{kjil} = \mu + \beta_i + \alpha(\beta)_{ji} + \delta(\beta\alpha)_{kji} + \varepsilon_{kjil}$$

Where:

Y_{kjil}, the response variable for the *k*th accession in the *j*th tester in the *i*th group in the *l*th replicate of the check;

μ, overall mean;

β_i, effect of the *i*th group;

$a(\beta)_{ji}$, effect of the *j*th tester in the *i*th group

$d(\beta\alpha)_{kji}$, effect of the *k*th accession in the *j*th tester in the *i*th group

ε_{kjil}, experimental error for the *k*th accession in the *j*th tester in the *i*th group in the *l*th replicate of the check.

As usual, $\varepsilon_{kjil} \sim N(0,\Sigma)$. Σ was modeled with an autoregressive model in row and column. Perhaps other spatial models can be used; for this example, we will consider accession as a random effect.

The SAS code to analyze plant height with the model above is the following:

```
PROC GLIMMIX data = Maize PLOTS = ALL;
class group tester acc;
model plh = group tester(group);
random acc(group*tester);
random residual/subject = intercept type = SP(POWA)(row col);
output out = TMP1 resid = resid;
covtest/wald;
run;
```

Due to the presence of missing data, SAS requires the inclusion of the column and row variables in the instruction CLASS, defining them as classification factors. This allows SAS correctly order the rows and columns. Results presented in Table 6 show significance levels of all the effects considered in the model.

With a code similar to the used in Example 1, we produce the different variograms whose results are presented in Fig. 5 and 6. The only necessary changes are to use the corresponding output file with residuals and to adjust the scale of the axis and the level instruction. The multidirectional variogram in Fig. 5 shows more or less flat pattern except for some angles at larger lags. It must be remembered that for larger lags, the number of observations is limited to estimate an autocorrelation parameter then it is recommended not to take in account lags with few observations. The one-directional variograms show a similar pattern than the directional variogram. Sometimes it is not easy to see whether a variogram is flat or not, then it would be recommendable to run several models and then compare the different variograms, even when the comparisons of variograms is not an easy task. For more details about the use and interpretation of variograms, Stefanova et al., (2014) is a good reference.

The analysis of grain yield per hectare did not fit for spatial model as shown in Table 7. The autoregressive term for column is not significant and for row, it is zero. We could not find an explanation for this after checking the original data, residuals and analyzed different models. These models included the adjusted of the spatial model just in one direction, row or column. In this case, it is recommended to use another model to analyze the data. Polynomial regression in the row and column directions (including interaction) (Federer, 1956; Federer et al., 1997) is another option. Splines or moving averages can also be used (Velazco et al., 2017). The separable autoregressive model has shown to be a parsimonious model that produces good results in the majority of cases (Grondona et al., 1996), and it is the standard spatial model used in Australia for wheat and in CIMMYT for wheat and maize trials.

Following, results of two additional models are presented in which the row and column effects are estimated. First, random linear regression effects (following Federer et al., 1997) is presented, and then random classification effects. In both cases, row and column effects were considered as random, as another way to model these effects compared with Example 1 in which row and column effects were considered as fixed effects. Usually it is recommended to use random effect for row and column as well as for block in an incomplete block design because inter block-information can be recovered. Recovery of inter-block information is more adequate when the block variance is small, compared to

TABLE 6. Main results for the analysis of plant height, example 2.

Covariance parameters				
Cov parm	Estimate	Standard error	*Z* value	*P* > *Z*
Acc(Group*Tester)	248.93	29.0353	8.57	< 0.0001
SP(POWA) Row	0.2039	0.08762	2.33	0.0200
SP(POWA) Col	0.3999	0.07503	5.33	< 0.0001
Residual	146.30	21.2196	6.89	< 0.0001

Test Type 3 of fixed effects				
Effect	Num DF	Den DF	*F* value	*P* > *F*
Group	2	531	5.69	0.0036
Tester(Group)	9	531	2.17	0.0228

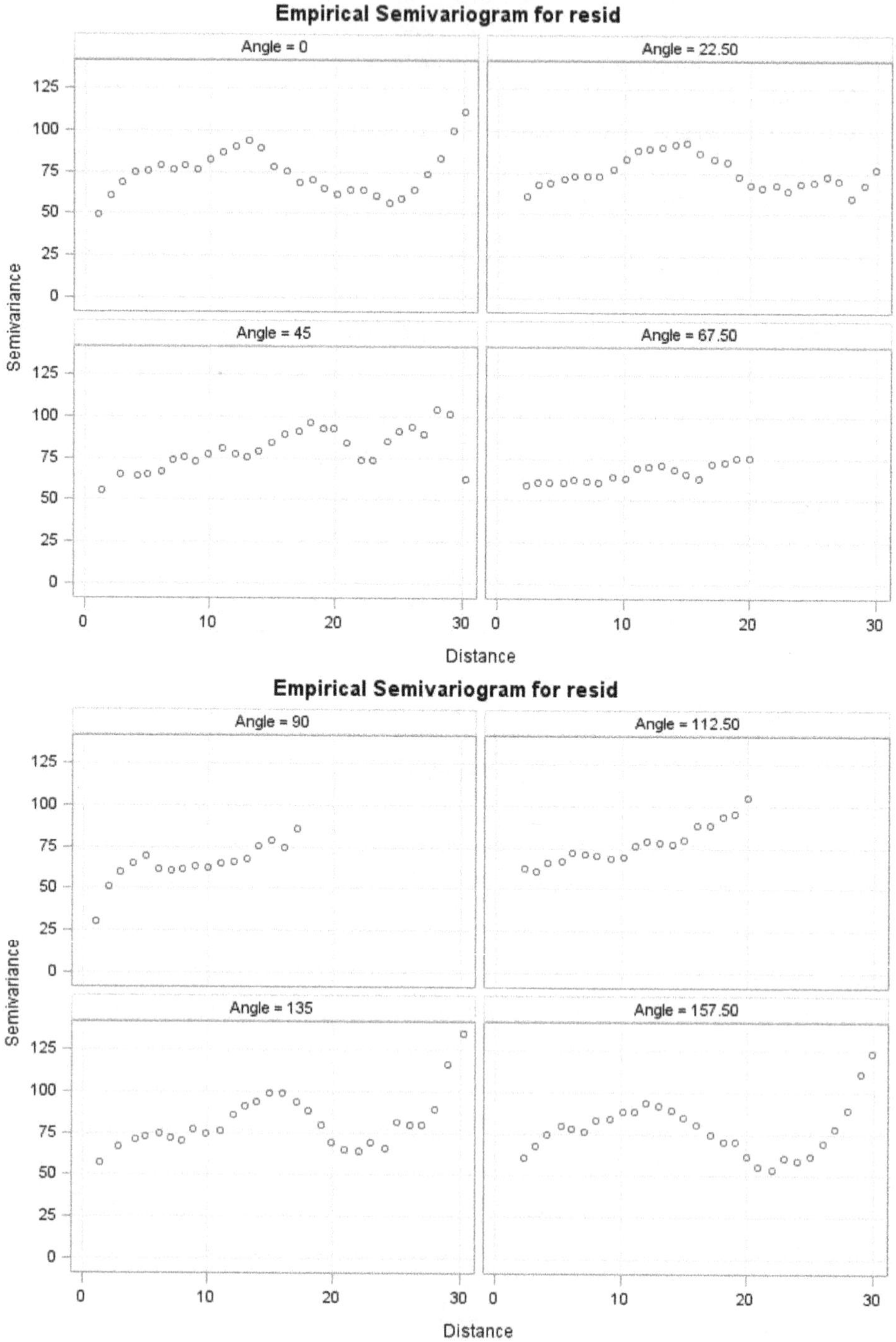

FIG. 5. Variogram of the residuals after fitting the spatial model at several directions, example 2. A 0° angle corresponds to row direction while a 90° angle corresponds to column direction.

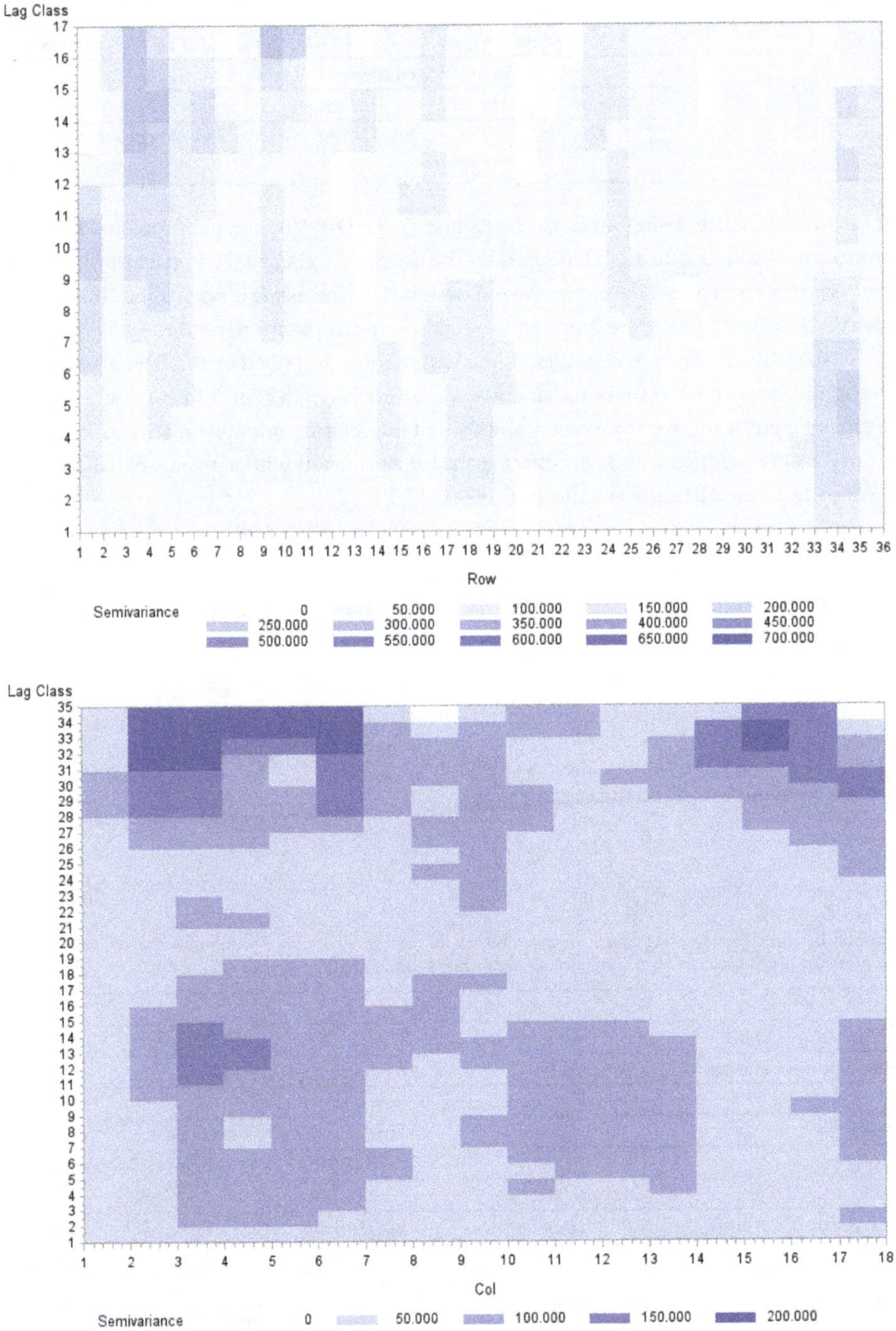

FIG. 6. One directional variogram in column direction for each row (above) and in row direction for each column (below) for spatial model, example 2.

TABLE 7. Main results for the spatial analysis of grain yield in an un-replicated experiment, example 2.

Covariance parameters				
Cov parm	Estimate	Standard error	*Z* value	*P* > *Z*
Acc(Group*Tester)	0.7044	0.09317	7.56	< 0.0001
SP(POWA) Row	0.07615	0.07844	0.97	0.3316
SP(POWA) Col	0	.	.	.
Residual	0.6290	0.07198	8.74	< 0.0001

Test Type 3 of fixed effects				
Effect	Num DF	Den DF	*F* value	*P* > *F*
Group	2	529	1.66	0.1902
Tester(Group)	9	529	0.73	0.6860

0the error variance. Using random effects also can be useful to avoid problems with the confusion of effects that are introduced by the design; this is usual when an optimization procedure is used to get the experimental design. This last issue is not a good thing and it should be solved while the experimental design is in the planning phase.

Tables 8 and 9 show the results for the two models described above. The AIC statistic is slightly lower for the first model; however, linear row effect and linear column effect were not significant. For this reason, the second model is a preferable model in which column effect is significant and row effect is almost significant with a *p*-value of 0.0755. SAS code to produce the results in Tables 8 and 9:

```
PROC GLIMMIX data = Maize;
class group tester acc;
model pgrhah = group tester(group);
random row col;
random acc(group*tester);
covtest/wald;
run;
PROC GLIMMIX data = Maize;

class group tester acc row col;
model pgrhah = group tester(group);
random row col;
random acc(group*tester);
covtest/wald;
run;
```

TABLE 8. Main results using linear random effects in row and column direction for the spatial analysis of grain yield in an un-replicated experiment, example 2.

Fit statistics				
-2 Res Log Likelihood			1884.93	
AIC (smaller is better)			1892.93	

Covariance parameters				
Cov Parm	Estimate	Standard error	*Z* value	*P* > *Z*
Row	0.000079	0.000135	0.59	0.2787
Col	0.001251	0.001864	0.67	0.2510
Acc(Group*Tester)	0.6956	0.08945	7.78	< 0.0001
Residual	0.5928	0.06797	8.72	< 0.0001

Test Type 3 of Fixed Effects				
Effect	Num DF	Den DF	*F* Value	*P* > *F*
Group	2	529	1.73	0.1790
Tester(Group)	9	529	0.85	0.5667

TABLE 9. Main results using row and column random effects for the analysis of grain yield in an unreplicated experiment, example 2.

Fit statistics				
-2 Res Log Likelihood			1887.80	
AIC (smaller is better)			1895.80	
Covariance parameters				
Cov parm	Estimate	Standard error	Z Value	P > Z
Row	0.03253	0.02265	1.44	0.0755
Col	0.05981	0.03115	1.92	0.0274
Acc(Group*Tester)	0.7125	0.08987	7.93	< 0.0001
Residual	0.5354	0.06863	7.80	< 0.0001
Test Type 3 of fixed effects				
Effect	Num DF	Den DF	F value	P > F
Group	2	529	1.84	0.1597
Tester(Group)	9	529	0.81	0.6105

Summary

Modeling residual spatial variability has shown to be effective in controlling sources of variation that the experimental design does not. In general, when the residuals are spatially modeled, the heritability (repeatability) and the precision with which a genotypic value is estimated are larger. This is a reason why spatial analysis of experiments is a good practice, especially in early stages of plant breeding programs. This practice is especially important for augmented designs, in which some genotypes are tested only in one plot; spatial modeling provides a way to separate the plot effect from the genotypic effect. However, this methodology introduces some challenges; solutions to those challenges are still under discussion. The two main challenges are to determine which experimental design is adequate when spatial analysis is going to be performed and how to decide which model is fine regarding the objective of our work. In recent years, some papers were published trying to answer these questions, but not complete agreement among scientists exists. Maybe the reason is that each trial is a different problem and only general recommendations can be provided. Models which perform well in some situations may not in others.

As a general recommendation, when possible, different models should be tested, starting with the model that mimics the experimental design. Then use the spatial model. Modeling spatial variability has a major effect in improving the results; however, other extra effects can be included as mentioned in the examples. The different models can be evaluated and compared by using fit statistics such as AIC and BIC, the semivariogram and other statistics as the heritability and the standard error of the difference between two genotypes.

Key Learning Points

- The experimental design cannot control small-scale effects neither some effect generated after the experiment is in the field.
- Modeling residual variability by spatial models is a method that provides advantages in field experimentation, producing more precision in the estimation of treatment (genotypic) effects and a better control of the noise.

- Besides standard statistics to compare different models (AIC, heritability), the semivariogram is probably the most important tool to compare models.
- Any effect that was assumed at the design–stage should be included in the model at the analysis stage, at least initially.

Review Questions

1. Explain with your own words what "spatial variability analysis" means
2. What are the different models that can be used to model the residual spatial variability?
3. Why does modeling the residual spatial variability usually have larger precision than a standard analysis in which spatial variability is not modeled?
4. Is it possible to model spatial variability of an experiment with five treatments in two replicates? Justify your answer.
5. If you are analyzing an experimental design; and you want to perform spatial analysis, are there any changes in the assumptions? If so, mention them.
6. What are the advantages of using spatial analysis?

Exercises

1. Replicate the analysis of the two examples presented above.
2. Perform the residual analysis of Example 1, including the option PLOTS = ALL in the *model* sentence. Do you think that the assumptions of the model are correct?
3. Perform the analysis of Example 2 considering row and column effects as random. Are there any changes in the results?

References

Al-Marshadi, A.H. 2014. Selecting the covariance structure in mixed model using statistical methods calibration. Journal of Mathematics and Statistics. 10(3):309–315. doi:10.3844/jmssp.2014.309.315

Brownie, C., Bowman, D. T. and Burton, J.W. 1993. Estimating spatial variation in analysis of data from yield trials: A comparison of methods. Agronomy Journal 85:1244c1253. doi:10.2134/agronj1993.00021962008500060028x

Burgueño, J., A. Cadena, J. Crossa, M. Banziger, A.R. Gilmour, and B. Cullis. 2000. User's guide for spatial analysis of field variety trials using ASREML. CIMMYT, Mexico-Veracruz, Mexico.

Casler, M.D. 2018. Blocking principles for biological experiments. In: B. Glaz and K.M. Yeater, editors, Applied statistics in the agricultural, biological, and environmental sciences. ASA, CSSA, SSSA, Madison, WI.

Cox, D.R. 2009. Randomization in the design of experiments. International Statistical Review 77(3):415–429. doi:10.1111/j.1751-5823.2009.00084.x

Cullis, B.R., and A.C. Gleeson. 1991. Spatial analysis of field experiments-an extension to two dimensions. Biometrics 47:1449–1460. doi:10.2307/2532398

Cullis, B.R., A. Smith, and N.E. Coombes. 2006. On the design of early generation variety trials with correlated data. Journal of Agricultural, Biological, and Environmental Statistics 11(4):381. doi:10.1198/108571106X154443

Federer, W.T. 1956. Augmented (or hoonuiaku) designs. Hawaiian Planters'. Record (Washington) LV(2):191–208.

Federer, W.T., E.A. Newton, and N.S. Altman. 1997. Combining standard block analyses with spatial analyses under a random effects model. In: T.G. Gregoire, D.R. Brillinger, P.J. Diggle, E. Russek-Cohen, W.G. Warren, and R.D. Wolfinger, editors, Modelling longitudinal and spatially correlated data. Lecture notes in statistics. Vol. 122. Springer, New York. doi:10.1007/978-1-4612-0699-6_33

Fisher, R.A. 1925. Statistical methods for research workers. Oliver and Boyd, Edinburgh.

Gilmour, A.R., B.R. Cullis, and A.P. Verbyla. 1997. Accounting for natural and extraneous variation in the analysis of field experiments. J. Agric. Biol. Environ. Stat. 2:269–293. doi:10.2307/1400446

Gleeson, A.C., and B.R. Cullis. 1987. Residual maximum likelihood (REML) estimation of a neighbour model for field experiments. Biometrics 43:277–288. doi:10.2307/2531812

Grondona, M.O., J. Crossa, P.N. Fox, and W.H. Pfeiffer. 1996. Analysis of variety trials using two-dimensional separable ARIMA processes. Biometrics 52:763–770. doi:10.2307/2532916

Gurka, M.J. 2006. Selecting the best linear mixed model under REML. Am. Stat. 60(1):19–26. doi:10.1198/000313006X90396

Hinkelmann, K., and O. Kempthorne. 2008. Design and analysis of experiments. Volume 1:, Introduction to experimental design, 2nd ed. Wiley, Hoboken, NJ.

Hu, X., and J. Spilke. 2009. Comparison of various spatial models for the analysis of cultivar trials. N. Z. J. Agric. Res. 52(3):277–287. doi:10.1080/00288230909510512

Kehel, Z., D.Z. Habash, S.A. Gezan, S.J. Welham, and M.M. Nachit. 2010. Estimation of spatial trend and automatic model selection in augmented designs. Agron. J. 102(6):1542–1552. doi:10.2134/agronj2010.0175

Kempton, R.A., J.C. Seraphin, and A.M. Sword. 1994. Statistical analysis of two-dimensional variation in variety yield trials. J. Agric. Sci. 122:335–342. doi:10.1017/S0021859600067253

Lin, B., Z. Pang, and J. Jiang. 2013. Fixed and random effects selection by REML and pathwise coordinate optimization. J. Comput. Graph. Stat. 22(2):341–355.

Lin, C.S., and G. Poushinsky. 1983. A modified augmented design for an early stage of plant selection involving a large number of test lines without replication. Biometrics 39:553–561. doi:10.2307/2531083

Martin, R.J. 1996. Spatial experimental design. In: S. Ghosh, C.R. Rao, editors, Handbook of statistics. Volume 13. Design and analysis of experiments. 1st edition. Elsevier. doi:10.1016/S0169-7161(96)13017-0

Nguyen, N.K., and E.R. Williams. 1993. An algorithm for constructing optimal resolvable row-column designs. Aust. J. Stat. 35:363–370. doi:10.1111/j.1467-842X.1993.tb01344.x

Papadakis, J. S. 1937. Méthode statistique pour des expériences sur champ. Bulletin de l`Institute d`Amélioration des Plantes à Salonique. Thessalonique 23.

Piepho, H.P., J. Möhring, M. Pflugfelder, W. Hermann, and E.R. Williams. 2015. Problems in parameter estimation for power and AR(1) models of spatial correlation in designed field experiments. Comm. In: Biometry abd. Crop Sci. 10(1):3–16.

Richter, C., B. Kroschewski, H.P. Piepho, and J. Spilke. 2015. Treatment comparisons in agricultural field trials accounting for spatial correlation. J. Agric. Sci. 153:1187–1207. doi:10.1017/S0021859614000823

Rivera, G. 1975. Métodos alternativos para el análisis de un número grande de variedades. Tesis de Maestria, Colegio de Postgraduados. Montecillos, Mexcio.

Robinson, G.K. 1991. That BLUP is a good thing: The estimation of random effects. Stat. Sci. 6(1):15–32. doi:10.1214/ss/1177011926

SAS Institute, Inc. 2015. SAS 9.4 procedures guide. SAS Institute, Inc., Cary, NC.

Stefanova, K.T., Smith, A.B., and Cullis, B.R. 2014. Enhanced diagnostics for the spatial analysis of field trials. JABES 14:392–409.

Stringer, J.K., A.B. Smith, and B.R. Cullis. 2008. Spatial analysis of agricultural field experiments. In: K. Hinkelmann and O. Kempthorne. Design and analysis of experiments, Volume 1, Introduction to experimental design, 2nd ed. Wiley, Hoboken, NJ.

Stroup, W.W. 2018. Analysis of non-Gaussian data. In: B. Glaz and K.M. Yeater, editors, Applied statistics in the agricultural, biological, and environmental sciences. ASA, CSSA, SSSA, Madison, WI.

Thöle, H., C. Richter, and D. Ehler. 2013. Strategy of statistical model selection for precision farming on-farm experiments. Precis. Agric. 14:434–449. doi:10.1007/s11119-013-9306-9

Velazco, J.G., M.X. Rodriguez-Alvarez, M.P. Boer, D.R. Jordan, P.H.C. Eilers, M. Malosetti, and F. van Eeuwijk. 2017. Modelling spatial trends in sorghum breeding field trials using a two-dimensional P-spline mixed model. Theor. Appl. Genet. 130(7):1375–1392. doi:10.1007/s00122-017-2894-4

Wang, S., P.X. Song, and J. Zhu. 2010. Doubly regularized reml for estimation and selection of fixed and random effects in linear mixed-effects models. The University of Michigan Department of Biostatistics Working Paper Series, Paper 89. University of Michigan School of Public Health. http://biostats.bepress.com/umichbiostat/paper89 (verified 3 Jan. 2018).

Wang, J., and G.B. Schaalje. 2009. Model selection for linear mixed models using predictive criteria. Commun. Stat. Simul. Comput. 38(4):788–801. doi:10.1080/03610910802645362

Williams, E.R., and H.P. Piepho. 2013. A comparison of spatial designs for field variety trials. Australian and New Zealand Journal of Statisitcs 55(3):253–258. doi: 10.1111/anzs.12034

Wood, T.B., and F.J.M. Stratton. 1910. The interpretation of experimental results. J. Agric. Sci. 3:417–440. doi:10.1017/S0021859600001210

Wilkinson, G.N., S.R. Eckert, T.W. Hancock, and O. Mayo. 1983. Nearest neighbour analysis of field experiments. Journal of Royal Statistical Society Series B 45:151–211.

CHAPTER 13: AUGMENTED DESIGNS-EXPERIMENTAL DESIGNS IN WHICH ALL TREATMENTS ARE NOT REPLICATED

Juan Burgueño,* José Crossa, Francisco Rodríguez, and Kathleen M. Yeater

In the research planning stages of a project, sometimes replication of treatments is impractical, prohibitively expensive, or impossible. Perhaps the researcher does not have enough material to fully randomize and replicate, or there are physical space and/or time limitations, or too many barriers to complete the data collection required for a fully replicated experiment to answer the proposed research question. In these scenarios, it can be advantageous to use unreplicated or augmented experimental designs. These designs reduce costs with an acceptable loss of precision. Other (though not all) important consequences are: (i) not all treatments are compared with the same precision; (ii) experimental error is partially estimated; (iii) in factorial experiments, some interaction effects cannot be estimated and some are confounded; (iv) standard experimental designs can (and should) be modified to deal with unreplicated treatments; and v) the statistical analysis approach is usually different to the usual analysis of variance.

INTRODUCTION

Statistical methodologies help scientists reduce the cost of experimentation through the use of an adequate sample size and appropriate statistical analyses that allow suitable inferential processes.

Early in the last century, experimental evaluation of genotypes was performed without replicates due to lack of resources and perhaps due to the absence of well-established statistical methods and procedures. In the 1920s, Fisher established the principles of experimental design (Fisher, 1925), which included replication, randomization, and blocking. However, many plant breeders continued evaluating genotypes without using replicates. In the best of cases, they used systematic checks to make better comparisons between genotypes. Checks were also used to calculate

---

Abbreviations: BLUE, best linear unbiased estimators; BLUP, best linear unbiased predictor; SED, standard error of the difference; SEM, standard error of the mean.

J. Burgueño, J. Crossa, and F. Rodríguez, CIMMYT, Biometrics and Statistics Unit, Mexico DF, Mexico; K.M. Yeater, USDA-ARS, Plains Area - Office of the Director, Fort Collins, CO; USDA-ARS, Fort Collins, CO. *Corresponding author (jburgueno@cgiar.org)

doi:10.2134/appliedstatistics.2016.0005

Applied Statistics in Agricultural, Biological, and Environmental Sciences
Barry Glaz and Kathleen M. Yeater, editors

the experimental error. It was Federer in 1956 who called attention to this and proposed augmented designs (see Federer et al. (1975), and Federer, 1993, 1998).

These new experimental (augmented) designs proposed by Federer provided a valid estimation of experimental error because treatments are randomly allocated, as opposed to the standard method in which checks were allocated systematically. With a valid estimation of experimental error, comparisons between cultivars can be done following Fisher's principles. Response surface designs are a class of augmented designs that were first used in the chemical industry (Box and Draper, 1987) to improve processes. However, they are not called "augmented designs"; instead, incomplete factorial designs and other names are used. In the chemical industry, the cost of an experimental unit is often much higher than the cost of an experimental unit in agronomic or biological research. Also, most of the confounding factors were controlled by incomplete factorial designs so the experimental error could be estimated with acceptable precision and without bias even if some treatments were not replicated. Most industrial designs continue to focus on optimizing the process. For example, regression models are used to analyze response surface experiments to obtain the best combination of factors that optimize the response (See Chapter 6 by Richter and Piepho, 2018 for a discussion of linear regression and Chapter 15 by Miguez et al., 2018 for a discussion of nonlinear regression). Response surface methodology is the name of the analytical approach used to analyze these experiments. In the extreme case of factorial designs, analytical technical approaches were proposed to analyze experiments in which treatments were not replicated (Montgomery, 2011; Box et al., 2005). Those ideas were used in agricultural experimentation and augmented designs such as those employed in Plan Puebla (Turrent-Fernández and Laird, 1975; Laird, 1987) were created. In this context, "augmented" had a different meaning: a standard factorial experiment is augmented with axial points located outside the factorial model's space, whether those axial points are replicated or not. When the points are not replicated (and sometimes none of the factorial design points are replicated), we have an augmented design. Usually only the central point is replicated in these kinds of experiments, where the central point is an experimental point in which each factor has the mean level of the levels assigned to that factor.

At this point, we should clarify our terminology because, as is often the case, several terms are used to express the same meaning or the same term is used with different meanings. On one side, we have the treatment design and on the other, we have the experimental design. The former is the way in which treatments in an experiment are defined and selected; the latter is the random way in which treatments are assigned to experimental units. Experimental designs such as split-plot designs or augmented designs as used by Federer (Federer, 1956; Federer et al., 1975) are suitable for specific treatment designs. This has led to misinterpretations such as thinking that a factorial is a kind of experimental design, when in fact it is a type of treatment design. Augmented design (Federer, 1956), central composite design (Montgomery, 2011), p-rep design (Cullis et al., 2006; Williams et al., 2011; Moehring et al., 2014), and unreplicated design are some names of experimental designs that share the condition that some treatments are not replicated.

In this chapter, we will discuss experiments in which some treatments are tested only once and others are replicated several times. Since in the literature there is no agreement on a general name for these kinds of experiments, in this chapter, we will use the term unreplicated experiments or, as Williams et al. (2011) did, the term *unreplicated trials* to indicate those kinds of experiments. Unreplicated experiments are conducted with different treatment designs and experimental designs. Our main focus will be on agricultural experiments where the research question is not necessarily a specific hypothesis; rather the focus would be in the realm of development or discovery research. In these experiments, we feel that due to differences in experimental conditions, particularly in field research, more attention is needed on the proper use of non-replicated experiments to provide better control on sources of experimental error.

Experimental Designs for Experiments in Which All Treatments are Not Replicated

Different experimental designs were developed for experiments in which all treatments are not replicated (we will refer to these as unreplicated designs). Different analytical approaches have also been suggested along with these different experimental designs. In 1956, Federer, while working on sugarcane (*Saccharum officinarum* L.) improvement, proposed the augmented design with the idea of augmenting basic standard designs (completely randomized design, randomized complete block design, lattice design) with other treatments which were impossible to replicate. The main advantage of using these designs is that they allow an appropriate measurement of experimental error to compare treatments, as opposed to the commonly used completely unreplicated and systematic design that does not provide an unbiased estimation of experimental error. To analyze these experiments, Federer proposed the standard analysis of variance for the original design considering the particular way in which replicated treatments were randomized.

After Federer, several unreplicated designs were proposed with different degrees of success. Rivera (1975), Lin and Poushinsky (1983), and Besag and Kempton (1986) proposed experimental designs in which one or more checks were located systematically in the field. The Lin and Poushinsky (1983) design used two checks, one located systematically and the other randomly. Federer and Raghavarao (1975) and Federer et al. (1975) presented an optimal augmented block design and an optimal augmented row and column design for estimating certain comparisons between unreplicated treatments. Cullis et al. (2006) introduced the partially replicated design (p-rep) in which checks were replaced by a set of replicated genotypes. They also introduced a method to allocate treatments in the field that optimized the analysis when a spatial model was used. Most of these designs are focused on early generation evaluation of genotypes where the available seed is not enough to replicate the treatments. However, the same ideas can be used in other kinds of experiments when replications is difficult, such as fertilizer experimentation, factorial experiments with one factor being date of application, and doses of different products.

Unreplicated experiments can be classified using at least two different criteria that produce different kinds of analyses. One criterion is the kinds of treatments that are replicated in the experiment; another is how treatments are allocated in the field, that is, randomly or systematically. Since unreplicated designs can be superimposed on different experimental designs, there are many different combinations of experimental designs and unreplicated experiments.

Replicated treatments can be checks or treatments with which, for the most part, we already have substantial experience. They are included in the experiments as controls, for example, treatments without fertilization or adapted cultivars. On the other hand, we can replicate some of the treatments of most interest about which we know less (for example, new cultivars) and thereby reduce the cost of testing less useful treatments. Usually we need to introduce a dummy variable in an analysis that compares checks against unreplicated treatments; then comparisons between checks (if there is more than one) and between unreplicated treatments are performed. Following the second criterion, when treatments are allocated randomly, the analysis can be performed with the usual ordinary least squares method (analysis of variance) or any of the available maximum likelihood methods (mixed models). The idea behind systematically allocating checks is that experimental error can be modeled or controlled (polynomial spatial model). It is important to have a good estimation of this model. When analyzing unreplicated experiments, two important issues must be considered: (i) we need to have a procedure that allows for statistically comparing treatments, and (ii) we need to adjust the values of unreplicated treatments. Since treatments are affected by the experimental units in which they are tested, and there is only one experimental unit for each unreplicated treatment, there is a confounding effect between the treatment and the experimental unit. However, we need to estimate the treatment effects as independently of other factors as possible. Based on a model that can be adjusted by the checks, it is possible to estimate the expected result in each experimental unit and then extract this effect from its value. Different models have been proposed for making this adjustment, some of these are a moving average (Papadakis, 1937), regression models (Rivera, 1975), and spatial models (Cullis et al., 2006).

Allocation of checks in the field can be done in two ways; if checks are allocated randomly, it is a good idea to adjust the values for spatial variability. However, randomization does not ensure that the estimation of spatial variability will be optimum. The alternate is to allocate checks in an optimal method, as proposed by Federer and Raghavarao (1975), Rivera (1975), Besag and Kempton (1986) and Cullis et al. (2006), depending on the model that will be used for the analysis.

An interesting design that can be used as a base design to construct an augmented design is the Sudoku design (Saba and Sinha, 2014). This is a design in which block structure controls the variability in three ways, columns, rows and "squares", a square is a set of rows and columns resembling the traditional game of Sudoku. In this design every repeated treatment is replicated in each row, column and square. Ensuring that every check is replicated in every row, column, and square allows a good distribution of the checks in the field. This is important to model the spatial variability. Due to the restriction in the distribution of checks, Sudoku designs are not as flexible as an augmented complete block design. However, nonsquare Sudoku designs can be built offering more flexibility.

Estimating the Experimental Error

We will discuss some considerations regarding the estimation of the experimental error when planning an unreplicated experiment. Experimental variability can be estimated only with the information provided for the replicated treatments except when unreplicated treatments have a random effect. For this reason, a key criterion for selecting checks is that the checks will provide a good estimate of experimental error. Ideally, the replicated treatments (checks) should respond to the environmental conditions in the experimental units, without interacting with the environment. A genotype that produces similar yields regardless of the environment is not a good check because it will cause experimental error to be biased downward. A genotype that interacts with the environment is also not a good check because it will produce a biased experimental error. The final estimation of the experimental error depends on the set of checks and genotypes used in the experiments as well as the analytical model. The bias produced by the checks will depend on the kind of interaction that the genotype has with the experimental unit (plot). In an ideal experiment, we assume that there is no interaction between treatment and plot or treatment and block; however, there is some degree of interaction, especially in large experiments. Using a fertilizer experiment as an example, if zero fertilizer dose is used as a check, the natural availability of nutrients in the soil will affect results and we can expect more variability and a higher experimental error than if we use a high level of fertilization as the check. In the last case, the effect of natural N availability in the soil will often produce only a marginal effect on experimental error. When unreplicated treatments have a random effect, the problem may be less important because it is possible to separate the experimental unit effect from the random effect of the treatment since experimental error is calculated from both plots with replicated treatments and unreplicated treatments. However, the experimental error includes the variability between experimental units and the interaction between the experimental plot and the treatment assigned to it. The last component cannot be estimated and will be confounded with the effect of the treatment. Some results have shown that when treatments are considered random, selection of checks is important to obtain a good estimation of the experimental error (Burgueño, 2005).

Another important question is how many checks and how many replicates of each check should be used. More replicates will produce higher precision, but at a higher cost, especially if checks are not treatments of interest. Issues to consider include the number of treatments, number of experimental units, arrangement of the plots in the field, the different kinds of treatments, checks, no checks, and characteristics of the treatments. Related to these issues, one must also consider the cost of one experimental unit and the total budget available for the experiment (or set of experiments). Based on all this information, researchers make a decision directed at obtaining the desired precision at an acceptable cost. Most of the time, making this decision is difficult because we usually do not know the experimental variability, the real effect of treatments, and the effects of other factors affecting the experimental units, such as climate and its interaction with the treatments. General guidelines can be found in the literature (Sahagun-Castellanos, 1985; Burgueño, 2005; Payne, 2006; Eshetie, 2011; Aguade, 2013; Casler, 2015); however, every experiment should be considered as a unique case.

Burgueño (2005) simulated different experimental designs for a uniform experiment in which only one genotype was sown and found that the number of checks did not affect estimation of experimental error; however, this must be interpreted with care. Sometimes a particular check may not perform well under trial conditions, which is why using only one check is not recommended. Using more than one check allows testing the analysis in a cross-validation scheme using one check and testing against the other. As is usual in problems of sample size, increasing the percentage of experimental units dedicated to checks has a large impact on precision, particularly when we increase this percentage from 1 to 5% or even 10%. However, as increases in the percentage of experimental units with checks compared with the overall sample size exceeds 10%, the gain is marginal. Considering all this, a general recommendation is to use between 5 and 10% of the experimental units for checks and two or three checks are usually sufficient. The number of replicates of each check is determined by the number of check plots and number of checks in the experiment, but at least four replicates are recommended.

Under special experimental conditions, such as experiments under drought or other stresses, the experimental error is higher and using a higher percentage of experimental units with checks is recommended, as is including some specific checks for the particular stress being examined. Based on our experience with testing maize (*Zea mays* L.) under drought stress, we assign approximately 15% of our experimental units to checks. The final decision is a compromise between experimental conditions, the desired precision, costs, and available budget. A guide is only that: a guide. Each experiment must be considered under its own conditions.

Precision

Any unreplicated experiment will have lower precision than any replicated experiment, no matter how sophisticated the design or analysis of the unreplicated experiment. This is clear in the formula for the standard error of the mean (SEM) and the standard error of the difference between two treatments i and j (SED_{ij}) for a completely randomized design or a complete block design:

$$SEM_i = \sqrt{\frac{MSE}{r_i}}, \quad SED_{ij} = \sqrt{MSE\left(\frac{1}{r_i} + \frac{1}{r_j}\right)}$$

where MSE is the mean square error, or the residual variance, and r the number of replicates.

These simple formulas are obtained from the analysis of variance of a completely randomized design or a randomized complete block design and also are valid for a completely randomized unreplicated experiment where treatment and block effects are fixed and errors are independent. For other designs or in cases with a different structure of the residuals, those formulas are not valid. The r factor appears when you estimate the SEM or the SED and affects the SEM and the SED. Not only is the SEM larger for treatments in which $r = 1$; different pairs of treatments also have different SED values depending on the number of replicates of the treatments under consideration; this is especially important in p-rep designs where some tested genotypes are replicated more than others and the researcher needs to decide which treatments have more replicates than the others. For incomplete block designs, it is well known that treatments in the same block are compared with more precision than treatments in different blocks; the same is true for augmented incomplete block designs. Finally, it is necessary to consider that modeling spatial variability affects the SED regarding the relative position in the field of the unreplicated treatments.

There are two considerations related to the precision and planning of the experiment: one is the precision required for comparing different treatments and the other is that if some treatments produce more variable results, then it is more difficult to estimate their effects. Again, a good example is a treatment in which the fertilizer dose is zero. Because this null treatment is probably more affected by natural conditions in the experimental units, it is appropriate to have more replicates. Also, if the researcher is interested only in comparing all the treatments against the check by Dunnett's test, then with more replicates of the check, there is less probability of Type 1 family error. That is, there is a lower probability of rejecting at least one true null hypothesis by error in a set of comparisons (family) (Keselman, 2015).

In plant breeding, treatments are usually genotypes derived from several crosses as well as offspring of the same cross (i.e., the male and female parents). Offspring from the same cross can be expected to have less difference than individuals from different crosses. Therefore, detecting statistical differences between related progeny is more difficult, so more precision is required. A possible consideration when using an incomplete block design is to put individuals from the same cross together in the same block to increase the precision of the comparison between them. However, some issues must be considered. First, block effect will be confused with family effect and the role of checks is more relevant. Second, family comparison is done with lower precision and for overall selection response, this may not be the best arrangement, as shown by Piepho et al. (2006). The question the breeder has to answer is: what is more relevant, comparisons between or within families? The same idea can be applied to other kinds of treatments in which differential precision is necessary.

A different problem is when you are interested in estimating the genetic variance between and within families in a progeny experiment and you are thinking of using a partially replicated design. In this case, we recommend that the replicated treatments be randomly selected from different families to have a good estimation of variance within each family. Selecting replicated treatments in this way will also help produce a good estimation of the experimental error.

As stated previously, there is a loss of precision when using unreplicated experiments. Therefore, these experiments should only be used when conducting a replicated experiment is not feasible. This kind of design was originally developed to test a large number of genotypes in the early stages of breeding programs in which the low amount of available seed and the large number of test genotypes make a standard replicated experiment impossible to perform. Unreplicated experiments have the advantage of testing more genotypes. They also are useful for multienvironmental trials. With moderate-to-high positive correlated environments, it could be better to use one replicate in more environments than a replicated design in less environments. This in turn allows the researcher to select the best individual from a larger population tested in several environments. Therefore, the pressure of selection is increased and a greater genetic gain can be obtained. The loss of precision in an unreplicated experiment can be compensated by the gain in selection pressure.

Spatial Adjustment

As mentioned above, the value obtained for any particular unreplicated treatment is affected directly by the conditions of the experimental unit to which it was assigned. If the experimental unit has good conditions for the trait of interest, then a treatment will probably have a higher value than the actual effect of the treatment. In industrial research, unlike in agronomic field research, the effect of the experimental unit is not important because most of the factors that produce the experimental unit are controlled: temperature, doses, humidity, etc. In field research, soil conditions, the microclimate, pest infestation levels, and other factors can affect the results of the experiment in unexpected ways. In field experiments, some factors, such as irrigation or fertilization, can be somewhat controlled. However, even in these cases, the soil in some experimental units may have better water-holding capacity or higher natural nutrient levels. Still other factors can be even more confounded by the environment. Due to the differences in the magnitude of the effects of the experimental units in industrial experiments compared with field experiments, there are at least two different developments in the design and analysis of unreplicated experiments. These differences can also be seen in different areas of biology. In experiments with microorganisms performed in growth chambers under controlled nutrient, temperature, and humidity conditions (among other factors), the experimental error is almost under control, and the number of factors that contribute to experimental error are reduced, contrasting with experiments with large animals growing in their natural conditions.

In the special case of field experiments, the soil is the main source of experimental error and it provides a substantial source of error. However, in most cases, the effect of the soil follows a spatial pattern. The knowledge of the spatial pattern can benefit the research.

Spatial analysis in field experiments was introduced by Papadakis (1937). Early spatial analysis models considered adjusting the values of each plot based on the values of its neighboring plots. A recent review of nearest neighbor methods can be found in Piepho et al. (2008). Currently, autoregressive models have prevailed among researchers as the usual method for analyzing unreplicated experiments (see Chapter 12 on Spatial Analysis by Burgueño, 2018). While these methods can also be used in replicated experiments, their use is highly recommended for unreplicated field experiments.

Despite all of these methods of adjustment, unreplicated experiments still have important shortcomings. In particular, the interaction between the experimental unit and the treatment located in it cannot be estimated separately, and if it is high, the effect of the treatment will be biased.

Not all traits are affected in the same way under experimental conditions. For example, spatial models can accommodate effects associated with soil characteristics or with the wind if it has a dominant direction. In the case of a disease, if its natural inoculum has a spatial distribution, it can be modeled. However, sometimes the inoculum is found in patches or in large areas relative to the experimental field area, and the common models cannot model it. Even if an important effect can be modeled, caution must be exercised when considering the use of an unreplicated experiment. Unreplicated experiments are very useful; they can reduce the cost of experimentation and provide acceptable precision, but only in some cases. For traits such as yield or plant height and other traits affected mainly by soil conditions, this kind of experiment, together with the use of an appropriate adjusting model, is useful and cost effective. But we need to think twice before using an unreplicated experiment for evaluating lodging, diseases, and other pests.

Analysis

An augmented design can be analyzed in much the same way as any experimental design. First, we decide which model is appropriate for our experiment; preferably the model is selected before initiating the experiment. The model will follow the randomization procedure which creates the augmented design. The randomization (the model) can consider block effects and other sources of experimental error such as plot size (if for example, the experiment is a split-plot design). What is special in an augmented design is the presence of two kinds of treatments, unreplicated treatments (that are being tested) and replicated treatments such as checks or controls. There are at least three cases to consider when deciding how to parameterize the treatment effects, as expressed in the following equations. For ease of explanation, we are stating overall treatment effect only; any design effects and the experimental error are omitted:

(P1) $\theta_i = \alpha_i \quad i=1,2,...,C,C+1,C+2,...,C+T$

(P2) $\theta_{ij} = \tau_i + \beta(\tau)_{ij} \quad i=1,2,...,C,C+1$ and $j=1,2,...,T$ if $i=C+1$

(P3) $\theta_{ij} = \delta_i + \gamma(\delta)_{ij} \quad i=1,2; j=1,2,...,C$ if $i=1$ and $j=1,2,...,T$ if $i=2$

where

θ, the overall treatment effect

α_i, the ith effect indexed from 1 to C for checks and from $C+1$ to $C+T$ for unreplicated treatments

τ_i, the ith effect of a dummy variable indexed from 1 to C for checks and $C+1$ for the average of unreplicated treatments

$\beta(\tau)_{ij}$, the jth effect of unreplicated treatments nested in the ith τ effect

δ_i, the ith effect of a dummy variable indexed as 1 for checks and as 2 for unreplicated treatments

$\gamma(\delta)_{ij}$, the jth effect of treatments nested in the ith δ effect

C, the number of checks

T, the number of unreplicated treatments

$C+T$, the number of treatments

In the first parameterization (P1), all treatments are considered in the same way and the relative effects of all treatments are estimated. In P2 and P3, we nested the treatment effect, with the difference being that in P2, we grouped unreplicated treatments in one level at the first hierarchy, whereas in P3 we included a dummy factor with two levels that compares the average of checks against the average of unreplicated treatments. From the ANOVA table, we have different tests in the three cases; for P1, the null hypothesis is that all treatments have the same effect whether they were replicated or not. In both P2 and P3, the ANOVAs test two different hypotheses. For P2, the first hypothesis is that there are no differences between checks and the average of unreplicated treatments considered as another level of the effect; the second hypothesis tests differences between unreplicated treatments only.

Parameterization 2 tests differences between the average of checks and the average of unreplicated treatments first and then the effects of the different checks nested in Level 1 of the dummy factor, and the effect of unreplicated treatments nested in the second level of the dummy factor. These last two hypotheses can be tested together or separately. Most statistical software packages produce an overall test that tests both hypotheses simultaneously default; however, separate tests can also be performed in most statistical software. In this last case (P3), the equation for this parameterization is $y_{ij} = \delta_i + \gamma(\delta)_{1j} + \gamma(\delta)_{2j}$ $i = 1,2$, $j = 1,2,\ldots C$ if $i = 1$ and $j = 1,2,\ldots,T$ if $i = 2$ it will not be used in the examples. The three parameterizations produce the same differences between treatments, whether they were replicated or not, and with the same standard error. Selection among the three parameterizations should be based on the kind of comparison that will be made. Sometimes for ease, the selected parameterization can be adjusted when it includes design factors, especially designs with more than one experimental error, such as split plot designs.

Another important consideration when choosing among the three parameterizations occurs during consideration of an unreplicated effect as a random effect. Usually, in an experiment with some unreplicated treatments, replicated and unreplicated treatment differs in their characteristics; usually unreplicated treatments are commercial cultivars or hybrids while replicated treatments are from a population of selection. Most plant breeders are interested in estimating the genetic variances of their populations. Checks are usually commercial genotypes with a different genetic background than the other genotypes being tested. If checks are included together with the tested genotypes and P1 is used, then the estimate of the genetic variance will be biased because unlike unreplicated genotypes, commercial cultivars (replicated checks) are considered as individuals of the same population. Therefore, P1 will simultaneously estimate the genetic variance between unreplicated treatments, between checks, and between checks and unreplicated treatments. Parameterizations 2 and 3 are preferable in these situations. In P2, we have different treatments only in Group C+1, and one variance component is estimated for unreplicated treatments. In P3, the variance between checks as well as the variance between unreplicated treatments can be estimated. Considering this variation in the effect of unreplicated treatments, we will analyze the data in five different ways, including three different parameterizations and two of them (P2 and P3) with the variant of considering fixed (P2-f and P3-f) or random effects (P2-r and P3-r).

It should be noted that when unreplicated treatments are considered as a specific set of levels, they are treated as fixed effects and do not contribute to the estimation of experimental error. However, when an effect is analyzed as a random effect, a residual component can be estimated from the plots with unreplicated treatments. This is possible because replicated treatments provide an initial estimate of the experimental error. This will affect the residual error and consequently the testing of hypotheses.

Example 1: Completely Randomized Design with Unreplicated Treatments

In this example, we use the three parameterizations presented and discuss how results differ among them. We provide the codes for the analysis and the results obtained with SAS software V9.4 (SAS Institute Inc., 2015).

SAS software has several procedures and libraries for analyzing linear models. The most complete procedure in SAS software is GLIMMIX, which covers all the models that can be analyzed in other procedures, such as ANOVA, GLM, and MIXED. The GLIMMIX procedure is a general procedure able to deal with unbalanced data, mixed models, nonlinear models, and several error distributions. In this section, we will cover only the analyses of augmented designs of a normal response distribution with independent error.

The following is a simple artificial example of a completely randomized design with eight treatments, two checks (C1 and C2) replicated three times and six unreplicated treatments (G3 to G8). Due to the small number of data points, asymptotic results are not completely valid. The data file has four columns; the first three identify the treatments and the fourth column is the response variable. The first column (*t*) uses a number for each treatment (G1-G8), the second column (d1) identifies all the unreplicated treatments with the same number (e.g., 3) and the third column (d2) uses number '1' for checks and '2' for unreplicated treatments. Columns d1 and d2 are dummy variables and are useful for estimating the different models.

```
data a;
input t $ d1 d2 y;
cards;
C1 1 1 22
C1 1 1 21
C1 1 1 24
C2 2 1 29
C2 2 1 26
C2 2 1 28
G3 3 2 28
G4 3 2 32
G5 3 2 27
G6 3 2 38
G7 3 2 35
G8 3 2 28
;
* Model P1-f;
PROC GLIMMIX;
class t;
model y = t;
lsmeans t;
estimate "Ch"      t 1-1;
estimate "UnT"     t 0  0  1-1;
estimate "Ch-UnT"  t 1  0-1;
run;
* Model P2-f;
PROC GLIMMIX;
class t d1;
model y = d1 t(d1);
lsmeans d1 t(d1);
estimate "Ch"     d1 1-1;
estimate "UnT"                    t(d1) 0  0  1-1;
estimate "Ch-UnT" d1 1  0-1  t(d1) 1 0-1;
run;
* Model P3-f;
PROC GLIMMIX;
class t d2;
model y = d2 t(d2);
lsmeans d2;
lsmeans t(d2)/slice = d2;
estimate "Ch"                t(d2) 1-1;
estimate "UnT"               t(d2) 0  0  1-1;
```

```
estimate "Ch-UnT" d2 1-1 t(d2) 1  0-1;
run;
* Model P2-r;
PROC GLIMMIX;
class t d1;
model y = d1/ddf = 4;
random t(d1)/s;
covtest "t(d1 = 3)" general 1 0;
lsmeans d1;
* Means (BLUP);
estimate "Ch 1"          intercept 1 d1 1          | t(d1) 1;
estimate "Ch 2"          intercept 1 d1 0 1        | t(d1) 0 1;
estimate "Mean of UnR"   intercept 1 d1 0 0 1 1 1 1 1 1 1;
estimate "G3"            intercept 1 d1 0 0 1 | t(d1) 0 0 1;
estimate "G4"            intercept 1 d1 0 0 1 | t(d1) 0 0 0 1;
estimate "G5"            intercept 1 d1 0 0 1 | t(d1) 0 0 0 0 1;
estimate "G6"  intercept 1 d1 0 0 1 | t(d1) 0 0 0 0 0 1   ;
estimate "G7"  intercept 1 d1 0 0 1 | t(d1) 0 0 0 0 0 0 1  /e;
estimate "G8"  intercept 1 d1 0 0 1 | t(d1) 0 0 0 0 0 0 0 1/e;
* Comparisons;
estimate "C1-C2" d1 1  -1    | t(d1) 1-1      / group 1 1;
estimate "G3-G4"              | t(d1) 0 0  1-1/ group 0 0 1 e;
estimate "C1-G3" d1 1  0-1   | t(d1) 1 0-1    / group 1 0 1;
estimate "C2-G3" d1 0  1-1   | t(d1) 0 1-1    / group 0 1 1;
run;
*Model P3-r;
PROC GLIMMIX;
class  d2 t;
model y = d2/ ddf = 4;
random  t(d2)/group = d2 s;
lsmeans d2/e;
estimate "int" intercept 1;
* Means (BLUP);
estimate "C1" intercept 1 d2 1   | t(d2) 1/group 1;
estimate "C2" intercept 1 d2 1   | t(d2) 0 1/group 1;
estimate "G1" intercept 1 d2 0 1 | t(d2) 0 0 1/group 0 1;
estimate "G2" intercept 1 d2 0 1 | t(d2) 0 0 0 1/group 0 1;
estimate "G3" intercept 1 d2 0 1 | t(d2) 0 0 0 0 1/group 0 1;
estimate "G4" intercept 1 d2 0 1 | t(d2) 0 0 0 0 0 1/group 0 1;
estimate "G5" intercept 1 d2 0 1 | t(d2) 0 0 0 0 0 0 1/group 0 1;
estimate "G6" intercept 1 d2 0 1 | t(d2) 0 0 0 0 0 0 0 1/group 0 1;
* Comparisons;
estimate "C1-C2"           | t(d2) 1-1      /e group 1;
estimate "G3-G4"           | t(d2) 0  0  1-1/e group 0 1;
estimate "C1-G3" d2 1-1  | t(d2) 1  0-1    /e group 1 1;
covtest "t(d2 = 1)" general 1  0;
covtest "t(d2 = 2)" general 0  1;
run;
```

Results of the five analyses are presented in Tables 1 to 3; the complete results obtained with SAS are presented in Appendix A1.

The first three analyses, which consider only fixed effects, are equivalent, in the sense that they produce the same results for the same tests; the same is true for mixed models (P2-r) and (P3-r), in which we find the same results for the variance component of unreplicated treatments. All analyses produce the same residual error variance estimate due to the equivalence among them (the residual error appears once at the bottom of Table 1); consequently, the standard errors of the least squares means (Table 2) are the same for all the analyses. Given that the variance components are the same in the two last analyses, the standard errors for the predicted values are the same in those two analyses. Least squares means are best linear unbiased estimators (BLUE) of the mean of the levels of fixed factors, whereas for random effects, we used the best linear unbiased predictor (BLUP).

TABLE 1. Test of fixed and random effects from Example 1 (the residual error was the same for all models).

	Type III tests of fixed effects					Covariance parameters			
Model	Effect	Num DF ‡	Den DF ‡	*F* value	*P* > *F*	Cov parm	Estimate	Standard error	*P*-Value LRT†
P1-f	t	7	4	16.1	0.0088	–	–	–	–
P2-f	d1	2	4	34.9	0.0029	–	–	–	–
	t(d1)	5	4	8.5	0.0295	–	–	–	–
P3-f	d2	1	4	51.6	0.0020	–	–	–	–
	t(d2)	6	4	10.1	0.0210	–	–	–	–
	t(d2 = 1)	1	4	18.3	0.0129				
	t(d2 = 2)	5	4	8.5	0.0295				
P2-r	d1	2	4	2.0	0.2504	t(d1 = 3)	17.53	12.67	0.0216
P3-r	d2	1	4	3.9	0.1213	t(d2 = 1)	13.44	20.12	0.0163
						t(d2 = 2)	17.53	12.67	0.0216
						Residual	2.33	1.65	

† Likelihood ratio test using a mixture of Chi-square distributions.

‡ NUM/DEN, Numerator and denominator degrees of freedom, respectively.

TABLE 2. Best linear unbiased estimates (BLUEs, Least squared means) and best linear unbiased predictors (BLUPs) for fixed and random effects of the five models used to analyze data from Example 1.

			P1-f		P2-f		P3-f		P2-r		P3-r	
d2	d1	t	BLUE	Standard error	BLUE	Standard error	BLUE	Standard error	BLUP	Standard error	BLUP	Standard error
1	—	-	—	—	—	—	25.0	0.62	—	—	25.0	2.67
1	1	C1	22.3	0.88	22.3	0.88	22.3	0.88	22.3	0.88	22.5	0.87
	2	C2	27.7	0.88	27.7	0.88	27.7	0.88	27.7	0.88	27.5	0.87
2	3	-	—	—	31.3	0.62	31.3	0.62	31.3	1.82	31.3	1.82
2	3	G3	28.0	1.53	28.0	1.53	28.0	1.53	28.4	1.45	28.4	1.45
		G4	32.0	1.53	32.0	1.53	32.0	1.53	31.9	1.45	31.9	1.45
		G5	27.0	1.53	27.0	1.53	27.0	1.53	27.5	1.45	27.5	1.45
		G6	38.0	1.53	38.0	1.53	38.0	1.53	37.2	1.45	37.2	1.45
		G7	35.0	1.53	35.0	1.53	35.0	1.53	34.6	1.45	34.6	1.45
		G8	28.0	1.53	28.0	1.53	28.0	1.53	28.4	1.45	28.4	1.45

Note that there are fewer significant differences between treatments based on BLUP results. For example, between Treatments G5 and G6, the difference between BLUE values is 11 and the difference between BLUP values is 9.7. In addition, it can be seen that BLUPs have a lower range than BLUEs (27.5–37.2 or Blups vs. 27.0–38.0 for Blues). This is due to the shrinkage effect, or regression to the mean, which results from considering a treatment as a random effect. A given BLUP has larger standard errors than its corresponding BLUE. Therefore, compared with BLUE, there are fewer significant differences between unreplicated BLUP values because their values are smaller and their standard errors are larger. For a more in-depth discussion on the use of BLUPs, see Robinson (1991). Also, Ticona-Benavente and da Silva Filho (2015) compare the use of BLUEs and BLUPs in breeding. The analysis that should be used depends on the researcher's hypothesis and on the means of the treatments. If we need a comparison against each check, then BLUE is appropriate. If unreplicated treatments are a random sampling of a population, then BLUP should be used. If the checks come from a different population than the unreplicated treatments, and

TABLE 3. Results from Example 1 of comparisons between two replicated treatments (C1 vs. C2), two unreplicated treatments (G3 vs. G4) and a check vs. an unreplicated treatment (C1 vs. G3).

	Contrast between	Estimate	Standard error	DF	$P > \|t\|$
Fixed models, P1-f, P2-f and P3-f	Checks	-5.33	1.247	4	0.0129
	Unreplicated treatments	-4.00	2.160	4	0.1377
	Check– Unreplicated treatment	-5.67	1.764	4	0.0325
Mixed model, P2-r	Checks	-5.33	1.247	4	0.0129
	Unreplicated treatments	-3.53	2.029	4	0.1569
	Check– Unreplicated treatment	-6.06	1.698	4	0.0234
Mixed model, P3-r	Checks	-5.04	1.213	4	0.0142
	Unreplicated treatments	3.53	2.029	4	0.1569
	Check– Unreplicated treatment	-5.91	5.817	4	0.3669

unreplicated treatments are random, then a dummy variable such as d1 needs to be included. These questions are a subset of the large number of important questions that the researcher needs to answer. before selecting a model.

For mixed models P2-r and P3-r, the standard error is a function of the number of replicates and of the variance components.

Table 3 shows the results of different comparisons between effects. Since analyses P1-f, P2-f, and P3-f all consider treatment as a fixed factor, all of them produce the same results, while analyses P2-r and P3-r produce different results between them, and between them and the first models. The results presented correspond to the comparisons between Treatments C1 and C2 (checks), G3 and G4 (unreplicated treatments), and between Treatments C1 and G3 (check– unreplicated treatment). In fixed models, the standard error for a comparison is a function of the number of replicates; checks can be compared with each other more precisely than one vs. another unreplicated treatment.

In some cases, the check effect should be considered as a fixed effect while the genotypic effect should be considered as random. One situation in which it is difficult to consider the check effect as random is when a low number of checks are used, that is, two to three checks. In this case, the estimation of the variance between checks is a poor estimate due to the few degrees of freedom associated with it.

SAS has some limitations with models in which a nested effect is a random effect. This produces the large standard error to compare a check with an unreplicated treatment in model P3-r. SAS also reports different degrees of freedom for the denominator regarding the effects included in the estimation. The option DDF = 4 was included to always obtain the same degree of freedom. These problems occur because of the way SAS codes the levels of a nested factor and its capabilities to build the vector of coefficients for an estimation. The option "e" after the slash in the ESTIMATE instruction, print the levels of the factor and the coefficients used in the ESTIMATE statement.

To estimate the difference between C1 and G3, we used the following instruction,

```
estimate "C1-G3" d2 1 -1   | t(d2) 1  0 -1 /e group 1 1;.
```

In it, d2 1 -1 compares the two levels of d2; t(d2) 1 0 -1 compares the effects of C1 and G3 while each of the instruction "group 1 1" after the slash is, for each level of d2, the weight of the coefficient of the random effects, that is, t(d2) 1 0 -1. In Table 4, it can be seen that t(d2) = (G3,1) and t(d2) = (C1,2) are included in the estimation because SAS applies the coefficients for t(d2) at each level of d2. These last two effects (level

TABLE 4. Coefficients to estimate the comparison between a check with an un-replicated treatment.

	Coefficients for estimate C1-G3			
Effect	t	d2	Group	Row1
INTERCEPT				
D2		1		1
D2		2		-1
T(D2)	C1	1	D2 1	1
T(D2)	C2	1	D2 1	
T(D2)	G3	2	D2 1	-1
T(D2)	G4	2	D2 1	
T(D2)	G5	2	D2 1	
T(D2)	G6	2	D2 1	
T(D2)	G7	2	D2 1	
T(D2)	G8	2	D2 1	
T(D2)	C1	1	D2 2	1
T(D2)	C2	1	D2 2	
T(D2)	G3	2	D2 2	-1
T(D2)	G4	2	D2 2	
T(D2)	G5	2	D2 2	
T(D2)	G6	2	D2 2	
T(D2)	G7	2	D2 2	
T(D2)	G8	2	D2 2	

combination) are not present in the data, but SAS included them in the model with effects of zero. Therefore, the estimation of the BLUP is right; however, the standard error is not.

To solve those problems, Piepho et al. (2006) proposed the use of a switch variable. This variable is a dummy variable with two levels; one for the level of interest and the other for the other levels in the nested factor. It is necessary to include as many switch variables as random effects are nested. In the case of the model P3-r, two switch variables are needed, one for d2 = 1 and the other for d2 = 2.

The following code shows how this approach works. First the new variables are created.

```
data a;
set a;
if d2 = 1 then switch1 = 1; else switch1 = 0;
switch2 = 1-switch1;
run;
PROC GLIMMIX;
class t d2;
model y = d2/noint solution;
random switch1*t switch2*t/s;
lsmeans d2/diff;
estimate "Ch" | switch2*t 1-1;
estimate "UnT" | switch1*t 0 0 1-1;
estimate "Ch-UnT" d2 1-1 | switch2*t 1       switch1*t 0 0-1/e;
estimate "Ch1" d2 1    0 | switch2*t 1;
estimate "Ch2" d2 1    0 | switch2*t 0 1;
estimate "UnT1"        d2 0    1 | switch1*t 0 0 1;
estimate "UnT2"        d2 0    1 | switch1*t 0 0 0 1;
covtest "switch1*t" general 1 0; covtest "switch2*t" general 0 1;
run;
```

Main results of this code are presented below.

Covariance parameter estimates		
Cov parm	**Estimate**	**Standard error**
switch1*t	17.5333	12.6726
switch2*t	13.4444	20.1208
Residual	2.3333	1.6499

Solutions for fixed effects						
Effect	**d2**	**Estimate**	**Standard error**	**DF**	***t* Value**	**Pr > |*t*|**
d2	1	25.0000	2.6667	4	9.37	0.0007
d2	2	31.3333	1.8196	4	17.22	< 0.0001

Type III Tests of Fixed Effects				
Effect	**Num DF**	**Den DF**	***F* Value**	***P* > *F***
d2	2	4	192.20	0.0001

Coefficients for estimate Ch-UnT			
Effect	***t***	**d2**	**Row1**
d2		1	1
d2		2	-1
switch1*t	C1		
switch1*t	C2		
switch1*t	G3		-1
switch1*t	G4		
switch1*t	G5		
switch1*t	G6		
switch1*t	G7		
switch1*t	G8		
switch2*t	C1		1
switch2*t	C2		
switch2*t	G3		
switch2*t	G4		
switch2*t	G5		
switch2*t	G6		
switch2*t	G7		
switch2*t	G8		

Estimates					
Label	**Estimate**	**Standard error**	**DF**	***t* value**	***P* > |*t*|**
Ch	-5.0417	1.2126	4	-4.16	0.0142
UnT	-3.5302	2.0294	4	-1.74	0.1569
Ch-UnT	-5.9123	1.6916	4	-3.50	0.0250
Ch1	22.4792	0.8698	4	25.84	< 0.0001
Ch2	27.5208	0.8698	4	31.64	< 0.0001
UnT1	28.3915	1.4508	4	19.57	< 0.0001
UnT2	31.9217	1.4508	4	22.00	< 0.0001

d2 Least squares means					
d2	**Estimate**	**Standard error**	**DF**	***t* value**	**$P > \|t\|$**
1	25.0000	2.6667	4	9.37	0.0007
2	31.3333	1.8196	4	17.22	< 0.0001

Tests of covariance parameters based on the restricted likelihood					
Label	**DF**	**-2 Res Log Like**	**ChiSq**	**$P >$ ChiSq**	**Note**
switch1*t	1	58.1353	4.09	0.0216	MI
switch2*t	1	58.6165	4.57	0.0163	MI

This approach avoids the use of the DDF options as well the option group, and it produces correct results.

Example 2: Augmented Design with Systematic Checks in a Latin-Square Arrangement

This experiment was designed to phenotype 576 accessions from CIMMYT's Germplasm Bank under the MasAgro Biodiversity project financed by the Mexican Ministry of Agriculture. Due to a lack of seed for planting replicates and the large number of accessions, we used an augmented design. In this example, we will discuss how to create a balanced and useful unreplicated experiment using a Latin-square design. The analysis of the experiment and its results are discussed in Chapter 12 (Burgueño, 2018) on Spatial Analysis.

An experimental design has at least two main goals aimed at optimizing statistical tests comparing treatments: one is to obtain a good estimate of the mean of the treatments (BLUE or BLUP) and the second is to make a good estimation of the experimental error. Augmented designs have an additional challenge: to adjust the means of unreplicated treatments for effects due to spatial variability rather than being due to treatment effects. Spatial variability is the major source of experimental error in a field design. It is controlled by experimental design and an accurate and unbiased estimate can be obtained by using replication and proper randomization. Mathematical and statistical models usually can be used to estimate spatial variability. Due to the fixed patterns of these models, replicated treatments allocated randomly will not necessarily provide the best adjustment. Geostatistics research has shown that a regular pattern of sampling points is a better source of information than randomized points. Based on these concepts, it is thought that systematic checks in the field can improve the estimation of spatial variability and is one way of justifying the use of systematic designs (Cressie, 1991).

Our experimental field had 36 ranges (perpendicular to sowing direction) and 18 rows (parallel with sowing direction) which resulted in a total of 648 plots. Due to the characteristics of the field and the environmental conditions, together with limited resources (seed, work force), we considered that 72 (11.1%) plots were sufficient for checks. Therefore, we had 576 plots remaining that we could use for 576 accessions. Seventy-two is convenient for the number of checks because it allowed us to allocate two checks in each range and four in each row. The 72 check plots were distributed systematically in the field starting in the left bottom corner and the rest in a knight's chess move pattern. Figure 1 shows the pattern of the checks in the field.

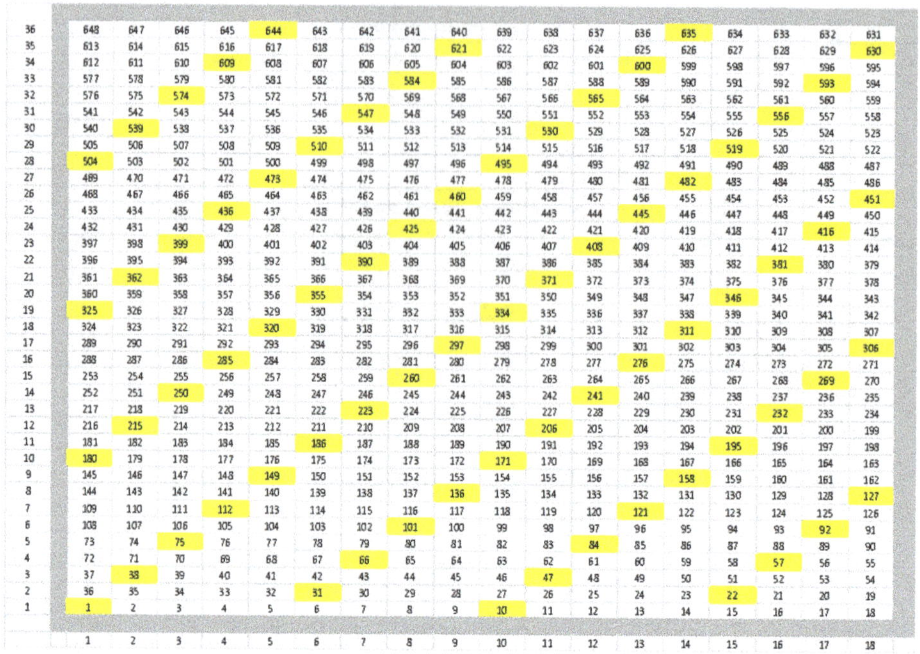

FIG. 1. Pattern of checks in a systematic augmented design comprised of 36 ranges and 18 rows. Checks are marked in yellow. Numbers inside the field are plot numbers.

We used four checks in this experiment: two standard checks plus a drought-resistant and a drought-susceptible genotype. The experiment was conducted under well-watered and drought conditions. To randomize the four checks in the check plots, we considered the two rows with check plots in the same range. For example, in Rows 1 and 10, checks plots are in Ranges 1, 10, 19, and 28, while in Rows 6 and 15, check plots are in Ranges 2, 11, 20, and 29 (see Fig. 2). In this way, we created an incomplete 4 × 2 Latin-square design (Fig. 2). The procedure followed was: (i) sorting

FIG. 2. First 12 ranges and four rows of the field layout after sorting by checks position in range and row. Numbers inside a cell are plot numbers. Different color cells represent the four different checks.

36	648	647	646	645	644	643	642	641	640	639	638	637	636	635	634	633	632	631
35	613	614	615	616	617	618	619	620	621	622	623	624	625	626	627	628	629	630
34	612	611	610	609	608	607	606	605	604	603	602	601	600	599	598	597	596	595
33	577	578	579	580	581	582	583	584	585	586	587	588	589	590	591	592	593	594
32	576	575	574	573	572	571	570	569	568	567	566	565	564	563	562	561	560	559
31	541	542	543	544	545	546	547	548	549	550	551	552	553	554	555	556	557	558
30	540	539	538	537	536	535	534	533	532	531	530	529	528	527	526	525	524	523
29	505	506	507	508	509	510	511	512	513	514	515	516	517	518	519	520	521	522
28	504	503	502	501	500	499	498	497	496	495	494	493	492	491	490	489	488	487
27	469	470	471	472	473	474	475	476	477	478	479	480	481	482	483	484	485	486
26	468	467	466	465	464	463	462	461	460	459	458	457	456	455	454	453	452	451
25	433	434	435	436	437	438	439	440	441	442	443	444	445	446	447	448	449	450
24	432	431	430	429	428	427	426	425	424	423	422	421	420	419	418	417	416	415
23	397	398	399	400	401	402	403	404	405	406	407	408	409	410	411	412	413	414
22	396	395	394	393	392	391	390	389	388	387	386	385	384	383	382	381	380	379
21	361	362	363	364	365	366	367	368	369	370	371	372	373	374	375	376	377	378
20	360	359	358	357	356	355	354	353	352	351	350	349	348	347	346	345	344	343
19	325	326	327	328	329	330	331	332	333	334	335	336	337	338	339	340	341	342
18	324	323	322	321	320	319	318	317	316	315	314	313	312	311	310	309	308	307
17	289	290	291	292	293	294	295	296	297	298	299	300	301	302	303	304	305	306
16	288	287	286	285	284	283	282	281	280	279	278	277	276	275	274	273	272	271
15	253	254	255	256	257	258	259	260	261	262	263	264	265	266	267	268	269	270
14	252	251	250	249	248	247	246	245	244	243	242	241	240	239	238	237	236	235
13	217	218	219	220	221	222	223	224	225	226	227	228	229	230	231	232	233	234
12	216	215	214	213	212	211	210	209	208	207	206	205	204	203	202	201	200	199
11	181	182	183	184	185	186	187	188	189	190	191	192	193	194	195	196	197	198
10	180	179	178	177	176	175	174	173	172	171	170	169	168	167	166	165	164	163
9	145	146	147	148	149	150	151	152	153	154	155	156	157	158	159	160	161	162
8	144	143	142	141	140	139	138	137	136	135	134	133	132	131	130	129	128	127
7	109	110	111	112	113	114	115	116	117	118	119	120	121	122	123	124	125	126
6	108	107	106	105	104	103	102	101	100	99	98	97	96	95	94	93	92	91
5	73	74	75	76	77	78	79	80	81	82	83	84	85	86	87	88	89	90
4	72	71	70	69	68	67	66	65	64	63	62	61	60	59	58	57	56	55
3	37	38	39	40	41	42	43	44	45	46	47	48	49	50	51	52	53	54
2	36	35	34	33	32	31	30	29	28	27	26	25	24	23	22	21	20	19
1	1	2	3	4	5	6	7	8	9	10	11	12	13	14	15	16	17	18
	1	2	3	4	5	6	7	8	9	10	11	12	13	14	15	16	17	18

FIG. 3. Final layout of an augmented design with systematic check plots in a Latin-square arrangement. Different colors represent the four different checks.

the field layout by the position of the check plots so that the check plots are side by side forming a square; (ii) randomizing five 4 × 4 Latin-squares and deleting the last two columns of the last square to have 18 columns (the same number of rows as in the field) with four rows; and (iii) assigning each column of the Latin-square design to each row (with check plots) in the field (Fig. 2). Finally the row and ranges are sorted in their original order.

By following this procedure, we ensured that the number of replicates of each check was almost the same, that we had each of the four different checks in each field row, that there were two different checks in each range in the field, and that the order of checks in the field had some balance. The final results of the design can be seen in Fig. 3. One can see intuitively that this layout of our checks was better in accounting for spatial variability than a random layout.

Example 3: Building and Analyzing an Augmented Design Using the Augmented Complete Block Design in R (Rodríguez et al., 2017) (ACBD-R)

With this example, we introduce the ACBD-R (program developed in the Biometrics and Statistics Unit of CIMMYT. Although there are commercial and free software products to analyze augmented designs, this development was done with the idea to have a free, easy, and simple way to create and analyze augmented complete block designs. It is also capable of generating Augmented Sudoku designs.

Augmented designs can be generated easily by first generating a standard design, then adding plots with the unreplicated treatments and finally randomizing replicated and unreplicated treatments within blocks. To our knowledge, just one R library (DiGGer) and two web platforms (Abhishek et al., 2004; Morejón Rivera et al., 2016) are available specifically designed to create and to analyze augmented designs.

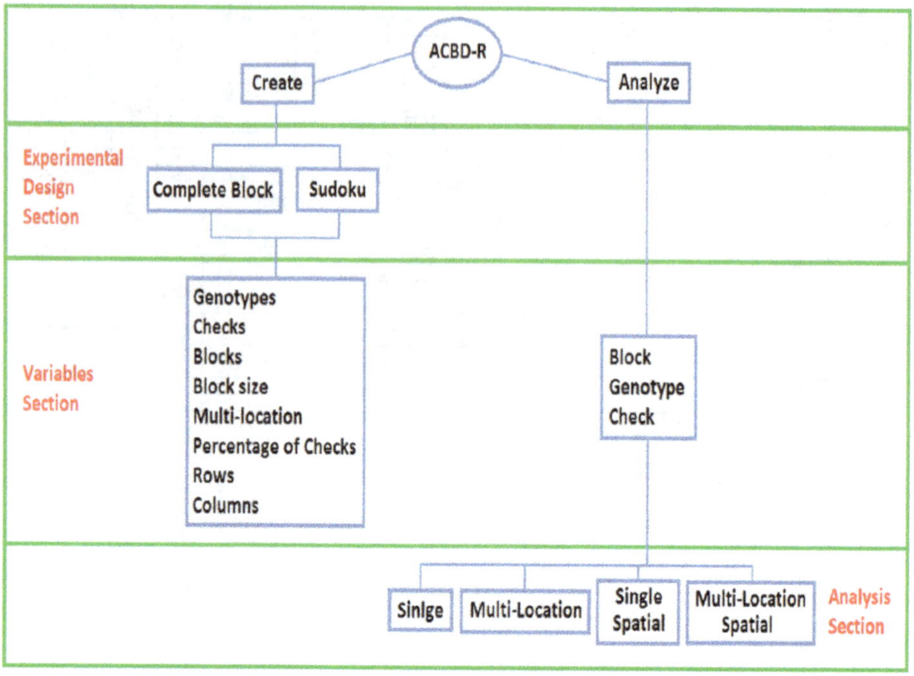

FIG. 4. Flow diagram showing the logical process that ACBD-R follows to generate and analyze augmented designs.

DiGGer was developed to generate optimized designs to be analyzed with a spatial model, but it requires a priori estimation of the expected correlation, which is not usually known. For SAS users, a good reference is the paper of Piepho (2015) which describes how to optimize different experimental designs including augmented and partially replicated experimental designs. A word of caution if using the SAS procedure OPTEX to optimize the design. This procedure does not check for balance and it usually produces an experimental design in which one or more treatments is replicated twice in one block and is not included in another block.

After opening the ACBD-R program, three options are offered, Create, Analyze and Help to create a design, and analyze a design. There is also an option for Help. Figure 4 shows the flow diagram for creating an experimental design or for analyzing data. The program is able to generate and analyze multienvironment experiments.

The following parameters define an augmented complete block design: i) the number of unreplicated treatments, ii) the number of checks per block, iii) the number of blocks, and iv) the block size, assuming the same block size for all blocks. Three of these four parameters must be defined in the program to generate a design. For an augmented Sudoku design, in addition to the number of replicated and unreplicated treatments, it is necessary to define the number of rows and columns per square. Also, instead of specifying the number of checks, the user must specify the percentage of desired checks. A summary of the available design is shown by the program. Field size (field features) can also be defined to introduce the spatial row column location of the plots in the field (Fig. 5 and 6).

In Fig. 7, the window to introduce the parameters for the analysis of an augmented complete block design is shown. After selecting the data file in .csv format

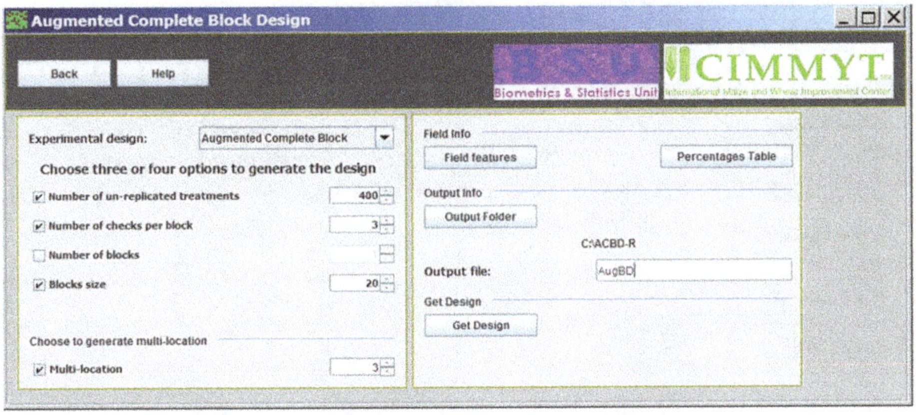

FIG. 5. Window options to generate an augmented complete block design.

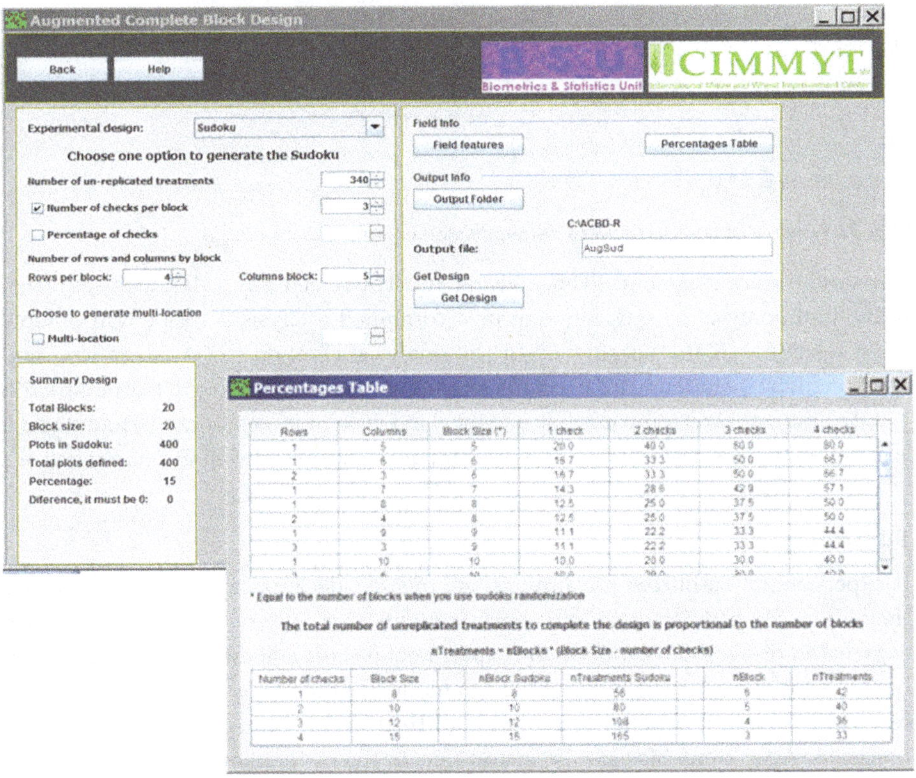

FIG. 6. Window options to generate an augmented Sudoku design.

and clicking on Open file icon, the file is displayed at the bottom of the window. The file format is the standard format with one row per observation (plot) and one column per each design identification (location, plot, block, entry, check, row, and column) and then the traits (GYG, grain yield; AD, anthesis date; and ASI, flowering interval). There is also a box to select if there is more than one environment and another box to select for a spatial or nonspatial analysis. If spatial analysis is selected, then the variables that define the grid in the field must be chosen. At the end, the

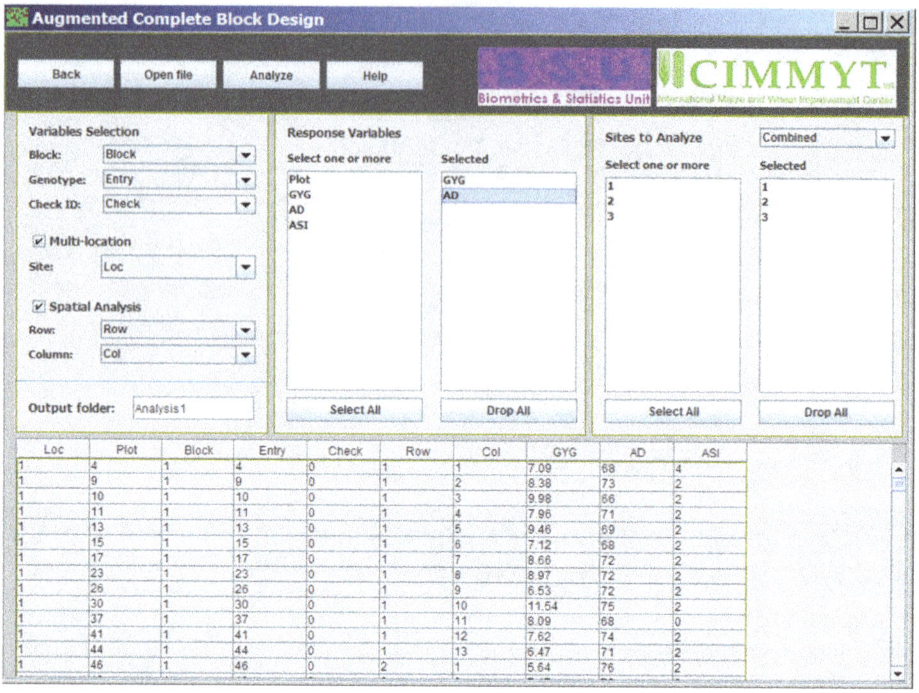

FIG. 7. Window options to analyze an augmented complete block design.

traits and the environments to be analyzed must be selected as well if the user wants individual analysis by environment or a combined analysis across environments. After selecting all the parameters of the analysis, click on the Analyze icon and results will be included in an Excel file located in the selected Output (Appendix A2).

The program is accompanied by a manual in which more details can be found and it is under continuous development. The program can be downloaded from data.cimmyt.org.

Summary

If experimental resources are limited, augmented designs can be useful because they allow for comparisons between treatments that may not be replicated. Also, augmented designs can estimate the experimental error and adjust unreplicated values if a spatial model has been applied to model spatial variability. However, when resources permit, it is better to replicate all treatments because doing so improves precision. Also, some bias can be introduced in the estimation of the means of unreplicated treatments when spatial variability is modeled for unreplicated treatments. An unreplicated treatment is under the effect of the plot to which it was assigned, and although the statistical model is able to model the spatial variability, some effects may not be estimated and bias can be introduced. Additionally, new problems in planning the design are introduced; distribution of replicated and unreplicated plots in the field, number of replicated treatments, and the number of replicates for replicated treatments. The question to answer before using an augmented design (and any other experimental design) is: based on the objectives of our experiment, do we expect our design to provide sufficient precision and confidence in the results? If the answer is yes, then you can save resources and have good results from your experiment. If the answer is no, then it may be more appropriate to increase costs and use replication.

Key Learning Points

- Although augmented designs reduce precision compared with experiments in which all treatments are replicated, augmented designs can be useful when resources are limited.
- A useful augmented design should provide an unbiased estimate of experimental error which can be used to conduct valid statistical tests, and adjust the values for unreplicated treatments through spatial modeling.
- Usually, replicated and unreplicated treatments must be managed differently in the analytical model.
- A model-based approach can be used for modeling spatial variability in systematic designs producing valid tests.

Review Questions

1. Define augmented design.
2. What is the relationship between an augmented design and classical designs like randomized complete block or incomplete block designs?
3. Why does an augmented design have a lower precision than a complete block design (if conducted in the same experimental field)?
4. Which analytical methods can adjust the values for unreplicated treatments?
5. Is it possible in a systematic design to estimate experimental error and conduct valid statistical tests?
6. What are the advantages and disadvantages of an augmented design?

Exercises

1. Download the ACBD-R program from data.cimmyt.org and run the analysis of the examples provided with the program using it. Run the analysis using the codes of SAS presented in Example 1 and compare the results.
2. Run the analysis of Example 1 in this chapter using the program ACBD-R.

References

Abhishek, R., Prasad, R and Gupta, V.K. 2004. Computer aided construction and analysis of augmented designs. Journal of Indian Society of Agricultural Statistics, 57 (Special volume), 320-344.

Aguade, A. 2013. Construction and analysis of augmented and modified augmented designs. Lambert Academic Publishing, Saarbrücken, Germany.

Besag, J.E., and R.A. Kempton. 1986. Statistical analysis of field experiments using neighbouring plots. Biometrics 42:231–251. doi:10.2307/2531047

Box, G.E.P., and N.R. Draper. 1987. Empirical model building and response surfaces. John Wiley & Sons, New York.

Box, G.E.P., J.S. Hunter, and W.G. Hunter. 2005. Statistics for experimenters. 2nd ed., Wiley Interscience, Hoboken, NJ.

Burgueño, J. 2005. Diseños experimentales no repetidos. Tesis de Doctorado, Colegio de Postgraduados. Montecillos, Estado de México.

Burgueño, J. 2018. Spatial Analysis of field experiments. In: B. Glaz and K.M. Yeater, editors, Applied statistics in the agricultural, biological, and environmental sciences. ASA, CSSA, SSSA, Madison, WI.

Casler, M.D. 2015. Fundamentals of experimental design: Guidelines for designing successful experiments. Agron. J. 107:692–705. doi:10.2134/agronj2013.0114

Cressie, N. 1991. Statistics for spatial data. John Wiley & Sons, New York.

Cullis, B.R., A.B. Smith, and N.E. Coombes. 2006. On the design of early generation variety trials with correlated data. J. Agric. Biol. Environ. Stat. 11(4):381–393. doi:10.1198/108571106X154443

Eshetie, A. 2011. Construction and analysis of augmented and modified augmented designs. Master of Science Thesis. Addis Ababa University, School of Graduate Students, College of Natural Sciences, Department of Statistics, Addis Ababa, Ethiopia.

Federer, W.T. 1956. Augmented (or hoonuiaku) designs. Hawaii. Plant. Rec. LV(2):191–208.

Federer, W.T., R.C. Nair, and D. Raghavarao. 1975. Some augmented row-column designs. Biometrics 31:361–373. doi:10.2307/2529426

Federer, W.T., and D. Raghavarao. 1975. On augmented designs. Biometrics 31(1):29–35. doi:10.2307/2529707

Federer, W.T. 1993. Statistical design and analysis for intercropping experiments. Volume I. Two Crops. Springer-Verlag, Heidelberg, Berlin. doi:10.1007/978-1-4613-9305-4

Federer, W.T. 1998. Recovery of interblock, intergradient, and intervariety information in incomplete block and lattice rectangle designed experiments. Biometrics 54:471–481. doi:10.2307/3109756

Fisher, R.A. 1925. Statistical methods for research workers. Oliver and Boyd, Edinburgh, U.K.

Keselman, H.J. 2015. Per family or familywise type I error control: "Eethet, eyether, neether, nyther, let's call the whole thing off! J. Mod. Appl. Stat. Methods 14(1):24–37. doi:10.22237/jmasm/1430453100

Miguez, F., S. Archontoulis, H. Dokoohaki. 2018. Nonlinear regression models and applications. In: B. Glaz and K.M. Yeater, Applied statistics in the agricultural, biological, and environmental sciences. ASA, CSSA, SSSA, Madison, WI.

Montgomery, D.C. 2011. Design and analysis of experiments. 8th ed. Wiley & Sons, New York.

Moehring, J., E.R. Williams, and H.P. Piepho. 2014. Efficiency of augmented p-rep designs in multi-environmental trials. Theor. Appl. Genet. 127(5):1049–1060. doi:10.1007/s00122-014-2278-y

Morejón Rivera, R., Cámara, F. A., Jiménez, D. E. and Díaz, S. H. 2016. SISDAM: Web application for processing data according to a modified augmented design. Cultivos tropicales, Vol. 37(3):153-164.

Laird, R.J. 1987. Técnicas de campo para experimentos con fertilizantes. Folleto Técnico No. 1. Instituto Nacional de Investigaciones Forestales y Agropecuarias. Tecomán, Colima.

Lin, C.S., and G. Poushinsky. 1983. A modified augmented design for an early stage of plant selection involving a large number of test lines without replication. Biometrics 39(3):553–561. doi:10.2307/2531083

Papadakis, J.S. 1937. Méthode statistique pour des expériences sur champ. Bulletin de l'Institut d'Amélioration des Plantes à Salonique. Thessalonique 23.

Payne, R.W. 2006. New and traditional methods for the analysis of unreplicated experiments. Crop Sci. 46:2476–2481. doi:10.2135/cropsci2006.04.0273

Piepho, H.P., E.R. Williams, and M. Fleck. 2006. A note on the analysis of designed experiments with complex treatment structure. HortScience 41(2):446–452.

Piepho, H.P., C. Richter, and E.R. Williams. 2008. Nearest neighbor adjustment and linear variance models in plant breeding trials. Biom. J. 50(2):164–189.

Piepho, H.P. 2015. Generating efficient designs for comparative experiments using SAS procedure OPTEX. Communications in Biometry and Crop Science 10:96–114.

Richter, C. and H.P. Piepho. 2018. Linear regression techniques. In: B. Glaz and K.M. Yeater, Applied statistics in the agricultural, biological, and environmental sciences. ASA, CSSA, SSSA, Madison, WI.

Rivera, G. 1975. Métodos alternativos para comparación de un número grande de variedades usando funciones de tendencia. Tesis de Maestria, Colegio de Postgraduados. Montecillos, Estado de México, Mexico.

Robinson, G.K. 1991. That BLUP is a good thing: The estimation of random effects. Stat. Sci. 6(1):15–32. doi:10.1214/ss/1177011926

Rodríguez, F., G. Alvarado, A. Pacheco, and J. Burgueño. 2017. ACBD-R. Augmented complete block design with R for windows. Version 3.0, International Maize and Wheat Improvement Center, Mexico-Veracruz, Mexico. http://hdl.handle.net/11529/10855 (verified 5 Jan. 2017).

Saba, M.F.A., and B.K. Sinha. 2014. SuDoKu as an experimental design- beyond the traditional latin square design. Statistics and Applications. 12(1&2):15–20.

Sahagun-Castellanos, J. 1985. Efficiency of augmented designs for selection. Paper 7882. Retrospective Theses and Dissertations. Iowa State University, Ames, IA.

SAS Institute Inc. 2015. SAS/STAT 14.1 User's Guide. SAS Institute Inc., Cary, NC.

Ticona-Benavente, C.A., and D.F. da Silva Filho. 2015. Comparison of BLUE and BLUP/REML in the selection of clones and families of potato (Solanum tuberosum). Genet. Mol. Res. 14(4):18421–18430. doi:10.4238/2015.December.23.30

Turrent-Fernández, A., and R.J. Laird. 1975. La matriz experimental Plan Puebla, para ensayos sobre prácticas de producción de cultivos. Agrociencia 19:117–143.

Williams, E., H.P. Piepho, and D. Whitaker. 2011. Augmented p-rep designs. Biom. J. 53(1):19–27. doi:10.1002/bimj.201000102

CHAPTER 14: MULTIVARIATE METHODS FOR AGRICULTURAL RESEARCH

Kathleen M. Yeater* and María B. Villamil

Agronomic research often involves measurement and collection of multiple response variables in an effort to understand the more complex nature of the system being studied. Multivariate statistical methods encompass the simultaneous analysis of all variables measured on each experimental or sampling unit. Many agronomic research systems studied are, by their very nature, multivariate; however, most analyses reported are univariate (i.e., analysis of one response at a time). The objective of this chapter is to use a hands-on approach to familiarize the researcher with a set of common applications of multivariate methods and techniques for the agronomic sciences: principal components analysis, multiple regression, and discriminant analysis. We use an agronomic data set, a subset of the data collected for the "Yield Challenge" program, established by the Illinois Soybean Association in collaboration with researchers from the University of Illinois in 2010. We provide the reader with a field guide to serve as a taxonomical key, a list of relevant references for each technique, a road map to our work, as well as R code to follow as we explore the data to assess quality and suitability for multivariate analyses. We also provide SAS code. The chapter illustrates how multivariate methods can capture the concept of variability to better understand complex systems. Important considerations along with advantages and disadvantages of each multivariate tool and their corresponding research questions are examined.

Most data collected in agriculture are multivariate. To fully understand such data, we need to analyze the variables simultaneously. Multivariate statistics can be applied to a broad variety of research questions in agriculture, providing essential tools for detailed description of data, identification of patterns, enhanced prediction of events, as well as empirical testing of complex theoretical ideas. As our research questions grow in complexity to find the answers within agriculturally managed systems, the availability of statistical software that can handle the intricacy of large multivariate

Abbreviations: CART, classification and regression tree; CDA, canonical discriminant analysis; CEC, cation-exchange capacity; Comp., principal component; CRD, crop reporting district; DA, discriminant analysis; DV, dependent variable; GLM, general linear model; IV, independent variable; LD, linear discriminant; MANOVA, multivariate analysis of variance; MR, multiple regression; NASS, National Agriculture Statistics Service; PCA, principal components analysis; SCN, soybean cyst nematode; SOM, soil organic matter; YC, yield challenge.

Kathleen M. Yeater, USDA-ARS Plains Area, Office of the Director, 2150 Centre Ave., Blg. D, Ste. 300, Fort Collins, CO, 80526. María B. Villamil, University of Illinois, Department of Crop Sciences, N-323 Turner Hall, 1102 S. Goodwin Ave., Urbana, IL, 61801 (villamil@illinois.edu). *Corresponding author (kathleen.yeater@ars.usda.gov).

doi:10.2134/appliedstatistics.2015.0083

Applied Statistics in Agricultural, Biological, and Environmental Sciences
Barry Glaz and Kathleen M. Yeater, editors

data is ever-growing, and as such, the applications of multivariate statistics become more suitable to agronomic and biological research.

The goal of this chapter is to use a hands-on approach that provides a conceptual introduction and familiarizes the reader with a set of multivariate procedures directly applicable in the agronomic sciences. We use real agronomic data to demonstrate three specific multivariate methods: principal components analysis (PCA), multiple regression (MR), and discriminant analysis (DA). Underlying statistical theory is not addressed, but remains uncompromised with our approach. Our experience working with students and researchers has taught us that the hands-on approach allows for a level of understanding of the underlying theory that makes these complex mathematical concepts accessible and even enjoyable without the need to refer to written formulas. Students and practitioners interested in the theoretical background of multivariate methodology are referred to the books by Sharma (1996), Johnson (1998), Johnson and Wichern (2002), and Tabachnick and Fidell (2012), among others, and to the open access article of Yeater et al. (2015) for a recent and brief introduction to the subject.

We created a field guide of the relevant methodology (Table 1) and collected examples of publications in agricultural journals for each of the methodologies listed, along with main procedures and functions for SAS (SAS Institute) and R, two mainstream statistical software programs that enable users to obtain the output of these methodologies (Table 2). Once we start using our specific example data, we encourage the reader to follow along using our road map (Fig. 1) for the multivariate exploration and analysis of our example data set. The data set itself is available for download as an online supplementary file. We use R coding as we move through the chapter, and provide relevant SAS code.

Questions and Methods: A Field Guide

Let us be clear: statistical analysis can never replace a good research question and a solid experiment. No matter how complicated the analysis, if there is no good scientific inquiry guiding the exploration, there will never be a good scientifically sound story. Yet it is helpful to recognize that the analysis of multivariate data involves a stage of data exploration (or data mining). Data exploration allows us to familiarize ourselves with the data we have so painstakingly gathered and also lets the data "talk," allowing it to show what additional patterns and nonrandom structures and relationships deserve our attention and might justify further research and explanation. We are not referring to a meaningless "fishing expedition," but rather to a close understanding of the data, its structure, and its potential as a source of scientific answers and further questions. This stage is characterized by an emphasis on the visualization of data and the lack of any associated stochastic model, so questions on significance at this stage are rarely relevant (Everitt, 2005). Once we have well-defined hypotheses, multivariate techniques allow us to extract the relevant information contained in the data set to statistically test our hypotheses (Sharma, 1996). The take-home message is that you need to be flexible and pragmatic in your approach and gather sufficient knowledge and skills to correctly apply the selected tools to

TABLE 1. A field guide of multivariate techniques in relation to research questions and goals when using the most common types of variables found in agricultural data sets.†

Research question	Goal	Dependent variables	Independent variables	Considerations	Statistical Technique	Desired output?
What is the structure of the data set?	Data reduction; remove redundancy	None or ≥1 latent	>2 cont. observed	Empirical	PCA	Create linear combinations of observed IV that might or might not represent latent variable(s)
	Uncover theoretical constructs	>2 cont. observed	≥1 latent	Theoretical	FA	Create linear combinations of observed variables to represent latent variables
		≥1 cont. observed and/or latent	≥1 cont. observed and/or latent	Theoretical Do not assume any functionv		Create linear combinations of observed and latent IVs to represent observed and latent DVs
	Explore relationships among variables	1 cont.	≥2 cont.	Assume linear relation	MR	Create a linear combination of IVs to predict DV. Explain DV variation based on IVs
		1 dich.	≥2 cont. and/ or cat.	Assume logistic relation	LR	Create a linear combination of the log of the odds to explain/ predict group membership
		1 dich.	≥2 cont. and/ or cat.	Do not assume any function	CART	Create a decision tree
		≥2 cont.	≥2 cont.	Assume linear relation between 2 sets of observations	CCA	Maximally correlate a linear combination of IVs with a linear combination of DVs
		≥2 cat.	≥2 cont. and/ or cat.	Do not assume linear relation or symmetry between 2 data sets	CCPA	Explain the distribution of observations in terms of a collection of variables (i.e. species in environments)
		none	≥2 cont.	Homogenous no pre-specified groups based on distance	CA	Identify a set of groups that minimizes within group variation and maximizes between group variation
How can I segregate pre-specified groups?	Identify significant differences among groups	1 disc. with ≥2 groups	≥2 cont.	Parametric tests	MANOVA DA	Determine reliability of group mean differences Create linear combinations of IVs that maximize group differences
	Describe major differences among groups	1 disc. with ≥2 groups	≥2 cont.	Assume linear discriminant function	DA	Identify the linear combinations of IVs that maximize group differences
		1 dich. only	≥2 cont.	Assume logistic discriminant function	LR	Identify the linear combination of the log of the odds that defines each group
		1 disc. with ≥2 groups	≥2 cont. and/ or cat.	Do not assume any function	CART	Identify IVs that contribute to a decision rule creating tree branches
	Predict group membership of future observations	1 disc. with ≥2 groups	≥2 cont.	Linear classification function	DA	Create linear combinations of IVs that maximize group differences
		1 dich. only	≥2 cont.	Logistic classification function	LR	Create a linear combination of the log of the odds of group membership
		1 disc. with ≥2 groups	≥2 cont. and/ or cat.	Decision tree classifier	CART	Decision tree for group membership

† Abbreviations: CA, cluster analysis; CART, classification and regression trees; cat., categorical variable; CCA, canonical correlation analysis; CCPA, canonical correspondence analysis; cont., continuous variable; DA, discriminant analysis; dich., dichotomous variable; disc., discrete variable; DV, dependent variable; FA, factor analysis; IV, independent variable; LR, logistic regression; MANOVA, multivariate analysis of variance; MR, multiple regression; PCA, principal component analysis; SEM, structural equation modeling.

TABLE 2. Previously listed abbreviations of statistical techniques with their full name, main SAS procedures, and R functions and packages along with several references for each technique published in agricultural related journals.

Statistical technique	Abbrev.	SAS procedures	R function {package}	Examples of publications from agriculture related journals
Principal component analysis	PCA	proc princomp	princomp{stats} prcomp{stats}	Martin et al., 2007; Sawchik and Mallarino, 2007; Yeater et al., 2015
Factor analysis	FA	proc factor	factanal{stats}	Shukla et al., 2006; So et al. 2009; Villamil et al., 2012a
Structural equation modelling	SEM	proc calis;	sem{sem}	Sorice et al., 2012; Smith et al., 2014; Kane et al. 2015
Multiple linear regression	MR	proc reg; proc glm	lm{stats}	Hao and Kravchenko, 2007; Westphal et al. 2009; Jacobson et al., 2011
Logistic regression	LR	proc logistic	glm{stats} with family = binomial	Pike et al., 2010; Villamil et al., 2012a; Schutte et al., 2014
Classification and regression trees	CART	proc tree	rpart{stats} ctree{party}	Roel and Plant, 2004; Williams et al., 2009; Zheng et al., 2009
Canonical correlation analysis	CCA	proc cancor	cancor{stats}	Tan et al., 2003; Martin et al. 2005; Liu et al., 2008
Canonical correspondence analysis	CCPA	proc corresp	corresp{MASS}; mca{MASS}	Didden et al. 1994; Albrecht 2003; Smukler et al., 2008
Cluster analysis	CA	proc fastclus; proc varclus	kmeans{MASS} hclust{MASS}	Ping et al., 2005; Goidts et al., 2009; DeDecker et al., 2014
Multivariate ANOVA	MANOVA	proc glm with manova statement	lm{stats}; manova{car}	Ping et al., 2005; Villamil et al. 2008; Taylor and Whelan, 2011; Yeater et al., 2015
Discriminant analysis	DA	proc discrim; proc candisc	lda{MASS}	Yeater et al., 2004; Villamil et al. 2008; Senthilkumar et al. 2009; Yeater et al., 2015

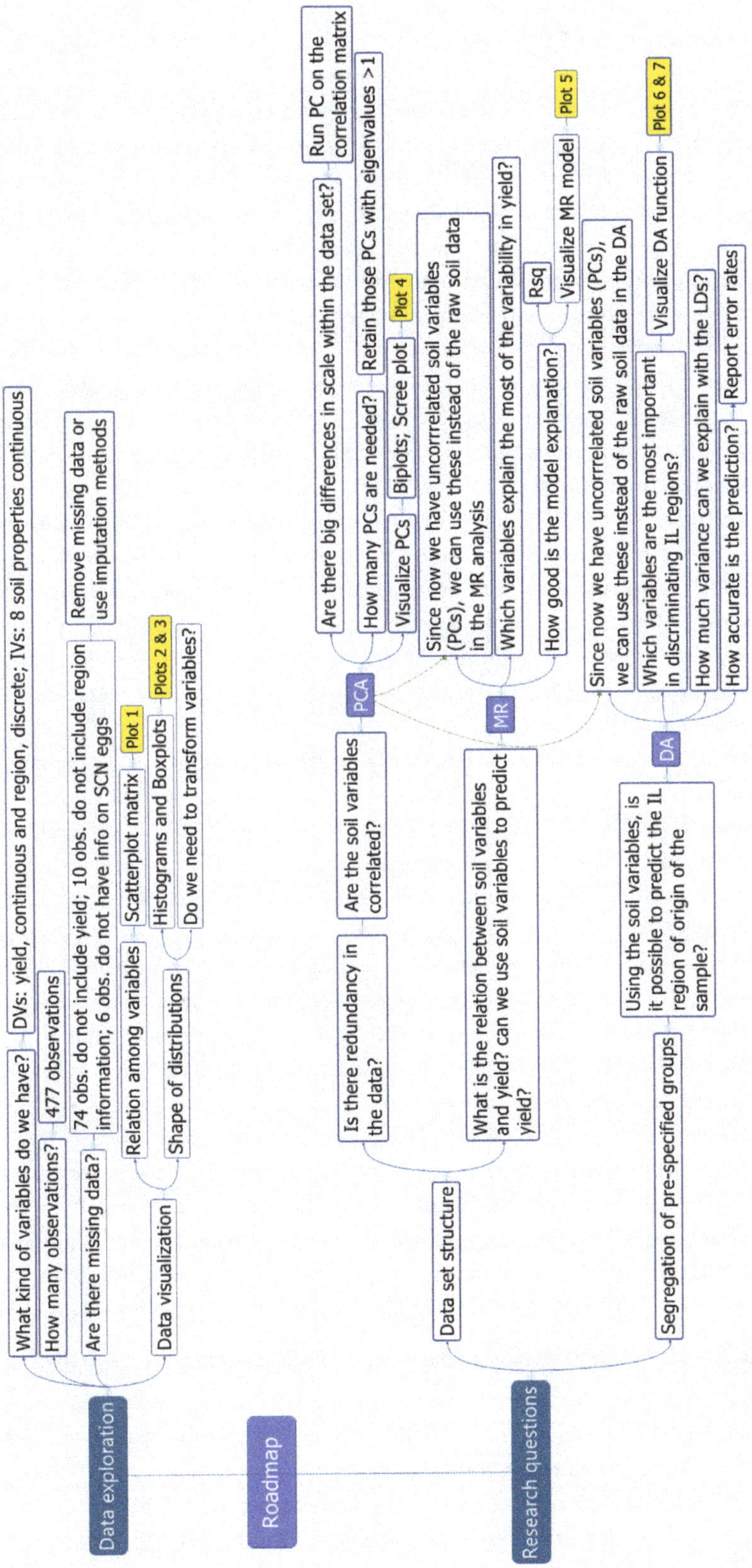

FIG. 1. Roadmap for multivariate exploration and analyses of our example data set.

your multivariate data to extract the relevant information that will help you find the solution to the stated problem.

Our field guide (Table 1) works as a taxonomical key, and it is based on the premise that we have a question that needs answering and that we have the data set(s) with certain characteristics that will allow us to achieve that understanding using specific statistical techniques, each with its own considerations, assumptions, and goals. The order of columns in Table 1 is thus not arbitrary, directing us to first focus on the general question (i.e., do we want to explore the structure of the data set, or do we want to be able to separate groups) and what we want to achieve with the statistical technique (i.e., do we want to remove redundancy—multicollinearity—in the data set, or explore the relationship among variables). This train of thought—and work—is accompanied by an acknowledgment of the kind of data available, that is, the type and number of dependent variables (DVs) and independent (IVs) variables in the data set(s), the assumptions that will need to be met, and other considerations for each statistical technique, linked to its abbreviation and particular result(s). The field guide continues in Table 2, where we expand those abbreviations by introducing the full name of the statistical technique and main procedures to run this analysis in SAS/STAT software, as well as functions within packages (i.e., lm[stats]) to work within the R environment. Most importantly for this hands-on approach to learning, for each of the methods listed, we have included three to four peer-reviewed publications with clear examples on the use of the specific technique in agricultural related research.

As stated, each of these multivariate analyses requires adherence to specific considerations, such as the rules of multivariate normality, linearity, and homoscedasticity for the use of PCA, MR, and linear DA that we cover in this chapter, yet linear regression and classification and regression tree (CART) analyses, for example, have far fewer assumptions than the aforementioned techniques. All procedures we are discussing (PCA, MR, DA) in addition to multivariate analysis of variance (MANOVA), which is a multivariate general linear model (GLM), have their beginnings in the same linear model. Linearity of variables is very important in these applications. Linearity is the assumption of a straight line fit between variables (all pairs have a linear relationship with each other). Additivity is also important to multivariate GLM, as one set of variables may be predicted from another set of variables; therefore the effect of the variables within the data set are additive in the prediction equation. These are just examples to help us explain that, although we describe the preparation of our data for our selected multivariate analysis, there are many important topics for each technique that are beyond the scope of this chapter. Therefore, we invite the reader to examine our references for greater depth and explanations.

Begin Data Analysis by Understanding the Nature of the Data

Due to the effect that characteristics of the data may have on the results, it is important to first consider the nature of the data, keeping in mind the assumptions that need to be met to use multivariate technique(s) of choice. Assessment and resolution of the specific challenges that a given data set presents are necessary to address during the data exploration stage to ensure a robust statistical evaluation. Tabachnick and Fidell (2012) provided an in-depth and appropriate data screening sequence that we have summarized in Table 3. The order of the steps is important because early decisions influence later decisions. For example, if the data are both nonnormal and there are outliers, the questions of whether to delete values (or not) or transform the data (or not) are addressed. If transformation is decided on first, will there be fewer outliers? Or if the outliers are deleted, might there be fewer variables with nonnormality? The researcher should consider the approaches and identify the data tuning that will best result in normality, linearity, and homoscedasticity, so as to meet the assumptions and appropriateness of the selected statistical approaches. This data "pretreatment" is a major challenge in functional genomics research and, although not our typical data set in agronomy, we encourage you to examine the work of van den Berg et al. (2006), which presents a clear and useful comparison of the pros and cons of different pretreatments (e.g., centering, scaling, and transformations) for clarifying the biological information and thus, the interpretability, of large "-omics" data sets.

Example Data, Exploration, and Multivariate Applications

The example data set used in this chapter is a subset of the data gathered for the "Yield Challenge" (YC) program, established by the Illinois Soybean Association (www.ilsoy.org, accessed 2 Feb. 2016) in collaboration with researchers from the

TABLE 3. Recommended data screening sequence for multivariate analysis of data (adapted from Tabachnick and Fidell, 2012).

1. Descriptive statistics—Inspect input values for accuracy
• Means and standard deviations
• Coefficient of variation, minimum, maximum, summary statistics
2. Missing data—Evaluate amount and sources—Is the "missingness" concentrated in observation(s) or variable(s)?
3. Verify independence of variables—Identify orthogonality
• Assess bivariate correlations—May foretell multicollinearity; large, inflated correlations may be due to redundancy of variable measures; small, deflated correlations may be due to restricted ranges
4. Identify nonnormal variables
• Skewness and kurtosis, Q–Q probability plot
• Transform, scale or center variables, if necessary and appropriate
• Assess multivariate results before and after pretreatments for example using the %multnorm macro in SAS (Khattree and Naik, 1999, 2000) or functions and plots in the MVN package (Korkmaz et al., 2015) within the R environment.
• Identify outliers—Univariate outliers, multivariate outliers
5. Evaluate pairwise—Correlation plots for nonlinearity and heteroscedasticity
6. Evaluate variables for multicollinearity and singularity
7. Check for spatial autocorrelation, if appropriate

University of Illinois in 2010. Details on experimental set-up, sampling methods, and initial data collection and analyses are provided in Davis et al. (2012) and Villamil et al. (2012). The YC subset contains soil properties and crop yield data collected in 2010 and 2011 from a total of 477 plots (*n*). Soil samples from each plot were sent to a commercial lab for baseline characterization of sites at planting. Soil variables in the subset included cation-exchange capacity (CEC), extractable nutrients (P, K, Ca, Mg), pH, and soil organic matter (SOM), along with soybean cyst nematode (SCN) egg counts. Soybean yield (kg ha^{-1}) was measured using the producer's commercial combine to harvest a minimum area of 0.8 ha from each plot and adjusted to 12% moisture content. Yield results were grouped according to the nine USDA National Agriculture Statistics Service (NASS) state crop reporting districts (CRDs) of Illinois, which were further combined into regions representative of IL: North (districts 1 and 2), Central (3 to 5), and South (6 to 9).

Our Road Map

We will use R (version 3.2.1) coding to accompany our road map (Fig. 1 built with XMind 6.0, http://www.xmind.net/), yet you can find the SAS (version 9.4, SAS/STAT 13.2) code provided in the boxes in this chapter to get similar results (some differences exist in output due to differences of precision of input values between R and SAS). For brevity, graphics coding for SAS is only included for the DA example. Briefly, after exploring our data and assessing its quality and suitability for multivariate analysis, we will: (i) evaluate soil variables to remove redundancy using PCA, (ii) explore the relationship between soybean yields and soil properties at time of planting using MR, and (iii) create a rule to identify the region of origin of the soil sample based on the lab data using DA. The data set (mvYCdata.csv) is available as a supplementary file and we recommend you download it to a folder in your desktop and change the directory for your R session ("Change dir..." in the File tab). Please pour yourself a good cup of coffee (mate or tea are also recommended) and follow us in our multivariate exploration and analysis of this data using the road map in Fig. 1.

Data Exploration

Let's begin by opening the file and checking that you have what we promised. We use the pound symbol (#) to add comments throughout, as we would do within the R console (this is equivalent to the /* prompt in SAS). See Box 1 for an example of equivalent SAS coding.

```
> YC<-read.csv("mvYCdata.csv", header=T, na.string='.')
#Here we are just stating the type and name of the data set, and that
the first row is for variable names and missing values are coded as
dots
> attach(YC)
> names(YC)
[1] "Yield"   "Region"   "pH"     "SOM"     "P_"     "K"    "Ca"   "Mg"
[9] "CEC"     "SCN"
```

In this data frame (our current data set) we have 477 observations of 10 variables. We have one categorical variable (factor "Region" with three levels), and the rest are all continuous numerical variables though some of them (Yield and SCN take integer numbers). We don't have to look at the data set itself to get this information; we

BOX 1.

```
/*infile mvYCdata.csv using favorite method*/
/*subset*/
/*remove scneggs = 12000 outlier*/
Data Ycallsubset;
Set mvYCdata;
if Region = '.' then delete;
if scneggs > 10000 then delete;
run;

Proc Print data=Ycallsubset;
run;
```

can use the str()function, which displays a breakdown of the structure of a data frame.

```
> str(YC)#to check what the data set looks like:
'data.frame':    477 obs. of 10 variables:
 $ Yield : int  4391 5030 4694 4923 3625 4499 4439 3632 3981 4351...
 $ Region: Factor w/3 levels "Central","North",..: 1 1 1 1 3 3 1 1 2
2...
 $ pH    : num  6 6.3 6.7 6.6 6.8 6.7 6.9 6.6 7.5 6.5...
 $ SOM   : num  3.4 3.4 3.2 3 2.9 3 3.5 3.4 3.8 3.9...
 $ P_    : num  31.5 16.5 34.5 27 40 32 40 27.5 36 26.5...
 $ K     : num  170 148 170 182 147...
 $ Ca    : num  2303 2454 2326 2470 2050...
 $ Mg    : num  464 504 244 268 191...
 $ CEC   : num  20 19.4 15 16.5 12.6 13.1 20.6 20.4 32 25.4...
 $ SCN   : int  40 80 0 0 360 0 280 200 280 240...
```

An important point to make here: note the number of observations we have in this data set. The general recommendation of minimum sample size for multivariate analysis falls into two categories. One category states that absolute number of observations (n) is necessary, while the other says that the observation/variable (n/p) ratio is important. These recommendations have been reviewed in the literature for the application of factor analysis and PCA (Arrindell and van der Ende, 1985; Velicer and Fava, 1998; MacCallum et al., 2001; Osborne and Costello, 2004). The results of Osborne and Costello (2004) indicate an interaction between the two, where the optimum outcomes occurred in analyses with large n and high ratios, agreeing with a heuristic argument where n should be at least five times p for DA, and n greater than 10 times p for hypothesis-driven approaches like MANOVA. The number of observations in a study is always directly related to power. Thus, we want to observe "real" differences when they exist so if we want to explore and analyze eight variables, we will require at least 80 observations.

Going back to our newly created YC object, we use the function summary to extract the basic info for each variable as follows:

```
> summary(YC)
     Yield           Region             pH              SOM               P_
 Min.   :2091    Central:288     Min.   :5.10    Min.   :1.00    Min.   :  8.5
 1st Qu.:4082    North  : 87     1st Qu.:6.10    1st Qu.:3.00    1st Qu.: 33.5
 Median :4499    South  : 92     Median :6.40    Median :3.30    Median : 47.6
 Mean   :4477    NA's   : 10     Mean   :6.44    Mean   :3.24    Mean   : 65.9
 3rd Qu.:4920                    3rd Qu.:6.70    3rd Qu.:3.80    3rd Qu.: 81.0
```

```
Max.   :5945                              Max.   :7.90   Max.   :4.90   Max.   :475.7
NA's   :74
K              Ca              Mg              CEC            SCN
Min.   :  56   Min.   :  794   Min.   :  72   Min.   : 6.3   Min.   :   0
1st Qu.: 142   1st Qu.: 1895   1st Qu.: 248   1st Qu.:13.2   1st Qu.:  40
Median : 202   Median : 2707   Median : 409   Median :16.4   Median :  80
Mean   : 256   Mean   : 3421   Mean   : 476   Mean   :16.9   Mean   : 180
3rd Qu.: 306   3rd Qu.: 4571   3rd Qu.: 640   3rd Qu.:20.0   3rd Qu.: 200
Max.   :1465   Max.   :13346   Max.   :1605   Max.   :40.0   Max.   :3000
NA's   :6
```

See Box 2 for an example of equivalent SAS coding.

We can easily note that among the soil variables there are only six missing data points (NA's) in SCN; we need to find out which observations are those and remove them, since the functions we will be using next do not work with missing data (error messages will appear):

```
> subset(YC, is.na(SCN))
    Yield  Region  pH SOM     P_     K     Ca      Mg  CEC SCN
336  5118   North 6.8 4.4   67.6 311.2  8712 1604.7 29.9  NA
361  5158 Central 6.7 3.1  187.7 927.0  3174  825.7 13.5  NA
371  4291 Central 5.9 3.6   53.3 382.1  5220  855.4 22.0  NA
432  4775   North 7.7 3.5   54.2 231.2 13346  809.6 37.1  NA
458  5501 Central 6.6 3.2  101.9 301.1  5765  637.5 19.0  NA
470  3914 Central 5.9 3.8   47.6 328.5  5036  663.1 20.2  NA
```

Missing data can occur from many reasons. Some of the specific "missingness" in our example data include: samples that were inadvertently uncollected from the field or mistakenly tossed before assay, equipment malfunction during assay, or survey responses that were unanswered for an unknown reason. The seriousness of the missing data is dependent on the reasons behind the missingness—is there a proportion of NA's in the data set, and are there "patterns" to this missingness that could be attributable to some immeasurable event, the latter being the more serious issue because they affect the generalizability of the results (Tabachnick and Fidell, 2012).

Imagine as an example you are conducting a survey on soybean yield and agronomic factors, yet several producers with low-yielding crops refuse to report yield (this is now a nonrandom distribution of missing values). If these responses (observations) with NA's on yield are simply deleted, the sample values on the agronomic aspects are distorted, thus the need for a method to estimate these NA's instead. The decision of how to handle missing data is important, so please refer to any of the multivariate analyses books listed in the references. Several statistical software tools have developed procedures and packages to assist the researcher with this issue (i.e., SAS Proc MI and MIANALYZE; R mitools, mvnmle, and other packages). Notice however, these tools are mostly developed for large sample size and normally distributed variables.

BOX 2.

```
Proc Univariate data = Ycallsubset normal plot;
var Yield pH SOM P_ K Ca Mg CEC SCN;
run;
```

Going back to our YC data example, we have enough observations (477) and only a few (6) random NA's in the SCN variable, so we are comfortable with removal of these NA's and proceeding with our analyses. Thus, we remove the vector of observations with NA's in the SCN variable from the original data YC, and we call this new object newYC:

```
> newYC<-YC[-c(336,361,371,432,458,470),]

> dim(newYC) #just checking the new dimensions here and as you can
see, it worked, the data set newYC has 471 obs instead of the original
477

[1] 471  10

# we can also check the specific variable SCN for NA's with the
function summary and using the '$' symbol..

> summary(newYC$SCN)

  Min. 1st Qu.  Median    Mean 3rd Qu.    Max.
  0.0    40.0    80.0   180.4   200.0  3000.0
```

We are now creating a data set that only has the soil variables so we can explore their characteristics,

```
> soil<-newYC[,3:10]
> str(soil)
'data.frame':   471 obs. of  8 variables:
 $ pH : num  6 6.3 6.7 6.6 6.8 6.7 6.9 6.6 7.5 6.5...
 $ SOM: num  3.4 3.4 3.2 3 2.9 3 3.5 3.4 3.8 3.9...
 $ P_ : num  31.5 16.5 34.5 27 40 32 40 27.5 36 26.5...
 $ K  : num  170 148 170 182 147...
 $ Ca : num  2303 2454 2326 2470 2050...
 $ Mg : num  464 504 244 268 191...
 $ CEC: num  20 19.4 15 16.5 12.6 13.1 20.6 20.4 32 25.4...
 $ SCN: int  40 80 0 0 360 0 280 200 280 240...
```

Visualization Step

See Box 3 for an example of equivalent SAS coding.

```
> pairs(soil)#basic scatterplot matrix
```

Figure 2 is an example of a scatterplot matrix. The variables are written in a diagonal line from top left to bottom right. Then each variable is plotted against each other. By default, the boxes on the upper right-hand side of the whole scatterplot are mirror images of the plots on the lower left-hand side. For example, the top square in the second column is an individual scatterplot of pH and SOM, with SOM as the x axis and pH as the y axis. This same plot is replicated in the first square in the second

BOX 3.

```
/*scatterplot correlation matrix*/
ods graphics on;/*ODS GRAPHICS statement options are available with SAS 9.2 and
later versions*/
proc corr data=Ycallsubset plots(Maxpoints=none)=matrix(histogram nvar=all);
var SOM ph P_ K Ca Mg CEC SCN;
run;
ods graphics off;
```

row. In this scatterplot, it's probably safe to say that there is no correlation between pH and SOM because the line of fit is flat (dots are just all over the place). Yet there seems to be strong and positive correlations between P and K, Ca and Mg, and CEC with all of them. In these situations, high x values correspond to high y values of each of those variable pairs. This is an indication of multicollinearity and a good justification for using PCA to remove redundancy in the data. We can corroborate this with the cor function in R:

```
> cor(soil)
       pH    SOM     P_      K      Ca      Mg    CEC     SCN
pH   1.000 -0.094 0.126 0.0877  0.32  0.047  0.17  0.1498
SOM -0.094  1.000 0.090 0.2098  0.31  0.405  0.52 -0.0638
P_   0.126  0.090 1.000 0.7764  0.39  0.277  0.16  0.0279
K    0.088  0.210 0.776 1.0000  0.52  0.392  0.25  0.0095
Ca   0.315  0.310 0.393 0.5169  1.00  0.692  0.75 -0.0605
Mg   0.047  0.405 0.277 0.3922  0.69  1.000  0.69 -0.1227
CEC  0.166  0.516 0.159 0.2496  0.75  0.687  1.00 -0.0805
SCN  0.150 -0.064 0.028 0.0095 -0.06 -0.123 -0.08  1.0000
```

The boxplot of the variables in the soil data set shown in Fig. 3 indicates that the variables differ in magnitude and that they have different variances, also.

```
> boxplot(soil)
```

A recent package in R called MVN, created by Korkmaz et al. (2015), helps us to solve the question of univariate and multivariate normality for this soil data set:

```
> library(MVN)
```

We use the uniNorm function to get the Shapiro–Wilks test of univariate normality along with the descriptive stats, including skewness and kurtosis for each variable:

```
> uniNorm(soil, type="SW", desc=TRUE)
$'Descriptive Statistics'
      n     Mean  Std.Dev Median   Min     Max    25th    75th   Skew Kurtosis
pH  471    6.440    0.451    6.4   5.1     7.9    6.10    6.70  0.430    0.454
SOM 471    3.232    0.815    3.3   1.0     4.9    3.00    3.80 -0.949    0.772
P_  471   65.611   53.542   46.5   8.5   475.7   33.25   80.90  2.948   13.337
K   471  254.264  178.529  198.0  56.0  1465.3  141.00  303.55  2.529    9.373
Ca  471 3376.774 2017.432 2691.0 794.5 12364.8 1892.00 4501.60  1.363    1.704
Mg  471  470.351  284.975  405.5  72.0  1567.3  245.85  633.05  1.133    1.044
CEC 471   16.786    5.259   16.3   6.3    40.0   13.10   19.85  0.590    0.482
SCN 471  180.382  321.029   80.0   0.0  3000.0   40.00  200.00  4.999   33.319

$'Shapiro-Wilk's Normality Test'
   Variable Statistic   p-value Normality
1   pH         0.9814         0    NO
2   SOM        0.9187         0    NO
3   P_         0.7305         0    NO
4    K         0.7705         0    NO
5   Ca         0.8696         0    NO
6   Mg         0.9053         0    NO
7   CEC        0.9766         0    NO
8   SCN        0.5045         0    NO
```

Visualization Step

None of these variables seem to meet the normality assumptions, so let's take a look at their histograms with the uniPlot function while we conduct the multivariate

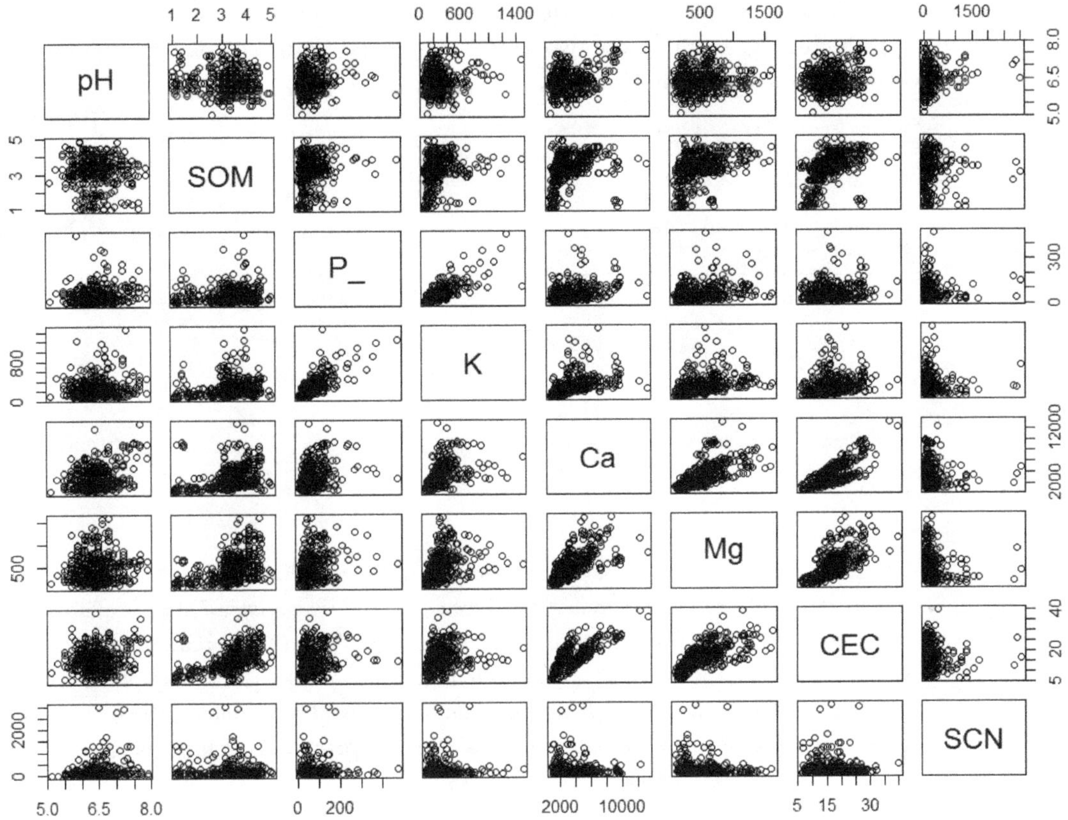

FIG. 2. Scatterplot matrix of the variables in the soil data set.

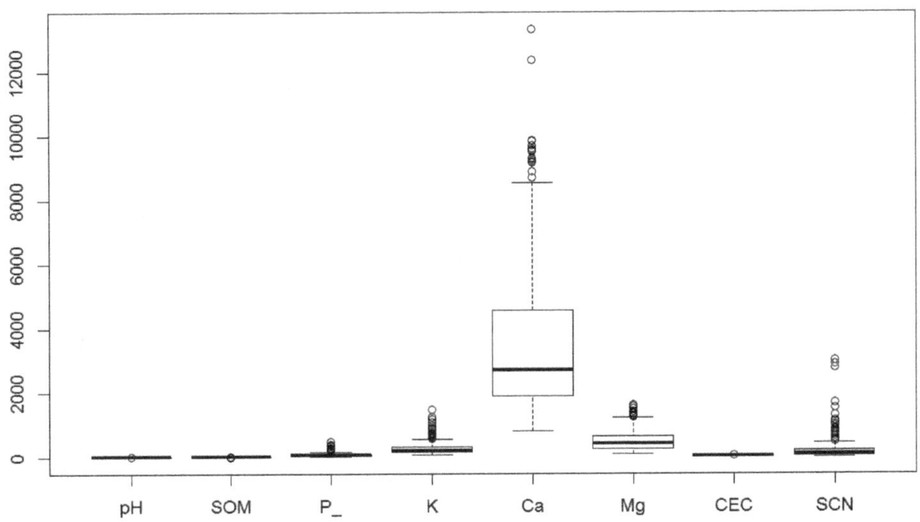

FIG. 3. Boxplots of the variables in the soil data set.

normality test and build the corresponding Chi-squared Q–Q plot and put them all together in Fig. 4, with these two lines of command in R, See Box 4 for an example of equivalent SAS coding:

```
> uniPlot(soil, type="histogram")
> result<-mardiaTest(soil, qqplot=T)
#to get all the plots within the same frame ☺

> result #to see the result of the MV normality test for the soil data
Mardia's Multivariate Normality Test
-----------------------------------
data: YC.nmd
g1p              : 65.56
chi.skew         : 5146
p.value.skew     : 0
g2p              : 175.6
z.kurtosis       : 82
p.value.kurt     : 0
chi.small.skew: 5186
p.value.small : 0
Result              : Data are not multivariate normal.
-----------------------------------
```

As we suspected, a visual inspection indicates that some of the variables in our data set are not normal. *Should we transform the variables?* It is probably not necessary in this case since we are going to use PCA on the correlation matrix that will standardize the variables by default. (See the manuscript on data pretreatment of van den Berg et al., 2006.) If we do not standardize the variables, those variables with the highest variances will dominate the first principal components regardless of the covariance structure of the variables.

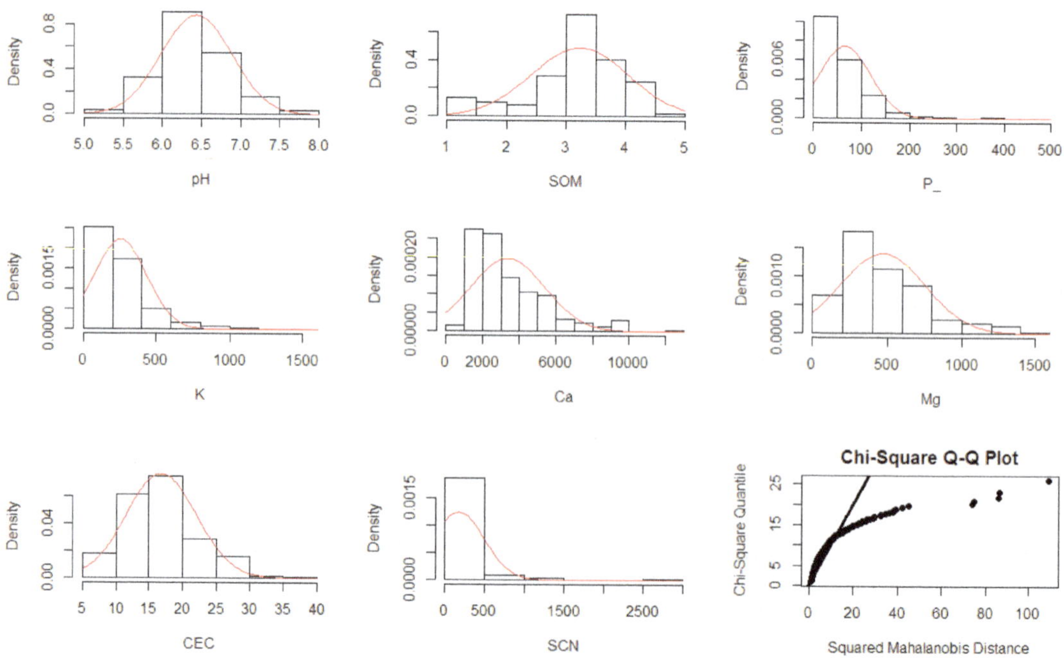

FIG. 4. Histograms of the variables in the soil data set and multivariate Chi-squared Q–Q plot to test multivariate normality.

BOX 4.

```
/*The following macro must be downloaded and saved on user's machine*/
/*Then call the macro using the %inc statement*/
%inc "<location of the MULTNORM macro>"
%multnorm(data=Ycallsubset, var=pH SOM P_ K Ca Mg CEC SCN, plot=mult)
```

So, let's continue with the MV analyses based on research questions.

1. Principal Components Analysis—Removing redundancy

Question: Do we really need all these variables? Can we reduce the number of variables and still explain a good proportion of the initial variability of the data set?

Principal components analysis enables us to avoid problems of multicollinearity by compiling the information into a new—hopefully smaller—set of uncorrelated variables. Principal components analysis uses all of the variables in the data set and determines weights (coefficients) in the new linear construct or loading factors of the eigenvectors based on the contribution of variability and correlation to the axis, the principal component (Comp.). The original observations are transformed into score values for each Comp. of significant importance, but let's do it so you'll see what we are talking about. We will be using several tools from the MASS package in R (Venables and Ripley, 2002), so we make this accessible in our session with the function library, See Box 5 for an example of equivalent SAS coding:

```
> library(MASS) #Probably one of the most useful R packages out there

> soil.pc<-princomp(soil, cor=T, scores=T) #the function 'princomp' conducts PCA
on the correlation matrix and extracts the scores with the specifications added

> summary(soil.pc)
Importance of components:
                        Comp.1 Comp.2 Comp.3 Comp.4  Comp.5  Comp.6  Comp.7  Comp.8
Standard deviation      1.8043 1.1976 1.0748 0.9509 0.76507 0.54102 0.47889 0.37868
Proportion of Variance 0.4069 0.1793 0.1444 0.1130 0.07317 0.03659 0.02867 0.01792
Cumulative Proportion  0.4069 0.5862 0.7306 0.8437 0.91682 0.95341 0.98208 1.00000

> round(soil.pc$sdev^2, 2) #The following are the eigenvalues of the correlation
matrix, R only gives us the standard deviation, so we need to square the
components:
Comp.1 Comp.2 Comp.3 Comp.4 Comp.5 Comp.6 Comp.7 Comp.8
3.26   1.43   1.16   0.90   0.59   0.29   0.23   0.14
```

The summed total variance of our eight components equals 8. Our results indicate that only the first three components (eigenvalues > 1) are needed to describe the space since Comp.1 accounts for ~41% (3.26/8) of the total variance while Comp.2 and Comp.3 explain an additional ~18% (1.43/8) and ~14% (1.16/8), respectively. With these three components we explain about 73% of the total variance, which can be seen in Fig. 5, the scree plot.

Visualization Step

```
> screeplot(soil.pc, main="")
```

To see which variables are responsible for the explanatory ability of the components, we check the loadings and pay attention to those with absolute

BOX 5.

```
/*PCA 472 observations, 8 variables*/
/*without outlier*/
ods graphics on;
proc princomp data=Ycallsubset out=prins plots=all;
var ph SOM P_ K Ca Mg CEC SCN;
run;
ods graphics off;

/*Note that this code creates all possible default plots and tables*/
/*Tables and Plots of interest include:
Eigenvalues of the Correlation Matrix contain Eigenvalues, Proportion, and
Cumulative Proportion for each of the 8 components,
Eigenvectors contain the Loading values for each of the 8 components,
Scree Plot and Variance Explained plot are equivalent to same-named
visualization step in R*/

/*We are creating an output file called prins, which includes the new PC scores,
we will use prins data set as input file for MR and DA*/
```

values higher than 0.4 (as an example). The "cutoff" or selection of important or influential variables is relative in comparison to the other loading values within a principal component (Tabachnick and Fidell, 2012). Though a nice thing to have, interpretability is not a necessary aspect of the principal components:

```
> loadings(soil.pc)
Loadings:
    Comp.1 Comp.2 Comp.3 Comp.4 Comp.5 Comp.6 Comp.7 Comp.8
pH  -0.128 -0.284  0.699  0.377 -0.451  0.243
SOM -0.295  0.390 -0.116 -0.453 -0.702  0.114        -0.186
P_  -0.323 -0.548 -0.316        -0.110         0.660 -0.213
K   -0.384 -0.448 -0.322                      -0.584  0.446
Ca  -0.491         0.194  0.155  0.218 -0.357 -0.329 -0.643
Mg  -0.451  0.220                0.413  0.754
CEC -0.445  0.332  0.218         0.106 -0.480  0.323  0.545
SCN        -0.326  0.457 -0.790  0.239

               Comp.1 Comp.2 Comp.3 Comp.4 Comp.5 Comp.6 Comp.7 Comp.8
SS loadings     1.00   1.00   1.00   1.00   1.00   1.00   1.00   1.00
Proportion Var  0.12   0.12   0.12   0.12   0.12   0.12   0.12   0.12
Cumulative Var   .12   0.25   0.38   0.50   0.62   0.75   0.88   1.00
```

While Comp.1 relates to CEC and the amounts of Ca and Mg primarily, Comp.2 explanatory ability is due to the relation among P and K and possibly SOM, whereas Comp.3 seems to highlight the important relation between pH of the soil and SCN egg counts in the sample.

The PCA thus reduced the dimensionality of the data set from eight (correlated) variables to three uncorrelated linear combinations (Comp.1 to Comp.3), which can be used for further analyses, with limited loss of information. So, next we will take the scores from the first three PCs out and create a new data frame mvpc to use in MR (on yield) and DA (among regions) as promised:

```
> mvpc<-data.frame(soil.pc$scores[,1:3])
> str(mvpc) #we use this function as a fast check
'data.frame':   471 obs. of 3 variables:
$ Comp.1: num  0.43 0.439 1.045 0.964 1.55...
```

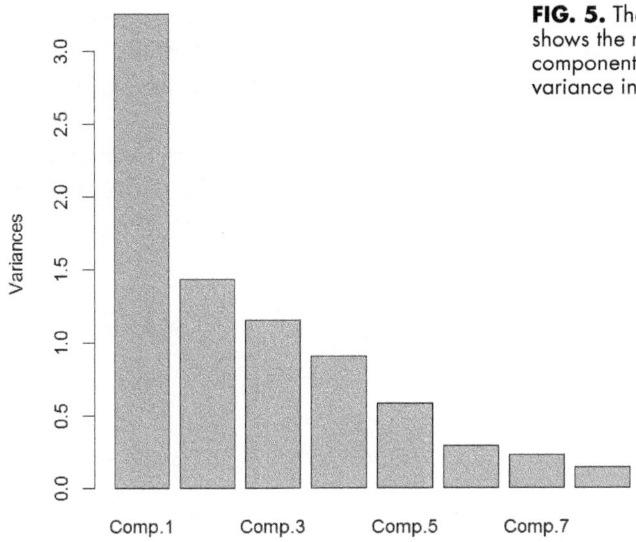

FIG. 5. The screenplot on the right, shows the relative importance of each component in explaining the total variance in the soil data set.

```
$ Comp.2: num  1.254 1.224 0.237 0.366 -0.528...
$ Comp.3: num  -0.521 0.117 0.312 0.284 0.907...
```

But before that, we also need to bring back the first two columns of the newYC data set that contain Yield and Region; we will call this new object ID:

```
> ID<-newYC[,1:2]

> str(ID) #just checking...
'data.frame':   471 obs. of 2 variables:
$ Yield: int  4391 5030 4694 4923 3625 4499 4439 3632 3981 4351...
$ Region: Factor w/3 levels "Central","North",..: 1 1 1 1 3 3 1 1 2
2...
```

And bind this new file to the previously created mvpc that contains the scores for the three principal components in a new object that we call mvYCidpc:

```
> mvYCidpc<-cbind(ID, mvpc)

> str(mvYCidpc) #checking again
'data.frame':   471 obs. of 5 variables:
$ Yield: int  4391 5030 4694 4923 3625 4499 4439 3632 3981 4351...
$ Region: Factor w/3 levels "Central","North",..: 1 1 1 1 3 3 1 1 2
2...
$ Comp.1: num  0.43 0.439 1.045 0.964 1.55...
$ Comp.2: num  1.254 1.224 0.237 0.366 -0.528...
$ Comp.3: num  -0.521 0.117 0.312 0.284 0.907...
```

So now to our second question of interest:

2. Question: How do these soil properties at planting impact the seed yield response?

To investigate this, we carry out a multiple regression (MR) of the previously obtained principal components on soybean yield. Multiple regression analyses are an extension of simple linear (bivariate) regression methods where several IVs (principal components) are combined to predict a value on a DV (soybean yield) for each observation in our study. One first needs to investigate the relationship between the DV and IVs, and then assess the importance of each of the IVs to the relationship. Potentially, a covariates analysis could be achieved where inquiries are made by adding some critical variable(s) to the developing prediction equation, thereby comparing the inclusion and exclusion of IVs to develop a best-fit prediction model for the given DV. Our following approach is an assessment of the potential importance of the principal components on the soybean seed yield response. See Box 6 for an example of equivalent SAS coding.

```
> MR.fit<-lm(Yield~Comp.1+Comp.2+Comp.3, data = mvYCidpc) #function
'lm' performs the MR using Yield as the response variable

> summary(MR.fit) #the essentials are extracted with this universal
function
Call:
lm(formula = Yield ~ Comp.1 + Comp.2 + Comp.3, data = mvYCidpc)

Residuals:
    Min      1Q  Median      3Q     Max
-2200.7  -393.7    14.6   428.0  1594.7
Coefficients:
               Estimate Std. Error t value Pr( > |t|)
(Intercept)      4466.9       31.6  141.19  <2e-16 ***
Comp.1            -11.6       17.3   -0.67   0.5022
Comp.2             25.9       26.1    0.99   0.3222
Comp.3            -84.6       28.6   -2.95   0.0033 **
---
Signif. codes:  0 '***' 0.001 '**' 0.01 '*' 0.05 '.' 0.1 ' ' 1

Residual standard error: 624 on 393 degrees of freedom
(74 observations deleted due to missingness)
Multiple R-squared:  0.0237,    Adjusted R-squared:  0.0163
F-statistic: 3.19 on 3 and 393 DF, p-value: 0.0238
```

The variable Comp.3 was the only significant contributor to the multiple regression equation ($p = 0.0033$). The regression for the grain yield had an explanatory R^2 of 0.0237 and an adjusted r^2 of 0.0163 (model $p = 0.0238$). This model only explains less than 3% of the yield variability represented with this YC data set (sad but true!). Since we are familiar with these data coming from the Yield Challenge Program, the results do not surprise us: most of the practices that aimed to increase yields were agronomic practices and inputs applied during the growing season, so the characteristics of the soil at planting alone did not have much of an impact within this system (see Davis et al., 2012; Villamil et al., 2012).

Visualization Step

```
> par(mfrow = c(2,2)) #to place ALL 4 plots in the same page
> plot(MR.fit) #to evaluate model assumptions
```

BOX 6.

```
/*use prins data set created from prior Principal Components Analysis*/
ods graphics on;
proc reg data=prins;
model yield = prin1 prin2 prin3/selection=none;/*comparable to R default*/
run;quit;
ods graphics off;
```

In Fig. 6, on the top left, we can see dots in a cloud kind of scatter for the residuals vs. predicted values, which indicates the model is appropriate and our assumptions are reasonably met. Visual inspections of the Q–Q plot, the scale–location plot, and the scatterplot of residuals vs. leverage indicate that there is a group of observations that seem influential. However, given the absolute lack of explanatory power of our model and our understanding of the origin of the data set, we can safely conclude that soil properties are not contributing to grain yield within these yield challenge plots, probably due to the high homogeneity and high inherent soil fertility of our soil samples (most soils are Mollisols).

So now to our third question of interest:

3. Question: Are these soil properties really different among these regions? Or in another way of thinking about this: Is it possible to create a rule to identify where the sample comes from knowing the lab results?

In our example data set, we have three pre-identified geographic regions based on state CRDs. With these known groups to the data, we can better define the relationship between the response variables and this classification variable using a (linear) discriminant analysis (DA) procedure. This method is an extension of the univariate regression analysis and ANOVA and is analogous to multiple regression analysis and MANOVA. A linear combination of response variables is used to describe or predict the behavior of categorical/dependent variable(s) in regression methodology. This procedure is extended in DA by utilizing these predefined group categories and their known relationships to the response variables. General DA provides a classification of the observations into groups, which can describe how well group membership can be predicted. The classification can be used to predict group membership of additional samples for which group membership is unknown. The key characteristics of DA entail that the variation among groups is maximized while variation within groups is minimized, and that the dimensionality of a multivariate data set is reduced into a smaller set of new variables, now called canonical functions, with a minimum loss of information. In canonical DA (CDA), an extension of MANOVA and closely related to canonical correlation analysis, the newly derived canonical variates summarize between-group variation and provide a simultaneous test describing which variables best account for group differences.

The goal of CDA is to test and describe the relationships among two or more groups based on a set of discriminating variables. Canonical DA involves deriving the linear combinations (i.e., canonical functions) of the variables that will discriminate the best (i.e., maximize the variation) among the predefined groups. For each sampled observation, a canonical score is calculated for each canonical variate, and the group centroid can be used to identify the most typical location of an observation from a given group. A comparison of group centroids indicates how far apart the groups are along the variate being tested. But enough explanation, let's do it so you can see what we are after here.

A little more data organization and cleaning is required before we can proceed. We will drop the variable Yield for the DA analysis and we also need to remove a few (10) missing values (NA's) present within the Region variable in the object that we created last, mvYCidpc.

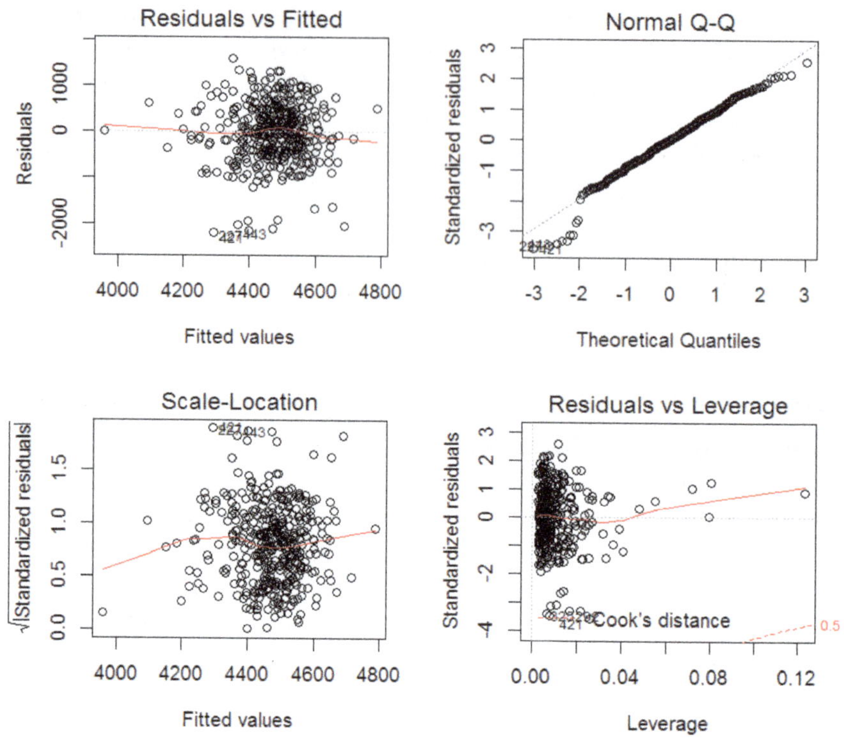

FIG. 6. Model fit assessment tools.

```
> YCrpc<-mvYCidpc[,2:5] #to drop 'Yield'

> YCrpc.nmd<-na.omit(YCrpc) #remove NA's
> dim(YCrpc.nmd) #checking again that there are now 10 obs less
[1] 461   4
```

Both steps will ensure a smooth work of the function lda{MASS} that we will use next (it should still be available in your console, but if you took a break and closed R, you might need to reload the package into your current session).

Since we are trying to create a rule to discriminate among regions, it will be useful to know how many observations per region we have in the data set:

```
> summary(YCrpc.nmd$Region)
Central    North    South
284        85        92
```

This simple step also tells us that the levels for the variable Region are not ordered in a logical manner so we do that using the function ordered in R,

```
> YCrpc.nmd$Region<-ordered(YCrpc.nmd$Region,levels=c("North","Centra
l","South"))
```

And now, to the actual DA, See Box 7 for an example of equivalent SAS coding.

```
> YCrpc.nmd.lda<-lda(YCrpc.nmd [,-1], YCrpc.nmd [,1])
> YCrpc.nmd.lda #to see the results, ask for it;-)

Call:
lda(YCrpc.nmd[,-1], YCrpc.nmd[, 1])

Prior probabilities of groups:
 North Central   South
```

```
0.1844  0.6161  0.1996

Group means:
          Comp.1    Comp.2    Comp.3
North    -0.7215  0.05106 -0.25419
Central -0.3478  0.13892  0.02236
South    1.5831 -0.51757  0.13851

Coefficients of linear discriminants:
          LD1       LD2
Comp.1  0.5746 -0.06054
Comp.2 -0.3915 -0.46707
Comp.3  0.1749 -0.75879
Proportion of trace:
  LD1    LD2
0.9689 0.0311
```

In our DA example, the first three principal components (Comp.1 to 3) are used to discriminate and classify among the three IL Regions. There are only two linear discriminants (LDs) ($p - 1$; number of groups minus one), which account for 100% of the among-groups variance, the first LD (LD1) housing the majority with 97% of the total variance explained (see output above). Notice here the difference with the components of a PCA; LDs are still linear combinations of original variables, but rather than explain as much variance as possible with each sequential LD, instead they are drawn to maximize the DIFFERENCE between groups (i.e., Regions) along that new variable. Rather than a similarity matrix, LD analysis (and MANOVA) use a comparison matrix of between and within groups' sum of squares and cross-products. Each LD is the linear combination of the independently measured variables and is orthogonal to the others (Vaylay and van Santen, 2002). The coefficients of the linear discriminants reflect the variance that the observed variables share with the LD and can be interpreted as the relative contribution of each variable to each LD function (Cruz-Castillo et al., 1994; Yeater et al., 2004). Thus, from the above output, we can see that LD1 is dominated by a large absolute loading from Comp.1 (0.57) followed by an important contribution from Comp.2 (0.39). To remind, Comp.1 is largely explained by Ca, Mg, and CEC while Comp.2 represents the influence of SOM, P, and K.

Visualization Step

The graphical representation of the LD information is useful in understanding the separation of a priori known IL Regions. Let's see our LD functions in action with a simple scatter in Fig. 7 obtained with the following commands,

```
> par(mfrow=c(1,1))#to return to the default values (1 plot per page)
the parameters of the graph page

> plot(YCrpc.nmd.lda, abbrev=TRUE, cex=0.7, ylim=c(-6,6),
xlim=c(-6,6))
> abline(h=0, v=0, lty=4)
```

Figure 7 represents LDs 1 and 2 on the x and y axes, respectively, with each replicate observation presented in the grouped scatter plot with Region groups as labels. A closer look at the loadings of the measured variables (Comp.1 to Comp.3) into each LD provides a more thorough interpretation. Thus, we create a parallel coordinates plot for Comp.1 to Comp.3 (also from the MASS package) for each region

BOX 7.

```
/*discriminant analysis*/
ods graphics on;
proc discrim data=prins pool=yes crossvalidate canonical distance
out=discrim_onprins_out;
priors prop;
class Region;
var prin1 prin2 prin3;
run;
ods graphics off;
/*Prior probabilities of groups are in Class Level Information,
Group means as described in R are most equivalent to the Class Means on
Canonical Variables (although not equivalent in values necessarily),
the Pooled Within Canonical Structure table is recommended to describe the
canonical variates (again the values are not equivalent to R output),
the Proportion of trace (from R) is the Proportion of the Eigenvalues of each
Canonical Correlation,
the table of error rates and percent of observation classified is located within
the Cross-validation Summary using Linear Discriminant Function*/

/*graphics code created by SAS ODS Graphics Designer*/
/*SAS 9.2 and newer required*/
/*This code creates a scatterplot of the first two canonical variates, similar
to PLOT6*/
proc template;
define statgraph Graph;
dynamic _CAN1A _CAN2A _REGION;
begingraph/designwidth=892 designheight=732;
   layout lattice/rowdatarange=data columndatarange=data rowgutter=10
columngutter=10;
      layout overlay;
        scatterplot x=_CAN1A y=_CAN2A/group=_REGION name='scatter';
        discretelegend 'scatter' /opaque=false border=true halign=left
valign=bottom displayclipped = true across=1 order=rowmajor location=inside;
      endlayout;
    endlayout;
endgraph;
end;
run;

proc sgrender data=DISCRIM_ONPRINS_OUT template=Graph;
dynamic _CAN1A="CAN1" _CAN2A="CAN2" _REGION="REGION";
run;
```

(Fig. 8) where the LD group means (centroid values for each Region) are plotted and the components standardized (0–1 scale):

```
> parcoord(YCrpc.nmd.lda$means, col=c("gray0", "gray30","gray60"),
lwd=2, lty=c(3,2,1), var.label=T); legend(1.35,1,c("North","Central"
, "South"),bg="white",cex=1,col=c("gray0", "gray30","gray60"), lwd=2,
lty=c(3,2,1))
```

A simple way to interpret this plot is by looking at it as if it were a table; this one has three columns with the data from Comp.1, Comp.2, and Comp.3. Now imagine each of these columns being mapped into a vertical axis in Fig. 8 with values ordered from minimum at the bottom and maximum at the top. Remember that instead of plotting every single data value we are plotting the group centroid, the LD mean value for each region, for each of our variables Comp.1, Comp.2, and Comp.3. The centroids

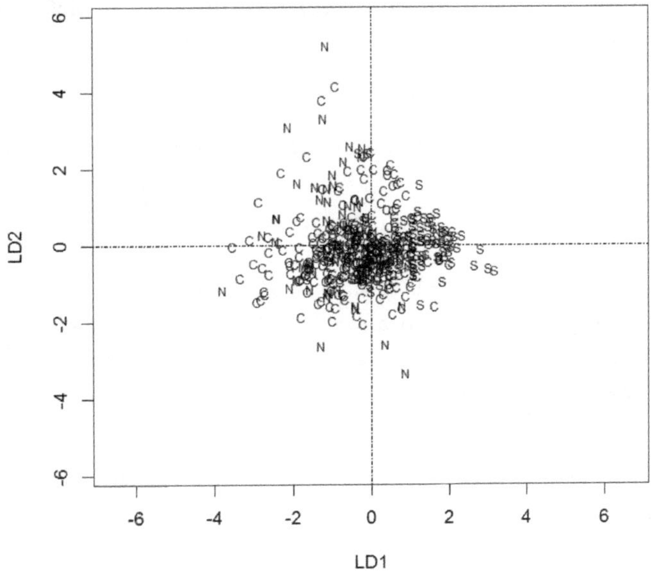

FIG. 7. Scatterplot of the linear discriminant (LD) scores for the first two linear discriminants (LD1 and LD2) showing each observation presented in the grouped scatter plot with Regions groups as labels.

are connected when they belong to the same region, creating the characteristic set of lines that makes this plot one of the most famous visualization techniques in multivariate analysis (Inselberg, 2009). This plot shows that Comp.1 and Comp.2 in our LD provide a nice separation of the North and Central regions versus the Southern IL region, and Comp.3 contributes to the discrimination of samples from North IL. How well does our LD achieve this? That goes to our last, but not least step of the analysis: we need to see how well our new discriminant function performs in the real world by providing an estimation of error rates associated with the DA using cross-validation (Johnson and Wichern, 2002).

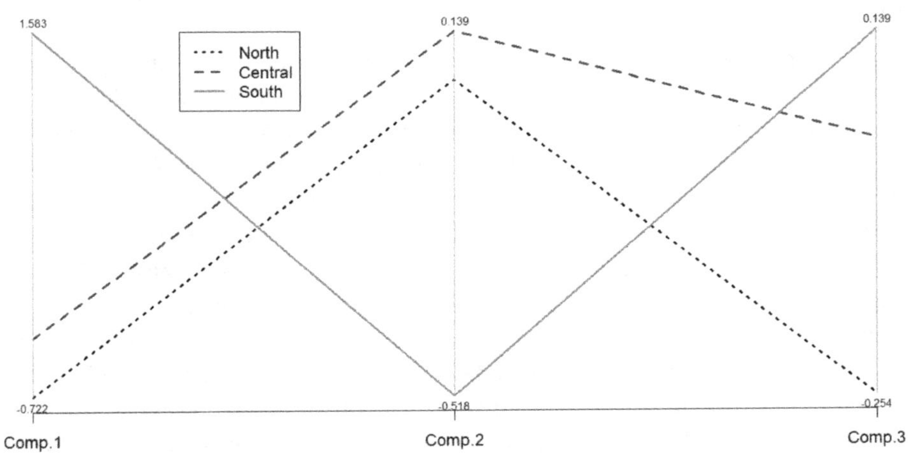

FIG. 8. Parallel coordinates plot for the mean values of the linear discriminant (LD) functions for each Illinois region.

We first use the generic function predict to create a new object YCrpc.ld that contains the predictions generated by our previous DA as follows,

```
> YCrpc.ld<-predict(YCrpc.nmd.lda)
```

Now, we re-run our DA but specifying CV = T, which requests crossvalidation of our predictions,

```
> YCrpc.nmd.ldcv<-lda(YCrpc.nmd [,-1], YCrpc.nmd [,1], CV=T)
```

With this method, the sample of data is partitioned into subsets such that the analysis is initially performed on a single subset (training set), while the other subset(s) (testing set) are retained for subsequent use in confirming and validating the initial analysis. Now we can create a table with both true and predicted values for our DA:

```
> table(true=YCrpc.nmd$Region, pred=YCrpc.nmd.ldcv$class)
         pred
true       North Central South
  North        2      81     2
  Central      2     271    11
  South        0      45    47
```

The above allows us to calculate our error rates. We obtained almost perfect characterization of the Central IL region with 95% (271 correctly classified/284 total Central observations) of the samples correctly classified, yet our discriminant rule failed badly when trying to classify samples from Southern and Northern IL, showing 51% (47 correctly classified/92 total) and 2% (2 correctly classified/85 total) correct classification, respectively—something we need to understand better. These results reflect two critical aspects of our original data set: high homogeneity and fertility of the soil samples submitted (i.e., low variability) and almost three times more samples submitted from Central IL than from other regions (i.e., dissimilar group size). High variability within the original data set and similar group sizes tend to result in better performance of multivariate methods.

Summary and Final Remarks

The concepts we have presented here with our field guide of relevant multivariate methods introduce the researcher to a logical application of these procedures in the agronomic sciences. In approaching a multivariate study, as our hands-on example also shows, the practical implementations of these approaches will most likely involve the inclusion of several methods of multivariate analysis, as well as univariate tests, to better understand the relationships between variables and their relevance to the complexities that are being researched. It is important to begin this process with proper examination of the potential research questions, specifying these a priori, allowing us to more efficiently design a study that properly examines the research questions. Before data collection and analysis, it is advantageous to understand the forth-coming data structure and to familiarize oneself with available software tools and applications that will be applied. Addressing adequate sample size, missingness of observations, and deciding on the appropriate measured variables all aid in optimizing information. The higher dimensionality of the multivariate analyses options can be difficult to interpret and are not a panacea to determining the underlying complexities in the system under research. As our example revealed, PCA assisted in describing the variability of the soil data. While MR results did not reveal a relationship between our soil parameters and the soybean yield, our knowledge of the data origins allowed us to interpret this statistical finding within the agronomical context. The DA among Illinois regions led to a moderately effective characterization of samples from central Illinois based on our PCA variables and allowed us to better understand the intricacies of our data set.

The amount of data and the complexity of the data sets collected in agrobiological systems are ever-increasing, and we believe multivariate approaches will become the norm rather than the exception when investigating and publishing systems-relevant information in our research areas. We hope that our hands-on strategy has given you an opportunity to think and approach your questions and data sets differently and has encouraged you to start using these powerful tools.

Key Learning Points

- Guided by a good research question, MA encompass the simultaneous analysis of all variables measured on each experimental unit.
- Analysis of multivariate data involves a stage of data exploration with an emphasis on visualization that allows for a greater understanding of your data.
- Your research question and the characteristics of the resulting data set will help you identify the most appropriate MA technique to extract the relevant information.
- Each MA technique requires adherence to specific considerations, such as the rules of multivariate normality, linearity, and homoscedasticity for the use of PCA, MR, and linear DA that we cover in this chapter, yet others have far fewer assumptions.
- A good data screening sequence will help you address the challenges that your data set presents to conduct the technique of your choice.
- Explore. Play. Enjoy.

Review Questions

True or False—*Hint:* Consult Table 1 to choose your answers.

1. We can explore relationships among variables in a data set with MR, CART, or even with cluster analysis among other tools.
2. The main difference between PCA and FA is that the PCA tries to uncover theoretical constructs underlying the data set.
3. SEM and CART are examples of flexible modern techniques that can integrate categorical and continuous variables.
4. MANOVA, PCA, and DA are examples of techniques that require multivariate normality and homogeneity of variances.
5. When the goal of the MA is to predict group membership, then DA, LR, or

CART are *not* good choices.

6. The desired result of applying MANOVA to a data set is to create linear combinations of variables that maximize our group differences.
7. No previous knowledge of group membership is required when we explore the data set with CA.
8. If the dependent variable in the data set only takes two values, a LR is the most appropriate technique to describe differences among the two possible outcomes.
9. If your data set does not meet the requirements of multivariate normality you cannot apply any of the MA techniques.
10. You can use canonical correspondence analysis to explore the distribution of several plant species on different environments characterized by topography, moisture level, aspect, etc.

Exercises

For the following exercises we will be working with the iris data set available from R by typing:

```
> library(datasets)
> iris
```

Once the data set is available on your workspace:

1. Investigate the structure of the data set and create a scatterplot matrix of the variables Sepal.Length, Sepal.Width, Petal.Length, Petal.Width. What can you infer from these results?
2. Explore the possibility of successfully separating the Iris species based on the sepal and petal variables listed. *Hint:* You will need the MASS library.
3. Calculate the error rates associated with the classification of each of the Iris species and provide an interpretation of your results.

References

Albrecht, H. 2003. Suitability of arable weeds as indicator organisms to evaluate species conservation effects of management in agricultural ecosystems. Agric. Ecosyst. Environ. 98:201–211. doi:10.1016/S0167-8809(03)00081-1

Arrindell, W.A., and J. van der Ende. 1985. An empirical test of the utility of the observations-to-variables ratio in factor and components analysis. Appl. Psychol. Meas. 9:165–178. doi:10.1177/014662168500900205

Cruz-Castillo, J.G., S. Ganeshanandam, B.R. MacKay, G.S. Lawes, C.R.O. Lawako, and D.J. Woolley. 1994. Applications of canonical discriminant analysis in horticultural research. HortScience 29:1115–1119.

Davis, V.M., L.S. Kull, and J.A. Nelson. 2012. Development of a team-based on-farm learning program while challenging soybean growers to increase yield. J. Extension 50 (online source). http://www.joe.org/joe/2012august/rb5.php (accessed 10 Aug. 2015)

DeDecker, J.J., J.B. Masiunas, A.S. Davis, and C.G. Flint. 2014. Weed management practice selection among Midwest U.S. Organic growers. Weed Sci. 62:520–531. doi:10.1614/WS-D-13-00138.1

Didden, W.A.M., J.C.Y. Marinissen, M.J. Vreekenbuijs, S.L.G.E. Burgers, R. Defluiter, M. Geurs, and L. Brussaard. 1994. Soil mesofauna and macrofauna in 2 agricultural systems- factors affecting population-dynamics and evaluation of their role in carbon and nitrogen dynamics. Agric. Ecosyst. Environ. 51:171–186. doi:10.1016/0167-8809(94)90042-6

Everitt, B.S. 2005. An R and S-plus companion to multivariate analysis. Springer-Verlag, London.

Goidts, E., B.A.S.v. Wesemael, and K. Van Oost. 2009. Driving forces of soil organic carbon evolution at the landscape and regional scale using data from a stratified soil monitoring. Glob. Change Biol. 15:2981–3000. doi:10.1111/j.1365-2486.2009.01940.x

Hao, X., and A.N. Kravchenko. 2007. Management practice effects on surface soil total carbon: Differences along a textural gradient. Agron. J. 99:18–26. doi:10.2134/agronj2005.0352

Inselberg, A. 2009. Parallel coordinates: Visual multidimensional geometry and its applications. Springer, New York.

Jacobson, L.M., M.B. David, and L.E. Drinkwater. 2011. A spatial analysis of phosphorus in the Mississippi river basin. J. Environ. Qual. 40:931–941. doi:10.2134/jeq2010.0386

Johnson, D.A., and R.A. Wichern. 2002. Applied multivariate statistical analysis. Prentice Hall, Upper Saddle River, NJ.

Johnson, D.E. 1998. Applied multivariate methods for data analysts. Brooks/Cole Publishing Co., Pacific Grove, CA.

Kane, D.A., S.S. Snapp, and A.S. Davis. 2015. Ridge tillage concentrates potentially mineralizable soil nitrogen, facilitating maize nitrogen uptake. Soil Sci. Soc. Am. J. 79:81–88. doi:10.2136/sssaj2014.07.0273

Khattree, R., and D.N. Naik. 1999. Applied multivariate statistics with SAS® software. 2nd ed. SAS Institute Inc., Cary, NC.

Khattree, R., and D.N. Naik. 2000. Multivariate data reduction and discrimination with SAS® software. SAS Institute Inc., Cary, NC.

Korkmaz, S., D. Goksuluk, and G. Zararsiz. 2015. MVN: An R package for assessing multivariate normality. https://cran.r-project.org/web/packages/MVN/vignettes/MVN.pdf (accessed 21 July 2015)

Liu, J.G., E. Pattey, M.C. Nolin, J.R. Miller, and O. Ka. 2008. Mapping within-field soil drainage using remote sensing, dem and apparent soil electrical conductivity. Geoderma 143:261–272. doi:10.1016/j.geoderma.2007.11.011

MacCallum, R.C., K.F. Widaman, K.J. Preacher, and S. Hong. 2001. Sample size in factor analysis: The role of model error. Multivariate Behav. Res. 36(4):611–637. doi:10.1207/S15327906MBR3604_06

Martin, N.F., G. Bollero, and D.G. Bullock. 2005. Associations between field characteristics and soybean plant performance using canonical correlation analysis. Plant Soil 273:39–55. doi:10.1007/s11104-004-6639-1

Martin, N.F., A.G. Bollero, and D.G. Bullock. 2007. Relationship between secondary variables and soybean oil and protein concentration. Trans. ASABE 50:1271–1278. doi:10.13031/2013.23638

Osborne, J.W., and A.B. Costello. 2004. Sample size and subject to item ratio in principal components analysis. Practical Assess. Res. Eval. 9(11). http://PAREonline.net/getv.asp?v=9&n-11 (accessed 27 Nov. 2015).

Pike, A.C., T.G. Mueller, A. Schorgendorfer, J.D. Luck, S.A. Shearer, and A.D. Karathanasis. 2010. Locating eroded waterways with United States geological survey elevation data. Agron. J. 102:1269–1273. doi:10.2134/agronj2010.0132

Ping, J.L., C.J. Green, K.F. Bronson, R.E. Zartman, and A. Doermann. 2005. Delineating potential management zones for cotton based on yields and soil properties. Soil Sci. 170:371–385. doi:10.1097/01.ss.0000169904.56743.75

Roel, A., and R.E. Plant. 2004. Factors underlying yield variability in two California rice fields. Agron. J. 96:1481–1494. doi:10.2134/agronj2004.1481

Sawchik, J., and A.P. Mallarino. 2007. Evaluation of zone soil sampling approaches for phosphorus and potassium based on corn and soybean response to fertilization. Agron. J. 99:1564–1578. doi:10.2134/agronj2006.0151

Schutte, B.J., A.S. Davis, S.A. Peinado, and J. Ashigh. 2014. Seed-coat thickness data clarify seed size-seed-bank persistence trade-offs in *Abutilon theophrasti* (Malvaceae). Seed Sci. Res. 24:119–131. doi:10.1017/S0960258514000099

Senthilkumar, S., A.N. Kravchenko, and G.P. Robertson. 2009. Topography influences management system effects on total soil carbon and nitrogen. Soil Sci. Soc. Am. J. 73:2059–2067. doi:10.2136/sssaj2008.0392

Sharma, S. 1996. Applied multivariate techniques. John Wiley & Sons, Hoboken, NJ.

Shukla, M.K., R. Lal, and M. Ebinger. 2006. Determining soil quality indicators by factor analysis. Soil Tillage Res. 87:194–204. doi:10.1016/j.still.2005.03.011

Smith, R.G., A.S. Davis, N.R. Jordan, L.W. Atwood, A.B. Daly, A.S. Grandy, M.C. Hunter, R.T. Koide, D.A. Mortensen, P. Ewing, D. Kane, M. Li, Y. Lou, S.S. Snapp, K.A. Spokas, and A.C. Yannarell. 2014. Structural equation modeling facilitates transdisciplinary research on agriculture and climate change. Crop Sci. 54:475–483. doi:10.2135/cropsci2013.07.0474

Smukler, S.M., L.E. Jackson, L. Murphree, R. Yokota, S.T. Koike, and R.F. Smith. 2008. Transition to large-scale organic vegetable production in the Salinas Valley, California. Agric. Ecosyst. Environ. 126:168–188. doi:10.1016/j.agee.2008.01.028

So, Y.F., M.M. Williams, J.K. Pataky, and A.S. Davis. 2009. Principal canopy factors of sweet corn and relationships to competitive ability with wild-proso millet (*Panicum miliaceum*). Weed Sci. 57:296–303. doi:10.1614/WS-08-158.1

Sorice, M.G., J.R. Conner, U.P. Kreuter, and R.N. Wilkins. 2012. Centrality of the ranching lifestyle and attitudes toward a voluntary incentive program to protect endangered species. Rangeland Ecol. Manage. 65:144–152. doi:10.2111/REM-D-10-00144.1

Tabachnick, B.G., and L.S. Fidell. 2012. Using multivariate statistics. 6th ed. Pearson, Boston, MA.

Tan, Z.X., R. Lal, N.E. Smeck, F.G. Calhoun, R.M. Gehring, and B. Parkinson. 2003. Identifying associations among soil and site variables using canonical correlation analysis. Soil Sci. 168:376–382 doi:10.1097/01.ss.0000070912.55992.d5

Taylor, J.A., and B.M. Whelan. 2011. Selection of ancillary data to derive production management units in sweet corn (*Zea mays* var. Rugosa) using MANOVA and an information criterion. Precis. Agric. 12:519–533. doi:10.1007/s11119-010-9195-0

van den Berg, R.A., H.C. Hoefsloot, J.A. Westerhuis, A.K. Smilde, and M.J. van der Werf. 2006. Centering, scaling, and transformations: Improving the biological information content of metabolomics data. BMC Genomics 7:142. doi:10.1186/1471-2164-7-142

Vaylay, R., and E. van Santen. 2002. Application of canonical discriminant analysis for the assessment of genetic variation in tall fescue. Crop Sci. 42:534–539. doi:10.2135/cropsci2002.0534

Velicer, W.F., and J.L. Fava. 1998. Effects of variable and subject sampling on factor pattern recovery. Psychol. Methods 3:231–251. doi:10.1037/1082-989X.3.2.231

Venables, W.N., and B.D. Ripley. 2002. Modern applied statistics with S. Springer, New York.

Villamil, M.B., M. Alexander, A.H. Silvis, and M.E. Gray. 2012a. Producer perceptions and information needs regarding their adoption of bioenergy crops. Renew. Sustain. Energy Rev. 16:3604–3612. doi:10.1016/j.rser.2012.03.033

Villamil, M.B., V.M. Davis, and E.D. Nafziger. 2012. Estimating factor contributions to soybean yield from farm field data. Agron. J. 104:881–887. doi:10.2134/agronj2012.0018n

Villamil, M.B., F.E. Miguez, and G.A. Bollero. 2008. Multivariate analysis and visualization of soil quality data for no-till systems. J. Environ. Qual. 37:2063–2069. doi:10.2134/jeq2007.0349

Westphal, A., L.J. Xing, R. Pillsbury, and T.J. Vyn. 2009. Effect of tillage intensity on population densities of *Heterodera glycines* in intensive soybean production systems. Field Crops Res. 113:218–226. doi:10.1016/j.fcr.2009.05.009

Williams, M.M., A.S. Davis, T.L. Rabaey, and C.M. Boerboom. 2009. Linkages among agronomic, environmental and weed management characteristics in North American sweet corn. Field Crops Res. 113:161–169. doi:10.1016/j.fcr.2009.05.005

Yeater, K.M., G.A. Bollero, D.G. Bullock, A.L. Rayburn, and S. Rodriguez-Zas. 2004. Assessment of genetic variation in hairy vetch using canonical discriminant analysis. Crop Sci. 44:185–189. doi:10.2135/cropsci2004.0185

Yeater, K.M., S.E. Duke, and W.E. Riedell. 2015. Multivariate analysis: Greater insights into complex systems. Agron. J. 107(2):799–810. doi:10.2134/agronj14.0017

Zheng, H., L. Chen, X. Han, X. Zhao, and Y. Ma. 2009. Classification and regression tree (CART) for analysis of soybean yield variability among fields in northeast china: The importance of phosphorus application rates under drought conditions. Agric. Ecosyst. Environ. 132:98–105. doi:10.1016/j.agee.2009.03.004

CHAPTER 15: NONLINEAR REGRESSION MODELS AND APPLICATIONS

Fernando Miguez,* Sotirios Archontoulis, and Hamze Dokoohaki

Nonlinear statistical models continue to be relevant because they have the advantages of simplicity, flexibility, and parsimony when used in the analysis and description of agricultural data. Often nonlinear models can be easier to interpret than other alternatives. These models have been applied to data ranging from leaf-level to landscape-level processes, such as photosynthesis, biomass accumulation, soil CO_2 efflux, and temperature responses. Our objective is to provide a multistep framework that helps with the application of these models with specific considerations in each step. Even though objectives and applications may vary, common priorities should be set when fitting nonlinear models: obtaining acceptable parameter estimates, a good model fit, and meeting standard assumptions of statistical models. The following steps are considered: (i) choose candidate models, (ii) set starting values, (iii) fit models, (iv) check convergence and parameter estimates, (v) find the "best" model among competing models, (vi) check model assumptions (residual analysis), and (vii) calculate statistical indexes and confidence intervals. In particular, we emphasize the first step (choose candidate models) by providing an extensive library of nonlinear functions (77 equations with the associated parameter interpretations) and application examples in agriculture. We hope that this contribution will clarify some of the difficulties and confusion with using nonlinear models.

Despite the wide availability of techniques for modeling the functional relationship between a dependent variable and independent variable(s), traditional nonlinear models continue to be relevant because they have the advantage of having meaningful parameters and being parsimonious. Recent developments in machine learning (such as neural networks) can provide good predictive abilities, but fail to shed light on the underlying processes. Still, one significant challenge in using nonlinear models is related to choosing the most appropriate model. We often ask the following questions: Which is the best model to describe our data? Which is the best statistical index to judge the goodness of fit? How do we choose among competing models? There are no simple answers to these questions. In this chapter, we attempt to provide agrono-

---

Abbreviations: AIC, Akaike information criterion; BIC, Bayesian information criterion; DOY, day of year; LRT, likelihood ratio test; QQ, quantile–quantile.

F. Miguez, Iowa State University, Dep. of Agronomy, 2104 Agronomy Hall, Ames, IA 50011; S. Archontoulis (sarchont@iastate.edu), H. Dokoohaki (hamzed@iastate.edu). *Corresponding author (femiguez@iastate.edu).

doi:10.2134/appliedstatistics.2016.0003

Applied Statistics in Agricultural, Biological, and Environmental Sciences
Barry Glaz and Kathleen M. Yeater, editors

mists with a general framework on how to approach these questions appropriately. Our specific objectives are: (i) to provide a succinct overview of nonlinear models and to develop guidelines to understand the family of functions used in agricultural applications; (ii) to describe techniques for modifying nonlinear models and how to cope with multiple nonlinear models; (iii) to discuss key methodological issues related to parameter estimation, model performance, and comparisons; and (iv) to demonstrate step-by-step analysis of experimental data using a nonlinear regression model. The chapter follows the structure of the flow diagram illustrated in Fig. 1. We start with the definition of nonlinear regression models and discuss their main advantages and disadvantages. Then we present 77 nonlinear functions with references to applications in agriculture. The last sections offer an updated overview of methodologies used to fit models, choose starting values, assess goodness of fit, select the right models, and evaluate residuals. Finally we provide an example of fitting a nonlinear model based on experimental data of biomass growth over time (Danalatos et al., 2009).

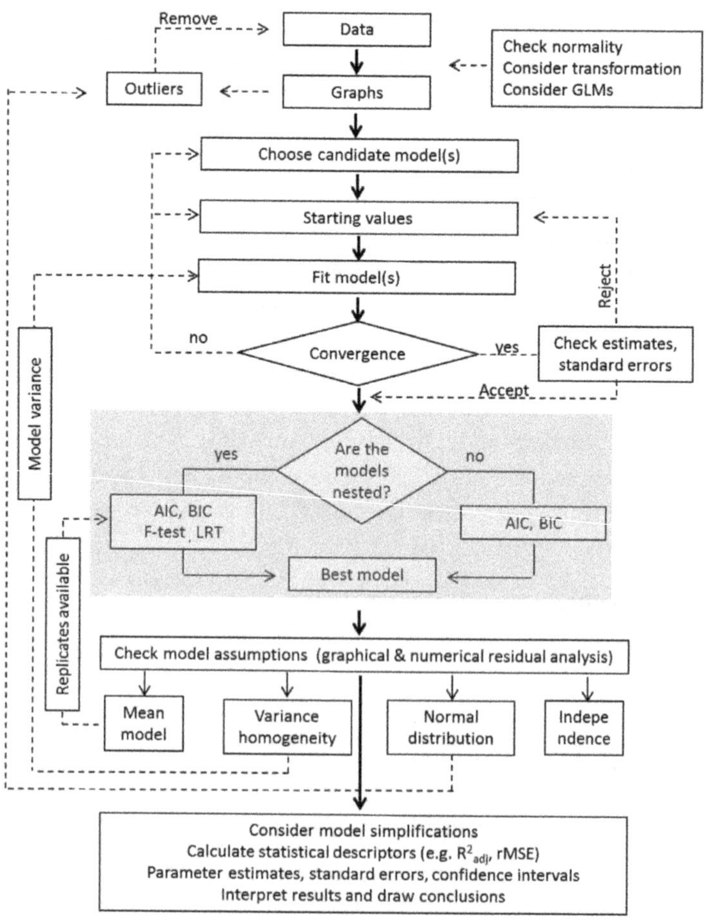

FIG. 1. Suggested workflow in the nonlinear regression analysis. Thick arrows indicate major steps, thin arrows indicate substeps and dashed arrows indicate feedback in nonlinear regression. The shaded part is optional and can be ignored in simple cases. GLM, generalized linear models; LRT, likelihood ratio test; AIC, Akaike information criterion; BIC, Bayesian information criterion.

Nonlinear Regression Model

Definition

In general, statistical models used in agricultural applications can be described with the following notation:

$$y = f(x, \theta) + \varepsilon \qquad [1]$$

where y is the response variable, f stands for function but can also be thought of as the *model* or *regression function*, x are the inputs, θ denotes the parameters to be estimated, and ε is the error. Each parameter can be evaluated for whether it is linear: if the second derivative of the function with respect to a parameter is not equal to zero, then the parameter is nonlinear. Thus, a given function (f) can have a mix of linear and nonlinear parameters.

Why Should We Use Nonlinear Models?

The main benefits of nonlinear models are parsimony, interpretability, and prediction (Bates and Watts, 2007). In general, nonlinear models are capable of accommodating a vast variety of patterns, although each individual nonlinear model can be less flexible than linear models (i.e., polynomials) in terms of the variety of data they can describe. However, nonlinear models appropriate for a given application can be more parsimonious (i.e., there will be fewer parameters involved) and more easily interpretable. Interpretability comes from the fact that the parameters can be associated with a biologically meaningful process. For example, one of the most widely used nonlinear models is the logistic equation (Eq. [2.1] in Table 1). This model describes the pervasive "S"-shaped growth curve. The parameters have a clear meaning (see Table 1) and units associated with their definition. The asymptotic parameter (y_{asym}) has units equal to the response variable (y), the inflection point (t_m) has units equal to the independent variable (t), and the parameter that determines the steepness of the curve (k) has units equal to t. A competing cubic or higher order polynomial model used to describe the same data would have the disadvantages that more parameters will be needed (than just three) and that the parameters will not be easily interpretable (Pinheiro and Bates, 2000). For example, what would be the interpretation of the parameters associated with the x^3, x^4, and x^5 in a five degree polynomial? In addition, when x is a large number the value of x^5 can be extremely large, making the parameter associated with it extremely small and possibly numerically unstable.

The other benefit of using nonlinear regression models is that their predictions tend to be more robust than competing polynomials, especially outside the range of observed data (i.e., extrapolation). Although extrapolation is not considered to be a good practice, when it is necessary to do so, nonlinear models based on functional relationships can have acceptable performance. Nonlinear regression models, however, come at a cost. Their main disadvantages are that they can be less flexible than competing linear models and that generally there is no analytical solution for estimating the parameters. The first point underscores that the choice of the nonlinear model is crucial, thus making it tempting to try a large library of functions and

TABLE 1. Nonlinear regression models. For example fits, see figures in the supplemental material online.

Eq.	Name/reference: form	Schematic graph/	changing parameter	Parameter meaning
Group I: Exponential				
[1.1]	Exponential decay: $Y = Y_o e^{-kt}$	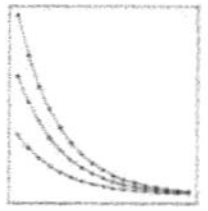	Y_o	Y is the response variable (e.g., soil organic matter), t is the explanatory variable (e.g., time), Y_o is the initial or the maximum Y value, k is a rate constant determines the steepness of the curve.
[1.2]	Exponential rise: $Y = Y_o(1 - e^{-kt})$		Y_o	
Group II: Sigmoid functions				
[2.1]	Logistic (Verhulst, 1838): $Y = \frac{Y_{asym}}{1+e^{-k(t-t_m)}}$	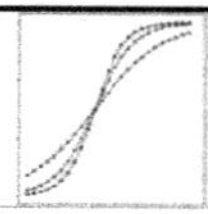	k	Y is the response variable (e.g., biomass), t is the explanatory variable (e.g., time), Y_{asym} or Y_{max} is the asymptotic or the maximum Y value, t_m is the inflection point at which the growth rate is maximized, k controls the steepness of the curve, v deals with the asymmetric growth (if $v = 1$, then Richards equation becomes logistic), a and b are parameters determine the shape of the curve, t_e is the time when $Y = Y_{asym}$, t_c is the critical time for a switch-off to occur (e.g., critical photoperiod), n is a parameter determines the sharpness of the response.
[2.2]†	Richards (1959): $Y = \frac{Y_{asym}}{\left[1+ve^{-k(t-t_m)}\right]^{\frac{1}{v}}}$	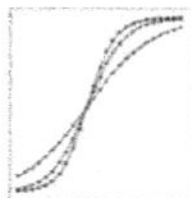	k	
[2.3]	Gompertz (1825): $Y = Y_{asym}e^{-e^{-k(t-t_m)}}$	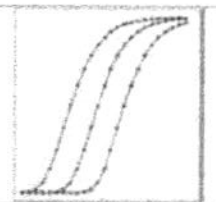	t_m	
[2.4]	Weibull (1951): $Y = Y_{asym}\left(1-e^{-at^b}\right)$	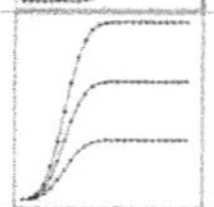	Y_{asym}	
[2.5]‡	Beta (Yin et al., 2003b): $Y = Y_{max}\left(1+\frac{t_e-t}{t_e-t_m}\right)\left(\frac{t}{t_e}\right)^{\frac{t_e}{t_e-t_m}}$	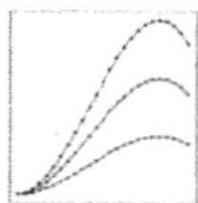	Y_{max}	

Eq.	Name/reference: form	Schematic graph/ changing parameter	Parameter meaning
[2.6]§	Hill (switch-off) function: $Y = \frac{t_c^n}{t_c^n + t^n}$	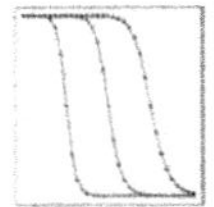 t_c	
Group III: Photosynthesis			
[3.1]	Blackman (1905): $Y = \min(Y_{asym}, aI) - R_d$	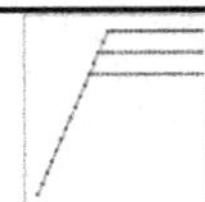 Y_{asym}	Y is the response variable (net photosynthesis), I is the explanatory variable (irradiance), Y_{asym} is the asymptotic Y value, a is the initial slope of the curve at low I levels, R_d is the dark respiration, θ is a dimensionless curvature parameter when $\theta = 1$ Eq. [3.4] is equivalent to Eq. [3.1] and when $\theta \to 0$ Eq. [3.4] is equivalent to Eq. [3.3].
[3.2] ¶	Asymptotic exponential: $Y = Y_{asym}\left(1 - e^{\frac{aI}{Y_{asym}}}\right) - R_d$	 Y_{asym}	
[3.3]¶	Rectangular hyperbola: $Y = \frac{aIY_{asym}}{Y_{asym} + aI} - R_d$	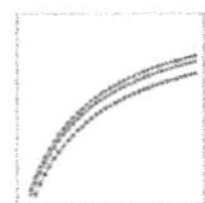 R_d	
[3.4] ¶,#	Non rectangular hyperbola: $Y = \frac{Y_{asym} + aI - \sqrt{(Y_{asym} + aI)^2 - 4\theta aIY_{asym}}}{2\theta} - R_d$	 Y_{asym}	
[3.5]	Modified logistic (Sinclair and Horie, 1989): $Y = Y_{asym}\left[\frac{2}{1+e^{-k(N-N_{min})}} - 1\right]$	 N_{min}	Y is the response variable (light saturated net photosynthesis), N is the explanatory variable (leaf nitrogen), Y_{asym} is the asymptotic Y value, k determines the curvature of the curve, N_{min} is the N value at or below which $Y = 0$.

Table continued.

TABLE 1. Continued.

Eq.	Name/reference: form	Schematic graph/ changing parameter		Parameter meaning
[3.6]††	Farquhar et al. (1980): $Y=\min\left[\frac{V_{\text{cmax}}\left(C_i-\Gamma^*\right)}{C_i+K_{\text{mc}}\left(1+O/K_{\text{mo}}\right)},\frac{J\left(C_i-\Gamma^*\right)}{4C_i+8\Gamma^*}\right]-R_{\text{day}}$		V_{cmax}	Y is the response variable (net photosynthesis), C_i is the explanatory variable (intercellular CO_2 concentration), V_{cmax} is the maximum carboxylation capacity, Γ^* is the CO_2 compensation point in the absence of R_d, K_{mc} and K_{mo} are Michaelis–Menten coefficients of Rubisco for CO_2 and O_2, respectively, O is the partial pressure of oxygen (= 210 mbar), J is the photosystem II electron transport rate, R_{day} is the dark respiration occurring in the light.
Group IV: Temperature dependencies				
[4.1]	Van't Hoff (1898) (known as Q_{10} function): $Y=Q_{10}^{\frac{T-T_{\text{ref}}}{10}}$		T_{ref}	Y is the response variable (e.g., respiration), T is the explanatory variable (temperature), T_{ref} is a reference temperature, at which $Y=1$, Q_{10} is the factor by which the rate of a process (respiration) increases for each 10°C temperature increase, E is the activation energy determines the increase in temperature response, R is the universal gas constant (= 8.314 J K^{-1} mol^{-1}), D is the deactivation energy that determines the decrease in the temperature response, S is the entropy term that determines the transition state of the curve, E_o is an activation-energy-like parameter that is temperature-adjusted, T_x is a fitted temperature parameter in K, T_{min} is the base or minimum temperature for $Y=0$.
[4.2]	Arrhenius (1889): $Y=e^{\left[\frac{E}{R}\left(\frac{1}{T_{\text{ref}}-273}-\frac{1}{T-273}\right)\right]}$	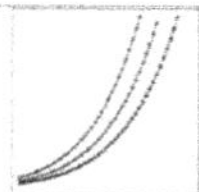	T_{ref}	
[4.3]‡‡	Modified Arrhenius: $Y=e^{\left[\frac{E}{R}\left(\frac{1}{T_{\text{ref}}-273}-\frac{1}{T-273}\right)\right]}\left\{\frac{1+e^{\left[\frac{(T_{\text{ref}}+273)S-D}{(T_{\text{ref}+273})R}\right]}}{1+e^{\left[\frac{E}{R}\frac{D}{(T+273)R}\right]}}\right\}$	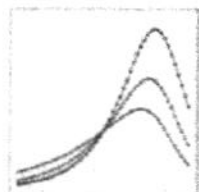	E	
[4.4]	Lloyd and Taylor (1994): $Y=e^{E_o\left(\frac{1}{T_{\text{ref}}-273-T_x}-\frac{1}{T-273-T_x}\right)}$	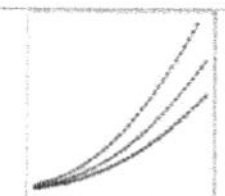	T_{ref}	

Eq.	Name/reference: form	Schematic graph/ changing parameter		Parameter meaning
[4.5]	Ratkowsky et al. (1982): $Y = \frac{(T - T_{min})^2}{(T_{ref} - T_{min})^2}$	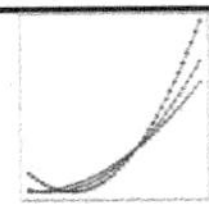	T_{min}	
Group V: Peak or bell shape curves				
[5.1]	Beta (Yin et al. 1995): $Y = \left[\left(\frac{T_c - T}{T_c - T_o}\right)\left(\frac{T - T_b}{T_o - T_b}\right)^{\frac{T_o - T_b}{T_c - T_o}}\right]^c$	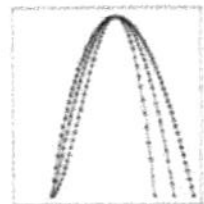	T_{ref}	Y is the response variable (e.g., rate of development), T is the explanatory variable (temperature), T_o is the optimum temperature for maximum Y, T_b is the base or minimum temperature for $Y = 0$ T_c is the ceiling or maximum temperature for $Y = 0$ c is a curvature parameter (default $c = 1$).
[5.2]§§	Bell curve: $Y = Y_{asym} e^{\left[a(X - X_o)^2 + b(X - X_o)^3\right]}$	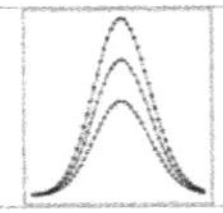	Y_{asym}	Y is the response variable, X is the explanatory variable, Y_{asym} is the asymptotic maximum Y value, X_o is the position of the center of the peak (Y_{asym}), a and b are coefficients controlling the width of the bell.
[5.3]	Gaussian function: $Y = Y_{asym} e^{\left[-0.5\left(\frac{X - X_o}{b}\right)^2\right]}$	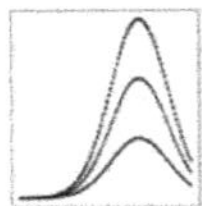	Y_{asym}	
Group VI: Others nonlinear equations				

Table continued.

TABLE 1. Continued.

Eq.	Name/reference: form	Schematic graph/ changing parameter		Parameter meaning
[6.1]	Power: $Y = aX^b$	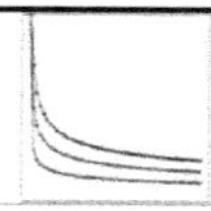	a	Y is the response variable, X is the explanatory variable, a and b parameters defines the shape of the curve and the magnitude of the Y value.
[6.2]	Modified hyperbola: $Y = \frac{aX}{1+bX}$	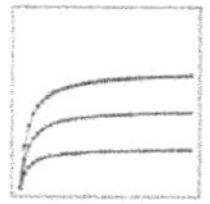	a	
[6.3]	Michaelis–Menten: $Y = -\frac{\mu X}{X + C_{sat}}$	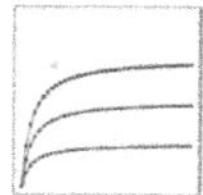	C_{sat}	Y is the response variable (e.g., denitrification rate), X is the explanatory variable, the substrate (e.g., NO_3), μ is the rate constant, C_{sat} is the half-saturation constant.
[6.4]	Rational function: $Y = \frac{a_1 X^{a_2}}{1 + a_3 X^{a_4}}$		a_1	Y is the response variable, X is the explanatory variable, a_1 and a_3 are parameters defining the magnitude of the Y value, a_2 and a_4 are parameters defining the shape of the curve; if $a_2 = a_4 + 1$, then equation shows a near-linear response; if $a_2 = a_4$, then equation becomes hyperbola; if $a_2 < a_4$, then equation takes a bell-shape; if $a_2 > a_4$ or $a_4 = 0$, the, equation becomes exponential; if $a_2 = 0$, then equation becomes exponential decay.
[6.5]	Ricker curve: $Y = a_1 X e^{-a_2 X}$	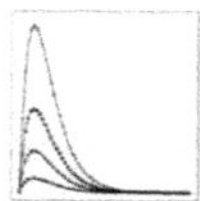	a_1	Y is the response variable, X is the explanatory variable, a_1 and a_2 parameters control both the height and the width of the skewed to the right "bell."

† The maximum growth rate for the Richards equation is given in Birch (1999).
‡ The maximum growth rate for the Beta equation is given in Yin et al. (2003b).
§ Cited in Amaducci et al. (2008).
¶ Cited in Goudriaan (1979).
# Goudriaan (1979) and Johnson et al. (2010) used the nonrectangular hyperbola to describe photosynthetic rate response to CO_2.
†† This is simplified version of the Farquhar model. For more details see references in the text.
‡‡ The optimum temperature for this equation is given in Fig. 2.
§§ Cited in Hammer et al. (2009).
Cited in Bril et al. (1994).

choose the model with lowest error. However, it is almost always better to choose a model based on whether it has been used successfully in similar applications and has biologically meaningful parameters (e.g., Table 1).

The lack of analytical solutions has two practical consequences. First, a numerical method needs to be used to find estimates for the parameters, and this implies that convergence of the algorithm needs to be checked (Fig. 1). A lack of convergence often results from the second consideration, which is that these numerical methods require starting values. Choosing a model with biologically meaningful parameters makes the process of choosing starting values easier as the starting values can usually be easily determined from visual inspection of the data (more on this later).

Typical Nonlinear Models and Application Examples

Choosing among competing models for an application is not always a simple task. We have developed a reference table as a guideline to understand the family of functions used in agricultural applications. Table 1 presents 27 common nonlinear equations and Tables 2 through 7 supplements these with 45 additional equations. We classified the equations into six groups. All equations have been used in agricultural applications, and most of the parameters have a meaningful interpretation. The variety of equations in Table 1 reflects the fact that one equation does not suit all processes.

Group I: Exponential

The exponential decay and exponential models, which give rise to maximum functions (Eq. [1.1] and [1.2], Table 1) have been used in a wide spectrum of soil and plant science applications. Both equations have been commonly used to describe light and nitrogen vertical distributions within plant canopies (Monsi and Saeki, 2005), nitrous oxide emission response to nitrogen fertilizer (e.g., Hoben et al., 2011), cumulative soil respiration (e.g., Gillis and Price, 2011), residue or soil organic matter decomposition (Thorburn et al., 2001), photoperiodic sensitivity (e.g., Wang and Engel, 1998), temperature or moisture responses on nitrification (e.g., Ma and Shaffer, 2001), and water infiltration rate (Horton, 1940), and general first-order kinetics processes. It is a simple equation with one major unknown, the rate constant (k), which is also termed *extinction coefficient* in crop physiology or decomposition rate constant in soil science. Also, the ratio log(2)/k is of importance in soil science because it denotes the mean residence time (e.g., soil organic matter). Equation [1.1] provided the starting point to develop case-specific nonlinear functions. Yin et al. (2000, 2003a) established a nonlinear function to describe leaf area index development as a function of canopy nitrogen content (Eq. [2.8], Table 2). Johnson et al. (2010) developed a flexible nonlinear function for the protein (nitrogen) vertical distribution within plant canopies (Eq. [2.9], Table 2). In soil science, Andren and Paustian (1987) and Gillis and Price (2011) extended Eq. [1.1] to better describe decomposition of straw residue and biochar, respectively (Eq. [2.5], [2.6], and [2.7], Table 2). Lastly, Eq. [1.1], as well as expolinear functions that combine linear and exponential phases (namely Eq. [3.15] and [3.16], Table 3) have also been applied to describe initial parts of growth curves, but not the entire growth curves because the growth profile often reaches an asymptotic value.

TABLE 2. Equations derived from the exponential equation. Numbering continues from Group I of the Table 1.

Eq.	Name and/or reference	Form	Parameter meaning
[2.3]	Parallel or double first order	$Y = fY_o e^{-k_1 t} + (1-f)Y_o e^{-k_2 t}$	Y is the response variable; t is the explanatory variable (e.g., time); Y_o is the initial or maximum Y value; k, k_1 and k_2 are rate constants determines the steepness of the curve (units: time^{-1}); f, f_1, and f_2 are fractions (range: 0–1) that determine the size of each pool; b is a rate constant for the linear component; t_i is the infection point for the logistic component (units: time); t_d is the distance from infection point to 3/4 maximum (units: time).
[2.4]	Triple first order	$Y = f_1 Y_o e^{-k_1 t} + f_2 Y_o e^{-k_2 t} + (1-f_1-f_2)Y_o e^{-k_3 t}$	
[2.5]	Consecutive first order (Andren and Paustian, 1987)	$Y = fY_o e^{-k_1 t} + (1-f)Y_o\left[e^{-k_2 t} + \frac{k_2}{k_1-k_2}\left(e^{-k_2 t} - e^{-k_1 t}\right)\right]$	
[2.6]	First order plus linear (Gillis and Price, 2011)	$Y = Y_o\left(1-e^{-kt}\right) + bt$	
[2.7]	First order plus logistic (Gillis and Price, 2011)	$Y = Y_o f\left(1-e^{-kt}\right) + \frac{Y_o(1-f)}{1+e^{-\frac{t-t_i}{t_d}}}$	
[2.8]	Leaf area index nitrogen limited equation Yin et al. (2000, 2003a)	$Y_{LAI} = \frac{1}{k_n}\ln\left(1+\frac{k_n X_N}{n_b}\right)$	Y_{LAI} *and* Y_N are the response variables, leaf area index and nitrogen content, respectively; X_N and X_{LAI} are the explanatory variables, canopy nitrogen content ,and leaf area index, respectively; k_n is the rate constant and determines the steepness of the curve (also called nitrogen extinction coefficient); n_b is the minimum value of nitrogen at or below which leaf photosynthesis under saturated light conditions is zero (usually n_b ranges from 0.25 to 0.80; see review by Archontoulis et al., 2012); Y_o is the initial or maximum Y value, c is a dimensionless coefficient.
[2.9]	Protein or nitrogen distribution within plant canopies (Johnson et al., 2010)	$Y_N = Y_o - \left(Y_o - n_b\right)\left(1-e^{-k_n X_{LAI}}\right)^c$	

Eq.	Name and/or reference	Form	Parameter meaning
[2.10]	Photoperiodic sensitivity (Wang and Engel, 1998)	$Y = e^{-P_{sen}\left(X - P_c\right)}$	Y is the response variable, X is the explanatory variable (photoperiod), P_{sen} is the photoperiod sensitivity coefficient being positive for short-days plants and negative for long days plants, P_c is the critical day-length, around 11 h for short days and 18 h for long days.
[2.11]	Model for water infiltration (Horton, 1940)	$Y = Y_c + (Y_o - Y_c)e^{-kt}$	Y is the response variable (infiltration capacity, mm/h), t is the explanatory variable (time, hours), k is the rate constant representing the rate of decrease, Y_c indicates a final or equilibrium capacity (mm/h) Y_o indicates the initial infiltration capacity (mm/h).

TABLE 3. Additional sigmoid type functions. Numbering continues from Group II of the Table 1.

Eq.	Name and/or reference	Form	Parameter meaning and application
Different version of the logistic equation†			
[3.7]	5-parameter	$Y = Y_{\text{low}} + \dfrac{Y_{\text{asym}}}{\left[1 + e^{-k(t - t_{\text{m}})}\right]^{c}}$	Y is the response variable (e.g., biomass), t is the explanatory variable (e.g., time), Y_{asym} is the asymptotic Y value, Y_{low} is the initial Y value, t_{m} is the infection point at which growth rate is maximum, k determines the steepness of the curve, c is a shape parameter.
[3.8]	4-parameter	$Y = Y_{\text{low}} + \dfrac{Y_{\text{asym}}}{1 + e^{-k(t - t_{\text{m}})}}$	
[3.9]	2-parameter	$Y = \dfrac{1}{1 + e^{-k(t - t_{\text{m}})}}$	
[3.10]	1-parameter	$Y = \dfrac{1}{1 + e^{-(t - t_{\text{m}})}}$	
Different versions of the beta function‡			
[3.11]	5-parameter; for cases when initial phase is important (Yin et al., 2003b)	$Y = Y_{\text{b}} + \left(Y_{\text{max}} - Y_{\text{b}}\right)\left(1 + \dfrac{t_{\text{e}} - t}{t_{\text{e}} - t_{\text{m}}}\right)\left(\dfrac{t - t_{\text{b}}}{t_{\text{e}} - t_{\text{b}}}\right)^{\frac{t_{\text{e}} - t_{b}}{t_{\text{e}} - t_{\text{m}}}}$	Y is the response variable, t is the explanatory variable (e.g., time), Y_{max} is the maximum Y value, t_{m} is the infection point at which growth rate is maximized, t_{e} is the t value (time) when $Y = Y_{\text{max}}$, t_{b} is the t value when growth starts, Y_{b} is the Y value at time t_{b}.

Eq.	Name and/or reference	Form	Parameter meaning and application
[3.12]	2-parameter; for cases when maximum growth rate achieved at the beginning of growth (Yin et al., 2003a)	$Y = Y_{\text{max}}\left(2t_e - t\right)\frac{t}{t_e^{\,2}}$	
[3.13]	2-parameter; for cases when maximum growth achieved toward the end of growth period (Yin et al., 2003a)	$Y = Y_{\text{max}}\left(3t_e - 2t\right)\frac{t^2}{t_e^3}$	
[3.14]	3-parameter; Applied to describe leaf senescence; (Yin et al., 2009)	$Y = Y_{\text{max}}\left[1 - \left(1 + \frac{t_e - t}{t_e - t_m}\right)\left(\frac{t}{t_e}\right)^{\frac{t_e}{t_e - t_m}}\right]$	
		Expolinear functions for plant growth analysis	
[3.15]	Expolinear (Goudriaan and Monteith, 1990)	$Y = \frac{c_m}{r_m}\log\left[1 + e^{r_m(t - t_o)}\right]$	Y is the response variable (e.g., biomass), t is the explanatory variable (time), c_m is the maximum growth rate in the linear phase, r_m is the maximum relative growth rate in the exponential phase, t_o is the time when growth begins, Y_{max} is the maximum Y value.
[3.16]	Symmetrical expolinear (Goudriaan, 1994)	$Y = \frac{c_m}{r_m}\log\left[\frac{1 + e^{r_m(t - t_o)}}{1 + e^{r_m\left(t - t_o - \frac{Y_{\text{max}}}{c_m}\right)}}\right]$	

Table continued.

TABLE 3. Continued.

Eq.	Name and/or reference	Form	Parameter meaning and application
Additional generic sigmoid functions			
[3.17]	Levakovic (cited in Zeide, 1993)	$Y = Y_{asym}\left(\frac{t^d}{\beta + t^d}\right)^c$	Y is the response variable (e.g., biomass), t is the explanatory variable (time), Y_{asym} is the asymptotic Y value, β and c determine the start of growth and the inflection point, d controls the shape of the curve.
[3.18]	Korf (cited in Zeide, 1993)	$Y = Y_{asym}e^{-\beta t^{-c}}$	
[3.19]	Special Von Bertalanffly, (1938)§	$Y = Y_{asym}\left[1 - e^{-k(t - t_o)}\right]$	Y is the response variable (e.g., biomass), t is the explanatory variable (time), Y_{asym} is the asymptotic maximum Y value, t_o is the time when growth begins, t_{50} is the time when $Y = 1/2$ of Y_{asym}, k controls the steepness of the curve, v controls the asymmetry and the shape of the curve (for $v = 1$, 2.20 becomes 2.19).
[3.20]	Generalized Von Bertalanffly, (1938)§	$Y = Y_{asym}\left[1 - e^{-k(t - t_o)}\right]^v$	
[3.21]	Log-logistic distribution (Ritz et al., 2013)	$Y = \frac{Y_{asym}}{1 + \exp\left[k\left\{\log(t) - \log(t_{50})\right\}\right]} = \frac{Y_{asym}}{1 + (t/t_{50})^k}$	
Inverse sigmoid functions used to quantify water retention curves¶			
[3.22]	van Genuchten (1980)	$\theta_h = \theta_r + \frac{\theta_s - \theta_r}{\left[1 + (ah)^n\right]^{1 - \frac{1}{n}}}$	θ_ψ is the volumetric water content (mm/mm), h is the matric potential (kPa), θ_r is the residual water content (mm/mm), θ_s is the saturated water content (mm/mm), a is the inverse of the air entry suction ($a > 0$), n is a measure of the poor-size distribution ($n > 1$).

Eq.	Name and/or reference	Form	Parameter meaning and application
[3.23]	Gardner (1958)	$\theta_h = \theta_r + \dfrac{\theta_s - \theta_r}{1+(ah)^n}$	
[3.24]	Brooks and Corey (1964)	$\theta_h = \theta_r + \dfrac{\theta_s - \theta_r}{(ah)^n}$	
[3.25]	Campbell (1974)	$\theta_h = \theta_s (ah)^n$	

† The 3-parameter logistic equation is given in Table 1 (see Eq. [2.1]).

‡ For additional versions of the beta growth function see Yin et al. (2003b).

§ For a comparison of the two von Bertalanffy (1938) models see Urban (2002).

¶ The R package HydroMe includes the listed water retention equations (see Omuto and Gumbe, 2009).

TABLE 4. Additional equations for photosynthesis. Numbering continues from Group III of the Table 1.

Eq.	Name and/or Reference	Form	Parameter meaning & application
[4.7]	Modified rectangular hyperbola proposed to describe photosynthetic reduction at over-saturated irradiance levels (e.g., photorespiration) (Ye and Zhao, 2010)	$Y = aI\frac{1-\beta I}{1+\gamma I} - R_d$	Y is the response variable (net photosynthesis), I is the explanatory variable (irradiance), a is the slope of the curve when I equals zero, β and γ are coefficients with no clear (biological) meaning.
[4.8]	Modified hyperbola used to describe light saturated photosynthesis response to nitrogen (del Pozo and Dennett, 1999)	$Y = Y_{asym}\frac{N - N_{min}}{N + k}$	Y is the response variable (light saturated net photosynthesis), N is the explanatory variable (leaf nitrogen), Y_{asym} is the asymptotic maximum Y value, k determines the curvature of the curve, N_{min} is the N value at or below which $Y = 0$.
[4.9]	Weibull type sigmoid curve used to assess water stress effects on photosynthesis (Vico and Porporato, 2008)	$Y = Y_{asym}e^{\left[-\left(\frac{-\Psi_L}{d}\right)^b\right]}$	Y is the response variable (e.g., carboxylation rate), Ψ_L is the explanatory variable (water potential), Y_{asym} is the asymptotic Y value, b and d determine the shape of the curve.

TABLE 5. Additional equations for temperature dependencies. Numbering continues from Group IV of the Table 1.

Eq.	Name and/or reference	Form	Parameter meaning & application
[5.6]	Power (e.g., Shibu et al., 2006)	$Y=\left(\frac{T}{T_o}\right)^c$	Y is the response variable, T is the explanatory variable (temperature), T_o is the optimum temperature for maximizing Y, c is a shape factor (usually $c = 2$), a determines the magnitude of the Y value, b determines the shape of the curve.
[5.7]	Logarithmic (e.g., Foereid et al., 2011)	$Y=a+b\log(T)$	
[5.8]	Exponential (e.g., Shibu et al., 2006)	$Y=ae^{bT}$	
[5.9]†	Modified exponential (O'Connell, 1990)	$Y=e^{\left[a+bT\left(1-0.5\frac{T}{T_o}\right)\right]}$	
[5.10]	Bell-shaped 3-parameter; Used to describe photosynthesis response to temperature (Johnson et al., 2010)	$Y=\left[\frac{(1+c)T_o-T_{min}-cT}{(1+c)T_o-T_{min}-cT_{ref}}\right]\left(\frac{T-T_{min}}{T_{ref}-T_{min}}\right)^c$	Y is the response variable (e.g., photosynthesis), T is the explanatory variable (temperature), T_{ref} is a reference temperature, at which $Y = 1$, T_{min} is the minimum temperature for $Y = 0$, T_o is the optimum temperature for maximum Y, c is a shape factor.

† Eq. [5.9] follows an increasing and decreasing pattern like the modified Arrhenius equation (see Eq. [4.3] in Table 1, Fig. 2). It was used by Kirschbaum (1995) to describe soil organic matter decomposition as a function of temperature.

TABLE 6. Additional peak or bell shape equations. Numbering continues from Group V of the Table 1.

Eq.	Name and/or reference	Form	Parameter meaning and application
[6.4]†	Lorentzian 3-parameter	$Y = \dfrac{a}{1+\left(\dfrac{X-X_o}{b}\right)^2}$	Y is the response variable, X is the explanatory variable, Y_o is the initial Y value, X_o is the position of the center of the peak, a and b are coefficients controlling the height and width of the peak, c determines the number of peaks ($0 < c < 1$ for one peak).
[6.5] †	Lorentzian 4-parameter	$Y = Y_o + \dfrac{a}{1+\left(\dfrac{X-X_o}{b}\right)^2}$	
[6.6]‡	Lognormal 3-parameter	$Y = ae^{-0.5\left[\dfrac{\log\left(\dfrac{X}{X_o}\right)}{b}\right]^2}$	
[6.7]	Pseudo–Voigt 4-parameters	$Y = a\left[\dfrac{c}{1+\left(\dfrac{X-X_o}{b}\right)^2}+(1-c)e^{-0.5\left(\dfrac{X-X_o}{b}\right)^2}\right]$	

† Condition $b > 1$.

‡ Condition $0 < b < 1$.

TABLE 7. Additional equations. Numbering continues from Group VI of the Table 1.

Eq.	Name and/or reference	Form	Parameter meaning & application
[7.6]†	Holling type IV (e.g., Bolker, 2008)	$Y = \frac{Y_{asym} X^2}{b + cX + X^2}$	Y is the response variable, X is the explanatory variable, Y_{asym} is the asymptotic Y value, b and c define the position of the peak and the shape of the curve, Y_o is the initial or maximum Y value.
[7.7]‡	Shepherd (e.g., Bolker, 2008)	$Y = \frac{Y_{asym} X}{b + X^c}$	
[7.8]‡	Hassell (e.g., Bolker, 2008)	$Y = \frac{Y_{asym} X}{(b + X)^c}$	
[7.9]§	2 parameter exponent (e.g., Ratkowsky, 1993)	$Y = Y_o b^X$	
[7.10]	Monod kinetics	$Y = \frac{\mu X}{X + C_{sat}}$	Y is the response variable (e.g., mineralization rate), X is the explanatory variable, the substrate (e.g., OC), μ is the rate constant, C_{sat} is the half-saturation constant, B is the rate of microorganisms (e.g., size of microbial biomass), if $B = 1$, then Monod becomes M–M equation.

† If $c < 0$ the equation takes a peak form; the peak occurs at $X = -2b/c$. This function is called rational function because expresses the ratio of two polynomial functions. It can also take a S-shape (see Bolker, 2008).

‡ If $c > 0$, then equation takes a peak form.

§ If $b > 1$, then Y increases; if $b < 1$ then Y decreases.

Group II: Sigmoid Curves

Sigmoid curves (mathematical functions having an "S" shape) are another important group of nonlinear models. These models are often applied to describe plant height, weight, leaf area index, or seed germination as a function of time, nitrogen application rate, or herbicide dose (e.g., Gan et al., 1996). Sigmoid equations are also used as 0–1 modifiers in process-based models to incorporate the effect of soil moisture or pH effects on soil nitrogen transformation processes (e.g., McGechan and Wu, 2001) and also as a switch-off function in studies assessing plant photoperiodic sensitivity (e.g., Amaducci et al., 2008). Table 1 (Group II) presents common sigmoid functions, and Table 3 provides additional sigmoidal equations, providing increased flexibility (e.g., when maximum growth or inflection point is achieved at the start or at the end of the growth period). Additional equations can be found in Zwietering et al. (1990), Zeide (1993), Leduc and Goelz (2009), and in many statistical textbooks or software manuals (e.g., SigmaPlot, JMP, TableCurve). In general, the suitability of a sigmoid equation to estimate maximum rate of increase or optimum X level for maximizing the Y value is an important research objective (Birch, 1999; Yin et al., 2003b). Each function has its advantages and disadvantages (for a discussion, see Birch, 1999; Yin et al., 2003b), and it is up to the researcher to select the most appropriate one to fit the experimental data. The logistic equation (Eq. [2.1], Table 1) describes symmetric growth having an inflection point at 1/2 of final size (i.e., asymptote). The Gompertz equation has an inflection point that is controlled by its asymptotic value and is at about one-third ($1/e = 0.3679$), while others like Richards or Weibull or Beta have more flexibility in dealing with asymmetric growth (the inflection point can be at any X value). Having a flexible inflection point is another important feature of a sigmoid curve. For that, Birch (1999), for example, modified the logistic equation (Eq. [2.1], Table 1) to deal with asymmetric growth by adding an extra shape parameter. When growth is known to decrease after a certain period of time, then the β growth function (Eq. [2.5], Table 1) might be a better option. On the other hand, Eq. [2.5] (Table 1) might not predict accurately initial growth, and, in cases when the initial phase should be considered, different versions of the β function should be used (see Eq. [3.11], Table 3). It is important to note that most sigmoid equations assume an initial Y value close to zero at time zero, which is reasonable in most cases; that is, at planting, biomass weight is close to zero.

Group III: Process Specific: Photosynthesis Examples

Photosynthesis is an important biological process involved in plant growth, and its rate is influenced by irradiance, temperature, nitrogen availability, vapor pressure deficit, and CO_2 concentration. Although there are detailed biochemistry models that have been proposed and used (e.g., von Caemmerer and Farquhar, 1981), there is a place for simpler models, especially when the main interest is in comparing different groups or treatments. Different nonlinear functions have been developed to describe photosynthesis response to different environmental variables. Functions to describe photosynthesis response to irradiance have been researched the most (Jassby and Platt, 1976; Goudriaan, 1979). All equations assume that dark respiration (R_d) is independent of light level. Among the equations presented in Table 1, the Blackman (Eq. [3.1]) is the simplest one, and the asymptotic exponential (Eq. [3.2]) and the nonrectangular hyperbola (Eq. [3.4]) are the most common. The rectangular hyperbola (Eq. [3.3], also termed the *Michaelis–Menten equation*) is used less frequently because it suggests that saturation is reached too fast. Currently, discussion on the photosynthetic capacity (Y_{asym}) and efficiency (a) of different plant species is based on the comparison of nonlinear regression estimates; for this, caution should be exercised because similar estimates from different equations can result in different responses (Fig. 2). Equation [3.2] is a simple three-parameter equation widely used in light driven process-based models like SUCROS and Hybrid-maize (Goudriaan and van Laar, 1994; Yang et al., 2004). Equation [3.4] offers more flexibility, and it is more accurate than Eq. [3.2] at the cost of one extra parameter (i.e., θ, curvature parameter). When θ = 1, Eq. [3.4] becomes equivalent to the Blackman equation, and when θ approaches zero, Eq. [3.4] becomes the rectangular hyperbola (Eq. [3.3]). Equation [3.4] comprises the reference equation when the biochemical model of Farquhar et al. (1980) or Collatz et al. (1992) is used in modeling studies. New equations are still being developed and tested (e.g., Eq. [4.7], Table 4).

Photosynthesis response to CO_2 has been quantified empirically using a nonrectangular hyperbola (Goudriaan, 1979; Johnson et al., 2010) and mechanistically using a biochemical model (Farquhar et al., 1980). The biochemical model is based on the Michaelis–Menten kinetics for substrate limited growth and the law of minimum between carboxylation and electron transport rates (Eq. [3.6], Table 1). Although its computation is laborious, this model is widely accepted. For more details on that model, we refer to the original publications (Farquhar et al., 1980; von Caemmerer and Farquhar, 1981) and to model application studies (Medlyn et al., 2002; Archontoulis et al., 2012). Photosynthesis response to leaf N, which is strongly related to Rubisco content, can be modeled using a modified logistic equation

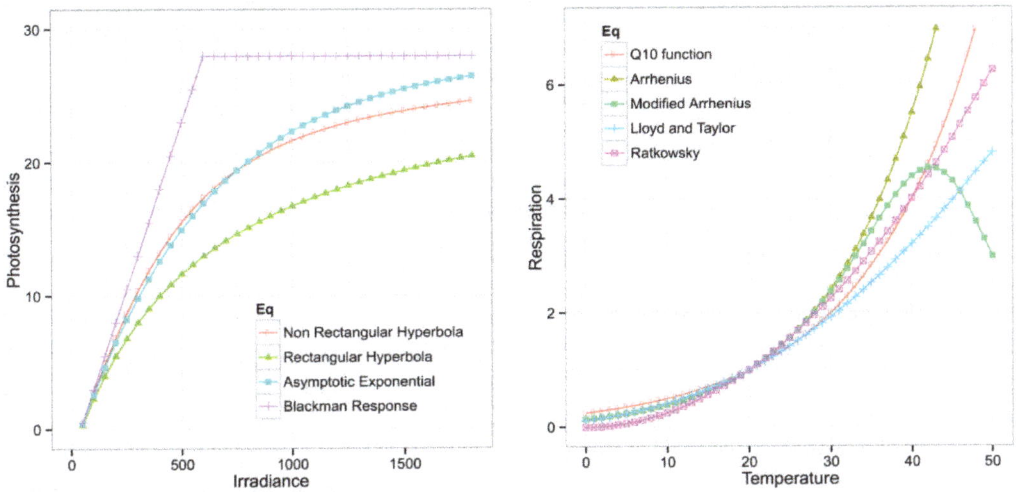

FIG. 2. Nonlinear models for describing photosynthesis response to irradiance (left panel) and respiration response to temperature (right panel). Equations and parameter meaning are given in Table 1. The following parameter values were used for these plots: Y_{asym} = 30 μmol CO_2 $m^{-2}s^{-1}$, a = 0.05 mol CO_2 mol^{-1} photons, θ = 0.7 and R_d = 2 μmol CO_2 $m^{-2}s^{-1}$; Q_{10} = 2, T_{ref} = 20°C, R = 8.314 (J K^{-1} mol^{-1}), E = 65000 (J mol^{-1}), S = 650 (J K^{-1} mol^{-1}), D = 207000 (J mol^{-1}), E_o = 350 (K), T_x = 225 (K), and T_{min} = 0°C. Note that at the reference temperature of 25°C, respiration takes the value of 1. The optimum temperature for the modified Arrhenius equation is: $T_{opt} = D/[S - R\ln(E/D - E)]$ = 42.3°C.

proposed by Sinclair and Horie (1989, see Eq. [3.5], Table 1), but there are alternatives (Eq. [4.8], Table 4). Photosynthesis response to water stress is usually described by sigmoid functions at the leaf level. For instance, Vico and Porporato (2008) utilized a Weibull-type curve (Eq. [4.9], Table 4). Photosynthesis response to vapor pressure deficit has been described by an exponential decay function (e.g., Osorio et al., 2006), but usually more sophisticated approaches are used (Collatz et al., 1992; Yin and Struik, 2009). Photosynthesis response to temperature is discussed below.

Group IV: Plant Responses to Temperature

A multitude of nonlinear regression models have been proposed and tested for modeling temperature dependence of various soil and plant processes (Lloyd and Taylor, 1994; Katterer et al., 1998; Davidson et al., 2006; Shibu et al., 2006; Portner et al., 2010). These include power, logarithmic, exponential, sigmoid, and bell-shape functions (Tables 1 and 5). The van't Hoff or Q_{10} function (i.e., Q_{10} is the factor by which the rate of a process increases for each 10°C temperature increase) is used in many studies, particularly in those addressing leaf or soil respiration rates. A common range for Q_{10} is between 1.4 and 4.9 (a value of 1 indicates no temperature effect) (Tjoelker et al., 2001; Atkin et al., 2005). In the Arrhenius equation, the Q_{10} term has been replaced by the activation energy. Both equations are equivalent, producing similar temperature responses (Fig. 2), but it should be noted that both Q_{10} and E coefficients are temperature range dependent. Usually, narrow temperature measurement ranges result in high and sometimes unrealistic Q_{10} or E estimates. Lloyd and Taylor (1994) noticed limitations of these two functions (i.e., the rate of reaction is not constant across temperatures) and developed a new equation known as Lloyd and Taylor (see Eq. [4.4], Table 1) to fit extensive literature data.

The above temperature functions describe a monotonic increase (Fig. 2). Most likely the rate of a process increases up to an optimum temperature and then drops (due to the lack of appropriate data, the drop is not always apparent). New equations or modifications of existing models have been developed to account for this. For example, the modified Arrhenius function, when compared with the standard Arrhenius equation, includes an additional two-parameter term (see D and S in Eq. [4.3], Table 1, Fig. 2) to capture the decline in the rate of a process at high temperatures (e.g., electron transport rate). If one of the two additional parameters is set to zero, then Eq. [4.3] becomes Eq. [4.2] (Table 1). Practitioners should be aware that this equation is "fragile" and requires careful parameterization (Medlyn et al., 2002; Archontoulis et al., 2012). Johnson et al. (2010) argued that temperature functions based on activation energy of chemical reactions are complex and difficult to apply routinely. They used a modified β function to describe photosynthesis response to temperature (Eq. [5.10], Table 5). Kirschbaum (1995) used a modified exponential temperature function (Eq. [5.9], Table 5) that provides a peak pattern to fit soil organic matter decomposition data. Additional (but difficult to interpret) peak temperature response functions are reported in Portner et al. (2010).

Group V: Bell Curve

In addition to temperature dependencies on photosynthesis, the bell-shape or peak functions are applied in agricultural sciences to describe the rate of phenological development as a function of temperature (e.g., Yin et al., 1995), the size of a leaf as a function of its rank in a plant (e.g., Hammer et al., 2009), or soil moisture effects on N_2O emissions (e.g., Rafigue 2011). Table 1 lists three commonly used equations (Eq. [5.1–5.3]). More application examples and different types of bell-curve equations can be found in Ma and Shaffer (2001) and in Table 6. In process-based simulation models, researchers have approximated a bell-shape response (viz. rate of development) with a two, three, or four segment (broken) linear regression model (e.g., APSIM; Keating et al., 2003). Typically, these segmented models should be fit using nonlinear methods as well.

Group VI: Others

In allometric studies, the relationships that exist between the growth rates of different plant components are quantified by means of regression analysis. Given the large variability that exists among plant species and plant components, numerous nonlinear models have been utilized to quantify relationships between plant processes, including power (Eq. [6.1], Table 1, e.g., plant N% vs. biomass weight), hyperbolic (Eq. [6.2], Table 1), and sigmoid curves (e.g., Eq. [2.1], Table 1). For application examples, see Vega et al. (2000), Vega and Sadras (2003), and Archontoulis et al. (2010). The Michaelis–Menten equation (Eq. [6.3], Table 1) is well-known and routinely applied to quantify the rate of a process (i.e., denitrification) that is dependent on the substrate (i.e., NO_3). In contrast, Eq. [6.4] is not as common in agronomy, but it appears to be very flexible taking many forms, from linear to exponential to bell-shaped. For example, it was applied to model temperature effects on soil nitrogen mineralization (Bril et al., 1994). The last equation in our reference Table 1 is the Ricker function (Eq. [6.5]), an option for hump-shaped patterns that are skewed to the right (Bolker, 2008).

Manipulating Nonlinear Functions

Sometimes there is a need to modify a "standard" nonlinear function to fit a set of data. This has led to the development of numerous versions of a standard equation (e.g., Birch, 1999; Tsoulakis, 2001; Tables 2, 3, and 4). Using the simplest form of the Michaelis–Menten hyperbolic function (see Eq. [7.1] in Fig. 3), we illustrate simple modification techniques. Equation [7.1] starts at zero when $x = 0$ and increases up to an asymptotic value of 1 as x increases. We can change the horizontal scale of this function by multiplying the variable x by a constant parameter, b, which is called a scale parameter (Bolker, 2008). If $b > 1$, then y saturates faster and if $0 < b < 1$, then y saturates more slowly (Eq. [7.2] in Fig. 3). We can change the vertical scale of the function by introducing a new parameter, a (Eq. [7.3] in Fig. 3). In this case, the asymptote moves from 1 to a. We can shift the whole curve to the right or the left by subtracting or adding a new parameter, c to the x variable (Eq. [7.4] in Fig. 3), which is called the location parameter (Bolker 2008). Similarly we can shift the whole curve up or down by adding or subtracting a new constant value, d, (Eq. [7.5] in Fig. 3). Lastly we can replace x with x^k, where k is a shape parameter, and then the equation can take many forms (exponential, sigmoid, etc.). A close example to the last modification is Eq. [2.6] in Table 1. When we modify nonlinear functions, we should add parameters that have meaningful interpretations.

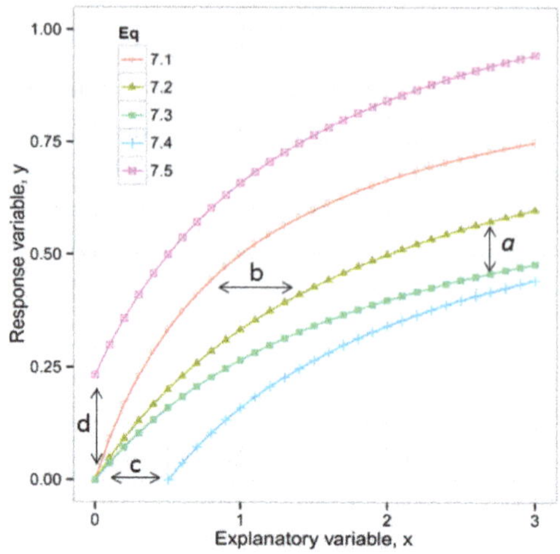

(#)	Equation	Parameters
7.1	$Y = \frac{x}{1+x}$	-
7.2	$Y = \frac{bx}{1+bx}$	b=0.5
7.3	$Y = a\frac{bx}{1+bx}$	b=0.5, a=0.8
7.4	$Y = a\frac{b(x-c)}{1+b(x-c)}$	b=0.5, a=0.8, c=0.5
7.5	$Y = a\frac{b(x-c)}{1+b(x-c)} + \mathrm{d}$	b=0.5, a=0.8, c=0.5, d=0.5

FIG. 3. Example of a nonlinear model modification. Starting with Eq. [7.1], the parameters *a*, *b*, *c*, and *d* were added step-by-step to Eq. [7.1], resulting in four new equations: Eq. [7.2], [7.3], E [7.4], and [7.5]. Horizontal or vertical arrows in the figure panel indicate how additional parameters affect the model.

Combining Nonlinear Functions

Sometimes a single equation is not enough to describe a set of experimental data, and researchers combine more than one equation to fit the data. The Farquhar model of photosynthesis is such an example because it combines two nonlinear models with the law of minimum. The use of the law of minimum is a common technique in modeling to simulate a process (e.g., biochar and tillage effects on soil bulk density, Archontoulis et al., 2016). However, the most common example that fits in this category comes from soil fertility, where researchers fit a quadratic (linear parameters) or quadratic-plus-plateau models to yield vs. nitrogen fertilizer experimental data (Sawyer et al., 2006). However, the estimation of the exact point at which the transition occurs (the optimum N rate, which is of great interest) is usually associated with uncertainty, especially when there are few points to fit the curve (Bachmaier, 2012).

When nonlinear functions are extended or combined to describe a phenomenon, it is important to recognize that there is an upper limit in the number of parameters that can be estimated from standard nonlinear regression analysis. This depends on the complexity of the model and the number of data points. For example, to fit growth curves with three parameters, we need at least four data points. When a process is described by a combination of nonlinear models (e.g., Farquhar model of photosynthesis or generic simulation crop models) then a step-wise parameterization method is usually applied. For application examples, see Miguez et al. (2009) and Archontoulis et al. (2012).

Fitting Nonlinear Models

Presently there are many statistical software packages available for fitting nonlinear models (e.g., SAS, R, JMP, GenStat, MatLab, Sigmaplot, OriginLab, SPSS). Nonlinear parameter estimates can be obtained using different methods (Bates and Watts, 2007); the most common methods are (i) nonlinear least squares, which minimizes the sum of squared error between observations and predictions, and (ii) the maximum likelihood method, which seeks the probability distribution parameters that make the observed data most likely. For non-normal data such as binomial or counts, generalized linear models should be used (Gbur et al., 2012). Most problems encountered during the use of standard nonlinear regression software functions are due to a poor choice of competing models, equation, or starting values (Fig. 1). The choice of estimation method can affect the parameter estimates (Ruppert et al., 1989), but in general, estimates from least square and from maximum likelihood methods tend to differ only when the data are not normally distributed and are approximately the same when the data follow a normal distribution (Myung, 2003).

Choosing Starting Values

All procedures for nonlinear parameter estimation require initial values. The choice of values influences the convergence of the estimation algorithm. In the worst case, there is no convergence, and in the best case, convergence is reached after a few iterations (Ritz and Streibig, 2005). There is no standard procedure for selecting initial estimates. These are five common methods:

1. If the model has parameters with biological meaning then use information from the literature.
2. Use graphical exploration (see examples).
3. Transform the nonlinear model into a linear one. For instance, log-transformation of Eq. [1.1] yields a linear equation ($\log Y = Y_0 - kt$) in which rough estimates of the parameter values can be easily obtained by linear regression. This method is recommended for getting initial estimates and to detect deviations from linearity. These estimates can be used as starting values (Bolker, 2008), but may also be used as final estimates when the variance model is appropriate for log Y (Ruppert et al., 1989). For more transformation examples, see Zeide (1993), Singh (2006), and Portner et al. (2010).
4. When no clear guidelines exist for choosing starting values, the recommendation is to use a grid search or "brute-force" approach (e.g., PROC NLIN in SAS or nls2 package in R). This grid search can be done by generating an extensive coverage of possible parameter values (and their combinations) and then evaluating the model at each one of these parameter combinations. The numerical method can then be used, starting with the combination that resulted in the best fit (lower mean squared error). The hope is that a sufficiently extensive coverage of the parameter space will result in a combination of parameters that will result in an adequate fit.
5. Use prespecified algorithms: this approach is specific to a given equation and can be used to calculate starting values for a given dataset (e.g., Pinheiro and Bates, 2000; Ritz and Streibig, 2008).

Check Algorithm Convergence

After the initial attempt at fitting a nonlinear model, we recommend that algorithm convergence is evaluated (Fig. 1). Convergence is achieved when a measure (such as the relative offset, change in LOSS function, or maximum change among parameter estimates; Bates and Watts, 2007) is below a certain threshold value (i.e., 10^{-5}), indicating that the algorithm has found a "best" solution (Fig. 1). If convergence is not achieved, the most likely problems are the poor choice of starting values or that the selected model is not well suited to describe the data. If convergence is achieved, the next step is to evaluate whether the parameter estimates are within a reasonable range. This requires not only evaluating the point estimates but also their standard errors. Unusually large standard errors are a sign of convergence problems, even if convergence was apparently achieved in the previous step. If no problems were encountered up to this point, the analysis can continue by simplifying the model and assessing model assumptions.

Evaluating Model Assumptions

When we are dealing with one model, the next step is to evaluate key model assumptions: normally distributed errors, independent errors, and homogeneous variance for the errors (Fig. 1). This step and the following steps are not unique to nonlinear models but are common to all linear models. Substantial deviations from the assumptions could result in bias (inaccurate estimates) and/or distorted standard errors (Ritz and Streibig, 2008). Violations of these assumptions can be detected from an analysis of the residuals by means of graphical procedures and formal statistical tests. For a more complete analysis, see Ritz and Streibig (2008). Briefly, to check whether the distribution of the measurement errors follows normality, the standardized residual plot is commonly applied (Pinheiro and Bates, 2000). Outliers and many extreme values are common causes for deviations from normality (Fig. 1). Heterogeneity of variance can be detected by looking at the plot of the fitted values over the residuals (absolute residuals: raw residuals stripped of the negative sign; or standardized residuals: raw residuals scaled by the variance; see Example 1). When the residual errors show a trend (e.g., increasing variability as the explanatory variable increases), this can be addressed by modeling the variance as a function of the independent variable or the fitted values (Fig. 1). If variance heterogeneity is ignored, the parameter estimates might not be influenced much, but this may result in severely misleading confidence and prediction intervals (Carroll and Ruppert, 1988). The residuals are assumed to be independent, and when this assumption is violated this is visually evident in a plot of correlations of residuals against "lag" (or units of separation in time or space). Typically, variables measured in time on the same subject (e.g., plant, animal or soil sample) tend to result in autocorrelated residuals which need to be accounted for by modeling the variance covariance matrix of the residuals.

Model Selection Criteria

When dealing with multiple models, the question is how to find the best one among competing models. Depending on the structure of the models, different statistical criteria can be used to find the "best" model: F test, Akaike information criterion (AIC), Bayesian information criterion (BIC), or likelihood ratio test (Zucchini, 2000; Burnham and Anderson, 2002; Hoffmann, 2005; Ritz and Streibig, 2008; Lewis et al., 2011). When models are *nested* (one model is a special case of another), the above criteria are applicable (Fig. 1). When models are *non-nested* (models having different structures, e.g., Eq. [2.1] vs. Eq. [2.5]), typically the AIC and the BIC criteria are used

(Fig. 1). However, from a practical point of view, one model might be preferred over another based on interpretability and specific objectives. There needs to be a balance between statistical model performance and how effectively the model answers research questions.

For two nested models, one with two parameters (reduced, e.g., Eq. [3.9] in Table 3) and one with four parameters (full, Eq. [3.8] in Table 3) to check whether the addition of the two extra parameters has a statistically significant contribution on the model performance, we can use the following F test:

$$F = \frac{\left(SS_{full} - SS_{reduced}\right)/\left(df_{full} - df_{reduced}\right)}{SS_{full}/df_{full}}$$

where SS_{full} and $SS_{reduced}$ are the regression model sum of squares for the full and reduced model, respectively, and df_{full} and $df_{reduced}$ are the degrees of freedom for the full and the reduced model, respectively. The p value can be calculated at $df_{full} - df_{reduced}$ (numerator), $n - p$ (denominator), where p is the number of parameters for the full model and n is the number of observations, and a decision can be made. This test is sometimes referred to as extra-sum-of-squares or multiple partial F test (Hoffmann, 2005; Ritz and Streibig, 2008), and it is computed when the least squares method is used to fit the data (see section on fitting nonlinear models). When the maximum likelihood method is used to fit the data, then the likelihood ratio test (LRT) statistic is computed to compare nested models:

$$\mathrm{LRT} = 2\log\left(L_{full} / L_{reduced}\right) = 2\left(\log L_{full} - \log L_{reduced}\right)$$

$$Q = 2\log\left(L_{full} / L_{reduced}\right) = 2\left(\log L_{full} - \log L_{reduced}\right)$$

where L_{full} and $L_{reduced}$ are the likelihood functions for the full and reduced model, respectively. These functions are closely related to the residual sum of squares (see Eq. [2.4] in Ritz and Streibig, 2008). It is assumed that the LRT statistic is approximately χ^2–distributed with $df_{full} - df_{reduced}$ degrees of freedom (for details, see Ritz and Streibig, 2008). Another approach for model selection involves calculating the AIC and BIC values for each model (i) separately:

$$\mathrm{AIC}_i = -2\log L_i + 2p_i$$

$$\mathrm{BIC}_i = -2\ \log L_i + p_i \log n$$

where L_i and p_i are the likelihood and the number of parameters for each model, and n is the number of observations. For both statistical criteria, a smaller value indicates a preferable model. The BIC criteria differs from the AIC only in the second term which depends on n. As n increases, the BIC favors simpler models (fewer parameters). This explains why sometimes the decision made on the basis of AIC and BIC indexes disagree. More information about these indexes is provided in Burnham and Anderson (2002). Note that LRT, AIC, and BIC are all designed to compare the performance of models that have been fitted to data via maximum likelihood estimation.

Goodness of Fit

To assess the goodness of fit, there are many different graphical methods and numerical indexes that highlight different features of the data and the model. Graphical comparisons provide a quick visual assessment of the goodness of fit (see examples below, Fig. 4). Numerical statistical indexes such as R^2, bias, mean squared error, root mean square error (RMSE), modeling efficiency, concordance correlation, and others (Wallach, 2006) can be used as quantitative summaries to assess the goodness of fit. Some indexes measure the absolute error (includes units), and some others measure the relative error (unitless). Depending on the data type, a combination of these indexes can be used. For example, the relative term is more meaningful than the absolute when comparing errors from different data sets such as biomass, yield, phenology, and nitrate leaching. The following equations are commonly used in model evaluation (for a comprehensive review of those we refer to Wallach, 2006):

$$\text{Bias} = \frac{1}{n}\sum_{i=1}^{n}\left(Y_i - \hat{Y}_i\right)$$

$$R^2 = 1 - \frac{\text{SS}_{\text{residual}}}{\text{SS}_{\text{total}}} = 1 - \frac{\sum_{i=1}^{n}(Y_i - \hat{Y}_i)^2}{\sum_{i=1}^{n}(Y_i - \hat{Y}_i)^2 + \sum_{i=1}^{n}(\hat{Y}_i - \bar{Y})^2}$$

$$\text{RMSE} = \sqrt{\frac{\text{SS}_{\text{residual}}}{n-p}}$$

$$\text{ME} = 1 - \frac{\text{SS}_{\text{residual}}}{\text{SS}_{\text{model}}} = 1 - \frac{\sum_{i=1}^{n}(Y_i - \hat{Y}_i)^2}{\sum_{i=1}^{n}(\hat{Y}_i - \bar{Y})^2}$$

where bias, R^2, RMSE, and ME (modeling efficiency) are numerical statistical indexes; n is the number of data points; Y_i and $\hat{Y}_i$ are the observed and predicted values; $\bar{Y}$ is

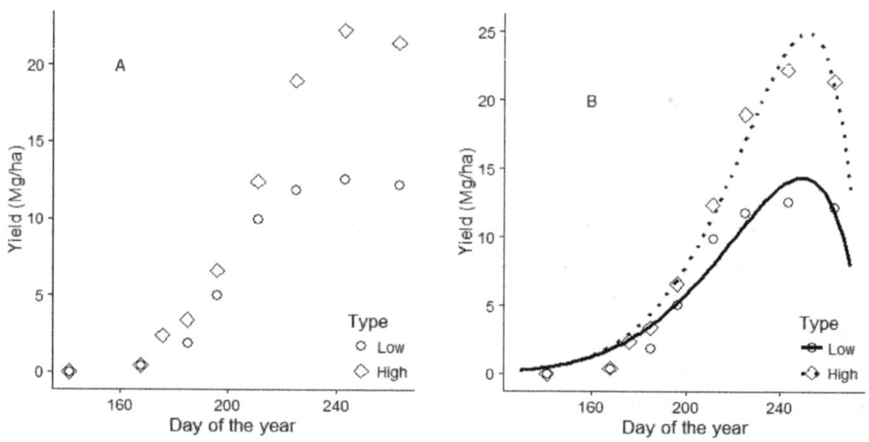

FIG. 4. (A) Observed yield during the growth season. (B) Predicted yield (line) versus observed (point) for both agricultural inputs

the mean observed value; p is the number of model parameters; $SS_{residual}$, SS_{model}, and SS_{total} are the sum of squares for the residual, regression model and the total, respectively.

Although often used, R^2 does not represent a good metric of model performance for nonlinear models. It has several limitations (e.g., it does not account for the number of parameters); therefore, we recommend that other measures of agreement (or combinations) should be used (Wallach, 2006). For nonlinear models, specifically, the main limitation of R^2 is that the full model does not necessarily include the simpler model with one single parameter, as is the case with linear models (see Exercise 2).

The numerical statistical descriptors indicate the average performance of the model across the sample. When the variability is not constant throughout the sample (e.g., biomass increase over time), then statistical indexes do not capture the fact that the uncertainty is not the same at different magnitudes of the response variables (see the extended example below). We should often be concerned with the predictive ability of the model. For this, cross-validation techniques can be used, and the mean squared error of prediction is an appropriate measure for goodness of fit (Wallach, 2006).

To illustrate the principles presented so far, we will now discuss examples of applications of nonlinear models with increasing level of complexity.

Example 1: Maize Biomass Accumulation

Goal: Investigate the effect of agricultural input level on maximum yield of maize using the beta growth function.

Fitting Procedure

The dataset used in this example contains plant biomass (Mg ha^{-1}) for different days of the year during the growing season and for different levels of agricultural inputs. The dataset is called "sm" and it is also part of the "nlraa" R package [available in R-forge and can be install through R command line by: install.packages("nlraa", repos = "http://RForge.R-project.org")]. As a first step we plot the observed data and confirm, visually, that the pattern is well represented by the function of choice. According to Fig. 4, the observed biomass for both low and high agricultural inputs follow a sigmoidal pattern (i.e., S-shape), making the Beta function a suitable choice. One of its main characteristics is that it allows for a decline in growth at later stages.

In the R command line this dataset can be loaded (the "nlraa" package is needed).

```
data(sm)
```

In the beta growth function (Eq. [2.5] in Table 1) there is one predictor (time or t) and one response variable (Y) while there are three coefficients that need to be estimated. In this example, Y is the response variable which stands for plant biomass or yield, while t represents time measured as day of the year (DOY).

Using the "nls" function, we can estimate the parameters of a nonlinear model using the least-squares method. The first argument to the "nls" function is the "formula" with the response variable on the left-hand side and the function on the right-hand side, separated by "$\sim$": ResponseResponse $\sim f$ (*predictor*).

The following is an example of how this function can be used:

```
data(sm)
# filtering the dataset for Maize, different agricultural inputs and
one specific block
# Maize-Block1-Low and high inputs
Maize.B1.low<-sm[(sm$Block==1 & sm$Crop=='M' & sm$Input==1),]
Maize.B1.high<-sm[(sm$Block==1 & sm$Crop=='M' & sm$Input==2),]
# Taking the day of year as the x value and the Yield as the Y value
# Low input
X.low<-Maize.B1.low$DOY
Y.low<-Maize.B1.low$Yield
#High inputs
X.high<-Maize.B1.high$DOY
Y.high<-Maize.B1.high$Yield
# Non linear regression for the low input data - B1=Block 1
reg.low<-nls(Y.low ~ bgf(X.low, w.max, t.e, t.m),data=Maize.B1.low,
start=list(w.max=13,t.e=230,t.m=180))

# Non linear regression for the High input data - B1=Block 1
reg.high<-nls(Y.high ~ bgf(X.high, w.max, t.e, t.m),data=Maize.
B1.high, start=list(w.max=20,t.e=240,t.m=210))
```

From visual inspection we can determine that the highest value for "high input" is about 20 to 25 Mg ha^{-1}, and the highest value for "low input" is about 13 Mg ha^{-1} (Fig. 4). These would be sensible starting values for Y.max parameter. Again, visually, we can estimate the time at which half of the maximum yield is achieved as approximately DOY 220 for both treatments. Similarly, the time at which maximum biomass is attained is about DOY 240.

Input	Y_{max}	t_e	t_m
Low	14.4	249.12	220.35
High	25	251.26	225.78

By using the "summary" function, we can obtain more detailed information about the standard error and *p* value for each coefficient. These *p* values are not typically informative, as they are a test of the null hypothesis that the parameter is equal to zero (strictly, the probability of observing the data that was observed given that the null hypothesis is true). This is almost always true in this type of single curve fitting, but it can be more informative as we evaluate treatment effects.

```
summary(reg.low)
##
## Formula: Y.low ~ bgf(X.low, Y.max, t.e, t.m)
##
## Parameters:
## Estimate Std. Error t value Pr(>|t|)
## Y.max 14.40 2.21 6.5 0.0013 **
## t.e 249.12 3.27 76.1 7.4e-09 ***
## t.m 220.35 4.75 46.4 8.8e-08 ***
## ---
## Signif. codes: 0 '***' 0.001 '**' 0.01 '*' 0.05 '.' 0.1 ' ' 1
##
## Residual standard error: 1.5 on 5 degrees of freedom
##
## Number of iterations to convergence: 12
## Achieved convergence tolerance: 3.74e-06
```

The fitted model is shown in Fig. 4 B.

Checking Model Assumptions

The first assumption that should be evaluated is whether the regression function is reasonable. This is best done visually by confirming that the function follows the data to a reasonable level. Although linear models have a mean for residuals numerically close to zero, this is not always the case in nonlinear models and checking that the mean of the residuals is "close" to zero is a good first step. This deviation, called bias, is slightly higher than we would like, which suggests that the regression function is not adequately predicting the data.

```
##Mean of the residual
mean(resid(reg.high))
mean(resid(reg.low))
```

Input	Low	High
Mean of residual	−0.23	−0.27

Residual Analysis

Analyzing the residuals of a fitted model can provide additional insight. Is there any clear pattern? Are deviations increasing as the fitted values increase? Do the residuals appear normal? The skill of visual analysis of residuals is not always intuitive, and it is developed through practice.

At first glance, Fig. 5 does not show any major issues with the residuals, but looking more closely, there seems to be a pattern in which model predictions of low yields are overestimated, and predictions of high yields are underestimated. One exception is that the highest predicted yield was overestimated. The deviations are not substantial, so there are no clear "outliers." Residuals are also not increasing with the fitted values (a common issue), suggesting that the assumption of equal variance is reasonable. With these few data points the assumption of normality is difficult to test, but the plot of residuals shows a similar spread in terms of negative and positive residuals, again, not showing a sign of concern.

A normal quantile–quantile (QQ) plot (Fig. 6) helps us evaluate whether the assumption of normality has been met to a reasonable degree. The closer the points are to the straight line, the more normal is their distribution. The QQ plots in this example show a more right skewed distribution for both low and high agricultural

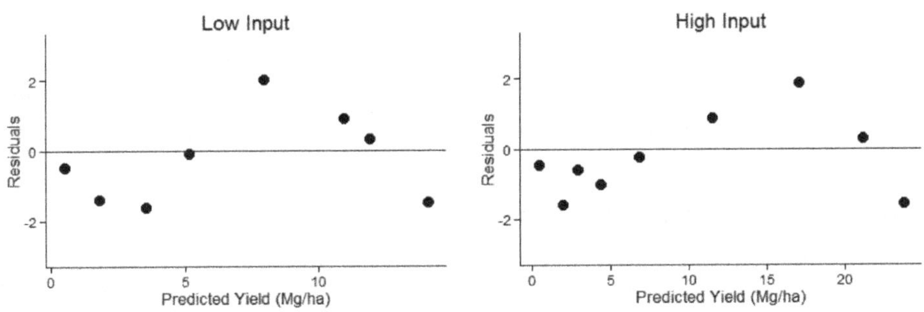

FIG. 5. Standardized residuals against predicted yield for high and low agricultural input fits.

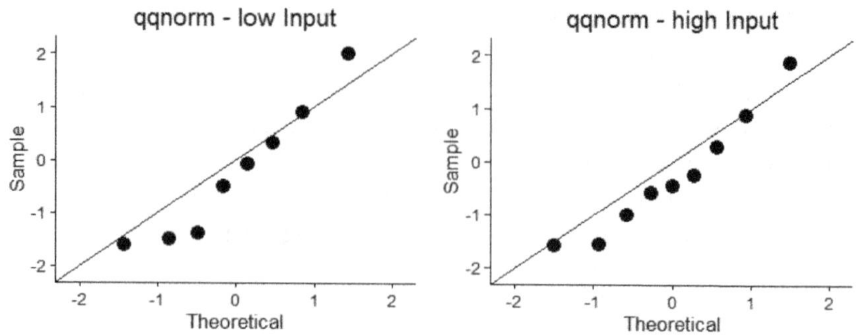

FIG. 6. Normal quantile–quantile (QQ) plot of residuals resulted from fitting process for both agricultural inputs.

inputs. It is also worthwhile to mention that the normality of the residuals also can be tested using more quantitative methods like the Shapiro–Wilk normality test.

Statistical Descriptors

Statistical indexes of model agreement can be calculated using equations in the goodness of fit section above. The first one is the RMSE, which gives an idea of the average deviation in the same units as the response variable. The normalized version of RMSE (NRMSE) shows the overall error relative to the mean of the observed response variable in percent. The bias is included here for completeness, but it was previously addressed. The last index (R^2) suggest a good agreement between the model and the data, but in this specific instance are not very useful. In general they do not provide diagnostic information, unless their values are extremely low.

Input	RMSE	NRMSE(%)	Bias	R^2
Low	1.22	18.05	−0.23	0.95
High	1.1	11.29	−0.27	0.99

The beta growth function was not able to fully capture the initial phase of growth and this was more clearly presented in the graph of residuals. This can be addressed by modifying the function by delaying the growth in the model at the cost of two extra parameters. The modified version of the Beta function (Eq. [3.11] in Table 3) allows for two offsets in both the x and y axes, and they are called t_b and Y_b, respectively. These two parameters (t_b and Y_b) can represent the planting date and the biomass at sowing date, respectively (Yin et al., 2003b). These parameters might not need to be estimated, since they are typically known and can be treated as fixed for a given application. These additional parameters are not necessary if the time scale is redefined as days since planting, but it might be preferred if planting date is an imposed treatment in a given study.

Comparing Alternative Models

GOAL: Compare S-shaped growth patterns models for the dataset introduced in the example for the maize, first block, and high agricultural input.

Answer

As was previously stated, the first step in fitting a nonlinear model is to visually inspect the data (Fig. 7). In the previous example, we found that by using the original beta growth function, some of the assumptions appeared to be not valid. In this exercise, we fit different sigmoid functions, including the modified version of Beta to find the model that best represents the observed yield.

Using the "data" function, we import the data set from the "nlraa" package and subsequently extract the desired predictor (DOY) and response variable (yield in Mg/ha) for the Maize and the first block. In the previous example, we explained how to use the "nls" function to fit our nonlinear model to data. The following block of code makes other uses of the nls function.

```
data(sm)
Maize.B1<-sm[(sm$Block==1 & sm$Crop=='M' & sm$Input==2),]
#Taking the day of year as the x value and yield as the Y value
X<-Maize.B1$DOY
Y<-Maize.B1$Yield
############### Nonlinear Regression for Gompertz function
reg.Gompertz=nls(Y~(Y.asim)*exp(-exp(-k*(X-t.m))),data=Maize.B1.low,
start=list(Y.asim=12.1,k=0.08,t.m=240.5))
############### Nonlinear Regression for Logistic function
reg.Logistic=nls(Y~(Y.asim)/(1+exp(-k*(X-t.m))),data=Maize.B1.low,
start=list(Y.asim=12.1,k=0.08,t.m=240.5))
############### Nonlinear Regression for Richards function
reg.Richards=nls(Y~rich(X, t.m, k, v, Y.asim),data=Maize.B1.low,
start=list(Y.asim=14.1,k=0.07,t.m=200,v=0.6))
############### Nonlinear Regression for Beta function
reg.Beta=nls(Y~SSbgf(X, Y.asim, t.e, t.m),data=Maize.B1.low,
start=list(Y.asim =10.1,t.e=250,t.m=220.1))
############### Nonlinear Regression for Beta function 2
reg.Beta2=nls(Y~bgf2(X, w.max, w.b, t.e, t.m, t.b),data = Maize.
B1.low, start=list(w.max=24,t.e=252,t.m = 215,w.b = 0,t.b = 141))
############### Nonlinear Regression for Weibull function
reg.Weibull=nls(Y~weibull((X-140),a = a, b = b, Yo=Y.asim),data=Maize.
B1.low, start=list(Y.asim=21.0,a=0.0000009,b = 3.3))
```

In the above code, the last four "nls" functions used predefined functions which produced cleaner code.

As a result of fitting different nonlinear growth models (Fig. 8), the following coefficients were obtained. Each model produced different values for their respective parameters but the mean of the residuals (bias) is more revealing in terms of how

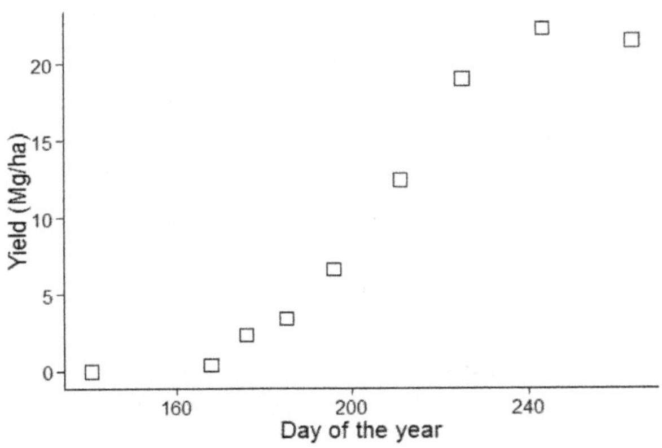

FIG. 7. Observed maize yield during the growth season for the first block and high agricultural input.

FIG. 8. Simulated yield versus observed for different models and high agricultural inputs

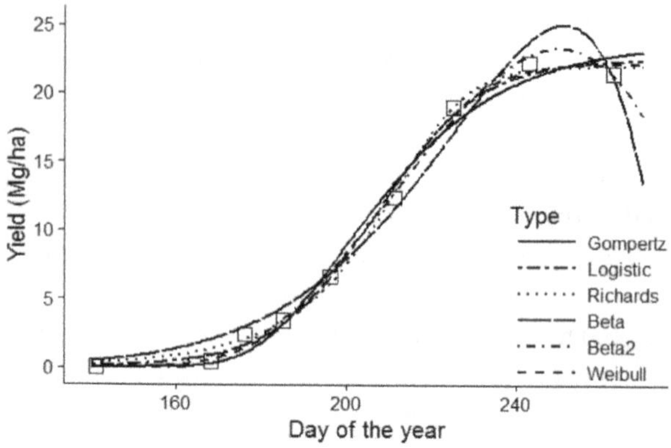

well each model fits the data. The parameter Y.asim presented in the table below is the equivalent to Y.max in beta growth functions.

Model	Y.asim	k	t.m	Mean of residuals
Gompertz	23.78	0.04	199.49	0.12
Logistic	22.52	0.08	206.87	0.02
Richards	21.98	0.13	212.10	–0.06
Beta	25.00	–	225.78	–0.27
Beta2	23.27	–	210.77	-1.265×10^{-8}
Weibull	22.12	–	–	0.08

Checking Model Assumptions

Residuals Mean

The mean of residuals is presented in the last column of the above table. The closest value to zero is given by the Beta and Logistic functions while the original Beta function has the largest absolute value.

Homogeneous Variance

The homogeneity of the variance assumption is investigated using a scatter plot (Fig. 9) between the residual and predicted response variable (Yield). An even distribution of points around the zero line is desirable and typical patterns that deviate from this might appear as a funnel shape (i.e., increasing scatter as the predicted response variable increases).

Normality

The normality of the residuals is more difficult to assess given the small sample size. The QQ plot of residuals for the different models shows a reasonable agreement with a normal distribution (Fig. 10), and, as it was stated earlier, the residuals are considered to follow a normal distribution the closer they are to the 1:1 line.

Based on the figures and interpretation presented above, among all the models, the modified Beta, Logistic, Richards, and Weibull show the least bias, and they meet standard assumptions. On the other hand, the Gompertz and the original

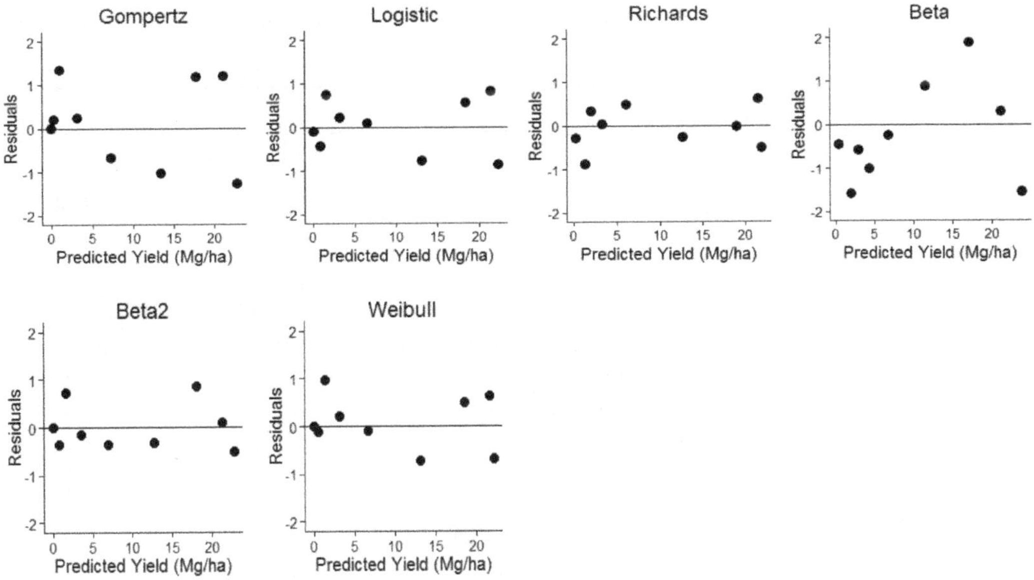

FIG. 9. Standardized residuals against predicted yield for all candidate models.

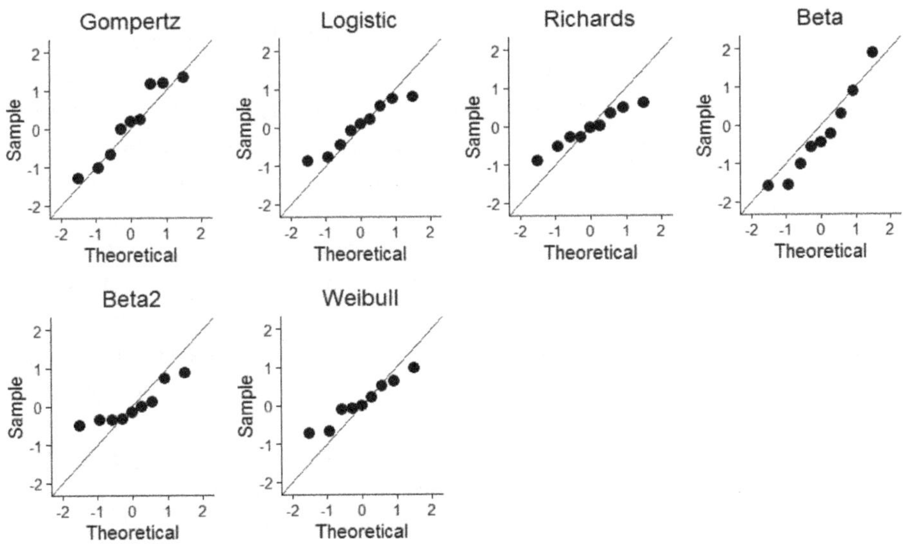

FIG. 10. Normal quantile–quantile (QQ) plot of residuals resulted from fitting process for all candidate models.

Beta function had higher mean of residuals, which are an indication of the pattern in the residuals.

Statistical Descriptors

The statistical descriptors are presented for the Logistic, Richards, Beta2, and Weibull models. From these models the Beta2 and Richards performed better than the Logistic and Weibull. For example, based on the bias, the Logistic and Weibull models underestimated the observed values, while the other two models tended to overestimate.

Model	RMSE	NRMSE (%)	Bias	R^2	AIC	BIC
Logistic	0.59	6.06	−0.024	0.99	32.23	33.02
Richards	0.46	4.69	0.064	0.99	21.62	22.60
Beta2	0.46	4.74	1.265×10^{-8}	0.99	23.70	24.88
Weibull	0.54	5.60	−0.081	0.99	22.52	23.31

This example illustrates that the original Beta function was modified to capture maize growth more effectively at the cost of having two extra parameters. There is commonly a tradeoff between the complexity of the models and their performance, but ultimately the goals of the analysis should dictate the choice of the model.

Example 2: Extended Application

The following extended example builds on the previous simpler examples, but with the emphasis shifted to the comparison among treatments. The example follows the workflow illustrated in Fig. 1.

DATA: We use data from Danalatos et al. (2009), which represent destructive measurements, taken in 2008 of aboveground biomass accumulation over time for three crops: fiber sorghum (*F*), sweet sorghum (*S*), and maize (*M*) growing in a deep fertile loamy soil of central Greece under two management practices: high and low input conditions. High refers to weekly irrigation (to match 100% of maximum evapotranspiration) and application of 200 kg N ha^{-1}, and low input refers to biweekly irrigation (~50% of maximum evapotranspiration) and application of 50 kg N ha^{-1}. The experiment was a 2 × 3 factorial completely randomized in four blocks. For more details, see Danalatos et al. (2009). With such data, many questions are possible. Here we will concentrate on three: (i) What is the maximum biomass accumulated by these crops? (ii) At what point in time was this biomass achieved? and (iii) Are there significant treatment effects and/or interactions? This requires us to statistically determine the effects of crop type on the function parameters and also the effect of input level (i.e., high or low). These questions are approachable through the use of a nonlinear model that captures the regression function and the structure of the data.

GRAPHS: Visually (Fig. 11), we can see that the sorghums have higher biomass and that the maximum biomass occurs later in the season. No outliers have been detected at this point. However, without a statistical analysis it is difficult to make sound statements based solely on data visualization.

CHOOSE CANDIDATE MODEL: Danalatos et al. (2009) analyzed the data using the beta growth function (Yin et al., 2003b; Eq. [2.5] in Table 1). That model was selected because it captures the decline of biomass toward the end of the growing season (Fig. 11). Also the parameters have clear meaning and are suitable to answer the research questions.

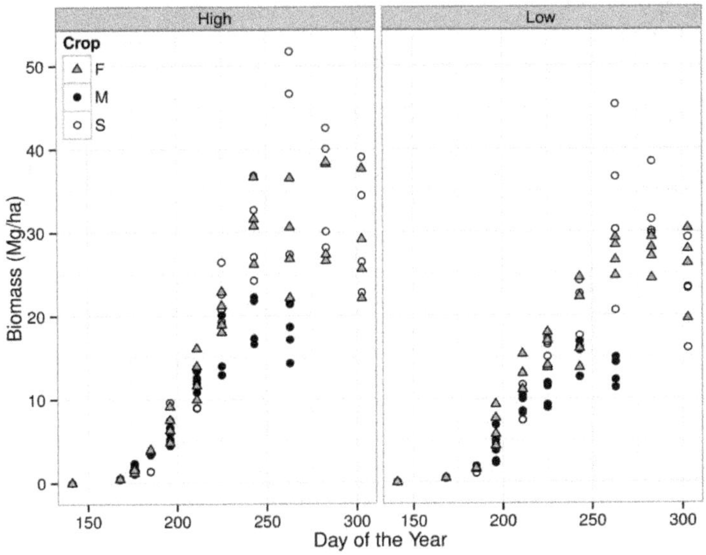

FIG. 11. Biomass accumulation with time for three crops: maize (*M*), fiber sorghum (*F*), and sweet sorghum (*S*) collected in Greece in 2008. The two panels represent the level of agricultural input: high and low.

STARTING VALUES: Since for this function parameters have straightforward interpretations, starting values can be found by visual inspection of Fig. 11. However, in this example, we used a prespecified algorithm that chooses the initial starting values automatically.

FIT MODEL AND CONVERGENCE: Model fit was conducted in the R package using the nonlinear least squares method (nls function, see the "nlraa" R package for details). There are three crops, two levels of agronomic input, and four blocks, which results in 24 possible combinations (experimental units). The model was fit to every experimental unit and apparent convergence was obtained for only 10 experiment units. This suggests that some modifications are needed (see section below on revising the regression model). Checking model assumptions can be useful for diagnosing the problem.

REVISING THE REGRESSION MODEL: This can be addressed in part by using a modified beta growth function (see Eq. [3.11] in Table 3), which is designed to capture more efficiently the initial phase of growth at the cost of two extra parameters. Equation [3.11] allows for an offset in the x axis (t_b) orientation and an offset in the y axis orientation (y_b). We do not fit these parameters, but rather keep them fixed; t_b is the planting date at 141 DOY, and y_b is the biomass weight at sowing, which is zero. As a first step, the fitting process was repeated as above with starting values determined visually as 30 for w.max (same as Y.max), 280 for t.e and 240 for t.m. Apparent convergence was obtained for all the experiment units. The final revised model was fitted to the entire dataset, but at this step the model included the effect of crop type, agronomic input level, and the interaction for each parameter. Full details are available in the vignette of the "nlraa" package.

CHECK MODEL ASSUMPTIONS: Visual inspection of the standardized residuals (Fig. 12) was used to evaluate the assumptions of appropriate regression function and normally distributed errors with homogeneous variance. In this case, a concentration of

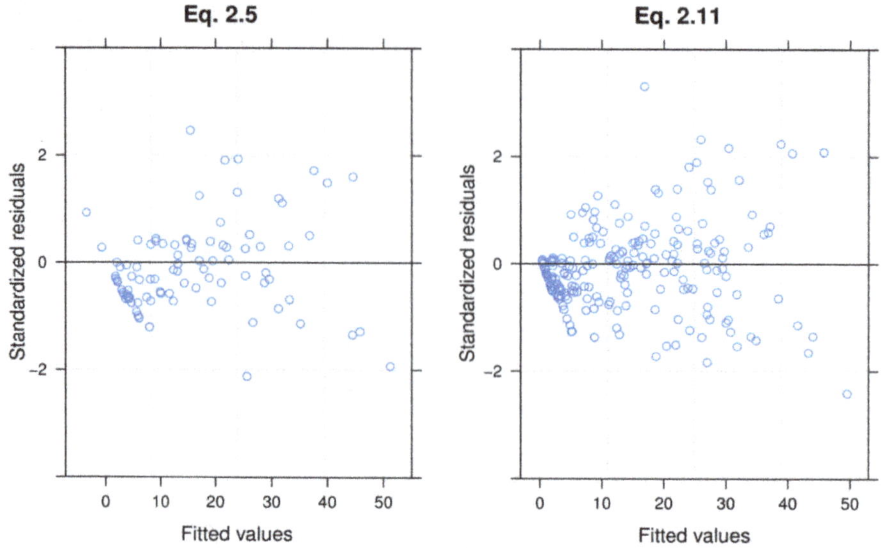

FIG. 12. Standardized residuals from individually fits to all experimental units. Left panel: Eq. [2.5] in Table 1; and right panel: Eq. [2.11] in Table 2. During the fitting procedure, Eq. [2.5] converged in only 10 out of the 24 experimental units, thus the fewer points in the left panel.

points stand out at low fitted values, which indicates overprediction (i.e., bias) at low values. Equation [3.11] (modified beta growth function) alleviated the overprediction at low values, but the bias did not disappear completely. A major argument for choosing Eq. [3.11] over Eq. [2.5] is that convergence was achieved for all experimental units (24 for Eq. [3.11] vs. only 10 for Eq. [2.5]).

VARIANCE HOMOGENEITY: The residual variance was modeled with a power function, and different power parameters were used for the three crops. This function is: $s^2(v) = s^2 * |v|^{2*\theta}$, where v is the covariate (the fitted values in this case), and θ depends on the crop (0.7, 0.86, and 0.89 for maize, fiber, and sweet sorghum respectively).

PARAMETER ESTIMATES AND STANDARD ERRORS: Table 8 illustrates the estimates of the model, the corresponding standard errors, and the p values. These values are final and account for modeling the residual variance. The interaction between agronomic input and crop species was significant for the maximum biomass weight (Y_{max}) and the final growth time point (t_e) (Table 8). The model could be further simplified by dropping the interaction and the effect of the input level on time of maximum biomass (t_m) (Table 9). The final model, which takes into account the interaction between crop level and management input level, has 18 parameters, which are used to describe the regression function and four parameters that describe the variance function (see the vignette in the "nlraa" package for details).

CALCULATE STATISTICAL DESCRIPTORS: Regarding the numerical summaries described in the section "Goodness of Fit," biomass has small initial values and high values at the end of the season (Fig. 13). In this case, use of the average RMSE of 4.1 Mg ha^{-1} might be misleading because it overestimates the error at the initial stages (biomass of 0–10 Mg ha^{-1}) and underestimates the error at advanced stages (biomass of 30–40 Mg ha^{-1}).

TABLE 8. Estimates with standard errors in parenthesis of the beta growth model (Eq. [3.11] in Table 3) used to fit the biomass data reported by Danalatos et al. (2009). Y_{max} is the maximum biomass (Mg ha^{-1}); t_m is the day of the year when the crop growth rate is maximized; and t_e is the day of the year when biomass is maximized. P values < 0.05 indicate significant effect of input levels (high or low) for every crop species and parameter.†

Maize			Fiber sorghum			Sweet sorghum		
High	Low	*p*	High	Low	*p*	High	Low	*p*
21.2 (0.99)	15.4 (2.27)	<0.001	38.6 (2.24)	31.8 (5.18)	0.02	43.2 (2.83)	33.9 (6.48)	0.01
215.7 (1.30)	217.1 (3.33)	0.5	234.5 (1.61)	235.8 (4.11)	0.61	239.4 (1.53)	240.0 (3.81)	0.79
248.0 (1.79)	248.8 (4.51)	0.76	277.2 (2.03)	279.4 (5.38)	0.5	278.6 (1.99)	279.0 (4.87)	0.89

† Note: in Eq. [3.11] the parameters Y_b, biomass weight at sowing and t_b, sowing date, were fixed at 0 Mg ha^{-1} and 141 d of the year, respectively.

TABLE 9. Test of fixed effects for the parameters in the model which include the effect of Crop, Management Input, and the interaction using the modified beta growth function. See Table 1, Eq. [2.5] for details.

Term	*F* value	*P* value
Y_{max}	11601	< 0.0001
Y_{max}.Crop	902	< 0.0001
Y_{max}.Input	441	< 0.0001
Y_{max}. (Input × Crop)	34	< 0.0001
t_e	23863	< 0.0001
t_e.Crop	90	< 0.0001
t_e.Input	33	< 0.0001
t_e. (Input × Crop)	59	< 0.0001
t_m	127944	< 0.0001
t_m.Crop	139	< 0.0001
t_m.Input	1	0.4017
t_m.(Input × Crop)	0	0.9667

INTERPRET RESULTS AND DRAW CONCLUSIONS: According to model predictions, the maximum estimated biomass was obtained for sweet sorghum under high inputs and this crop reached a total of 43 Mg ha^{-1} on DOY 279 (Fig. 14; Table 8). On the other extreme, maize reached its maximum biomass under high input of 21 Mg ha^{-1} on DOY 248. The maximum biomass and the time when it was reached were significantly affected by the Crop and Input interaction. In practice, the specific objectives can guide the additional analysis to be conducted such as treatment differences, their significance level (p value) and/or having a model capable of producing robust predictions within the range of observed values.

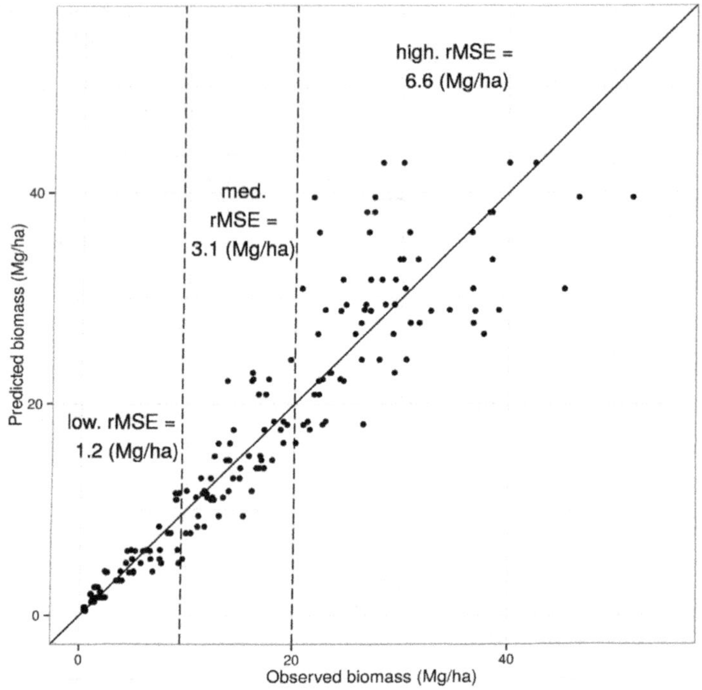

FIG. 13. Observed versus predicted biomass values. The RMSE was used as a measure of goodness of fit, but given that the variability is increasing along with the biomass weight, three RMSE values were calculated for ranges indicated by vertical dashed lines (0–10 10–20, and >20 Mg ha^{-1}).

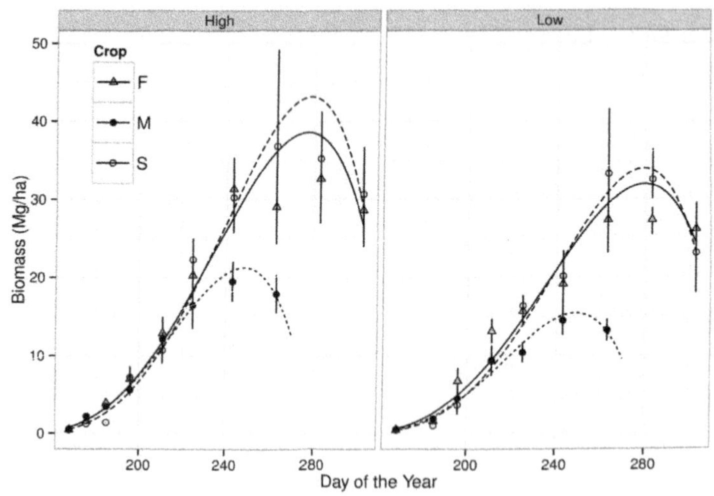

FIG. 14. Observed data and fit for the final model. Vertical bars indicate confidence interval of observations.

Summary

The most critical step which distinguishes nonlinear models from linear models is that the choice of the main function is critical and this can be difficult without appropriate guidance. Here, we present an extensive library of nonlinear functions (77 equations with the associated parameter meaning) and the typical applications, which, we hope, will make the task of choosing candidate models easier. Our review of nonlinear equations is incomplete, as there are countless numbers of potential functions (Ratkowsky, 1990) and their ad hoc modifications. We also contribute a suggested workflow (Fig. 1) that should provide the necessary structure to avoid common errors in the use of nonlinear regression models.

Key Learning Points

- Nonlinear models have the advantages of having meaningful parameters and being parsimonious (few parameters are needed).
- Choosing an appropriate model for a given application is crucial, and both statistical arguments and application specific interpretability should be considered.
- Nonlinear models do not have analytical solutions, and iterative algorithms should be used.
- Model convergence is highly improved by appropriate choice of starting values.
- Graphical analyses are important to evaluate whether the model is correct on average.
- As with linear models the assumptions of the residuals should be evaluated: independence, homogeneity of variance and normality.

Review Questions

1. Which of these functions is not used for data that follows a sigmoidal (S-shape) pattern?
   (a) Logistic
   (b) Gompertz
   (c) Richards
   (d) Weibull
   (e) Arrhenius
2. Which of these functions is typically used for temperature response data?
   (a) Logistic
   (b) Rectangular hyperbola
   (c) Gompertz
   (d) Arrhenius
   (e) Weibull
3. In the beta growth function as described by Yin et. al. (2003b), the parameter tm represents:
   (a) Maximum biomass attained
   (b) Time at which maximum biomass is attained
   (c) Time at which maximum growth rate is attained
   (d) Time at which growth starts
   (e) Time at which growth ends
4. In most photosynthesis functions the parameter Rd
   (a) Maximum rate of photosynthesis
   (b) Dark respiration
   (c) Curvature
   (d) Irradiance level
   (e) Quantum efficiency
5. In the exponential decay function the parameter k cannot be greater than 1.

(a) True
(b) False

Exercises

Exercise 1: Example of Derivatives of Linear Models

The first derivative of a quadratic model can be used to calculate the slope at that point

$$y = \beta_0 + \beta_1 x + \beta_2 x^2$$

where y is the response variable, β_0 is the intercept, β_1 is the slope associated with the independent variable x, and β_2 is the parameter associated with the quadratic term (x^2). The derivative of this function with respect to x is

$$\frac{dy}{dx} = \beta_1 + 2\beta_2 x$$

However, if we calculate the partial derivative with respect to each parameter we have

$$\frac{dy}{d\beta_1} = x$$

and

$$\frac{dy}{d\beta_2} = x^2$$

The second partial derivatives for both of these equations will be exactly zero. Therefore, this is considered a linear model.

1.A. Show that the second derivative of a simple exponential decay function with respect to its parameter is not zero.

1.B. Identify the linear and nonlinear parameters in the following model. Show your work and justify.

$$y = \beta_0 + \beta_1 \exp\left(-\frac{x}{\theta}\right)$$

Exercise 2: Full and Reduced Models

It is common to compare a full model to a simpler model. With linear models, it is common to compare a full model to a reduced model with one single parameter (intercept). This is the basis for the R^2

$$R^2 = \frac{\text{SSE}(\text{reduced}) - \text{SSE}(\text{full})}{\text{SSE}(\text{reduced})}$$

where SSE is the sum of squares of the error. For a simple linear regression:
Full model:

$$y_i = \beta_0 + \beta_1 x_i$$

Reduced model:

$$y_i = \beta_0$$

It is important to recognize that we can obtain the reduced model by setting β_1 equal to zero. For a model with just one parameter (intercept) the least-square estimate is imply the mean ($\bar{y}$). The R^2 for this model is

$$R^2 = 1 - \frac{\sum_i \left[y_i - \left(\hat{\beta}_0 + \hat{\beta}_1 x_i\right)\right]^2}{\sum_i \left(y_i - \bar{y}\right)^2}$$

where $\hat{\beta}_0$ and $\hat{\beta}_1$ are the least-square estimates for β_0 and β_1.

For nonlinear models, the *reduced* model with one single parameter (intercept) is not always a subset of the *full* model. For this to be true, the intercept only model should be a result of setting one or more of the parameters of the full model to a fixed value (typically zero). This is why the R^2 is often not appropriate for nonlinear models.

2.A. Provide examples of nonlinear models which can be reduced to a single parameter (intercept). Show your work.

2.B. Provide examples of nonlinear models which cannot be reduced to a single parameter. Show your work.

References

Amaducci, S., M. Colauzzi, G. Bellocchi, and G. Venturi. 2008. Modelling post-emergent hemp phenology (*Cannabis sativa* L.): Theory and evaluation. Eur. J. Agron. 28:90–102. doi:10.1016/j.eja.2007.05.006

Andren, O., and K. Paustian. 1987. Barley straw decomposition in the field: A comparison of models. Ecology 68:1190–1200. doi:10.2307/1939203

Archontoulis, S.V., P.C. Struik, X. Yin, L. Bastiaans, J. Vos, and N.G. Danalatos. 2010. Inflorescence characteristics, seed composition, and allometric relationships predicting seed yield in the biomass crop *Cynara cardunculus*. GCB Bioenergy 2:113–129.

Archontoulis, S.V., I. Huber, F.E. Miguez, P.J. Thorburn, N. Rogovska, and D.A. Laird. 2016. A model for mechanistic and system assessments of biochar effects on soils and crops and trade-offs. GCB Bioenergy. doi:10.1111/gcbb.12314

Archontoulis, S.V., X. Yin, J. Vos, N.G. Danalatos, and P.C. Struik. 2012. Leaf photosynthesis and respiration of three bioenergy crops in relation to temperature and leaf nitrogen: How conservative are biochemical model parameters among crop species? J. Exp. Bot. 63:895–911. doi:10.1093/jxb/err321

Arrhenius, S. 1889. Ueber die Reaktionsgeschwindigkeit bei der Inversion von Rohrzucker durch Sauren. Z. Phys. Chem. 4:226–248.

Atkin, O.K., D. Bruhn, V.M. Hurry, and M.G. Tjoelker. 2005. The hot and the cold: Unraveling the variable response of plant respiration to temperature. Funct. Plant Biol. 32:87–105. doi:10.1071/FP03176

Bates, D.M., and D.G. Watts. 2007. Nonlinear regression and its applications. Wiley Series in Probability and Statistics. Wiley, New York.

Bachmaier, M. 2012. Sources of inaccuracy when estimating economically optimum N fertilizer rates. Agric. Sci. 3:331–338

Birch, C.P.D. 1999. A new generalized logistic sigmoid growth equation compared with the Richards growth equation. Ann. Bot. (Lond.) 83:713–723. doi:10.1006/anbo.1999.0877

Blackman, F.F. 1905. Optima and limiting factors. Ann. Bot. (Lond.) 19:281–295.

Bolker, B.M. 2008. Ecological models and data in R. Princeton Univ. Press, Princeton, NJ.

Bril, J., H.G. Van Faassen, and H. Klein Gunnewiek. 1994. Modeling N20 emission from grazed grassland. Report 24 of the DLO Research Institute for Agrobiology and Soil Fertility (AB-DLO). AB-DLO, Haren, Germany.

Brooks, R.H., and A.T. Corey. 1964. Hydraulic properties of porous medium. Hydrology Paper 3. Colorado State University, Fort Collins.

Burnham, K.P., and D.R. Anderson. 2002. Model selection and multimodel inference: A practical information—Theoretical approach. 2nd ed. Springer-Verlag, New York.

Campbell, G.S. 1974. A simple method for determining unsaturated conductivity from moisture retention data. Soil Sci. 117:311–314. doi:10.1097/00010694-197406000-00001

Carroll, R.J., and D. Ruppert. 1988. Transformations and weighting in regression. Chapman and Hall, New York. doi:10.1007/978-1-4899-2873-3

Collatz, G.J., M. Ribas-Carbo, and J.A. Berry. 1992. Coupled photosynthesis stomatal conductance model for leaves of C4 Plants. Aust. J. Plant Physiol. 19:519–538. doi:10.1071/PP9920519

Danalatos, N.G., S.V. Archontoulis, and K. Tsiboukas. 2009. Comparative analysis of sorghum versus corn growing under optimum and under water/nitrogen limited conditions in central Greece. In: Proceeding of the 17th European Biomass Conference, Hamburg, Germany, p. 538–544.

Davidson, E.A., I.A. Janssens, and Y. Luo. 2006. On the variability of respiration in terrestrial ecosystems: Moving beyond Q10. Glob. Change Biol. 12:154–164. doi:10.1111/j.1365-2486.2005.01065.x

del Pozo, A., and M.D. Dennett. 1999. Analysis of the distribution of light, leaf nitrogen, and photosynthesis within the canopy of *Vicia faba* L. at two contrasting plant densities. Aust. J. Agric. Res. 50:183–189. doi:10.1071/A94105

Farquhar, G.D., S. von Caemmerer, and J.A. Berry. 1980. A biochemical model of photosynthetic CO_2 assimilation in leaves of C3 species. Planta 149:78–90. doi:10.1007/BF00386231

Foereid, B., J. Lehmann, and J. Major. 2011. Modeling black carbon degradation and movement in soil. J. Plant Soil 345:223. doi:10.1007/s11104-011-0773-3

Gan, Y., E.H. Stobbe, and C. Njue. 1996. Evaluation of selected nonlinear regression models in quantifying seedling emergence rate of spring wheat. Crop Sci. 36:165–168. doi:10.2135/cropsci1996.0011183X003600010029x

Gardner, W. 1958. Some steady state solutions of the unsaturated moisture flow equation with application to evaporation from a water table. Soil Sci. 85:228–232. doi:10.1097/00010694-195804000-00006

Gbur, E.E., W.W. Stroup, K.S. McCarter, S. Durham, L.J. Young, M. Christman, M. West, and M. Kramer. 2012. Analysis of generalized linear mixed models in the agricultural and natural resources sciences. ASA, CSSA, and SSSA, Madison, WI. doi:10.2134/2012.generalized-linear-mixed-models

Gillis, J.D., and G.W. Price. 2011. Comparison of a novel model to three conventional models describing carbon mineralization from soil amended with organic residues. Geoderma 160:304–310. doi:10.1016/j.geoderma.2010.09.025

Gompertz, B. 1825. On the nature of the function expressive of the law of human mortality, and on a new mode of determining the value of life contingencies. Philos. Trans. R. Soc. 115:513–583. doi:10.1098/rstl.1825.0026

Goudriaan, J. 1979. A family of saturation type curves, especially in relation to photosynthesis. Ann. Bot. (Lond.) 43:783–785.

Goudriaan, J. 1994. Using the expolinear growth equation to analyze resource capture. In: J.L. Monteith, R.K. Scott, and M.H. Unsworth, editors, Resource capture by crops. Nottingham Univ. Press, Nottingham, UK. p. 99–110.

Goudriaan, J., and J.L. Monteith. 1990. A mathematical function for crop growth based on light interception and leaf area expansion. Ann. Bot. (Lond.) 66:695–701.

Goudriaan, J., and H.H. van Laar. 1994. Modelling potential crop growth processes. Kluwer Academic Publ., Dordrecht, The Netherlands. doi:10.1007/978-94-011-0750-1

Hammer, G.L., Z. Dong, G. Mc, A. Lean, C. Doherty, J. Messina, C. Schussler, S. Zinselmeier, S. Paszkiewicz, and M. Cooper. 2009. Can Changes in Canopy and/or Root System Architecture Explain Historical Maize Yield Trends in the U.S. Corn Belt? Crop Sci. 49:299–312. doi:10.2135/cropsci2008.03.0152

Hoben, J.P., R.J. Gehj, N. Millar, P.R. Graces, and G.P. Robertson. 2011. Nonlinear nitrous oxide (N2O) response to nitrogen fertilizer in on-farm corn crops of the US Midwest. Glob. Change Biol. 17:1140–1152. doi:10.1111/j.1365-2486.2010.02349.x

Hoffmann, J.P. 2005. Linear regression analysis: Assumptions and applications. SPSS version. Available at: https://www.researchgate.net/profile/John_Hoffmann/publication/228896673_Linear_Regression_Analysis_Assumptions_and_Applications/links/00b7d51cb759e51bce000000.pdf (accessed 21 Sept. 2016).

Horton, R.E. 1940. An approach towards a physical interpretation of infiltration capacity. Soil Sci. Soc. Am. Proc. 5:227–237.

Jassby, A.D., and T. Platt. 1976. Mathematical formulation of the relationship between photosynthesis and light for phytoplankton. Limnol. Oceanogr. 21:540–547. doi:10.4319/lo.1976.21.4.0540

Johnson, I.R., J.H.M. Thornley, J.M. Frantz, and B. Bugbee. 2010. A model of canopy photosynthesis incorporating protein turnover distribution through the canopy and its acclimation to light, temperature and CO_2. Ann. Bot. (Lond.) 106:735–749. doi:10.1093/aob/mcq183

Katterer, T., M. Reichstein, O. Andren, and A. Lomander. 1998. Temperature dependence of organic matter decomposition: A critical review using literature data analyzed with different models. Biol. Fertil. Soils 24:258–262.

Keating, B.A., P.S. Carberry, G.L. Hammer, M.E. Probert, M.J. Robertson, D. Holzworth, N.I. Huth, J.N.G. Hargreaves, H. Meinke, Z. Hochman, G. McLean, K. Verburg, V. Snow, J.P. Dimes, M. Silburn, E. Wang, S. Brown, K.L. Bristow, S. Asseng, S. Chapman, R.L. McCown, D.M. Freebairn, and C.J. Smith. 2003. An overview of APSIM, a model designed for farming systems simulation. Eur. J. Agron. 18:267–288. doi:10.1016/S1161-0301(02)00108-9

Kirschbaum, M.U.F. 1995. The temperature dependence of soil organic matter decomposition, and the effect on global warming on soil organic science. Soil Biol. Biochem. 27:753–760. doi:10.1016/0038-0717(94)00242-S

Leduc, D., and L. Goelz. 2009. A height-diameter curve for longleaf pine plantations in the Gulf Coastal Plain. South. J. Appl. For. 33:164–170.

Lewis, F., A. Butler, and L. Gilbert. 2011. A unified approach to model selection using the likelihood ratio test. Methods Ecol. Evol. 2:155–162. doi:10.1111/j.2041-210X.2010.00063.x

Lloyd, J., and J.A. Taylor. 1994. On the temperature dependence of soil respiration. Funct. Ecol. 8:315–323. doi:10.2307/2389824

Ma, L., and M.J. Shaffer. 2001. A review of carbon and nitrogen processes in nine U.S. soil nitrogen dynamic models. In: M.J. Shaffer, L. Ma, and S. Hansen, editors, Modeling carbon and nitrogen dynamics for soil management. Lewis Publ., Boca Raton, FL, p. 55–103.

McGechan, M.B., and L. Wu. 2001. A review of carbon and nitrogen processes in European soil nitrogen dynamic model. In: M.J. Shaffer, L. Ma, and S. Hansen, editors, Modeling carbon and nitrogen dynamics for soil management. Lewis Publ., Boca Raton, FL. p. 103–171.

Medlyn, B.E., D. Loustau, and S. Delzon. 2002. Temperature response of parameters of a biochemically based model of photosynthesis. I. Seasonal changes in mature maritime pine (*Pinus pinaster* Alt.). Plant Cell Environ. 25:1155–1165. doi:10.1046/j.1365-3040.2002.00890.x

Miguez, F.E., X. Zhu, S. Humphries, G.A. Bollero, and S.P. Long. 2009. A semimechanistic model predicting growth and production of the bioenergy crops Miscanthus x giganteus: Description, parameterization and validation. GCB Bioenergy 1:282–296. doi:10.1111/j.1757-1707.2009.01019.x

Monsi, M., and T. Saeki. 2005. On the factor light in plant communities and its importance for matter production. Ann. Bot. (Lond.) 95:549–567. doi:10.1093/aob/mci052

Myung, I.J. 2003. Tutorial on maximum likelihood estimation. J. Math. Psychol. 47:90–100. doi:10.1016/S0022-2496(02)00028-7

O'Connell, A.M. 1990. Microbial decomposition (respiration) of litter in eucalypt forests of south-western Australia: An empirical model based on laboratory incubations. Soil Biol. Biochem. 22:153–160. doi:10.1016/0038-0717(90)90080-J

Omuto, C.T., and L.O. Gumbe. 2009. Estimating water infiltration and retention characteristics using a computer program in R. Comput. Geosci. 35:579–585. doi:10.1016/j.cageo.2008.08.011

Osorio, M.L., E. Breia, A. Rodrigues, J. Osorio, X. Le Roux, F.A. Daudet, I. Ferreira, and M.M. Chaves. 2006. Limitations to carbon assimilation by mild drought in nectarine tress growing under field conditions. Environ. Exp. Bot. 55:235–247. doi:10.1016/j.envexpbot.2004.11.003

Pinheiro, J.C., and D.M. Bates. 2000. Mixed-effects models in S and S-plus. Springer-Verlag, New York. doi:10.1007/978-1-4419-0318-1

Portner, H., H. Bugmnann, and A. Wolf. 2010. Temperature response functions introduce high uncertainty in modeled carbon stocks in cold temperature regimes. Biosciences 7:3669–3684.

Rafigue, R. 2011. Measurements and modeling of nitrous oxide emissions from Irish grasslands. Ph.D. diss. National University of Ireland, Cork.

Ratkowsky, D.A., J. Olley, T.A. McMeekin, and A. Ball. 1982. Relationship between temperature and growth rate of bacterial cultures. J. Bacteriol. 149:1–5.

Ratkowsky, D.A. 1990. Handbook of nonlinear regression models. Marcel Dekker, New York, 241 pp.

Ratkowsky, D.A. 1993. Principles of nonlinear regression modeling. J. Ind. Microbiol. 12:195–199. doi:10.1007/BF01584190

Richards, F.J. 1959. A flexible growth function for empirical use. J. Exp. Bot. 10:290–301. doi:10.1093/jxb/10.2.290

Ritz, C., C.B. Pipper, and J.C. Streibig. 2013. Analysis of germination data from agricultural experiments. Eur. J. Agron. 45:1–6. doi:10.1016/j.eja.2012.10.003

Ritz, C., and J.C. Streibig. 2005. Bioassay analysis using R. J. Stat. Softw. 12:1–21. doi:10.18637/jss.v012.i05

Ritz, C., and J.C. Streibig. 2008. Nonlinear regression with R. Springer, New York.

Ruppert, D., N. Cressie, and R.J. Carroll. 1989. A transformation/weighting model for estimating Michaelis–Menten parameters. Biometrics 45:637–656. doi:10.2307/2531506

Sawyer, J., E. Nafziger, G. Randall, L. Bundy, G. Rehm, and B. Joern. 2006. Concepts and rationale for regional nitrogen rate guidelines for corn. PM 2015. Iowa State Univ. Ext., Ames.

Sinclair, T.R., and T. Horie. 1989. Leaf nitrogen, photosynthesis, and crop radiation use efficiency: A review. Crop Sci. 29:90–98. doi:10.2135/cropsci1989.0011183X002900010023x

Shibu, M.E., P.A. Leffelaar, H. van Keulen, and P.K. Aggarwal. 2006. Quantitative description of soil organic matter dynamics—A review of approaches with reference to rice-based cropping systems. Geoderma 137:1–18. doi:10.1016/j.geoderma.2006.08.008

Singh, P. 2006. Modeling crop production systems—Principles and application. Science Publishers, Inc., Enfield, NH.

Thorburn P.J., M.E. Probert, and F.A. Robertson. 2001. Modelling decomposition of sugar cane surface residues with APSIM-Residue. Field Crops Res. 70:223–232.

Tjoelker, M.G., L. Oleksyn, and P. Reich. 2001. Modelling respiration of vegetation: Evidence for a general temperature-dependence Q10. Glob. Change Biol. 7:223–230. doi:10.1046/j.1365-2486.2001.00397.x

Tsoulakis, A. 2001. Analysis of logistic growth models. Res. Lett. Inf. Math. Sci. 2:23–46.

Urban, J.H. 2002. Modeling growth of different development stages in bivalves. Mar. Ecol. Prog. Ser. 238:109–114. doi:10.3354/meps238109

van Genuchten, M.Th. 1980. A closed-form equation for predicting the hydraulic conductivity of unsaturated soils. Soil Sci. Soc. Am. J. 44:892–898. doi:10.2136/sssaj1980.03615995004400050002x

van 't Hoff, J.H. 1898. Lectures on theoretical and physical chemistry. Part 1. Chemical dynamics. Edward Arnold, London.

Vega, C.R.C., and V.O. Sadras. 2003. Size dependent growth and development of inequality in maize, sunflower and soybean. Ann. Bot. (Lond.) 91:795–805. doi:10.1093/aob/mcg081

Vega, C.R.C., V.O. Sadras, F.H. Andrade, and S.A. Uhart. 2000. Reproductive allometry in soybean, maize and sunflower. Ann. Bot. (Lond.) 85:461–468. doi:10.1006/anbo.1999.1084

Verhulst, P.F. 1838. A note on population growth. Corresp. Math. Phys. 10:113–121.

Vico, G., and A. Porporato. 2008. Modelling C3 and C4 photosynthesis under water-stressed conditions. Plant Soil 313:187–203. doi:10.1007/s11104-008-9691-4

von Bertalanffy, L. 1938. A quantitative theory of organic growth (inquiries on growth laws. ii). Human Biology. 10(2):181–213.

von Caemmerer, S., and G.D. Farquhar. 1981. Some relationships between the biochemistry of photosynthesis and the gas exchange of leaves. Planta 153:376–387. doi:10.1007/BF00384257

Wallach, D. 2006. Evaluating crop models. In: D. Wallach, D. Makowski, and J.W. Jones, editors, Working with dynamic crop models—Evaluations, analysis, parameterization, and applications. Elsevier, New York. p. 11–53.

Weibull, W. 1951. A statistical distribution function of wide applicability. J. Appl. Math. 18:293–297.

Wang, E., and T. Engel. 1998. Simulation of phenological development of wheat crops. Agric. Syst. 58:1–24. doi:10.1016/S0308-521X(98)00028-6

Yang, H.S., A. Dobermann, J.L. Lindquist, D.T. Walters, T.J. Arkebauer, and K.G. Cassman. 2004. Hybrid-maize—a maize simulation model that combines two crop modeling approaches. Field Crops Res. 87:131–154. doi:10.1016/j.fcr.2003.10.003

Ye, Z., and Z. Zhao. 2010. A modified rectangular hyperbola to describe the light-response curve of photosynthesis of *Bidens pilosa* L. grown under low and high light conditions. Front. Agric. China 4:50–55. doi:10.1007/s11703-009-0092-0

Yin, X., M.J. Kroff, G. McLean, and R.M. Visperas. 1995. A nonlinear model for crop development as a function of temperature. Agric. For. Meteorol. 77:1–16. doi:10.1016/0168-1923(95)02236-Q

Yin, X., A.D.H.C.M. Schapendonk, M. Kroff, M. van Oijen, and P.S. Bindraban. 2000. A Generic equation for nitrogen-limited leaf area index and its application in crop growth models for predicting leaf senescence. Ann. Bot. (Lond.) 85:579–585. doi:10.1006/anbo.1999.1104

Yin, X., E.A. Lantinga, A.D.H.C.M. Schapendonk, and X. Zhong. 2003a. Some quantitative relationships between leaf area index and canopy nitrogen content and distribution. Ann. Bot. (Lond.) 91:893–903. doi:10.1093/aob/mcg096

Yin, X., J. Goudriaan, E.A. Lantinga, J. Vos, and J.H.J. Spiertz. 2003b. A flexible sigmoid function of determinate growth. Ann. Bot. 91:361–371 (erratum 91:753). doi:10.1093/aob/mcg029

Yin, X., and P.C. Struik. 2009. C3 and C4 photosynthesis models: An overview from the perspective of crop modelling. NJAS Wagening. J. Life Sci. 57:27–38. doi:10.1016/j.njas.2009.07.001

Yin, X., W. Guo, and J.H. Spiertz. 2009. A quantitative approach to characterize sink–source relationships during grain filling in contrasting wheat genotypes. Field Crops Res. 114:119–126. doi:10.1016/j.fcr.2009.07.013

Zeide, B. 1993. Analysis of growth equations. For. Sci. 39:594–616.

Zucchini, W. 2000. An Introduction to Model Selection. J. Math. Psychol. 44:41–61. doi:10.1006/jmps.1999.1276

Zwietering, M.H., I. Jongenburger, F.M. Rombouts, and K. van't Riet. 1990. Modeling bacterial growth curve. Appl. Environ. Microbiol. 56:1875–1881.

CHAPTER 16: ANALYSIS OF NON-GAUSSIAN DATA

Walter W. Stroup

Abstract

The term non-Gaussian data refers to response variables assumed not to have a normal distribution. In this chapter, the more technically correct "Gaussian distribution" is used instead of the more common term "normal" distribution. This is to dispel the misimpression that non-Gaussian data are somehow abnormal. In many areas of plant science, non-Gaussian data are the rule, not the exception. Few areas of statistics have changed as recently and as extensively in terms of what is considered best practice for data analysis. The purpose of this chapter is to provide plant science students, researchers, reviewers, and technical editors with an overview of the changes that have occurred, an introduction to the essential features of the generalized linear mixed model (GLMM) – now considered the standard tool analyzing for non-Gaussian data – and to present examples of the setup, implementation, and interpretation of GLMM analyses of non-Gaussian data.

> The fact is, Martin, you can do nothing till you know a little mathematics. If you are not going to be a cookbook bacteriologist, like most of them, you must be able to handle some fundamentals of science. ... Then how can you make progress if you do not know physical chemistry, and how can you know much physical chemistry without much mathematics? from Sinclair Lewis, *Arrowsmith (1925)*.

The chapter concerns analysis of data for data that do not meet assumption of normality – data that are increasingly common in plant research. Some of the methods in this chapter are an order of magnitude more difficult than most statistical methods for the plant sciences. The learning curve is steep. Yet the spirit of what the fictional aspiring medical researcher Martin Arrowsmith was told by his mentor is alive and well: your research will be substantially more meaningful and have greater impact if you can analyze and

---

Abbreviations: Analysis of Variance, ANOVA; Generalized Linear Model, GLM; Generalized Linear Mixed Model, GLMM; Linear Model, LM; Linear Mixed Model, LMM

Key Words: Generalized Linear Mixed Model, probability distribution, likelihood, Gaussian, binomial, Poisson, negative binomial, analysis of variance, linear predictor, link function, fixed effect, random effect, data and model scale, inverse link, conditional and marginal inference, PROC GLIMMIX

Department of Statistics, University of Nebraska, Lincoln 68583-0963, *Corresponding author (wstroup1@unl.edu)

doi:10.2134/appliedstatistics.2015.0081

Applied Statistics in Agricultural, Biological, and Environmental Sciences
Barry Glaz and Kathleen M. Yeater, editors

interpret data appropriately, regardless of whether your data are normally distributed.

Classical analysis of variance (ANOVA) and regression require the assumption that the observations are independent and normally distributed with constant variance. Mixed model methods for repeated measures and spatial data relax the assumption of independence but assume normality. The "normal" – or, more properly, the Gaussian – distribution characterizes variability for data that are continuous and spread symmetrically about the mean. While this is appropriate for many response variables in the plant sciences – e.g. height, weight, yield, etc. – many other response variables do not plausibly fit the Gaussian framework. Examples include continuous percent data (e.g. leaf area affected), discrete percent data (number of seeds that germinate out of the number planted), counts (number of weeds in a plot), ratings and classifications, and time-to-event data. In fact, in some areas, response variables of this type are the rule, not the exception: data that satisfy the assumption of normality would actually be highly ab-normal in these disciplines. To avoid the tempting but erroneous idea that "normal" data are in some sense good and "non-normal" data are somehow defective, in this chapter we avoid the label "normal" in favor of the more technically correct term *Gaussian*.

Few areas of statistical analysis relevant to plant scientists have undergone changes as recent and as extensive as those for non-Gaussian data. For those who took their statistical methods course prior to 2005, the methods that were taught for non-Gaussian data are already seriously dated. Even now, those who teach these courses are scrambling to update their non-Gaussian sections so that what is taught will be consistent with our contemporary understanding of best practices for the analysis of non-Gaussian data, as well as the design of experiments when non-Gaussian response variables are expected to be the primary criteria of interest. The purpose of this chapter is to provide students and instructors with a description of best practices for analyzing non-Gaussian plant science data and with relevant examples for implementing these analyses.

This chapter is divided into three sections. The first provides a review of the history of statistical methods for non-Gaussian data over the past century. This is included so that students (and advisors whose statistical training is pre-2005 and reviewers and technical editors with whom researchers who use these methods will inevitably have to interact) will understand why the changes in best practices are so recent, so drastic, and so important. The second section of this chapter provides an introduction to key features of the generalized linear mixed model (GLMM), the fundamental tool for non-Gaussian data analysis. The third section will present five examples of the analysis of non-Gaussian data drawn from representative experiments that should be familiar to plant scientists. The material for the first two sections draws heavily from Stroup's 2014 open access *Agronomy Journal* article, "Rethinking the Analysis of Non-Normal Data in Plant and Soil Science." The introduction to the GLMM is necessarily a brief survey, and the examples, while representative, obviously cannot be exhaustive. References will be given to direct interested readers to more complete materials covering the GLMM.

A Brief History of Non-Gaussian Data Analysis in the Plant Sciences

To help readers understand the paradigm shift occurring in applied statistics, we begin with a brief history. Fisher and Mackenzie (1923) published the first use of ANOVA for experimental data. Fisher's *Statistical Methods for Research Workers* (1925) and *The Design of Experiments* (1935) established the template for statistics in agricultural research and rapidly became the template for experimental research in general. Yates (1940) introduced the recovery of inter-block information – a crucial precursor to mixed model methodology – and was a leader in extending Fisher's work to complex experiments, notably split-plots. Bartlett (1947) introduced transformations for non-normal data within the ANOVA framework. Eisenhart (1947), Henderson (1953, 1963), and Harville (1976, 1978) did seminal work essential to modern mixed models. Searle (1971) and Graybill (1976) integrated matrix algebra with linear model theory. This, along with the development of computers that could be programmed by written instructions made modern statistical software (e.g. SAS) possible. SAS PROC GLM, a comprehensive linear model package, was introduced in 1976, and quickly became the standard ANOVA and regression package at North American agricultural research facilities. Despite its success, the limitations of PROC GLM and similar software were apparent from the outset. Specifically, the "general" linear model theory upon which these packages were based were ill-suited to mixed-effects models – notably multi-level, split-plot type experiments – and non-Gaussian data.

Nelder and Wedderburn (1972) introduced generalized linear models, a major departure in approaching non-Gaussian data. Whereas transformations altered the data to meet ANOVA assumptions, Nelder and Wedderburn extended the linear model basis of ANOVA and regression to accommodate more plausible probability assumptions about the data. In 1982, the USDA-supported University Statisticians of Southern Experiment Stations, the group responsible for developing SAS, initiated a regional project to address PROC GLM's shortcomings. The project publication (S-198 Regional Project, 1989) along with Laird and Ware (1982) brought mixed model methods to the attention of larger research communities including agriculture. Prior to 1982, awareness of mixed models was confined to a few highly specialized applications. By the early 1990s mixed model methods were considered mainstream. Liang and Zeger (1986) were similarly instrumental in expanding awareness and applicability of generalized linear models. In 1992, SAS introduced PROC MIXED, which implemented mixed model analysis for normally-distributed data, and PROC GENMOD, which implemented fixed-effects-only generalized linear models for non-normal data. Breslow and Clayton (1993) and Wolfinger and O'Connell (1993) published seminal papers integrating mixed model and generalized linear model theory and methods. The next decade saw intense development of GLMM theory and methods, and concurrent growth in computing technology. Nearly all scientists soon had computers in their offices that were capable of carrying out these new, more complex analyses that were becoming more mainstream.

By the mid-2000s, practical GLMM software began to appear. SAS PROC GLIMMIX was introduced in 2005. Several GLMM packages in R – GLMPQL, GEE, LME4, etc. – appeared as well. This was a watershed moment for statistical analysis. For the first time, useable software existed to implement the full range of statistical models explicitly intended to accommodate both complex experiments (primarily a mixed model issue) and non-Gaussian data (primarily a generalized linear model issue). The extensive development of theory and methodology over previous decades became available to researchers in accessible form. This explains why the entire question of how to do statistical analysis of non-Gaussian data from experimental research is now being reassessed. Statistics, like all other disciplines, is dynamic; it is not a fixed set of unchanging rules passed down from Fisher and company.

Why do these changes matter? Stroup (2013a) presented a comprehensive description of the distribution of observations and the consequences of various estimation and inference strategies for non-Gaussian data obtained from designs typical of research in the biological sciences – blocked designs, split-plot experiments, etc. Stroup (2014) presented examples from agronomic experiments to illustrate the problems with standard ANOVA or regression, with or without transformation, and why the GLMM is the method of choice. To summarize briefly, standard ANOVA and regression, with or without transformation, gives inaccurate estimates of treatment means and their standard errors, misleading confidence intervals, and, in general, underpowered tests of hypotheses. GLMMs were developed to address these issues. The main impediments to the use of GLMMs prior to the mid 2000s were the unavailability of sufficient computing power and lack of useable software. Now that computing power and available software are non-issues, the GLMM is clearly the method of choice for non-Gaussian data.

Introduction to GLMM – Key Concepts and Requirements for Implementation

The section is divided into five subsections. In the next section, an overview of probability distributions, focusing on essentials for the analysis of non-Gaussian data, is presented. "What Would Fisher Do?" introduces a technique Stroup (2013b) developed as a tool to facilitate extracting an appropriate model and strategy for data analysis from a description of a design and its objectives. "A Gentle Introduction to the GLMM" provides an overview of the essential features of the GLMM, and why they matter. "Key Issues" describes three issues unique to the analysis of non-Gaussian data and how they affect modeling decisions and interpretation. "GLIMMIX Syntax Basics" lists the basic commands required to compute GLMM analysis using SAS PROC GLIMMIX.

Some Things You Need to Know About Probability Distributions

All observed data, especially biological data, are assumed to arise from a combination of **systematic** and **random variation**. The randomized block design provides an example familiar to plant scientists in which we characterize observation as the sum of treatment effects + block effects + residual. **Residual**, also called **error**, is a catchword for “everything else.” Treatment effects are systematic; residual effects are random. For reasons discussed later, block effects lie somewhere in the middle. A **probability distribution** is a mathematical description of how we think the random component of observations, vary. We characterize probability distributions by a **density** or **distribution plot**, a list of **parameters**, and a distribution or **density function**.

The most familiar probability distribution, and the basis of classical statistical methods beginning with Fisher's ANOVA and associated regression analysis, is the "normal" or, more properly, the Gaussian distribution. The density plot is the familiar bell-shaped curve, shown in Fig. 1. The parameters for the Gaussian distribution are the mean and the variance, or alternatively the standard deviation. In statistical writing, the mean is usually denoted μ, the variance σ^2, and the standard deviation σ. Figure 1 shows a Gaussian distribution with a mean of 100 and standard deviation of 10.

The Gaussian distribution applies to variables that are continuous, distributed symmetrically about their mean, and are, in theory, unbounded between plus or minus infinity. In practice, as you can see on Fig. 1, almost all of the distribution lies within $\pm 3\sigma$ of the mean, or within a range of 6σ. This is why the Gaussian distribution can be used for variables such as height, weight and yield: they are strictly non-negative but they are continuous, symmetrically distributed, and their 6σ range lies well to the right of zero.

The density function is $f(y)=\sqrt{1/(2\pi\sigma^2)}e^{-\frac{(y-\mu)^2}{2\sigma^2}}$. The shorthand notation for the Gaussian distribution is $Y \sim N(\mu,\sigma^2)$. Most statistical methods textbooks pay little attention to the density function because if you limit attention to classical ANOVA and regression, and you are a consumer of statistics rather than a statistical theorist, the density function plays a background, essentially invisible role. However, when we start to consider contemporary methods for non-Gaussian data, we need to pay attention to the density function.

The two most common non-Gaussian distributions, and the two distributions that illustrate the essentials for doing analysis with non-Gaussian data, are for **discrete proportions**, a.k.a. binomial data and for count data. Let's start with binomial data.

Discrete proportions arise from "yes-no" data – the plant is alive or dead, infected or healthy, stalk lodging is present or it is not, the seed germinates or it does

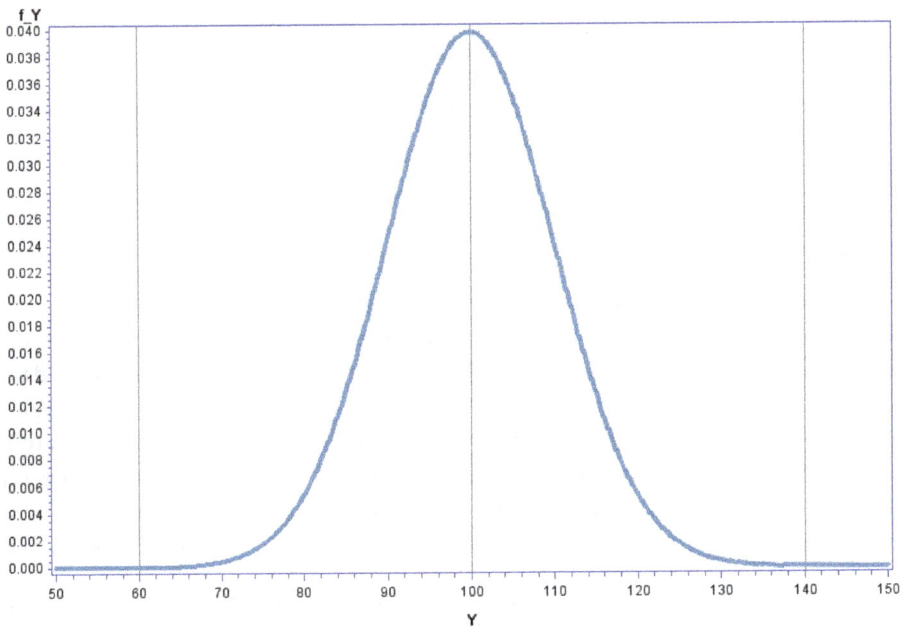

FIG. 1. Gaussian Probability Density Plot, Mean 100, Standard Deviation 10.

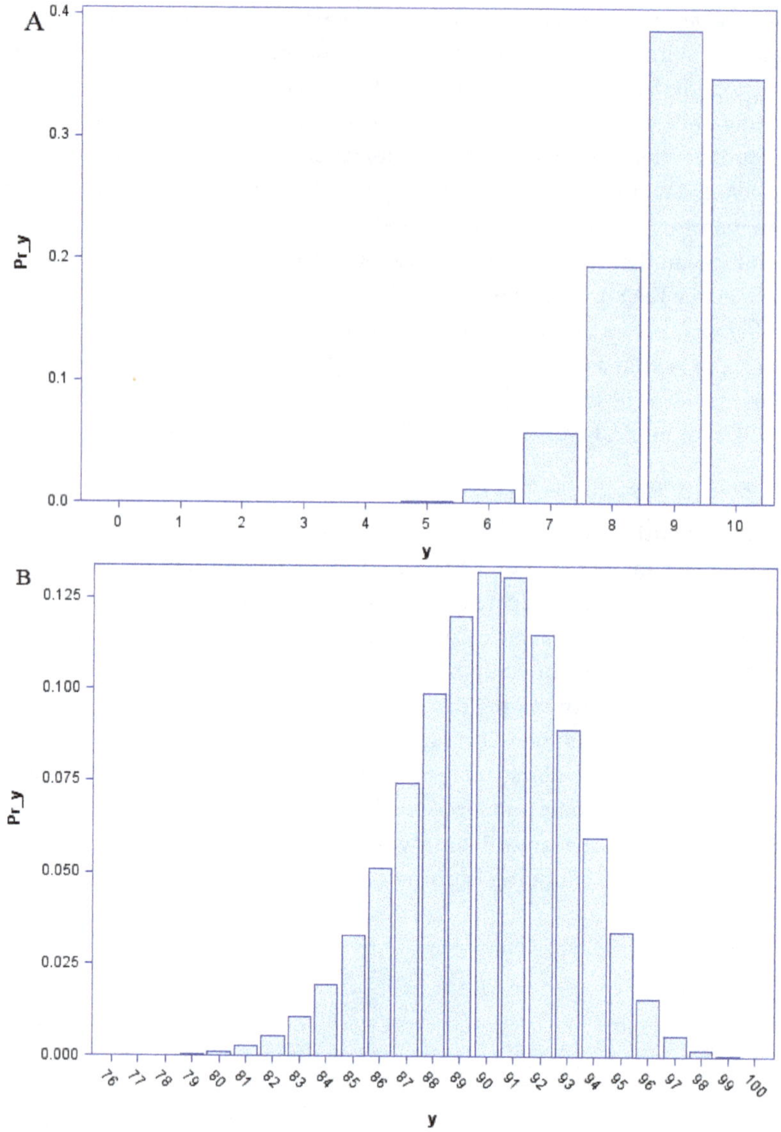

FIG. 2. Binomial distribution plots, $P = 0.9$. A. $N = 10$. B. $N = 100$.

not. In each experimental unit, e.g. a plot, N observations are taken. For example, N plants are observed, and of these, Y show the response or characteristic of interest, called a "success" in statistical jargon. Because N can vary among plots, most data analysts use the sample proportion, defined as $pct = Y/N$, as the response variable.

Formally, Y has a **binomial** distribution, written $Y \sim \text{Binomial}(N, p)$, where p denotes the probability that an observation drawn at random has the characteristic of interest. Note that p must be greater than 0 and less than 1. The distribution function gives the probability of exactly y "successes" out of N observations, and is written Probability $(Y=y)=f(y)=\binom{N}{y}p^{y}(1-p)^{N-y}$. Figure 2 shows two plots of the binomial distribution, the first when $N = 10$ and $p = 0.9$, the second when $N = 100$ and $p = 0.9$.

In multi-treatment experiments, p_i denotes the probability for the ith treatment. Analysis focuses on estimating p_i for each treatment and testing the equality of p_i among treatments. The expected value of the binomial random variable is Np and its variance is $Np(1-p)$. Notice that the mean and variance of the Binomial depend on N and p only. If you know the mean, you also know the variance. With the Gaussian distribution, estimates of the mean and variance require distinct calculations; with the binomial, a single calculation, the estimate of p, determines both the mean and variance. As we will see below, this crucially changes the meaning of "residual" in the analysis of binomial data – a fact that is greatly underappreciated in the literature of classical statistical methods.

For large N, the binomial distribution approximates the Gaussian – the approximation becomes more accurate as N increases. You can see this in Fig. 2; with $N = 10$ the distribution is not even close to a bell-shaped curve, whereas with $N = 100$ it is. However, there is no universal agreement about what qualifies as "large." Textbooks give rules of thumb for when it is appropriate to use the "normal approximation to the binomial" ranging from $Np \geq 5$ to $Np \geq 10$ if $p < 0.5$, or $N(1-p) \geq 5$ to $N(1-p) \geq 10$ if $p > 0.5$. Even when the normal approximation is accurate, for multi-treatment experiments, unequal p_i guarantees unequal variance, violating a key ANOVA assumption. From the Bartlett (1947) through the appearance of GLMM software, the standard way to deal with this problem was to use a **variance stabilizing transformation**. The standard variance-stabilizing transformation for proportion data is the arc sine square root transformation, that is, $\sin^{-1}\left(\sqrt{pct}\right)$, where pct is the sample proportion, Y/N, as defined above. This is also known as the angular transformation. However, as we shall see later in this section, neither the normal approximation nor the angular transformation qualify as best practices for analyzing binomial data compared to the GLMM alternative.

Count data are discrete, non-negative, integer-valued, and typically have right-skewed distributions. Two commonly used distributions for count data are the **Poisson** and the **negative binomial**. In classical probability theory, counts imply a Poisson distribution. The Poisson distribution function is Probability $(Y=y)=f(y)=e^{-\lambda}\lambda^{y}/y!$, where λ denotes the mean. Figure 3 shows two Poisson distributions, the first with a mean $\lambda = 2$ and the second with mean $\lambda = 10$.

You can see that as counts increase, the Poisson approximates the normal distribution – the larger the count, the better the approximation. Poisson variables have equal variance and mean, that is mean $= \lambda =$ variance. By definition, this means Poisson counts violate the equal variance assumption of ANOVA. For smaller counts, data analysts are often advised to use log or square root transformations. However, assuming Poisson-distributed counts is problematic for biological count data. Requiring mean and variance to be equal is a rigid and usually unrealistic assumption. Based on biological theory and accumulated experience, the negative binomial provides a more realistic distribution. Negative binomial random variables have mean $\lambda > 0$, and variance $\lambda + \phi\lambda^2$, where $\phi > 0$ is called the *scale parameter.* Notice that, unlike the normal, whose mean, μ, and variance, σ^2, are distinct entities, the

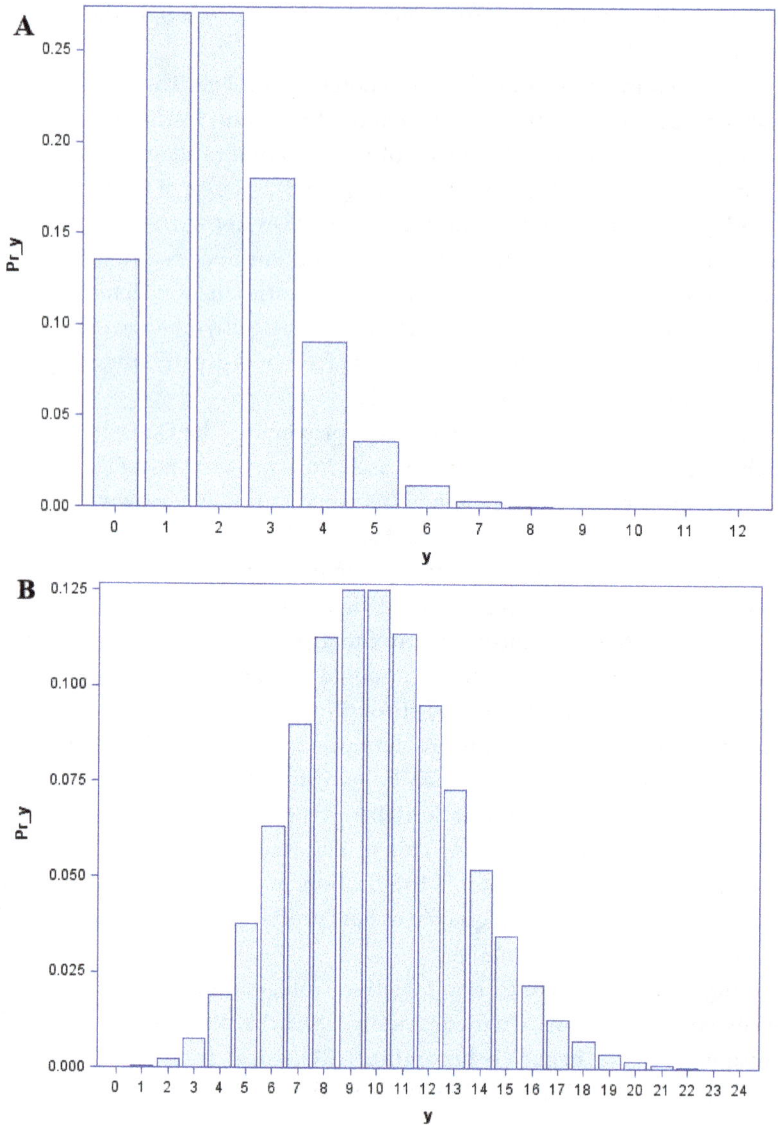

FIG. 3. Poisson distribution plots. A. λ = 2. B. λ = 10.

negative binomial's variance depends on the mean and an additional scale term. Figure 4 shows plots of the negative binomial distribution for $\lambda = 2$ and $\lambda = 10$, both with scale parameter $\phi = 0.5$ (a typical value for count data in the plant sciences).

The distribution of the negative binomial is equivalent to a Poisson with the mean λ multiplied by a term with a Gamma distribution that reflects additional variability at the experimental unit level. For this reason, the negative binomial distribution is sometimes referred to as a Poisson-Gamma. This will be explained in more detail in subsequent sections. The important point for our purposes in this chapter is that the logic of the GLMM developed for working with the Poisson applies equally to the negative binomial.

Other distributions of interest in the plant sciences, although less common than binomial and count data distributions, include the beta (for continuous proportions), the gamma (for time to event and non-negative, right-skewed continuous data), and the log-normal (for continuous, right-skewed data). Detailed discussions of these distributions can be found in Stroup (2013a) and in textbooks on statistical distribution theory. We will give essential characteristics of these distributions in "Gentle Introduction to the GLMM", our introduction to the GLMM.

What Would Fisher Do?

Analysis begins with a statistical model, a description of the impact of experimental factors and random variation on the observed response. For designed experiments, the statistical model and the ANOVA table are intimately linked. The ANOVA table is a good place to start to understand GLMMs, how they differ from traditional ANOVA models, and how to set up and work with GLMMs. **Accurately translating the design and the distribution of the observations into a model is an *essential* and *crucial* step in the analysis of non-Gaussian data.** Doing so for non-Gaussian data differs in subtle but important ways from setting up the model for analysis of Gaussian data.

Stroup (2013b) introduced an exercise called WWFD ("What Would Fisher Do?") to help students and data analysts work through steps leading from description of design to ANOVA table to model. The procedure was based on Fisher's comments following Yates (1935) Complex Experiments. Fisher said that any experiment could be described in terms of its "topographical" and "treatment" components, the former being the physical elements such as blocks, experimental units – what Federer (1955) and later Milliken and Johnson (2009) would call the "experiment design." By writing an ANOVA for the topographical design, another for the treatment design, then integrating them, Fisher suggested that the appropriate analysis would then be apparent. Stroup's WWFD adapted Fisher's thought process by showing how one translates the integrated ANOVA into a GLMM.

The specifics of this process will be illustrated for each of the examples presented in the examples section of this chapter. We begin in this section with a generic description of the essential feature of the WWFD ANOVA process using a randomized complete block design with 8 blocks, 2 treatments, and therefore 2 experimental units per block with treatments randomly assigned to experimental units, one unit per block per treatment. Thus the sources of variation for the topographical ANOVA are blocks and units within blocks; the sources of variation for the treatment ANOVA are treatment and whatever is left, a term Fisher called "parallels." Table 1 shows the topographical, treatment and combined ANOVA tables.

The placement of the rows in the topographic and treatment ANOVAs is important. Sources of variation in the treatment ANOVA are always placed in the line immediately above the line in the topographical ANOVA corresponding to the unit to which they were randomly assigned. Here, "treatment" was randomly assigned to "unit(block)." Then one "slides" the sources of variation to the right to obtain the combined ANOVA. Notice the name of the source of variation in the last line of the combined ANOVA: "unit(block) | treatment." Read this as, "unit within block after accounting for treatment." Also notice that its 7 degrees of freedom result from the

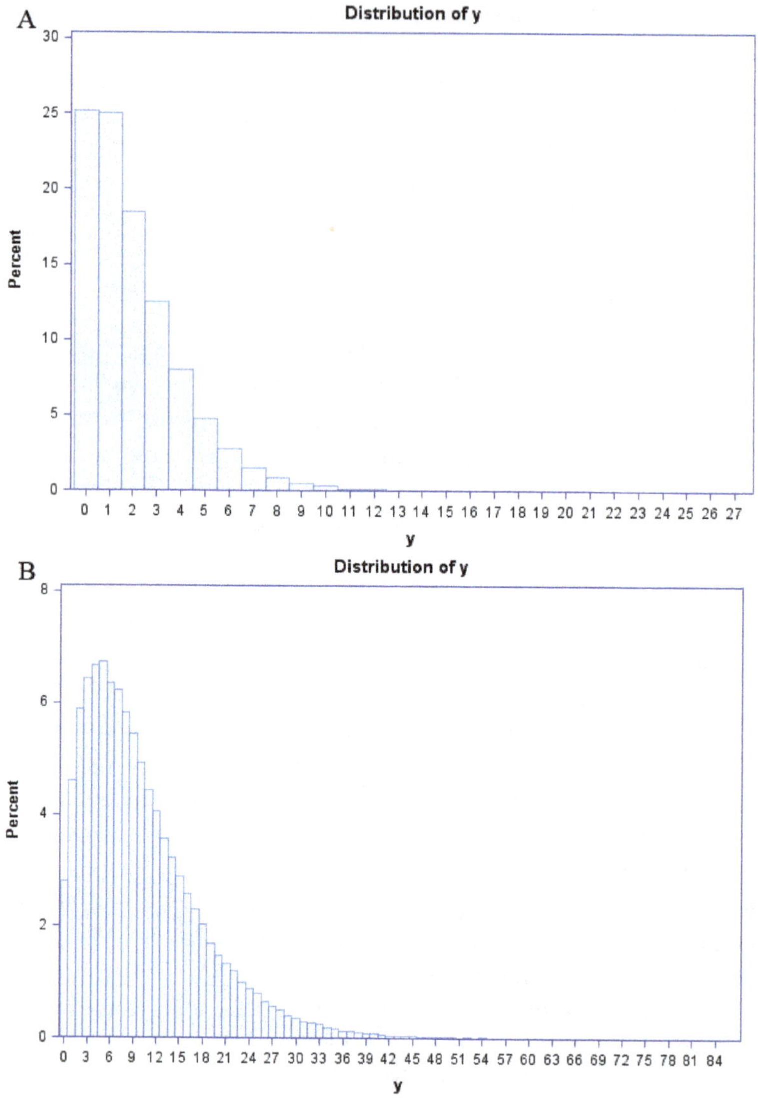

FIG. 4. Negative binomial distribution plots. Scale parameter $\phi = 0.5$. A. $\lambda = 2$. B. $\lambda = 10$.

TABLE 1. WWFD ANOVAs for Randomized Complete Block Design.

Topographical		Treatment		Combined	
Source	df	Source	df	Source	df
Block	7			Block	7
		Treatment	1	Treatment	1
unit(block)	8	"parallels"	14	unit(block) | Treatment a.k.a. "residual" a.k.a. Block × Treatment	8-1=7
TOTAL	15	TOTAL	15	TOTAL	15

original df for experimental units in the topographical ANOVA minus the df of the treatments applied to those units. While these line placement and degree of freedom protocols seem obvious with a simple, single-factor design such as the RCBD, disciplined application of WWFD rules greatly facilitates defining sensible models for arbitrarily complex designs.

Traditionally, this leads to an equation read as "observation = overall mean + treatment + block + error," where, aside from "overall mean" each term in the model corresponds to a line in the ANOVA table. This model is written in statistical notation as $y_{ij} = \mu + \tau_i + b_j + e_{ij}$, where y_{ij} denotes the observation on the ith treatment and j^{th} block, and each line in the combined ANOVA implies a corresponding term on the right-hand side: "block" implies b_j; "treatment" implies τ_i; and "unit(block) | treatment" implies e_{ij}. Statistical texts commonly refer to e_{ij} as the "residual" or "random error" term, assumed to be independent and have a Gaussian distribution with mean 0 and variance σ^2.

This equation provides the standard justification for using the tools of analysis of variance – sums of squares, mean squares, F-tests, etc. Provided, that is, that the Gaussian assumption is plausible. However, if the data are not Gaussian, we have a problem.

The GLMM takes a different approach. We still use the ANOVA sources of variation, but in a way that is more adaptable for data that do not fit classical assumptions. First, we identify the unit of observation and the most plausible probability distribution of the observations taken on those units. Then we need to decide which sources of variation in the combined ANOVA have probability distributions – in modeling language, which sources are fixed and which are random – and, for the random model effects, what probability distributions are reasonable to assume. Finally, we need to specify the **linear predictor** – the GLMM generalization of the model equation used for classical ANOVA and regression.

For example, for normally distributed data, write the distribution of the observations as $y_{ij} \sim N(\mu_{ij}, \sigma^2)$, read "the observations (y_{ij}) have a normal distribution with mean μ_{ij} and variance σ^2." Write the linear predictor as $\mu_{ij} = \mu + \tau_i + b_j$. Notice that the linear predictor describes how the observation's mean is affected by the sources of variation, block and treatment, listed in the ANOVA. This is called the **probability distribution** formulation of the model; the traditional approach in the previous paragraph is called the **model equation** formulation.

To complete the model, we need to decide if the treatment effects, τ_i are **fixed** or **random**, and do the same for the block effects, b_j. To decide if an effect is fixed or random, ask the following questions:

- if we repeated this experiment, or if someone else did an identical experiment, e.g. to see to what extent are our results reproducible, would the same levels of the effect necessarily be used?
- is there anything special about the levels of this effect used in this experiment?
- is it sufficient that the levels used in this experiment be representative of levels that could have been used?

If the answer to these questions are no, the same levels need not be used, no, there is nothing special about the levels used, and yes, it is sufficient that they be representative, then the effect is random. Its levels form a sample from a larger population, and, being observations on that sample, they have a probability distribution.

TABLE 2. ANOVA Sources of Variation and Associated Distributions for Randomized Block Design with Binomial Data.

Source of Variation	df	Distribution
block	7	$b_j \sim NI(0, \sigma_B^2)$
treatment	1	NA - τ_i are fixed effects
unit(block) | treatment = unit of observation	7	$y_{ij} \mid b_j \sim Binomial(N_{ij}, p_{ij})$

On the other hand, if the same levels would be used and there is something special about those levels and those levels only, then the effect is fixed. In this example, block effects are random – the blocks used in the experiment merely need to be typical and representative blocks – whereas treatment effects are fixed, because we are doing the experiment specifically to compare the two treatments under study. No other treatments will do. To complete the model, we specify the distribution of the block effects. The usual assumption is $b_j \sim NI(0, \sigma_B^2)$, read in words as "block effects are normally and independently distributed with mean zero and variance σ_B^2."

For normally distributed data, these two formulations – model equation and probability distribution – are simply different ways of expressing the same thing. Readers may ask, "If it doesn't matter, why add a new complication?" The answer: because the probability distribution approach accommodates non-Gaussian data, whereas the model equation does not.

To illustrate, suppose the data from this randomized complete block design are binomial. We would them view the combined ANOVA in Table 2.

If we use a classical model equation, we need to ask how each term in the model would be interpreted. Recall that the equation is $\mu + \tau_i + b_j + e_{ij}$. Once we have estimates of each term in the model equation, we use them to estimate the mean. For binomial data, the mean we are estimating is p_{ij} assuming, as is usual, that we use the sample percent, y_{ij}/N_{ij} as the response variable. With binomial data, once we know the mean, we also know the variance, $p_{ij}(1-p_{ij})/N$. This is unlike Gaussian data, where the estimated mean and the variance are distinct terms. Also notice that in the Gaussian case, the information from treatment and block are clearly used to estimate the mean and the information from the unit of observation line of the ANOVA corresponds to the e_{ij} term in the model and is clearly used to estimate the experimental unit variance, σ^2. With binomial data, there is no σ^2 to estimate, e_{ij} is therefore a meaningless term, and the model equation is nonsense. The same would be true of Poisson data, where the mean and variance are identical. The negative binomial is even more complex. There is no distinct variance to estimate, but the variance, $\lambda(1+\lambda\phi)$, depends on the mean *and* a *scale* parameter. In addition to the mean, the scale parameter must be estimated.

In addition to the residual term being meaningless for non-Gaussian data, it turns out, as we shall see below, that the "mean" we obtain from the model equation is not really an estimate of the mean we want. Clearly, non-Gaussian data call for a different approach. For this, we now turn to the next section, a brief, users' introduction to the GLMM.

Gentle Introduction to the GLMM

A statistical model has two aspects: it is a mathematical description of how we think the data arose and it provides a template for analyzing and interpreting the data. Often, data analysts focus on the latter and lose track of the former, an example being the use of the model equation (a template for analysis) which does not accurately reflect the process that gave rise to the data (the sources of variation and their distributions in the ANOVA). The key to an accurate and informative analysis is to align both aspects of the model. This is the fundamental principle motivating the generalized linear mixed model.

The following is intended as a "what consumers need to know" guide. For the rigorous, "gloves off" presentation of GLMM theory, see Stroup (2013b). For most users of statistics, full immersion in GLMM theory is unrealistic and unnecessary. What readers of this chapter need is a general understanding of the main issues and how contemporary statistical science approaches them. Some GLMM intuition is essential to appropriately analyzing non-Gaussian data and interpreting the results.

The GLMM begins with the WWFD ANOVA table sources of variation and associated distributions, as illustrated by Table 2. The GLMM must resolve two questions raised by this table in order to obtain an informative analysis. The first question is: once we write a linear predictor, what does it actually make sense for it to estimate in the context of the distribution of the observed data? The second question, which is intimately connected to the first, is: if the source of variation and their distributions provide an accurate mathematical description of how the data arose, what does the distribution of the data we can actually observe really look like?

Let's start with the first question. With Gaussian data, the linear predictor estimates the mean, For data from a blocked design, the mean, μ_{ij}, is estimated by the linear predictor $\mu+\tau_i+b_j$. For non-Gaussian data, it usually does not make sense to use the linear predictor to directly estimate the mean. For binomial data, it makes more sense for the linear predictor to estimate $\log\left[p_{ij}/\left(1-p_{ij}\right)\right]$, called the **logit**. For count data, it makes more sense for the linear predictor to estimate the log of the mean, that is, $\log\left(\lambda_{ij}\right)$. The reason comes from the probability distribution function. In contemporary statistics, estimation comes from maximizing the **likelihood** – for the blocked design, this is equivalent to finding the values of the treatment and block effects that maximize the probability distribution of the data. As a matter of terminology, for estimation, the distribution function is called the likelihood: the difference is that when we look at the function as a probability distribution, we assume we know the parameter and we want to know how the observations, Y, behave; for estimation, we have the data and our question concerns the parameters, that is, "what values of the parameter are most consistent with the data we have observed?" Because it is much easier to work with the log of the function, and whatever maximizes the log also maximizes the function, the **log likelihood** is the primary tool for modern statistical analysis.

Start with the binomial. Recall that probability of getting y_{ij} "yes" responses out of N observations is given by binomial distribution function $\binom{N}{y_{ij}} p_{ij}^{y_{ij}}\left(1-p_{ij}\right)^{N-y_{ij}}$.

The natural log of the distribution, the binomial log likelihood, after some strategic rearranging, is $\log\binom{N}{y_{ij}} + y_{ij}\log\left(\frac{p_{ij}}{1-p_{ij}}\right) + N\log\left(1-p_{ij}\right)$.

The term $y_{ij}\log\left(\frac{p_{ij}}{1-p_{ij}}\right)$ is particularly important: in the log likelihood, whatever is multiplied by y_{ij} is called the **natural** (or **canonical) parameter.** The natural parameter is always a function of the mean. For the binomial, the mean is Np_{ij} and the natural parameter is $\log\left(\frac{p_{ij}}{1-p_{ij}}\right)$. In categorical data, $\log\left(\frac{p_{ij}}{1-p_{ij}}\right)$ is also called the **log odds.** In modeling theory it is called the **logit.**

Why is this important? Without belaboring the underlying theory, the natural parameter is a better candidate for regression and ANOVA-like models that the mean itself. Following this idea, write the linear predictor as $\eta_{ij} = \log\left(\frac{p_{ij}}{1-p_{ij}}\right) = \eta + \tau_i + b_j$. In generalized linear model terminology, η_{ij} is called the **link function.** It is the function that "links" the mean to the linear predictor. The **inverse link** expresses the mean in terms of the linear predictor. For the logit, the inverse link is $p_{ij} = 1/\left(1+e^{-\eta_{ij}}\right)$. Notice that we use η rather than μ to denote the intercept of the linear predictor. This is because η is the value of the link function when all other model effects are zero – it is *not* the "overall mean." Using notation other than μ helps us avoid possible confusion and misstatement.

For count data, we use the same logic, but use the Poisson distribution. Recall that the distribution function is $e^{-\lambda_{ij}}\lambda_{ij}^{y_{ij}}/\left(y_{ij}\right)!$. The log likelihood, again with some strategic rearranging, is $y_{ij}\log\left(\lambda_{ij}\right) - \lambda_{ij} - \log\left(y_{ij}!\right)$. The natural, or canonical, parameter is $\log\left(\lambda_{ij}\right)$ and therefore the linear predictor for count data from a randomized block design would be written $\eta_{ij} = \log\left(\lambda_j\right) = \eta + \tau_i + b_j$, where $\eta_{ij} = \log\left(\lambda_j\right)$ is the link function. Once you estimate the model, you would obtain the estimates of the mean using the count data inverse link, $\lambda_{ij} = e^{\eta_{ij}}$, which, for the blocked design, would be $e^{\eta+\tau_i+b_j}$.

To summarize, the GLMM must be specified in probability distribution form. The required elements of a GLMM are

- The **linear predictor.** This will follow from the sources of variation in the combined column of the WWFD ANOVA table
- The **distribution** of all **random** effects in the linear predictor
- The **distribution** of the **observations, conditional on the random model effects**
- The **link function**

For binomial data from a blocked design, the model is thus

- Linear predictor: $\eta_{ij} = \eta + \tau_i + b_j$ (but see discussion in the "Key Issues" section, What to Include in the Linear Predictor)
- Distribution of block effects: $b_j \sim NI\left(0, \sigma_B^2\right)$
- Distribution of observations: $y_{ij} \mid b_j \sim Binomial\left(N_{ij}, p_{ij}\right)$
- Logit link: $\eta_{ij} = \log\left[p_{ij}/\left(1-p_{ij}\right)\right]$

Note that when unit-level distribution, $y_{ij} \mid b_j$ for the blocked design, is Gaussian, the link function is often not mentioned because it is equal to the unit-level mean

– in GLMM notation, $\eta_{ij} = \mu_{ij}$. This is called the **identity link**. This matters to agronomic researchers mainly when describing the statistical analysis used, for example in the "Materials and Methods" section of a thesis or journal article. If the data are Gaussian, the identity link is understood.

The second question we need to address in our "gentle introduction to the GLMM," is how are the data we can actually observe distributed and what does this say about how we interpret an analysis using the GLMM. How does this interpretation compare with the interpretation of a "traditional" statistical analysis?

To illustrate, let us explore further our binomial example. The observations taken on each experimental unit are assumed to have a Binomial distribution with N_{ij} "yes-no" observations (e.g. seeds that either do or do not germinate) and probability p_{ij} of a "yes" response (e.g. the seed germinates) on any given observation. Suppose $N_{ij} = 100$ for all experimental units, and $p = 0.9$, the distribution shown in Fig. 2b. The mean for the ij^{th} unit is thus Np_{ij} = 100 × 0.9 = 90 and the variance is $Np_{ij}(1-p_{ij}) = \sqrt{100 \times 0.9 \times (1-0.9)} = 3$. Figure 5 shows the binomial distribution with a Gaussian distribution having with same mean and standard deviation superimposed. You can see that the two distributions are very similar.

Inspection of Fig. 5 reveals why, with large N, the normal approximation to the binomial reassures and tempts data analysts. However, the reassurance is only apparent. Life is not so simple. Our model tells us that we have two distributions that account for our observations. One is the binomial, the other is the distribution of the block effects.

What happens when p_{ij} is affected by a block effect that has a normal distribution? Figure 6 shows the distribution of the resulting observations. This is the distribution of the data that we actually observe. It has two features that are crucial to understanding the logic of analyzing non-Gaussian data.

First, even though the normal approximation is convincing for the binomial distribution at the experimental unit level, and the distribution of the block effects is normal by assumption, the resulting distribution of the observations is not even

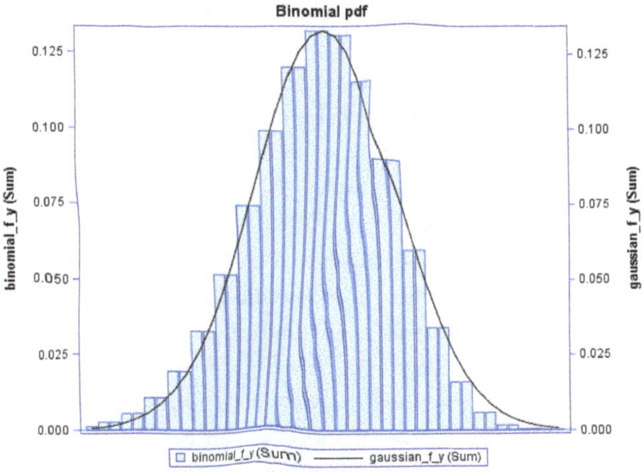

FIG. 5. Gaussian (normal) Approximation of the Binomial for p = 0.9, N = 100.

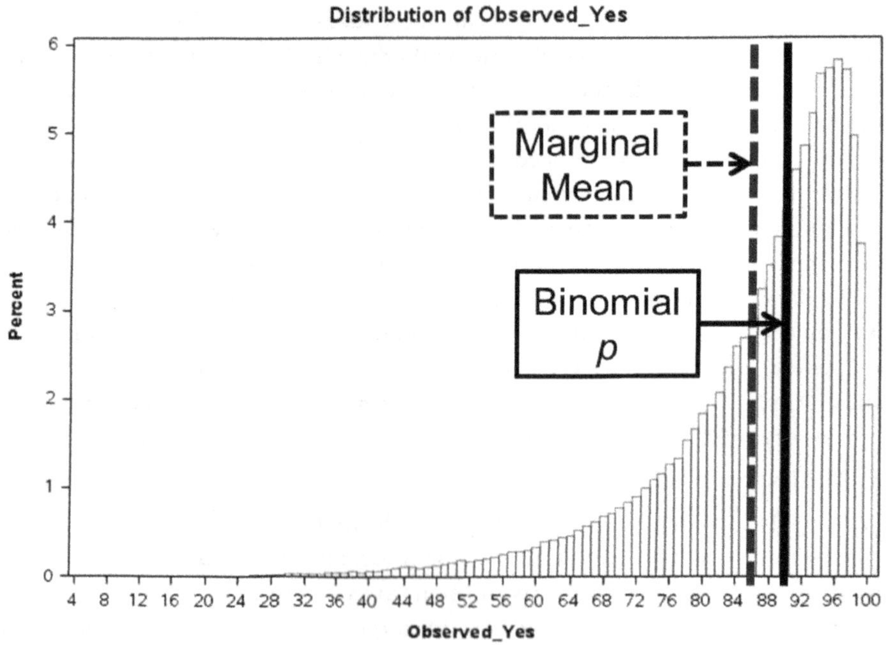

FIG. 6. Actual distribution of binomial (p = 0.9, N = 100). Observations from blocked design.

remotely normal. In this case it is strongly left-skewed. In general, for $p_{ij} > 0.5$ the distribution of the observations will be left-skewed; for $p_{ij} < 0.5$ the distribution will be right-skewed. The skewness increases as the probability approaches 0 or 1. The observations are symmetrically distributed *only* when $p_{ij} = 0.5$ for all treatments – not likely in practice.

The second important feature of the distribution shown in Fig. 6 is that its mean is *not* equal to the binomial probability $p = 0.9$. For the distribution and block variance shown in Fig. 6, the mean is 86.6. Consequently, the mean sample proportion is 0.866. This is important because with standard ANOVA – the *model equation* approach defined earlier – yields an unbiased estimate of the mean of the distribution of the observations. If your goal is to estimate the binomial probability, $p = 0.9$, you will not get it using ANOVA. Analysis of variance will give you an unbiased estimate of 0.866. A similar set of illustrations could be developed for the count data. In fact, the discrepancy between the sample mean of the observations and λ, the actual mean you want to estimate, tends to be even greater for count data. As we will see in examples later in this chapter, the sample mean tends to substantially overestimate λ.

Formally, there are three distributions relevant to the analysis of experimental data. Two of them follow from sources of variation identified by the WWFD ANOVA exercise: the distribution at the unit level (the binomial distribution in our example); and the distribution of the random model effects (blocks in our example).

The block and unit-given-block distributions follow from the sources of variation listed in the ANOVA. However, neither can be observed directly. The *only* distribution we can actually observe is called the *marginal distribution* of the observations. The important thing for users to remember is that when we say we have "binomial" data, we are referring to the distribution of the observations conditional on the ij^{th} unit. However the distribution of the *observed* data – the marginal distribution – is emphatically not binomial.

The reason this is a non-issue with Gaussian data is that if the random effects distribution and the unit-given-random-effects distribution are Gaussian, the resulting marginal distribution is Gaussian as well. However, this only happens when the unit-of-observation level distribution, that is, the distribution of $y_{ij} \mid b_j$ is Gaussian. For all other data – binomial, counts, continuous proportions, time-to-event, etc. – the marginal distribution of the observed data quite different. Our usual intuitions can betray and mislead.

Important note to Advisors: Certainly, researchers – including your graduate students – in the early stages of analyzing data, should plot the raw data, look both for anomalies and at the big picture – what biological story the plants, through the data, tell us. However, do not be surprised when hand-calculated sample means of non-Gaussian data do not agree with estimates obtained from GLMM analysis. The whole point of this discussion is that the sample means are misleading, and it is well-established that the GLMM results are more accurate, and give a far better picture of what the data tell us than computationally much simpler methods taught in pre-computer days. This is not because incorrect methods were consciously taught back then. In many cases, the theory was there but the ability to do anything with it was not – we simply lacked the tools we now have at our disposal. In other cases, the theory had not yet caught up with the nuances of analyzing non-Gaussian data. Please allow and encourage your students to use contemporary best practices for non-Gaussian data. This appeal goes to reviewers and technical editors as well.

The fundamental problem of analyzing non-Gaussian data, especially with designs most commonly used in agronomic research, is that what we want to estimate or test – e.g. treatment effects on p_{ij} for binomial data – involves parameters of distributions we cannot directly observe. What GLMM analysis does that standard ANOVA and regression cannot do is provide a way to extract the information we want – about the effects listed from the WWFD ANOVA exercise – from the observations we have, where these effects are camouflaged in a complex marginal distribution.

This concludes our introduction to the GLMM. Before concluding this section, however, a brief note on acronyms that are likely to cause confusion, especially when someone who took statistics before 2008 and someone who learns statistics using this textbook attempt to communicate.

In contemporary statistical terminology, any model specified with a linear predictor – whether in model equation or probability distribution form – is technically a GLMM. However, linear models are usually referred to by the following acronyms:

- **LM:** "linear model" – unit-level distribution is Gaussian; fixed-effect-only model, that is, no random model effects.
- **LMM:** "linear mixed model" – unit-level-given-random effects distribution is Gaussian; linear predictor contains random model effects. Blocked designs where block effects are considered random (see above discussion) and all designs with split-plot features require random model effects; all repeated measures in time or space require mixed model methodology.

TABLE 3. Mean, Variance and Usual Link Function for Common Distributions.

Distribution	Mean	Variance	Link
Normal	μ	σ^2	μ (identity)
Negative Binomial	λ	$\lambda+\phi\lambda^2$	$\log(\lambda)$
Poisson	λ	λ	$\log(\lambda)$
Binomial proportion	p	$p(1-p)/N$	logit=$\log\left(\frac{p}{1-p}\right)$ † probit=$\Phi^{-1}(p)$
Beta	μ	$\frac{\mu(1-\mu)}{1+\varphi}$	$\log\left(\frac{\mu}{1-\mu}\right)$
exponential	μ	μ^2	$\log(\mu)$ ‡
Gamma	μ	μ/φ	$\log(\mu)$ ‡

† Natural parameter is Logit

‡ Natural parameter is $1/\mu$ but rarely used as link function; log is better choice

- **GLM:** "generalized linear model" – unit-level distribution is non-Gaussian; fixed-effects-only linear predictor.
- **GLMM:** "generalized linear mixed model" – unit-level-given-random-effect distribution is non-Gaussian; linear predictor contains random effects.

Table 3 summarizes essential information about distributions most common in plant science research. Use this table as a guide to deciding what distribution makes sense, commonly used link functions, commonly used notation to denote the mean, and the variance. The last item is especially useful in deciding whether the last line in the WWFD ANOVA should or should not be included in the linear predictor. More about this in the next subsection.

Before moving on, a comment about the acronym GLM is in order, because it is guaranteed to confuse researchers raised on ANOVA via SAS. From its introduction in 1976 through the mid-1990s, PROC GLM was the preeminent SAS procedure for analyzing experimental data. At the time it was introduced, GLM meant "general linear model." Don't confuse "general" and "generalized" – they have very different meanings in statistical science. The "general" linear model is "general" in the sense that it can accommodate any linear predictor, provided the data are normally distributed and all model effects are fixed. By 1976 standards, that was "general." By 2015 standards it is not. The "general" linear model cannot fully accommodate random model effects nor can it accommodate non-Gaussian data (except via transformations, which do not accommodate non-Gaussian data so much as they attempt to force data to appear Gaussian instead). The contemporary acronym for the "general linear model" is LM. SAS PROC GLM can compute the analysis of LMs but not GLMs.

In SAS, PROC MIXED is specifically intended for LMMs. Many researchers still use PROC GLM for certain mixed models but this is emphatically discouraged – see Littell, et al. (2006) for a full treatment of this subject. The generalized linear mixed model (GLMM) integrates mixed and generalized linear models. PROC GLIMMIX was developed for GLMMs. Notice that all linear models are special cases of the GLMM: the GLM is a GLMM with no random model effects; the LMM is a GLMM with normally-distributed data; the LM is a GLMM with no random effects and normally-distributed data. For this reason, software intended to implement GLMMs can implement any linear model. For most data analyses, PROC GLIMMIX has effectively replaced PROCs GLM, MIXED, and GENMOD. On the other hand, PROC GLM cannot implement LMMs, GLMs, or GLMMs.

Key Issues – Data Versus Model Scale; Conditional Versus Marginal Model

Three crucial issues are raised by GLMs and GLMMs that are likely to be unfamiliar to those new to generalized models. These are

- the *model* scale versus *data* scale
- what terms to include in the linear predictor and what terms must not be included
- *conditional* versus *marginal* inference and, as a consequence, *conditional* versus *marginal* models

These are discussed briefly below. Examples (provided later) will further illustrate relevant aspects of these issues. Interested readers are referred to Gbur, et al. (2012) and Stroup (2013b) for more in-depth discussions.

Model Versus Data Scale

For GLMs and GLMMs, all statistical analysis occurs in terms of the link function. For example, with the binomial example above, analysis is in terms of the logit. The test of equal treatments implies testing for equal τ_i and hence equal logits. Ask for the "mean" and you get the mean logit. This is fine for testing – tests using GLMMs are demonstrably more accurate than standard ANOVA, with or without transformation – but the mean logit or the difference between two mean logits is not the stuff of understandable reports. Instead, results and discussion of research ought to include estimates of the probabilities, p_i, for each treatment, obtained using the inverse link. Estimates of the mean logit are examples of the *link* or *model* scale. The estimate converted to a probability is an example of the data scale.

Notice that with Gaussian data, there is no data scale-model scale distinction. The identity link guarantees that the estimate of the mean using the model is in fact an estimate of the ith treatment mean on the data scale. Until Nelder and Wedderburn introduced GLMs in 1972, there was no reason for the data/model scale distinction to occur to anyone. Because statistical methods textbooks still focus on traditional ANOVA, data/model scale issues have yet to come to their attention.

What to Include in the Linear Predictor

For Gaussian data, the hard part of writing the model is getting the WWFD ANOVA right. The model equation follows. The linear predictor is defined by each line of the combined ANOVA except the last.

With non-Gaussian data, writing the linear predictor requires more care. One must consider the last line of the combined ANOVA and the variance implied by the response variable's distribution. To illustrate, consider the randomized block example with Gaussian data as opposed to the same design with binomial data. For Gaussian data, there is no ambiguity about the linear predictor. It consists of an intercept plus a term for each line in the ANOVA except the last, for example, $\mu + \tau_i + b_j$ for the randomized block design. We used these terms to estimate the means, μ_{ij}. The last line of the ANOVA gives us MS(residual), which we use as the estimate of the variance, σ^2.

What if we borrow the normal-data linear predictor and use it for binomial data? Using the linear predictor $\eta + \tau_i + b_j$ means not accounting for any variation that may be attributable to the last line of the ANOVA. This may be a problem. Why? It is reasonable to assume that the ij^{th} experimental unit has unique characteristics not fully explained by the treatment and block effects alone. In the normal case, experimental unit uniqueness is implicit: e_{ij} appears in the model equation to account for it and σ^2 measures it. But, unlike the normal, the binomial does not have a separate variance term. Once you estimate p_{ij} you automatically have the mean and the variance. If the binomial model merely borrows the linear predictor from the normal model, experimental unit uniqueness is ignored. This results in the model accounting for less variation than may be actually present in the data. This is one form of **overdispersion**, the modeling term for the observed variance exceeding the variance accounted for by the model. Overdispersion results in downward biased standard errors, meaning confidence intervals are too narrow, and upward biased test statistics, meaning type I error rates are inflated, often severely.

The solution to this problem lies in accurately understanding the WWFD combined ANOVA. How this plays out depends on the specifics of the distribution. Examples in the next section will show how this issue is dealt with in the context of several designs and distributions.

Conditional and Marginal

Earlier in this section, we explored the distinction between the conditional and marginal distributions of the data. Stroup (2014) presented three examples showing how ANOVA on the untransformed data estimated treatment means that were different from those estimated by the corresponding GLMM. The distribution of the observations from the blocked design with binomial data shown in Fig. 6 showed two different understandings of "the mean:" the actual p_i, and the marginal mean. Inference applicable to p_i is called **conditional** inference, and is the inference that is usually the appropriate focus when working with a GLMM. Inference on the marginal mean is, as the name implies, called **marginal** inference. The distinction between conditional and marginal inference exists for all non-Gaussian data. Working effectively with non-Gaussian data requires understanding the difference.

The conditional, or GLMM estimate and the marginal estimate address two potentially useful, but distinctly different, questions. In the binomial case, the GLMM estimate says, "If I take an average member of the population – which means a member of the population whose block effect $b_j = 0$ – what is the estimated binomial probability?" The marginal mean addresses the question, "If I average over all the members of the population, what is the mean proportion?" Because the marginal distribution is highly skewed, another way to think of this is to consider the mean and median. Because the distribution of block effects is symmetric, an estimate of the probability at $b_j = 0$ is closely related to the median. Household income provides a useful analogy, because income data tend to be strongly right-skewed: median income accurately characterizes a typical household; mean income does not provide a useful characterization of a typical household, but it does accurately measure the amount of money in the overall economy. Conditional inference: median of a skewed distribution. Marginal inference: mean of a skewed distribution. Which is right? It depends. What is the question?

Because the normal distribution is symmetric, the conditional-marginal issue does not arise with LMs and LMMs. With GLMs and GLMMs, researchers must decide which understanding of "expected value" best addresses a study's objectives.

GLIMMIX Syntax Basics

Before computers generally available to plant science researchers reached the level of a typical high-end desktop or laptop PC circa 2005, GLMM analysis would have been considered inconceivable. Generalized Linear Mixed Model analysis requires software unavailable prior to 2000. As of this writing, the most capable software available to agricultural researchers for GLMMs is the current version SAS PROC GLIMMIX, released in essentially its current form in 2008. Many – but not all – of the analyses discussed in the next section can be implemented with other software, notably the LME4 package in R. However, these packages lack the flexibility and range of options of GLIMMIX. Some of the examples in the next section simply cannot be done with R, as of the time this chapter is written. Period. For this reason, for the examples presented, use PROC GLIMMIX.

In this section, we present the essential syntax required to run a GLIMMIX program. Note that because all linear models are special cases of the GLMM, any analysis from a simple one-way ANOVA or linear regression through a complex GLMM can be implemented with PROC GLIMMIX. The essential syntax is as follows

- PROC statement. The mandatory statement is PROC GLIMMIX.
  The most commonly used option is METHOD=LAPLACE or METHOD=QUADRATURE. For Gaussian data, PROC GLIMMIX uses the widely-accepted standard algorithm for LMMs, restricted maximum likelihood (REML), as the default method. For non-Gaussian data, PROC GLIMMIX uses pseudo-likelihood, an adaptation of REML for GLMMs. As the name implies, pseudo-likelihood uses an approximation of the GLMM likelihood, not the true likelihood. For non-Gaussian GLMMs, the true likelihood is defined by a complex integral ("intractable" in mathematician's language) that requires numerical approximation. METHOD=LAPLACE and METHOD=QUADRATURE are both called integral approximation methods. The LAPLACE option uses a Laplace integral approximation. The QUADRATURE option uses adaptive Gauss-Hermite quadrature – "adaptive" means that the algorithm determines the optimal number of quadrature points, then computes the estimating equations. Quadrature is capable of greater accuracy than Laplace, although the difference is small in the first four examples in this chapter. Quadrature is also more computationally intense; the more complex the

random effects part of the model, the greater the discrepancy between Laplace and quadrature. There are pros and cons to pseudo-likelihood and integral approximation. When you want to use information criteria to compare covariance structures, or if you want to test variance components, e.g. test homogeneity of variance, you need to work with the true likelihood. That is, you need to use either LAPLACE or QUADRATURE. On the other hand, for typical GLMMs used by plant scientists, of which the examples in this chapter are representative, the integral approximation methods have an inflated type I error rate when testing treatment effects, whereas pseudo-likelihood produces more accurate *p*-values and confidence bounds. Examples in section III will show when and why to use the methods available in PROC GLIMMIX.

- CLASS statement. Syntax is identical to other SAS PROCs (GLM, MIXED, etc.). If a model effect is a classification variable, it needs to be named in the CLASS statement. If it is a regression variable, and you want to estimate the coefficients of a regression equation, it must not be named in the CLASS statement.
- MODEL statement. This statement is in the form of RESPONSE = <list of fixed effects>. The syntax is identical to PROC MIXED, with one exception. If a response variable is given as Y/N, where Y is the variable that identifies the number of "successes" out of N observations in the unit of observation, then GLIMMIX treats the response as having a binomial distribution and uses the logit link. All other response variables are assumed to be Gaussian with an identity link unless otherwise specified. For example, MODEL Y=TRT; will use the identity link with a Gaussian distribution, whereas MODEL Y=TRT / D=POISSON; will use the log link with the Poisson distribution.
- RANDOM statement. This statement lists the random effects in the linear predictor. The syntax is identical to PROC MIXED. There are variations on the way random effects are specified that are explained where relevant in the examples in the next section.
- RANDOM _RESIDUAL_ statement. This replaces the REPEATED statement in MIXED, but serves essentially the same purpose. It is mainly, but not exclusively, used for repeated measures experiments and spatial data. Example 3 in the next section shows a repeated measures example. Gbur et al. (2012) present more advanced examples drawn from agriculture and natural resource science.

Options for inference on treatment means, contrasts, and so forth use familiar commands from GLM, MIXED and other SAS PROCs. The primary option that is important for non-Gaussian data is ILINK. As the acronym implies, this option implements the inverse link. All inference, by default, occurs on the model, or link, scale. If you also want data scale inference, you must ask for it. For example, the statement LSMEANS TRT; for binomial data will obtain the mean logit for each treatment. LSMEANS TRT / ILINK; obtains the mean logits and the corresponding estimate of p_i for each treatment.

The Examples

This section contains five examples showing increasingly complex designs with data from non-Gaussian distributions typical of the plant sciences. Obviously, no textbook can present an exhaustive collection of examples that cover every possibility readers may encounter. These examples are intended to cover enough of a range so that you have the essential ideas in place to allow you to analyze non-Gaussian data.

Each example is organized as follows. Consider these to be essential steps that you should always follow. Do not skip steps. Most of the problems people have implementing GLMM analysis results from either glibly applying methods they have learned for Gaussian data or skipping the steps role-modeled here.
The steps for each example are

- Description of the study and its objectives
- Visualization of the plot plan
- What would Fisher Do? Construction of the combined ANOVA
- Which sources of variation are fixed? Which are random?
- Write the implied model
- Write the essential SAS GLIMMIX statements
- Implement the analysis and interpret relevant output

Example 1 – A Randomized Complete Block Design with Binomial Data

Description of the Experiment and its Objectives: These data were collected to compare seed germination rates for two plant varieties. The experiment design was a randomized complete block with eight blocks. For each variety within each block, 100 seeds were planted, and the number of seeds that germinated was recorded. The number of seeds planted is denoted N in the data set accompanying this example; the response variable, the number of seeds that germinate, is denoted Y.

Step 1. Visualization.

In this case, you can visualize this experiment using a standard plot plan for a complete block design, shown as follows. Each cell below corresponds to a plot within a block and is the experimental unit with respect to variety.

	Block							
Variety	1	2	3	4	5	6	7	8
1								
2								

Step 2: WWFD ANOVA.

The units in the **experiment design** (what Fisher called the "topographical aspect") are block and plot within block. The **treatment design** consists of the two varieties.

Each variety is randomly assigned to a plot within each block, denoted plot(block) below. Hence plot is the **experimental unit**. The resulting WWFD ANOVA is as follows.

Experiment Design		Treatment Design		Combined	
Source	df	Source	df	Source	df
block	7			block	7
		variety	1	variety	1
plot(block)	8	"parallels"	14	plot(block) | variety	8-1=7
TOTAL	15	TOTAL	15	TOTAL	15

Variety appears on the line immediately above plot(block). This signifies that plot(block) is the experimental unit with respect to variety. In the combined table, the source of variation for experimental unit is denoted plot(block)|treatment. Read this as "plot within block after accounting for variety." Notice that its degrees of freedom result from the 8 df for plot(block) minus one df used to account for the variety effect.

Important: in classical ANOVA, the last line is often called "residual" or "error." In "Introduction to GLMM", we saw that this is appropriate for Gaussian data, but not for non-Gaussian data. Non-Gaussian data does not have a variance parameter, σ^2. However, we do need to include unit-level variability over and above what is accounted for by block and treatment effect. The next step shows how you proceed when the response variable is binomial.

Step 3. Which Sources are Random? What are the Associated Distributions?

The blocked design in this example is similar to the example used in "Introduction to GLMM" to introduce key modeling concepts. Most of the decisions required for this study were discussed in the previous section. However, reviews never hurt, so here is a brief review.

A source of variation is *random* if the following criteria apply: the levels observed in the experiment are representative of a larger population; there is nothing special about the levels used in the experiment; any representative set could have been used. These criteria always apply to the experimental units – they are by definition a representative sample of the experiment's target population. These criteria also apply to blocks – it is difficult to imagine an experiment for which the blocks observed are unique, and no other blocks could conceivably be used to do the research. On the other hand, the two varieties observed are included in the experiment because they are the varieties we wish to study. Given the experiment's purpose, no other varieties will do. Hence variety is a fixed effect.

Typically, the Gaussian distribution provides a reasonable approximation of block effects. That is, it is reasonable to assume that variation among blocks is distributed symmetrically about the mean response. Generalized Linear Mixed Model theory assumes normality for random model effects. Generalized Linear Mixed Model software such as PROC GLIMMIX requires the normality assumption. If you assume block effects do not have a Gaussian distribution, you must use software beyond the scope of this chapter. Fortunately, GLMM analysis is robust to the assumption of normally distributed block effects, meaning that block effects can deviate substantially from normality with little substantive impact on the accuracy of the analysis. The normality assumption for blocks is rarely a problem.

For variation at the experimental unit level, there are two prevalent GLMM approaches. The first is to assume a plot(block)|variety random effect with a Gaussian distribution. This is called a logit-normal GLMM. The second is to assume that the probability of a seed germinating varies among experimental units according to a Beta distribution. This is called a beta-binomial model.

Before continuing with this example, a brief introduction to the Beta distribution. The Beta distribution is used for random variables that are bounded by 0 and 1. Two common uses of the Beta distribution in plant science are i) characterizing the distribution of probability, e.g. when it varies as it does in this example, and ii) for modeling continuous proportions, e.g. percent of leaf area affected, which does not arise from a Y "successes" out of N "trials" process. A rigorous description of the Beta distribution is beyond the scope of this chapter. Interested readers are referred to textbooks on mathematical statistics such as Casella and Berger (2002) or GLMM textbooks such as Stroup (2013b). For our purposes, it is sufficient to know that the Beta distribution can be characterized by its mean, denoted μ, which must be between 0 and 1, and a scale parameter, denoted Φ. The scale parameter determines the shape and the variance of the distribution. Figure 7 shows the density function of the Beta distribution with mean $\mu = 0.2$ for selected values of Φ ranging from 5 to 100. Notice that when Φ is small, the distribution is highly skewed and widely spread. As Φ increases, the distribution becomes more symmetrical and the variance decreases.

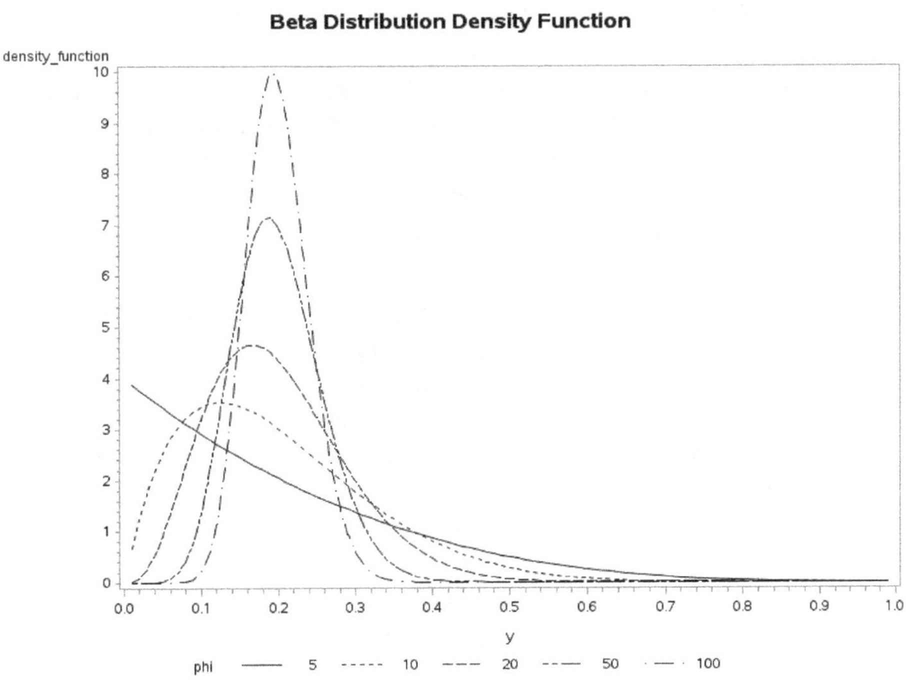

FIG. 7. Beta density function for mean = 0.2 and various scale parameter (phi) values.

If the number of seeds planted is equal for all plots, you can implement either the logit-normal or beta-binomial model using PROC GLIMMIX. If there are unequal number of seeds per experimental unit, the PROC GLIMMIX will provide accurate results for the logit-normal model, but not for the beta-binomial. For this reason, this example continues below, illustrating the logit-normal model. Before continuing, two important notes.

First, if you are in doubt about which model to use, you should consult a qualified statistician. With unequal number of seeds per experimental unit, the beta-binomial requires advanced software that is well beyond the scope of this chapter.

The second note is extremely important. Inexperienced researchers often record only the percent data, and fail to keep track of the number of seeds per plot on which the percent is based. When this happens, you have no choice but to use the beta distribution. However, because you do not know if the seeds per plot are equal or not, you have no guarantee that the results of your analysis will be accurate. If your response variable is binomial, it is essential to make sure you know how many seeds per plot were used to determine percent germination.

We now continue with the remaining steps.

Step 4: Write the Resulting Model.

The required elements of a statistical model are

- the **response variable** and its distribution
- the **link function** associate with the response variable distribution
- the **linear predictor**, typically the sum of the sources of variation in the combined WWFD ANOVA
- the distributions assumed for the random effects in the linear predictor

For this experiment, the resulting model is

- response variable: Y_{ij} = number of seeds that germinate for the *i*th treatment in the *j*th block. The distribution is $Y_{ij} \sim Binomial\left(N_{ij}, p_{ij}\right)$, for which N_{ij} is the number of seeds planted, and p_{ij} is the probability that a given seed germinates in the *i*th treatment in the *j*th block.
- Link function. The most common link function for binomial data is the **logit**, denoted $\eta_{ij} = \log\left[p_{ij}/\left(1-p_{ij}\right)\right]$
- Linear predictor. Using the WWFD ANOVA, the linear predictor is logit = block + variety + plot(block) | variety, or, in formal statistical notation, $\eta_{ij} = \eta + \nu_i + b_j + u_{ij}$, where η denotes an intercept, ν_i denotes the variety effect, b_j denotes the block effect, and u_{ij} denotes the experimental unit (plot(block) | variety) effect. Note two conventions of statistical notation: first, fixed effects are denoted with Greek letter and random effects with Latin letters; second, fixed effects appear first in the model, followed by random effects.
- Distributions. The block effects are assumed $b_j \sim NI\left(0, \sigma_B^2\right)$, where $\sim NI$ means "normally and independently distributed." Unit effects are assumed $u_{ij} \sim NI\left(0, \sigma_U^2\right)$.

Step 5: PROC GLIMMIX statements.

The SAS PROC GLIMMIX statements that implement the model described above are as follows:

```
proc glimmix data=RCBD_binomial;
 class Block Variety;
 model Y/N=Variety;
 random intercept Variety / subject=Block;
 lsmeans Variety / ilink diff oddsratio cl;
 estimate 'Odds: Variety 1' intercept 1 Variety 1 0 / exp cl;
 estimate 'Odds: Variety 2' intercept 1 Variety 0 1 / exp cl;
```

The PROC and CLASS statements should be familiar to any plant science researcher with previous SAS experience.

The MODEL statement gives the response variable to the left of the equal sign and a list of the *fixed* effects to the right. Like all SAS linear model software (e.g. PROC GLM and PROC MIXED) the intercept (η) is included by default; VARIETY defines the effect denoted v_i above. The binomial response variable is specified by Y/N, where Y is the variable in the data set that gives the number of seeds that germinated and N gives the number of seeds planted in a given plot. Any variable specified in the Y/N format is treated as having a binomial distribution. The default link function is the logit.

The RANDOM statements in PROC GLIMMIX and PROC MIXED use identical syntax. From the model definition in Step 4, the conventional way to write this statement is RANDOM BLOCK BLOCK*VARIETY. Note that BLOCK*VARIETY uniquely specifies an experimental unit, and is therefore the way you define the plot(block)|variety effect so that SAS understands it. What is shown in the SAS statements above is a computationally more efficient, and, especially for non-Gaussian models, preferable form. SAS reads RANDOM INTERCEPT / SUBJECT=BLOCK as RANDOM BLOCK, and RANDOM VARIETY / SUBJECT=BLOCK as RANDOM BLOCK*VARIETY, but instead of having work with one large matrix that includes all blocks, it works with much smaller matrices one block at a time. The results are the same, but the computing requirements are greatly reduced. This data set is small, so the advantage of using the SUBJECT specification is negligible, but with large data sets and complex models, failure to use the SUBJECT option can result in SAS being unable to compute the analysis.

The LSMEANS and ESTIMATE statements allow you to compute estimated means and differences. In model terms, LSMEAN for variety gives you an estimate of $\eta + v_i$ for the ith variety. This is an estimate of the log-odds, i.e. $\log\left[p_i/(1-p_i)\right]$, where p_i is the probability of a seed of the ith variety germinating. The ILINK option converts the log-odds estimate to an estimate of p_i – in the language of the previous section, ILINK converts the **model scale** estimate (log-odds) to the **data scale** estimate (probability). The DIFF option, as the name implies, obtains the difference between the log-odds of varieties and CL obtains confidence intervals for all estimates. ODDSRATIO obtains a statistic frequently of interest in categorical data analysis. In categorical data, the odds-ratio is defined as $\left[p_1/(1-p_1)\right]/\left[p_2/(1-p_2)\right]$, that is, the odds of a variety 1 seed germinating divided by the odds of a variety 2 seed germinating. The difference in the log-odds is $\log\left[p_1/(1-p_1)\right]-\log\left[p_2/(1-p_2)\right]=$ $\log\left[p_1/(1-p_1)\right]/\left[p_2/(1-p_2)\right]$. Taking e^{diff} gives you the odds-ratio. Additionally, the EXP option for the ESTIMATE statement gives you the odds for each variety. That is,

$e^{\eta+v_i} = e^{\log[p_i/(1-p_i)]} = p_i/(1-p_i)$. Use this if you desire to obtain and report estimates of the odds.

Relevant Output

The above PROC GLIMMIX statements produce the following items.

Convergence criterion (PCONV=1.11022E-8) satisfied.

If you do not see "convergence criterion satisfied," you will not see any other useable output. In addition, you should check the SAS LOG to make sure that there are no warnings. Often, convergence criteria will be technically satisfied, but there are nonetheless symptoms that the results of the analysis are not to be trusted.

Fit Statistics

-2 Res Log Pseudo-Likelihood	48.31
Generalized Chi-Square	13.69
Gener. Chi-Square / DF	0.98

The "-2 Res Log Pseudo-Likelihood" and "Generalized Chi-Square" are of little intrinsic interest, but the "Gener. Chi-Square/DF" is an important diagnostic. With Gaussian data, the "Generalized Chi-Square" is equal to the sum of square for residual, and hence "Gener. Chi-Square/DF" = MS(residual) = the experimental unit variance. In general, all distributions have a mean parameter and a scale parameter. For the Gaussian distribution, the variance is its scale parameter, and it must be estimated. For the binomial, the scale parameter is 1 – the variance of the binomial is proportional to $p(1-p)$, so if you know the mean, you know that variance without any additional estimate. "Gener. Chi-Square/DF" is an estimate of the scale parameter, so if you really have a valid model for your binomial data, "Gener. Chi-Square/DF" should be equal, or at least close, to 1. If it isn't, this would imply that the variance is proportional to $\Phi p(1\text{-}p)$ where Φ denotes the scale parameter estimated by "Gener. Ci-Sq/DF". If the generalized chi-square suggests that $\Phi \neq 1$, this is evidence that your model is not properly specified and your analysis should not be trusted. For example, if you failed to include the unit effect, BLOCK*VARIETY, as a random effect, "Gener. Chi-Square/DF" = 7.05 – strong evidence of a poorly specified model.

Covariance Parameter Estimates

Cov Parm	Subject	Estimate	Standard Error
Intercept	Block	0.5201	0.6037
Variety	Block	0.8335	0.4926

Here, the word "Covariance" is misleading. These are the estimates of the variance components. Under the column labelled ESTIMATE, the row labelled "*Intercept* Block" gives you the estimated block variance, $\hat{\sigma}_B^2 = 0.52$, and the row labelled "*Variety* Block" gives you the estimate of the unit, a.k.a. plot(block)|variety, variance, $\hat{\sigma}_U^2 = 0.83$. Although GLIMMIX prints a column of standard errors for the estimates, these are "asymptotic" standard errors and are notoriously inaccurate unless your sample size is at least in the thousands. With eight blocks, they are virtually meaningless.

Type III Tests of Fixed Effects

Effect	Num DF	Den DF	F Value	P > F
Variety	1	7	6.75	0.0355

Read this output as you would any other F-value and associated *p*-value. This is a test statistic used to test $H_0 : p_1 = p_2$, the null hypothesis that the probability of a seed germinating is equal for each variety. Here, you have evidence to reject this hypothesis with $P < 0.05$. Notice that sums of square and mean squares have no meaning in the context of non-Gaussian data and hence do not appear in the output. Reviewers should not insist on sums of square or mean squares because, as noted in Stroup (2014), if you run an analysis that gives you sums of square and mean squares, the resulting F-test, as well as the estimates of the p_i, will be incorrect.

Variety Least Squares Means

Variety	Estimate	Standard error	DF	t Value	P > |t|	Alpha	Lower	Upper	**Mean**	**Standard Error Mean**	**Lower Mean**	**Upper Mean**
1	2.5502	0.4405	7	5.79	0.0007	0.05	1.5087	3.5917	**0.9276**	**0.02958**	**0.8189**	**0.9732**
2	1.2698	0.4232	7	3.00	0.0199	0.05	0.2691	2.2704	**0.7807**	**0.07245**	**0.5669**	**0.9064**

The values under ESTIMATE are estimates of the log-odds for each variety. They are typically of little interest to plant science researchers. The values in the last four columns, here shown in boldface, are the only Least Squares Means output of interest. MEAN gives the estimates of the germination probabilities, i.e. for variety 1, $\hat{p}_1 = 0.93$ and for variety 2, $\hat{p}_2 = 0.78$. The next columns give the standard error of the $\hat{p}_i$ and the lower and upper 95% confidence bounds.

Differences of Variety Least Squares Means

Variety	_Variety	Estimate	Standard error	DF	t Value	P > |t|	Alpha	Lower	Upper	**Odds ratio**	**Lower confidence limit for odds ratio**	**Upper confidence limit for odds ratio**
1	2	1.2805	0.4927	7	2.60	0.0355	0.05	0.1154	2.4455	**3.598**	**1.122**	**11.537**

As with the Least Squares Mean output, the estimates are on the log-odds scale, and are typically not of interest. Notice that the *p*-value is identical to that of the F-test for variety, because there are only two varieties. The last three columns, shown here in boldface, are columns of potential interest, as they show the estimated odds-ratio and its lower and upper confidence limits. An odds-ratio of one would mean that germination is equally likely for each variety. The odds-ratio of 3.6 means that the odds of a seed germinating are 3.6 times as great for variety 1 as for variety 2. This is statistically significant because even the lower confidence bound of this estimate is greater than one. You can add the option ODDS to the LSMEANS statement to obtain the odds for each variety.

Variety Least Squares Means

Label	Estimate	Standard Error	DF	t Value	P > |t|	Alpha	Lower	Upper	**Odds**	**Lower Odds**	**Upper Odds**
Odds: Variety 1	2.5502	0.4405	7	5.79	0.0007	0.05	1.5087	3.5917	**12.8102**	**4.5210**	**36.2973**
Odds: Variety 2	1.2698	0.4232	7	3.00	0.0199	0.05	0.2691	2.2704	**3.5601**	**1.3088**	**9.6836**

As noted earlier, the "Estimate" given here is of the log-odds – generally of little interest. The only items here of potential interest are the "Exponentiated Estimate" and associated lower and upper confidence bounds, shown here in boldface. These give the estimated odds, that is, $\hat{p}_i/(1-\hat{p}_i)$ for each variety.

Example 2 – Split-Plot Experiment with Count Data

Description of the Experiment and its Objectives: Data were collected at three center pivot irrigation sites to study the effect of four seed mixes at three irrigation levels on plant species regeneration. Each center pivot unit was set so that each of the three irrigation levels were applied in concentric rings. Irrigation levels were randomly assigned to the inner circle, the middle ring, or the outer ring. Each site was also divided into quarters; the four seed mixes were assigned at random so that each mix was planted to one quarter at each field. Key species were targeted. Species count, i.e. the number of plants of targeted species, was the response variable of interest. This example focuses on the analysis for one of the species considered to be desirable.

Step 1. Visualization.

The following schematic shows the experiment design for this study.

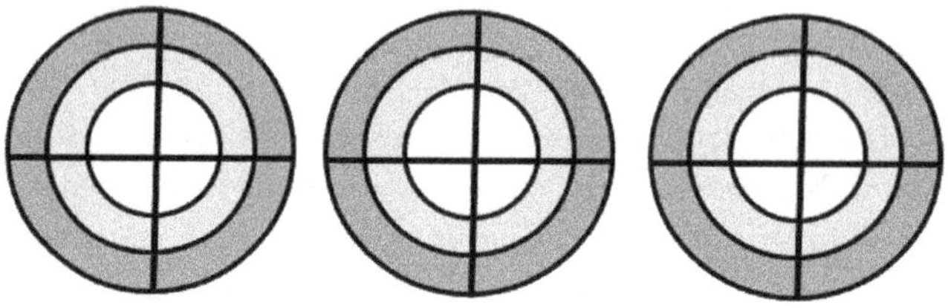

Each circle represents a center-pivot unit. The concentric rings show the placement of irrigation levels. Note that these are the experimental units with respect to irrigation level. The solid lines at right angles show the boundaries between areas of the field planted to each mix. These quarter-fields are the experimental units with respect to seed mix. Finally, the experimental unit with respect to irrigation × mix treatment combinations is the intersection of irrigation rings and mix quarter-fields.

While this design is a member of the split-plot family, to be technically correct, it should be referred to as a **strip–split-plot** or, more simply, as a **strip-plot.** A **split-plot** occurs when the experimental units for one factor are sub-units of the experimental units for another factor. Here, the irrigation and mix experimental units are not sub-units of one another. However, the irrigation × mix experimental unit is the intersection of the irrigation and mix units. This is the defining characteristic of a **strip-plot** experiment. See Milliken and Johnson (2009) for additional information about split-plot and strip-plot experiments.

Step 2. WWFD ANOVA.

The units in the experiment design are center pivot sites, concentric rings within sites, quarter-fields within sites, and the ring × quarter-field intersections within site. The treatment design consists of two factors: irrigation (three levels) and seed mix (four mixes). All twelve factorial combinations appear in the experiment. The resulting WWFD ANOVA appears below.

Experiment Design		Treatment		Combined	
Source	df	Source	df	Source	df
Site	2			Site	2
		Irrigation	2	Irrigation	2
Circle(Site)	6			Circle(Site) | Irrig = Site × Irrig	6-2=4
		Mix	3	Mix	3
Quarter(Site)	9			Quarter(Site) | Mix = Site × Mix	9-3=6
		Irrigation × Mix	6	Irrigation × Mix	6
Circle × Quarter (Site)	18	"parallels"	24	Circle×Quarter(Site) | Irrig×Mix = Site×Irrig×Mix	18-6=12
TOTAL	**35**	**TOTAL**	**35**	**TOTAL**	**35**

Notice two key features of this table. First, because the treatment design is a factorial, treatment effects are specified by main effects (of irrigation and mix) and interaction. These appear in the row immediately above their corresponding experimental unit. Second, the experimental units with respect to each treatment effect have two names in the **combined** sources of variation list. The first is their literal, technical description. The second is the way you need to specify them so that statistical software recognizes the effect you intend. In other words, the variables Site, Irrigation and Mix appear in the data set you use to analyze these data.

Technically, the experimental unit with respect to irrigation is "circle within site after accounting for irrigation," denoted *Circle(Site)|Irrig*. However, if you write "Circle(Site)|Irrig" in your PROC GLIMMIX program, SAS will return an error message, because "Circle" is not a variable in the data set. On the other hand, if you specify the site and the level of irrigation, that is, $Site \times Irrig$, this uniquely identifies the circle within site for a given irrigation level. It is important to recognize this equivalence.

Important: Many writers inexperienced with mixed model concepts will say that the model includes estimates of the $Site \times Irrig$, $Site \times Mix$ and $Site \times Irrig \times Mix$ interaction effects. This is not true, and hence inaccurate and misleading writing. $Site \times Irrig$, etc. are programming devices to get software to understand what random effects – error terms with respect to each treatment effect – need to be estimated, but they are not interaction effects. The effect being estimated is *Circle(Site)|Irrig*. The programming device required to specify this effect is to describe it to the software as $Site \times Irrig$.

Step 3. Which Sources are Random? What are the Associated Distributions?

Applying the criteria for random and fixed effects described in Step 3 of Example 1, sites and all experimental units with respect to each treatment factor are random effects. All treatment factors – irrigation, mix and their interaction – are fixed. Notice a general principle: effects that appear in the experiment design columns of the WWFD ANOVA are usually random effects. This is because they are usually blocking criteria or experimental units understood to represent the target population to which the results of the experiment are supposed to apply. On the other hand, effects that appear in the treatment design column are usually fixed, because they are *the* factors under study.

Two thought processes guide deciding distributions to be assumed for random effects. One is for blocking criteria and experimental units – a.k.a. "error terms" – for whole plot treatment factors. The other is for the source of variation that corresponds to the unit at which observations are taken. In this case, *Site* is the blocking criterion, and, because this is a strip-plot design, *circle(site)|irrig* and *quarter-field(site)|mix* are both considered whole plot error terms. The smallest unit in the source of variation list, *circle×quarter-field(site)|irrig×mix*, is the unit on which observations are taken, often referred to as the unit-of-observation level effect.

For the blocking and whole plot error effects, their distributions are typically well-approximated by the Gaussian distribution. Also, because random assignment is a given in a properly designed experiments, blocking and whole plot error effects can be assumed to be independent.

For the unit of observation, the data are counts, and therefore the assumed distribution must be one that is plausible for count data. Stroup (2014) presented several ways in which plant science count data have been analyzed. The one that has emerged as the method of choice for biological count data is to assume a negative binomial distribution. This will be the focus of the rest of this example.

Aside: The negative binomial is equivalent to assuming that the *circle × quarter-field(site)|irrig × mix effect* – the split plot error term – has a Gamma distribution, and the counts, conditional on the block, whole plot, and split plot effects, have a Poisson distribution. It is not essential that you know this, but it will be of interest to readers with a background in mathematical statistics and probability theory.

Step 4. Write the Resulting Model.

The required elements of the model are as follows.

- **Response variable:** Y_{ijk} = species count for the ith irrigation level, j^{th} seed mix, k^{th} center pivot site. The distribution is negative binomial. Its formal specification is given below under **distributions**.
- **Link function:** Count data use the log link, $\eta_{ijk} = \log(\lambda_{ijk})$, where λ_{ijk} denotes the mean species count for the ith irrigation level, j^{th} seed mix, k^{th} center pivot site.
- **Linear Predictor:** Using the WWFD ANOVA, the linear predictor is intercept + irrigation + mix + irrigation×mix + site + site×irrigation + site×mix. Note that the "site×..." terms are not interactions, but whole plot error terms. Also, the unit of observation level effect does not appear in the linear predictor. This source of variation is used to estimate the scale parameter of the negative binomial distribution, defined below under **distributions**.

In formal statistical notation, the linear predictor is written $\eta_{ijk} = \eta + \alpha_i + \gamma_j + \alpha\gamma_{ij} + s_k + s\alpha_{ik} + s\gamma_{jk}$, where α refers to irrigation effects, γ refers to seed mix effects, and s refers to site effects.

- **Distributions:**
  - Site: $s_k \sim NI(0, \sigma_S^2)$

- Irrigation whole plot: $s\alpha_{ik} \sim NI\left(0,\sigma_{SI}^2\right)$
- Seed mix whole plot: $s\gamma_{jk} \sim NI\left(0,\sigma_{SM}^2\right)$
- Observation: $Y_{ijk} \mid s_k, s\alpha_{ik}, s\gamma_{jk} \sim Negative\ Binomial\left(\lambda_{ijk},\phi\right)$, where λ_{ijk} denotes the mean seed count, as defined above, and ϕ denote the scale parameter. The distribution of the observations in a GLMM is *always* conditional on the random model effects. The variance of negative binomial counts depends on both the mean and the scale parameter – specifically, the variance is $\lambda_{ijk} + \phi\lambda_{ijk}^2$. Note that the variance of count data is not constant; it increases as the mean increases. This explains why standard ANOVA, which assumes constant variance, is not suitable for count data. The negative binomial variance, on the other hand, accurately models the observed variance typically seen in biological count data.

Step 5. PROC GLIMMIX statements.

The SAS PROC GLIMMIX statements that implement the model described above are as follows:

```
proc glimmix data=ctr_pivot_strip_plot;
 class site irrig mix;
 model count=irrig|mix / d=negbin;
 random intercept irrig mix / subject=site;
 lsmeans irrig*mix / ilink plot=meanplot(sliceby=mix join ilink);
```

The syntax in the MODEL statement IRRIG|MIX is SAS shorthand for IRRIG MIX IRRIG*MIX. That is, it specifies the main effects of irrigation and mix and their interaction. Response variables that are not specified using the Y/N syntax used in Example 1 are assumed to have a Gaussian distribution unless otherwise defined using the D= option in the MODEL statement. Here, D=NEGBIN alerts GLIMMIX to analyze COUNT as a negative binomial random variable. The syntax for the RANDOM statement used the same strategy as in Example 1. In this case, the random sources of variation are SITE and the two whole plot error terms (understood by GLIMMIX as SITE*IRRIG and STE*MIX, or equivalently, as shown here, IRRIG MIX / SUBJECT=SITE). The LSMEANS statement shown here is a first step: the PLOT statement constructs an interaction plot, and the ILINK option obtains estimates of the expected counts for each irrigation × mix combination. These allow you to get an overview of the results of the experiment.

When you run the statements above, you will get the following.

Iteration History

Iteration	Restarts	Subiterations	Objective Function	Change	Max Gradient
0	0	7	83.33913367	2.00000000	0.000042
1	0	7	82.33807513	1.47941203	2.08E-6
...	...	...	...	...	...
17	0	0	82.887789443	0.00000089	9.215E-6
18	0	0	82.887789328	0.00000046	9.151E-6
19	0	0	82.887789384	0.00000021	9.185E-6

Did not converge.

Notice the column with the heading CHANGE. The GLMM estimation procedure is iterative. The CHANGE value shows the amount by which estimates changed from the previous iteration. When the change is less than 10^{-8}, the difference between successive estimates is considered sufficiently negligible that successive iterations will not produce consequential improvement in the accuracy of estimated model effects. The default number of iterations in GLIMMIX is 20. Notice that CHANGE moves steadily toward 10^{-8}, but does not quite reach it. The fix is to add an additional statement to increase the number of iterations. The revised program is

```
proc glimmix data=ctr_pivot_strip_plot;
 class site irrig mix;
 model count=irrig|mix / d=negbin;
 random intercept irrig mix / subject=site;
 lsmeans irrig*mix / ilink plot=meanplot(sliceby=mix join ilink);
 nloptions maxiter=50;
```

This directs GLIMMIX to raise the iteration limit to 50. You can raise the limit even higher if the CHANGE information in the iteration history tells you that the procedure is making progress. With negative binomial data, setting the iteration limit to MAXITER=5000 is not unheard of – iterations take a minute fraction of a second and even 5000 iterations can be computed in just a few minutes.

The results now appear as follows.

Iteration History

Iteration	Restarts	Subiterations	Objective Function	Change	Max Gradient
0	0	7	83.33913367	2.00000000	0.000042
...	...	...	...	...	...
23	0	0	82.887789376	0.00000001	9.148E-6

Convergence criterion (PCONV=1.11022E^{-8}) satisfied.

If you see "convergence criterion satisfied" and MAX GRADIENT less than 10^{-3}, you can move on to interpreting subsequent output. Here, MAX GRADIENT is roughly equal to 9×10^{-6}.

The next diagnostic appears in the FIT STATISTICS.

Fit Statistics	
-2 Res Log Pseudo-Likelihood	82.89
Generalized Chi-Square	23.49
Gener. Chi-Square / DF	0.98

The "-2 Res Log Pseudo-Likelihood" and "Generalized Chi-Square" have no intrinsic meaning (at least none of interest here) but the "Generalized Chi-Square/DF" in theory should be 1 and in practice should not appreciably exceed 1. There is no exact rule of thumb for exactly what "appreciably exceeds 1" means, but here, chi-square/DF = 0.98 tells you that you may proceed with the analysis.

The interaction plot appears as follows.

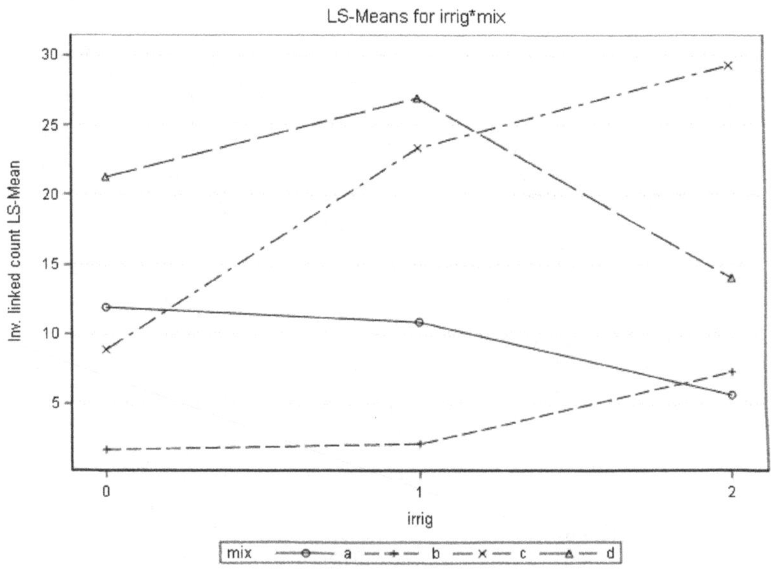

Depending on the specifics of the researchers' objectives, there are several possibilities for pursuing the analysis at this point.

What follows are merely examples of how one might pursue the analysis. These are not intended to be prescriptive not exhaustive. They are simply meant to illustrate inferences researchers might consider, how to implement them in the context of a GLMM, and how to read and interpret the GLIMMIX output.

Start with the GLMM adaptation of the ANOVA table to assess overall tests of main effects and interaction. The relevant output is

Type III Tests of Fixed Effects

Effect	Num DF	Den DF	*F* Value	*P* > *F*
irrig	2	4	0.35	0.7213
mix	3	6	2.88	0.1253
irrig*mix	6	12	2.36	0.0964

While none of the effects are statistically significant at the $P < 0.05$ level, the irrigation × mix interaction has 6 numerator degrees of freedom and a p-value < 0.10. From the interaction plot, you can see that changes in species count over irrigation level are visually quite different for seed mix "C" relative to the others, and less obviously, but perhaps importantly, for mix "D." In addition, it appears that relative differences in expected seed counts vary over irrigation level: mix "D" yields relatively high expected counts at low and intermediate irrigation level, whereas mix "C" does better at higher irrigation levels, and mixes "A" and "B" do poorly regardless of irrigation. In addition, irrigation level appears to have a negligible effect on expected counts for mixes "A" and "B."

The following PROC GLIMMIX statements include post-processing options that create estimates and test-statistics to address comparisons of interest given in the above paragraph.

```
proc glimmix data=ctr_pivot_strip_plot;
 class site mix irrig;
 model count=irrig|mix / d=negbin;
```

```
random intercept irrig mix / subject=site;
lsmeans irrig*mix / ilink slice=mix slicediff=irrig;
contrast 'linear irrig, mix c' irrig 1 0 -1 mix*irrig 0 0 0  0 0 0  1 0 -1  0 0 0;
contrast 'quad irrig, mix c' irrig 1 -2 1 mix*irrig 0 0 0  0 0 0  1 -2 1  0 0 0;
contrast 'linear irrig, mix d' irrig 1 0 -1 mix*irrig 0 0 0  0 0 0  0 0 0  1 0 -1;
contrast 'quad irrig, mix d' irrig 1 -2 1 mix*irrig 0 0 0  0 0 0  0 0 0  1 -2 1;
contrast 'irrig x mix a vs b' mix*irrig 1 0 -1  -1 0 1  0 0 0  0 0 0,
                              mix*irrig 1 -2 1  -1 2 -1 0 0 0  0 0 0;
contrast 'irrig x mix c vs d' mix*irrig  0 0 0  0 0 0 1 0 -1  -1 0 1,
                              mix*irrig  0 0 0  0 0 0 1 -2 1  -1 2 -1;
contrast 'linear irrig x mix c vs d' mix*irrig  0 0 0  0 0 0 1 0 -1  -1 0 1;
contrast 'quad irrig x mix c vs d'   mix*irrig  0 0 0  0 0 0 1 -2 1  -1 2 -1;
contrast 'irrig x mix a&b vs c' mix*irrig 1 0 -1  1 0 -1  -2 0 2  0 0 0,
                                mix*irrig 1 -2 1  1 -2 1 -2 4 -2  0 0 0;
contrast 'linear irrig x mix a&b vs c' mix*irrig 1 0 -1  1 0 -1  -2 0 2  0 0 0;
lsmestimate mix*irrig 'irrig 1 v 3 mix a&b' 1 0 -1  1 0 -1 0 0 0 0 0 0,
                      'irrig 1 v 3 mix c' 0 0 0  0 0 0  1 0 -1  0 0 0 ,
                      'irrig 1 v 3 x mix a&b vs c' 1 0 -1  1 0 -1  -2 0 2  0 0 0/
                       exp divisor=2,1,2 cl;
contrast 'irrig x mix a&b vs d' mix*irrig 1 0 -1  1 0 -1  0 0 0  -2 0 2,
                                mix*irrig 1 -2 1  1 -2 1  0 0 0 -2 4 -2;
contrast 'quad irrig x mix a&b vs d' mix*irrig 1 -2 1  1 -2 1  0 0 0 -2 4 -2;
lsmestimate mix*irrig 'irrig 0, mix a & c' 1 0 0  0 0 0  1 0 0  0 0 0,
                      'irrig 0, mix d vs a&c' 1 0 0  0 0 0  1 0 0  -2 0 0,
                      'irrig 0, mix d vs b' 0 0 0  1 0 0  0 0 0  -1 0 0 /
                      divisor=2,2,1 exp cl;
lsmestimate mix*irrig 'irrig 1, mix c & d' 0 0 0  0 0 0  0 1 0  0 1 0,
                      'irrig 1, mix a vs c&d' 0 2 0  0 0 0  0 -1 0  0 -1 0,
                      'irrig 1, mix b vs c&d' 0 0 0  0 2 0  0 -1 0  0 -1 0 /
                      divisor=2,2,2 exp cl;
lsmestimate mix*irrig 'irrig 2, mix a & b' 0 0 1  0 0 1  0 0 0  0 0 0,
                      'irrig 2, mix c vs a&b' 0 0 1  0 0 1  0 0 -2  0 0 0,
                      'irrig 2, mix d vs a&b' 0 0 1  0 0 1  0 0 0  0 0 -2 /
                      divisor=2,2,2 exp cl;
nloptions maxiter=50;
```

The SLICE option in the LSMEANS statement obtains statistics to test the irrigation effect for each MIX. The SLICEDIFF option obtains pairwise comparisons between each MIX within each level of irrigation. Notice that the order of MIX and IRRIG in the CLASS statement. This was done to make the order of the coefficients easier (at least in the opinion of the program's author). If you reverse the order to IRRIG MIX, you will need to reorder the coefficients in every contrast in the program accordingly. If you have any doubt about the order, print the LSMEANS and follow the order of the treatment combinations in the output.

The SLICE results are as follows:

Tests of Effect Slices for Mix*irrig Sliced by Mix

mix	Num DF	Den DF	*F* Value	*P* > *F*
a	2	12	0.68	0.5254
b	2	12	2.29	0.1439
c	2	12	1.90	0.1925
d	2	12	0.55	0.5919

Superficially, these results suggest a lack of evidence that irrigation level has any effect on expected species count for any seed mix. However, we can partition these effects into their linear and quadratic (or "non-linear" if you prefer). For more details on linear and quadratic contrasts, see Chapter 7 (Vargas et al., 2018) in this text. The program has statements to do this for seed mixes "C" and "D," labelled "linear irrig, mix c," "quad irrig, mix c," and so forth for mix "D." Seed mixes "A" and

"B" are not shown here in the interest of space but readers can follow the template for the "C" and "D" contrasts. The results are:

Contrasts

Label	Num DF	Den DF	F Value	P > F
linear irrig within mix c	1	12	3.44	0.0884
quad irrig within mix c	1	12	0.45	0.5133
linear irrig within mix d	1	12	0.44	0.5214
quad irrig within mix d	1	12	0.67	0.4283

The test of no linear irrigation effect for mix "C" has a p-value=0.0084, moderate evidence of a linear effect. For mix "D" despite visual evidence of a quadratic irrigation effect, the test statistics provide no formal support. This is our first encounter with a common feature of count data: recall that the variance increases as a function of the square of the mean $\left(\lambda+\phi\lambda^2\right)$. As a result, the larger the estimated count, the more difficult it is to show differences to be statistically significant. Like many experiments, this one was planned using conventional wisdom appropriate for Gaussian response variables. This is a recipe for an underpowered experiment if the primary response variable of interest is a count. Stroup (2013b) and Gbur et al. (2012) describe power analysis and design planning strategies appropriate for experiments with non-Gaussian data. The researchers responsible for this example would have benefited.

Additional CONTRAST statements addresses additional questions about the irrigation effect for different mixes suggested by the interaction plot and the SLICE results. The contrasts labeled "irrig by mix a vs b" and "irrig by mix c vs d" compare the overall irrigation effects for mix "A" and "B" and for mix" C" and "D" respectively. Notice that each contrast consists of two comparisons, one per degree of freedom. The first degree of freedom is defined on the linear contrast, the second on the quadratic. The same strategy is employed to compare the irrigation effect averages over mixes "A" and "B" first with the irrigation effect over "C" and then with the irrigation effect over mix "D." Results are as follows:

Contrasts

Label	Num DF	Den DF	F Value	P > F
irrig x mix a vs b	2	12	3.95	0.0482
irrig x mix c vs d	2	12	2.57	0.1177
irrig x mix a&b vs c	2	12	1.05	0.3791
irrig x mix a&b vs d	2	12	1.18	0.3410

Despite the fact that there is little visual evidence that the irrigation effect differs for mixes "A" and "B," the p-value for this test is < 0.05. This is an artifact of the small counts – and hence the small variance – associated with these two mixes. On the other hand, obvious differences in irrigation effect between mix "C" and "D" and between each of these mixes and the average irrigation effect for mixes "A" and "B" do not show up as significant. These contrasts can be decomposed into their linear and quadratic components. Selected contrast appear in the program above and their results are as follows:

Contrasts

Label	Num DF	Den DF	F Value	P > F
linear irrig x mix c vs d	1	12	5.11	0.0432
quad irrig x mix c vs d	1	12	0.02	0.9042
linear irrig x mix a&b vs c	1	12	1.54	0.2389
quad irrig x mix a&b vs d	1	12	0.94	0.3504

The obvious visual difference between the irrigation effect for mix "C" and for mix "D" is that mix "C" shows a strong upward linear trend over irrigation levels whereas mix "D" does not. The "linear irrig x mix c vs d" contrast confirms this. On the other hand, the strong linear irrigation effect for "C" versus the essentially nonexistent linear trend for mixes "A" and "B" receives no formal support from the contrast's p-value of just less than 0.24. Similarly, the difference between an apparent quadratic irrigation effect for mix "D" and the lack of an irrigation effect for mixes "A" and "B" has an associated p-value of 0.35. The results may be artifacts of an underpowered experiment, the high variance associated with the counts for mixes "C" and "D," the possibility that what appears on the interaction plot is more apparent than real, or a combination of all of the above.

Turning now from tests of contrasts to estimates of mean differences, begin with SLICEDIFF results, shown below.

Simple Effect Comparisons of mix*irrig Least Squares Means By irrig

Simple Effect Level	mix	_mix	Estimate	Standard Error	DF	t Value	P > |t|	Alpha	Lower	Upper
irrig 0	a	b	1.9700	0.9289	12	2.12	0.0554	0.05	-0.05383	3.9938
irrig 0	a	c	0.2987	0.8430	12	0.35	0.7292	0.05	-1.5380	2.1354
irrig 0	a	d	-0.5822	0.8272	12	-0.70	0.4949	0.05	-2.3845	1.2200
irrig 0	b	c	-1.6713	0.9278	12	-1.80	0.0968	0.05	-3.6927	0.3501
irrig 0	b	d	-2.5522	0.9142	12	-2.79	0.0163	0.05	-4.5442	-0.5603
irrig 0	c	d	-0.8809	0.8255	12	-1.07	0.3069	0.05	-2.6796	0.9177
irrig 1	a	b	1.6582	0.9113	12	1.82	0.0938	0.05	-0.3274	3.6438
irrig 1	a	c	-0.7629	0.8276	12	-0.92	0.3748	0.05	-2.5660	1.0402
irrig 1	a	d	-0.9079	0.8256	12	-1.10	0.2930	0.05	-2.7067	0.8908
irrig 1	b	c	-2.4211	0.8961	12	-2.70	0.0192	0.05	-4.3735	-0.4687
irrig 1	b	d	-2.5662	0.8944	12	-2.87	0.0141	0.05	-4.5148	-0.6175
irrig 1	c	d	-0.1450	0.8072	12	-0.18	0.8604	0.05	-1.9038	1.6137
irrig 2	a	b	-0.2636	0.8682	12	-0.30	0.7667	0.05	-2.1553	1.6282
irrig 2	a	c	-1.6581	0.8479	12	-1.96	0.0742	0.05	-3.5055	0.1892
irrig 2	a	d	-0.9183	0.8549	12	-1.07	0.3039	0.05	-2.7808	0.9443
irrig 2	b	c	-1.3946	0.8257	12	-1.69	0.1170	0.05	-3.1936	0.4045
irrig 2	b	d	-0.6547	0.8326	12	-0.79	0.4469	0.05	-2.4688	1.1594
irrig 2	c	d	0.7399	0.8109	12	0.91	0.3795	0.05	-1.0270	2.5067

The "Estimate" is the estimated difference between the log counts of the mixes given in the MIX and _MIX columns when they are applied in conjunction with the irrigation level shown in the "Simple Effect Level" column. For example, the first row gives the estimate and associated statistics for $\log(\lambda_{0A}) - \log(\lambda_{0B})$, the difference between the log count for mix "A" with irrigation level 0 and mix "B" with irrigation

0. This table gives estimates, standard errors, 95% confidence interval lower and upper bounds, and test statistics (t- and p-values) for all pairwise simple effect comparisons between mixes at each irrigation level.

This table contains our second encounter with count data's mean-variance relationship. At irrigation level 0, the estimated counts for mixes "B" and "D" are $\hat{\lambda}_{0B} = 1.7$ and $\hat{\lambda}_{0D} = 21.3$ respectively – a difference of 19.6. At irrigation level 2, the estimated counts for mixes "A" and "C" are $\hat{\lambda}_{2A} = 5.6$ and $\hat{\lambda}_{2C} = 29.4$ respectively – a difference of 23.8. Yet the p-value for the difference between mixes "B" and "D" at irrigation 0 is 0.0163, whereas the p-value for the difference between "A" and "C" at irrigation 2 is 0.0742. Smaller mean counts, and their associated differences, have lower variances, and hence are estimated with greater precision and are tested with greater power. If you expect that you will have larger counts, you need to plan accordingly, using precision and power analysis methods shown in Stroup (2013b).

Expressing differences in term of log counts may not be the best way to communicate results. However, you can use the result that the difference between log counts is equal to the log of the ratio of the counts to re-express estimated differences in more understandable form. For example, $\log(\lambda_{0A}) - \log(\lambda_{0B}) = \log(\lambda_{0A}/\lambda_{0C})$ and hence $e^{\log(\lambda_{0A}/\lambda_{0C})} = \lambda_{0A}/\lambda_{0C}$, the ratio of the expected count of mix "A" relative to mix "C" at irrigation level 1. You can add the following statement to the GLIMMIX program

```
ods output slicediffs=simple_effects;
```

This creates a data set containing the SLICEDIFF output. Then add the following statements:

```
data simple_effects;
 set simple_effects;
 ratio=exp(estimate);
 ratio_lower=exp(lower);
 ratio_upper=exp(upper);
 std_error_ratio=exp(estimate)*stderr;
proc print data=simple_effects;
 var SimpleEffectLevel mix _mix ratio std_error_ratio ratio_lower
ratio_upper;
run;
```

Notice that the lower and upper confidence interval bounds use the same exponentiation as the estimate. The standard error uses the **delta rule**, discussed in Stroup (2013b, 2014) and Gbur et al (2012). The output appears as

Obs	SimpleEffectLevel	mix	_mix	ratio	std_error_ratio	ratio_lower	ratio_upper
1	irrig 0	a	b	7.17063	6.66054	0.94760	54.2614
2	irrig 0	a	c	1.34810	1.13640	0.21482	8.4600
3	irrig 0	a	d	0.55865	0.46210	0.09213	3.3873
4	irrig 0	b	c	0.18800	0.17442	0.02490	1.4192
5	irrig 0	b	d	0.07791	0.07123	0.01063	0.5710
6	irrig 0	c	d	0.41439	0.34209	0.06859	2.5035
7	irrig 1	a	b	5.24996	4.78442	0.72081	38.2379
8	irrig 1	a	c	0.46632	0.38590	0.07684	2.8298
9	irrig 1	a	d	0.40336	0.33300	0.06676	2.4371
10	irrig 1	b	c	0.08882	0.07959	0.01261	0.6258
11	irrig 1	b	d	0.07683	0.06871	0.01095	0.5393
12	irrig 1	c	d	0.86498	0.69821	0.14901	5.0212

Obs	SimpleEffectLevel	mix	_mix	ratio	std_error_ratio	ratio_lower	ratio_upper
13	irrig 2	a	b	0.76830	0.66707	0.11587	5.0945
14	irrig 2	a	c	0.19050	0.16152	0.03003	1.2083
15	irrig 2	a	d	0.39921	0.34126	0.06199	2.5710
16	irrig 2	b	c	0.24794	0.20472	0.04102	1.4985
17	irrig 2	b	d	0.51960	0.43261	0.08469	3.1879
18	irrig 2	c	d	2.09562	1.69937	0.35809	12.2641

This output is especially useful in research were the relative count of between pairs of counts is of primary interest.

In many cases, comparisons beyond simple pairwise comparisons are of interest. For example, at irrigation level 2, mixes "A" and "B" have similar expected counts noticeably lower than the other mixes, especially mix "C." However, none of the pairwise differences are statistically significant. You can use the fact that contrasts that test differences between groups of means – where the grouping makes sense – can be tested with greater power than simple pairwise tests. The LSMESTIMATE statement is the method of choice for obtaining estimates and tests for such contrasts. The statement

```
lsmestimate mix*irrig 'irrig 2, mix a & b' 0 0 1  0 0 1  0 0 0  0 0 0,
                      'irrig 2, mix c vs a&b' 0 0 1  0 0 1  0 0 -2  0 0 0,
                      'irrig 2, mix d vs a&b' 0 0 1  0 0 1  0 0 0  0 0 -2
                             / divisor=2,2,2 exp cl;
```

gives the following output.

Least Squares Means Estimates

Effect	Label	Estimate	Standard Error	DF	*t* Value	*P* > |*t*|	Alpha	Lower	Upper
mix*irrig	irrig 2, mix a & b	1.8522	0.6175	12	3.00	0.0111	0.05	0.5069	3.1976
mix*irrig	irrig 2, mix c vs a&b	-1.5263	0.7154	12	-2.13	0.0542	0.05	-3.0852	0.03249
mix*irrig	irrig 2, mix d vs a&b	-0.7865	0.7236	12	-1.09	0.2984	0.05	-2.3630	0.7900

Least Squares Means Estimates

Effect	Label	Exponentiated estimate	Exponentiated lower	Exponentiated upper
mix*irrig	irrig 2, mix a & b	6.3741	1.6601	24.4736
mix*irrig	irrig 2, mix c vs a&b	0.2173	0.04572	1.0330
mix*irrig	irrig 2, mix d vs a&b	0.4554	0.09414	2.2034

The first line gives the average estimated log count and associated statistics for combined mix "A" and "B" given irrigation level 2. The second line gives the difference between the log counts of mix "C" and the combined mixes "A and B." The third gives the log count difference between mix "D" and mixes "A" and "B" given irrigation 2. Exponentiating the log count gives the estimated count; exponentiating the difference, as we saw earlier, gives you the estimated ratio of expected counts. These appear in the lower box.

Other, similar, contrasts implemented using the LSMESTIMATE statements appear in the GLIMMIX program. In the interest of space, these are not discussed here, but are left as an exercise to the reader to run and interpret.

Example 3 – A Multi-Location (Blocked) Design with Categorical (Multinomial) Data

Description of the Experiment and its Objectives: These data were collected to compare three poinsettia varieties for subjective, aesthetic qualities. Twelve growers participated in the study. Each grower had plants from each of the three varieties; the number of plants for a given variety at a given grower ranged from 10 to 31. At each location, volunteers with training in rating ornamentals assigned a score to each plant on a scale of 1 to 3, where "1" meant "superior quality – suitable for sale at a boutique florist," "2" meant "excellent quality – suitable for sale in most commercial settings, and "3" meant "marginally acceptable," e.g. suitable for "discounted for immediate sale" only.

Important Note: The data in this example are ordinal categorical data. That is, the ratings – "superior," "excellent," and "acceptable" – have an obvious order, but the numbers used as numeric codes – 1, 2 and 3 – have no intrinsic meaning. In has been common in biological research to treat rating data as if the numeric code meant something and was suitable for classical analysis of variance. However, this approach is prone to giving insufficiently informative and potentially misleading results. For example, a mean rating of 2 can result from half the plants receiving a "1" and half the plants receiving a "3." However, it can also result from all of the plants receiving a "2," or from a myriad of other combinations of "1," "2" or "3." The means are all 2 but the underlying reality and economic consequences are likely to be very different. Generalized Linear Mixed Models allow these data to be analyzed using ordinal logit models, which greatly increase the accuracy and useable information available from the analysis. This example focuses on correctly implementing the ordinal logit analysis. This should be considered the "gold standard" for this type of data.

Step 1. Visualization.

As with Example 1, you can visualize this trial by using a standard plot plan for a randomized complete block design, shown as follows. Within each cell (i^{th} variety × j^{th} grower) combination) there are n_{ij} plants to be rated, where n_{ij} is between 10 and 31.

	Grower											
Variety	1	2	3	4	5	6	7	8	9	10	11	12
1												
2												
3												

Step 2: WWFD ANOVA.

The units in the **experiment design** (what Fisher called the "topographical aspect") are grower and cell within grower. The **treatment design** consists of the three varieties. Varieties are randomly assigned to cells within grower, denoted cell(grower) below. Hence cell(grower) is the **experimental unit**. The resulting WWFD ANOVA is as follows.

Experiment Design		Treatment Design		Combined	
Source	df	Source	df	Source	df
grower	11			grower	11
		variety	2	variety	2
cell(grower)	24	"parallels"	33	cell(grower) | variety	24-2=22
TOTAL	**35**	**TOTAL**	**35**	**TOTAL**	**35**

Variety appears on the line immediately above cell(grower). This signifies that cell(grower) is the experimental unit with respect to variety. In the combined table, the source of variation for experimental unit is denoted cell(grower)|variety, read as "cell within grower after accounting for variety." Notice that although each cell contains n_{ij} plants and rating scores (observations) are at the plant level, the actual observation to be used in the analysis is y_{ijk}, the number of plants of the ith variety at the jth grower that receive rating k. This observation occurs at the cell level.

Important: as with Example 1, the last line in the combined ANOVA should *not* be written as "residual." The next steps will show how you proceed when the response variable is a multiple category rating.

Step 3. Which Sources are Random? What are the Associated Distributions?

Applying the criteria given in Example 1, the GROWER effect is random because the growers in this trial represent a population of growers. In theory, any representative group of growers could have been used in this trial, and the results are presumed to be applicable to all growers similar to this study's participants. On the other hand, VARIETY effects are fixed because the three varieties in this trial are specifically of interest to the researcher.

Step 4: Write the Resulting Model.

For this experiment, the resulting model is

- Response variable. Y_{ijk} = number of plants in rating category k for the ith variety at the jth grower. The distribution is $Y_{ijk} \sim Multinomial\left(N_{ij}, p_{ij1}, p_{ij2}, p_{ij3}\right)$, where N_{ij} is the number of plants, and $\left\{p_{ij1}, p_{ij2}, p_{ij3}\right\}$ are the probabilities that a given plant belongs to rating category 1, 2 or 3 respectively for the ith variety at the jth grower. Note that the sum of the three probabilities must equal one, that is $\sum_{k=1}^{3} p_{ijk} = 1$.
- Link function. The link function of choice for ordinal multinomial data is the **cumulative logit**. More correctly, the link functions are cumulative logits – if there are k rating categories, there will be $k-1$ link functions, one per boundary between adjacent categories, illustrated as follows:

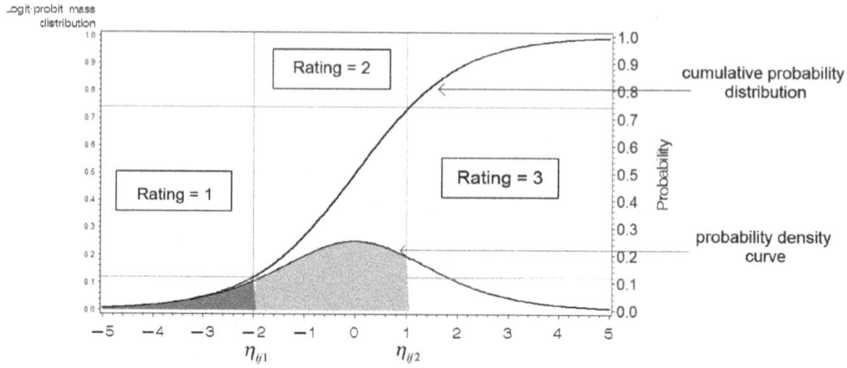

The darkest shaded area on the left represents rating category 1, the area in the center represents category 2, and the unshaded area on the right represents category

3. These three areas under the **probability density curve** correspond to the probabilities $\{p_{ij1}, p_{ij2}, p_{ij3}\}$ for the three categories. The cumulative logits, η_{ij1} and η_{ij2} model the boundary between categories 1 and 2 and between categories 2 and 3, respectively. For each boundary, project a line upwards. At the point where it intersects the **cumulative probability distribution**, project a line across to the y-axis, labelled **probability**. The probability projected from the lower logit, η_{ij1}, is the probability of a variety *i* plant at location *j* having rating 1. The probability projected from the upper logit, η_{ij2}, is the sum of the probabilities of a variety *i* plant at location *j* having rating 1 or 2. This is called the cumulative probability. As you can see, the cumulative logit model focuses on estimating the probabilities $\{p_{ij1}, p_{ij2}, p_{ij3}\}$ and testing the variety effect on them. This is more accurate, informative and useful than focusing on the mean of a numeric code that is arbitrary and has no intrinsic meaning.

Formally, the cumulative links are defined as:
$\eta_{ij1} = \log\left[p_{ij1}/\left(1-p_{ij1}\right)\right] = \log\left[p_{ij1}/\left(p_{ij2}+p_{ij3}\right)\right]$, and
$\eta_{ij2} = \log\left\{\left(p_{ij1}+p_{ij2}\right)/\left[1-\left(p_{ij1}+p_{ij2}\right)\right]\right\} = \log\left[\left(p_{ij1}+p_{ij2}\right)/p_{ij3}\right]$.

- Linear predictor. Using the WWFD ANOVA, the linear predictor has the form logit = grower + variety + cell(grower) | variety. Models for multinomial data are unique in that there must be one linear predictor for each link function. For the cumulative logit model, the linear predictors are $\eta_{ij1} = \eta_1 + v_i + g_j + u_{ij}$, and $\eta_{ij2} = \eta_2 + v_i + g_j + u_{ij}$ where η_1 and η_2 denote the intercepts for the lower and upper link functions, respectively, v_i denotes the variety effect, g_j denotes the grower effect, and u_{ij} denotes the experimental unit (cell(grower) | variety) effect. Notice that $\eta_1 < \eta_2$, and the variety, grower and unit effects are the same for both link functions. This guarantees that the boundary estimated by the lower link will always be less than the boundary estimated by the upper link. Otherwise, the results of the analysis would be incoherent.
- Distributions. The grower effects are assumed $g_j \sim NI\left(0, \sigma_G^2\right)$, experimental unit effects are assumed $u_{ij} \sim NI\left(0, \sigma_U^2\right)$.

Step 5: PROC GLIMMIX Statements.

The statements that implement the model described above are as follows:

```
proc glimmix data=Variety_Trial;
 class Rating Grower Variety;
 model Rating=Variety / d=multinomial solution;
 random intercept Variety / subject=Grower;
 weight N_of_Plants;
```

Notice that categorical response variables, in this case RATING, must be included in the CLASS statement along with the treatment factor, VARIETY, and the blocking criterion, GROWER. The WEIGHT statement is required if the data set gives the number of plants (N_of_Plants) in each rating category for each grower-variety combination. If there is one line in the data set per plant, do not include the WEIGHT statement.

The MODEL and RANDOM statement are similar to those used in Example 1. These statements follow a form common to all data sets from blocked designs with categorical data. The only differences between these statements and Example 1 are how the response variable and distribution are specified. In Example 1 there were only two response categories, germinated or did not germinate. Hence the Y/N syntax and the default use of the binomial distribution and logit link. With three or more response categories, you name the response variable, in this case RATING, and you must name the distribution, in this case D=MULTINOMIAL. By default, PROC GLIMMIX regards all response variables not in the Y/N format as Gaussian unless there is a D= statement that says otherwise. The default link function for the multinomial distribution is the cumulative logit, so you do not need to also specify the link function.

Note to Plant Breeders: Geneticists who work with categorical data often use "Threshold" models. A threshold model is otherwise known as a cumulative probit multinomial GLMM. You can implement it by adding the option LINK=CPROBIT immediately after D= MULTINOMIAL in the MODEL statement.

Additional statements for estimates and tests are given below. Their rationale will be more intelligible after first considering preliminary output from the SOLUTION option in the model statement.

Relevant Output I: Solution

The first output of interest shows the SOLUTION for the fixed effects, that is, the estimates of the intercept and variety effects.

Solutions for Fixed Effects

Effect	Rating	Variety	Estimate	Standard Error	DF	*t* Value	*P* > |*t*|
Intercept	1		-1.1862	0.5960	11	-1.99	0.0720
Intercept	2		2.7756	0.5970	11	4.65	0.0007
Variety		1	0.1655	0.6095	22	0.27	0.7885
Variety		2	1.1907	0.6105	22	1.95	0.0640
Variety		3	0	.	.	.	.

The ESTIMATE column for INTERCEPT 1 and INTERCEPT 2 give the boundaries on the logit scale between rating categories 1 and 2 and rating categories 2 and 3 respectively. Notice that GLIMMIX sets the last VARIETY effect equal to zero, so INTERCEPT 1 and INTERCEPT 2 give the boundaries on the logit scale for VARIETY 3. The VARIETY 1 and 2 effects are estimates of how far up or down the boundaries are shifted. Adding the VARIETY 1 or 2 effect to INTERCEPT 1 and INTERCEPT 2 gives you the boundaries between categories on the logit scale for varieties 1 and 2. You can translate all of these logits into probability estimates using the ESTIMATE statement and ILINK option. These are shown below.

Additional Post-Processing. Once you have the fixed effects estimates, you can move on to the statistics that are generally of the greatest interest with multinomial data: what are the estimated probabilities of each rating category for each treatment, and is there evidence of a variety effect on these probabilities? The following statements allow you to address these questions.

```
estimate 'variety 1 superior' intercept 1 0 variety 1 0 0 / ilink;
estimate 'variety 1 excellent or better' intercept 0 1 variety 1 0 0 / ilink;
estimate 'variety 2 superior' intercept 1 0 variety 0 1 0 / ilink;
estimate 'variety 2 excellent or better' intercept 0 1 variety 0 1 0 / ilink;
estimate 'variety 3 superior' intercept 1 0 variety 0 0 1 / ilink;
```

```
estimate  'variety 3 excellent or better' intercept 0 1 variety 0 0 1 / ilink;
contrast '1 v 3' variety 1 0 -1;
contrast '2 v others' variety -1 2 -1;
```

The ESTIMATE statements define the logits for the boundaries for each variety. For example, the first statement obtains $\hat{\eta}_1 + \hat{v}_1$, the boundary between category 1 ("superior" rating) and category 2 ("excellent"). The second ESTIMATE statement obtains $\hat{\eta}_2 + \hat{v}_1$, the boundary between category 2 and category 3 ("adequate") – the cumulative logit for categories 1 and 2. The rest of the ESTIMATE statements define the analogous boundaries for varieties 2 and 3. The INLINK option translates these logit estimates into probabilities. The output is

Estimates

Label	Estimate	Standard error	DF	*t* Value	P > |*t*|	Mean	Standard error mean
variety 1 superior	-1.0206	0.5955	22	-1.71	0.1006	0.2649	0.1160
variety 1 excellent or better	2.9411	0.5971	22	4.93	< 0.0001	0.9498	0.02845
variety 2 superior	0.004495	0.5963	22	0.01	0.9941	0.5011	0.1491
variety 2 excellent or better	3.9662	0.5986	22	6.63	< 0.0001	0.9814	0.01092
variety 3 superior	-1.1862	0.5960	22	-1.99	0.0591	0.2339	0.1068
variety 3 excellent or better	2.7756	0.5970	22	4.65	0.0001	0.9413	0.03296

The ESTIMATE column gives estimates on the logit scale. The MEAN column gives estimates on the probability scale. For variety 1, the MEAN of 0.2649 gives the estimated probability of a variety 1 plant having a "superior" rating; in the second row, the MEAN of 0.9498 gives the estimate of a variety 1 plant having a rating of "excellent" or "superior. Thus, the estimated probability of a plant having an "excellent" rating is 0.9498 – 0.2649 = 0.6849, and the estimated probability of a variety 1 plant have an "adequate" rating is 1 – 0.9498 = 0.0502. You can see that for varieties 1 and the, the breakdown is approximately 25-70-5 superior-excellent-adequate, whereas for variety 2 it approximately 50-48-2. Variety 2 appears to have a much higher proportion of "superior" plants and a correspondingly lower proportion of "excellent" or "adequate" plants.

The CONTRAST statements allow you to test this formally. The CONTRAST output is

Contrasts

Label	Num DF	Den DF	*F* Value	*P* > *F*
1 v 3	1	22	0.07	0.7885
2 v others	1	22	4.39	0.0479

The results show no evidence of a different between varieties 1 and 3 ($p > 0.79$) but evidence that the set of probabilities for variety 2 differs from the average of the other two varieties ($p < 0.05$).

How much difference does the cumulative logit GLMM make? Suppose these data were analyzed using the numeric ratings in an analysis of variance. The mean ratings are 1.82, 1.65 and 1.86 respectively for varieties 1, 2 and 3. The p-value for the "1 vs 3" contrast is 0.78 and the p-value for the "2 vs others" contrast is 0.06. The mean ratings defy intelligible interpretation and the tests of hypotheses for variety differences lose statistical power – the inevitable result of making an indefensible assumption regarding the distribution of the response variable.

Analysis treating numeric ratings as if their distribution was Gaussian for data with response variables that are categories appears in the literature mainly because until useable GLMM software such as PROC GLIMMIX became available, there was no truly viable alternative. However, as this example illustrates, focusing on means of numerically coded ratings, and failing to focus on proportions in the various rating categories, entails reduced power and loss of valuable information. While plant scientists pre-2008 had an excuse, contemporary researchers with categorical data should use contemporary methods to analyze their data.

Example 4 – A Repeated Measures Experiment with Count Data

Description of the Experiment and its Objectives: These data were collected to compare microbe counts at increasing soil depth for three tillage management types. The experiment design consisted of four blocks with three plots each. Tillage types (labeled A, B and C in the data set) were randomly assigned to plots within each block. At each plot, a soil probe capable of taking readings at five equally spaced depths was used to obtain microbe counts. In theory, the counts should have decreased with increasing depth, but the rate of decrease for the particular microbe and the tillage types under study was only vaguely understood prior to this study.

Important Note: By definition, experiments with observations taken at planned intervals in space or over time are called **repeated measures** experiments. The data from such experiments are often referred to as **longitudinal data**, although this term is more commonly used in the medical and social sciences than in the agricultural and natural resource sciences. The repeated measures in this example result from observing the experimental unit for tillage – the plot – with a single core at multiple depths. Another form of repeated measures common in the plant sciences involves measurements on the same experimental unit – e.g. plot in a field or pot in a greenhouse – at planned times over the growing season. The approach to analysis is identical. If the repeated measure is time, replace depth in this example by time. Chapter 10 (Gezan and Carvalho, 2018) covers repeated measures in more detail.

Step 1. Visualization. The following shows the layout of a block in example 4.

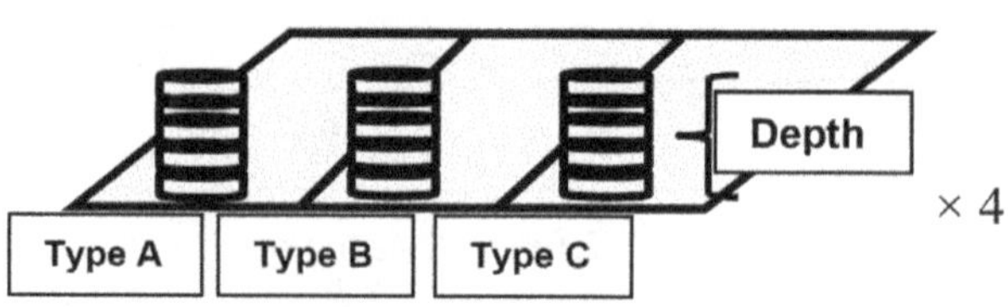

Obviously, the order of tillage types would be randomized in each block – the order shown above is for illustration only. Each cone in the stack of five in each plot represents a depth at which the microbe counts are taken.

Step 2: WWFD ANOVA.

The units in the **experiment design** are block, plot within block, and soil depth within plot. The **treatment design** consists of three tillage types and soil depth. Plot within block is the experimental unit with respect to tillage type. Soil depth within plot is the experimental unit with respect to the treatment factor depth. In repeated measures experiments, the unit on which repeated measures are taken, in this case depth, is both an element of the experiment design and a factor in the treatment design. Some users prefer to give this source of variation a different name in the experiment design and treatment design lists, to avoid confusion. Doing so is not necessary, provided you keep the distinction in mind as you construct the combined ANOVA. The resulting WWFD ANOVA is as follows.

Experiment Design		Treatment Design		Combined	
Source	df	Source	df	Source	df
block	3			block	3
		type	2	type	2
plot(block)	8			plot(block) | type = block×type	8-2=6
		depth	4	depth	4
		type×depth	8	type×depth	8
soil depth(plot)	48	"parallels"	45	soil depth(plot) | type×depth = block×depth(type)	48-4-8=36
TOTAL	59	TOTAL	59	TOTAL	59

The line appearing immediately below TYPE in the combined ANOVA is "plot within block after accounting for type." Equivalently, specifying the block and the type, i.e. block×type, uniquely identifies the plot and provides the appropriate way to translate this source of variation into language understandable to PROC GLIMMIX or equivalent software. In repeated measures jargon, block×type is also called the **between subjects** error term. It is also called the **whole plot error**, because the ANOVA for repeated measures bears a superficial resemblance to the split-plot ANOVA. However, as will become clear in Step 4 below, the superficial resemblance is just that – superficial. There is more complexity in modeling repeated measures. As with previous examples, the last line in the combined ANOVA should *not* be written as "residual." It is technically "soil depth within plot after accounting for type and the treatment factor depth," and is specified in GLIMMIX program statements as block×depth(type). In repeated measures jargon, it is called the **within subjects** error term.

Step 3. Which Sources are Random? What are the Associated Distributions?

The terms that originate in the experiment design ANOVA, and appear in their "after account for" the associated treatment factor form in the combined ANOVA, correspond to entities that represent larger populations. Therefore, these sources of variation will be represented by random effects in the model. Typically, the block and block×type effects are assumed to be independent and have a Gaussian distribution. The within subjects – block×depth(type) – term has a more complex distribution because taking adjacent measurements in time or space on the same plot induces potential correlation that the model must take into account. This will be explained below in more detail. Finally, the observations are counts, so the assumed distribution must account for the mean-variance relationship with count data. In the repeated measures setting, the Poisson is the distribution of choice.

Step 4: Write the Resulting Model.

For this experiment, the resulting model is

- Response variable. Y_{ijk} = microbe counts for the ith tillage type at the jth soil depth in the kth block. The distribution is

  $Y_{ijk} \mid random\ effects\,(see\ linear\ predictor) \sim Poisson(\lambda_{ijk})$
- Link function. The standard link function for count data is $\eta_{ijk} = \log(\lambda_{ijk})$.
- Linear predictor. Using the WWFD ANOVA, the linear predictor has the form log = block + type + block×type + depth + type×depth + block×depth(type). In formal statistical notation, write the model as $\eta_{ijk} = \eta + \tau_i + \delta_j + \tau\delta_{ij} + r_k + b_{ik} + w_{ijk}$, where τ denotes type effects, δ denotes depth effects, r denotes block effects, b denotes between subjects (block×type) effects, and w denotes within subjects [block×depth(type)] effects.
- Distributions. The block effects are assumed $r_k \sim NI(0, \sigma_R^2)$. Between subjects effects are assumed $b_{ik} \sim NI(0, \sigma_B^2)$. Within subjects effects must account for correlations between observations on different depths at the same plot. To do this, write the distribution of the within subjects terms for a given plot (block×type) as a group:

$$\begin{bmatrix} w_{i1k} \\ w_{i2k} \\ w_{i3k} \\ w_{i4k} \\ w_{i5k} \end{bmatrix} \sim N\left(\begin{bmatrix} 0 \\ 0 \\ 0 \\ 0 \\ 0 \end{bmatrix}, \begin{bmatrix} \sigma_1^2 & & & & \\ \sigma_{21} & \sigma_2^2 & & & \\ \sigma_{31} & \sigma_{32} & \sigma_3^2 & & \\ \sigma_{41} & \sigma_{42} & \sigma_{43} & \sigma_4^2 & \\ \sigma_{51} & \sigma_{52} & \sigma_{53} & \sigma_{54} & \sigma_5^2 \end{bmatrix}\right)$$

This is called a **multivariate normal distribution**. As with any block or "error" effect, the mean for each element is zero. The variance of the within subject effect at the jth time is σ_j^2 for $j = 1,2,3,4,5$; the covariance between any pair of terms at times j and j' is denoted $\sigma_{j'j}$. The form written above is called an **unstructured** covariance model, because it allows the variance at each time and covariance between every pair of times to be different. In most repeated measures experiments, this is an overly detailed and inefficient way to model correlation among repeated measures. In practice, the model must account for the possibility that observations closer together tend to be more alike than observations farther apart. The simplification that often adequately accounts for this is a **first-order autoregressive** covariance model, usually referred to by its acronym **AR(1)**. Write the AR(1) model as

$$\begin{bmatrix} w_{i1k} \\ w_{i2k} \\ w_{i3k} \\ w_{i4k} \\ w_{i5k} \end{bmatrix} \sim N\left(\begin{bmatrix} 0 \\ 0 \\ 0 \\ 0 \\ 0 \end{bmatrix}, \sigma^2 \begin{bmatrix} 1 & & & & \\ \rho & 1 & & & \\ \rho^2 & \rho & 1 & & \\ \rho^3 & \rho^2 & \rho & 1 & \\ \rho^4 & \rho^3 & \rho^2 & \rho & 1 \end{bmatrix}\right)$$

Notice that ρ denotes the correlation between within subject effects at adjacent depths, ρ^2 denotes correlation two units of depth apart, etc. If the depths are not equally spaced, a variation on the AR(1), called as first-order spatial power model, simply sets the correlation to ρ^d, where d is the distance between a given pair of observations.

If all of the correlations are zero, the distribution of the within subjects effects reduces to $w_{ijk} \sim NI\left(0, \sigma^2\right)$. This is identical to the model for a split-plot experiment. The problem with using a split-plot model to analyze repeated measures is that if there is correlation and it is not accounted for, test statistics are biased upward, and confidence interval widths are biased downward. On the other hand, if the correlation structure is over-modeled, for example the unstructured structure is used when the AR(1) would suffice, power suffers and confidence interval widths will be biased upward.

Step 5: PROC GLIMMIX statements.

Following the discussion in the preceding paragraph, the first step in repeated measures analysis is to decide how complex the covariance model needs to be in order to account for correlation between observations at different depths in the same plot. Textbooks such as Stroup (2013) and Gbur, et al (2012) give a complete account of how to do this and the underlying theory. Here, we consider an abridged version of this process. This idea is to use the simplest covariance model that adequately accounts for within subject correlation. Do this by computing the **Akaike information criterion** (AICC) for all covariance structures deemed plausible prior to analyzing the data. In this example we show the unstructured, the AR(1) and the split-plot model. In practice, there are other models you should also consider. See Stroup (2013b) or Gbur et al. (2012).

Run the following GLIMMIX programs:

```
proc glimmix data=RptM_counts method=laplace;
 class Block Type Depth;
 model count=Type|Depth / d=Poisson;
 random intercept / subject=Block;
 random Depth / subject=Block*Type type=UN;
proc glimmix data=RptM_counts method=laplace;
 class Block Type Depth;
 model count=Type|Depth / d=Poisson;
 random intercept Type / subject=Block;
 random Depth / subject=Block*Type type=AR(1);
proc glimmix data=RptM_counts method=laplace;
 class Block Type Depth;
 model count=Type|Depth / d=Poisson;
 random intercept Type / subject=Block;
 random Depth / subject=Block*Type;
```

Note the use of METHOD=LAPLACE. To obtain valid information criteria, you need to use an integral approximation method. You could use METHOD=QUADRATURE, but it can be much slower than LAPLACE and will produce similar results for experiments of this type. The first GLIMMIX program computes the unstructured model, the second the AR(1) model, and the third the split-plot model. In practice, TYPE=UN in conjunction with non-Gaussian data, often fails to run. If this happens, replace TYPE=UN with TYPE=CHOL. For the purposes of obtaining fit statistics, they are equivalent. Ignore all of the output from each of these runs except the FIT STATISTICS output – specifically, the AICC. The results appear as

Fit Statistics	UN	AR(1)	Split-Plot
-2 Log Likelihood	499.95	525.54	532.64
AIC (smaller is better)	557.95	563.54	568.64
AICC (smaller is better)	**615.95**	**582.54**	**585.33**
BIC (smaller is better)	540.15	551.88	557.60
CAIC (smaller is better)	569.15	570.88	575.60
HQIC (smaller is better)	518.90	537.96	544.40

The AICC statistics are shown in boldface. Interpret them just as directed in the output: "smaller is better." Don't overthink it. Smaller is better. Period. The AICCs are: AR(1), 582.5; split-plot, 585.3; unstructured, 616.0. Subsequent analysis should use the AR(1) covariance model.

At this point, the analysis follows standard protocol for a factorial treatment design with one qualitative factor (TYPE) and one quantitative factor (DEPTH). There is no one-size-fits-all way to do this – specifics can vary depending on the experiment's particulars. In this example, we illustrate one possible approach: an initial GLIMMIX run to obtain an overview, and then specific focus on regression over depths and possible effects of TYPE on those regressions.

Overview:

```
proc glimmix data=RptM_counts;
 class block type depth;
 model count=type|depth / d=poisson;
 random intercept type / subject=block;
 random depth / subject=block*type type=ar(1);
 lsmeans type*depth / plot=meanplot(sliceby=type ilink join);
 ods select meanplot;
```

Notice that we now remove METHOD=LAPLACE and use the default pseudo-likelihood method to complete the analysis. We do this because pseudo-likelihood provides better type I error control and more accurate inferential statistics. The output of primary interest at this point is the MEANPLOT, shown as follows:

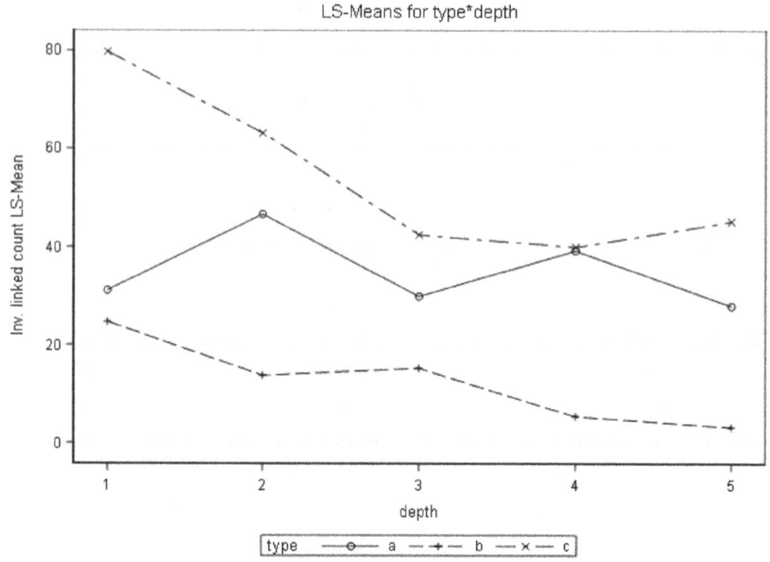

Most evident from the plots are trends over depth that differ (at least visually) for each tillage type. Tillage type "A" does not appear to have a consistent pattern of change. This is often characterized as a linear trend with a zero slope. Tillage type "B" appears to have a linear trend decreasing with depth. Tillage type "C" shows a possible nonlinear trend, in that mean counts decline linearly over depths 1 through 3, then show no further decrease at depths 4 and 5. With this in mind, a useful strategy is to fit a linear regression model over depth with separate slopes for each tillage type. The model should include a term for lack of fit, in order to test for nonlinear trends such as the suspected pattern for tillage type "C".

The next step in the analysis is to pursue this regression strategy. Specifically, the linear predictor $\eta+\tau_i+\beta_i D+\tau\delta_{ij}$ allows you to fit a linear regression over depth for each type – where β_i denotes the slope for the ith tillage type and D is the depth (i.e. $D=1,2,3,4,5$) – and test lack of fit from linear regression – where $\tau\delta_{ij}$ denotes the lack of fit term. The following GLIMMIX statements allow you to compute this model and the appropriate estimates and tests associated with it.

```
proc glimmix;
 class block type depth;
 model count=type type*d type*depth / ddfm=kr2 d=poisson htype=1;
 random intercept type / subject=block;
 random depth / subject=block*type type=ar(1);
 nloptions maxiter=50;
```

Notice that the CLASS and RANDOM statements must be kept identical to those in the previous GLIMMIX run. In addition, you must create a new variable, D, and *not* include it in the CLASS statement, so GLIMMIX can compute it as a linear regression. The value of D in this case is identical to DEPTH; the difference is that the variable DEPTH is treated as a CLASS variable, allowing you to estimate lack of fit (TYPE*DEPTH) and define the repeated measures covariance structure with the second RANDOM statement. The term TYPE*D in the MODEL statement defines the separate linear regression for each TYPE. HTYPE=1 instructs GLIMMIX to test TYPE*D (the linear component of the depth effect) first, then test TYPE*DEPTH afterwards. This makes TYPE*DEPTH a test of whatever depth effect remains over and above the linear effect – in other words, by definition, lack-of-fit. Finally, the NLOPTIONS statement overrides the GLIMMIX default limit of 20 iterations. This run actually required 28 iterations to converge. This is typical of complex GLMMs. The author has analyzed data sets that required as many as 2500+ iterations. There is no upper limit on the MAXITER override.

Relevant output:

Type I Tests of Fixed Effects

Effect	Num DF	Den DF	F Value	P > F
type	2	6.192	2.29	0.1804
d*type	3	8.956	3.98	0.0469
type*depth	9	29.33	0.65	0.7494

The result for TYPE*DEPTH, F=0.65, p>0.74, tells you that there is no evidence of lack of fit from linear regression over depth. This allows you to move to the next step: drop the lack-of-fit term (TYPE*DEPTH) from the model and focus on the estimated slopes for linear regression over depth and whether there is evidence that they differ by type.

The GLIMMIX statements are

```
proc glimmix;
 class block type depth;
 model count=type type*d / noint d=poisson solution ddfm=kr2;
 random intercept type / subject=block;
 random depth / subject=block*type type=ar(1);
 contrast 'slope a vs c' type*d 1 0 -1;
 contrast 'slope a&c vs b' type*d 1 -2 1;
 contrast 'intercept a vs b' type 1 -1 0;
 contrast 'intercept a&b vs c' type 1 1 -2;
```

The NOINT option causes $\eta + \tau_i$ to be combined into a single intercept coefficient for each TYPE. In other words, the linear predictor is $\beta_{0i} + \beta_{1i}D$ where β_{0i} and β_{1i} denote the intercept and slope for the linear regression over depth for the ith type.

Relevant output:

Solutions for Fixed Effects

Effect	type	Estimate	Standard Error	DF	*t* Value	P > |*t*|
type	a	3.5742	0.7139	17.27	5.01	0.0001
type	b	3.7468	0.7198	18.02	5.21	< 0.0001
type	c	4.4687	0.7086	16.91	6.31	< 0.0001
d*type	a	-0.03441	0.1319	7.907	-0.26	0.8009
d*type	b	-0.4999	0.1424	9.681	-3.51	0.0059
d*type	c	-0.1471	0.1299	7.637	-1.13	0.2917

The ESTIMATE column shows the regression coefficient estimates for each type. For example, the regression equation for type A is 3.57 – 0.034 × *D*. Note that this give you an estimate of the log count at depth D; if you want an estimate of the mean count at depth D you must use the inverse link function, for example $e^{3.57-0.034D}$. You can see that the slopes for type A and C appear to be relatively similar, but each appears to differ compared with the slope for type B, whereas the intercept for type C appears to be noticeable greater than the intercepts for types A and B.

Relevant output:

Contrasts

Label	Num DF	Den DF	*F* Value	P > *F*
slope a vs c	1	7.772	0.37	0.5600
slope a&c vs b	1	9.027	5.80	0.0392
intercept a vs b	1	12.9	0.03	0.8652
intercept a&b vs c	1	12.52	0.89	0.3630

There is sufficient evidence that the linear change over depth for type B differs from the linear change over depth for types A and B ($p < 0.04$), but insufficient evidence to conclude any other differences in the regression for each type. Recall from the MEANPLOT that type B is the type with the smallest mean counts; visually, type C had generally greater mean counts and appeared to have a much greater intercept and slope. However, recall also that with counts, the variance increases with the mean: it is easy to show differences between small counts but showing differences between larger counts is more difficult and requires more replication. This study should have received more attention at the time it was designed if the apparent difference between type C and the other types was considered important and researchers wanted sufficient power and precision to draw credible conclusions about depth effects. Designs that work for Gaussian data do not necessarily work for non-Gaussian data. If you know that your primary response variable will be a count, use precision and power analysis strategies as presented in Stroup (2013b).

Given these results, the researcher has one of two options for a final report. One option would be to simply report the regression equations for each type from the output above, along with the contrasts showing that there is no evidence of a statistically significant difference among any of the intercepts but a significant difference between the slope for type B versus the other two types. Alternatively, you could combine estimates where there is no evidence of a difference. That is, estimate a common regression equation for types A and C, and estimate an equation for type B with the same intercept but a different slope.

If you choose the latter option, the SAS statements are

```
data pool_a_c;
 set RptM_counts;
 if type='a' or type='c' then pooled_type='ac';
 if type='b' then  pooled_type='_b';
proc glimmix;
 class block type pooled_type depth;
 model count=pooled_type*d / d=poisson solution ddfm=kr2;
 random intercept type / subject=block;
 random depth / subject=block*type type=ar(1);
```

Relevant Output:

Solutions for Fixed Effects

Effect	pooled_type	Estimate	Standard error	DF	*t* Value	*P* > |*t*|
Intercept		3.9402	0.4314	6.19	9.13	< 0.0001
d*pooled_type	_b	-0.5249	0.1338	18.49	-3.92	0.0010
d*pooled_type	ac	-0.08048	0.09323	11.87	-0.86	0.4051

The estimated intercept for all three types is 3.94. The estimated slope for type B is -0.52 and for types A and C is -0.08. Again, keep in mind that these regression equations estimate the *log* count for a give depth. If you want to predicted count, you must apply the inverse link.

Which of the final two options to use, separate regressions or combined, depends on the context of the study and the consequences of each decision. If the differences that appear in the MEANPLOT are important, it may be better to leave the regression estimates separate, and use the analysis as an argument for a better-designed follow-up study. On the other hand, if the differences in the MEANPLOT are not large enough to be scientifically consequential, combining the regression equations may tell the most accurate story about depth effects for the tillage types in this study.

Example 5: A Complex Study with Split-Plot and Repeated Measures Design Features and Binomial Data

This example is included as a final, but in the author's experience, essential, illustration of how to accurately translate a study design into a model. Most "modeling problems" are really design-to-model interface issues. While GLMMs can accommodate an unprecedented combination of design and data distribution options, with flexibility comes greatly increased ways for things to go wrong. Complex designs exacerbate opportunities for trouble.

The most important aspect of this example involves getting the model and associated GLIMMIX CLASS, MODEL, and RANDOM statements right. Once those are in place, post-processing – that is, use of LSMEANS, CONTRAST, ESTIMATE and LSMESTIMATE statements – is no different from any other factorial treatment design. Therefore, in this example, we focus exclusively on the design description, visualization of the plot plan, the WWFD ANOVA, and the GLIMMIX CLASS, MODEL and RANDOM statements. Along the way, we will pay attention not only to the "correct" model, but also to typical mistakes readers are likely to be tempted to make.

Description of the Study and Objectives

The study used four locations. Each location was divided into two sections. In one section the first half of the study was initiated in year 1; in the other, the final half of the study was initiated the following year. Each section was divided into four blocks. Each block consisted of two whole plot experimental units. The whole plot treatment factor was planting date, with two dates. Within each block, one whole plot experimental unit was assigned to each plant date. Each whole plot experimental unit consisted of 10 split plot experimental units. The split plot treatment design was a factorial with 2 sod suppression × 5 seeding rate combinations. The 10 split plot experimental units were assigned at random to the 10 treatment combinations. In each plot, 40 segments were observed; the response variable was the number of segments containing a seedling out of the 40 segments observed. Hence, the response variable was binomial. Establishment was measured at two times: "pre," i.e. during the year of planting; and "post," i.e. four growing seasons after initial planting.

Thus, the treatment design is a 2 × 5 × 2 × 2 factorial (planting date × seedling rate × sod suppression × pre-post observation time) and the experiment design is a split plot with repeated measures. In discussing the objectives, it became clear that the researchers were also interested in seeing if treatment effects differed by location. Including locations, the treatment design is actually a 4 × 2 × 5 × 2 × 2 factorial (adding the four locations) and the experiment design is a split-split plot with repeated measures.

The objectives of the study were to assess planting date, seedling rate and sod suppression effects on percent of seedlings emerging, to estimate possible location effect on treatment differences, and to evaluate establishment survival four years "post" planting.

Step 1: Plot Plan Visualization

The following table was used in consulting sessions to visualize the experiment and treatment designs. The areas labelled "Year 1" and "Year 2" at each location are adjacent physical entities called "section" in the description above and the WWDF ANOVA below. The "obs pre" refers to observations taking the initial year of planting; "obs post" refers to observations taken on the same plots four years later.

Location 1								
	Year 1				Year 2			
	Block 1,1		Block 1,4		Block 2,1		Block 2,4	
	PD 1	PD 2	PD 1	PD 2	PD 1	PD 2	PD 1	PD 2
	2 S × 5 Rate combinations randomized to 10 subplots in [PD(block)]	ditto	2 S×5 Rate randomized to 10 subplots	ditto	2 S × 5 Rate randomized to 10 subplots	ditto	2 S × 5 Rate randomized to 10 subplots	ditto
Obs Pre								
Obs Post								

Location 4								
	Year 1				Year 2			
	Block 1,1		Block 1,4		Block 2,1		Block 2,4	
	PD 1	PD 2	PD 1	PD 2	PD 1	PD 2	PD 1	PD 2
	2 S×5 Rate combinations randomized to 10 subplots in [PD(block)]	ditto	2 S× 5 Rate randomized to 10 subplots	ditto	2 S × 5 Rate randomized to 10 subplots	ditto	2 S × 5 Rate randomized to 10 subplots	ditto
Obs Pre								
Obs Post								

†PD, planting date; S, suppression.

Step 2: WWFD ANOVA

Location is clearly a physical entity that is part of the experimental material prior to the assignment of treatments. Locations used in the study are sampled from a target population, and are not randomly assignment. Years are also in some sense sampled, but the experiment is constructed by randomly assigning year to one of each location's two sections. Ordinarily, any source of variation that originates in the **experiment** column would not appear in the **treatment** column. However, the researchers are interested in location effects on treatment differences, i.e. "location'..." effects, so these must appear in the **treatment** column. This is typical of multi-location experiments. On the other hand, an effect originating in the **treatment** column would ordinarily have all possible interactions with other factors that also appear subsequently in the **treatment** column. However, in this case, year acts as more of a replicate that as a treatment factor.

Finally, notice the terms in the **combined** column. Those in **bold** are random model effects. **These must be limited to actual physical entities in the experiment design.** This is important, because data analysts are often tempted to include every interaction involving a random factor, whether it corresponds to a physical entity or not. In this case, the temptation is to include all "year ×..." interaction terms. Resist this temptation: it is improper modeling and will almost surely cause the estimation procedure to fail. If you are unlucky, the estimation procedure will not fail, and you will be stuck with trying to interpret nonsense results.

Experiment		Treatment		Combined	
Source	df	Source	df	Source	df
Location (Loc)	3			Loc	3
		Year	1	Year	1
Section(Loc)	4			**Loc × Year***	4-1=3
Block(Section)	24			**Block(Year × Loc)***	24
		Planting Date (PD)	1	PD	1
		PD×Loc	3	PD×Loc	3
WholePlot(Block)	32			**PD × Block(Year × Loc)***	32-4=28
		Suppression (S)	1	S	1
		Rate	4	Rate	4
		S × Rate	4	S × Rate	4
		Loc × S	3	Loc × S	3
		Loc × Rate	12	Loc × Rate	12
		Loc × S × Rate	12	Loc × S × Rate	12
		PD × S	1	PD × S	1
		PD × Rate	4	PD × Rate	4
		PD × S × Rate	4	PD × S × Rate	4
		PD × Loc × S	3	PD × Loc × S	3
		PD × Loc × Rate	12	PD × Loc × Rate	12
		PD × Loc × S × R	12	PD × Loc × S × Rate	12
SubPlot(WholePlot)	576			**PD × S × R × Blk(Year × Loc)***	576-72=504
		Time (T)	1	T	1
		Loc × T	3	Loc × T	3
		PD × T	1	PD × T	1
		Loc × PD × T	3	Loc × PD × T	3
		S × T	1	S × T	1
		Rate × T	4	Rate × T	4
		S × Rate × T	4	S × Rate × T	4
		Loc × T	3	Loc × T	3
		Loc × S × T	3	Loc × S × T	3
		Loc × Rate × T	12	Loc × Rate × T	12
		Loc × S × Rate × T	12	Loc × S × Rate × T	12
		PD × S × T	1	PD × S × T	1
		PD × Rate × T	4	PD × Rate × T	4
		PD × S × Rate × T	4	PD × S × Rate × T	4
		Loc × PD × S × T	3	Loc × PD × S × T	3
		Loc × PD × Rate × T	12	Loc × PD × Rate × T	12
		Loc × PD × S × Rate × T	12	Loc × PD × S × Rate × T	12
Obs(Plot)	640	"parallels"	1216	**Time × PD× S × Rate(Blk × Year × Loc)***	640-83=557
TOTAL	1279	TOTAL	1279	TOTAL	1279

* Terms shown in **bold** in the **Combined** sources of variation column are random effects corresponding to blocking criteria and whole or split plot experimental units. It is possible that the variance of these effects may not be these same for each year of planting. GLIMMIX program below shows how to obtain separate variance estimates for each year.

Resulting SAS PROC GLIMMIX statements

```
proc glimmix data=CombininedInitPost method=laplace;
class  time year location block date suppression rate;
model Y/N=location|date|suppression|rate|time/ dist=binomial;
```

*The following is the random statement originally attempted. Note that it has no chance of yielding a solution because there are too many terms that involve year. In addition, many of these terms do not correspond to physical entities.

```
*random year year*location block(year*location)
date*block(year*location) suppression*rate*date*block(year*location);
```

The following follows from WWFD and is therefore what should be done.
Note that the random part of this model is best written as two RANDOM statements.

```
random intercept block  /subject=location group=year;
random intercept suppression*rate Time*suppression*rate /
subject=date*block*location group=year;
parms (1.2)(0.001)(0.001)(0.3)(1.1)(.001)(1.2)(0.3)(0.1)(0.9);
nloptions maxiter=250 absfconv=1e-2;
covtest 'year diff?' homogeneity;
run;
```

The option GROUP=YEAR in the RANDOM statements fits a separate set of variance estimates for each year of planting. This potentially makes sense given the uniqueness of each growing season. However, because there are only two years and the design is already complex, estimating separate variance terms for each year is computationally demanding. The ABSFCONV option relaxes the criterion for convergence – arguably well past the point of responsible statistical analysis. The COVTEST statement produces a test statistic to test the equality of the two sets of variances. This option can only legitimately be used in conjunction with METHOD=LAPLACE or METHOD=QUADRATURE. Here, LAPLACE is the only viable alternative, because the RANDOM structure is far too complex for the quadrature method to run. Unless there is overwhelming evidence that the variances should be separate by year, the GROUP=YEAR option should not be used for the final analysis.

The relevant output is the COVTEST result:

Tests of Covariance Parameters
Based on the Likelihood

Label	DF	-2 Log Like	ChiSq	$P >$ ChiSq	Note
year diff?	6	8068.57	6.41	0.3788	DF

The p-value of 0.3788 means that there is no statistically significant evidence that the variance terms for the two years are different. Given this result, you would remove the GROUP=YEAR options from both random statements and proceed with the analysis. On the other hand, if there is evidence that covariance parameters differ by year, simply retain the GROUP=YEAR options and proceed with the analysis using the heterogeneous variance model. With mixed model methodology, heterogeneous variance models are well-defined and legitimate to use. The only reason to drop the GROUP=YEAR option is that the homogeneous variance model is more efficient if there is evidence that the assumption of homogeneous variance is satisfied.

In the interest of space, the further details are left to the reader. Suffice it to say that analysis at this point proceeds like any other complex, multi-factor experiment.

As a note of interest, the COVTEST result took 52 hours of CPU time and an additional 40 hours of post-processing time to obtain. This on a reasonably high-end 2014-vintage desktop PC. Your results will vary depending on your computer, but if you run models of this complexity, expect it to take a LONG time!

Key Learning Points

- Know the required elements of a generalized linear mixed model.
- Know the difference between the model scale and data scale.
- Know the difference between a conditional GLMM and its target of inference, and a marginal GLMM and its target of inference.
- Know when to use the PROC GLIMMIX default and when to override it with an integral approximation method (LAPLACE or QUADRATURE).
- Know how to translate the experiment design and treatment design sources of variation into an appropriate and "sensible" model.
- Know what overdispersion is, what diagnostic to use to recognize it, why it occurs, and what to do about it if there is evidence of overdispersion.
- For logit models, know what the model scale LSMEANS and DIFF estimate, and what their data scale analogs estimate (be careful).
- For count data models that use the log link (Poisson and negative binomial), know what the model scale LSMEANS and DIFF estimate, and what their data scale analogs estimate (be careful).
- For repeated measures with non-Gaussian data, know how to account for within-subject correlation. Be sure you know how doing so differs from accounting for correlation in Gaussian models.
- Know how to select an appropriate covariance structure for repeated measures with non-Gaussian data.

Review Questions

1. The combined WWFD ANOVA for Example 1 was

Source	df
block	
variety	
plot(block) I variety	
TOTAL	

   a. How do you read the term "plot(block) I variety"?

   b. How do you determine the DF for this term?

2. True or False: For the combined ANOVA given in (1), the following GLIMMIX statements should be fine for any binomial data

```
proc glimmix;
class block variety;
model y/n=variety;
random block; (or random intercept / subject=block; )
```

3. True or False: For the combined ANOVA given in (1), the following GLIMMIX statements would be better than (2) for binomial data

```
proc glimmix;
class block variety;
model y/n=variety;
random block plot(block*variety);
  (or random intercept plot*variety / subject=block; )
```

4. When you are working with binomial data

   a. How is the sample proportion for each experimental unit defined?

   b. True or False: It is okay to just record the sample proportion for each experimental unit.

   c. If you answered FALSE to (b), what should you record (i.e. include in the data)?

5. True or False: if the model is $\alpha_i + b_j$, α is a random effect and b is a fixed effect.

6. True or False: use the following SAS statements if you want to estimate a separate linear regression equation for each treatment (TRT) over level of irrigation (IRRIG).

```
proc glimmix;
 class trt irrig;
 model y = trt trt*irrig;
```

If you said FALSE, modify the statements so that they will estimate the desired terms.

7. What is the "link function of choice" for a GLMM with ordinal multinomial data?

8. Which of the following is ordinal multinomial data?

   a. Leaf shape: round, oval, clover-leaf, heart-shaped

   b. Damage level: intact, minor damage, moderate damage, completely destroyed.

9. True or False: in a repeated measures experiment, the repeated measures must be equally spaced.

10. Does your answer in (9) change depending on whether the repeated measures are in time or in space?

11. True or False: if you want to compare different covariance structures for repeated measures experiments with binomial or count data, you must use a RANDOM...RESIDUAL statement.

12. We know that the usual strategy for comparing repeated measures covariance structures is to use information criteria. True or False: to obtain information criteria to compare different covariance structures for repeated measures experiments with binomial or count data, you must use either the METHOD=LAPLACE or METHOD=QUAD option.

13. True or False: once you choose a covariance structure for repeated measures binomial or count data, you should use METHOD=LAPLACE to complete the analysis.

14. True or False: the negative binomial is the distribution of choice for repeated measures count data.

15. True or False: a repeated measures experiment can legitimately be called

a "split-plot in time" only if you can assume that the within-subject correlation is zero and hence the within subject effects are ~

16. True or False: the "split-plot-in-time" analysis is equivalent to a repeated measures analysis with a compound symmetry (CS) covariance structure.
17. True or False: In a GLM or GLMM model with a logit link, the model scale estimate of the LSMEANS is the log of the odds and the model scale estimate of the LSMEAN difference (DIFF) is the log of the odds-ratio.
18. True or False: In a GLM or GLMM model with a logit link, the data scale estimate of the LSMEANS is the odds and the data scale estimate of the LSMEAN difference (DIFF) is the odds-ratio.
19. True or False: In a GLM or GLMM model with a log link, the data scale estimate of the LSMEANS is the mean count and the data scale estimate of the LSMEAN difference (DIFF) is the ratio of the mean counts.
20. True or False: Use a binomial distribution for a sample proportion (i.e. Y/N) and a negative binomial distribution for a continuous proportion.

REFERENCES

Bartlett, M.S. 1947. The use of transformations. Biometrics 3: 39-52.

Breslow, N.E. and D.G. Clayton. 1993. Approximate inference in generalized linear mixed models. J. Amer. Statist. Assoc. 88: 9-25.

Casella, G. and R.L. Berger. 2002. Mathematical statistics, 2nd ed. Pacific Grove, CA: Duxbury.

Eisenhart, C. 1947. The assumptions underlying analysis of variance. Biometrics 3: 1-21.

Federer. W.T. 1955. Experimental design. New York: MacMillan.

Fisher, R.A. 1925. Statistical methods for research workers. Oliver and Boyd, Edinburgh, U.K.

Fisher, R.A. 1935. The design of experiments. Oliver and Boyd, Edinburgh, U.K.

Fisher, R.A. and W.A. Mackenzie. 1923. Studies in crop variation II: the manurial response of different potato varieties. J. Agric. Sci. 13:311-320.

Gbur, E.E., W.W. Stroup, K.S. McCarter, S. Durham, L.J. Young, M. Christman, M. West and M. Kramer. 2012. Generalized linear mixed models in the agricultural and natural resources sciences. American Society of Agronomy, Madison, WI.

Gezan, S.A., and M. Carvalho. 2018. Analysis of repeated measures for the biological and agricultural sciences. In: B. Glaz and K.M. Yeater, editors, Applied statistics in the agricultural, biological, and environmenta sciences. ASA, CSSA, SSSA, Madison, WI.

Graybill, F.A. 1976. Theory and application of the linear model. Duxbury, North Sciuate. MA.

Harville, D.A. 1976. Extensions of the Gauss-Markov theorem to include the estimation of random effects. Annals of Statistics 4: 384-395.

Harville, D.A. 1978. Maximum likelihood approaches to variance component estimation and to related problems. J American Statist Assoc 72: 320-338.

Henderson, C.R. 1953. Estimation of variance and covariance components. Biometrics 9: 226-252.

Henderson, C.R. 1963. Selection index and expected genetic advance. In: W.D. Hanson and H.F. Robinson, Statistical genetics and plant breeding. 141-163, National Academy of Science - National Research Council Publication 982.

Laird, N.M. and J.H. Ware, 1982. Random-effects models for longitudinal data. Biometrics 38: 963-973.

Lewis, S. 1925. Arrowsmith. Paperback printing 2008. Penguin, New York.

Liang, K.-Y. and S.L. Zeger. 1986. Longitudinal data analysis using generalized linear models. Biometrika. 73: 13-22.

Littell, R.C., G.A. Milliken, R.D. Wolfinger, W.W. Stroup and O. Schabenberger. 2006. SAS for Mixed Models, 2nd ed. SAS Institute, Cary, NC.

Milliken, G.A. and D.E. Johnson. 2009. Analysis of messy data, Vol. 1, 2nd Ed. Chapman and Hall, New York.

Nelder, J.A. and R.W.M. Wedderburn. 1972. Generalized linear models. J Royal Statist Soc A 135: 370-384.

S-189 Regional Project, Various Authors. 1989. Applications of mixed models in agriculture and related disciplines, Southern Cooperative Series Bulletin No.343. Louisiana Agricultural Experiment Station, Baton Rouge, LA.

Searle, S.R. 1971. Linear models. New York: Wiley.

Stroup, W.W. 2013a. Non-normal data in agricultural experiments. Proceedings of the 25th Annual Conference on Applied Statistics in Agriculture. Kansas State University, Manhattan, KS.

Stroup, W.W. 2013b. Generalized linear mixed models. Boca Raton, FL. CRC Press.

Vargas, M., B. Glaz, J. Crossa, and A. Morgounov. 2018. Analysis and interpretation of interactions of fixed and random effects. In: B. Glaz and K.M. Yeater, editors, Applied statistics in the agricultural, biological, and environmental sciences. ASA, CSSA, SSSA, Madison, WI.

Wolfinger, R.D. and M. O'Connell. 1993. Generalized linear mixed models: a pseudo-likelihood approach. J. Statist. Computation and Simulation. 4: 233-243.

Yates, F. 1935. Complex experiments. J Royal Statistical Society, Supplement, 2: 181-223.

Yates. F. 1940. The recovery of inter-block information in balanced incomplete block designs. Annals of Eugenics 10: 317-325.

APPENDIX A.1 AUTHOR SUPPLEMENTS

Table of Contents

Introduction: The Story of a Grumpy Ox A3
Barry Glaz

CHAPTER 1. Errors in Statistical Decision-Making A4
Kimberly Garland-Campbell

CHAPTER 2. Analysis of Variance and Hypothesis Testing A10
Marla S. McIntosh

CHAPTER 3. Blocking Principles for Biological Experiments A15
Michael D. Casler

CHAPTER 4. Power and Replication – Designing Powerful Experiments A26
Michael D. Casler

CHAPTER 5. Multiple Comparison Procedures: The Ins and Outs A32
David J. Saville

CHAPTER 6. Linear Regression Techniques A35
Christel Richter and Hans-Peter Piepho

CHAPTER 7. Analysis and Interpretation of Interactions of Fixed and Random Effects A52
Mateo Vargas, Barry Glaz, Jose Crossa, and Alex Morgounov

CHAPTER 8. The Analysis of Combined Experiments A80
Philip M. Dixon, Kenneth J. Moore, and Edzard van Santen

CHAPTER 9. Analysis of Covariance A83
Kevin S. McCarter

CHAPTER 10. Analysis of Repeated Measures for the Biological and Agricultural Sciences A88
Salvador A. Gezan and Melissa Carvalho

CHAPTER 11. The Design and Analysis of Long-term Rotation Experiments A92
Roger William Payne

CHAPTER 12. Spatial Analysis of Field Experiments A117
Juan Burgueño

CHAPTER 13. Augmented Designs—Experimental Designs in which all Treatments are Not Replicated A118
Juan Burgueño, Jose Crossa, and Kathleen M. Yeater

CHAPTER 14. Multivariate Methods for Agricultural Research A125
Kathleen M. Yeater and María B. Villamil

CHAPTER 15. Nonlinear Regression Models and Applications A128
Fernando Miguez, Sotirios Archontoulis, and Hamze Dokoohaki

CHAPTER 16. Analysis of Non-Gaussian Data A130
Walter W. Stroup

Online appendices are available at https://dl.sciencesocieties.org/publications/additional-appendices-app-stats.

INTRODUCTION: THE STORY OF A GRUMPY OX

Barry Glaz

Review Questions

1. Alpha and Beta rejected a H_o based on results with $p = 0.0498$ and did not reject when $p = 0.0501$. Were these good decisions?

a. Yes, we live and die by 0.05.

b. They should have rejected the H_o in both cases.

c. They should have accepted the H_o in both cases.

d. Had they also considered the effects of a Beta error, it is extremely likely that they would have either rejected or accepted the H_o in both cases.

Answer: a is technically correct in that we currently live and die by 0.05, but doing so is killing us. The best answer is d.

2. Rho did not understand statistics. You should ignore a significant interaction if at least one main effect is significant.

a. True

b. False

Answer: b. False. The grumpy ox had this one right.

3. Alpha and Beta's research on maximizing the morning snack of the oxen employees of Delta Oxlines should follow up with higher rates of coffee and donut.

a. True

b. False

Answer: a. True. Since oxcart-pulling distance increased at the highest rates of coffee and donut (1000 ml and 500 g, respectively), Alpha and Beta should conduct more research to find the rate at which the response to coffee and donut is maximized.

4. What is δαϱν in English?

a. The equation of a complex mixed model.

b. A Type 5 error.

c. The ox fraternity in The Wondrous Land.

d. darn.

Answer: d. darn.

CHAPTER 1: ERRORS IN STATISTICAL DECISION-MAKING

Kimberly Garland-Campbell

Software code

For hypothetical experiment, critical *F* values and beta values were calculated as below:

SAS

We created a dataset named ALPHA with two variables; the first variable is also named ALPHA and is a range of levels for Type 1 error from zero to 0.9999. The second variable is named FVALUE and is the *F* value associated with the effect that we would like to obtain (1.5, 4, and 9.285). The following code creates a new dataset named ERROR with the variables from the ALPHA dataset plus the following variables: PROB, NUMDF, DENDF, NONCENT, FCRIT, POWER, BETA, and AVEERROR, where PROB is the probability of alpha error, NUMDF is the numerator degrees of freedom for the effect, DENDF is the denominator degrees of freedom associated with the experimental error, NONCENT is the noncentrality parameter, FCRIT is the critical *F* value for tests of a significant difference in effects, POWER is the power of the test and BETA is 1-POWER or the Type 2 error associated with the test. AVEERROR is the average of the alphas and the beta errors for the various scenarios. The resulting ERROR dataset can be exported into a spreadsheet using Export Wizard in the File menu, or copied from the PROC PRINT statement. The ERROR dataset contains the data used in Table 1 and in Figures 1-3.

```
DATA ERROR;  SET ALPHA;
PROB=(1-ALPHA);
NUMDF=7;
DENDF=14;
NCP=NUMDF*FVALUE;
FCRIT=FINV(PROB, NUMDF, DENDF,0);
POWER=1-PROBF(FCRIT,NUMDF,DENDF,NCP);
BETA=1-POWER;
AVEERROR=(ALPHA+BETA)/2;
PROC PRINT DATA=ERROR;
RUN:
```

In R:

```
Error<-read.csv("alpha.csv")
Error$prob <- (1-(alpha))
Error$numdf<-7
Error$dendf<-14
Error<-transform(Error,ncp=(numdf*Fvalue))
Error<-transform(Error,fcrit=qf(alpha,numdf,dendf))
Error<-transform(Error,power=1-pf(fcrit,numdf,dendf,ncp))
Error<-transform(Error,beta=1-power)
Error<-transform(Error,aveterror=(alpha+beta)/2)
print(Error)
```

Yates Oat Experiment:

Data are available from the 'agridat' package in R as yates.oat.csv.

In this dataset, the x value is the columns and the y variable is the rows in the experimental design. The cultivar effect is named GEN and the manure effect is named NITRO.

```
SAS code using Proc Mixed:

PROC SORT DATA=YATES;
BY GEN NITRO BLOCK;
RUN;

PROC MIXED DATA=YATES METHOD=REML COVTEST PLOTS=ALL;
```

*This statement identifies the dataset, requests REML analysis and full residuals plots. The COVTEST statement requests tests of the covariance parameters;

```
TITLE 'YATES OAT SPLIT PLOT ANALYSIS';
```

* The title statement is an identifier and can be modified as needed;

```
CLASS GEN NITRO BLOCK;
```

*The class statement identifies the nitrogen and block variables as factors rather than nominal variables. In this analysis, nitrogen is considered to be a factor (categorical) variable, but it is conceivable that it would be a nominal (quantitative) variable in another type of analysis;

```
MODEL YIELD=GEN NITRO GEN*NITRO;
```

*The genotype and nitrogen treatments are fixed effects and replications are random so they are specified as such in the RANDOM statement;

```
RANDOM BLOCK BLOCK*GEN;
```

* The random statement specifies the two random effects for linear trend;

```
PROC MIXED DATA=YATES METHOD=REML COVTEST PLOTS=ALL;
TITLE 'YATES OAT SPLIT PLOT WITH LINEAR TREND';
CLASS GEN NITRO BLOCK X;
```

*The *X* variable is added to the classification statements to detect trends along the columns and is also added to the fixed effects below;

```
MODEL YIELD=GEN NITRO GEN*NITRO X;
RANDOM BLOCK BLOCK*GEN;
LSMEANS X GEN NITRO GEN*NITRO/ E CL;
RUN;
```

Similar models can be tested using R as described in the documentation for the agridat package: (agridat.pdf. pp. 318-319). A few additional pieces of code have been added to the code included in the agridat.pdf documentation for reasons described below. The code below is for the dataset available in the agridat package as well as for a new dataset containing the new randomization.

```
library(agridat)
require(lattice)
require(lme4)
require(lsmeans)
require(lucid)
```

#yield of oats in a split block experiment with four nitrogen levels, three oat cultivars, six blocks, total of 72 plots

import data from agridat package:

```
data("yates.oats")
dat<-yates.oats
```

#Plots the general layout of the data yields
#Includes the experimental design in the plots with the yield data.

```
desplot(yield ~ x*y, data=dat, outl=block, text=gen, col=nitro,
        cex=1, main="Yield Data for original Yates.oats Dataset")
```

#The plot shows that there is a linear gradient across the field.
Right-half of each block has lower yield.
Conduct split-plot analysis using lmer function in the 'lme4' package in R
#Have to note nitro as factor or the program will evaluate it with 1 df as a numerical variable

```
oatsp <- lmer(yield ~ factor(nitro) * gen + (1|block/gen), data=dat)
```

#print out all effects

```
summary(oatsp)
```

#print summary of fixed effects

```
anova(oatsp)
```

#print out fixed and random effects for model

```
fixef(oatsp)
ranef(oatsp)
```

#print out variance components for model

```
vc(oatsp)
```

#obtain lsmeans for the main effects and print them

```
lsmnit <- lsmeans(oatsp, "nitro")
lsmgen <- lsmeans(oatsp, "gen")
print(lsmnit)
print(lsmgen)
```

#plot residuals

```
qqmath(ranef(oatsp))
```

#Rerun model with a linear effect to correct for the linear trend.
Add a linear trend for column to the split-plot arrangement

```
oatsplin <- lmer(yield ~ x + factor(nitro) * gen + (1|block/gen),
data=dat)
summary(oatsplin)
anova(oatsplin)
fixef(oatsplin)
ranef(oatsplin)
```

#The residual variance is reduced

```
vc(oatsplin)
```

#The means do not change

```
lsmnitlin <- lsmeans(oatsplin, "nitro")
lsmgenlin <- lsmeans(oatsplin, "gen")
print(lsmnitlin)
print(lsmgenlin)
```

#The residuals plot has a better fit when the linear trend is accounted for

```
qqmath(ranef(oatsplin))
```

Review Questions, Answers:

True or False:

1. The central *F* distribution is calculated based on the numerator and error degrees of freedom.

TRUE. The *F* distribution depends on the ratio between the treatment or numerator and the error or denominator degrees of freedom.

2. Type 1 error should always be controlled below 5% whenever possible.

FALSE. While controlling α error below 0.05 or 5% is common, it is not always the most desirable option given the relative importance of α and β errors and the constraints to the experimental design.

3. The noncentrality parameter is associated with the effect size.

TRUE. The noncentrality parameter, λ, is determined by the effect size, the size of the experiment, and the unexplained experimental error.

4. Experiments should always be designed to obtain the minimum average error.

FALSE. This is a bit of a trick question. If money, space, time, and other similar constraints did not impact experimental design then this statement would be true, but because these all do impact our ability to conduct research, tradeoffs are required.

5. Effect sizes can be divided into those that measure differences between groups and those that measure association.

TRUE. While there are many statistics that can be used to measure effect size, this is a nice way to group them.

6. When spatial variation is discovered after the experiment is conducted, it will have to be included in the unexplained error for the experiment.

FALSE. Major trends can be detected and removed and other types of mixed model design that model covariance among experimental errors can be used to model spatial variation. But it is a good idea to anticipate spatial and temporal variation and block for it during the experimental design phase.

7. An experiment with a good deal of power will be associated with a lower probability of false positives.

FALSE. Actually, more power will have a lower probability of false negatives (β) and may or may not impact false positive error (α) depending on the experimental design.

8. The null hypothesis test is a valid approach to agronomic and environmental research.

TRUE. While criticized, the null hypothesis test still provides a good framework for decision making in agronomic and environmental research as long as the assumptions are met as described in this and other chapters in this book.

Solutions to Exercises:

Question 1:

In SAS:

```
PROC MIXED DATA=YATES METHOD=REML COVTEST PLOTS=ALL;
TITLE 'YATES OAT RCB ANALYSIS';
CLASS GEN NITRO BLOCK;
MODEL YIELD=GEN NITRO GEN*NITRO;
RANDOM BLOCK;
LSMEANS GEN NITRO GEN*NITRO/ E CL;
RUN;
```

In R:

```
oatrcb <- lmer(yield ~  factor(nitro) * gen + (1|block), data=dat)
summary(oatrcb)
anova(oatrcb)
```

```
fixef(oatrcb)
ranef(oatrcb)
vc(oatrcb)
```

#The means do not change

```
lsmnit <- lsmeans(oatrcb, "nitro")
lsmgen <- lsmeans(oatrcb, "gen")
print(lsmnit)
print(lsmgen)
qqmath(ranef(oatrcb))
```

Question 2:

In SAS:

```
PROC MIXED DATA=YATES METHOD=REML COVTEST PLOTS=ALL;
TITLE 'YATES OAT RCB ANALYSIS WITH LINEAR TREND';
CLASS BLOCK GEN NITRO X;
MODEL YIELD=X GEN NITRO GEN*NITRO;
RANDOM BLOCK;
LSMEANS X GEN NITRO GEN*NITRO/ E CL;
RUN;
```

In R:

```
oatrcblin <- lmer(yield ~ x + factor(nitro) * gen + (1|block/gen),
data=dat)
summary(oatrcblin)
anova(oatrcblin)
fixef(oatrcblin)
ranef(oatrcblin)
```

#The residual variance is reduced

```
vc(oatrcblin)
```

#The means do not change

```
lsmnitlin <- lsmeans(oatrcblin, "nitro")
lsmgenlin <- lsmeans(oatrcblin, "gen")
print(lsmnitlin)
print(lsmgenlin)
```

#The residuals plot has a better fit when the linear trend is accounted for

```
qqmath(ranef(oatrcblin))
```

QUESTION 3:

In SAS or R:

Same as above except replace *x* with *y*. (R is case sensitive so use lower case).

Question 4:

Notes:

Grand mean = 104.

25% of grand mean = 26.

Standard error of a difference can be calculated as $(2^*MSE/24)^{1/2}$ for Geno; $(2^*MSE/18)^{1/2}$ for Nitro and $(2^*MSE/6)^{1/2}$ for their interaction. The *t* value for each of these can be calculated as 26/SED for each effect and the *F* value for the contrast as t^2. The following code can then be used to figure out experimental design parameters that work for each effect. An example for geno in the RCB analysis is below for alpha=0.05.

IN SAS:

```
DATA ERROR;  SET ALPHA;
TITLE 'ALPHA ERROR FOR GEN';
MSE=254.2; *MSE is from mixed model analysis of dataset;
SED=SQRT((2*MSE)/24);
```

```
ALPHA=0.05;
FVALUE=(26/SED)*(26/SED);
PROB=(1-ALPHA);
NUMDF=2;
DENDF=60;
NCP=NUMDF*FVALUE;
FCRIT=FINV(PROB, NUMDF, DENDF,0);
POWER=1-PROBF(FCRIT,NUMDF,DENDF,NCP);
BETA=1-POWER;
AVEERROR=(ALPHA+BETA)/2;
PROC PRINT DATA=ERROR;
RUN;
```

In R:

```
Error<-read.csv("alpha.csv")
Error$alpha<-0.05
Error$prob <- (1-(alpha))
Error$numdf<-2
Error$dendf<-60
Error$sed<-(sqrt(2*254/24))
Error$t<-(26/Error$sed)
Error$Fvalue<-(Error$t)*(Error$t);
Error<-transform(Error,ncp=(numdf*Fvalue))
Error<-transform(Error,fcrit=qf(alpha,numdf,dendf))
Error<-transform(Error,power=1-pf(fcrit,numdf,dendf,ncp))
Error<-transform(Error,beta=1-power)
Error<-transform(Error,aveterror=(alpha+beta)/2)
str(Error)
```

CHAPTER 2: ANALYSIS OF VARIANCE AND HYPOTHESIS TESTING

Marla S. McIntosh

Supplement 1. Statbean Data.

Loc	Blk	Mulch	Ca_Trt	pH	Ca	Pct_Total	Trt_No
Central	1	0	0	5.58	765.604	5	1
Central	1	0	G1X	5.91	1072.395	5.93	2
Central	1	0	G2X	5.55	884.1182	13.33	3
Central	1	0	L1X	6.07	976.6795	7.86	4
Central	1	0	L2X	6.07	1030.769	12.86	5
Central	2	0	0	5.35	484.6031	29.23	1
Central	2	0	G1X	5.55	700.2342	22.5	2
Central	2	0	G2X	5.6	944.7611	24.44	3
Central	2	0	L1X	6.27	1651.486	15.38	4
Central	2	0	L2X	5.97	1159.154	24.29	5
Central	3	0	0	5.34	352.0018	28.7	1
Central	3	0	G1X	5.18	350.7162	34.55	2
Central	3	0	G2X	5.93	1041.485	25.83	3
Central	3	0	L1X	5.55	554.0077	22.61	4
Central	3	0	L2X	6.13	1397.836	18.33	5
West	1	0	0	5.89	1152.135	17.14	1
West	1	0	G1X	6.16	1730.585	10.71	2
West	1	0	G2X	5.66	1269.248	7.86	3
West	1	0	L1X	6.63	2309.693	25	4
West	1	0	L2X	6.52	1764.905	21.43	5
West	2	0	0	7.1	2673.196	28.89	1
West	2	0	G1X	6.89	2127.684	49.29	2
West	2	0	G2X	7.12	2806.581	40.83	3
West	2	0	L1X	7.24	2414.204	46.43	4
West	2	0	L2X	7.31	3146.425	23.57	5
West	3	0	0	6.76	2116.01	28.15	1
West	3	0	G1X	6.74	2522.883	22.96	2
West	3	0	G2X	6.77	1868.224	32.59	3
West	3	0	L1X	6.94	2513.69	30.37	4
West	3	0	L2X	6.97	2452.25	24.35	5
East	1	0	0	3.85	71.59624	0.77	1
East	1	0	G1X	3.93	84.90909	2.14	2
East	1	0	G2X	3.88	51.67797	0	3
East	1	0	L1X	4.12	94.65318	0.74	4
East	1	0	L2X	4.44	294.2328	2.96	5
East	2	0	0	4.21	153.2803	35.71	1
East	2	0	G1X	4.08	225.5879	16.43	2
East	2	0	G2X	4.18	242.5236	6.4	3
East	2	0	L1X	4.43	360.1202	38.57	4
East	2	0	L2X	4.76	465.0364	47.86	5

East	3	0	0	3.98	33.1985	0	1
East	3	0	G1X	4.09	79.34783	0	2
East	3	0	G2X	3.99	101.7181	0	3
East	3	0	L1X	3.92	22.76089	2.22	4
East	3	0	L2X	4.12	83.36722	1.6	5
Central	1	1	0	6.05	893.019	2.86	6
Central	1	1	G1X	5.81	930.887	4.29	7
Central	1	1	G2X	5.87	1203.349	9.29	8
Central	1	1	L1X	6.08	1084.533	6.67	9
Central	1	1	L2X	6.12	1450.82	9.29	10
Central	2	1	0	5.53	588.843	4.44	6
Central	2	1	G1X	5.69	823.8247	3.2	7
Central	2	1	G2X	5.57	860.6557	3.33	8
Central	2	1	L1X	5.7	863.8875	5.22	9
Central	2	1	L2X	5.78	817.8222	3.57	10
Central	3	1	0	5.47	646.8615	3.75	6
Central	3	1	G1X	6.09	1521.4	4.76	7
Central	3	1	G2X	5.08	548.4234	16.67	8
Central	3	1	L1X	6.08	1605.782	5.45	9
Central	3	1	L2X	6.29	1496.161	7.83	10
West	1	1	0	6.48	1799.348	21.54	6
West	1	1	G1X	6.42	1593.126	13.33	7
West	1	1	G2X	5.76	1171.701	14.62	8
West	1	1	L1X	6.58	2200.379	38.46	9
West	1	1	L2X	6.84	2782.864	24	10
West	2	1	0	7.02	2211.675	27.86	6
West	2	1	G1X	7.1	2497.6	51.43	7
West	2	1	G2X	7.15	2718.435	58.57	8
West	2	1	L1X	7.25	2562.135	61.43	9
West	2	1	L2X	7.28	2625.38	37.14	10
West	3	1	0	6.92	1977.072	40	6
West	3	1	G1X	6.67	1532.194	38.4	7
West	3	1	G2X	6.58	1954	19.05	8
West	3	1	L1X	6.93	2139.155	30	9
West	3	1	L2X	7.03	2470.402	41.6	10
East	1	1	0	4.03	12.2972	0	6
East	1	1	G1X	3.97	45.73493	0	7
East	1	1	G2X	3.92	97.56944	0	8
East	1	1	L1X	4.06	35.53188	1.48	9
East	1	1	L2X	4.12	56.58363	1.6	10
East	2	1	0	4.33	187.6108	15	6
East	2	1	G1X	4.21	305.4299	11.43	7
East	2	1	G2X	4.13	375.1605	17.14	8
East	2	1	L1X	4.67	416.7319	22.86	9
East	2	1	L2X	4.76	450	25	10
East	3	1	0	3.95	9.765314	0	6
East	3	1	G1X	3.91	173.9345	0	7
East	3	1	G2X	3.75	402.1609	1.6	8
East	3	1	L1X	4.21	123.3071	0.69	9
East	3	1	L2X	4.14	210.8767	1.48	10

Supplement 2. SAS code - PROC MIXED for pH - by location (with and without contrast statements) and combined over locations.

Note: The data used for this program is provided as an Excel file 'Statbean Data.xlsx' in Supplement 1 and as a SAS dataset 'statbean.sas7bdat' in Supplement 6.

Replace the Data statement to use the dataset in Supplement 1.

```
data anova.statbean; set anova.statbean;
run;
proc sort; by loc;
run;
Title 'Statbean Data';
proc print;
run;
Title 'Mixed pH ANOVA by location without contrasts';
proc mixed data=anova.statbean plots=residualpanel method=type3; by
loc;
class Blk Mulch Ca_Trt;
model pH=Mulch Ca_Trt Mulch*Ca_Trt;
random Blk;
lsmeans Mulch Ca_Trt Mulch*Ca_Trt;
run;
Title 'Mixed pH ANOVA combined locations';
proc mixed data=anova.statbean plots=residualpanel method=type3;
class Loc Blk Mulch Ca_Trt;
model pH=Loc|Mulch|Ca_Trt;
random Blk(Loc);
lsmeans Loc|Mulch|Ca_Trt;
run;
Title 'Mixed pH ANOVA by location with contrasts';
proc mixed data=anova.statbean plots=residualpanel method=type3; by
loc;
class Blk Mulch Ca_Trt;
model pH=Mulch Ca_Trt Mulch*Ca_Trt;
random Blk;
lsmeans Mulch Ca_Trt Mulch*Ca_Trt;
contrast "Main Effect-Gypsum Linear" Ca_Trt -1 0 1 0 0/E;
contrast "Main Effect-Gypsum Quadratic" Ca_Trt 1 -2 1 0 0/E;
contrast "Main Effect-Lime Linear" Ca_Trt -1 0 0 0 1/E;
contrast "Main Effect-Lime Quadratic" Ca_Trt 1 0 0 -2 1/E;
contrast "Main Effect-Lime vs Gypsum" Ca_Trt 0 1 1 -1 -1/E;
contrast "Interaction-Gypsum Linear*Mulch" Ca_Trt*Mulch -1 0 1 0 0 1 0 -1 0 0/E;
contrast "Interaction-Gypsum Quadratic* Mulch" Ca_Trt*Mulch 1 -2 1 0 0 -1 2 -1 0 0/E;
contrast "Interaction-Lime Linear*Mulch" Ca_Trt*Mulch -1 0 0 0 1 1 0 0 0 -1/E;
contrast "Interaction-Lime Quadratic*Mulch" Ca_Trt*Mulch 1 0 0 -2 1 -1 0 0 2 -1/E;
contrast "Interaction-Lime vs Gypsum*Mulch" Ca_Trt*Mulch 0 1 1 -1 -1 0 -1 -1 1 1/E;
run;
```

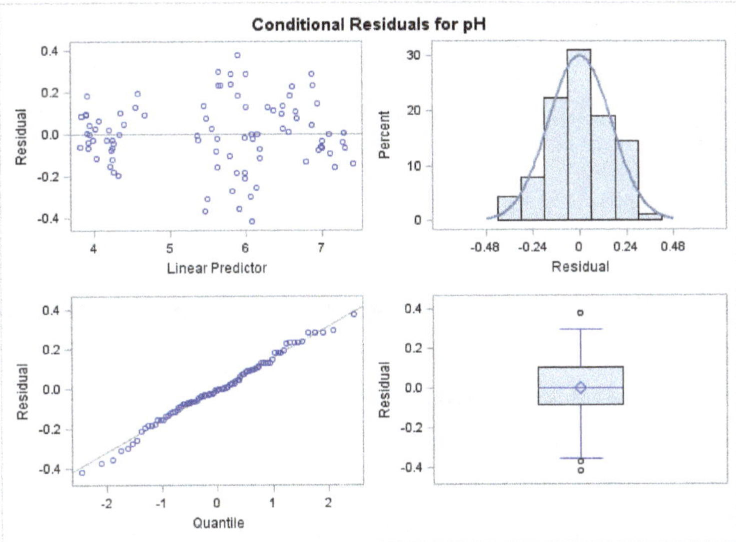

Supplement 3. Provided in the electronic supplemental materials.

Supplement 4. SAS code for ANOVA for pH, Ca, and Pct_Total

```
data anova.statbean; set anova.statbean;
proc sort; by loc;
run;
Title 'Statbean Data';
proc print;
run;
Title 'Mixed pH ANOVA by location';
proc mixed data=anova.statbean plots=residualpanel method=type3; by
loc;
class Blk Mulch Ca_Trt;
model pH=Mulch Ca_Trt Mulch*Ca_Trt;
random Blk;
lsmeans Mulch Ca_Trt Mulch*Ca_Trt;
run;
Title 'Mixed pH ANOVA combined locations';
proc mixed data=anova.statbean plots=residualpanel method=type3;
class Loc Blk Mulch Ca_Trt;
model pH=Loc|Mulch|Ca_Trt;
random Blk(Loc);
lsmeans Loc|Mulch|Ca_Trt;
run;
Title 'Mixed Calcium ANOVA by location';
proc mixed data=anova.statbean plots=residualpanel method=type3; by
loc;
class Blk Mulch Ca_Trt;
model Ca=Mulch Ca_Trt Mulch*Ca_Trt;
random Blk;
lsmeans Mulch Ca_Trt Mulch*Ca_Trt;
run;
Title 'Mixed Calcium ANOVA combined locations';
proc mixed data=anova.statbean plots=residualpanel method=type3;
class Loc Blk Mulch Ca_Trt;
model Ca=Loc|Mulch|Ca_Trt;
random Blk(Loc);
lsmeans Loc|Mulch|Ca_Trt;
run;
Title 'Mixed Percent total ANOVA by location';
proc mixed data=anova.statbean plots=residualpanel method=type3; by
loc;
class Blk Mulch Ca_Trt;
```

```
model Pct_total=Mulch Ca_Trt Mulch*Ca_Trt;
random Blk;
lsmeans Mulch Ca_Trt Mulch*Ca_Trt;
run;
Title 'Mixed Percent total ANOVA combined locations';
proc mixed data=anova.statbean plots=residualpanel method=type3;
class Loc Blk Mulch Ca_Trt;
model Pct_total=Loc|Mulch|Ca_Trt;
random Blk(Loc);
lsmeans Loc|Mulch|Ca_Trt;
run;
```

Supplement 5 provided in the electronic supplemental materials.

Answers to Review Questions

1. True
2. False
3. False. The F-value numerator is the treatment MS which includes both Var(Treatment) and Var(Residual). See the EMS.
4. False, the Type 1 error rate is a probability set by the researcher.
5. True

CHAPTER 3: BLOCKING PRINCIPLES FOR BIOLOGICAL EXPERIMENTS

Michael D. Casler

Example 1. Conduct a linear mixed model ANOVA from an augmented design.

Problem: Augmented designs are unbalanced, specifically with reference to test treatments that are typically unreplicated.

Solution: Residual or error variances must be estimated from replicated treatments, which should be arranged in a manner that also allows estimation and removal of some spatial variation within the experimental area. Estimates for unreplicated treatments are then adjusted for spatial variation.

Example: Twenty-one soybean cultivars were evaluated in an augmented design, with four cultivars arranged in three randomized complete blocks and the other 17 cultivars each represented in only one of the three blocks (Scott and Milliken, 1993). Each column of the data set below represents one block.

SAS Code: The following code gives an ANOVA with a separate *F* test for check cultivars and test cultivars. It also provides adjusted cultivar means and standard errors for the four check cultivars and the 17 test cultivars. Note that we are using a trick that will allow SAS Proc Mixed to compute a separate *p* value for check cultivars, which are replicated, and for test cultivars, which are not replicated. The trick is to recode the cultivar number into two sets of numbers. The first set, *c*, codes the checks and has $c = 0$ for all the test cultivars. The second set, *x*, codes the test cultivars and has $x = 0$ for the check cultivars. The ANOVA model is then set up with two terms: check cultivars and test cultivars nested within check cultivars.

```
options nocenter;
data a; input entryno entry$ y1 y2 y3; datalines;
1     Sibley      4098  4060  4283
1     Sibley        .     .   3952
2     Hardin      4020  4414  3571
3     Weber       4440  3835  4154
4     Kato        3860  3865  3674
5     TEgg        2169    .     .
6     Harlon      3250    .     .
7     Rampage     3807    .     .
8     Steele      4068    .     .
9     Vinton      3871    .     .
10    Vinton81    3838    .     .
11    BSR101        .   4244    .
12    Norsoy        .   3290    .
13    WBlack        .   3019    .
14    Mandarin      .   3506    .
15    Hark          .   4384    .
16    Coles         .   4148    .
17    Hodgson78     .     .   4167
18    Lakota        .     .   4023
19    Mandan507     .     .   2435
20    Bert          .     .   4595
```

```
21   Leslie       .     .    3957
;
data b; set a;
yield=y1; block=1; output;
yield=y2; block=2; output;
yield=y3; block=3; output;
drop y1-y3; run;
data c; set b;
x=entryno; if x<5 then x=0;
c=entryno; if c>4 then c=0;
proc mixed; class block x c;
model yield = c x(c);
random block;
lsmeans x(c);
run;
```

SAS Output: The output below contains the mixed models ANOVA and the least squares means for all 21 cultivars.

```
The Mixed Procedure

                  Model Information
Data Set                        WORK.C
Dependent Variable              yield
Covariance Structure            Variance Components
Estimation Method               REML
Residual Variance Method        Profile
Fixed Effects SE Method        Model-Based
Degrees of Freedom Method       Containment

             Class Level Information
Class     Levels     Values
block          3     1 2 3
x             18     0 5 6 7 8 9 10 11 12 13 14 15
                     16 17 18 19 20 21
c              5     0 1 2 3 4

            Dimensions
Covariance Parameters               2
Columns in X                       27
Columns in Z                        3
Subjects                            1
Max Obs per Subject                30
          Number of Observations
Number of Observations Read                 66
Number of Observations Used                 30
Number of Observations Not Used             36

                      Iteration History
Iteration     Evaluations     -2 Res Log Like       Criterion
        0               1        130.48103794
        1               1        130.48103794      0.0000000

                   Convergence criteria met.

The Mixed Procedure
Covariance Parameter

      Estimates
Cov Parm      Estimate
block                0
Residual         68856
           Fit Statistics
-2 Res Log Likelihood           130.5
AIC (Smaller is Better)          132.5
AICC (Smaller is Better)         133.1
```

BIC (Smaller is Better) 131.6

Type 3 Tests of Fixed Effects

Effect	Num DF	Den DF	F Value	Pr > F
c	4	7	3.61	0.0667
x(c)	16	7	6.50	0.0090

Least Squares Means

Effect	x	c	Estimate	Standard Error	DF	t Value	Pr > |t|
x(c)	5	0	2169.00	262.41	7	8.27	< 0.0001
x(c)	6	0	3250.00	262.41	7	12.39	< 0.0001
x(c)	7	0	3807.00	262.41	7	14.51	< 0.0001
x(c)	8	0	4068.00	262.41	7	15.50	< 0.0001
x(c)	9	0	3871.00	262.41	7	14.75	< 0.0001
x(c)	10	0	3838.00	262.41	7	14.63	< 0.0001
x(c)	11	0	4244.00	262.41	7	16.17	< 0.0001
x(c)	12	0	3290.00	262.41	7	12.54	< 0.0001
x(c)	13	0	3019.00	262.41	7	11.51	< 0.0001
x(c)	14	0	3506.00	262.41	7	13.36	< 0.0001
x(c)	15	0	4384.00	262.41	7	16.71	< 0.0001
x(c)	16	0	4148.00	262.41	7	15.81	< 0.0001
x(c)	17	0	4167.00	262.41	7	15.88	< 0.0001
x(c)	18	0	4023.00	262.41	7	15.33	< 0.0001
x(c)	19	0	2435.00	262.41	7	9.28	< 0.0001
x(c)	20	0	4595.00	262.41	7	17.51	< 0.0001
x(c)	0	1	4098.25	131.20	7	31.24	< 0.0001
x(c)	0	2	4001.67	151.50	7	26.41	< 0.0001
x(c)	0	3	4143.00	151.50	7	27.35	< 0.0001
x(c)	0	4	3799.67	151.50	7	25.08	< 0.0001

Results and Conclusions: Check cultivars differed from each other with a p value of 0.07, while the test cultivars had $p < 0.01$. The check cultivars have 4 df in the numerator because this term is testing differences among five means: Sibley, Hardin, Weber, Cato, and the mean of all 17 test cultivars. Least squares means allow the researcher to choose the best test cultivars for further, more advanced, testing. Least squares means are adjusted for block effects, but not for spatial variation on a finer scale. Note that there are three standard error values, each one corresponding to $r = 1$, $r = 4$, or $r = 5$ experimental units per cultivar (262, 151, and 131, respectively).

Example 2. Make a logical and objective decision regarding whether or not random design effects should be retained in the final model for publication purposes.

Problem: Modern mixed models analysis is often taught in a manner that encourages researchers to used reduced models, containing only those terms that are important. This practice results in pooling random design effects with residual effects. There are three philosophies that can be employed in pooling when it is clear that a design component is small or nonsignificant: always pool, never pool, or pool using an objective decision tool that seeks to avoid Type 2 errors.

Solution: The example below will illustrate how to employ a likelihood ratio test to quantify a p-value for a random design component, then set a decision rule for "calling" that term significant or nonsignificant, with the final result to either include or exclude that term form the model. The methodology is based on the concepts and philosophy of Carmer et al. (1969) but using modern likelihood ratio tests, rather than F tests.

Example: The data are percentage survivorship of 14 Italian ryegrass (*Lolium multiflorum* Lam.) cultivars planted in factorial combination with three seeding rates (200,

400, and 800 seeds m^{-2}). The experiment was designed as a randomized complete block with four replicates and randomized with seeding rates as whole plots and cultivars as subplots.

SAS Code: There are two blocks of data in the SAS code below. The first consists of the field map in 24 rows × 7 columns: rows are identified by row number, rep number, and seeding rate, while the data in each column are the cultivar numbers (1 through 14). The second block of data consists of percentage survivorship for the 24 × 7 grid. The two blocks of data are merged together by row number, prior to conducting the analyses of variance.

```
options nocenter;
data a; input row rep rate x1-x7; datalines;
 1      4     40      1      11      13      9      4      7     10
 2      4     40      5       8       2      6      3     12     14
 3      4     80      3      11       6      1     10      7      8
 4      4     80     14       9       4     12      5      2     13
 5      4     20      4       8       2     11      1     10     12
 6      4     20      3       9      14      7      6     13      5
 7      3     80      4       5       2      3     10      6      1
 8      3     80      8      12       7     14      9     11     13
 9      3     40      5       7       9      8      4     14     11
10      3     40      1       2       3     10     13     12      6
11      3     20     11       5       4      3     13      7      1
12      3     20      6      14       9      8      2     10     12
13      2     40     13       9       2     12      6      8      5
14      2     40     11      14      10      3      4      1      7
15      2     20     12      13       7      1     11     10      4
16      2     20      6      14       5      9      3      8      2
17      2     80     14       1       9     12      2      3      6
18      2     80      4      11       8      5      7     10     13
19      1     80      5       2       8      9      3     14     11
20      1     80     12       1      13     10      7      4      6
21      1     20     11       1       8     13     14      7     10
22      1     20      5       3       6      4      9      2     12
23      1     40     10       1      12     11      4      8      2
24      1     40      6       5      14     13      3      7      9
;
data aa; set a;
cultivar=x1; col=1; output;
cultivar=x2; col=2; output;
cultivar=x3; col=3; output;
cultivar=x4; col=4; output;
cultivar=x5; col=5; output;
cultivar=x6; col=6; output;
cultivar=x7; col=7; output;
drop x1-x7;
proc sort; by row col; run;
data b; input row x1-x7; datalines;
 1     50      5      5      5      5      5      5
 2      5     50     40      5      5      5     10
 3      5      5      5     40      5     10     25
 4     50      5     60      5     20     45      5
 5      5     20      5      5     20      5      5
 6      5      5     10      5      5      5      5
 7     20      5     70      5      5     10     50
 8     40      5      5     60     10     15      5
 9      5     10      5     80     50     50      5
10     30     80      5      5     10      5      5
11      5      5     20      5      5     10     40
12      5      5      5     40     30      5      5
13      5      5     40      5      5     30      5
14      5     15      5      5     50     30     20
15      5      5      5     15      5      5     35
```

```
16     5    15     5     5     5    25    50
17    20    75    10     5    60     5     5
18    30     5    60    10    30     5     5
19     5    80    75     5     5    30    15
20     5    60    15     5    40    55    10
21     5    10    20     5     5     5     5
22     5     5     5    10     5    40     5
23     5    25     5     5    40    30    65
24     5     5    40     5     5     5     5
;
data bb; set b;
gc95=x1; col=1; output;
gc95=x2; col=2; output;
gc95=x3; col=3; output;
gc95=x4; col=4; output;
gc95=x5; col=5; output;
gc95=x6; col=6; output;
gc95=x7; col=7; output;
drop x1-x7;
proc sort; by row col; run;
data c; merge aa bb; by row col;
libname arfs '~/arfs';
proc mixed cl; class rep rate cultivar;
model gc95 = rate|cultivar;
random rep rep*rate;
lsmeans cultivar;
run;
proc mixed cl; class rep rate cultivar;
model gc95 = rate|cultivar;
random rep;
lsmeans cultivar;
run;
```

SAS Output: The output consists of mixed models analysis results and least squares means for cultivars for two different models. According to the model, least squares means for seeding rates and the cultivar × seeding rate interaction would also be important output, but have been left out here for brevity. The first model is the full model that includes three random effects: Blocks, Error(a), and Error(b) of the split-plot randomization. The second model is based on the visual observation that Error(a) is very small any may not actually be significant. The second model is identical to the first, except that it excludes Error(a), collapsing that term into Error(b), effectively treating this analysis as a simple randomized complete block without the split-plot randomization restriction.

```
The Mixed Procedure

                  Model Information
Data Set                        WORK.C
Dependent Variable              gc95
Covariance Structure            Variance Components
Estimation Method               REML
Residual Variance Method        Profile
Fixed Effects SE Method         Model-Based
Degrees of Freedom Method       Containment

              Class Level Information
Class         Levels    Values
rep                4    1 2 3 4
rate               3    20 40 80
cultivar          14    1 2 3 4 5 6 7 8 9 10 11 12 13
                        14
```

```
              Dimensions
Covariance Parameters               3
Columns in X                       60
Columns in Z                       16
Subjects                            1
Max Obs per Subject               168

           Number of Observations
Number of Observations Read                168
Number of Observations Used                168
Number of Observations Not Used              0

                      Iteration History
Iteration    Evaluations    -2 Res Log Like       Criterion
        0              1      1013.63658070
        1              1      1011.54388284      0.00000000
                    Convergence criteria met.

            Covariance Parameter Estimates
Cov Parm    Estimate   Alpha     Lower     Upper
rep          2.2003     0.05    0.2619   1.377E10
rep*rate     4.3932     0.05    0.7764   33250 <--- This is Error(a)
Residual     108.39     0.05   85.2069    142.56

             Fit Statistics
-2 Res Log Likelihood            1011.5
AIC (Smaller is Better)          1017.5
AICC (Smaller is Better)         1017.7
BIC (Smaller is Better)          1015.7

           Type 3 Tests of Fixed Effects
                     Num      Den
Effect                DF       DF     F Value     Pr > F
rate                   2        6       13.53     0.0060
cultivar              13      117       29.20     < 0.0001
rate*cultivar         26      117        1.96     0.0082

                             Least Squares Means
                                         Standard
Effect     cultivar    Estimate         Error       DF     t Value     Pr > |t|
cultivar      1        37.0833         3.1541      117      11.76      < 0.0001
cultivar      2        50.4167         3.1541      117      15.98      < 0.0001
cultivar      3         5.0000         3.1541      117      1.59        0.115
cultivar      4        31.6667         3.1541      117      10.04      < 0.0001
cultivar      5        6.6667         3.1541       117       2.11      0.0367
cultivar      6        5.8333         3.1541       117       1.85     0.0669
cultivar      7        12.5000         3.1541      117       3.96      0.0001
cultivar      8        41.2500         3.1541      117      13.08      < 0.0001
cultivar      9        5.8333          3.1541      117       1.85     0.0669
cultivar     10        5.0000          3.1541      117       1.59      0.1156
cultivar     11        6.6667          3.1541      117       2.11      0.0367
cultivar     12        5.0000          3.1541      117       1.59      0.1156
cultivar     13        6.2500          3.1541      117       1.98     0.0499
cultivar     14        25.8333          3.1541      117       8.19    < 0.0001

The Mixed Procedure
                  Model Information
Data Set                       WORK.C
Dependent Variable             gc95
Covariance Structure           Variance Components
Estimation Method              REML
Residual Variance Method       Profile
Fixed Effects SE Method       Model-Based
Degrees of Freedom Method      Containment

              Class Level Information
Class        Levels    Values
```

```
rep                 4     1 2 3 4
rate                3     20 40 80
cultivar           14     1 2 3 4 5 6 7 8 9 10 11 12 13
                          14

            Dimensions
Covariance Parameters                  2
Columns in X                          60
Columns in Z                           4
Subjects                               1
Max Obs per Subject                  168

          Number of Observations

Number of Observations Read                  168
Number of Observations Used                  168
Number of Observations Not Used                0

                        Iteration History
Iteration    Evaluations     -2 Res Log Like        Criterion
        0              1       1013.63658070
        1              1       1012.20561338       0.00000000
                      Convergence criteria met.

              Covariance Parameter Estimates
Cov Parm       Estimate       Alpha         Lower         Upper
rep              3.5932        0.05        0.7115       3947.53
Residual         111.39        0.05       88.0562        145.45

             Fit Statistics
-2 Res Log Likelihood              1012.2
AIC (Smaller is Better)             1016.2
AICC (Smaller is Better)            1016.3
BIC (Smaller is Better)             1015.0

          Type 3 Tests of Fixed Effects
                     Num      Den
Effect               DF       DF     F Value      Pr > F
rate                  2      123       20.64     < 0.0001
cultivar             13      123       28.41     < 0.0001
rate*cultivar        26      123        1.91     0.0103

                              Least Squares Means
                                        Standard
Effect       cultivar      Estimate       Error       DF      t Value        Pr > |t|
cultivar       1          37.0833        3.1907      123      11.62         < 0.0001
cultivar       2          50.4167        3.1907      123      15.80         < 0.0001
cultivar       3           5.0000        3.1907      123       1.57         0.1197
cultivar       4          31.6667        3.1907      123       9.92         < 0.0001
cultivar       5           6.6667        3.1907      123       2.09         0.0387
cultivar       6           5.8333        3.1907      123       1.83        0.0699
cultivar       7          12.5000        3.1907      123       3.92         0.0001
cultivar       8          41.2500        3.1907      123      12.93         < 0.0001
cultivar       9           5.8333        3.1907      123       1.83        0.0699
cultivar      10           5.0000        3.1907      123       1.57         0.1197
cultivar      11           6.6667        3.1907      123       2.09         0.0387
cultivar      12           5.0000        3.1907      123       1.57         0.1197
cultivar      13           6.2500        3.1907      123       1.96         0.0524
cultivar      14          25.8333        3.1907      123       8.10         < 0.0001
```

Results and Conclusions: The first step in interpreting the results is to conduct a likelihood ratio test of Error(a), in other words, testing the null hypothesis Ho:σ^2_a=0, where σ^2_a is the Error(a) covariance component (estimated as 4.39 in the first mixed models ANOVA). The formula for this test is to compute the difference between the –2RLL (residual log likelihood) values and divide them by the difference in the number of covariance parameters. This becomes: (1012.2 – 1011.5)/(3 – 2) = 0.7. This value is tested as a χ^2 variate with 3 – 2 = 1 df, resulting in $p = 0.40$. Using the Carmer et al. (1969) guideline, that the null hypothesis should not be rejected unless $p < 0.50$, we would *not* reject this null hypothesis. In other words, we would *not* reduce the model; we would use the first mixed models analysis above as our final result, assuming all other diagnostics and results are correct. Note the change in denominator df between the two analyses, from 117 to 123—this is correctly indicating that the second estimate of residual variance is the pooled Error(a) and Error(b) value with 117 + 6 = 123 df. Also, note that the standard error of a cultivar mean actually went up after pooling, from 3.15 to 3.19—this is another sign that pooling, or model reduction, is a bad idea in this case.

Example 3. Review Exercise. Arranging blocks and blocking patterns for future experiments.

Problem: Decisions on exactly how to arrange blocks in many field experiments are very difficult, because there is often little information available to determine if there are gradients and the patterns of any gradients. This is especially true on agricultural experiment stations, which are often located on sites that have a uniform visual appearance.

Solution: Conduct retrospective analyses of previous experiments on a given site to determine the size, scale, and direction of spatial variation.

Example: Step 1. Consider a randomized complete block experiment that has been completed on a given site. In this example, the experiment was designed with 12 rows and 6 columns. Conduct an ANOVA on the data and output the residuals from the ANOVA. The blocking pattern of the previous experiment is not important, because the residuals are adjusted for both block and treatment effects, representing only pure spatial variation present in the field, at the scale of the experimental unit. In this case, we added the grand mean back to each residual, so that the data appears in the original units of measurement (Mg ha^{-1} of plant biomass). Step 2. Create dummy variables that can be used to simulate different blocking patterns on this site. Use each dummy variable to conduct a simulated one-way ANOVA with two sources of variation: among dummy blocks and within dummy blocks. Choose a design pattern with a high *F* ratio for dummy blocks (low within-block variance).

SAS Code: The following data are residuals + grand mean from a field experiment conducted in 6 columns × 12 rows, so that treatment and block variation has already been removed. The SAS code creates several combinations of rows and columns, after which it computes a one-way ANOVA for 22 different blocking arrangements and sizes. Comparison of the residual variances across the different blocking arrangements and sizes can help a researcher better understand the scale and dimensions of spatial variation within a field and plan better blocking schemes for future experiments.

```
options nocenter;
data a; input row y1 y2 y3 y4 y5 y6;
datalines;
1        7.85      7.86      9.13      9.63     10.52     10.24
2       10.08     10.78     15.18     11.66     12.57     13.60
3       14.41     10.19     16.11     13.53     12.66     14.37
```

```
4       14.18     9.77    13.85    12.99    13.09    13.38
5        9.87     6.24     7.27     9.68    13.91     9.62
6       11.63    14.00    15.34    10.99    13.04    13.21
7       11.39    12.33    18.57    18.65    12.79    12.10
8       19.20    13.75    11.47     9.43    17.94     9.26
9       19.03    13.59    11.52    14.09    16.19    12.21
10      11.31     5.52    11.40     9.71    13.51    10.65
11       8.23    13.27    14.51    13.16    17.15    16.81
12      12.59    11.27    11.44    13.58    15.69    13.87
;
data b; set a;
yield=y1; col=1; output;
yield=y2; col=2; output;
yield=y3; col=3; output;
yield=y4; col=4; output;
yield=y5; col=5; output;
yield=y6; col=6; output;
drop y1-y6; run;
data c; set b;
r1=row;
r2=int((row+1)/2);
r3=int((row+2)/3);
r4=int((row+3)/4);
r6=int((row+5)/6);
c1=col;
c2=int((col+1)/2);
c3=int((col+2)/3);
proc glm; class r1 c1; model yield = r1*c1;
proc mixed; class r2 c1; model yield = r2*c1;
proc mixed; class r3 c1; model yield = r3*c1;
proc mixed; class r4 c1; model yield = r4*c1;
proc mixed; class r6 c1; model yield = r6*c1;
proc mixed; class c1; model yield = c1;
proc mixed; class r1 c2; model yield = r1*c2;
proc mixed; class r2 c2; model yield = r2*c2;
proc mixed; class r3 c2; model yield = r3*c2;
proc mixed; class r4 c2; model yield = r4*c2;
proc mixed; class r6 c2; model yield = r6*c2;
proc mixed; class c2; model yield = c2;
proc mixed; class r1 c3; model yield = r1*c3;
proc mixed; class r2 c3; model yield = r2*c3;
proc mixed; class r3 c3; model yield = r3*c3;
proc mixed; class r4 c3; model yield = r4*c3;
proc mixed; class r6 c3; model yield = r6*c3;
proc mixed; class c3; model yield = c3;
proc mixed; class r1; model yield = r1;
proc mixed; class r2; model yield = r2;
proc mixed; class r3; model yield = r3;
proc mixed; class r4; model yield = r4;
proc mixed; class r6; model yield = r6; run;
```

SAS Output: The output from this SAS code results in one GLM ANOVA, which provides the variance among the raw yield values. This value of 8.63 is the estimated residual variance expected from the use of a completely randomized design on this site. The remaining output consists of the results from 22 mixed models ANOVAs, only the first of which is shown below. The critical item here is the Residual Covariance Parameter estimate of 9.0313. The remaining output shown below consists of the "Residual" line from all 22 Proc Mixed ANOVAs. Each of the 22 Proc Mixed ANOVAs provides a residual variance that would be expected for a different blocking design. For example, the first one corresponds to blocks that contain two experimental units in an arrangement of 2 rows × 1 column, while the last one corresponds to blocks that contain 36 experimental units in an arrangement of 6 rows × 6 columns.

```
The GLM Procedure
              Class Level Information

Class         Levels    Values

r1                12    1 2 3 4 5 6 7 8 9 10 11 12
c1                 6    1 2 3 4 5 6

Number of Observations Read          72
Number of Observations Used          72

The GLM Procedure
Dependent Variable: yield

                                 Sum of
Source               DF          Squares     Mean Square    F Value    Pr > F

Model                71      612.7471875       8.6302421          .         .
Error                 0        0.0000000               .
Corrected Total      71      612.7471875

R-Square     Coeff Var      Root MSE    yield Mean
1.000000             .             .      12.49458

Source               DF        Type I SS     Mean Square    F Value    Pr > F
r1*c1                71      612.7471875       8.6302421          .         .

The Mixed Procedure
                          Model Information
Data Set                     WORK.C
Dependent Variable           yield
Covariance Structure         Diagonal
Estimation Method            REML
Residual Variance Method     Profile
Fixed Effects SE Method     Model-Based
Degrees of Freedom Method    Residual

             Class Level Information
Class    Levels    Values
r2            6    1 2 3 4 5 6
c1            6    1 2 3 4 5 6

            Dimensions
Covariance Parameters           1
Columns in X                   37
Columns in Z                    0
Subjects                        1
Max Obs per Subject            72

          Number of Observations
Number of Observations Read              72
Number of Observations Used              72
Number of Observations Not Used           0

Covariance Parameter
      Estimates
Cov Parm      Estimate
Residual        9.0313
            Fit Statistics
-2 Res Log Likelihood          206.3
AIC (Smaller is Better)        208.3
AICC (Smaller is Better)       208.5
BIC (Smaller is Better)        209.9

         Type 3 Tests of Fixed Effects
               Num      Den
```

Effect	DF	DF	F Value	Pr > F
r2*c1	35	36	0.91	0.6093

Residual	9.0313
Residual	8.0655
Residual	9.0267
Residual	8.0114
Residual	8.1919
Residual	5.1116
Residual	8.0005
Residual	7.5977
Residual	8.6949
Residual	8.0895
Residual	8.3881
Residual	6.6952
Residual	7.89
Residual	7.8557
Residual	8.7369
Residual	8.1976
Residual	8.5585
Residual	6.5478
Residual	7.8181
Residual	8.0338
Residual	8.7223
Residual	8.1588

Results and Conclusions: The best way to visualize these results is to organize them into a graph. The residual variances above were matched up with two additional columns of data, the number of rows within the blocks and the number of columns within the blocks. The graph below illustrates the relationships. Clearly, the best three blocking scenarios were based on a single row across 2, 3, or 6 columns. All scenarios with multiple rows were relatively inefficient, indicating that most of the spatial variation was in the direction of rows. These results indicate that future experiments in this field should make all attempts to block out the variability associated with rows.

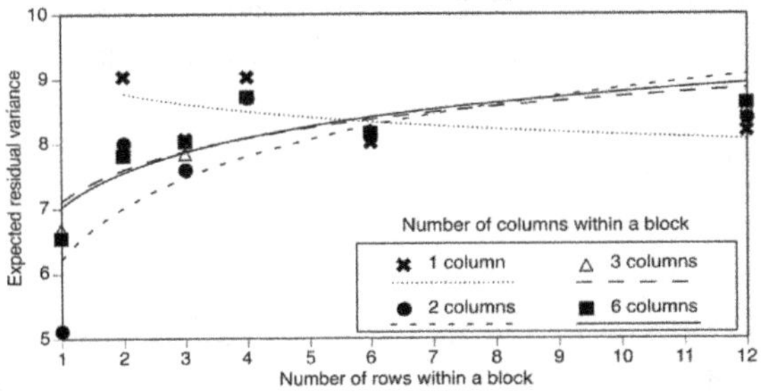

CHAPTER 4: POWER AND REPLICATION—DESIGNING POWERFUL EXPERIMENTS

Michael D. Casler

EXERCISE #1.

Predict the power of a future hypothetical experiment using the probability distribution method.

Solution

Using estimates of residual variances and covariance parameter estimates of other random factors estimated from previous experiments, it is possible to predict the power of future experiments under a wide range of design scenarios. From these predictions, it is then possible to make an intelligent assessment and comparison of different designs and choose an optimal design that balances statistical power with financial cost.

Example

Consider a proposed completely randomized design in which treatments are replicated as shown in Figure 1C and, additionally, multiple sampling (observational) units are created within each experimental unit, from which one data point is collected on each observational unit. This could equally apply to experiments in the field, glasshouse, laboratory, or benchtop. Replicates are nested within treatments and sampling units are nested within experimental units. The goal is to detect a difference between treatment means of 5% at a Type 1 error rate of $\alpha = 0.05$. Prior estimates of experimental error (5) and sampling error (10) are available (covariance parameter estimates of random effects from previous experiments); these values are defined in the "parms" statement.

SAS code

The SAS code below computes the expected power for the design described above with $r = 4$ replicates per treatment and $s = 2$ sampling units per experimental unit. It first creates the representative data set with treatment means of 95 and 100, following which Proc GLIMMIX is used to compute the F-ratio within the representative data set. The final set of computations generates the non-centrality parameter, assuming a normal distribution, which then leads to the power computation. Any of the input parameters can be changed to allow investigation and comparison of different designs.

```
options nocenter;
data a; input trt y;
*CREATE A REPRESENTATIVE DATA SET WITH 4 REPS AND 2 SAMPLES PER EXPTL UNIT;
do rep=1 to 4 by 1;
do samples=1 to 2 by 1;
output;
```

```
end;
end;
*CREATE TREATMENT MEANS WITH THE DESIRED DETECTION VALUE;
datalines;
1 95
2 100
run;
*COMPUTE THE NON-CENTRALITY PARAMETER;
data b; set a;
proc glimmix; class trt rep;
model y = trt;
random rep(trt);
*INPUT PARAMETER ESTIMATES FOR EXPTL AND SAMPLING ERRORS;
parms (5)(10) / hold=1,2;
ods output tests3=power_terms;
*COMPUTE POWER OF THE TEST;
data power; set power_terms;
alpha=0.05;
ncparm=numdf*Fvalue;
F_critical=finv(1-alpha, numdf, dendf, 0);
power=1-probf(F_critical, numdf, dendf, ncparm);
proc print;
run;
```

SAS output

The output consists of one run of Proc GLIMMIX, including all relevant diagnostic and estimation information. The last line is the result of the "proc print" statement, printing out the results of the computations made after obtaining the GLIMMIX output.

```
The GLIMMIX Procedure

                    Model Information
Data Set                        WORK.B
Response Variable               y
Response Distribution           Gaussian
Link Function                   Identity
Variance Function               Default
Variance Matrix                 Not blocked
Estimation Technique            Restricted Maximum Likelihood
Degrees of Freedom Method       Containment
  Class Level Information
Class     Levels    Values
trt            2    1 2
rep            4    1 2 3 4
Number of Observations Read            16
Number of Observations Used            16
          Dimensions
G-side Cov. Parameters          1
R-side Cov. Parameters          1
Columns in X                    3
Columns in Z                    8
Subjects (Blocks in V)          1
Max Obs per Subject            16

          Parameter Search

                              Objective
   CovP1        CovP2          Function

  5.0000      10.0000      66.284236398

           Optimization Information

Optimization Technique         Dual Quasi-Newton
Parameters in Optimization     2
Equality Constraints           2
Lower Boundaries               2
Upper Boundaries               2
Fixed Effects                 Profiled
Starting From                  Data
```

```
                                Iteration History
                                      Objective                      Max
Iteration   Restarts   Evaluations     Function          Change  Gradient
    0           0           4         66.284236398                   0

      Convergence criterion (ABSGCONV=0.00001) satisfied.

             Fit Statistics
-2 Res Log Likelihood              66.28
AIC  (smaller is better)           66.28
AICC (smaller is better)           66.28
BIC  (smaller is better)           66.28
CAIC (smaller is better)           66.28
HQIC (smaller is better)           66.28
Generalized Chi-Square              0.00
Gener. Chi-Square / DF              0.00

 Covariance Parameter Estimates
                        Standard
Cov Parm    Estimate       Error
rep(trt)      5.0000           .
Residual     10.0000           .

        Type III Tests of Fixed Effects
              Num       Den
Effect        DF        DF     F Value     Pr > F
trt             1         6       5.00     0.0067

                  Num
Obs  Effect  DF  DenDF  FValue  ProbF   alpha  ncparm  F_critical  power
1    trt     1   6      5.00    0.0667  0.05   5       5.98738     0.46741
```

Results and Conclusions

The predicted power for this future hypothetical design is 0.47. Increasing the number of replicates or samples would increase the predicted power, a result that can easily be investigated by repeated runs of this code, simply by changing the values in the two "do" statements.

EXERCISE #2.

Solution

Expanding on Example #1, we broaden the SAS code to a number of different designs using a SAS macro.

Example

Consider the proposed completely randomized design in which treatments are replicated as shown in Figure 1C and, additionally, multiple sampling (observational) units are created within each experimental unit, from which one data point is collected on each observational unit (Example #1). Replicates are nested within treatments and sampling units are nested within experimental units. The goal remains to detect a difference between treatment means of 5% at a Type 1 error rate of $\alpha = 0.05$. Prior estimates of experimental error (5) and sampling error (10) are available. We wish to predict the power for six different designs with $r = 4$, 5, or 6 replicates and $s = 2$ or 3 sampling units per replicate.

SAS code

The SAS code below computes the expected power for the design described above for all six variations of the numbers of replicates per treatment and sampling units per experimental unit. It first creates a macro titled "one". The macro then creates two new variables: "obsv" is the number of sampling units per experimental unit and "repl" is the number of replicates. These two values are allowed to vary with maxima of "obsmax" and "repmax", respectively. The line that reads "%one(3,6);" is the place to set the upper limits for the investigation. In this example, we have chosen to vary the number of replicates from 4 to 6 and the number of sampling units from 2 to 3. The remainder of the code is identical to that in Example #1.

```
options nocenter;
%macro one(obsmax,repmax);
data a;
%do obsv=2 %to &obsmax;
group1=&obsv;
%do repl=4 %to &repmax;
group2=&repl;
do obs=1 to &obsv by 1;
do rep=1 to &repl by 1;
do trt=0 to 1 by 1;
output; end; end; end;
%end; %end;
%mend one;
%one(3,6);              /*    <--- change values here     */
run;
proc sort; by group1 group2;
data b; set a; by group1 group2;
if trt=0 then y=95;
if trt=1 then y=100;
run;
proc glimmix; class trt rep; by group1 group2;
model y = trt;
random rep(trt);
parms (5)(10) / hold=1,2;
ods output tests3=power_terms;
data power; set power_terms;
alpha=0.05;
ncparm=numdf*Fvalue;
F_critical=finv(1-alpha, numdf, dendf, 0);
power=1-probf(F_critical, numdf, dendf, ncparm);
proc print; run;
```

SAS output

The output below is abbreviated by eliminating all the proc glimmix output, which is necessary only for the purpose of checking the SAS run for errors. The output below is the result of the "proc print" statement at the end of the SAS code, printing all the calculated parameter estimates for the six design scenarios, shown under the headings "group1" (number of sampling units) and "group2" (number of replicates).

Obs	group1	group2	Effect	NumDF	DenDF	FValue	ProbF	alpha	ncparm	F_critical	power
1	2	4	trt	1	6	5.00	0.0667	0.05	5.00	5.98738	0.46741
2	2	5	trt	1	8	6.25	0.0369	0.05	6.25	5.31766	0.59308
3	2	6	trt	1	10	7.50	0.0209	0.05	7.50	4.96460	0.69494
4	3	4	trt	1	6	6.00	0.0498	0.05	6.00	5.98738	0.53734
5	3	5	trt	1	8	7.50	0.0255	0.05	7.50	5.31766	0.67085
6	3	6	trt	1	10	9.00	0.0133	0.05	9.00	4.96460	0.77140

Results and Conclusions

The output allows direct comparisons of designs that are both statistically and logistically efficient. It shows that there are multiple ways to achieve an expected level of power, e.g. $r = 6$ replicates and $s = 2$ sampling units is roughly equivalent to $r = 5$ replicates and $s = 3$ sampling units. The results can be expanded to a wider range of values and used to graphically display the design comparisons as shown in Figure 2, in which $r = 3$ to 20 and $s = 3$ to 20.

EXERCISE #3.

Solution:

The exercise is similar to Exercise #2, but the design differs, providing another illustration for conducting power analyses.

Example

The experiments described in Casler (1998; 2013) were used to obtain the following estimates of random factors: blocks (0), treatment ′ location interaction (0.02), and residual variance (2.0). Power was predicted for a randomized complete block design, but the random block effect was assumed to be zero, based on previous estimates from (Casler, 1998). The desired detection limit was set to 5%, with representative treatment means of 9.5 and 10 with a Type 1 error rate of $\alpha = 0.05$.

SAS code

The SAS code below computes the expected power for the design described above for 25 variations of the numbers of replicates per location and number of locations. It first creates a macro entitled "two". The macro then creates two new variables: "locn" is the number of locations and "repl" is the number of replicates. These two values are allowed to vary with maxima of "locmax" and "repmax", respectively. The line that reads "%two(6,8)" is the place to set the upper limits for the investigation. In this example, we have chosen to vary the number of replicates from 4 to 8 and the number of locations from 2 to 6. The remainder of the code is similar to that in Examples #1 and 2.

```
options nocenter;
%macro two(locmax,repmax);
data a;
%do locn=2 %to &locmax;
group1=&locn;
%do repl=4 %to &repmax;
group2=&repl;
do loc=1 to &locn by 1;
do rep=1 to &repl by 1;
do trt=0 to 1 by 1;
output;
end;
end;
end;
%end;
%end;
%mend two;
%two(6,8);            /*   <--- change here     */
run;
proc sort; by group1 group2;
```

```
data b; set a; by group1 group2;
if trt=0 then y=9.5;
if trt=1 then y=10;
run;
proc glimmix; class loc trt rep; by group1 group2;
model y = trt;
random trt*loc;
parms (0.02)(0.2) / hold=1,2;
ods output tests3=power_terms;
data power;
set power_terms;
alpha=0.05;
ncparm=numdf*Fvalue;
F_critical=finv(1-alpha, numdf, dendf, 0);
power=1-probf(F_critical, numdf, dendf, ncparm);
proc print;
run;
```

SAS output

The output below is abbreviated by eliminating all the proc glimmix output, which is necessary only for the purpose of checking the SAS run for errors. The output below is the result of the "proc print" statement at the end of the SAS code, printing all the calculated parameter estimates for the six design scenarios, shown under the headings "group1" (number of locations) and "group2" (number of replicates).

Obs	group1	group2	Effect	Num DF	DenDF	FValue	ProbF	alpha	ncparm	F_critical	power
1	2	4	trt	1	2	3.57	0.1994	0.05	3.5714	18.5128	0.20180
2	2	5	trt	1	2	4.17	0.1780	0.05	4.1667	18.5128	0.22463
3	2	6	trt	1	2	4.69	0.1628	0.05	4.6875	18.5128	0.24407
4	2	7	trt	1	2	5.15	0.1514	0.05	5.1471	18.5128	0.26082
5	2	8	trt	1	2	5.56	0.1425	0.05	5.5556	18.5128	0.27539
6	3	4	trt	1	4	5.36	0.0816	0.05	5.3571	7.7086	0.42377
7	3	5	trt	1	4	6.25	0.0668	0.05	6.2500	7.7086	0.47726
8	3	6	trt	1	4	7.03	0.0569	0.05	7.0312	7.7086	0.52112
9	3	7	trt	1	4	7.72	0.0499	0.05	7.7206	7.7086	0.55749
10	3	8	trt	1	4	8.33	0.0447	0.05	8.3333	7.7086	0.58799
11	4	4	trt	1	6	7.14	0.0369	0.05	7.1429	5.9874	0.60896
12	4	5	trt	1	6	8.33	0.0278	0.05	8.3333	5.9874	0.67420
13	4	6	trt	1	6	9.37	0.0222	0.05	9.3750	5.9874	0.72379
14	4	7	trt	1	6	10.29	0.0184	0.05	10.2941	5.9874	0.76214
15	4	8	trt	1	6	11.11	0.0157	0.05	11.1111	5.9874	0.79231
16	5	4	trt	1	8	8.93	0.0174	0.05	8.9286	5.3177	0.74472
17	5	5	trt	1	8	10.42	0.0121	0.05	10.4167	5.3177	0.80635
18	5	6	trt	1	8	11.72	0.0090	0.05	11.7187	5.3177	0.84922
19	5	7	trt	1	8	12.87	0.0071	0.05	12.8676	5.3177	0.87980
20	5	8	trt	1	8	13.89	0.0058	0.05	13.8889	5.3177	0.90214
21	6	4	trt	1	10	10.71	0.0084	0.05	10.7143	4.9646	0.83824
22	6	5	trt	1	10	12.50	0.0054	0.05	12.5000	4.9646	0.88893
23	6	6	trt	1	10	14.06	0.0038	0.05	14.0625	4.9646	0.92097
24	6	7	trt	1	10	15.44	0.0028	0.05	15.4412	4.9646	0.94192
25	6	8	trt	1	10	16.67	0.0022	0.05	16.6667	4.9646	0.95608

Results and Conclusions

The output allows any researcher to make direct comparisons of designs that are both statistically and logistically efficient. The results can be expanded to a wider range of values and used to graphically display the design comparisons as shown in Figure 2 for $l = 2$ to 6 locations and $r = 4$ to 20 replicates.

CHAPTER 5: MULTIPLE COMPARISON PROCEDURES: THE INS AND OUTS

David J. Saville

Solutions

Exercise 1

Each of the main effect means for the "within row spacing" factor is an average of 12 data values, so each has an effective sample size of $n = 12$. The LSD(5%) which is appropriate for comparing the two main effect means is therefore

LSD(5%) = 2.131 x √[2 x 853,113 / 12] = 804

where 2.131 is the two-sided 5% critical value for the t distribution with the residual degrees of freedom (15) (this critical value for t_{15} can also be calculated in Excel by typing the formula "=tinv(0.05,15)" into any cell), and 853,113 is the residual mean square (which is also the "pooled variance estimate").

Exercise 2.

(a) When sorted into descending order, the treatment means are 7604, 7493, 6150, 5679, 4838 and 4192. We now search for "homogeneous" groups of means, and assign a letter to each such group.

We start our search with the largest mean, 7604. The second largest mean, 7493, differs from 7604 by only 111, which is less than the least significant difference (LSD(5%)=1392), so the two means do not differ significantly, so we include this mean of 7493 in a homogeneous group along with the first mean of 7604. The third largest mean, 6150, however, differs from 7604 by 1454, which is greater than the LSD(5%) of 1392, so the two means differ significantly, so we cannot include this mean of 6150 in a homogeneous group along with the first two means. Therefore our first homogeneous group consists of just the first two means, 7604 and 7493. To indicate this result, we assign the letter "a" to each of these means.

We now continue our search for homogeneous groups by forgetting about the largest mean, and examining the second largest mean, 7493, in relation to the remaining four means. Now the third largest mean, 6150, differs from 7493 by 1343, which is less than the LSD(5%) of 1392, so the two means do not differ significantly, so we include this mean of 6150 in a homogeneous group along with the mean of 7493. The fourth largest mean, 5679, however, differs from 7493 by 1814, which is greater than the LSD(5%) of 1392, so the two means differ significantly, so we cannot include this mean of 5679 in a homogeneous group along with the other two means. Therefore our second homogeneous group consists of just two means, 7493 and 6150. To indicate this result, we assign the letter "b" to each of these means.

Continuing our search, we forget about the two largest means, and examine the third largest mean, 6150, in relation to the remaining three means. The fourth largest mean, 5679, differs from 6150 by 471, which is less than the LSD(5%) of 1392, so the two means do not differ significantly, so we include this mean of 5679 in a homogeneous group along with the mean of 6150. The fifth largest mean, 4838, differs from 6150 by 1312, which is also less than the LSD(5%) of 1392, so the two means do not differ significantly, so we also include this mean of 4838 in a homogeneous group along with the other two means. The sixth largest mean, 4192, however, differs from 6150 by 1958, which is greater than the LSD(5%) of 1392, so the two means differ significantly, so we cannot include this mean of 4192 in a homogeneous group along with the other three means. Therefore our third homogeneous group consists of three means, 6150, 5679 and 4838. To indicate this result, we assign the letter "c" to each of these three means.

Continuing our search, we forget about the three largest means, and examine the fourth largest mean, 5679, in relation to the remaining two means. The fifth largest mean, 4838, differs from 5679 by 841, which is less than the LSD(5%) of 1392, so the two means do not differ significantly, so we include the mean of 4838 in a homogeneous group along with the mean of 5679. The sixth largest mean, 4192, however, differs from 5679 by 1487, which is greater than the LSD(5%) of 1392, so the two means differ significantly, so we cannot include the mean of 4192 in a homogeneous group along with the other two means. Therefore our fourth homogeneous group consists of two means, 5679 and 4838. We notice, however, that this fourth homogeneous group is included in the third homogeneous group (assigned the letter "c"), so we do not declare a fourth homogeneous group, and do not assign any more letters to these two means.

To complete our search, we compare the fifth largest mean, 4838, to the only remaining mean, 4192. These means differ by 646, which is less than the LSD(5%) of 1392, so the two means do not differ significantly, so we include the mean of 4192 in a homogeneous group along with the mean of 4838. Therefore our fourth homogeneous group consists of two means, 4838 and 4192. To indicate this result, we assign the letter "d" to each of these three means.

This completes the process. The final result, in terms of the sorted means, is:

7604	a
7493	ab
6150	bc
5679	c
4838	cd
4192	d

When re-sorted into the order of the treatments, this gives the same lettering as shown in Table 2.

Aside: In this example, the six treatments were equally replicated, so a single LSD(5%) could be used for comparing all pairs of treatment means. The above procedure can then be easily performed by computer. However, if the treatments had been unequally replicated, several different LSD(5%) values would have been required and the procedure is more complicated, with the result that attempts at generating an appropriate computer routine are not universally successful.

(b) For the within-row spacing of 5 cm, *yes*, there is a 5% significant difference between the "3 rows per bed" treatment mean (6150) and the "5 rows per bed" treatment mean (7604), since they do not have a letter in common (which reflects the fact that they differ by 1454, which is greater than the LSD(5%) of 1392). Similarly, for the within-row spacing of 10 cm, *yes,* there is a 5% significant difference between the "3 rows per bed" treatment mean (4192) and the "5 rows per bed" treatment mean (5679), since they do not have a letter in common (which reflects the fact that they differ by 1487, which is greater than the LSD(5%) of 1392).

For each within-row spacing (5 cm and 10 cm), *no,* there is no significant difference between the "4 rows per bed" treatment mean and either the "3 rows per bed" or "5 rows per bed" treatment means, since for all four pairwise comparisons the two means being compared have a letter in common (which reflects the fact that the two means always differ by less than the LSD(5%) of 1392).

(c) For each between-row spacing (3, 4, and 5 rows per bed), yes, there is a 5% significant difference between the "5 cm" treatment mean and the "10 cm" treatment mean? For example, the two "3 rows per bed" treatment means are 6150 and 4192 for 5 cm and 10 cm spacing, respectively, and these means differ significantly since they do not have a letter in common (which reflects the fact that they differ by 1958, which is greater than the LSD(5%) of 1392). For "4 rows per bed" the respective means again do not have a letter in common, and differ by 2655 (>1392). For "5 rows per bed" the respective means also do not have a letter in common, and differ by 1925 (>1392).

CHAPTER 6: LINEAR REGRESSION TECHNIQUES

Christel Richter and Hans-Peter Piepho

Appendix 1 (Refers to Example 1: Datafile EEL with variables LENGTH and WEIGHT)

Grey: Externally Studentized residuals with $|\hat{e}_i^{**}| > 2$. Framed: high leverage

Eel_no	Length (cm)	Weight (g)
1	33	108.6
2	34	114.1
3	36	120.4
4	36	128.6
5	37	137.5
6	39	144.2
7	39	148.3
8	40	152.4
9	41	160.5
10	42	166.4
11	42	165.9
12	42	162.8
13	43	179.0
14	43	172.1
15	44	178.0
16	45	189.7
17	46	194.9
18	46	184.8
19	46	191.8
20	46	189.7
21	47	202.6
22	47	198.9
23	47	198.2
24	48	209.6
25	49	212.1
26	51	224.5
27	51	224.7
28	51	228.3
29	51	221.6
30	52	231.7
31	53	246.2
32	54	247.5
33	55	254.8
34	58	275.0

```
DATA eel;
SET eel;
LABEL length='Length [cm]' weight='Weight [g]';
RUN;

ODS GRAPHICS ON;

TITLE 'Example 1: Generation of a template for Figure 1';
PROC TEMPLATE;
DEFINE STATGRAPH eel_temp;
BEGINGRAPH;
ENTRYTITLE "Weight and length of eels" /TEXTATTRS=(SIZE=11pt);
LAYOUT lattice/COLUMNS = 2 ROWS = 2 COLUMNWEIGHTS = (.8 .2) ROWWEIGHTS
= (.8 .2)
COLUMNDATARANGE = union ROWDATARANGE = union;
COLUMNAXES;
COLUMNAXIS /LABEL = "Length [cm]" GRIDDISPLAY = on
LABELATTRS=(SIZE=12)TICKVALUEATTRS=(SIZE=10);
COLUMNAXIS /LABEL = "" GRIDDISPLAY = on;
ENDCOLUMNAXES;
ROWAXES;
ROWAXIS /LABEL = "Weight [g]" GRIDDISPLAY = on LABELATTRS=(SIZE=12)
TICKVALUEATTRS=(SIZE=10);
ROWAXIS /LABEL = "" GRIDDISPLAY = on;
ENDROWAXES;
LAYOUT overlay;
SCATTERplot X = length Y = weight/MARKERATTRS=(COLOR=black SIZE=10
```

```
SYMBOL=circlefilled);
ENDLAYOUT;
BOXPLOT Y=weight/ORIENT = vertical;
BOXPLOT Y=length/ORIENT = horizontal;
ENDLAYOUT;
ENDGRAPH;
END;
RUN;
TITLE 'Example 1: Figure 1';
PROC SGRENDER DATA = eel TEMPLATE = eel_temp;
RUN; QUIT;

TITLE 'Example 1: Regression with Figure 3 and some further
representations with PROC REG';
PROC REG DATA=eel PLOTS(LABEL)=all;
MODEL weight = length /CLB;
OUTPUT OUT=eel_out P=yhat LCLM=lclm UCLM=uclm LCL=lcl UCL=ucl R=yresid
STUDENT=student RSTUDENT=rstudent H=h COOKD=cookd COVRATIO=covratio
DFFITS=dffits PRESS=PRESS;
RUN; QUIT;

TITLE 'Example 1: Regression with PROC MIXED';
PROC MIXED DATA=eel plots=all;
MODEL weight = length /s CL outp=eel_out_mixed;
RUN; QUIT;

TITLE 'Example 1: Regression with PROC GLM';
PROC GLM DATA=eel plots=all;
MODEL weight = length /SOLUTION CLPARM ;
OUTPUT OUT=eel_out_glm P=yhat LCLM=lclm UCLM=uclm LCL=lcl UCL=ucl
R=yresid STUDENT=student RSTUDENT=rstudent H=h COOKD=cookd
COVRATIO=covratio DFFITS=dffits PRESS=PRESS;
RUN; QUIT;

TITLE 'Example 1: Figure 4 Confidence ellipse';
PROC CORR DATA=eel PLOTS=SCATTER(ellipse=confidence alpha= 0.05);
VAR length weight;
RUN; QUIT;

TITLE 'Example 1: Figure 4 Prediction ellipse';
PROC CORR DATA=eel PLOTS=SCATTER(alpha= 0.05);
VAR length weight;
RUN; QUIT;

TITLE 'Example 1: Figure 6 and tests for normality';
PROC UNIVARIATE DATA= eel_out NORMAL;
QQPLOT student/NORMAL (MU=0 SIGMA=1) ODSTITLE='Q-Q Plot for Eel_
Weight';
VAR student;
RUN; QUIT;

TITLE 'Example 1: Figure 7';
DATA eel_out;
SET eel_out; IF student<-2 OR student>2 THEN studcrit="S obs.
no.="||trim(left(_N_));
IF studcrit ne '      ' THEN crit =length;
RUN; QUIT;

PROC SGPLOT DATA= eel_out NOAUTOLEGEND;
TITLE 'Internally studentized residual against length';
SCATTER X=length Y=student;REFLINE 0 2 -2;
SCATTER X=crit Y=student/ DATALABEL=studcrit
DATALABELATTRS=(Family=Arial SIZE=10
STYLE=Italic Weight=Bold) MARKERATTRS=(COLOR=black SIZE=8
SYMBOL=circlefilled);
REFLINE 0 2 -2;
RUN; QUIT;
```

Appendix 2 (Refers to Example 2: Datafile CALIBRATION with variables *C* and EXT)

Grey: Externally Studentized residuals with $\left|\hat{e}_i^{**}\right| > 2$. No high leverage.

Obs_no	Concentration (mmol/l)	Extinction
1	0.5	55.5
2	1.0	77.0
3	1.5	100.7
4	3.0	165.5
5	5.0	253.4

Regression analysis and Fig. 3 analogous to PROC REG of Example 1; for Fig. 2, the Template of Example 1 must be adjusted (among others: boxplot only for the extinction)

```
TITLE 'Example 2: Generation of concentration values as initial
values for the iteration';
DATA cali_start;
DO start=4.3 TO 4.6 BY 0.01;
OUTPUT; END;
RUN; QUIT;

TITLE 'Example 2: Iterative solution for CL and CU';
/*rootMSE and regression function from PROC REG; Cmean and SSC
from PROC UNIVARIATE*/
PROC MODEL DATA=cali_start OUT=result;
rootMSE=0.7320449; n=5; Cmean=2.2; SSC=13.3; ext=230;
EXT1=33.70602+43.9609*CL; /*regression function*/
EXT2=33.70602+43.9609*CU; /*regression function*/
eq.CL=EXT2- rootMSE *sqrt(1+1/n+(CL- Cmean)**2/
SSC)*tinv(0.975,n-2)-ext;
eq.CU=EXT1+ rootMSE *sqrt(1+1/n+(CU- Cmean)**2/
SSC)*tinv(0.975,n-2)-ext;
SOLVE CL CU;
RUN; QUIT;
```

Appendix 3 (Refers to Example 3 and Example 3 (modfied): Datafile FIBER with variables DAY, CONTENT, and BLOCK)

Grey: Externally Studentized residuals with $\left|\hat{e}_i^{**}\right| > 2$. No high leverage.

No. obs.	Day	Crude fiber content (g/kg)	Block	No. obs.	Day	Crude fiber content (g/kg)	Block
1	0	218	1	11	10	289	3
2	0	225	2	12	10	297	4
3	0	229	3	13	15	297	1
4	0	239	4	14	15	307	2
5	5	246	1	15	15	317	3
6	5	258	2	16	15	343	4
7	5	254	3	17	20	316	1
8	5	269	4	18	20	336	2
9	10	257	1	19	20	354	3
10	10	275	2	20	20	351	4

For Fig. 2, the Template of Example 1 must be adjusted; Fig. 3, 6, and 7, Table 2 A and D analogous to the code of Example 1

```
DATA fiber;
SET fiber;
LABEL content='Fiber Content [g/kg]' Day='Days after the first
cut';
RUN; QUIT;

TITLE 'Example 3: Table 2 B';
PROC REG DATA=fiber;
MODEL content=day / lackfit;
RUN; QUIT;

TITLE 'Example 3: Remark to explain the lack-of-fit';
DATA fiber;
SET fiber;
day2=day*day; day3=day2*day; day4=day2*day2;
RUN; QUIT;
PROC REG DATA=fiber;
MODEL content=day day2 day3 day4/lackfit ss1;
RUN; QUIT;

TITLE 'Example 3: Table 2 C with PROC GLM';
PROC GLM DATA=fiber;
CLASS day;
MODEL content=day;
RUN; QUIT;

TITLE 'Example 3: Table 2 C with PROC MIXED';
PROC MIXED DATA=fiber;
CLASS day;
MODEL content=day;
RUN; QUIT;

TITLE 'Example 3: Calculation of the means per day';
PROC MEANS DATA=fiber;
CLASS day;
VAR content;
OUTPUT OUT=fiber_mean_file MEAN=content_mean;
RUN; QUIT;

TITLE 'Example 3: Table 3';
PROC REG DATA=fiber_mean_file;
MODEL content_mean = day;
OUTPUT OUT=fiber_mean_out P=yhat STDP=stdp LCLM=lclmean UCLM=uclmean
STDR=stdr STDI=stdi LCL=lclind UCL=uclind;
RUN; QUIT;

TITLE 'Example 3 (modified): Table 18 A';
PROC GLM data=fiber;
CLASS block;
MODEL content =day block;
RUN; QUIT;

TITLE 'Example 3 (modified): Table 18 B';
PROC MIXED DATA=fiber;
/*Estimated function values and confidence intervals per fixed block are
given in fiber_fix*/
CLASS block;
MODEL content = day block /S CL OUTP=fiber_fix;
ESTIMATE 'b0+block1' int 1 block 1 /cl;
ESTIMATE 'b0+block2' int 1 block 0 1/cl;
ESTIMATE 'b0+block3' int 1 block 0 0 1/cl;
ESTIMATE 'b0+block4' int 1 block 0 0 0 1/cl;
RUN; QUIT;

TITLE 'Example 3 (modified): Table 19 and Table 20 (broad inference';
PROC MIXED DATA=fiber;
```

```
/*Confidence intervals (broad inference) are given in fiber_rand*/
CLASS block;
MODEL content =day /S CL ddfm=Kenwardroger(firstorder) OUTPM=fiber_rand;
RANDOM block /S;
RUN; QUIT;

TITLE 'Example 3 (modified): Table 20';
/*To have a common divisor all values in the estimate statement are
multiplied with 4 = number of blocks*/
PROC MIXED DATA=fiber;
/*Calculation of confidence intervals for the mean of all fixed blocks*/
CLASS block;
MODEL content =day block /S CL;
ESTIMATE 'day0'   int 4 block 1 1 1 1 day  0/divisor=4 cl;
ESTIMATE 'day5'   int 4 block 1 1 1 1 day 20/divisor=4 cl;
ESTIMATE 'day10'  int 4 block 1 1 1 1 day 40/divisor=4 cl;
ESTIMATE 'day15'  int 4 block 1 1 1 1 day 60/divisor=4 cl;
ESTIMATE 'day20'  int 4 block 1 1 1 1 day 80/divisor=4 cl;
RUN; QUIT;

TITLE 'Example 3 (modified): Table 20';
PROC MIXED DATA=fiber;
/*Calculation of confidence intervals for the mean of all random
blocks*/
CLASS block;
MODEL content =day /S CL ddfm=Kenwardroger(firstorder);
RANDOM block /S;
ESTIMATE 'day0'  int 4 day 0  | block 1 1 1 1 /divisor=4 cl;
ESTIMATE 'day5'  int 4 day 20 | block 1 1 1 1 /divisor=4 cl;
ESTIMATE 'day10' int 4 day 40 | block 1 1 1 1 /divisor=4 cl;
ESTIMATE 'day15' int 4 day 60 | block 1 1 1 1 /divisor=4 cl;
ESTIMATE 'day20' int 4 day 80 | block 1 1 1 1 /divisor=4 cl;
RUN; QUIT;
```

Appendix 4 (Refers to Example 4: Datafile GRASS with variables WGRASS and YIELD)

Grey: Externally Studentized residuals with $|\hat{e}_i^{**}| > 2$. Framed: high leverage

No. Obs.	Rx_i	Wind grass (number/m²) x_i	yield (g / plot) y_i	Ry_i	No. Obs.	Rx_i	Wind grass (number/m²) x_i	yield (g / plot) y_i	Ry_i
1	1.5	0	9310	48	27	27.5	99	6410	23
2	1.5	0	8460	43	28	27.5	99	6640	25.5
3	4	1	9770	52	29	29	100	6010	19
4	4	1	9320	49	30	30	101	6940	28
5	4	1	8620	44	31	31.5	102	4930	14
6	6	2	7850	36.5	32	31.5	102	7620	33
7	7.5	3	9520	51	33	33.5	132	5680	17
8	7.5	3	9080	46	34	33.5	132	7570	32
9	9	4	6340	22	35	35	145	5240	16
10	10	5	9340	50	36	36	152	6320	21
11	11	17	7940	39	37	37	161	3970	9
12	12	21	8730	45	38	38	167	4790	13
13	13	22	7870	38	39	39	197	5230	15
14	14	24	8160	41	40	40	243	5980	18
15	15	31	7700	35	41	41	250	7240	29
16	16	37	8120	40	42	42	251	2340	5
17	17	46	7850	36.5	43	43	258	4320	12
18	18	51	7530	31	44	44	268	2830	6
19	19.5	56	7630	34	45	45	288	3810	8
20	19.5	56	9120	47	46	46	305	4260	11

21	21.5	57	6640	25.5	47	47	311	4130	10
22	21.5	57	7490	30	48	48	337	3050	7
23	23	61	6280	20	49	49	901	1740	2
24	24	81	8450	42	50	50	927	1750	3
25	25	84	6920	27	51	51	1102	1980	4
26	26	88	6550	24	52	52	1204	540	1

For Fig. 2, the Template of Example 1 must be adjusted; for Fig. 3, 6, and 7 the code of Example 1 can be used

```
DATA grass;
SET grass; sqrt_wgrass=sqrt(wgrass); log_yield=log(yield);
LABEL wgrass='wind grass [number/plot]' yield='Yield [g/plot]'
sqrt_wgrass='sqrt(wind grass)';
RUN; QUIT;

TITLE 'Example 4: Spearman's correlation coefficient';
PROC CORR DATA=grass SPEARMAN;
VAR wgrass yield;
RUN; QUIT;

TITLE 'Example 4: Regression with logarithm of yield
(multiplicative errors after back-transformation)';
PROC REG DATA=grass;
MODEL log_yield=wgrass;
RUN; QUIT;

TITLE 'Example 4: Non-linear regression yield=a*exp(b*wgrass)
with additive errors';
PROC NLIN DATA=grass PLOTS=all;
PARMS  a=8000 b=-0.01;
MODEL yield=a*exp(b*wgrass);
RUN; QUIT;

TITLE 'Example 4: Non-linear regression
yield=a*exp(b*wgrass)+c with additive errors and Fig. 8';
PROC NLIN DATA=grass PLOTS=all;
PARMS a=7000 b=-0.01 c=1200;
MODEL yield=a*exp(b*wgrass)+c;
RUN; QUIT;
```

Appendix 5 (Refers to Example 5: Datafile SHAPE with input variables x1, x2, BLOCK, and YIELD)

```
TITLE 'Example 5: DATA input and calculation of the variables
TREATMENT, AREA, SHAPE_INDEX1, and SHAPE_INDEX2';
DATA space;
INPUT x1 x2 @@; treatment=x1*1000+x2; area=x1*x2;
shape_ind1=x1/x2; shape_ind2=(x1+x2)/2/sqrt(x1*x2);
DO block=1 TO 4; INPUT yield@@; OUTPUT; END;
DATALINES;
30 30 5.95 5.30 6.50 6.35
30 24 7.10 6.45 6.60 5.75
30 20 7.00 6.50 6.35 8.90
30 15 8.10 5.50 6.60 7.50
24 24 8.85 7.65 7.00 7.90
24 20 7.65 6.90 8.25 8.30
24 15 7.80 6.75 8.20 7.25
20 20 8.05 6.65 8.10 8.05
20 15 9.30 8.75 8.75 8.00
15 15 9.35 8.10 7.60 7.75
;TITLE 'Example 5: Table 6 A TO D';
PROC GLM DATA=space;
CLASS block treatment;
MODEL yield =block treatment;
```

```
RUN; QUIT;

PROC GLM DATA=space;
CLASS block treatment;
MODEL yield =block area treatment / SS1;
RUN; QUIT;

PROC GLM DATA=space;
CLASS block treatment;
MODEL yield =block area shape_ind1 treatment / SS1;
RUN; QUIT;
PROC GLM DATA=space;
CLASS block treatment;
MODEL yield =block area shape_ind2 treatment / SS1;
RUN; QUIT;

TITLE 'Example 5: Table 7';
PROC GLM DATA=space;
CLASS block;
MODEL yield =block area / SS1;
OUTPUT OUT=residual_B r=resi_B;
RUN; QUIT;

PROC GLM DATA=space;
CLASS block;
MODEL yield =block area shape_ind1 / SS1;
OUTPUT OUT=residual_C r=resi_C;
RUN; QUIT;

PROC GLM DATA=space;
CLASS block;
MODEL yield =block area shape_ind2 / SS1;
OUTPUT OUT=residual_D r=resi_D;
RUN; QUIT;

PROC SORT DATA= residual_B; BY treatment block; RUN;
PROC SORT DATA= residual_C; BY treatment block; RUN;
PROC SORT DATA= residual_D; BY treatment block; RUN;

DATA resi;
MERGE residual_B residual_C residual_D; BY treatment block;
KEEP treatment block yield resi_B resi_C resi_D;
RUN; QUIT;

PROC MEANS DATA=resi;
VAR yield resi_B resi_C resi_D;
BY treatment;
OUTPUT OUT=mwresi MEAN= mwyield mwresi_B mwresi_C mwresi_D;
RUN; QUIT;
```

Appendix 6 (Refers to Example 6: Datafile POTATO with variables SIZE and WEIGHT).

The datafile Potato.xls is available in the supplemental material.

```
DATA potato;
SET potato;
size2=size*size; size3=size*size*size;
size_reciprocal = 1/size;
label size='Size [mm]' weight='Weight [g]';
RUN; QUIT;

TITLE 'Example 6: Table 8 A and B';
/* Sequential approach with sequence x1→ x2 → x3; partially also
with PROC REG (with option ss1) and PROC MIXED (with option htype=1)
possible*/
```

```
PROC GLM DATA= potato;
MODEL weight = size size2 size3 /SOLUTION SS1;
RUN; QUIT;
/* Sequential approach with sequence x3→ x1 → x2*/
PROC GLM DATA= potato;
MODEL weight =size3 size size2 / SOLUTION SS1;
RUN; QUIT;
/* Partial approach does not depend on the sequence*/
PROC GLM DATA= potato;
MODEL weight =size size2 size3 / SOLUTION ;
RUN; QUIT;

TITLE 'Example 6: Table 8 C';
/*pcorr2 bases on the partial approach, pcorr1 on the sequential
approach*/
PROC REG DATA=potato;
MODEL weight =size size2 size3/ PCORR2 TOL;
RUN; QUIT;

TITLE 'Example 6: Table 9 A';
/*m123... m3 are labels for the different models with intercepts. The
fit criteria are in the potato_info_int file*/
PROC REG DATA= potato OUTEST= potato_info_int;
m123: MODEL weight = size size2 size3 / ADJRSQ AIC PRESS BIC SBC SSE;
m12:  MODEL weight = size size2 / ADJRSQ AIC PRESS BIC SBC SSE;
m13:  MODEL weight = size size3 / ADJRSQ AIC PRESS BIC SBC SSE;
m23:  MODEL weight = size2 size3 / ADJRSQ AIC PRESS BIC SBC SSE;
m1:   MODEL weight = size / ADJRSQ AIC PRESS BIC SBC SSE;
m2:   MODEL weight = size2 / ADJRSQ AIC PRESS BIC SBC SSE;
m3:   MODEL weight = size3 / ADJRSQ AIC PRESS BIC SBC SSE;
RUN; QUIT;

TITLE 'Example 6: Table 9 B';
/*m123... m3 are labels for the different models without
intercepts. The fit criteria are in the potato_info_noint file*/
PROC REG DATA= potato OUTEST= potato_info_noint;
m123: MODEL weight = size size2 size3 / NOINT ADJRSQ AIC PRESS
BIC SBC SSE;
m12:  MODEL weight = size size2 / NOINT ADJRSQ AIC PRESS BIC
SBC SSE;
m13:  MODEL weight = size size3 / NOINT ADJRSQ AIC PRESS BIC
SBC SSE;
m23:  MODEL weight = size2 size3 / NOINT ADJRSQ AIC PRESS BIC
SBC SSE;
m1:   MODEL weight = size / NOINT ADJRSQ AIC PRESS BIC SBC SSE;
m2:   MODEL weight = size2 / NOINT ADJRSQ AIC PRESS BIC SBC SSE;
m3:   MODEL weight = size3 / NOINT ADJRSQ AIC PRESS BIC SBC SSE;
RUN; QUIT;

TITLE 'Example 6: Table 12';
/*first step: PROC TRANSREG to get the LLtransreg-values*/
/*with log(size) as regressor and using lambda=0 */
PROC TRANSREG details DATA= potato PLOTS=all SS2 PBOXCOXTABLE
CL;
MODEL BOXCOX(weight /  LAMBDA=0 ) = log(size);
RUN; QUIT;

/*with log(size) as regressor and searching for the optimal
lambda */
PROC TRANSREG details DATA= potato PLOTS=all SS2 PBOXCOXTABLE
CL;
MODEL BOXCOX(weight /  LAMBDA=0 to 1 by 0.01 ) = log(size);
RUN; QUIT;

/*with size as regressor and using lambda=1/3 if it is in
confidence interval of the optimal lambda*/
PROC TRANSREG details DATA= potato PLOTS=all SS2 PBOXCOXTABLE CL;
```

```
MODEL BOXCOX (weight / CONVENIENT CLL =0.333333 lambda=0.3 TO 0.4 by
0.001) = IDENTITY(size);
RUN; QUIT;

/*second step: Calculation of -2LL=LLnew in Table 12 using the
LLtransreg-values of PROC TRANSREG; for the first row in Table
12: -1166.3, for the second row: -1157.06, and for the third row:
-1159.38*/
DATA calc;
LLtransreg=-1166.3;
n=524; PI=constant('PI'); LLnew=-2*(LLtransreg-n/2*log(2*PI)-(n-2)/2);
PROC PRINT DATA=calc;
RUN; QUIT;

/*for the fourth row in Table 12; with size3 as regressor, no
transformation*/
PROC MIXED data=potato method=ML;
MODEL weight=size3;
run;quit;

TITLE 'Example 6: Table 14';
/*For using the ML-method, replace method = REML (default) with method
= ML */
/*Weighted regression with 1/size*/
PROC MIXED DATA=potato METHOD=REML PLOTS=all;
MODEL weight = size3 /NOINT S CL DDFM=KenwardRoger(firstorder);
WEIGHT size_reciprocal;
RUN; QUIT;

/*Power-of-x model; due to convergence problems, initial values for
the covariance parameters are specified*/
PROC MIXED DATA=potato METHOD=REML PLOTS=all;
MODEL weight = size3 /NOINT S CL DDFM=KenwardRoger(firstorder);
REPEATED / LOCAL=exp(size);
PARMS (0.2) (1.5);
RUN; QUIT;

/*Power-of-mean model in two steps*/
ODS OUTPUT SolutionF=sf;
PROC MIXED DATA=potato;
MODEL weight = size3 /NOINT s;
RUN; QUIT;

PROC MIXED DATA=potato method=REML PLOTS=all;
MODEL weight = size3 /NOINT s;
REPEATED / LOCAL=pom(sf) DDFM=KenwardRoger(firstorder);
RUN; QUIT;

/*Model with individual variance per size; with Fig. 13 right*/
PROC MIXED DATA=potato METHOD=REML PLOTS=all;
MODEL weight = size3 / NOINT s DDFM=KenwardRoger(firstorder)
OUTPM=potato_out_ind residual;
REPEATED / GROUP=size;
RUN; QUIT;

/*Fig. 13*/
PROC REG DATA= potato;
MODEL weight = size3 /NOINT;
OUTPUT OUT=potato_out P=yhat R=yresid STUDENT=student
RSTUDENT=rstudent;
RUN; QUIT;

PROC SGPLOT DATA=potato_out;
TITLE 'Unweighted Regression';
LABEL size='size';
SCATTER x=size y=student;
RUN; QUIT;
```

```
PROC SGPLOT DATA= potato_out_ind;
TITLE 'Regression with individual variance per size';
LABEL size='size' ;
SCATTER x=size y=StudentResid;
RUN; QUIT;
```

Appendix 7 (Refers to Example 7: Datafile APPLE with variables VARIETY, YEAR1_4, and Year1_10)

Grey: Externally Studentized residuals with $|\hat{e}_i^{**}| > 2$. Framed: high leverage

Variety A Yield (kg/tree)				Variety B Yield (kg/tree)			
Year1_4	Year1_10	Year1_4	Year1_10	Year1_4	Year1_10	Year1_4	Year1_10
22	109	39	137	23	73	40	93
27	119	40	142	27	79	40	96
28	125	41	148	27	72	40	91
30	115	42	154	29	79	41	99
33	127	42	152	30	76	42	87
34	133	42	160	30	86	42	92
34	148	42	155	31	78	44	91
34	141	43	144	32	84	45	97
36	141	43	161	33	79	48	99
36	134	44	170	35	85	51	100
37	144	45	165	35	86	53	105
37	140	47	165	36	82	54	101
38	133	48	164	37	91	55	110
38	144	48	167	37	87	56	108
39	140	54	190	38	89	56	113

```
DATA apple;
SET apple;
LABEL year1_10='Yield [kg/tree] year 1 to 10' year1_4='Yield
[kg/tree] year 1 to 4';
RUN; QUIT;

TITLE 'Example 7: Table 15';
/*Before running the following procedures, the datafile must be
sorted by variety*/
PROC REG DATA=apple;
MODEL year1_10=year1_4 / CLB ADJRSQ;
BY variety;
OUTPUT OUT=apple_out P=yhat LCLM=lclmean UCLM=uclmean
STDR=stdr STDI=stdi  LCL=lclind UCL=uclind;
RUN; QUIT;

TITLE 'Example 7: Fig. 14';
PROC SGPLOT DATA=apple;
TITLE 'Cumulative yield after 10 years against after 4 years';
REG x=year1_4 y=year1_10 /group=variety clm cli
MARKERATTRS=(SIZE=9 SYMBOL=circle);
XAXIS LABEL ="yield [kg/tree] year 1 to 4 "
LABELATTRS=(SIZE=12)  VALUEATTRS=(SIZE=10);
YAXIS LABELATTRS=(SIZE=12) VALUEATTRS=(SIZE=10) VALUES=(50
100 150  200);
RUN; QUIT;

TITLE 'Example 7: Table 16 A';
PROC GLM DATA=apple; /*with PROC GLM*/
```

```
MODEL year1_10=year1_4 /SOLUTION CLPARM;
RUN; QUIT;
PROC MIXED DATA=apple; /*alternatively with PROC MIXED*/
MODEL year1_10=year1_4 /S CL;
RUN; QUIT;

TITLE 'Example 7: Table 16 B';
PROC GLM DATA=apple;/*with PROC GLM*/
CLASS variety;
MODEL year1_10=year1_4 variety variety*year1_4 /SOLUTION CLPARM;
ESTIMATE 'int+b01' intercept 1 variety 1;
ESTIMATE 'int+b02' intercept 1 variety 0 1;
ESTIMATE 'b1+b11' year1_4 1 year1_4*variety 1;
ESTIMATE 'b1+b12' year1_4 1 year1_4*variety 0 1;
ESTIMATE 'b01-b02' variety 1 -1;
ESTIMATE 'b11-b12' year1_4*variety 1 -1;
RUN; QUIT;
PROC MIXED DATA=apple; /*alternatively with PROC MIXED*/
CLASS variety;
MODEL year1_10=year1_4 variety year1_4*variety /S CL;
ESTIMATE 'int+b01' int 1 variety 1 / CL;
ESTIMATE 'int+b02' int 1 variety 0 1 / CL;
ESTIMATE 'b1+b11' year1_4 1 year1_4*variety 1 / CL;
ESTIMATE 'b1+b12' year1_4 1 year1_4*variety 0 1 / CL;
ESTIMATE 'b01-b02' variety 1 -1 / CL;
ESTIMATE 'b11-b12' year1_4*variety 1 -1 / CL;
RUN; QUIT;

TITLE 'Example 7: Table 16 C';
PROC MIXED DATA=apple PLOT=all ;
CLASS variety;
MODEL year1_10= year1_4 year1_4*variety/S CL RESIDUAL OUTP=apple_
OUT_c_varhom;
ESTIMATE 'b1+b11' year1_4 1 year1_4*variety 1;
ESTIMATE 'b1+b12' year1_4 1 year1_4*variety 0 1;
RUN; QUIT;

TITLE 'Example 7: Table 17 A and B';
/*Variance Homogeneity*/
PROC MIXED DATA=apple plot=all ;
CLASS variety;
MODEL year1_10= year1_4 year1_4*variety/S CL RESIDUAL OUTP=apple_
out_c_varhom;
RUN; QUIT;
/*Variance Heterogeneity*/
PROC MIXED DATA=apple plot=all ;
CLASS variety;
MODEL year1_10= year1_4 year1_4*variety/S CL RESIDUAL OUTP=apple_
out_c_varhet;
REPEATED / GROUP=variety;
RUN; QUIT;

TITLE 'Example 7: Figure 15 upper part';
PROC SGPLOT DATA=apple NOAUTOLEGEND;
TITLE 'Cumulative yield after 10 years against after 4 years';
REG x=year1_4 y=year1_10 /clm cli;
SCATTER x=year1_4 y=year1_10/ DATALABEL=variety
DATALABELATTRS=(Family=Arial SIZE=10 STYLE=Italic WEIGHT=Bold)
MARKERATTRS=(SIZE=7 SYMBOL=circlefilled);
XAXIS LABEL ="yield [kg/tree] year 1 to 4 " LABELATTRS=(SIZE=12)
VALUEATTRS=(SIZE=10);
YAXIS LABELATTRS=(SIZE=12) VALUEATTRS=(SIZE=10) VALUES=(50 100 150
200);
RUN; QUIT;

TITLE 'Example 7: Figure 15 lower part left';
PROC SGPLOT DATA=Apple_out_c_varhom;
```

```
TITLE 'Cumulative yield after 10 years against after 4 years';
SCATTER x = year1_4 y = year1_10 /GROUP=variety MARKERATTRS=(SIZE=9
symbol=circle);
BAND x=year1_4 LOWER=LOWER UPPER=UPPER/GROUP=variety
FILLATTRS=(TRANSPARENCY=0);
SERIES x=year1_4 y=pred/GROUP=variety LINEATTRS=(PATTERN=solid
THICKNESS=2);
XAXIS LABEL ="yield [kg/tree] year 1 to 4 " LABELATTRS=(SIZE=12)
VALUEATTRS=(SIZE=10);
YAXIS LABELATTRS=(SIZE=12) VALUEATTRS=(SIZE=10) VALUES=(50  100  150
200);
RUN;QUIT;

TITLE 'Example 7: Figure 15 lower part right';
PROC SGPLOT DATA= apple_OUT_c_varhom NOAUTOLEGEND;
TITLE 'Externally Studentized residuals';
SCATTER x=pred y=studentresid/ DATALABEL=variety
DATALABELATTRS=(Family=Arial SIZE=10 STYLE=Italic WEIGHT=Bold)
MARKERATTRS=(SIZE=7 SYMBOL=circlefilled) ; REFLINE 0;
XAXIS LABEL ="PREDICTED value year 1 TO 10 " LABELATTRS=(SIZE=12)
VALUEATTRS=(SIZE=10);
YAXIS LABELATTRS=(SIZE=12) VALUEATTRS=(SIZE=10) ;
RUN; QUIT;
```

Example 7, Figure 16 left and right analogous to Example 7, Figure 15 lower part with Datafile = apple_out_c_varhet

Appendix 8 (Refers to Example 8: Datafile AIRTEMP with variables TEMP and YEAR)

Grey: Externally Studentized residuals with $\left|\hat{e}_i^{**}\right| > 2$. No high leverage.

Year	Mean air temperature (°C)
1960	8.94
1961	9.35
1962	7.98
1963	7.79
1964	8.42
1965	8.02
1966	8.93
1967	9.50
1968	8.65
1969	7.78
1970	7.93
1971	8.89
1972	8.16
1973	8.53
1974	9.46
1975	9.47
1976	8.51
1977	9.05
1978	8.20
1979	8.00
1980	7.56
1981	8.42
1982	9.42
1983	9.48
1984	8.58
1985	8.00
1986	8.27

Year	Mean air temperature (°C)
1987	7.57
1988	9.57
1989	9.94
1990	10.11
1991	8.85
1992	9.66
1993	8.79
1994	9.85
1995	9.17
1996	7.34
1997	9.04
1998	9.46
1999	10.00
2000	10.41
2001	9.32
2002	9.82
2003	9.46
2004	9.04
2005	9.31
2006	9.89
2007	10.42
2008	10.06
2009	9.43
2010	8.02
2011	10.14
2012	9.27
2013	9.34

```
TITLE 'Example 8: Table 21, Figure 17 and Durbin-Watson test';
PROC REG DATA=airtemp PLOTS (LABEL)=all;
ID Year;
MODEL Temp=Year /DWPROB CLB CLI CLM ADJRSQ;
RUN; QUIT;

TITLE 'Example 8: Table 22';
PROC MIXED DATA=airtemp PLOTS=residualpanel(unpack) ;
MODEL Temp=Year /S CL DDFM=kr(firstorder);
REPEATED / TYPE=AR(1) SUBJECT=intercept;
RUN; QUIT;

TITLE 'Example 8: Analysis as AR(1)';
PROC AUTOREG DATA=airtemp /*AR(1) */;
MODEL Temp=Year /NLAG=1 DW=1 DWPROB;
OUTPUT OUT=auto_ar1 P=pred PM=predm R=rest LCL=lcl LCLM=lclm UCL=ucl
UCLM=uclm;
RUN; QUIT;

TITLE 'Example 8: Analysis as autoregressive model with backward
algorithm';
PROC AUTOREG DATA=airtemp /*Backward algorithm results in AR(4) */;
MODEL Temp=Year /NLAG=12 DW=12 BACKSTEP DWPROB;
OUTPUT OUT=auto_ar4 P=pred PM=predm R=rest LCL=lcl LCLM=lclm UCL=ucl
UCLM=uclm;
RUN; QUIT;

TITLE 'Example 8: Figure 18 at top left';
PROC SGPLOT DATA=auto_ar1;
TITLE 'Analysis by PROC AUTOREG';
BAND X=year LOWER=lcl UPPER=ucl/ TRANSPARENCY=0.6
LEGENDLABEL="Prediction interval" NAME="band1";
BAND X=year LOWER=lclm UPPER=uclm/ TRANSPARENCY =0.3
LEGENDLABEL="Confidence interval" NAME="band2";
SCATTER x=year y=temp/ MARKERATTRS=(SIZE=7 SYMBOL=circlefilled) ;
SERIES x=year y=predm / LINEATTRS=(COLOR=verydarkblue THICKNESS=2 ) ;
SERIES x=year y=pred / LINEATTRS=(COLOR=verydarkred THICKNESS =2 ) ;
XAXIS LABEL ="Year " LABELATTRS=(SIZE=12)  valueATTRS=(SIZE=10);
YAXIS LABEL ="Temperature [grd C]" LABELATTRS=(SIZE=12)
VALUEATTRS=(SIZE=10) ;
KEYLEGEND "band1" "band2" / LOCATION=inside
POSITION=bottomright;
RUN; QUIT;

/*Example 8: Figure 18 at top right analogous to Figure 18 at top left
with datafile = auto_ar4*/
TITLE 'Example 8: Figure 18 at bottom';
PROC NLIN DATA=airtemp PLOTS=all;
PARMS a=7 b=1 c=1960 d=11 e=0.03 ;
MODEL temp=a-b*sin((year-c)*2*3.14/d)+e*year;
OUTPUT OUT=NLIN PREDICTED=pred L95=lcl L95M=lclm U95=ucl U95M=uclm;
RUN; QUIT;

TITLE 'Example 8: Figure 19 ';
PROC NLIN DATA=airtemp PLOTS=all CONVERGEPARM=1E-7;
PARMS a=0.034 b=-60 Year0=1985 ;
IF (Year >Year0) THEN MODEL Temp=a*Year + b;
ELSE MODEL Temp=a*Year0+b;
RUN; QUIT;
```

Appendix 9.

Solutions to Exercises

1. The scatterplots of the data for the two levels 1 and 2 of Org give the impression that a quadratic function of N may be suitable to describe the relation. It is not clear whether this is also a suitable approach for the variant without organic fertilization.
2. Assuming a quadratic function and considering the fact that the design is a randomized complete block design, the coincidence test rejects the hypothesis of coincidence with a Type 1 error rate of 0.05

(F-value = (2552.716682-1962.3931)/(10-4)/5.826405 =16.886 and p value < 0.0001).

To obtain the intermediate results for the coincidence test we used PROC GLM with the following CLASS and MODEL statements:

```
CLASS org block;
MODEL yield = org block N N*org N*N N*N*org/s;
CLASS block;
MODEL yield = block N N*N /s;
```

Alternatively, a joint test for coincidence is possible using a contrast statement in PROC GLM:

```
CLASS block org;
MODEL yield=block org org*n org*N*N / SOLUTION;
CONTRAST 'coincidence' org 1 -1 0,
                       org 1 0 -1,
                       org*N 1 -1 0,
                       org*N 1 0 -1,
                       org*N*N 1 -1,
                       org*N*N 1 0 -1;
```

The estimates and tests for the intercepts averaged across the blocks can be obtained by the statements

```
ESTIMATE 'intercept Org=0' intercept 3 block 1 1 1 org 3 0 0/ DIVISOR=3;
ESTIMATE 'intercept Org=1' intercept 3 block 1 1 1 org 0 3 0/ DIVISOR=3;
ESTIMATE 'intercept Org=2' intercept 3 block 1 1 1 org 0 0 3/ DIVISOR=3;
```

Based on the common estimated residual variance $s^2 = 5.8264$, the results for all regression parameters are:

Parameter	Estimate	Standard Error	DF	t Value	P > |t|
		Org = 0			
Intercept	36.4467	1.3936	22	26.15	< 0.0001
b1	0.08258	0.05131	22	1.61	0.1218
b2	0.00013	0.000348	22	0.37	0.7115
		Org = 1			
Intercept	34.6407	1.3583	22	25.5	< 0.0001
b1	0.5711	0.04363	22	13.09	< 0.0001
b2	-0.00291	0.000279	22	-10.45	< 0.0001
		Org = 2			
Intercept	36.3637	1.3583	22	26.77	< 0.0001
b1	0.3741	0.04363	22	8.57	< 0.0001
b2	-0.00197	0.000279	22	-7.07	< 0.0001

All parameters of the two levels with organic fertilization (Org = 1 and 2) are significantly different from zero; the regression coefficients of the variant Org = 0 are not significantly different from zero. The reason for their non-significance may be that in this case a linear function is appropriate and the quadratic approach is overparametrized (the tests are based on the partial approach).

As a consequence, polynomial functions of different order need to be considered for the different levels of organic fertilization, a situation which we did not discuss in this chapter. The need for this somewhat more complicated analysis does not arise if for all three levels of Org the same function type (e.g., a linear or quadratic function) is appropriate. We demonstrate two approaches to take this in consideration. The first one separates the analysis in two parts: the analysis for Org = 0 and the joint analysis for Org = 1 and 2. The second approach does not separate the analysis for the three Org levels using a trick which has several advantages compared to the first approach.

3.1 Approach with separate analyses of Org = 0 and joint analysis of Org = 1 and 2

If the function type is not the same for all levels of a factor, the first idea may be to analyze the data separately for each level. If we would analyze the three organic variants separately and consider the block structure of the experiment, for each variant different block means would be estimated. The efficient analysis of the RCB design requires, however, that all variants are analysed jointly. Moreover, we know from the F-test that there are significant differences between blocks. To consider the same block effects for all variants, we estimate block effects based on a joint analysis of all treatments as block effect = (total mean – mean of all values of the corresponding block), subtract these estimates from all observed values and then analyze the corrected data without block effects separately for each function type (Org = 0 and Org =1 and 2). In doing so, we do not regress observed values but corrected values (observed values – block effects) on the N fertilization rates. Therefore the degrees of freedom of the residuals have to be corrected.

3.1.1 Regression for the variant without organic fertilization

Due to our supposition that a linear approach is better suited for Org = 0, we choose a sequential approach starting with the linear term followed by a quadratic. For a polynomial regression analysis with a linear and a quadratic term and nine observed values, the DF of the residuals would normally be N-3 = 6. To consider the correction of the observed values by the block effects (DF blocks = 2), we set the denominator DF to N-3-2 = 4. This can be achieved in PROC MIXED by

```
MODEL yield_corr= N N2 / df=4,4 HTYPE=1;
```

The corrected yield corresponds to the variable `yield _ corr`. The parameter estimates are the same as above in the joint analysis and $s^2 = 7.2204$. The sequential F-tests confirm our supposition that a linear function is better suited.

Effect	Numerator DF	Denominator DF	*F* Value	*P* > *F*
N	1	4	49.63	0.0021
N2	1	4	0.11	0.7534

Finally, the fitted linear regression function is the following (using $s^2 = 6.3057$):

Parameter	Estimate	Standard Error	DF	*t* Value	*P* > |*t*|
Intercept	36.3071	1.3971	5†	25.99	< 0.0001
b1	0.1012	0.01342	5	7.54	0.0007

†For the linear function, we set the denominator DF = N-2-2 = 5.

3. 1. 2 Regressions for the variants with organic fertilization

Again, we use the corrected values. The denominator DF have been set to N-6-2 = 24-6-2 = 16 because the model has 6 regression parameters. The residual variance is common for both variants and is estimated as $s^2 = 4.7144$. Again, the regression parameter estimates are the same as in the joint analysis of all three variants.

Parameter	Estimate	Standard Error	DF	*t* Value	*P* > |*t*|
		Org = 1			
intercept	34.6407	1.2218	16	28.35	< 0.0001
b1	0.5711	0.03924	16	14.55	< 0.0001
b2	-0.00291	0.000251	16	-11.62	< 0.0001
		Org = 2			
intercept	36.3637	1.2218	16	29.76	< 0.0001
b1	0.3741	0.03924	16	9.53	< 0.0001
b2	-0.00197	0.000251	16	-7.86	< 0.0001

The disadvantage of this approach with separation of the analysis by the function type is that the estimation of the residual variance is not based on all observed values and that degrees of freedom are lost for the tests. Therefore, we do not discuss further options under this approach (test of coincidence of parameters or variance heterogeneity for Org = 1 and 2) and recommend instead the following joint analysis of all three levels of Org.

3.2 Regressions based on a joint analysis for Org = 0, 1, and 2

We want to do a joint analysis fitting a linear regression for Org = 0 and a quadratic regression for Org = 1, 2. The trick to suppress the quadratic term for Org = 0 is based on the definition of an auxiliary variable denoted as *switch* (Piepho et al., 2006). *switch* is equal to 0 if Org = 0 and it is equal to 1 if Org = 1 or 2 so that the variable switch*org*N*N is equal to 0 if Org = 0 and it is equal to org*N*N if Org = 1 or 2.

With this new variable *switch*, we analyze the data with the following CLASS and MODEL statements in PROC GLM and test the coincidence:

```
CLASS block org;
MODEL yield=block org org*n switch*org*N*N / SOLUTION;
CONTRAST 'coincidence' org 1 -1 0,
                       org 1 0 -1,
                       org*N 1 -1 0,
                       org*N 1 0 -1,
                       switch*org*N*N 1 -1,
                       switch*org*N*N 1 0 -1;
```

Whether the three intercepts, the three linear terms, and the two quadratic terms (Org =1 and 2) coincide can be tested by

```
CONTRAST 'coincidence intercept' org 1 -1 0,
                                 org 1 0 -1;
CONTRAST 'coincidence linear'    org*N 1 -1 0,
                                 org*N 1 0 -1;
CONTRAST 'coincidence quadratic' switch*org*N*N 0 1 -1;
```

The results are the following:

Contrast	DF	Contrast SS	Mean Square	*F* Value	*P* > *F*
joint coincidence	5	1337.95021	267.590042	47.71	< 0.0001
coincidence intercept	2	6.072477	3.036238	0.54	0.5892
coincidence linear	2	778.172377	389.086188	69.37	< 0.0001
coincidence quadratic	1	33.27615	33.27615	5.93	0.023

We see that the three intercepts coincide, whereas the linear and quadratic terms do not. The corresponding estimates and tests are shown in the table below.

The intercepts have been calculated as averages across blocks and the tests are based on the common estimated residual variance $s^2 = 5.6086$:

Parameter	Estimate	Standard Error	DF	*t* Value	*P* > |*t*|
		Org = 0			
Intercept	36.3071	1.31758	23	27.56	< 0.0001
b1	0.1012	0.01266	23	7.99	< 0.0001
		Org = 1			
Intercept	34.6407	1.33269	23	25.99	< 0.0001
b1	0.5711	0.04280	23	13.34	< 0.0001
b2	-0.00291	0.000273	23	-10.65	< 0.0001
		Org = 2			
Intercept	36.3637	1.33269	23	27.29	< 0.0001
b1	0.3741	0.04280	23	8.74	< 0.0001
b2	-0.00197	0.000273	23	-7.21	< 0.0001

The estimated regression parameters are identical to those obtained by the first approach; the test results differ slightly due to differences in the residual variance used and differences in the DF.

The intercepts of the three levels of org do not differ significantly, so we fit a model with a common intercept. At first, we test with PROC MIXED whether variant-specific residual variances or a common one should be assumed for this model.

```
CLASS block org;
MODEL yield=block org*n switch*org*N*N / SOLUTION;
REPEATED/ GROUP = org;
```

The LR-test for the comparison of the two models indicates a better fit of the model with a common variance (p value = 0.1115). Therefore, we delete the repeated statement in PROC MIXED.

The common intercept in the mean of the blocks can be estimated and tested by the statement

```
ESTIMATE 'intercept ' intercept 3 block 1 1 1 /DIVISOR=3;
```

Finally, we obtain the following results assuming a common variance ($s^2 = 5.4028$):

Parameter	Estimate	Standard Error	DF	*t* Value	*P* > |*t*|
		Org = 0			
Intercept	35.7746	0.7523	25	47.55	< 0.0001
b1	0.1053	0.00943	25	11.17	< 0.0001
		Org = 1			
Intercept	35.7746	0.7523	25	47.55	< 0.0001
b1	0.5460	0.03472	25	15.73	< 0.0001
b2	-0.00279	0.00024	25	-11.47	< 0.0001
		Org = 2			
Intercept	35.7746	0.7523	25	47.55	< 0.0001
b1	0.3871	0.03472	25	11.15	< 0.0001
b2	-0.00203	0.00024	25	-8.35	< 0.0001

Reference

Piepho, H.P., E.R. Williams, and M. Fleck. 2006. A note on the analysis of designed experiments with complex treatment structure. HortScience 41:446–452.

CHAPTER 7: ANALYSIS AND INTERPRETATION OF INTERACTIONS OF FIXED AND RANDOM EFFECTS

Mateo Vargas, Barry Glaz, Jose Crossa, and Alex Morgounov

Appendix 1: SAS code for orthogonal polynomial contrasts and graphs SAS code for calculating the orthogonal polynomial contrasts

The following SAS code can be used for calculating the orthogonal polynomial coefficients for any contrast of interest and applied for the analyses of the example shown in Section 1. This procedure generates the correct coefficients for levels or rates that are equally or unequally spaced (as for the P rates we used). It is necessary to have installed the Interactive Matrix Language (**IML**) procedure. Our example here will show only how to calculate coefficients for P, because this is our only factor with quantitative rates or levels > 2. We show the code for calculating the contrasts for the main effect of P, and exclusively one two-way (from three), and one three-way (from three), and the unique four-way interactions with P, the extension to the other interactions is straightforward. We are including the four-way interaction contrast, even though this was not included in the reduced model. In Appendix 2, we show the complete code for calculating exhaustively all the possible contrasts involving P in the four-way analysis of variance

SAS macro program

```
**** Reading data: Data Wheat is available in attached CSV file ****;
Data Wheat;
      Infile "C:\Experiment 1 Data Wheat.CSV" dlm="," firstobs=3;
      Informat Soil$ 10.;
      Input Year Soil N P Rep Yield ;
Datalines;
Run;
**** Here begins the macro code for calculating the orthogonal
polynomial coefficients in an automatic way ****;
%Macro Coefficients;
Proc IML;
Plevels = {0, 50, 150, 250};
Coeff = Orpol(Plevels,3);
Ncoef = ncol(coeff)-1;
Call symputx("ncoef", ncoef);
      %do K = 1%to &ncoef;
            CoefGrade&K = t(Coeff[,&K+1]);
            PosCoefGrade&K = rowcat(char( CoefGrade&K ,15,10));
            NegCoefGrade&K = rowcat(char(-(CoefGrade&K),15,10));
            Call symputx("PosCoefGrade&K", PosCoefGrade&K);
            Call symputx("NegCoefGrade&K", NegCoefGrade&K);
      %end;
Run;
Title1 "Four-way ANOVA, decomposing df for P into three contrasts";
Proc GLIMMIX Data = Wheat;
      Class Year Soil N P Rep;
      Model Yield = Year | Soil | N | P;
      Random Rep(Year Soil);
**** P main effects contrasts ****;
```

```
Contrast "Linear P" P &PosCoefGrade1 ;
Contrast "Quadratic P" P &PosCoefGrade2 ;
Contrast "Cubic P" P &PosCoefGrade3 ;
**** Year × P two-way interaction contrasts ****;
Contrast "Linear Y*P" Year*P &PosCoefGrade1 &NegCoefGrade1 ;
Contrast "Quadratic Y*P" Year*P &PosCoefGrade2 &NegCoefGrade2 ;
Contrast "Cubic Y*P" Year*P &PosCoefGrade3 &NegCoefGrade3 ;
**** Year × Soil × P three-way interaction contrasts ****;
Contrast "Linear Y*S*P" Year*Soil*P
     &PosCoefGrade1 &NegCoefGrade1 &NegCoefGrade1 &PosCoefGrade1;
Contrast "Quadratic Y*S*P" Year*Soil*P
     &PosCoefGrade2 &NegCoefGrade2 &NegCoefGrade2 &PosCoefGrade2;
Contrast "Cubic Y*S*P" Year*Soil*P
     &PosCoefGrade3 &NegCoefGrade3 &NegCoefGrade3 &PosCoefGrade3;
**** Year × Soil × N × P four-way interaction contrasts ****;
Contrast "Linear Y*S*N*P" Year*Soil*N*P
     &PosCoefGrade1 &NegCoefGrade1 &NegCoefGrade1 &PosCoefGrade1
     &NegCoefGrade1 &PosCoefGrade1 &PosCoefGrade1 &NegCoefGrade1;
Contrast "Quadratic Y*S*N*P" Year*Soil*N*P
     &PosCoefGrade2 &NegCoefGrade2 &NegCoefGrade2 &PosCoefGrade2
     &NegCoefGrade2 &PosCoefGrade2 &PosCoefGrade2 &NegCoefGrade2;
Contrast "Cubic Y*S*N*P" Year*Soil*N*P
     &PosCoefGrade3 &NegCoefGrade3 &NegCoefGrade3 &PosCoefGrade3

     &NegCoefGrade3 &PosCoefGrade3 &PosCoefGrade3 &NegCoefGrade3;
Run;
%Mend;
     %Coefficients;
Run;
```

Brief explanation of preceding SAS code

In the **Plevels** statement it is necessary to express how many and which levels of the variable you wish to calculate. For instance, for the data set we used in Section 1 of this chapter, P has 4 unevenly spaced rates: 0, 50, 150, and 250 kg ha^{-1}. The **Orpol** function is used for obtaining the required coefficients; since there are three degrees of freedom, we can calculate three polynomial coefficients––linear, quadratic, and cubic––which are assigned to macro variables that are later used in the data step in the complete four-way ANOVA and contrasts.

In the contrast statements for the main effects, we need only positive values of the coefficients, but in the two-, three, and four-way contrasts, a combination of positive and negative coefficients are required, depending on the levels of each of the factors involved in those interactions. In Appendix 2, we show the procedure for obtaining the corresponding order of signs for each contrast. These positive and negative coefficients are assigned to the **PosCoefGrade** and **NegCoefGrade** macro variables, respectively, using the **Call Symput** function. The order of the polynomial contrasts is obtained using the **%do** - **%end** cycle. Note that the appropriate number and grade of the coefficients are automatically determined in the **ncoef = ncol(coeff)-1** statement. This program can be easily modified for different numbers and/or levels of factors.

It is important to mention that the order of the interaction coefficients in the CONTRAST statement depends on the order of the factors that are listed in the CLASS statement. If this is not the case the program may be executed without any errors but you may not get the intended contrast. This will be shown in a more detailed manner in the Exercise 1 of the Appendix 3.

Once we have the appropriate coefficients, the contrasts can be calculated using SAS procedures such as GLM, MIXED, or GLIMMIX.

SAS Code for Graphing Interactions

Graphing the two-way interaction Year × P

The following code is useful for graphing a two-way interaction, the generalization to a three-way or four-way interaction is straightforward as will be shown in Appendix 2.

```
**** Reading data: Data Wheat is available in attached CSV file ****;
Data Wheat;
      Infile "C:\Experiment 1 Data Wheat.CSV" dlm="," firstobs=3 ;
      Informat Soil$ 10. ;
      Input Year Soil N P Rep Yield ;
Datalines;
Run;
ODS select Covparms Tests3 Lsmeans LSMLines Meanplot; Title1 " ";
Title2 "Four-way ANOVA and LSD for Year x P Interaction using the
Confidence Interval" ;
Proc GLIMMIX Data = Wheat ;
      Class Year Soil N P Rep ;
      Model Yield = Year Soil N P
                    Year*Soil Year*N Year*P Soil*N Soil*P
                    Year*Soil*N Year*Soil*P Soil*N*P ;
                    Random Rep(Year Soil);
      LSMeans Year*P / Lines CL Plot = mean (sliceby = Year Join CL);
      ODS OUTPUT LSMeans = LSMeans ;
Run;
***** Generating the different curves to be used in graphing the LSD bars *****;
Data Graph;
      Set LSMeans ;
      HWCI=Estimate-Lower;
      LSD = Sqrt(2)*(HWCI);
      if year = 2007 then Y_07=Estimate;
      if year = 2008 then Y_08=Estimate;
      if year = 2007 then do;
          Yield1=Estimate; output;
          Yield1=Estimate - LSD; output;
          Yield1=Estimate + LSD; output;
      end;
      if year = 2008 then do;
Yield2=Estimate; output;
Yield2=Estimate -LSD;
output;
Yield2=Estimate +LSD;
output;
end;
Proc GPlot Data = Graph ;
      Plot (Y_07 Y_08)*P (Yield1 Yield2)*P / frame overlay
                         vaxis = axis1 haxis = axis2 nolegend;
Symbol1 v=dot cv=black h = 2.0 l=1 w=2 i=rq ci=black ;
Symbol2 v=dot    cv=red    h = 2.0   l=1 w=2  i=RQ        ci=Red ;
Symbol3                              l=1 w=2  i=hiloct    ci=black ;
Symbol4                              l=1 w=2  i=hiloct    ci=red ;
axis1 length = 4.5 in order = (0.8 to 2.4 by 0.4)
label=(f=Albany h=2.0 a=90 r=0 "Grain yield (Mg ha-1) ")
      value=(f=Albany
      h=2.0 ) offset = (1) minor=none;
axis2 length = 4.5 in order = (0 to 250 by 50)
      label=(f=Albany h=2.0 "P fertilizer rate (kg ha-1) ")
      value=(f=Albany h=2.0 ) offset = (3) minor=none;
Run;
```

Brief explanation of preceding SAS code

The output delivery system (ODS), ***ODS select Covparms Tests3 LSMLines Meanplot*** is for saving exclusively the useful information. ***ODS Output LSMeans = LSMeans*** statement is useful for creating a temporary file with only the information that will be needed later. For generating the different variables containing information for the regression lines associated with each combination of the Year × P interaction, we created the new variables **Y_07** and **Y_08** from the yield values. Similarly, cycles **if – then do – end** are used for obtaining the information needed in the upper and lower LSD bars. The codes can be adapted if, for example, four regression lines need to be depicted with their corresponding LSD error bars, as in the Soil × N × P interaction shown in Fig. 1.3 from Section 1 and as provided with code in Example 2 of Appendix 2, and so on to any number of regression lines needed.

If we are interested in the ANOVA for only the terms found to be significant in the final model (Table 1.2), i.e. the four main effects, five two-way, and three three-way interactions, without including the four-way interaction; all those terms should be included in the model statement when computing the correct LSD and/or the confidence interval. In the LSMeans statement, we have included only the interaction that is of interest for graphing the LSD values, thus simplifying the output.

The selected output includes only three sections: with the covariance parameters estimates and the Type III test of fixed effects we have obtained all the information shown in Table 1.2.

The GLIMMIX Procedure

Covariance Parameter Estimates

Cov Parm	Estimate	Standard Error
Rep(Year*Soil)	0.000222	0.000953
Residual	0.008593	0.002084

Type III Tests of Fixed Effects

Effect	Num DF	**Den DF**	***F* Value**	***P* > *F***
Year	1	4	1204.82	< 0.0001
Soil	1	4	556.51	< 0.0001
N	1	34	11.36	0.0019
P	3	34	37.75	< 0.0001
Year*Soil	1	4	125.30	0.0004
Year*N	1	34	9.45	0.0041
Year*P	3	34	6.40	0.0015
Soil*N	1	34	19.80	< 0.0001
Soil*P	3	34	13.72	< 0.0001
Year*Soil*N	1	34	33.31	< 0.0001
Year*Soil*P	3	34	7.94	0.0004
Soil*N*P	6	34	3.21	0.0131

From the Year*P Least Squares Means (lsmeans) table we can calculate the LSD value considering the confidence interval given by the upper and lower values, and then multiplying the half width of the interval (upper value minus the estimate or estimate minus lower value) by the square root of 2, as shown in the SAS code in the statements ***HWCI=Estimate-Lower; LSD = Sqrt(2)*(HWCI)***.

The estimates for the least squares means are then used for graphing the response curves by means of the ***GPLOT*** Procedure and using an interpolation method; the highest degree of the orthogonal polynomial contrast that was found to be significant in that particular two-way interaction, in this case for Year × P interaction in Table 1.2 was the quadratic contrast. This was performed with the code ***i=rq***, were ***i*** means interpolation and ***rq*** means Regression Quadratic, thus using the following statements and obtaining the graph shown below.

```
Symbol1 v=dot cv=black h = 2.0 l=1 w=2 i=rq ci=black ;
Symbol2 v=dot cv=red h = 2.0 l=1 w=2 i=rq ci=red ;
```

Year*P Least Squares Means

Year	P	Estimate	Standard Error	DF	t Value	P >|t|	Alpha	Lower	Upper
2007	0	1.6537	0.03361	34	49.20	< 0.0001	0.05	1.5854	1.7221
2007	50	1.8600	0.03361	34	55.34	< 0.0001	0.05	1.7917	1.9283
2007	150	2.0537	0.03361	34	61.10	< 0.0001	0.05	1.9854	2.1221
2007	250	2.0838	0.03361	34	62.00	< 0.0001	0.05	2.0154	2.1521
2008	0	0.9138	0.03361	34	27.19	< 0.0001	0.05	0.8454	0.9821
2008	50	1.0213	0.03361	34	30.38	< 0.0001	0.05	0.9529	1.0896
2008	150	1.0563	0.03361	34	31.43	< 0.0001	0.05	0.9879	1.1246
2008	250	1.1250	0.03361	34	33.47	< 0.0001	0.05	1.0567	1.1933

The next section of output includes the LSD grouping of the lsmeans which are calculated using the option ***Lines*** in the ***LSMeans*** statement as follows:

```
LSMeans Year*P / Lines CL Plot = mean (sliceby = Year Join CL);
```

T Grouping for Year*P
Least Squares Means (Alpha=0.05)

LS-means with the same letter are not significantly different.

Year	P	Estimate		
2007	250	2.0838		A
2007	150	2.0537		A
2007	50	1.8600		B
2007	0	1.6537		C
2008	250	1.1250		D
2008	150	1.0563	E	D
2008	50	1.0213	E	
2008	0	0.9138		F

Finally, the option ***Plot = mean (sliceby = Year Join CL)***, directly creates the following graph of means responses for each year (sliceby=year), joining the mean values using a straight line (Join) and including the 95% confidence limits (CL).

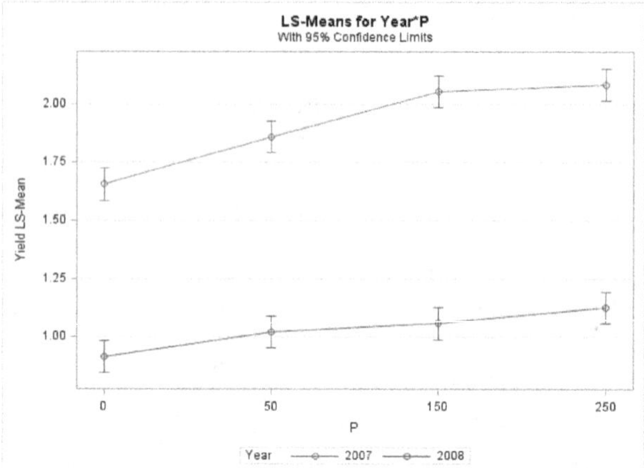

As you can see these graphs are directly and easily obtained using a single statement. However, the main limitation for this kind of graph is that you can only use straight lines as response curves and you cannot use a more sophisticated interpolation method as you can using the ***GPLOT*** procedure.

In Appendix 2, we show other examples for obtaining more complicated graphs, for three- and four-way interactions, and we will show that the extension from this two-way interaction case is straightforward.

Appendix 2: Generalization of SAS code programs shown in Appendix 1

Example 1: SAS macro code for calculating orthogonal polynomial coefficients

In this Appendix, we describe the complete SAS code for calculating all possible orthogonal polynomial contrasts involving P in the data set Wheat. This will include the contrasts for the main effect of P, the six two-way interactions, the four three-way interactions, and the four- way interaction. Although we recommended not including the four-way interaction in our reduced analysis in Section 1, as a resource for readers, we will include here the code for the contrasts associated with this interaction.

As stated in Appendix 1, it is necessary to have installed the Interactive Matrix Language (**IML**) procedure for calculating the orthogonal polynomial coefficients. In the contrast statements for the main effects, we need only positive values of the coefficients, but in the two-, three, and four-way contrasts, positive and negative coefficients are required depending on the levels of each of the factors involved in those interactions. For obtaining the order of signs of the coefficients in the corresponding contrast, one can use Table A2.1 (in this appendix) as an example. This table is only for the linear contrasts (CoefGrade1) where CoefGrade1 is a vector of length 1 × 4 (1 row, 4 columns), one coefficient value for each P rate.

One may use the coefficient -1 for Year 1 and the coefficient +1 for Year 2, and so on, for any factor: -1 for Black soil and +1 for Chesnut soil; -1 for the first rate of N (N0) and +1 for the second rate of N (N30). In fact, this selection of signs is what is used in a factorial experiment 2^k, where 2 is the number of levels of each factor and *k* is the number of factors. The order of positive and negative coefficients is not important because the significance of each contrast is based on sum of squares of differences among the levels tested. Therefore, the sign in these differences is cancelled by squaring them. However, if you are using the *Estimate* statement instead of the *Contrast* statement, the sign is important, because the interpretation would reverse. The significance of the comparison does not differ between the *Contrast* and *Estimate* statements. That is, the standard errors, the t values, and the p-values are exactly the same despite the choice of which level to consider as -1 or +1. Again, the significance of a *Contrast* or an *Estimate* statement, whichever you prefer to use, will be equivalent.

Table A2.1. Coefficients for the linear contrast for P main effect, two-way, three-way, and four-way interactions.

Effect	Factor 1	Coefficient	Factor 2	Coefficient	Factor 3	Coefficient	Phosphorus fertilizer rates
P main effect	P	1					PosCoefGrade1
Two-way Y × P Interaction	Year 1	1					PosCoefGrade1
	Year 2	-1					NegCoefGrade1
Three-way Y × Soil × P Interaction	Year 1	1	Black	1			PosCoefGrade1
			Chestnut	-1			NegCoefGrade1
	Year 2	-1	Black	1			NegCoefGrade1
			Chestnut	-1			PosCoefGrade1
Four-way Y × Soil × N × P Interaction	Year 1	1	Black	1	N_0	1	PosCoefGrade1
					N_{30}	-1	NegCoefGrade1
			Chestnut	-1	N_0	1	NegCoefGrade1
					N_{30}	-1	PosCoefGrade1
	Year 2	-1	Black	1	N_0	1	NegCoefGrade1
					N_{30}	-1	PosCoefGrade1
			Chestnut	-1	N_0	1	PosCoefGrade1
					N_{30}	-1	NegCoefGrade1

Therefore, using the sequence of signs shown in Table A2.1, the complete SAS code for estimating all the possible polynomial contrasts involving the P factor, is the following:

SAS macro program

```
**** Reading data: Data Wheat is available in attached CSV file ****;
Data Wheat;
     Infile "C:\Experiment 1 Data Wheat.CSV" dlm="," firstobs=3 ;
     Informat Soil$ 10. ;
     Input Year Soil N P Rep Yield ;
Datalines;
Run;
**** Here begins the macro code for calculating the orthogonal
polynomial contrasts in an automatic way ****;
%Macro Coefficients;
Proc IML;
     Plevels = {0, 50, 150, 250};
     Coeff = Orpol(Plevels,3);
     Ncoef = ncol(coeff)-1;
     Call symputx("ncoef", ncoef);
     %do K = 1%to &ncoef;
          CoefGrade&K = t(Coeff[,&K+1]);
          PosCoefGrade&K = rowcat(char( CoefGrade&K ,15,10));
          NegCoefGrade&K = rowcat(char(-(CoefGrade&K),15,10));
          Call symputx("PosCoefGrade&K", PosCoefGrade&K);
          Call symputx("NegCoefGrade&K", NegCoefGrade&K);
     %end;
Run;
Title1 "Four-way ANOVA, decomposing df for P into three contrasts";
Proc GLIMMIX Data = Wheat;
     Class Year Soil N P Rep;
     Model Yield = Year | Soil | N | P;
     Random Rep(Year Soil);
**** P main effects contrasts ****;
Contrast "P Linear " P &PosCoefGrade1 ;
Contrast "P Quadratic " P &PosCoefGrade2 ;
Contrast "P Cubic" P &PosCoefGrade3 ;
**** Year × P two-way interaction contrasts ****;
Contrast "Y × P Linear " Year*P &PosCoefGrade1 &NegCoefGrade1 ;
Contrast "Y × P Quadratic" Year*P &PosCoefGrade2 &NegCoefGrade2 ;
```

```
Contrast "Y × P Cubic" Year*P &PosCoefGrade3 &NegCoefGrade3 ;
**** Soil × P two-way interaction contrasts ****;
Contrast "S × P Linear" Soil*P &PosCoefGrade1 &NegCoefGrade1 ;
Contrast "S × P Quadratic" Soil*P &PosCoefGrade2 &NegCoefGrade2 ;
Contrast "S × P Cubic" Soil*P &PosCoefGrade3 &NegCoefGrade3 ;
**** N × P two-way interaction contrasts ****;
Contrast "N × P Linear" N*P &PosCoefGrade1 &NegCoefGrade1 ;
Contrast "N × P Quadratic" N*P &PosCoefGrade2 &NegCoefGrade2 ;
Contrast "N × P Cubic " N*P &PosCoefGrade3 &NegCoefGrade3 ;
**** Year × Soil × P three-way interaction contrasts ****;
Contrast "Y × S × P Linear" Year*Soil*P
      &PosCoefGrade1 &NegCoefGrade1 &NegCoefGrade1 &PosCoefGrade1 ;
Contrast "Y × S × P Quadratic" Year*Soil*P
      &PosCoefGrade2 &NegCoefGrade2 &NegCoefGrade2 &PosCoefGrade2 ;
Contrast "Y × S × P Cubic" Year*Soil*P
      &PosCoefGrade3 &NegCoefGrade3 &NegCoefGrade3 &PosCoefGrade3 ;
**** Year × N × P three-way interaction contrasts ****;
Contrast "Y × N × P Linear" Year*N*P
      &PosCoefGrade1 &NegCoefGrade1 &NegCoefGrade1 &PosCoefGrade1 ;
Contrast "Y × N × P Quadratic" Year*N*P
      &PosCoefGrade2 &NegCoefGrade2 &NegCoefGrade2 &PosCoefGrade2 ;
Contrast "Y × N × P Cubic" Year*N*P
      &PosCoefGrade3 &NegCoefGrade3 &NegCoefGrade3 &PosCoefGrade3 ;
**** Soil × N × P three-way interaction contrasts ****;
Contrast "S × N × P Linear" Soil*N*P
      &PosCoefGrade1 &NegCoefGrade1 &NegCoefGrade1 &PosCoefGrade1 ;
Contrast "S × N × P Quadratic" Soil*N*P
      &PosCoefGrade2 &NegCoefGrade2 &NegCoefGrade2 &PosCoefGrade2 ;
Contrast "S × N × P Cubic" Soil*N*P
      &PosCoefGrade3 &NegCoefGrade3 &NegCoefGrade3 &PosCoefGrade3 ;
**** Year × Soil × N × P four-way interaction contrasts ****;
Contrast "Y × S × N × P Linear" Year*Soil*N*P
      &PosCoefGrade1 &NegCoefGrade1 &NegCoefGrade1 &PosCoefGrade1
      &NegCoefGrade1 &PosCoefGrade1 &PosCoefGrade1 &NegCoefGrade1;
Contrast "Y × S × N × P Quadratic" Year*Soil*N*P
      &PosCoefGrade2 &NegCoefGrade2 &NegCoefGrade2 &PosCoefGrade2
      &NegCoefGrade2 &PosCoefGrade2 &PosCoefGrade2 &NegCoefGrade2;
Contrast "Y × S × N × P Cubic" Year*Soil*N*P
      &PosCoefGrade3 &NegCoefGrade3 &NegCoefGrade3 &PosCoefGrade3
      &NegCoefGrade3 &PosCoefGrade3 &PosCoefGrade3 &NegCoefGrade3;
Run;
%Mend;
%Coefficients;
Run;
```

This SAS code was discussed previously in Appendix 1. Here, we are only adding all possible contrasts.

The selected output for this complete four-way model

The covariance parameter estimates and the Type III tests of fixed effects were already shown in Table 1.1 of Section 1. We repeat that information here and we also show here the complete results for all the possible contrasts involving the P rates from the Experiment 1 data.

The GLIMMIX Procedure

Covariance Parameter Estimates

Cov Parm	Estimate	Standard Error
Rep(Year*Soil)	0.000292	0.000955
Residual	0.008036	0.002148

Type III Tests of Fixed Effects

Effect	Num DF	Den DF	*F* Value	*P* > *F*
Year	1	4	1204.82	< 0.0001

Soil	1	4	556.51	< 0.0001
N	1	28	12.15	0.0016
P	3	28	40.37	< 0.0001
Year*Soil	1	4	125.30	0.0004
Year*N	1	28	10.11	0.0036
Year*P	3	28	6.84	0.0013
Soil*N	1	28	21.17	< 0.0001
Soil*P	3	28	14.67	< 0.0001
N*P	3	28	1.49	0.2400
Year*Soil*N	1	28	35.62	< 0.0001
Year*Soil*P	3	28	8.50	0.0004
Year*N*P	3	28	0.65	0.5892
Soil*N*P	3	28	5.39	0.0047
Year*Soil*N*P	3	28	2.13	0.1183

Contrasts

Label	Num DF	Den DF	*F* Value	*P* > *F*
P Linear	1	28	106.69	< 0.0001
P Quadratic	1	28	12.60	0.0014
P Cubic	1	28	1.82	0.1886
Y × P Linear	1	28	15.21	0.0006
Y × P Quadratic	1	28	5.09	0.0321
Y × P Cubic	1	28	0.24	0.6285
S × P Linear	1	28	43.81	< 0.0001
S × P Quadratic	1	28	0.19	0.6635
S × P Cubic	1	28	0.00	0.9801
N × P Linear	1	28	2.59	0.1190
N × P Quadratic	1	28	0.48	0.4927
N × P Cubic	1	28	1.39	0.2491
Y × S × P Linear	1	28	24.96	< 0.0001
Y × S × P Quadratic	1	28	0.02	0.8787
Y × S × P Cubic	1	28	0.50	0.4835
Y × N × P Linear	1	28	0.82	0.3718
Y × N × P Quadratic	1	28	0.49	0.4902
Y × N × P Cubic	1	28	0.64	0.4307
S × N × P Linear	1	28	3.00	0.0943
S × N × P Quadratic	1	28	0.07	0.7866
S × N × P Cubic	1	28	13.09	0.0012
Y × S × N × P Linear	1	28	0.50	0.4870
Y × S × N × P Quadratic	1	28	0.01	0.9436
Y × S × N × P Cubic	1	28	5.90	0.0218

Generalization of SAS Code for Graphing Interactions

Example 2: Graphing the three-way interaction Year × Soil × P

Here we present an example for graphing a three-way interaction. For Fig. 1.2, we are interested in the ANOVA for all the terms which were significant, that is, the four main effects, five two-way, and three three-way interactions; thus all those terms should be included in the model statement when computing the correct LSD and/or the confidence interval. However, in the LSMeans statement, we have included only the interactions that are of interest for graphing the LSD values in order to simplify the output.

```
**** Reading data: Data Wheat is available in attached CSV file ****;
Data Wheat;
Infile "C:\Experiment 1 Data Wheat.CSV" dlm="," firstobs=3 ;
```

```
Informat Soil$ 10. ;
Input Year Soil N P Rep Yield ;
Datalines;
Run;
ODS select CovParms Tests3 LSMLines Meanplot;
Title2 "Four-way ANOVA and LSD for Year x Soil x P Interaction using
the Confidence Interval";
Proc GLIMMIX data = Wheat;
      Class Year Soil N P Rep ;
      Model Yield = Year Soil N P
                    Year*Soil Year*N Year*P Soil*N Soil*P
                    Year*Soil*N Year*Soil*P Soil*N*P;
      Random Rep(Year Soil);
LSMeans Year*Soil*P / Lines CL plot=mean(sliceby=Year*Soil Join CL);
ODS Output LSMeans = LSMeans;
Run;
***** Generating the different curves to be used in graphing the LSD
bars *****;
Data Graph;
      Set LSMeans ;
      HWCI = Estimate - Lower;
      LSD = Sqrt(2)*(HWCI);
if Year = 2007 and Soil = "black" then Y1S1 = Estimate ;
if Year = 2007 and Soil = "chestnut" then Y1S2 = Estimate ;
if Year = 2008 and Soil = "black" then Y2S1 = Estimate ;
if Year = 2008 and Soil = "chestnut" then Y2S2 = Estimate ;
if Year = 2007 and Soil = "black" then do;
      Yield1 = Estimate; output;
      Yield1 = Estimate - (LSD/2); output;
      Yield1 = Estimate + (LSD/2); output;
end;
if Year = 2007 and Soil = "chestnut" then do;
      Yield2 = Estimate; output;
      Yield2 = Estimate - (LSD/2); output;
      Yield2 = Estimate + (LSD/2); output;
end;
if Year = 2008 and Soil = "black" then do;
      Yield3 = Estimate; output;
      Yield3 = Estimate - (LSD/2); output;
      Yield3 = Estimate + (LSD/2); output;
end;
if Year = 2008 and Soil = "chestnut" then do;
      Yield4 = Estimate; output;
      Yield4 = Estimate - (LSD/2); output;
      Yield4 = Estimate + (LSD/2); output;
end;
Run;
** Graphics options for creating an enhanced meta-file **;
FILENAME Figure 'C:\Output\Figure 2.1, Y x S x P.EMF';
GOPTIONS DEVICE=SASEMF GSFNAME=Figure GSFMODE=Replace;
Proc Gplot data = Graph ;
      Plot (Y1S1 Y1S2 Y2S1 Y2S2)*P (Yield1 Yield2 Yield3 Yield4)*P /
           overlay frame vaxis = axis1 haxis = axis2 nolegend;
Symbol1 v=dot cv=black h = 2.0 l=1 w=2 i=rl ci=black ;
Symbol2 v=dot cv=blue h = 2.0 l=1 w=2 i=rq ci=blue ;
Symbol3 v=dot cv=green h = 2.0 l=1 w=2 i=rl ci=green ;
Symbol4 v=dot cv=red h = 2.0 l=1 w=2 i=rl ci=Red ;
Symbol5 l=1 w=2 i=hiloct ci=black;
Symbol6 l=1 w=2 i=hiloct ci=blue;
Symbol7 l=1 w=2 i=hiloct ci=green;
Symbol8 l=1 w=2 i=hiloct ci=Red;
axis1 length = 4.5 in order = (0.5 to 3.0 by 0.5)
      label=(f=Albany h=3.0 a=90 r=0 "Grain yield (Mg ha-1)")
      value=(f=Albany h=3.0) offset = (1) minor=none;
axis2 length = 7.0 in order = (0 to 250 by 50)
      label=(f=Albany h=3.0 "P fertilizer rate (kg ha-1)")
      value=(f=Albany h=3.0) offset = (3) minor=none;
Title1 f=Albany h=2.0 "Figure 1.2.- Year × Soil × P Interaction
response profiles";
Run;
```

Brief Description of SAS code

The **ODS select** ***CovParms Tests3 LSMLines Meanplot*** option is used to save the useful and needed output: the covariance parameters of estimates for random terms (***CovParms***); the Type III tests of fixed effects (***Tests3***); the least squares means and their t (LSD) grouping (***LSMLines***); and finally the mean response profiles for the three-way interaction (***Meanplot***). The SAS system is insensitive to small or capital letters, we use a mix of both only for emphasis.

The ***meanplot*** option or simply ***mean*** requests that the least squares means (lsmeans) be displayed. For example, in the line ***LSMeans Year*Soil*P / Lines CL plot=mean(sliceby=Year*Soil Join CL),*** the lsmeans response profiles are requested for the three- way Year × Soil × P interaction. The ***meanplot-options*** controls the display of the least square means; ***join*** or ***connect*** connects the lsmeans with lines; ***Sliceby=Year*Soil*** creates four response profiles coming from the two-way interaction Year × Soil at each P rate; and the ***CL*** code displays upper and lower confidence limits for the lsmeans. By default, 95% limits are drawn. The confidence levels can be changed with the **alpha**= option. In the next example, we will ask for the lsmeans, LSD grouping, and the response profiles for all the main effects, two-way, three-way, and the four-way interactions.

The second part of the SAS program is for calculating and graphing the response profiles for each mean using the ***GPLOT*** procedure, in order to include a different interpolation curve for each profile, rather than simply using lines joined in the default output of the ***GLIMMIX*** procedure.

Using information saved in the temporary file named ***LSMeans*** in the ODS Output ***LSMeans = LSMeans*** statement, we calculate first the half width of the confidence interval using the ***HWCI = Estimate – Lower*** statement, and finally we calculate the LSD value using the expression ***LSD = Sqrt(2)*(HWCI).***

In the next block of statements, using the first statement as an example: ***if Year = 2007 and Soil = "black" then Y1S1 = Estimate*** creates one response profile for the first Year × Soil combination (Year 1, Soil 1) and assigns the values to a new variable ***Y1S1,*** which will be used for graphing the corresponding response profiles, and so on for the other three Year × Soil combinations. Similarly the following ***if-then do-end*** cycle:

```
if Year = 2007 and Soil = "black" then do;
Yield1 = Estimate; output;
Yield1 = Estimate -(LSD/2); output;
Yield1 = Estimate + (LSD/2); output;
end;
```

is used for calculating the center, lower, and upper LSD bars to be included for each response profile at each P rate. This is similar for the other three Year × Soil combinations.

In the ***Plot (Y1S1 Y1S2 Y2S1 Y2S2)*P (Yield1 Yield2 Yield3 Yield4)*P / overlay*** statement for the ***GPLOT*** procedure we ask to simultaneously plot the four response profiles and the four LSD bars associated with each curve. Then, with the ***Symbol*** option, we can use a different interpolation method for each curve, using, for example, ***i=rl, i=rq,*** or ***i=rc,*** for a linear, quadratic or cubic regression, respectively. For drawing the LSD bars, we used the interpolation alternative ***i=hiloct.*** The generalization to a four-way interaction in which we need to include eight response profiles is straightforward from this example.

Complete Selected Output from this example

The results for the covariance parameter estimates and the Type III tests of fixed effects are the same here as those shown previously in Table 1.2 from Section 1, which corresponded to the reduced final model.

The GLIMMIX Procedure

Covariance Parameter Estimates

Cov Parm	Estimate	Standard Error
Rep(Year*Soil)	0.000222	0.000953
Residual	0.008593	0.002084

Type III Tests of Fixed Effects

Effect	Num DF	Den DF	*F* Value	*P* > *F*
Year	1	4	1204.82	< 0.0001
Soil	1	4	556.51	< 0.0001
N	1	34	11.36	0.0019
P	3	34	37.75	< 0.0001
Year*Soil	1	4	125.30	0.0004
Year*N	1	34	9.45	0.0041
Year*P	3	34	6.40	0.0015
Soil*N	1	34	19.80	< 0.0001
Soil*P	3	34	13.72	< 0.0001
Year*Soil*N	1	34	33.31	< 0.0001
Year*Soil*P	3	34	7.94	0.0004
Soil*N*P	6	34	3.21	0.0131

The following lsmeans and their LSD grouping were also shown in Table 1.4 of Section 1.

T Grouping for Year*Soil*P Least

Squares Means (Alpha=0.05)
LS-means with the same letter are not significantly different.

Soil	Year	P	Estimate		
black	2007	250	2.7075		A
black	2007	150	2.5600		B
black	2007	50	2.2075		C
black	2007	0	1.9475		D
chestnut	2007	150	1.5475		E
chestnut	2007	50	1.5125		E
chestnut	2007	250	1.4600	F	E
chestnut	2007	0	1.3600	F	G
black	2008	250	1.3050	H	G
black	2008	150	1.2225	H	
black	2008	50	1.1825	H	
black	2008	0	1.0375		I
chestnut	2008	250	0.9450	J	I
chestnut	2008	150	0.8900	J	K
chestnut	2008	50	0.8600	J	K
chestnut	2008	0	0.7900		K

The following graph showing the four response profiles for the Year × Soil × P interaction was obtained with the statement ***LSMeans Year*Soil*P / Lines CL plot=mean(sliceby =Year*Soil Join CL)***, already explained above.

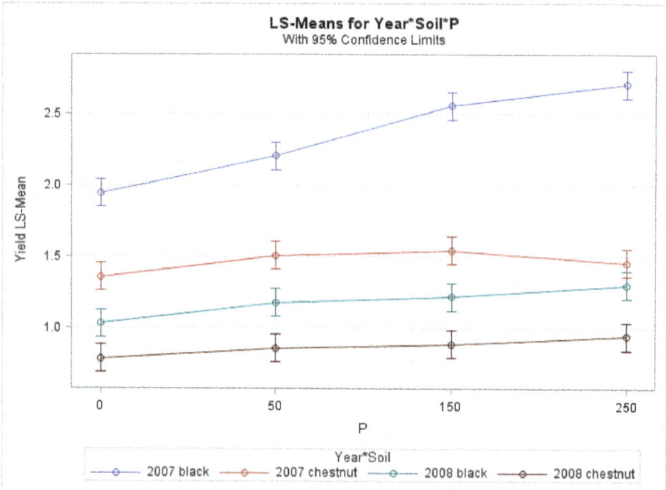

Finally, the next plot was obtained using the ***GPLOT*** procedure and corresponds to Fig. 1.2 shown previously in Section 1 when explaining the significant interactions found in the final model.

Figure 1.2.- Year x Soil x P

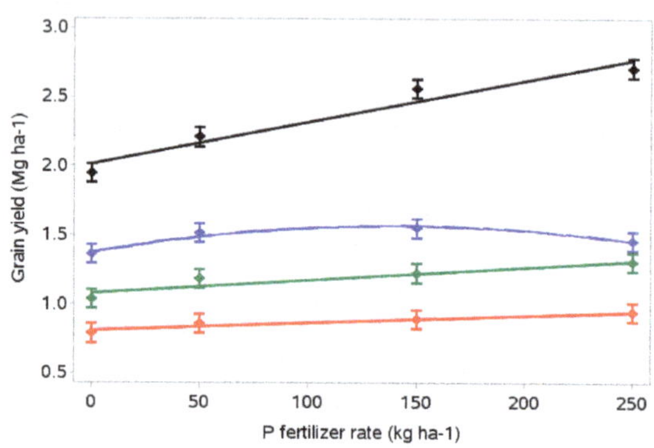

Example 3: Four-way ANOVA, Mean Plots, and LSD values for Main Effects and all Two-, Three-, and Four-Way Interactions.

```
**** Reading data: Data Wheat is available in attached CSV file ****;
Data Wheat;
      Infile "C:\Experiment 1 Data Wheat.CSV" dlm="," firstobs=3 ;
      Informat Soil$ 10. ;
      Input Year Soil N P Rep Yield ; Datalines;
Run;
ODS select CovParms Tests3 Lsmeans Diffs LSMLines Meanplot;
Title2 "Four-way ANOVA and mean plots for all main effects and all
interactions";
Proc GLIMMIX data = Wheat;
      Class Year Soil N P Rep;
      Model Yield = Year | Soil | N | P ;
      Random Rep(Year Soil);
      LSMeans Year | Soil | N | P / PDIFF Lines CL;
      LSMeans Year Soil N P / plot = mean (Join CL);
      LSMeans Year*Soil Year*N Year*P/plot=mean (sliceby=Year Join CL);
```

```
      LSMeans Soil*N Soil*P / plot = mean (sliceby = Soil Join CL);
      LSMeans N*P / plot = mean (sliceby = N Join CL);
      LSMeans Year*Soil*N Year*Soil*P/plot=mean (sliceby=Year*Soil Join CL);
      LSMeans Year*N*P / plot = mean (sliceby = Year*N Join CL);
      LSMeans Soil*N*P / plot = mean (sliceby = Soil*N Join CL);
      LSMeans Year*Soil*N*P / plot=mean (sliceby = Year*Soil*N Join CL);
      ODS output lsmeans = LSMEANS diffs=DIFFS tests3=DOF;
Run;
*** Strategy for Calculating the LSD using the Average Standard Error
of Differences ***;
Proc Sort data = DOF;
      By Effect;
Proc Sort data = DIFFS;
      By Effect;
Proc Means data=DIFFS mean noprint;
      By Effect;
      Output out = ASED mean = AvStdErr;
      Var StdErr;
Data LSD;
      Merge ASED DOF;
      By Effect;
      t = tinv(1-0.05 / 2, DenDF);
      LSD = t*AvStdErr;
      Drop NumDF ;
Run;
If AvStdErr = . then delete ;
Title2 "LSD calculated using the Average Standard Error of Differences"
;
Proc Print Data = LSD ;
      Var Effect AvStdErr DenDF t LSD;
Run;
```

Brief description of the SAS code.

Almost all the components of this SAS code were explained previously in this appendix; now we will describe only additional statements used. Firstly, because we are interested in the complete model and in all lsmeans, we are using the bar notation of SAS in both the ***Model*** and ***LSMeans*** statements. The new option ***PDIFF*** in the ***LSMeans*** statement is used for generating the least squares differences, their standard errors, t values, and p-values, which will be saved in the temporary file ***DIFFS*** with the statement ***ODS output lsmeans diffs=DIFFS tests3=DOF***, and later used for calculating the LSD value by using the average standard error of differences (ASED) obtained in the related ***Proc Means*** block of statements.

The temporary file ***DOF*** is generated for saving the information related to the degrees of freedom (NumDF, DenDF) for numerator and denominator, respectively, in the Type III tests of fixed effects for each term included in the model. The ***DenDF*** are used later in asking SAS for the accumulated ***t*** probability using the ***t = tinv(1 - 0.05 / 2, DenDF)*** statement, and finally calculating the LSD values for each effect through the expression ***LSD = t*AvStdErr***.

In addition to the description of the statements used for graphing the response profiles in Example 2 above, now we will provide a more detailed explanation of this code. In the line ***LSMeans Year Soil N P / plot = mean (join cl)***, the lsmeans response profiles are requested for all the main effects Year, Soil, N, and P, simultaneously.

The statement ***LSMeans Year*Soil Year*N Year*P / plot = mean (sliceby = Year Join CL)*** is used for simultaneously requesting the response profiles for the Year × Soil, Year × N, and Year × P interactions. **Sliceby**=fixed-effect specifies the Year effect by which to group the means in a single plot, and the levels for the Soil, N, and P effects to be drawn in the horizontal axis, because Year is a qualitative factor while

N and P are quantitative factors. Similarly, the statement ***LSMeans Soil*N Soil*P / plot = mean (sliceby = Soil Join CL)*** is used to draw the individual response profiles for each Soil at the N and P levels in the horizontal axis.

For separating means of three-way interactions, we can use a statement like the following: ***LSMeans Year*Soil*N Year*Soil*P / plot = mean (sliceby=Year*Soil Join CL)***, in which we are analyzing the three-way interactions Year × Soil × N and Year × Soil × P, creating four response profiles coming from the two-way interaction Year × Soil at each level of the N and P rates.

Finally, the statement ***LSMeans Year*Soil*N*P / plot = mean (sliceby = Year*Soil*N Join CL)*** is useful for drawing the eight response profiles for the three-way combination of the levels of factors Year, Soil, and N, all of them with two levels, and for each level of the P factor in the horizontal axis. Only one of the 15 graphs obtained with the provided SAS code are included in the SAS output (see Fig. 1.3, Section 1)

Partial Selected Output of SAS Code for Example 3

The covariance parameter estimates and Type III tests of fixed effects are the same as shown in the first example of this appendix. Instead of showing these again, we show the structure for the lsmeans, the differences, as well as the t grouping for the two-way Year × Soil interaction, as an example for visualizing the standard errors of the differences used for calculating the LSD values using the ASED.

Year*Soil Least Squares Means

Soil	Year	Estimate	Standard Error	DF	t Value	P > |t|	Alpha	Lower	Upper
black	2007	2.3556	0.02546	4	92.52	< 0.0001	0.05	2.2849	2.4263
chestnut	2007	1.4700	0.02546	4	57.74	< 0.0001	0.05	1.3993	1.5407
black	2008	1.1869	0.02546	4	46.62	< 0.0001	0.05	1.1162	1.2576
chestnut	2008	0.8712	0.02546	4	34.22	< 0.0001	0.05	0.8006	0.9419

Differences of Year*Soil Least Squares Means

Soil	Year	_Soil	_Year	Estimate	Standard Error	DF	t Value	P > |t|	Alpha	Lower	Upper
black	2007	chestnut	2007	0.8856	0.03601	4	24.60	< 0.0001	0.05	0.7857	0.9856
black	2007	black	2008	1.1687	0.03601	4	32.46	< 0.0001	0.05	1.0688	1.2687
black	2007	chestnut	2008	1.4844	0.03601	4	41.22	< 0.0001	0.05	1.3844	1.5843
chestnut	2007	black	2008	0.2831	0.03601	4	7.86	0.0014	0.05	0.1832	0.3831
chestnut	2007	chestnut	2008	0.5988	0.03601	4	16.63	< 0.0001	0.05	0.4988	0.6987
black	2008	chestnut	2008	0.3156	0.03601	4	8.77	0.0009	0.05	0.2157	0.4156

T Grouping for Year*Soil Least Squares Means (Alpha=0.05)
LS-means with the same letter are not significantly different.

Soil	Year	Estimate	
black	2007	2.3556	A
chestnut	2007	1.4700	B
black	2008	1.1869	C
chestnut	2008	0.8712	D

LSD calculated using the Average Standard Error of Differences

Obs	Effect	AvStdErr	DenDF	t	LSD
1	N	0.02241	28	2.04841	0.04591
2	N*P	0.04482	28	2.04841	0.09181
3	P	0.03169	28	2.04841	0.06492
4	Soil	0.02546	4	2.77645	0.07069
5	Soil*N	0.03318	28	2.04841	0.06796
6	Soil*N*P	0.06400	28	2.04841	0.13109
7	Soil*P	0.04574	28	2.04841	0.09369

8	Year	0.02546	4	2.77645	0.07069
9	Year*N	0.03318	28	2.04841	0.06796
10	Year*N*P	0.06400	28	2.04841	0.13109
11	Year*P	0.04574	28	2.04841	0.09369
12	Year*Soil	0.03601	4	2.77645	0.09997
13	Year*Soil*N	0.04752	28	2.04841	0.09734
14	Year*Soil*N*P	0.09089	28	2.04841	0.18619
15	Year*Soil*P	0.06520	28	2.04841	0.13355

Finally, following is the graph obtained for the three-way Soil × N × P interaction, like that shown in Fig. 1.3 from Section 1.

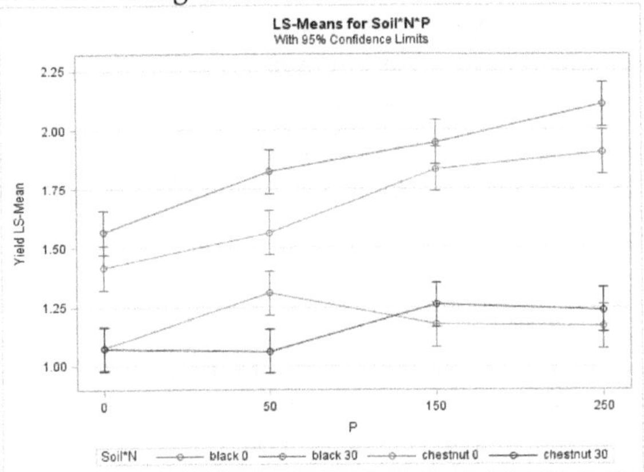

Appendix 3: Answers to Review Questions and Exercises

Review Questions: True or False

1. When the researcher can identify predetermined contrasts, then these and not the LSD should be used for mean separation. *True*
2. It is not important whether you include all of the interactions in your model, your *F*-test result for each effect will be identical in the model with and without the interactions. For example, if a study includes genotypes and compost rates, we can ignore the interaction of Genotype × Compost and each p-value for Genotype and Compost will be identical to each p-value for Genotype and Compost in the analysis that includes the Genotype × Compost interaction. *False*. By including interactions, the residual variance will change, therefore each term in the Analysis of Variance that included interactions will have a different F value compared with the analysis that did not include interactions.
3. In a study with three factors, the initial analysis must include main effects, two-way interactions, and the three-way interaction. If the three-way interaction is significant, then the researcher cannot properly discuss any main effect or two-way interaction. *False*. It is likely that the researcher will report results based on the analysis of the three-way interaction and we encourage researchers to follow this approach. However, if the significant *F* value for the three-way interaction is based largely on non-crossover interactions, then results for some main effects and some two-way interactions may provide similar information in a simpler format than the three-way interaction. When this is the case, the author may choose to present the data of the lower order effect as long as it is done so within the context

of the higher-order interaction. When presented with this situation, there is not a standard correct approach. The author must determine which approach best (most accurately and in the simplest format) makes the points he/she feels are important.

4. It is never appropriate to speculate about what causes a significant effect. *False*.

Researchers should carefully identify results that lead to logical speculation based on their subject-area knowledge. It is crucial however, that the author clearly identifies these comments as speculation.

5. A researcher conducts an experiment in each of two years. When analyzing such an experiment, it is always necessary that Year be considered as a random effect. *False*. Often when there are a low number of levels for what is often considered as a random effect, it is preferable to analyze it as a fixed effect.

6. The more levels a researcher has for an effect that is normally considered as a random effect (such as Year or Location), the more likely it would be useful to analyze it as a random effect. *True*

7. It is not important when making inferences whether an effect is fixed or random. *False*.

Inferences about a random effect should pertain to the population being tested. Inferences about fixed effects should pertain only to the specific effects (qualitative factor) or range of effects (quantitative factor) that were tested.

Exercise 1

Using the data corresponding to the example shown in Section 1 from this chapter, construct a SAS program that calculates in a step-wise fashion the F values for the main effects, 2-way, 3- way, and then the 4-way interactions as well as the single degree of freedom contrasts for the linear, quadratic, and cubic responses of GY to P for all effects involving P. Rather than using the method that we provided in Appendices 1 and 2 that automatically calculates the coefficients, in this exercise we ask that you insert the actual coefficients into each line of code.

Answer

The first step is to run Proc IML to obtain the correct coefficients. Since our unequally spaced rates of P are 0, 50, 150, and 250 kg P ha^{-1}, the following code will produce the correct regression coefficients.

```
Proc IML;
      Pcoeff=Orpol({0,50,150,250});
      Print Pcoeff;
Run;
```

The output from this program is the following:

```
                              Pcoeff
   0.5          -0.58585          0.4959593         -0.401004
   0.5          -0.325472        -0.280609           0.7518821
   0.5           0.1952834       -0.678681          -0.501255
   0.5           0.716039         0.4633304          0.1503764
```

The numbers in the second column in this table are the coefficients needed to calculate a linear regression for the four rates of P we input. The third column provides the coefficients for a quadratic regression, and the final column contains the coefficients needed to calculate a cubic regression. In the code we provided in Appendices 1 and 2, the numbers in the second column (linear) are the "PosCoefGrade1", the numbers in the third column are the "PosCoefGrade2", and the numbers in the final column are the "PosCoefGrade3". To obtain the NegCoefGrade1, NegCoefGrade2, and NegCoefGrade3, we multiplied each column by (-1).

Now that we have our linear, quadratic, and cubic regression coefficients, our next step will be to read in our data and print it out to check that it has been read in correctly.

```
**** Reading data: Data Wheat is available in attached CSV file ****;
Data Wheat;
      Infile "C:\Experiment 1 Data Wheat.CSV" dlm="," firstobs=3 ;
      Informat Soil$ 10. ;
      Input Year Soil N P Rep Yield ; Datalines;
Run;
```

The next step is to insert the Proc GLIMMIX code for this data with four fixed-effect factors. In this first set of code, there will be only main effects. The order of the variables in the Class statement is particularly important. For our SAS code that generates the coefficients, P must follow all other fixed variables. If Year, Soil, or N had different numbers of levels, then their order would also be important. Since they all have the same number of levels (two), their order is not important except that they must be listed prior to P.

```
ODS Select CovParms Tests3 Contrasts; Proc GLIMMIX data = Wheat;
      class Year Soil N P Rep ;
      model Yield = Year Soil N P ;

      random Rep(Year Soil);
Run;
```

Our code for the three (linear, quadratic, and cubic) contrasts for the main effect of P follow. We use the coefficients from Columns 2, 3, and 4, respectively for each of these contrasts. The code for the linear effect of P is:

```
contrast "Linear P" P -0.58585 -0.325472 0.1952834 0.716039;
```

The words "Linear P" comprise a user-chosen title to describe the effect being tested. Following the closed quotation marks is P which tells SAS that all calculations will be done on this effect. The linear coefficients follow the P and the ";" ends the statement. Similarly, we calculate the quadratic and cubic effects as follows:

```
contrast "Quadratic P" P 0.4959593 -0.280609 -0.678681 0.4633304;
contrast "Cubic P" P -0.401004 0.7518821 -0.501255 0.1503764;
```

We insert a run statement after the final contrast statement and combine our Proc GLIMMIX code from above.

```
ODS Select CovParms Tests3 Contrasts;
Proc GLIMMIX data = Wheat;
      class Year Soil N P Rep ;
      model Yield = Year Soil N P ;
      random Rep(Year Soil);
      contrast "Linear P" P -0.58585 -0.325472 0.1952834 0.716039;
      contrast "Quadratic P" P 0.4959593 -0.280609 -0.678681 0.4633304;
      contrast "Cubic P" P -0.401004 0.7518821 -0.501255 0.1503764;
Run;
```

In addition to the standard Proc GLIMMIX output, running this code will result in the following output for these three Contrast statements.

Contrasts

Label	Num DF	Den DF	*F* Value	*P* > *F*
Linear P	1	52	25.94	<0.0001
Quadratic P	1	52	3.06	0.0860
Cubic P	1	52	0.44	0.5093

Now, we need to delete the last run statement and add our SAS code for the two-way interactions. Everything except the new contrast statements follows:

```
ODS Select CovParms Tests3 Contrasts; Proc GLIMMIX data = Wheat;
class Year Soil N P Rep ;
model Yield = Year Soil N P
     Year*Soil Year*N Year*P Soil*N Soil*P N*P ;
random Rep(Year Soil);
contrast "Linear P" P -0.58585 -0.325472 0.1952834 0.716039; contrast
"Quadratic P" P 0.4959593 -0.280609 -0.678681 0.4633304; contrast
"Cubic P" P -0.401004 0.7518821 -0.501255 0.1503764;
```

Now we need to construct and add the contrast statements for all 2-way interactions involving P.

```
contrast "Year × P Linear" Year*P
     -0.58585 -0.325472 0.1952834 0.716039
      0.58585 0.325472 -0.1952834 -0.716039;
```

Because we are analyzing the linear effect of P for the Year × P interaction, in the first line of code, after the title of our interaction within quotation marks, we insert the code Year*P to direct SAS to apply the coefficients to this interaction. The first line of coefficients are from the second column of the Proc IML output because this is a linear effect. These coefficients correspond to the code "PosCoefGrade1" in Appendices 1 and 2. The second line of coefficients are the coefficients in the first line multiplied by (-1) and these correspond to "NegCoefGrade1". Note that the ";" follows the second line of coefficients. We could have put these three lines of code without carriage returns as follows:

```
contrast "Year × P Linear" Year*P
     -0.58585 -0.325472 0.1952834 0.716039
       0.58585 0.325472 -0.1952834 -0.716039;
```

which is equivalent to the statement

```
*&PosCoefGrade1 &NegCoefGrade1; when using the SAS Macro code;
```

However, we recommend using the separate lines to facilitate finding errors in code.

If we add all of our contrast statements for 2-way interactions involving P, then we will have the following code.

```
ODS Select CovParms Tests3 Contrasts;
Proc GLIMMIX data = Wheat;
      class Year Soil N P Rep ;
      model Yield = Year Soil N P
           Year*Soil Year*N Year*P Soil*N Soil*P N*P;
      random Rep(Year Soil);
      contrast "Linear P" P -0.58585 -0.325472 0.1952834 0.716039;
      contrast "Quadratic P" P 0.4959593 -0.280609 -0.678681 0.4633304;
      contrast "Cubic P" P -0.401004 0.7518821 -0.501255 0.1503764;
      contrast "Year × P Linear" Year*P
          -0.58585 -0.325472 0.1952834 0.716039
           0.58585 0.325472 -0.1952834 -0.716039;
* First line is the original linear coefficients from Proc IML output;
*Second line is the Proc IML linear coefficients multiplied by -1;
*Note that P must come after Soil in the Class statement for this to
calculate properly;
*&PosCoefGrade1 &NegCoefGrade1;
contrast "Year × P Quadratic" Year*P
     0.4959593 -0.280609 -0.678681 0.4633304
    -0.4959593 0.280609 0.678681 -0.4633304;
```

```
*&PosCoefGrade2 &NegCoefGrade2 ;
contrast "Year × P Cubic" Year*P
     -0.401004 0.7518821 -0.501255 0.1503764
      0.401004 -0.7518821 0.501255 -0.1503764;
*&PosCoefGrade3 &NegCoefGrade3 ;
contrast "Soil × P Linear" Soil*P
     -0.58585 -0.325472 0.1952834 0.716039
      0.58585 0.325472 -0.1952834 -0.716039;
contrast " Soil × P Quadratic" Soil*P
      0.4959593 -0.280609 -0.678681 0.4633304
     -0.4959593 0.280609 0.678681 -0.4633304;
contrast "Soil × P Cubic" Soil*P
     -0.401004 0.7518821 -0.501255 0.1503764
      0.401004 -0.7518821 0.501255 -0.1503764;
contrast "N × P Linear" N*P
     -0.58585 -0.325472 0.1952834 0.716039
      0.58585 0.325472 -0.1952834 -0.716039;
contrast "N × P Quadratic" N*P
      0.4959593 -0.280609 -0.678681 0.4633304
     -0.4959593 0.280609 0.678681 -0.4633304;
contrast "N × P Cubic" N*P
     -0.401004 0.7518821 -0.501255 0.1503764
      0.401004 -0.7518821 0.501255 -0.1503764;
```

As noted above, and for all other interactions, note that in the Class statement, P must have followed whatever fixed effect(s) that comprise the interaction, in this case Year, Soil, and N. This is because we calculated these coefficients as follows:

For P, we used the coefficients generated by Proc IML, and for Year we used +1, -1, for Year 1 and Year 2, respectively. Note that when you are using rates, such as 0, 50, 150, and 250, SAS will assign the coefficients to these rates from lowest to highest rate. For Year 1 or Year 2, SAS will assign the coefficients 1 -1 alphabetically so Year 1 will correspond to 1 and -1 to Year 2. Had the treatment been month, and the two months in question January and August, then 1 would have been assigned to August and -1 would have been assigned to January. The output follows for all of the main effects and two-way interaction contrasts that include P as one of the factors:

Contrasts

Label	Num DF	Den DF	*F* Value	*P* > *F*
Linear P	1	41	40.42	< 0.0001
Quadratic P	1	41	4.77	0.0347
Cubic P	1	41	0.69	0.4117
Year × P Linear	1	41	5.76	0.0210
Year × P Quadratic	1	41	1.93	0.1725
Year × P Cubic	1	41	0.09	0.7648
Soil × P Linear	1	41	16.60	0.0002
Soil × P Quadratic	1	41	0.07	0.7880
Soil × P Cubic	1	41	0.00	0.9877
N × P Linear	1	41	0.98	0.3280
N × P Quadratic	1	41	0.18	0.6710
N × P Cubic	1	41	0.52	0.4729

Note that now by adding the two-way interactions, the linear, quadratic, and cubic responses to the main effect of P changed because the denominator degrees of freedom decreased from 52 to 41.

Based on the order of variables in the Class statement, we calculated the 8 coefficients for the Year × P interaction using the following table.

Phosphorus fertilizer rates

Year treatment	Year coefficient	kg P ha⁻¹ 0	50	150	250	Sum of coefficients
		Coefficients for linear response to P				
		-0.58585	-0.325472	0.1952834	0.716039	0.0000004
		Coefficients for Year × P linear interaction				
Year 1	1	-0.58585	-0.325472	0.1952834	0.716039	0.0000004
Year 2	-1	0.58585	0.325472	-0.1952834	-0.716039	-0.0000004
Sum of coefficients	0	0.00000	0.000000	0.00000000	0.00000	0.0000000

The table above shows how we multiplied the P coefficients by the Year coefficients to obtain the coefficients used in the SAS code for the linear response to P in the Year × P interaction.

Note that the sum of the coefficients in the rows for Year (Year 1 and Year 2) do not sum exactly to 0. However, they are extremely close to 0. In order to ensure that the contrasts are calculated correctly, it is important not to round off the coefficients generated by Proc IML.

If the Class statement would have had P listed before Year, then we would have calculated our coefficients using the following table:

P fertilizer kg ha^{-1}	Coefficients for linear response to P	Year coefficients: Year 1 (1)	Year 2 (-1)
0	-0.5858500	-0.5858500	0.5858500
50	-0.3254720	-0.3254720	0.3254720
150	0.1952834	0.1952834	-0.1952834
250	0.7160390	0.7160390	-0.7160390
Sum of coefficients	0.0000004	0.0000004	0.0000004

From this table, we could have run the following program to calculate the linear response to P fertilizer for the Year × P Interaction: Notice that P is listed before Year in the Class statement below.

```
Proc GLIMMIX data = Wheat;
     class Soil N P Year Rep ;
     model Yield = Year Soil N P
          Year*Soil Year*N Year*P Soil*N Soil*P N*P ;
     random Rep(Year Soil);
     contrast " Year × P Linear" Year*P
         -0.5858500 0.5858500
         -0.3254720 0.3254720
          0.1952834 -0.1952834
          0.7160300 -0.7160300
Run;
```

Our output for the contrast Year × P Linear follows:

Contrasts

Label	Num DF	Den DF	*F* Value	*P* > *F*
Year × P Linear	1	45	5.76	0.0206

The output is the same as previously when P came after instead of before Year in the Class statement. The values can be calculated correctly either way. (Verify this on your own.) However, the researcher needs to be aware of the order of variables in the Class statement so he/she can properly order the coefficients in the SAS code.

As long as the coefficients are listed in an order that is in agreement with the variables in the Class statement, then the contrast will be calculated correctly.

Next, we calculate the coefficients for the three-way single degree of freedom interactions that include P, beginning with the Year × Soil × P interaction.

Beginning with Year and Soil, we assign the following coefficients:

Year 1 = +1

Year 2 = -1

Black Soil = +1

Chestnut Soil = -1

We have a simple table with +1 and -1 being the coefficients on the row and +1 and -1 being the coefficients on the column. Multiplying the Year × Soil coefficients we get the following 4 coefficients: 1, -1, -1, 1 shown in yellow in the following table.

Year	Year	Soil		
	Coefficients	Black	Chestnut	
		Soil coefficients		Sum of coefficients
		1	-1	0
Year 1	1	1	-1	0
Year 2	-1	-1	1	0
Sum of coefficients	0	0	0	0

The numbers shaded in yellow are the products of the Year × Soil coefficients, and the sums in the bottom row are of the two rows immediately above the bottom, and in the final column, the sums are of the two columns immediately to the left of the final column.

The coefficients 1, -1, -1, and 1 are then placed in the rows with the P linear coefficients forming the columns (we placed P after Year and Soil in the Class statement) to calculate the contrast for the linear response to P in the Year × Soil x P interaction.

Year × Soil	Coefficients for linear response of P linear				
Coefficients	-0.58585	-0.325472	0.1952834	0.71603	Sum of coefficients
1	-0.58585	-0.325472	0.1952834	0.716039	-0.0000004
-1	0.58585	0.325472	-0.1952834	-0.716039	0.0000004
-1	0.58585	0.325472	-0.1952834	-0.716039	0.0000004
1	-0.58585	-0.325472	0.1952834	0.716039	-0.0000004
Sum of coefficients	0.00000	0.000000	0.0000000	0.00000	0.0000000

The numbers shaded in yellow are the products of the coefficients for the linear response to P in the Year × Soil × P interaction. These are the numbers that we will use as coefficients in the SAS code for the Contrast statement that calculates the linear response to P in each of our three-way interactions. The coefficients for the quadratic and cubic responses to P are calculated similarly. The following code calculates linear, quadratic, and cubic responses to P in each three-way interaction involving P as a factor.

```
Proc GLIMMIX data=Wheat;
      class Year Soil N P Rep;
      model Yield=Year Soil N P
      Year*Soil Year*N Year*P Soil*N Soil*P N*P
      Year*Soil*N Year*Soil*P Year*N*P Soil*N*P;
random Rep(Year Soil);
**** Year × Soil × P three way interaction contrasts ****;
contrast "Year × Soil × P Linear" Year*Soil*P
      0.58585 0.325472 -0.1952834 -0.716039
     -0.58585 -0.325472 0.1952834 0.716039
     -0.58585 -0.325472 0.1952834 0.716039
      0.58585 0.325472 -0.1952834 -0.716039;
contrast "Year × Soil × P Quadratic" Year*Soil*P
```

```
        -0.4959593 0.280609 0.678681 -0.4633304
         0.4959593 -0.280609 -0.678681 0.4633304
         0.4959593 -0.280609 -0.678681 0.4633304
        -0.4959593 0.280609 0.678681 -0.4633304;
contrast "Year × Soil × P Cubic" Year*Soil*P
         0.401004 -0.7518821 0.501255 -0.1503764
        -0.401004 0.7518821 -0.501255 0.1503764
        -0.401004 0.7518821 -0.501255 0.1503764
         0.401004 -0.7518821 0.501255 -0.1503764;
contrast "Year × N × P Linear" Year*N*P
         0.58585 0.325472 -0.1952834 -0.716039
        -0.58585 -0.325472 0.1952834 0.716039
        -0.58585 -0.325472 0.1952834 0.716039
         0.58585 0.325472 -0.1952834 -0.716039;
contrast "Year× N × P Quadratic" Year*N*P
        -0.4959593 0.280609 0.678681 -0.4633304
         0.4959593 -0.280609 -0.678681 0.4633304
         0.4959593 -0.280609 -0.678681 0.4633304
        -0.4959593 0.280609 0.678681 -0.4633304;
contrast "Year × N × P Cubic" Year*N*P
         0.401004 -0.7518821 0.501255 -0.1503764
        -0.401004 0.7518821 -0.501255 0.1503764
        -0.401004 0.7518821 -0.501255 0.1503764
         0.401004 -0.7518821 0.501255 -0.1503764;
contrast "Soil × N × P Linear" Soil*N*P
         0.58585 0.325472 -0.1952834 -0.716039
        -0.58585 -0.325472 0.1952834 0.716039
        -0.58585 -0.325472 0.1952834 0.716039
         0.58585 0.325472 -0.1952834 -0.716039;
contrast "Soil × N × P Quadratic" Soil*N*P
        -0.4959593 0.280609 0.678681 -0.4633304
         0.4959593 -0.280609 -0.678681 0.4633304
         0.4959593 -0.280609 -0.678681 0.4633304
        -0.4959593 0.280609 0.678681 -0.4633304;
contrast "Soil × N × P Cubic" Soil*N*P
         0.401004 -0.7518821 0.501255 -0.1503764
        -0.401004 0.7518821 -0.501255 0.1503764
        -0.401004 0.7518821 -0.501255 0.1503764
         0.401004 -0.7518821 0.501255 -0.1503764;
Run;
```

The results for the contrasts of each three-way interaction follow.

Contrasts

Label	Num DF	Den DF	*F* Value	*P* > *F*
Year × Soil × P Linear	1	31	22.49	< 0.0001
Year × Soil × P Quadratic	1	31	0.02	0.8847
Year × Soil × P Cubic	1	31	0.45	0.5053
Year × N × P Linear	1	31	0.74	0.3955
Year × N × P Quadratic	1	31	0.44	0.5118
Year × N × P Cubic	1	31	0.58	0.4536
Soil × N × P Linear	1	31	2.70	0.1103
Soil × N × P Quadratic	1	31	0.07	0.7970
Soil × N × P Cubic	1	31	11.79	0.0017

To finish the exercise, it is necessary to calculate the three single degree of freedom contrasts for the four way interaction Year × Soil × N × P. For this set of contrasts, we will provide the detail for calculating the coefficients for the cubic response of P for this four-way interaction. From our calculations for the coefficients used in the three-way interactions, we know that the coefficients for the Year × Soil interaction are 1, -1, -1, and 1. To obtain the coefficients for the three-way Year × Soil × N interaction, we place these four coefficients as rows and for N, add 1 and -1 as columns, giving us the following table.

Year × Soil coefficients		N coefficients	Sum of coefficients
	1	-1	
1	1	-1	0

-1	-1	1	0
-1	-1	1	0
1	1	-1	0
Sum of coefficients	0	0	0

By adding the factor N which had two levels, we have evolved from four coefficients for the Year × Soil two-way interaction to eight coefficients for the Year × Soil × N three-way interaction. The coefficients which are the product of the rows and columns in the above table are 1, -1, -1, 1, -1, 1, 1, and -1. To obtain our coefficients for the four-way interaction, we build a new table using these 8 coefficients as the rows and (since we are calculating the cubic response to P rates) using the cubic coefficients for our rates of P generated previously by Proc IML as the columns. The results are in the following table.

Y × S × N coefficients	Cubic coefficients for P from Proc IML				Sum of coefficients
	-0.401004	0.7518821	-0.501255	0.1503764	-0.0000005
1	-0.401004	0.7518821	-0.501255	0.1503764	-0.0000005
-1	0.401004	-0.7518821	0.501255	-0.1503764	0.0000005
-1	0.401004	-0.7518821	0.501255	-0.1503764	0.0000005
1	-0.401004	0.7518821	-0.501255	0.1503764	-0.0000005
-1	0.401004	-0.7518821	0.501255	-0.1503764	0.0000005
1	-0.401004	0.7518821	-0.501255	0.1503764	-0.0000005
1	-0.401004	0.7518821	-0.501255	0.1503764	-0.0000005
-1	0.401004	-0.7518821	0.501255	-0.1503764	0.0000005

The coefficients necessary to calculate the significance of the Year × Soil × N × P Cubic interaction are those in yellow:

Getting back to the code we provided in Appendix 2, to calculate the significance of this interaction, we have the following groups:

```
-0.401004   0.7518821   -0.501255    0.1503764: "PosCoefGrade3"
 0.401004  -0.7518821    0.501255   -0.1503764: "NegCoefGrade3"
 0.401004  -0.7518821    0.501255   -0.1503764: "NegCoefGrade3"
-0.401004   0.7518821   -0.501255    0.1503764: "PosCoefGrade3"
 0.401004  -0.7518821    0.501255   -0.1503764: "NegCoefGrade3"
-0.401004   0.7518821   -0.501255    0.1503764: "PosCoefGrade3"
-0.401004   0.7518821   -0.501255    0.1503764: "PosCoefGrade3"
 0.401004  -0.7518821    0.501255   -0.1503764: "NegCoefGrade3"
```

We would calculate the linear (PosCoefGrad1" and "NegCoefGrade1") and quadratic coefficients (PosCoefGrad2" and "NegCoefGrade2") similarly, but using the second and third columns, respectively, from the Proc IML output instead of the fourth column which we used to generate these cubic coefficients.

The code for calculating the linear, quadratic, and cubic responses to P for the four-way interaction Year × Soil × N × P is the following:

```
Proc GLIMMIX data = Wheat;
class Year Soil N P Rep ;
model Yield = Year Soil N P
             Year*Soil Year*N Year*P Soil*N Soil*P N*P
             Year*Soil*N Year*Soil*P Year*N*P Soil*N*P
             Year*Soil*N*P;
random Rep(Year Soil);
contrast "Year × Soil × N × P Linear" Year*Soil*N*P
     -0.58585 -0.325472 0.1952834 0.716039
      0.58585 0.325472 -0.1952834 -0.716039
      0.58585 0.325472 -0.1952834 -0.716039
```

```
     -0.58585 -0.325472 0.1952834 0.716039
      0.58585 0.325472 -0.1952834 -0.716039
     -0.58585 -0.325472 0.1952834 0.716039
     -0.58585 -0.325472 0.1952834 0.716039
      0.58585 0.325472 -0.1952834 -0.716039;
contrast "Year × Soil × N × P Quadratic" Year*Soil*N*P
      0.4959593 -0.280609 -0.678681 0.4633304
     -0.4959593 0.280609 0.678681 -0.4633304
     -0.4959593 0.280609 0.678681 -0.4633304
      0.4959593 -0.280609 -0.678681 0.4633304
     -0.4959593 0.280609 0.678681 -0.4633304
      0.4959593 -0.280609 -0.678681 0.4633304
      0.4959593 -0.280609 -0.678681 0.4633304
     -0.4959593 0.280609 0.678681 -0.4633304;
contrast "Year × Soil × N × P Cubic" Year*Soil*N*P
     -0.401004 0.7518821 -0.501255 0.1503764
      0.401004 -0.7518821 0.501255 -0.1503764
      0.401004 -0.7518821 0.501255 -0.1503764
     -0.401004 0.7518821 -0.501255 0.1503764
      0.401004 -0.7518821 0.501255 -0.1503764
     -0.401004 0.7518821 -0.501255 0.1503764
     -0.401004 0.7518821 -0.501255 0.1503764
      0.401004 -0.7518821 0.501255 -0.1503764;
Run;
```

The output for the linear, quadratic, and cubic responses to P for the four-way interaction follows:

Contrasts

Label	Num DF	Den DF	*F* Value	*P* > *F*
Year × Soil × N × P Linear	1	28	0.50	0.4870
Year × Soil × N × P Quadratic	1	28	0.01	0.9436
Year × Soil × N × P Cubic	1	28	5.90	0.0218

The complete Proc GLIMMIX SAS code for calculating all two-way, three-way, and four-way interactions in the ANOVA as well as the linear, quadratic, and cubic responses to P for all contrasts involving P, can be easily obtained putting together the programs shown above. Note: *using Proc GLIMMIX, instead of the CONTRAST statement you can also use the LSMESTIMATE statement because it provides a mechanism for obtaining hypothesis tests among the least squares means. However, in contrast to the hypotheses tested with the ESTIMATE or CONTRAST statements, the LSMESTIMATE statement enables to form linear combinations of the least squares means, rather than linear combination of fixed-effects parameter estimates and/or random-effects solutions*

Exercise 2

Using the data set (Wheat) shown in Section 1 from this chapter, calculate the Mean Squared Error (MSE) for Rep (Year Soil) for the complete model using two approaches.

Approach 1.

Proc GLIMMIX provides the covariance parameter estimate (or variance) for Rep (Year Soil). Referring to Expected Mean Squares, we know the following:

```
MSE for Rep (Year Soil) = Residual variance + (P × N) × [Variance
(Rep (Year × Soil)]
Where P and N equal the number of levels of P and N, respectively.
P × N = 4 × 2 = 8
MSE for Rep (Year Soil) = Residual variance + (8) × [Variance (Rep
(Year × Soil)]
```

We know from the Proc GLIMMIX output that Variance (Rep (Year × Soil) = 0.000292 and we know that the residual variance is 0.008036. Thus,

```
MSE for Rep (Year × Soil) = 0.008036 + 8(0.000292)
MSE for Rep (Year × Soil) = 0.008036 + 0.002336
MSE for Rep (Year × Soil) = 0.010372
```

Approach 2.

We can also use our LSD output in Appendix 2, Example 3, from Proc GLIMMIX to calculate the MSE for Rep (Year Soil). One column of output is the average standard error of the differences between two means (ASED) which is labeled "AvStdErr" in the Proc GLIMMIX output. We know that the MSE for Rep (Year Soil) is used to calculate the standard error of the difference (SED) between two Year, Soil, or Year × Soil means, so we can use that formula in reverse.

Using the output for Year × Soil (we could also have used Year or Soil because, as stated above, the ASED for each of these was calculated from the same Mean Square as was used to calculate the LSD for Year × Soil).

The actual formula for the SED is SED = [2(MSE for Rep (Year Soil))/n](1/2) where n = number of observations for each Year × Soil mean. This number n is calculated from 2 reps, 4 rates of P, and 2 rates of N so n = 2 × 4 × 2 = 16 observations for calculating each Year × Soil mean. Substituting ASED (which is AvStdErr from the Proc GLIMMIX output) for SED, this gives us:

```
0.036011 = [2(MSE for Rep (Year Soil))/16]^(1/2)
```

Squaring both sides of the equation gives us:

```
0.0012967 = 2(MSE for Rep (Year Soil)/16
0.0012967 = MSE for Rep (Year Soil)/8
8 × 0.0012967 = MSE for Rep (Year Soil)
```

MSE for Rep (Year Soil) = 0.0103736, which is equal to 5 decimal places to the answer from our Approach 1.

Calculating LSD values

Now, using the residual variance (MSE) or the mean square of Rep (Year Soil) if appropriate, let's calculate the LSD for all effects and interactions and compare these LSDs with the LSDs calculated with Proc GLIMMIX, which are provided in the output in Appendix 2. Remember, the LSD is a mechanism for doing all possible t tests with a mean variance. However, the Pdiff option in SAS actually calculates all possible t tests and does so not based on a mean variance, but based on the actual pooled variance of the difference of the two means being tested. Thus, using the Pdiff option is a more precise way of carrying out the intentions of the LSD so we recommend, if the results differ, using the significance results from the Pdiff option rather than from the LSD. Still, when we publish, it is useful to provide measures of variability in our tables and figures. The LSD is a useful statistic to use for this purpose. While the actual probabilities may not be exactly equal, comparisons among every single pair of means for significance using the LSD will almost always yield the same significance results (that is, significant or not significant) as the Pdiff option. For rare instances when the Pdiff option identifies a comparison as significant or not significant and this determination is different from that of the LSD, the author can point this out and indicate that it is the result from the Pdiff options on which conclusions are based, but that the LSD value is presented because it is still a useful estimate of variation.

For our calculations, we need to remember that the general formula for calculating the LSD is for t (α, df) (and we will conduct our calculations at $\alpha = 0.05$),

```
LSD = t [(2 × variance)/n] (1/2)
```

Where n = the number of observations for each mean.

We will show the calculations for calculating the LSDs for comparing means of N, Soil, Year × Soil, and Year × N and results for all effects and interactions are provided in the table below. We need to recall that each effect had the following number of levels.

```
Year = 2
Soil = 2
N = 2
P = 4
Rep = 2
```

First for N.

The two N means were comprised of 2 × 2 × 4 × 2 = 32 observations (number of levels of Year, Soil, P, and Rep multiplied together), so n = 32.

Based on our Proc GLIMMIX output from Appendix 2, we see that the denominator df associated with N = 28. This tells us two things. First, the appropriate variance is the residual (error mean square) since there are 28 df associated with this value. Second, to find the appropriate t value, we need to use df = 28.

We see from the SAS output that the mean square error = 0.008036 and from a t table, we find that t = 2.048. (We can also use the following function in Excel to obtain our *t* value:

```
=tinv (0.05, 28) = 2.048407
```

We now have the information to calculate the LSD which is:

$$LSD = 2.048407\ [2 \times 0.008036)/32]^{(1/2)}$$

LSD = 0.045906. This compares with 0.04591 calculated by using the ASED estimated by Proc GLIMMIX.

LSD for comparing means of Soil.

This is equivalent to the LSD for comparing means of Year because there are two levels of Year and two levels of Soil, and to calculate the LSD for year or soil, the variance is the same and it is the MSE for Rep (Year Soil) = 0.010372 as calculated above.

As for N, there are 32 observations that make up each Soil mean (Year × N × P × Rep) = 2 × 2 × 4 × 2 = 32.

Unlike the example of calculating an LSD for comparing N means, where denominator df = 28, we see from our Proc GLIMMIX output that for Soil (or Year), the denominator df = 4. Thus, we must use the mean square for Rep (Year Soil), which we already calculated as 0.010372, in our LSD formula and we must use 4 df for finding our t value which is $t = 2.776445$. Thus,

```
LSD = 2.776445 [2 × 0.010372)/32] (1/2)
LSD = 2.776445 × 0.0254597
LSD = 0.0706903.
```

This compares with 0.07069 calculated in Appendix 2 by using the ASED estimated by Proc GLIMMIX.

LSD for comparing means of the Year × Soil interaction.

There are 16 observations that comprise each Year × Soil mean (Levels of N × P × Rep) = 2 × 4 × 2 = 16. We see that the denominator df = 4 for Year × Soil, thus we will use the mean square error of Rep (Year Soil) as the variance, which is 0.010372 and from the t table, we find that t = 2.77645. Thus,

```
LSD = 2.776445 × [2 × 0.010372/16] (1/2)
LSD = 2.776445 × 0.0360069
LSD = 0.09997129
```

which compares with 0.09997 calculated by using the ASED estimated by Proc GLIMMIX.

LSD for comparing means of the Year × N interaction.

There are 16 observations that comprise each Year × N mean (Levels of Soil × P × Rep) = 2 × 4 × 2 = 16. We see that the denominator df = 28 for Year × N, thus we will use the residual as the variance = 0.008036, and from the t table, we find that t = 2.048407. Thus,

```
LSD = 2.048407 [2 × 0.008036/16] (1/2)
LSD = 2.048407 × 0.0316938
LSD = 0.0649219
```

which compares with 0.06796 calculated in Appendix 2, Example 3 by using the ASED estimated by Proc GLIMMIX.

The table below shows that our LSD values were nearly identical to 5 decimal places, whether calculated in Appendix 2 by the ASED or here using the variance. In most cases, it should be fine to use the correct estimated variance in order to estimate the LSD. To be most certain of precision, it would be safest to use Proc GLIMMIX and the ASED method we provided in Appendix 2.

Effect	Variance	ASED	Den df†	Critical t‡	LSD by variance	LSD by ASED
Year (Y)	0.010372	0.02546	4	2.77645	0.07069	0.07069
Soil (S)	0.010372	0.02546	4	2.77645	0.07069	0.07069
Nitrogen (N)	0.008036	0.02241	28	2.04841	0.04590	0.04591
Phosphorus (P)	0.008036	0.02546	28	2.04841	0.06492	0.06492
Y × S	0.010372	0.03601	4	2.77645	0.09997	0.09997
Y × N	0.008036	0.03318	28	2.04841	0.06492	0.06796
Y × P	0.008036	0.04574	28	2.04841	0.09181	0.09369
S × N	0.008036	0.03318	28	2.04841	0.06422	0.06796
S × P	0.008036	0.04574	28	2.04841	0.09181	0.09369
N × P	0.008036	0.04482	28	2.04841	0.09181	0.09181
Y × S × N	0.008036	0.04752	28	2.04841	0.09181	0.09734
Y × S × P	0.008036	0.13355	28	2.04841	0.12984	0.13355
Y × N × P	0.008036	0.06400	28	2.04841	0.12984	0.13109
S × N × P	0.008036	0.06400	28	2.04841	0.12984	0.13109
Y × S × N × P	0.008036	0.09089	28	2.04841	0.18363	0.18629

† Denominator degrees of freedom from the ANOVA for the fixed effects
‡ Critical t value at 0.05/2 significance level and for Den df.

CHAPTER 8: THE ANALYSIS OF COMBINED EXPERIMENTS

Philip M. Dixon, Kenneth J. Moore, and Edzard van Santen

Answers to Review Questions, with short explanations

1. In this study, modeling environment effects and block effects as fixed effects or as random effects leads to the same inference about the difference between the two types of tillage.

True. The design is balanced. The only differences would arise if a random effect variance is estimated as 0.

2. If conclusions about the effect of tillage in a new location are desired, you should use narrow sense inference.

False. Narrow sense inference makes conclusions about the locations and years used in the study

3. Narrow sense conclusions about the effects of tillage usually have smaller standard errors than do broad sense conclusions.

True. The variance for broad sense conclusions includes an additional non-negative variance component.

4. To obtain narrow-sense conclusions, omit the treatment by environment interaction from the model.

False. You should model the treatment by environment interaction as a fixed effect. Omitting it from the model assumes no interaction and pools the interaction with the plot-plot variation.

5. To obtain broad-sense conclusions, model the treatment by environment interaction as a random effect.

True.

6. Broad-sense confidence intervals for the difference between two types of tillage will be based on T distributions with 18 degrees of freedom.

False. The degrees of freedom for the confidence interval will be the df of the treatment by environment interaction, which is (18 – 1)(3-1) = 34

7. The combined analysis across environments requires that plot-plot variation be pooled across environments.

False. Although error variances are often pooled, they do not need to be.

8. The combined analysis across environments requires that variation between blocks be pooled across environments.

False. Although block variances are often pooled, they do not need to be. If blocks are considered a fixed effect, pooling is not an issue.

9. The 18 environments are actually 6 locations, each studied for 3 years. Tillage effects are expected to vary somewhat among locations because of different soil

characteristics. Tillage effects are not expected to vary among years at the same location. In this case, subdividing the treatment by environment interaction will have minimal effect on the conclusions about the average treatment effect.

True. Although the location by treatment variance component is expected to be somewhat different from the year by treatment variance component, this appears to have minimal effect on conclusions about the average treatment effect.

10. Imagine that the three tillage treatments are three levels of some quantitative factor, e.g. amount of soil disturbance. The data for each environment could be analyzed using a regression model with a linear effect of soil disturbance. It is possible to construct a combined analysis of those regression models in all 18 environments.

True. The model for the combined analysis includes a random regression slope by environment interaction.

Chapter 8 Answers to Exercises

1. An experiment was conducted to assess the effect of a fungicide treatment on soybean yield (kg ha^{-1}). It was conducted as an on-farm strip-plot trial with six pairs of side-by-side strips of which one randomly received fungicide treatment. The experiment was repeated at eleven farms (environments). The data were extracted from a much larger dataset provided by the Iowa Soybean Association and are provided in the on-farm soybean dataset in the supplemental materials.

   a. Analyze the experiment separately for each environment.

   See SAS code – first block of code

   b. Evaluate the error variances to determine whether or not they may be considered homogeneous.

   See SAS code – second block of code using repeated statement and reml to compare variances across environments

Various approaches are possible, including Levene's Test, a Likelihood Ratio Test, and comparison of AIC, AICc, or BIC statistics. All are consistent and indicate that the model with separate variances for each location is superior to using a single pooled estimate. The inference is that variances are heterogeneous.

   c. Conduct a combined analysis of variance assuming environment and replication to be random factors and treatment as fixed.

   See SAS code – third block of code

The F ratio for treatment is 13.78. Since the probability associated with its occurrence is quite small (0.0038) the effect of fungicide treatment is considered to be significant.

   d. Interpret the results of the experiment with respect to the efficacy of fungicide treatment in improving soybean yield.

The mean difference between fungicide and untreated soybean over all environments was approximately 1.65 kg ha^{-1} (se = 0.44 kg ha^{-1}). To be economical on average for represented environments, the per-ha cost of application would need to be less than the market value of 1.65 kg of soybeans.

2. Antonio Mallorino at Iowa State University has studied corn response to P fertilization since 2002. The Prate.csv file contains 13 years of data from the SouthWest research farm.

The design is a RCBD with three blocks of five plots each. Four P rates (0, 28, 56, and 112 lb ac^{-1}) were used; the 0 level was replicated twice in each block. Blocks and plots can be considered independent across years. The response variable is yield in bu ac^{-1}.

Consider years to be a fixed factor and P rate (Prate) to be a continuous variable.

a. What sort of polynomial model is appropriate to describe yield response to Prate? Linear? quadratic? With one coefficient for all years or coefficients that differ among years?

See SAS code - first block of code. A reasonable model has a quadratic response to P fertilization, with different linear coefficients but a single quadratic term. The quadratic × year interaction is not significant, but when that term is omitted, the quadratic effect, the linear effect, and the linear × year interaction are all significant. There is no evidence of lack of fit in either the interaction or main effect terms.

Now consider years to be a random factor. Fit a quadratic model that allows the intercept and linear Prate coefficient to vary between years (but the quadratic coefficient is constant).

b. What is the equation that predicts yield as a function of Prate for a year not in the data set, e.g, 2015?

See SAS code - second block of code, Solution for Fixed Effects in the output.
Yield = 170.2 + 0.36 × Prate - 0.00214 × Prate^2.

c. What is the year-to-year variability in the linear Prate slope? Use the standard deviation to describe that variability.

See SAS code - second block of code. The slope variance is the UN(2,2) parameter in the SAS output. We want its square root, = 0.12.

d. Examine the residuals. Is it appropriate to use yield as the response variable, or should yield be transformed?

See SAS code - third block of code. A plot of residuals against predicted values indicates no need to transform the response variable. The plot shows no evidence of lack of fit and no change in residual variability with increases in predicted values.

e. Apply Levene's test to the residuals to assess whether the error variance differs among years.

See SAS code - fourth block of code. There is evidence of different error variances in different years. The p-value is 0.022.

f. Refit the model used in parts b-e with year-specific error variances. Do the answers to questions b and c change much?

See SAS code - fifth block of code. The answers are slightly different, but the differences are small. When fit with year-specific error variances, the equation is:

Yield = 170.3 + 0.35 × Prate - 0.00202 × Prate2
The variability in the slope between years is 0.11.

CHAPTER 9: ANALYSIS OF COVARIANCE

Kevin S. McCarter

Solutions to Selected Review Questions

1. T - Refer to discussion on pages 265 and 266
2. F - Refer to Examples, 1, 4, 5
3. F - Refer to Example 4
4. T - Refer to Example 4
5. T - Refer to Example 4, Key Learning Points
6. Define events A – F according to the corresponding descriptions in the statement of the question. The requested probabilities are then given below.

   a. Probabilities of events A – F, rounded to five decimal places:

      i. P(A) = 0.34372

      ii. P(B) = 0.47739

      iii. P(C) = 0.15588

      iv. P(D) = 0.02192

      v. P(E) = 0.00108

      vi. P(F) = 0.00001

   b. P(A or B) = 0.82110

   c. P(E or F) = 0.00109

   d. Yes. See Examples 1, 4, 5. Discussion left for the reader.

   e. Discussion left for the reader.
7. Discussion left for the reader.

Data Analysis Exercises 1 – 5

Analyses are left to the reader. See the SAS program provided in this appendix.

Sample SAS code for examples and exercises

```
dm "log;clear";
options nodate pageno=1;

ods html close;
ods html;

*Create the SAS datasets for the examples and exercises;

data example_1;
```

```
input grp $ y x @@;
datalines;
A 71.3 12.7 A 63.4 13.3 A 55.0  8.6 A 54.0  7.3 A 54.6  8.2
A 47.7  6.4 A 49.1  7.0 A 88.1 14.2 A 59.4  8.6 A 70.5 10.7
B 63.0  7.6 B 80.9 13.4 B 78.7 10.3 B 85.1 14.8 B 78.5 13.6
B 73.0 13.5 B 53.0  5.3 B 76.3  9.9 B 68.7  9.4 B 84.9 14.2
;
run;
data example_2;
input grp $ y x @@;
datalines;
A 69.2 10.8 A  59.4 10.7 A 70.2 13.1 A 52.3  6.6 A 61.0 9.6
A 73.9 13.4 A  57.1  7.3 A 64.9 10.0 A 68.2 13.8 A 75.1 14.8
B 89.9 17.4 B 101.3 21.2 B 73.2 13.4 B 96.4 20.1 B 86.4 17.6
B 74.8 14.8 B  81.2 17.2 B 97.3 20.2 B 99.4 21.9 B 79.3 13.9
;
run;

data example_3;
input grp $ y x @@;
datalines;
A 81.2 14.1 A 58.7  8.8 A 47.4  5.7 A 49.4  5.5 A 66.1  9.1
A 72.5 14.6 A 71.1 12.7 A 53.5  6.2 A 62.2  8.0 A 68.5 12.4
B 56.6 15.9 B 57.5 17.9 B 75.6 21.6 B 68.5 19.9 B 58.0 14.1
B 57.7 15.1 B 62.6 16.8 B 73.9 20.6 B 77.0 20.9 B 44.4 13.0
;
run;

data example_4;
input grp $ y x @@;
datalines;
A 55.1  8.2 A 67.1 11.0 A 73.6 14.8 A 64.6  9.3 A 76.4 13.8
A 45.5  5.3 A 47.3  6.4 A 57.4  6.6 A 78.5 12.9 A 61.9  8.4
B 61.7  9.5 B 56.3 11.3 B 54.2 12.0 B 68.0  9.8 B 58.5 12.6
B 59.2 10.5 B 60.8 10.7 B 68.2  9.8 B 52.7 13.1 B 78.5  6.4
;
run;
data exercise_1;
input grp $ y x @@;
datalines;
A  49.4  5.2 A  72.2 13.2 A 61.6  9.2 A  63.4 11.3 A  71.0 12.3
A  49.6  7.2 A  56.6  8.3 A 61.1  6.7 A  66.1  9.7 A  71.9 12.0
B  95.3 20.2 B  96.7 20.5 B 75.8 12.5 B 102.6 21.9 B  78.5 15.3
B 101.4 20.8 B 100.8 21.3 B 88.2 17.2 B  79.1 14.4 B 108.0 21.2
;
run;
data exercise_2;
input grp $ y x @@;
datalines;
A 56.7  7.0 A 46.3  5.7 A 71.2 14.6 A 69.0 11.9 A 59.7 10.8
A 59.8 10.2 A 60.1  8.1 A 66.8 10.8 A 72.5 14.5 A 57.6  7.9
B 70.3 10.5 B 78.0 12.8 B 63.1  9.4 B 57.1  6.6 B 68.7 11.3
B 70.4  9.5 B 73.3  9.4 B 61.7  6.8 B 73.7 11.9 B 77.4 11.6
;
run;

data exercise_3;
input grp $ y x @@;
datalines;
A 54.8  8.9 A 54.6  8.3 A 52.5 10.0 A 68.2 12.6 A 59.6  9.4
A 76.8 13.2 A 56.3  7.9 A 71.2 15.0 A 62.0  9.2 A 68.0 13.8
B 54.4 14.7 B 55.8 19.1 B 69.4 20.3 B 51.7 15.3 B 67.2 21.3
B 43.9 12.2 B 71.6 21.6 B 66.7 20.5 B 42.4 13.4 B 55.7 15.9
;
data exercise_4;
input grp $ y x @@;
datalines;
```

```
A 73.9 14.4 A 46.7  6.1 A 48.9  6.3 A 83.0 14.6 A 80.7 15.0
A 73.1 12.1 A 54.8  7.2 A 53.4  7.3 A 79.3 13.5 A 60.6 11.5
B 68.6  9.6 B 66.9  9.9 B 53.5  5.0 B 79.6 13.3 B 82.0 13.2
B 72.4 12.2 B 67.6  9.9 B 77.4 13.4 B 70.6 10.0 B 75.4 11.1
;
run;

data exercise_5;
input grp $ y x @@;
datalines;
A 70.9 12.6 A 71.8 13.5 A 61.0 10.9 A 71.4 13.9 A 56.9  6.4
A 55.8  9.1 A 62.1  7.4 A 80.8 13.2 A 54.8  8.8 A 50.8  5.6
B 52.4 12.3 B 57.9 10.7 B 68.9 11.1 B 62.7 11.6 B 62.4 10.2
B 78.5  6.5 B 55.0 14.8 B 79.2  6.8 B 82.6  5.5 B 51.6 13.7
;
run;
* Create SAS macro for performing analyses;

%macro analyze_data(DATASET);

footnote1 "Chapter 9 - Analysis of Covariance - Sample SAS Code for
Examples and Exercises";
footnote2 "Analysis of Dataset &DATASET";

* Print listing of dataset;

title1 "Listing of the Dataset &DATASET";
proc print data=&DATASET;
run;

* Calculate summary statistics and perform exploratory analysis;

title1 "Summary of the Dataset";
proc means data=&DATASET mean std;
var y x;
by grp;
run;

title1 "Boxplots of Response Variable Y for Each Group";
proc sgplot data=&DATASET;
vbox y / group=grp extreme ;
run;

title1 "Boxplots of Covariate X for Each Group";
proc sgplot data=&DATASET;
hbox x / group=grp extreme ;
run;

title1 "Scatter Plot of Response Y vs Covariate X";
proc sgplot data=&DATASET;
scatter y=y x=x / group=grp;
run;

* ANCOVA which forces the relationship between the response and the
covariate to be the same across treatment groups;

ods graphics on;

title1 "ANOVA To Compare Mean of Response Variable Y Across Groups";
proc mixed data=&DATASET;
class grp;
model y = grp ;
lsmeans grp / pdiff cl ;
run;

title1 "ANOVA To Compare Mean of Covariate X Across Groups";
proc mixed data=&DATASET;
```

```
class grp;
model x = grp ;
run;

title1 "ANCOVA To Compare Mean of Response Variable Y Across Groups,
Adjusting for X";
title2 "Parallel Slopes Model";
proc mixed data=&DATASET;
class grp;
model y = x grp ;
lsmeans grp / pdiff cl ;
run;
* Using PROC GLM to produce the ANCOVA plot, which is not produced by
PROC MIXED;

* Comment the following ODS SELECT statement out to see all of the
output from GLM;
* Leave it uncommented to output only the ANCOVA plot;

ods select ANCOVAPlot;

title1 "ANCOVA To Compare Mean of Response Variable Y Across Groups,
Adjusting for X";
title2 "Parallel Slopes Model";
proc glm data=&DATASET;
class grp;
model y = x grp ;
lsmeans grp / pdiff cl ;
run;
quit;

* ANCOVA which allows for the relationship between the response and
the covariate to differ across treatment groups. ;

ods graphics on;

title1 "ANCOVA To Compare Mean of Y Across Groups, Adjusting for X";
title2 "Non-Parallel Slopes";
proc mixed data=&DATASET;
class grp;
model y = x grp x*grp ;
lsmeans grp / pdiff cl ;
run;
* Using PROC GLM to obtain the ANCOVA plot;

* Comment the following ODS SELECT statement out to see all of the
output from GLM;
* Leave it uncommented to output only the ANCOVA plot;

ods select ANCOVAPlot;

title1 "ANCOVA To Compare Mean of Y Across Groups, Adjusting for X";
title2 "Non-Parallel Slopes";
proc glm data=&DATASET;
class grp;
model y = x grp x*grp ;
lsmeans grp / pdiff cl ;
run;
quit;

%mend;

* Choose dataset to analyze by uncommenting the appropriate line below. ;
* To uncomment a line, remove the asterisk at the beginning of the line. ;

%analyze_data(DATASET=example_1);
*%analyze_data(DATASET=example_2);
```

```
*%analyze_data(DATASET=example_3);
*%analyze_data(DATASET=example_4);

*%analyze_data(DATASET=exercise_1);
*%analyze_data(DATASET=exercise_2);
*%analyze_data(DATASET=exercise_3);
*%analyze_data(DATASET=exercise_4);
*%analyze_data(DATASET=exercise_5);
```

CHAPTER 10: ANALYSIS OF REPEATED MEASURES FOR THE BIOLOGICAL AND AGRICULTURAL SCIENCES

Salvador A. Gezan and Melissa Carvalho

Answers to True or False

1. Spatial correlation is a type of correlation that is present between observations that belong to the same experimental unit. (F)
2. If we have missing data, then repeated measures analysis can't be used. (F)
3. Combining all data from several time points into a single analysis will provide greater statistical power than analyzing every time point separately. (T)
4. For random effects, the statistical inferences are valid only for the levels that are considered in the corresponding factor. (F)
5. The compound symmetry (CS) structure is the simplest structure that can model some form of correlation. (T)
6. The AR(1) and ARH(1) structures do not need identical intervals between measurements. (F)
7. Comparing two models by using the residual log-likelihood (ReslogL) requires that the fixed effects between models are the same. (T)
8. The F- and t-tests from a repeated measures analysis are no longer valid tests because their degrees of freedom are incorrect. (F)
9. Linear mixed models can only be used on normally distributed response variables. (T)

GenStat

```
" Set working directory - change to location of your data file "
SET [WORKINGDIRECTORY='C:/.../CodeChapter']

" Read and display data from working directory "
FILEREAD [NAME='HEIGHT.TXT'; IMETHOD=read] FGROUPS=7(yes),no,no
FSPREADSHEET  Plot,Spp,Stk,Prep,Trt,Blk,Time,Initial,Ht

" Single point measurement analysis "
RESTRICT Plot,Spp,Stk,Prep,Trt,Blk,Time,Initial,Ht; CONDITION=Time.
EQ.1984
RESTRICT Plot,Spp,Stk,Prep,Trt,Blk,Time,Initial,Ht; CONDITION=Time.
EQ.1985
RESTRICT Plot,Spp,Stk,Prep,Trt,Blk,Time,Initial,Ht; CONDITION=Time.
EQ.1986
RESTRICT Plot,Spp,Stk,Prep,Trt,Blk,Time,Initial,Ht; CONDITION=Time.
EQ.1987
RESTRICT Plot,Spp,Stk,Prep,Trt,Blk,Time,Initial,Ht; CONDITION=Time.
EQ.1988
RESTRICT Plot,Spp,Stk,Prep,Trt,Blk,Time,Initial,Ht; CONDITION=Time.
```

```
EQ.1989
VCOMPONENTS [FIXED=Initial+Blk+Trt; FACTORIAL=9]
REML [PRINT=model,components,deviance,waldTests; FMETHOD=automatic;
MVINCLUDE=*; METHOD=AI;\
 MAXCYCLE=20] Ht;
VPLOT [RMETHOD=all] fittedvalues,normal,halfnormal,histogram
VPREDICT [PRINT=description,prediction,se,avesed] CLASSIFY=Trt;
LEVELS=*; PARALLEL=*
RESTRICT Plot,Spp,Stk,Prep,Trt,Blk,Time,Initial,Ht;

" Fitting different error structures with factor trt "
VCOMP [FIXED=Initial+Time+Time.Blk+Trt+Time.Trt; CADJUST=none;
FACTORIAL=9] Plot.Time;\
 CONSTRAIN=positive
VSTRUCTURE [Plot.Time] FACTOR=Time; MODEL=identity; ORDER=1;
HETEROGENEITY=none;      " ID error structure "
VSTRUCTURE [Plot.Time] FACTOR=Time; MODEL=uniform; ORDER=1;
HETEROGENEITY=none;       " CS error structure "
VSTRUCTURE [Plot.Time] FACTOR=Time; MODEL=ar; ORDER=1;
HETEROGENEITY=none;            " AR(1) error structure "
VSTRUCTURE [Plot.Time] FACTOR=Time; MODEL=diagonal; ORDER=1;
HETEROGENEITY=none;      " DIAG error structure "
VSTRUCTURE [Plot.Time] FACTOR=Time; MODEL=uniform; ORDER=1;
HETEROGENEITY=outside;    " CSH error structure "
VSTRUCTURE [Plot.Time] FACTOR=Time; MODEL=ar; ORDER=1;
HETEROGENEITY=outside;         " ARH(1) error structure "
VSTRUCTURE [Plot.Time] FACTOR=Time; MODEL=banded; ORDER=5;
HETEROGENEITY=outside;     " TOEPH error structure "
VSTRUCTURE [Plot.Time] FACTOR=Time; MODEL=unstructured; ORDER=*;
HETEROGENEITY=none; " UN error structure "
REML [PRINT=model,components,deviance,waldTests; MAXCYCLE=20;
FMETHOD=automatic; MVINCLUDE=explanatory,\
yvariate; METHOD=AI] Ht
VAIC [PRINT=deviance,aic,bic]

" Detailed model UN error structure - full treatment structure "
VCOMP [FIXED=Initial+Time+Time.Blk+Time*Spp*Stk*Prep; CADJUST=none;
FACTORIAL=17] Plot.Time;\
 CONSTRAIN=positive
VSTRUCTURE [Plot.Time] FACTOR=Time; MODEL=unstructured; ORDER=*;
HETEROGENEITY=none;  " UN error structure "
REML [PRINT=model,components,deviance,waldTests; MAXCYCLE=20;
FMETHOD=automatic; MVINCLUDE=explanatory,\
yvariate; METHOD=AI] Ht
VAIC [PRINT=deviance,aic,bic]
```

SAS

```
ods graphics on;

* Read data - change directory to location of your data file;
data HEIGHT;
  infile 'C:\...\CodeChapter\HEIGHT.TXT' firstobs=2 expandtabs;
  input Plot $ Spp $ Stk $ Prep $ Trt $ Blk $ Time $ Initial Ht;
run;
proc print data=HEIGHT (obs=20); run;

* Sorting data by Time;
proc sort data=HEIGHT;
  by Time;
run;

* Single point measurement analysis;
proc mixed data=HEIGHT plots=studentpanel;
  by Time;
  class Trt Blk Time;
  model Ht = Initial Blk Trt / ddfm=KR;
  lsmeans Trt / cl;
run;

* Fitting different error structures with factor trt;
proc mixed data=HEIGHT;
  class Plot Spp Stk Prep Trt Blk Time;
```

```
  model Ht = Initial Time Time(Blk) Trt Time*Trt / ddfm=KR;
  *repeated Time / subject=Plot type=VC;          * ID error
structure;
  *repeated Time / subject=Plot type=CS;          * CS error
structure;
  *repeated Time / subject=Plot type=AR(1);       * AR(1) error
structure;
  *repeated Time / subject=Plot type=TOEP;       * TOEP error structure;
  *repeated Time / subject=Plot type=VC group=Time; * DIAG error structure;
  *repeated Time / subject=Plot type=CSH;         * CSH error structure;
  *repeated Time / subject=Plot type=ARH(1);     * ARH(1) error structure;
  *repeated Time / subject=Plot type=TOEPH;      * TOEPH error structure;
  repeated Time / subject=Plot type=UN;           * UN error structure;
  lsmeans Time*Trt / slice=Time;
run;

* Detailed model UN error structure - full treatment structure;
proc mixed data=HEIGHT plots=studentpanel;
  class Plot Spp Stk Prep Trt Blk Time;
  model Ht = Initial Time Time(Blk) Spp|Stk|Prep Time*Spp Time*Stk
Time*Prep Time*Spp*Stk Time*Spp*Prep Time*Stk*Prep Time*Spp*Stk*Prep /
ddfm=KR residual outp=resid;
  repeated Time / subject=Plot type=UN r rcorr;
run;

ods graphics off;
```

R

```
# Read data - change directory to location of your data file;
rm(list=ls())
setwd("C:/Users/sgeza/Desktop/Repeatead/Paper_AGJournal_2016/
Revisions_Nov2016")

# Loading libraries
library(nlme)
library(lsmeans)

# Reading data
datasoy<-data.frame(Soybean[Soybean$Year=='1988',])
head(datasoy)

# Defining factors
datasoy$Plot<-as.factor(datasoy$Plot)
datasoy$Variety<-as.factor(datasoy$Variety)
datasoy$Time<-as.factor(datasoy$Time)
str(datasoy)

# Some EDA
boxplot(weight~Time,data=datasoy)
hist(datasoy$weight)
datasoy$logweight<-log(datasoy$weight)
boxplot(logweight~Time,data=datasoy)

# Obtaining subsets of the data by Time
T14<-datasoy[datasoy$Time==14,]

# Single point measurement analysis
modelSingle<-lm(logweight~Variety,data=T14)
summary(modelSingle)
anova(modelSingle)
plot(modelSingle)
lsmeans(modelSingle,~Variety)

# ID error structure
rstruct<-varIdent(form=~1)
rheter<-varIdent(form=~1)
modID<-gls(logweight~Time+Variety+Time:Variety,
           correlation=rstruct,weights=rheter,data=datasoy)
#summary(modID)
#anova(modID)
#plot(modID)
```

```
# CS error structure
rstruct<-corCompSymm(form=~Time|Plot)
rheter<-varIdent(form=~1)
modCS<-gls(logweight~Time+Variety+Time:Variety,
           correlation=rstruct,weights=rheter,data=datasoy)

# CSH error structure
rstruct<-corCompSymm(form=~Time|Plot)
rheter<-varIdent(form=~1|Time)
modCSH<-gls(logweight~Time+Variety+Time:Variety,
           correlation=rstruct,weights=rheter,data=datasoy)

# DIAG error structure
rstruct<-varIdent(form=~1)
rheter<-varIdent(form=~1|Time)
modDIAG<-gls(logweight~Time+Variety+Time:Variety,
           correlation=rstruct,weights=rheter,data=datasoy)

# AR(1) error structure
rstruct<-corAR1(form=~1|Plot)
rheter<-varIdent(form=~1)
modAR1<-gls(logweight~Time+Variety+Time:Variety,
           correlation=rstruct,weights=rheter,data=datasoy)

# ARH(1) error structure
rstruct<-corAR1(form=~1|Plot)
rheter<-varIdent(form=~1|Time)
modARH1<-gls(logweight~Time+Variety+Time:Variety,
           correlation=rstruct,weights=rheter,data=datasoy)

# US error structure (with extra output)
rstruct<-corSymm(form=~1|Plot)
rheter<-varIdent(form=~1|Time)
modUS<-gls(logweight~Time+Variety+Time:Variety,
           correlation=rstruct,weights=rheter,data=datasoy)

# Comparing US against ARH(1) using Likelihood ratio test
anova(modUS,modARH1)

# Output for selected model
#mod<-modID
#mod<-modCS
#mod<-modCSH
#mod<-modDIAG
#mod<-modAR1
mod<-modARH1
#mod<-modUS
output<-summary(mod)
anova(mod,type='marginal')   # Marginal ANOVA table
anova(mod,type='sequential') # Sequential ANOVA table
attr(output$apVar,"Pars")    # Variance components
(logL<-2*mod$logLik)         # log-likelihood value
(AIC<-output$AIC)            # AIC
(BIC<-output$BIC)            # BIC
```

CHAPTER 11: THE DESIGN AND ANALYSIS OF LONG-TERM ROTATION EXPERIMENTS

Roger William Payne

Appendix 1. Genstat commands to analyze the potato yields from the Westmaas experiment.

```
IMPORT [PRINT=*] 'wmpotato.xlsx'
" try various random models "
CAPTION 'Split-plot nested within years'; STYLE=meta
VCOMPONENTS [FIXED=Year*Rotation*Nitrogen] Year/Block/Wholeplot
REML [PRINT=components] Yield
VAIC [PRINT=deviance,aic,sic,dfrandom]
VRACCUMULATE [PRINT=*] 'Split-plot nested within years'
CAPTION 'Nested split-plot, different residual variance each year';\
  STYLE=meta
VCOMPONENTS [FIXED=Year*Rotation*Nitrogen; EXPERIMENT=Year]\
  Year/Block/Wholeplot
REML [PRINT=components] Yield
VAIC [PRINT=deviance,aic,sic,dfrandom]
VRACCUMULATE [PRINT=*] 'Nested split-plot meta analysis'
" Nested split plot: power-distance correlation structure over years"
VCOMPONENTS [FIXED=Year*Rotation*Nitrogen] Year/Block/Wholeplot
VARIATE Ycoord; VALUES=Year
VSTRUCTURE [TERM=Year.Block.Wholeplot; COORDINATES=Ycoord] power;\
  FACTOR=Year
REML [PRINT=*] Yield
VRACCUMULATE [PRINT=*] 'Nested split-plot and power distance'
" Nested split plot with different residual variance in each year
  and power-distance correlation structure over years "
VCOMPONENTS [FIXED=Year*Rotation*Nitrogen; EXPERIMENT=Year]\
  Year/Block/Wholeplot
VSTRUCTURE [TERM=Year.Block.Wholeplot; COORDINATES=Ycoord] power;\
  FACTOR=Year
REML [PRINT=*] Yield
VRACCUMULATE [PRINT=*]\
  'Nested split-plot meta analysis and power distance'
" Nested split plot with different residual variances
  and variance components "
FORMULA [VALUE=Block/Wholeplot] differentvcterms
VRMETA [EXPERIMENTSFACTOR=Year; RANDOM=Random] 77,79,80...88;\
  LOCALTERMS=differentvcterms
VCOMPONENTS [FIXED=Year*Rotation*Nitrogen; EXPERIMENTS=Year] #Random
REML [PRINT=*; MVINCLUDE=explanatory] Yield
VRACCUMULATE 'Meta analysis with different variance components'
" use split plot with different residual variance in each year "
CAPTION 'Nested split-plot, different residual variance each year';\
  STYLE=meta
VCOMPONENTS [FIXED=Year*Rotation*Nitrogen; EXPERIMENT=Year]\
  Year/Block/Wholeplot
REML [PRINT=wald] Yield
" drop unnecessary fixed terms "
VCOMPONENTS [FIXED=Year*Rotation*Nitrogen-Year.Rotation.Nitrogen;\
  EXPERIMENTS=Year] Year/Block/Wholeplot
REML [PRINT=wald; MVINCLUDE=explanatory; WORKSPACE=100] Yield
VCOMPONENTS [FIXED=Year*Rotation*Nitrogen - Year.Rotation.Nitrogen\
  - Rotation.Nitrogen; EXPERIMENTS=Year] Year/Block/Wholeplot
REML [PRINT=wald; MVINCLUDE=explanatory; WORKSPACE=100] Yield
VCOMPONENTS [FIXED=Year*Rotation*Nitrogen - Year.Rotation.Nitrogen\
  - Rotation.Nitrogen - Year.Nitrogen; EXPERIMENTS=Year]\
```

```
  Year.Block/Wholeplot
REML [PRINT=wald; MVINCLUDE=explanatory; WORKSPACE=100] Yield
" predicted means for Nitrogen,
  and for Year.Rotation with a summary of the sed's "
VDISPLAY [PRINT=means; PTERMS=Nitrogen+Year.Rotation]
" to print the means with all the sed's, set option PSE
VDISPLAY [PRINT=means; PTERMS=Year.Rotation; PSE=alldifferences]"
PEN 11...15; LINESTYLE=1,8,1,8,1;\
  CSYMBOL='crimson','darkviolet','violet','darkblue','royalblue';\
  CLINE='crimson','darkviolet','violet','darkblue','royalblue';\
  CFILL='crimson','darkviolet','violet','darkblue','royalblue';\
  SYMBOL='circle','heavycross','diamond','heavyplus','square'
VGRAPH [METHOD=linesandpoints] Year; GROUP=Rotation;\
  PENS=!(11...15); YTITLE='Yield (t ha~^{-1})';\
  TITLE='Estimated mean yield of potato for rotations and years'
```

Appendix 2. Genstat output for the potato yields from the Westmaas experiment.

```
IMPORT [PRINT=*] 'wmpotato.xlsx'
" try various random models "
CAPTION 'Split-plot nested within years'; STYLE=meta
```

Split-plot nested within years

```
VCOMPONENTS [FIXED=Year*Rotation*Nitrogen] Year/Block/Wholeplot
REML [PRINT=components] Yield
```

Estimated variance components

Random Term	Component	S.e.
Year.Block	1.407	1.403
Year.Block.Wholeplot	7.079	1.893

Residual variance model

Term	Model(order)	Parameter	Estimate	s.e.
Residual	Identity	Sigma2	5.195	0.701

```
VAIC [PRINT=deviance,aic,sic,dfrandom]
```

```
    Deviance                                950.43
    Akaike information coefficient          956.43
    Schwarz Bayes information coefficient   965.75
    d.f. of random model                    3
```

Note: omits constant, -log(det(X'X)), that depends only on the fixed model.

```
VRACCUMULATE [PRINT=*] 'Split-plot nested within years'
CAPTION 'Nested split-plot, different residual variance each year';\
   STYLE=meta
```

Nested split-plot, different residual variance each year

```
VCOMPONENTS [FIXED=Year*Rotation*Nitrogen; EXPERIMENT=Year]\
   Year/Block/Wholeplot
REML [PRINT=components] Yield
```

Estimated variance components

Random term	component	s.e.
Year.Block	2.12	1.17
Year.Block.Wholeplot	0.91	0.52

Residual model for each experiment
Experiment factor: Year

Experiment	Term Factor	Model(order)	Parameter	Estimate	s.e.
77.00	Residual	Identity	Variance	0.819	0.352
79.00	Residual	Identity	Variance	2.684	1.149
80.00	Residual	Identity	Variance	1.344	0.588
81.00	Residual	Identity	Variance	5.231	2.558
82.00	Residual	Identity	Variance	4.538	1.873
83.00	Residual	Identity	Variance	1.988	0.887
84.00	Residual	Identity	Variance	4.013	1.660
85.00	Residual	Identity	Variance	5.415	2.185
86.00	Residual	Identity	Variance	14.28	5.56
87.00	Residual	Identity	Variance	58.21	22.43
88.00	Residual	Identity	Variance	12.90	5.25

```
VAIC [PRINT=deviance,aic,sic,dfrandom]
```

```
Deviance                                     891.60
Akaike information coefficient               917.60
Schwarz Bayes information coefficient        957.98
d.f. of random model                         13
```

Note: omits constant, -log(det(X'X)), that depends only on the fixed model.

```
VRACCUMULATE [PRINT=*] 'Nested split-plot meta analysis'
" Nested split plot: power-distance correlation structure over years"
VCOMPONENTS [FIXED=Year*Rotation*Nitrogen] Year/Block/Wholeplot
VARIATE Ycoord; VALUES=Year
VSTRUCTURE [TERM=Year.Block.Wholeplot; COORDINATES=Ycoord] power;\
   FACTOR=Year
REML [PRINT=*] Yield
VRACCUMULATE [PRINT=*] 'Nested split-plot and power distance'
" Nested split plot with different residual variance in each year
   and power-distance correlation structure over years "
VCOMPONENTS [FIXED=Year*Rotation*Nitrogen; EXPERIMENT=Year]\
   Year/Block/Wholeplot
VSTRUCTURE [TERM=Year.Block.Wholeplot; COORDINATES=Ycoord] power;\
   FACTOR=Year
REML [PRINT=*] Yield
VRACCUMULATE [PRINT=*]\
   'Nested split-plot meta analysis and power distance'
" Nested split plot with different residual variances
   and variance components "
FORMULA [VALUE=Block/Wholeplot] differentvcterms
VRMETA [EXPERIMENTSFACTOR=Year; RANDOM=Random] 77,79,80...88;\
   LOCALTERMS=differentvcterms
VCOMPONENTS [FIXED=Year*Rotation*Nitrogen; EXPERIMENTS=Year] #Random
REML [PRINT=*; MVINCLUDE=explanatory] Yield
VRACCUMULATE 'Meta analysis with different variance components'
```

Accumulated summary of REML random models

	Deviance	AIC	SIC	Random d.f.
Split-plot nested within years	950.43	956.43	965.75	3
Nested split-plot meta analysis	891.60	917.60	957.98	13
Nested split-plot and power distance	950.38	958.38	970.80	4
Nested split-plot meta analysis and power distance	891.57	919.57	963.06	14
Meta analysis with different variance components	852.53	918.53	1021.02	33

Note: omits constant, -log(det(X'X)), that depends only on the fixed model.

```
" use split plot with different residual variance in each year "
CAPTION 'Nested split-plot, different residual variance each year';\
   STYLE=meta
```

Nested split-plot, different residual variance each year

```
VCOMPONENTS [FIXED=Year*Rotation*Nitrogen; EXPERIMENT=Year]\
   Year/Block/Wholeplot
REML [PRINT=wald] Yield
```

Tests for fixed effects

Sequentially adding terms to fixed model

Fixed term	Wald statistic	n.d.f.	F statistic	d.d.f.	F pr
Year	851.65	10	85.45	10.4	<0.001
Rotation	106.93	4	26.73	27.0	<0.001
Nitrogen	110.56	2	55.28	58.1	<0.001
Year.Rotation	154.93	40	3.75	22.8	<0.001
Year.Nitrogen	29.16	20	1.27	48.7	0.242
Rotation.Nitrogen	15.43	8	1.93	58.1	0.073
Year.Rotation.Nitrogen	108.45	80	1.19	40.6	0.271

Dropping individual terms from full fixed model

Fixed term	Wald statistic	n.d.f	F statistic	d.d.f.	F pr
Year.Rotation.Nitrogen	108.45	80	1.19	40.6	0.271

Message: denominator degrees of freedom for approximate F-tests are calculated using algebraic derivatives ignoring fixed/boundary/singular variance parameters.

```
" drop unnecessary fixed terms "
VCOMPONENTS [FIXED=Year*Rotation*Nitrogen-Year.Rotation.Nitrogen;\
   EXPERIMENTS=Year] Year/Block/Wholeplot
REML [PRINT=wald; MVINCLUDE=explanatory; WORKSPACE=100] Yield
```

Tests for Fixed Effects

Sequentially adding terms to fixed model

Fixed term	Wald statistic	n.d.f.	F statistic	d.d.f.	F pr
Year	796.16	10	79.79	10.7	<0.001
Rotation	91.94	4	22.98	32.8	<0.001
Nitrogen	83.70	2	41.85	128.4	<0.001
Year.Rotation	117.01	40	2.92	25.3	0.003
Year.Nitrogen	29.67	20	1.36	88.3	0.166
Rotation.Nitrogen	10.34	8	1.29	128.4	0.253

Dropping individual terms from full fixed model

Fixed term	Wald statistic	n.d.f.	*F* statistic	d.d.f.	F pr
Year.Rotation	117.01	40	2.92	25.3	0.003
Year.Nitrogen	29.67	20	1.36	88.3	0.166
Rotation.Nitrogen	10.34	8	1.29	128.4	0.253

Message: denominator degrees of freedom for approximate *F*-tests are calculated using algebraic derivatives ignoring fixed/boundary/singular variance parameters.

```
VCOMPONENTS [FIXED=Year*Rotation*Nitrogen - Year.Rotation.Nitrogen\
   - Rotation.Nitrogen; EXPERIMENTS=Year] Year/Block/Wholeplot
REML [PRINT=wald; MVINCLUDE=explanatory; WORKSPACE=100] Yield
```

Tests for fixed effects

Sequentially adding terms to fixed model

Fixed term	Wald statistic	n.d.f.	F statistic	d.d.f.	F pr
Year	783.23	10	78.49	10.8	<0.001
Rotation	89.66	4	22.42	34.0	<0.001
Nitrogen	83.72	2	41.86	136.7	<0.001
Year.Rotation	112.76	40	2.82	26.9	0.003
Year.Nitrogen	29.82	20	1.38	95.6	0.154

Dropping individual terms from full fixed model

Fixed term	Wald statistic	n.d.f.	F statistic	d.d.f.	F pr
Year.Rotation	112.76	40	2.82	26.9	0.003
Year.Nitrogen	29.82	20	1.38	95.6	0.154

Message: denominator degrees of freedom for approximate F-tests are calculated using algebraic derivatives ignoring fixed/boundary/singular variance parameters.

```
VCOMPONENTS [FIXED=Year*Rotation*Nitrogen - Year.Rotation.Nitrogen\
   - Rotation.Nitrogen - Year.Nitrogen; EXPERIMENTS=Year]\
   Year.Block/Wholeplot
REML [PRINT=wald; MVINCLUDE=explanatory; WORKSPACE=100] Yield
```

Tests for fixed effects

Sequentially adding terms to fixed model

Fixed term	Wald statistic	n.d.f.	*F* statistic	d.d.f.	F pr
Year	780.43	10	78.21	10.8	<0.001
Rotation	92.68	4	23.17	32.6	<0.001
Nitrogen	83.21	2	41.60	142.0	<0.001
Year.Rotation	117.47	40	2.93	27.0	0.002

Dropping individual terms from full fixed model

Fixed term	Wald statistic	n.d.f.	*F* statistic	d.d.f.	F pr
Nitrogen	83.21	2	41.60	142.0	<0.001
Year.Rotation	117.47	40	2.93	27.0	0.002

Message: denominator degrees of freedom for approximate F-tests are calculated using algebraic derivatives ignoring fixed/boundary/singular variance parameters.

```
" predicted means for Nitrogen,
  and for Year.Rotation with a summary of the sed's "
VDISPLAY [PRINT=means; PTERMS=Nitrogen+Year.Rotation]
```

Table of predicted means for Nitrogen

Nitrogen	N1	N2	N3
	43.37	45.03	45.65

Standard error of differences: 0.2585

Table of predicted means for Year.Rotation

Rotation Year	IIf	III	IIIf	IV	IVf
77	23.44	21.92	21.87	25.01	21.20
79	39.44	46.62	45.74	49.50	44.35
80	40.53	41.14	40.90	41.54	42.48
81	30.37	38.04	39.45	39.43	38.14
82	39.75	40.49	42.95	45.53	45.63
83	31.14	36.55	39.53	37.96	37.61
84	54.59	57.67	56.63	58.40	58.99
85	51.48	47.80	49.79	52.54	49.75
86	48.21	51.03	54.58	51.48	52.99
87	42.99	48.76	41.56	45.67	51.85
88	54.28	62.76	64.41	58.52	62.46

Standard errors of differences

Average	2.483
Maximum	3.686
Minimum	1.413

Average variance of differences: 6.366

Standard error of differences for same level of factor:

	Year	Rotation
Average:	1.978	2.524
Maximum:	3.686	3.527
Minimum:	1.413	2.074

Average variance of differences:

	Year	Rotation
	4.328	6.529

```
" to print the means with all the sed's, set option PSE
VDISPLAY [PRINT=means; PTERMS=Year.Rotation; PSE=alldifferences]"
PEN 11...15; LINESTYLE=1,8,1,8,1;\
  CSYMBOL='crimson','darkviolet','violet','darkblue','royalblue';\
 CLINE='crimson','darkviolet','violet','darkblue','royalblue';\
 CFILL='crimson','darkviolet','violet','darkblue','royalblue';\
  SYMBOL='circle','heavycross','diamond','heavyplus','square'
VGRAPH [METHOD=linesandpoints] Year; GROUP=Rotation;\
   PENS=!(11...15); YTITLE='Yield (t ha~^{-1})';\
  TITLE='Estimated mean yield of potato for rotations and years'
```

Appendix 3. ASReml-R commands to analyze potato yields from the Westmaas experiment.

```
#'## Initialize
library(asreml4)
library(asremlPlus)
library(ggplot2)
library(knitr)
#knitr::spin("Appendix1.v4.r")
options(width = 110)

#'## Load data, order it and create a 3 level nested Wholeplot factor
load(file = "wmpotato.rda")
```

```
wmpotato <- with(wmpotato, wmpotato[order(Year,Block,Wholeplot,Subplot), ])
wmpotato$NWholeplot <- factor(rep(1:5, each=3, times=22))

#'## Try various random models
#'### Split plot within years
model1.asr <- asreml(Yield ~ Year*Rotation*Nitrogen,
                     random = ~Year:Block/Wholeplot,
                     data = wmpotato)
summary(model1.asr)$varcomp
info.accumulate <- data.frame(Model = "Split-plot",
                              infoCriteria(model1.asr),
                              stringsAsFactors = FALSE)

#'### Split plot with different residual variance in each year
model2.asr <- asreml(Yield ~ Year*Rotation*Nitrogen,
                    random = ~Year:Block/Wholeplot,
                    residual = ~ idh(Year):Block:NWholeplot:Subplot,
                    data = wmpotato)
model2.asr <- update(model2.asr)
summary(model2.asr)$varcomp
info.accumulate <- rbind(info.accumulate,
                         data.frame(Model = "Split plot meta analysis",
                                    infoCriteria(model2.asr),
                                    stringsAsFactors = FALSE))

#'### Split plot with EXP structure over years
model3.asr <- asreml(Yield ~ Year*Rotation*Nitrogen,
                    random = ~Year:Block/Wholeplot,
                    residual = ~ exp(Year):Block:NWholeplot:Subplot,
                    data = wmpotato)
summary(model3.asr)$varcomp
info.accumulate <- rbind(info.accumulate,
                         data.frame(Model = "Split-plot and EXP",
                                    infoCriteria(model3.asr),
                                    stringsAsFactors = FALSE))

#'### Split plot: different residual variance in each year & EXP
structure
model4.asr <- asreml(Yield ~ Year*Rotation*Nitrogen,
                    random = ~Year:Block/Wholeplot,
                    residual = ~ exph(Year):Block:NWholeplot:Subplot,
                    data = wmpotato)
model4.asr <- update(model4.asr)
summary(model4.asr)$varcomp
info.accumulate <- rbind(info.accumulate,
                   data.frame(Model = "Split-plot meta analysis and EXP",
                                    infoCriteria(model4.asr),
                                    stringsAsFactors = FALSE))

dimnames(summary(model4.asr)$varcomp)

#'### Split plot with different residual variances and variance
components
model5.asr <- asreml(Yield ~ Year*Rotation*Nitrogen,
                    random = ~ idh(Year):Block/Wholeplot,
                    residual =   ~ idh(Year):Block:NWholeplot:Subplot,
                    data = wmpotato)
summary(model5.asr)$varcomp
vcnames <- rownames(summary(model5.asr)$varcomp)[1:22]
model5.asr <- setvarianceterms(model5.asr$call, terms = vcnames,
                 bounds = "U", initial=0.01, ignore.suffices = FALSE)
model5.asr <- update(model5.asr)
summary(model5.asr)$varcomp
info.accumulate <- rbind(info.accumulate,
     data.frame(Model = "Meta analysis with different variance components",
                        infoCriteria(model5.asr, bound.exclusions = "F"),
                        stringsAsFactors = FALSE))
```

```
#'### Accumulated summary of REML information criteria
info.accumulate
#'### Use split plot with different residual variance in each year
wald(model2.asr, denDF = "algebraic")

#'## Drop unnecessary fixed terms
model2a.asr <- asreml(Yield ~ Year*Rotation*Nitrogen-Year:Rotation:Nitrogen,
                      random = ~ Year:Block/Wholeplot,
                     residual = ~ idh(Year):Block:NWholeplot:Subplot,
                      data = wmpotato)
wald(model2a.asr, denDF = "algebraic")

model2b.asr <- asreml(Yield ~ Year*Rotation*Nitrogen-
Year:Rotation:Nitrogen -
                                       Rotation:Nitrogen,
                       random = ~ Year:Block/Wholeplot,
                     residual =   ~ idh(Year):Block:NWholeplot:Subplot,
                       data = wmpotato)
wald(model2b.asr, denDF = "algebraic")

model2c.asr <- asreml(Yield ~ Year*Rotation*Nitrogen-
Year:Rotation:Nitrogen -
                         Rotation:Nitrogen - Year:Nitrogen,
                       random = ~ Year:Block/Wholeplot,
                       residual =   ~ idh(Year):Block:NWholeplot:Subplot,
                       data = wmpotato)
wald(model2c.asr, denDF = "algebraic")

#'## Get predictions and plot
predict(model2c.asr, classify = "Nitrogen")$pvals
predYR <- predict(model2c.asr, classify = "Year:Rotation")$pvals
predYR

cols <- c('red','darkviolet','violet','darkblue','lightskyblue')
ggplot(data = predYR,
      aes(x =Year, y=predicted.value, colour=Rotation, linetype=Rotation,
       shape = Rotation)) + geom_point() + geom_line() +
       labs(y = "Yield") + scale_color_manual(values =
       cols) + scale_shape_manual(values = c(16,4,18,3,15))
```

Appendix 4. ASReml-R output from the analysis of potato yields from the Westmaas experiment.

Initialize

```
library(asreml4)
## Loading required package: Matrix
## Licensed to University of South Australia, serial number 402060331,
expires 31-jan-2018, 373 days.
library(asremlPlus)
library(ggplot2)
library(knitr)
#knitr::spin("Appendix1.v4.r")
options(width = 110)
```

Load data, order it and create a 3 level nested Wholeplot factor

```
load(file = "wmpotato.rda")
wmpotato <- with(wmpotato, wmpotato[order(Year,Block,Wholeplot,Subplot), ])
wmpotato$NWholeplot <- factor(rep(1:5, each=3, times=22))
```

Try various random models

Split plot within years

```
model1.asr <- asreml(Yield ~ Year*Rotation*Nitrogen,
                     random = ~Year:Block/Wholeplot,
                     data = wmpotato)
```

```
## Model fitted using the gamma parameterization.
## ASReml4 Beta-release 4.0.0.9005 Mon Jan 23 15:55:32 2017
##          LogLik        Sigma2     DF     wall     cpu
##  1      -341.227       10.0174    165 15:55:32    0.0
##  2      -332.725        7.9880    165 15:55:32    0.0
##  3      -326.355        6.4141    165 15:55:32    0.0
##  4      -324.012        5.5986    165 15:55:32    0.0
##  5      -323.596        5.2382    165 15:55:32    0.0
##  6      -323.590        5.1960    165 15:55:32    0.0
##  7      -323.590        5.1952    165 15:55:32    0.0
summary(model1.asr)$varcomp
##                        component std.error  z.ratio bound %ch
## Year:Block              1.407175 1.4026645 1.003216     P   0
## Year:Block:Wholeplot    7.079126 1.8929384 3.739755     P   0
## units(R)                5.195198 0.7005203 7.416198     P   0
info.accumulate <- data.frame(Model = "Split-plot",
                              infoCriteria(model1.asr),
                              stringsAsFactors = FALSE)
Split plot with different residual variance in each year
model2.asr <- asreml(Yield ~ Year*Rotation*Nitrogen,
                     random = ~Year:Block/Wholeplot,
                     residual = ~ idh(Year):Block:NWholeplot:Subplot,
                     data = wmpotato)
## Model fitted using the sigma parameterization.
## ASReml4 Beta-release 4.0.0.9005 Mon Jan 23 15:55:32 2017
##          LogLik        Sigma2     DF     wall     cpu
## 1      -325.726        1.0     165 15:55:32    0.0 (2 restrained)
##  2      -315.906            1.0    165 15:55:32    0.0
##  3      -307.104            1.0    165 15:55:32    0.0
##  4      -300.698            1.0    165 15:55:32    0.0
##  5      -295.773            1.0    165 15:55:32    0.0
##  6      -294.551            1.0    165 15:55:32    0.0
##  7      -294.313            1.0    165 15:55:32    0.0
##  8      -294.249            1.0    165 15:55:32    0.0
##  9      -294.216            1.0    165 15:55:32    0.0
## 10      -294.197            1.0    165 15:55:32    0.0
## 11      -294.187            1.0    165 15:55:32    0.0
## 12      -294.181            1.0    165 15:55:32    0.0
## 13      -294.178            1.0    165 15:55:32    0.0
## Warning in asreml(Yield ~ Year * Rotation * Nitrogen, random =
~Year:Block/Wholeplot, : Log-likelihood not
## converged
## Warning in asreml(Yield ~ Year * Rotation * Nitrogen, random =
~Year:Block/Wholeplot, : Some components
## changed by more than 1% on the last iteration.
model2.asr <- update(model2.asr)
## Model fitted using the sigma parameterization.
## ASReml4 Beta-release 4.0.0.9005 Mon Jan 23 15:55:32 2017
##          LogLik        Sigma2     DF     wall     cpu
##  1      -294.176            1.0    165 15:55:32    0.0
##  2      -294.176            1.0    165 15:55:32    0.0
##  3      -294.176            1.0    165 15:55:32    0.0
##  4      -294.175            1.0    165 15:55:32    0.0
##  5      -294.175            1.0    165 15:55:32    0.0
summary(model2.asr)$varcomp
##                                          component std.error  z.ratio bound %ch
## Year:Block                               2.1188142 1.1698023 1.811258     P 0.0
## Year:Block:Wholeplot                     0.9225528 0.5212748 1.769801     P 0.5
## Year:Block:Nwholeplot:Subplot(R)        1.0000000        NA       NA     F 0.0
## Year:Block:Nwholeplot:Subplot!Year_77   0.8192466 0.3522389 2.325826     P 0.0
## Year:Block:Nwholeplot:Subplot!Year_79   2.6812820 1.1480199 2.335571     P 0.0
## Year:Block:Nwholeplot:Subplot!Year_80   1.3431204 0.5869084 2.288467     P 0.0
## Year:Block:Nwholeplot:Subplot!Year_81   5.1776315 2.5260176 2.049721     P 0.5
## Year:Block:Nwholeplot:Subplot!Year_82   4.5354571 1.8726148 2.421991     P 0.0
## Year:Block:Nwholeplot:Subplot!Year_83   1.9832480 0.8836502 2.244381     P 0.1
## Year:Block:Nwholeplot:Subplot!Year_84   4.0128598 1.6612209 2.415609     P 0.0
## Year:Block:Nwholeplot:Subplot!Year_85   5.4153625 2.1867447 2.476449     P 0.0
## Year:Block:Nwholeplot:Subplot!Year_86  14.2810395 5.5608347 2.568147     P 0.0
## Year:Block:Nwholeplot:Subplot!Year_87  58.1502751 22.4010672 2.595871    P 0.0
## Year:Block:Nwholeplot:Subplot!Year_88  12.8700511 5.2352869 2.458328     P 0.1
info.accumulate <- rbind(info.accumulate,
                         data.frame(Model = "Split plot meta analysis",
                                    infoCriteria(model2.asr),
                                    stringsAsFactors = FALSE))
## Warning in infoCriteria.asreml(model2.asr): The following bound terms
```

```
were discounted:
##  Year:Block:NWholeplot:Subplot(R)
Split plot with EXP structure over years
model3.asr <- asreml(Yield ~ Year*Rotation*Nitrogen,
                     random = ~Year:Block/Wholeplot,
                   residual = ~ exp(Year):Block:NWholeplot:Subplot,
                     data = wmpotato)
## Model fitted using the gamma parameterization.
## ASReml4 Beta-release 4.0.0.9005 Mon Jan 23 15:55:32 2017
##           LogLik         Sigma2     DF      wall     cpu
##  1      -340.831        10.0380    165 15:55:32     0.0
##  2      -332.931         8.0998    165 15:55:32     0.0
##  3      -326.549         6.4773    165 15:55:32     0.0
##  4      -324.050         5.6183    165 15:55:32     0.0
##  5      -323.597         5.2428    165 15:55:32     0.0
##  6      -323.590         5.1963    165 15:55:32     0.0
##  7      -323.590         5.1952    165 15:55:32     0.0
## Warning in asreml(Yield ~ Year * Rotation * Nitrogen, random =
~Year:Block/Wholeplot, : Some components
## changed by more than 1% on the last iteration.
summary(model3.asr)$varcomp
##                         component std.error      z.ratio bound  %ch
## Year:Block        1.407182e+00 1.4026520 1.0032293744      P  0.0
## Year:Block:Wholeplot      7.079156e+00 1.8933937 3.7388719187      P  0.0
## Year:Block:NWholeplot:Subplot(R)     5.195228e+00 0.7006187 7.4151999388   P  0.0
## Year:Block:NWholeplot:Subplot!Year!pow 1.090783e-05 0.1010570 0.0001079374   U 58.6
info.accumulate <- rbind(info.accumulate,
                         data.frame(Model = "Split-plot and EXP",
                                    infoCriteria(model3.asr),
                                    stringsAsFactors = FALSE))
Split plot with different residual variance in each year and EXP structure
model4.asr <- asreml(Yield ~ Year*Rotation*Nitrogen,
                     random = ~Year:Block/Wholeplot,
                   residual = ~ exph(Year):Block:NWholeplot:Subplot,
                     data = wmpotato)
## Model fitted using the sigma parameterization.
## ASReml4 Beta-release 4.0.0.9005 Mon Jan 23 15:55:32 2017
##           LogLik          Sigma2     DF      wall     cpu
##  1      -326.106          1.0   165 15:55:32    0.0 (2 restrained)
##  2       -315.850           1.0    165 15:55:32    0.0
##  3       -307.051           1.0    165 15:55:32    0.0
##  4       -300.617           1.0    165 15:55:32    0.0
##  5       -295.655           1.0    165 15:55:32    0.0
##  6       -294.412           1.0    165 15:55:32    0.0
##  7       -294.167           1.0    165 15:55:32    0.0
##  8       -294.097           1.0    165 15:55:32    0.0
##  9       -294.060           1.0    165 15:55:32    0.0
## 10       -294.037           1.0    165 15:55:32    0.0
## 11       -294.024           1.0    165 15:55:32    0.0
## 12       -294.016           1.0    165 15:55:32    0.0
## 13       -294.012           1.0    165 15:55:32    0.0
## Warning in asreml(Yield ~ Year * Rotation * Nitrogen, random =
~Year:Block/Wholeplot, : Log-likelihood not
## converged
## Warning in asreml(Yield ~ Year * Rotation * Nitrogen, random =
~Year:Block/Wholeplot, : Some components
## changed by more than 1% on the last iteration.
model4.asr <- update(model4.asr)
## Model fitted using the sigma parameterization.
## ASReml4 Beta-release 4.0.0.9005 Mon Jan 23 15:55:32 2017
##           LogLik          Sigma2     DF      wall     cpu
##  1       -294.009           1.0    165 15:55:32    0.0
##  2       -294.009           1.0    165 15:55:33    0.0
##  3       -294.008           1.0    165 15:55:33    0.0
##  4       -294.007           1.0    165 15:55:33    0.0
##  5       -294.007           1.0    165 15:55:33    0.0
##  6       -294.007           1.0    165 15:55:33    0.0
summary(model4.asr)$varcomp
##                     component  std.error   z.ratio bound %ch
## Year:Block                                    2.12260198  1.1684830 1.8165450
P 0.0
## Year:Block:Wholeplot                 0.84472627  0.5064265  1.6680136     P 0.6
## Year:Block:NWholeplot:Subplot(R)   1.00000000     NA         NA          F 0.0
## Year:Block:NWholeplot:Subplot!Year!pow  0.06584539  0.1081466 0.6088532    U 0.2
## Year:Block:NWholeplot:Subplot!Year_77   0.81429437  0.3493945 2.3305874    P 0.0
```

```
## Year:Block:NWholeplot:Subplot!Year_79   2.68405000  1.1469765 2.3401091     P 0.0
## Year:Block:NWholeplot:Subplot!Year_80   1.34566016  0.5892322 2.2837521     P 0.0
## Year:Block:NWholeplot:Subplot!Year_81   5.19119457  2.5411514 2.0428513     P 0.5
## Year:Block:NWholeplot:Subplot!Year_82   4.51796149  1.8574883 2.4322961     P 0.0
## Year:Block:NWholeplot:Subplot!Year_83   2.11597580  0.9675121 2.1870277     P 0.1
## Year:Block:NWholeplot:Subplot!Year_84   4.13956058  1.7259741 2.3983910     P 0.0
## Year:Block:NWholeplot:Subplot!Year_85   5.48604076  2.2146976 2.4771060     P 0.0
## Year:Block:NWholeplot:Subplot!Year_86 14.41228320 5.5903544 2.5780625  P 0.0
## Year:Block:NWholeplot:Subplot!Year_87  58.23033685  22.3474570   2.6056807   P 0.0
## Year:Block:NWholeplot:Subplot!Year_88  13.12521093  5.3138979 2.4699780     P 0.1
info.accumulate <- rbind(info.accumulate,
                    data.frame(Model = "Split-plot meta analysis and EXP",
                               infoCriteria(model4.asr),
                               stringsAsFactors = FALSE))
## Warning in infoCriteria.asreml(model4.asr): The following bound terms
were discounted:
##   Year:Block:NWholeplot:Subplot(R)
dimnames(summary(model4.asr)$varcomp)
## [[1]]
##  [1] "Year:Block"                                  "Year:Block:Wholeplot"
##  [3] "Year:Block:NWholeplot:Subplot(R)"
"Year:Block:NWholeplot:Subplot!Year!pow"
##  [5] "Year:Block:NWholeplot:Subplot!Year_77"
"Year:Block:NWholeplot:Subplot!Year_79"
##  [7] "Year:Block:NWholeplot:Subplot!Year_80"
"Year:Block:NWholeplot:Subplot!Year_81"
##  [9] "Year:Block:NWholeplot:Subplot!Year_82"
"Year:Block:NWholeplot:Subplot!Year_83"
## [11] "Year:Block:NWholeplot:Subplot!Year_84"
"Year:Block:NWholeplot:Subplot!Year_85"
## [13] "Year:Block:NWholeplot:Subplot!Year_86"
"Year:Block:NWholeplot:Subplot!Year_87"
## [15] "Year:Block:NWholeplot:Subplot!Year_88"
##
## [[2]]
## [1] "component" "std.error" "z.ratio"   "bound"     "%ch"
Split plot with different residual variances and variance components
model5.asr <- asreml(Yield ~ Year*Rotation*Nitrogen,
                     random = ~ idh(Year):Block/Wholeplot,
                    residual =   ~ idh(Year):Block:NWholeplot:Subplot,
                     data = wmpotato)
## Model fitted using the sigma parameterization.
## ASReml4 Beta-release 4.0.0.9005 Mon Jan 23 15:55:33 2017
##          LogLik          Sigma2      DF      wall      cpu
##  1      -325.726           1.0     165 15:55:33     0.0 (13 restrained)
##  2      -298.707           1.0     165 15:55:33     0.0 (6 restrained)
##  3      -285.216           1.0     165 15:55:33     0.0 (6 restrained)
##  4      -278.707           1.0     165 15:55:33     0.0 (7 restrained)
##  5      -276.570           1.0     165 15:55:33     0.0 (8 restrained)
##  6      -275.950           1.0     165 15:55:33     0.0 (5 restrained)
##  7      -275.800           1.0     165 15:55:33     0.0 (1 restrained)
##  8      -275.795           1.0     165 15:55:33     0.0 (1 restrained)
##  9       -275.795            1.0     165 15:55:33     0.0
## 10       -275.795            1.0     165 15:55:33     0.0
summary(model5.asr)$varcomp
##                          component   std.error    z.ratio bound %ch
## Year:Block!Year_7     2.321580e+00   3.3536595 0.69225281     P   0
## Year:Block!Year_79    2.601849e+00   4.1605472 0.62536224     P   0
## Year:Block!Year_80    8.148340e-01   1.5000504 0.54320444     P   0
## Year:Block!Year_81    1.028062e+01  16.8785574 0.60909369     P   0
## Year:Block!Year_82    3.782552e-07          NA         NA     B   0
## Year:Block!Year_83    1.311193e-01   0.8770122 0.14950686     P   0
## Year:Block!Year_84    3.149464e+00   4.8633243 0.64759494     P   0
## Year:Block!Year_85    3.782552e-07          NA         NA     B   0
## Year:Block!Year_86    5.980740e-06          NA         NA     B   0
## Year:Block!Year_87    4.368026e-05          NA         NA     B   0
## Year:Block!Year_88    2.563949e-01   3.5621790 0.07197699     P   0
## Year:Block:Wholeplot!Year_77      3.782552e-07 NA          NA     B   0
## Year:Block:Wholeplot!Year_79  7.705295e-01  1.2528742   0.61500946   P   0
## Year:Block:Wholeplot!Year_80    7.394376e-01  0.8696140    0.85030549   P   0
## Year:Block:Wholeplot!Year_81    7.413468e+00  5.7598117    1.28710245   P   0
## Year:Block:Wholeplot!Year_82    2.701891e-01  1.3808788    0.19566460   P   0
## Year:Block:Wholeplot!Year_83  1.617967e+00  1.6054174    1.00781705    P   0
## Year:Block:Wholeplot!Year_84    5.980740e-06       NA        NA       B   0
## Year:Block:Wholeplot!Year_85 5.980740e-06   NA       NA       B   0
```

```
## Year:Block:Wholeplot!Year_86    1.481352e-05        NA        NA           B   0
## Year:Block:Wholeplot!Year_87   5.832377e+01   38.6155750   1.51036903   P   0
## Year:Block:Wholeplot!Year_88    7.262868e+00  7.3523383 0.98783110  P   0
## Year:Block:NWholeplot:Subplot(R)   1.000000e+00      NA      NA      F   0
## Year:Block:Nwholeplot:Subplot!Year_77 7.466666e-01  0.2822136 2.64575043 P  0
## Year:Block:Nwholeplot:Subplot!Year_79 2.718222e+00  1.2156259 2.23606798   P  0
## Year:Block:Nwholeplot:Subplot!Year_80 1.368230e+00  0.6118913 2.23606798    P  0
## Year:Block:Nwholeplot:Subplot!Year_81 2.158255e+00  0.9652010 2.23606798    P  0
## Year:Block:Nwholeplot:Subplot!Year_82 4.794074e+00  2.1439751 2.23606798    P  0
## Year:Block:Nwholeplot:Subplot!Year_83 1.855457e+00  0.8297855 2.23606798    P  0
## Year:Block:Nwholeplot:Subplot!Year_84 4.328390e+00  1.6359796 2.64574820    P  0
## Year:Block:Nwholeplot:Subplot!Year_85 5.397298e+00  1.9708165 2.73861005    P  0
## Year:Block:Nwholeplot:Subplot!Year_86 1.417377e+01 5.1755335 2.73860958   P  0
## Year:Block:Nwholeplot:Subplot!Year_87 8.108572e+00  3.6262636 2.23606798   P  0
## Year:Block:Nwholeplot:Subplot!Year_88 8.893420e+00  3.9772582 2.23606798   P  0
vcnames <- rownames(summary(model5.asr)$varcomp)[1:22]
model5.asr <- setvarianceterms(model5.asr$call, terms = vcnames, bounds = "U",
                               initial=0.01, ignore.suffices = FALSE)

## Model fitted using the sigma parameterization.
## ASReml4 Beta-release 4.0.0.9005 Mon Jan 23 15:55:33 2017
##          LogLik     Sigma2    DF     wall      cpu
##   1    -393.211       1.0    165 15:55:33    0.0 (18 restrained)
##   2    -367.175       1.0    165 15:55:33    0.0 (9 restrained)
##   3    -338.725       1.0    165 15:55:33    0.0 (7 restrained)
##   4    -303.755       1.0    165 15:55:33    0.0 (8 restrained)
##   5    -285.353       1.0    165 15:55:33    0.0 (7 restrained)

## Warning in asreml(fixed = Yield ~ Year * Rotation * Nitrogen, random =
~idh(Year):Block/Wholeplot, :
## Singularity in average information matrix

## 2 singularities in Average Information matrix
##   6    -278.943      1.0    165 15:55:33    0.0 (7 restrained)
##   7    -276.627      1.0    165 15:55:33    0.0 (6 restrained)
##   8    -275.985      1.0    165 15:55:33    0.0 (3 restrained)
##   9    -276.843      1.0    165 15:55:33    0.0 (2 restrained)
##  10    -277.981      1.0    165 15:55:33    0.0 (2 restrained)
##  11    -279.125      1.0    165 15:55:33    0.0 (2 restrained)
##  12    -280.273      1.0    165 15:55:33    0.0 (2 restrained)
##  13    -281.422      1.0    165 15:55:33    0.0 (2 restrained)

## Warning in asreml(fixed = Yield ~ Year * Rotation * Nitrogen, random =
~idh(Year):Block/Wholeplot, : Log-
## likelihood not converged

## Warning in asreml(fixed = Yield ~ Year * Rotation * Nitrogen, random =
~idh(Year):Block/Wholeplot, : Some
## components changed by more than 1% on the last iteration.

model5.asr <- update(model5.asr)

## Model fitted using the sigma parameterization.
## ASReml4 Beta-release 4.0.0.9005 Mon Jan 23 15:55:33 2017
##        LogLik    Sigma2  DF  wall     cpu
## 1     -282.573     1 .0   165 15:55:33   0.0 (2 restrained)

## Warning in asreml(fixed = Yield ~ Year * Rotation * Nitrogen, random =
~idh(Year):Block/Wholeplot, :
## Singularity in average information matrix

## 1 singularities in Average Information matrix
##  2   -282.543                1.0    165 15:55:33    0.0 (2 restrained)
##  3  -282.518                 1.0    165 15:55:33    0.0 (2 restrained)
##  4       -282.499            1.0    165 15:55:33    0.0 (2 restrained)
##  5       -282.487            1.0    165 15:55:33    0.0 (1 restrained)
##  6       -282.483            1.0    165 15:55:33    0.0 (1 restrained)
##  7       -282.483            1.0    165 15:55:33    0.0 (1 restrained)
summary(model5.asr)$varcomp
```

```
##                                           component      std.error   z.ratio     bound %c
## Year:Block!Year_77                        2.344716e+00   3.3536596  0.69915147  U   0.
## Year:Block!Year_79                        2.601849e+00   4.1605472  0.62536224  U   0.
## Year:Block!Year_80                        8.148340e-01   1.5000504  0.54320444  U   0.
## Year:Block!Year_81                        1.028062e+01  16.8785574  0.60909369  U   0.
## Year:Block!Year_82                        1.074595e-01   0.4935380 -0.21773306  U   0.
## Year:Block!Year_83                        1.311193e-01   0.8770122  0.14950686  U   0.
## Year:Block!Year_84                        3.152837e+00   4.8662813  0.64789447  U   0.
## Year:Block!Year_85                       -1.372373e-01   0.2891636 -0.47460095  U   0.
## Year:Block!Year_86                       -3.118268e+06   NA         NA          S   0.
## Year:Block!Year_87                       -3.176159e+00  16.4108486 -0.19354020  B   0.
## Year:Block!Year_88                        2.563949e-01   3.5621790  0.07197699  U   0.
## Year:Block:Wholeplot!Year_77             -1.619506e-01   0.1621616 -0.99869904  U   0.
## Year:Block:Wholeplot!Year_79              7.705295e-01   1.2528742  0.61500946  U   0.
## Year:Block:Wholeplot!Year_80              7.394376e-01   0.8696140  0.85030549  U   0.
## Year:Block:Wholeplot!Year_81              7.413468e+00   5.7598117  1.28710245  U   0.
## Year:Block:Wholeplot!Year_82              3.776488e-01   1.5691971  0.24066374  U   0.
## Year:Block:Wholeplot!Year_83              1.617967e+00   1.6054174  1.00781705  U   0.
## Year:Block:Wholeplot!Year_84             -2.360219e-02   1.1987019 -0.01968980  U   0.
## Year:Block:Wholeplot!Year_85             -6.354705e-01   1.3623072 -0.46646637  U   0.
## Year:Block:Wholeplot!Year_86             -9.913303e-01   3.8453095 -0.25780247  U   0.
## Year:Block:Wholeplot!Year_87              6.149949e+01  45.4102563  1.35430844  U   0.
## Year:Block:Wholeplot!Year_88              7.262868e+00   7.3523383  0.98783110  U   0.
## Year:Block:NWholeplot:Subplot(R)          1.000000e+00   NA         NA          F   0.
## Year:Block:NWholeplot:Subplot!Year_77     8.854815e-01   0.3959994  2.23606798  P   0.
## Year:Block:NWholeplot:Subplot!Year_79     2.718222e+00   1.2156259  2.23606798  P   0.
## Year:Block:NWholeplot:Subplot!Year_80     1.368230e+00   0.6118913  2.23606798  P   0.
## Year:Block:NWholeplot:Subplot!Year_81     2.158255e+00   0.9652010  2.23606798  P   0.
## Year:Block:NWholeplot:Subplot!Year_82     4.794074e+00   2.1439751  2.23606798  P   0.
## Year:Block:NWholeplot:Subplot!Year_83     1.855457e+00   0.8297855  2.23606798  P   0.
## Year:Block:NWholeplot:Subplot!Year_84     4.348626e+00   1.9447645  2.23606798  P   0.
## Year:Block:NWholeplot:Subplot!Year_85     6.170008e+00   2.7593116  2.23606798  P   0.
## Year:Block:NWholeplot:Subplot!Year_86     1.584683e+01   7.0869141  2.23606875  P   0.
## Year:Block:NWholeplot:Subplot!Year_87     8.108572e+00   3.6262636  2.23606798  P   0.
## Year:Block:NWholeplot:Subplot!Year_88     8.893420e+00   3.9772582  2.23606798  P   0.
```

```
info.accumulate <- rbind(info.accumulate,
                         data.frame(Model = "Meta analysis with
different variance components",
                                    infoCriteria(model5.asr, bound.
exclusions = "F"),
                                    stringsAsFactors = FALSE))
## Warning in infoCriteria.asreml(model5.asr, bound.exclusions =
"F"): The following bound terms were discounted:
##  Year:Block:NWholeplot:Subplot(R)
Accumulated summary of REML information criteria
info.accumulate
##Model                                            DF NBound AIC      BIC      logREML
## 1 Split-plot                                    3  0      653.1797 662.4975 -323.5899
## 2 Split plot meta analysis                      13 1      614.3496 654.7269 -294.1748
## 3 Split-plot and EXP                            4  0      655.1797 667.6035 -323.5899
## 4 Split-plot meta analysis and EXP              14 1      616.0131 659.4963 -294.0065
## 5 Meta analysis with different variance components 33 1  630.9657 733.4619 -282.4829
```

Use split plot with different residual variance in each year

```
wald(model2.asr, denDF = "algebraic")
## Model fitted using the sigma parameterization.
## ASReml4 Beta-release 4.0.0.9005 Mon Jan 23 15:55:33 2017
##          LogLik        Sigma2     DF     wall    cpu
##  1      -294.175         1.0    165 15:55:33    0.0
##  2      -294.175         1.0    165 15:55:33    0.0
##  3      -294.175         1.0    165 15:55:33    0.2
## $Wald
##
## Wald tests for fixed effects.
## Response: Yield
##
##                         Df denDF   F.inc       Pr
## (Intercept)              1  10.5 15300.0 0.000000
## Year                    10  10.4    85.4 0.000000
## Rotation                 4  27.0    26.7 0.000000
## Nitrogen                 2  58.2    55.3 0.000000
## Year:Rotation           40  22.7     3.7 0.000683
## Year:Nitrogen           20  48.7     1.3 0.241290
## Rotation:Nitrogen        8  58.2     1.9 0.072879
## Year:Rotation:Nitrogen  80  40.6     1.2 0.270253
##
## $stratumVariances
## NULL
```

Drop unnecessary fixed terms

```
model2a.asr <- asreml(Yield ~ Year*Rotation*Nitrogen-Year:Rotation:Nitrogen,
                     random = ~ Year:Block/Wholeplot,
                     residual = ~ idh(Year):Block:Nwholeplot:Subplot,
                         data = wmpotato)
## Model fitted using the sigma parameterization.
## ASReml4 Beta-release 4.0.0.9005 Mon Jan 23 15:55:34 2017
##          LogLik        Sigma2     DF     wall    cpu
##  1      -454.257         1.0    245 15:55:34    0.0
##  2      -436.705         1.0    245 15:55:34    0.0
##  3      -432.022         1.0    245 15:55:34    0.0
##  4      -429.833         1.0    245 15:55:34    0.0
##  5      -428.217         1.0    245 15:55:34    0.0
##  6      -426.955         1.0    245 15:55:34    0.0
##  7      -426.438         1.0    245 15:55:34    0.0
##  8      -426.321         1.0    245 15:55:34    0.0
##  9      -426.301         1.0    245 15:55:34    0.0
## 10      -426.297         1.0    245 15:55:34    0.0
## 11      -426.296         1.0    245 15:55:34    0.0
## 12      -426.296         1.0    245 15:55:34    0.0
wald(model2a.asr, denDF = "algebraic")
## Model fitted using the sigma parameterization.
## ASReml4 Beta-release 4.0.0.9005 Mon Jan 23 15:55:34 2017
##          LogLik        Sigma2     DF     wall    cpu
##  1      -426.296         1.0    245 15:55:34    0.0
##  2      -426.296         1.0    245 15:55:34    0.0
##  3      -426.296         1.0    245 15:55:34    0.0
## $Wald
##
## Wald tests for fixed effects.
## Response: Yield
##
##                    Df denDF   F.inc       Pr
## (Intercept)         1  10.7 14670.0 0.000000
## Year               10  10.7    79.8 0.000000
## Rotation            4  32.8    23.0 0.000000
## Nitrogen            2 128.4    41.9 0.000000
## Year:Rotation      40  25.3     2.9 0.002830
## Year:Nitrogen      20  88.3     1.4 0.166185
## Rotation:Nitrogen   8 128.4     1.3 0.252994
##
## $stratumVariances
## NULL
model2b.asr <- asreml(Yield ~ Year*Rotation*Nitrogen-Year:Rotation:Nitrogen
-
                                                  Rotation:Nitrogen,
                random = ~ Year:Block/Wholeplot,
```

```
                 residual =   ~ idh(Year):Block:NWholeplot:Subplot,
                     data = wmpotato)
## Model fitted using the sigma parameterization.
## ASReml4 Beta-release 4.0.0.9005 Mon Jan 23 15:55:34 2017
##          LogLik        Sigma2     DF     wall      cpu
##  1     -458.206          1.0     253 15:55:34     0.0 (1 restrained)
##  2     -452.705          1.0     253 15:55:34     0.0
##  3     -440.278          1.0     253 15:55:34     0.0
##  4     -433.140          1.0     253 15:55:34     0.0
##  5     -429.272          1.0     253 15:55:34     0.0
##  6     -427.727          1.0     253 15:55:34     0.0
##  7     -427.310          1.0     253 15:55:34     0.0
##  8     -427.236          1.0     253 15:55:34     0.0
##  9     -427.226          1.0     253 15:55:34     0.0
## 10     -427.225          1.0     253 15:55:34     0.0
## 11      -427.224          1.0    253 15:55:34     0.0
wald(model2b.asr, denDF = "algebraic")
## Model fitted using the sigma parameterization.
## ASReml4 Beta-release 4.0.0.9005 Mon Jan 23 15:55:34 2017
##          LogLik        Sigma2     DF     wall      cpu
##  1     -427.224          1.0     253 15:55:34     0.0
##  2     -427.224          1.0     253 15:55:34     0.0
##  3     -427.224          1.0     253 15:55:34     0.0
## $Wald
##
## Wald tests for fixed effects.
## Response: Yield
##
##                 Df denDF   F.inc        Pr
## (Intercept)      1  10.8 14460.0 0.000000
## Year            10  10.8    78.5 0.000000
## Rotation         4  34.0    22.4 0.000000
## Nitrogen         2 136.7    41.9 0.000000
## Year:Rotation   40  26.9     2.8 0.003045
## Year:Nitrogen   20  95.6     1.4 0.154287
##
## $stratumVariances
## NULL
model2c.asr <- asreml(Yield ~ Year*Rotation*Nitrogen-Year:Rotation:Nitrogen
-
                       Rotation:Nitrogen - Year:Nitrogen,
                     random = ~ Year:Block/Wholeplot,
                   residual =   ~ idh(Year):Block:NWholeplot:Subplot,
                     data = wmpotato)
## Model fitted using the sigma parameterization.
## ASReml4 Beta-release 4.0.0.9005 Mon Jan 23 15:55:34 2017
##          LogLik        Sigma2     DF     wall      cpu
##  1    -482.365         1.0      273 15:55:34     0.0 (1 restrained)
##  2     -473.437          1.0     273 15:55:34     0.0
##  3     -457.815          1.0     273 15:55:34     0.0
##  4     -449.143          1.0     273 15:55:34     0.0
##  5     -444.554          1.0     273 15:55:34     0.0
##  6     -443.052          1.0     273 15:55:34     0.0
##  7     -442.746          1.0     273 15:55:34     0.0
##  8     -442.702          1.0     273 15:55:34     0.0
##  9     -442.695          1.0     273 15:55:34     0.0
## 10     -442.694          1.0     273 15:55:34     0.0
## 11     -442.694          1.0     273 15:55:34     0.0
wald(model2c.asr, denDF = "algebraic")
## Model fitted using the sigma parameterization.
## ASReml4 Beta-release 4.0.0.9005 Mon Jan 23 15:55:34 2017
##          LogLik        Sigma2     DF     wall      cpu
##  1     -442.694          1.0     273 15:55:34     0.0
##  2     -442.694          1.0     273 15:55:34     0.0
##  3     -442.694          1.0     273 15:55:34     0.0
## $Wald
##
## Wald tests for fixed effects.
## Response: Yield
##
##                 Df denDF   F.inc          Pr
## (Intercept)      1  10.8 14360.0 0.00000000
## Year            10  10.8    78.2 0.00000001
## Rotation         4  32.6    23.2 0.00000000
## Nitrogen         2 142.0    41.6 0.00000000
```

```
## Year:Rotation 40  27.0     2.9 0.00223137
##
## $stratumVariances
## NULL
Get predictions and plot
predict(model2c.asr, classify = "Nitrogen")$pvals
## Model fitted using the sigma parameterization.
## ASReml4 Beta-release 4.0.0.9005 Mon Jan 23 15:55:34 2017
##            LogLik        Sigma2     DF     wall    cpu
##  1      -442.694           1.0    273 15:55:34    0.0
##  2      -442.694           1.0    273 15:55:34    0.0
##  3      -442.694           1.0    273 15:55:34    0.0
##
## Notes:
## - The predictions are obtained by averaging across the hypertable
##   calculated from model terms constructed solely from factors in
##   the averaging and classify sets.
## - Use 'average' to move ignored factors into the averaging set.
## - The simple averaging set: Year,Rotation
## - The ignored set: Block,Wholeplot
## ##
##   Nitrogen predicted.value std.error    status
## 1       N1          43.36754  0.402073 Estimable
## 2       N2          45.02538  0.402073 Estimable
## 3       N3          45.64905  0.402073 Estimable
predYR <- predict(model2c.asr, classify = "Year:Rotation")$pvals
## Model fitted using the sigma parameterization.
## ASReml4 Beta-release 4.0.0.9005 Mon Jan 23 15:55:34 2017
##            LogLik        Sigma2     DF     wall    cpu
##  1      -442.694           1.0    273 15:55:34    0.0
##  2      -442.694           1.0    273 15:55:34    0.0
##  3      -442.694           1.0    273 15:55:34    0.0
predYR
##
## Notes:
## - The predictions are obtained by averaging across the hypertable
##   calculated from model terms constructed solely from factors in
##   the averaging and classify sets.
## - Use 'average' to move ignored factors into the averaging set.
## - The simple averaging set: Nitrogen
## - The ignored set: Block,Wholeplot
##
## ##    Year Rotation predicted.value std.error    status
## 1    77      IIf         23.43704  1.485051 Estimable
## 2    77      III         21.91852  1.485051 Estimable
## 3    77     IIIf         21.87037  1.485051 Estimable
## 4    77       IV         25.01481  1.485051 Estimable
## 5    77      IVf         21.19630  1.485051 Estimable
## 6    79      IIf         39.44074  1.533421 Estimable
## 7    79      III         46.61852  1.533421 Estimable
## 8    79     IIIf         45.73704  1.533421 Estimable
## 9    79       IV         49.50370  1.533421 Estimable
## 10   79      IVf         44.35185  1.533421 Estimable
## 11   80      IIf         40.52593  1.448644 Estimable
## 12   80      III         41.13704  1.448644 Estimable
## 13   80     IIIf         40.89630  1.448644 Estimable
## 14   80       IV         41.54444  1.448644 Estimable
## 15   80      IVf         42.48148  1.448644 Estimable
## 16   81      IIf         30.37407  1.546351 Estimable
## 17   81      III         38.04444  1.546351 Estimable
## 18   81     IIIf         39.45185  1.546351 Estimable
## 19   81       IV         39.42963  1.546351 Estimable
## 20   81      IVf         38.13704  1.546351 Estimable
## 21   82      IIf         39.75185  1.594618 Estimable
## 22   82      III         40.48519  1.594618 Estimable
## 23   82     IIIf         42.94815  1.594618 Estimable
## 24   82       IV         45.53333  1.594618 Estimable
## 25   82      IVf         45.62963  1.594618 Estimable
## 26   83      IIf         31.14444  1.535587 Estimable
## 27   83      III         36.54815  1.535587 Estimable
## 28   83     IIIf         39.52963  1.535587 Estimable
## 29   83       IV         37.95556  1.535587 Estimable
## 30   83      IVf         37.61481  1.535587 Estimable
## 31   84      IIf         54.58889  1.634753 Estimable
## 32   84      III         57.67407  1.634753 Estimable
```

```
## 33   84      IIIf         56.63333  1.634753 Estimable
## 34   84        IV         58.40000  1.634753 Estimable
## 35   84       IVf         58.99259  1.634753 Estimable
## 36   85       IIf         51.48148  1.781401 Estimable
## 37   85       III         47.80000  1.781401 Estimable
## 38   85      IIIf         49.78519  1.781401 Estimable
## 39   85        IV         52.53704  1.781401 Estimable
## 40   85       IVf         49.75185  1.781401 Estimable
## 41   86       IIf         48.20926  2.132871 Estimable
## 42   86       III         51.02778  2.132871 Estimable
## 43   86      IIIf         54.57593  2.132871 Estimable
## 44   86        IV         51.48148  2.132871 Estimable
## 45   86       IVf         52.98704  2.132871 Estimable
## 46   87       IIf         42.98519  2.809408 Estimable
## 47   87       III         48.75556  2.809408 Estimable
## 48   87      IIIf         41.55556  2.809408 Estimable
## 49   87        IV         45.67407  2.809408 Estimable
## 50   87       IVf         51.85000  2.809408 Estimable
## 51   88       IIf         54.28333  1.917713 Estimable
## 52   88       III         62.76111  1.917713 Estimable
## 53   88      IIIf         64.41407  1.917713 Estimable
## 54   88        IV         58.52407  1.917713 Estimable
## 55   88       IVf         62.45556  1.917713 Estimable
cols <- c('red','darkviolet','violet','darkblue','lightskyblue')
ggplot(data = predYR,
       aes(x =Year, y=predicted.value, colour=Rotation, linetype=Rotation,
shape = Rotation)) +
  geom_point() + geom_line() + labs(y = "Yield") +
  scale_color_manual(values = cols) + scale_shape_manual(values = c(16,4,18,3,15)
```

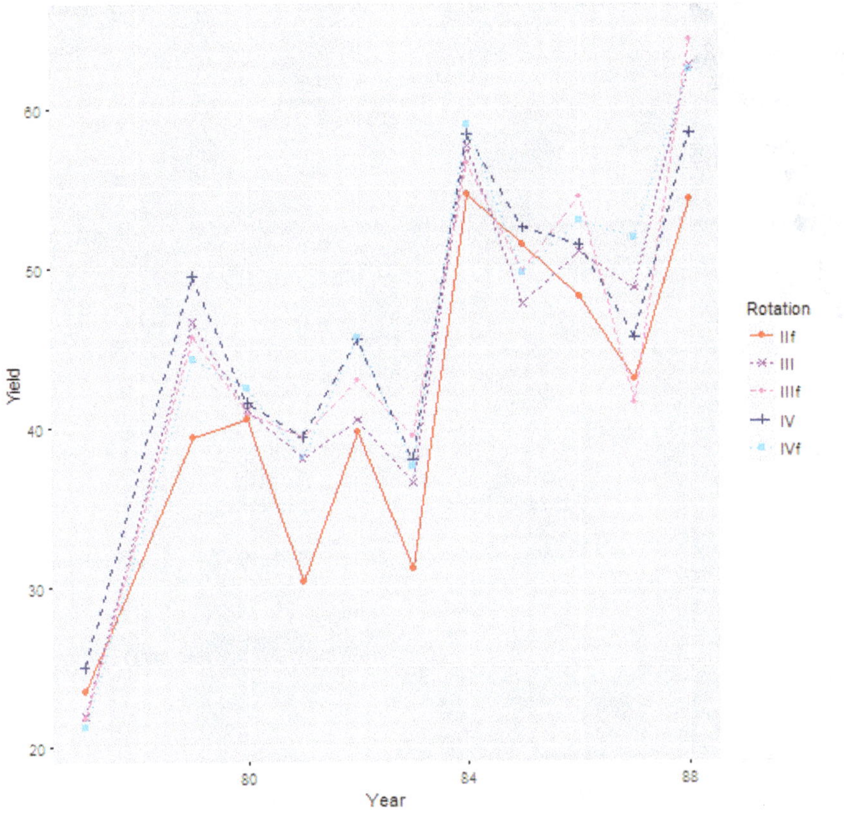

Appendix Fig. 1. plot of chunk unnamed-chunk-11

Appendix 5. SAS code and output from the analysis of potato yields from the Westmaas experiment (supplemental provided by Kathleen Yeater)

```
/*Try Various Random Models*/
/*Split plot within years*/
title1 'Split-plot nested within years';
Proc Mixed data=wmpotato method=REML covtest;*method=REML is default;
class Year_ Block_ Wholeplot_ Rotation_ Nitrogen_;
model Yield = Year_|Rotation_|Nitrogen_ / ddfm=kr;
random Block_ Block_*Wholeplot_ / subject=Year_;
ods select covparms fitstatistics;
run;
```

Covariance Parameter Estimates					
Cov Parm	Subject	Estimate	Standard Error	*Z* Value	*P* > *Z*
Block_	Year_	1.4072	1.4027	1.00	0.1579
Block_*Wholeplot_	Year_	7.0791	1.8929	3.74	< 0.0001
Residual		5.1952	0.7005	7.42	< 0.0001

Fit Statistics	
-2 Res Log Likelihood	950.4
AIC (Smaller is Better)	956.4
AICC (Smaller is Better)	956.6
BIC (Smaller is Better)	957.6

```
/*Nested split-plot, different residual variance each year*/
```

```
title1 'Nested split plot with different residual variance in each
year';
Proc Mixed data=wmpotato method=REML covtest;
class Year_ Block_ Wholeplot_ Rotation_ Nitrogen_;
model Yield = Year_|Rotation_|Nitrogen_ / ddfm=kr;
random Block_ Block_*Wholeplot_ / subject=Year_;
repeated / group=Year_;
ods select covparms fitstatistics;
run;
```

Covariance Parameter Estimates						
Cov Parm	Subject	Group	Estimate	Standard Error	Z Value	P > Z
Block_	Year_		2.1173	1.1783	1.80	0.0362
Block_*Wholeplot_	Year_		0.9103	0.8170	1.11	0.1326
Residual		Year_ 77	0.8185	0.3435	2.38	0.0086
Residual		Year_ 79	2.6840	1.1544	2.32	0.0100
Residual		Year_ 80	1.3442	0.5864	2.29	0.0109
Residual		Year_ 81	5.2395	4.1243	1.27	0.1020
Residual		Year_ 82	4.5379	1.8554	2.45	0.0072
Residual		Year_ 83	1.9885	0.9469	2.10	0.0179
Residual		Year_ 84	4.0134	1.6360	2.45	0.0071
Residual		Year_ 85	5.4145	2.1357	2.54	0.0056
Residual		Year_ 86	14.2836	5.4923	2.60	0.0047
Residual		Year_ 87	58.2196	23.1443	2.52	0.0059
Residual		Year_ 88	12.8988	5.5662	2.32	0.0102

Fit Statistics	
-2 Res Log Likelihood	891.6
AIC (Smaller is Better)	917.6
AICC (Smaller is Better)	920.0
BIC (Smaller is Better)	922.8

```
/*Nested split plot with AR1 structure over years*/
title1 'Nested split-plot and AR1';
title2 ' ';
Proc Mixed data=wmpotato method=REML covtest;
class Year_ Block_ Wholeplot_ Rotation_ Nitrogen_;
model Yield = Year_|Rotation_|Nitrogen_ / ddfm=kr;
random Block_ Block_*Wholeplot_ / subject=Year_ ;
repeated / subject=Year_ type=AR(1);
ods select covparms fitstatistics;
run;
```

Covariance Parameter Estimates					
Cov Parm	Subject	Estimate	Standard Error	Z Value	P Z
Block_	Year_	1.3777	1.3946	0.99	0.1616
Block_*Wholeplot_	Year_	6.7744	2.0190	3.36	0.0004
AR(1)	Year_	0.06903	0.1568	0.44	0.6598
Residual		5.4732	1.0339	5.29	< 0.0001

Fit Statistics	
-2 Res Log Likelihood	950.2
AIC (Smaller is Better)	958.2
AICC (Smaller is Better)	958.5
BIC (Smaller is Better)	959.8

```
/*Nested split plot with different residual variance in each year AND
AR1 structure*/
title1 'Nested split-plot with heterogeneous AR1 structure in each
year';
Proc Mixed data=wmpotato method=REML covtest;
class Year_ Block_ Wholeplot_ Rotation_ Nitrogen_;
model Yield = Year_|Rotation_|Nitrogen_ / ddfm=kr;
random Block_ Block_*Wholeplot_ / subject=Year_;
repeated / group=Year_ type=ARH(1);
```

```
ods select covparms fitstatistics;
run;
```

Covariance Parameter Estimates

Cov Parm	Subject	Group	Estimate	Standard Error	Z Value	P Z
Block_	Year_		2.1173	1.1783	1.80	0.0362
Block_*Wholeplot_	Year_		0.9101	0.8167	1.11	0.1326
Var(1)		Year_ 77	0.8185	0.3435	2.38	0.0086
ARH(1)		Year_ 77	0	.	.	.
Var(1)		Year_ 79	2.6840	1.1544	2.32	0.0100
ARH(1)		Year_ 79	0	.	.	.
Var(1)		Year_ 80	1.3442	0.5863	2.29	0.0109
ARH(1)		Year_ 80	0	.	.	.
Var(1)		Year_ 81	5.2402	4.1239	1.27	0.1019
ARH(1)		Year_ 81	0	.	.	.
Var(1)		Year_ 82	4.5380	1.8553	2.45	0.0072
ARH(1)		Year_ 82	0	.	.	.
Var(1)		Year_ 83	1.9885	0.9469	2.10	0.0179
ARH(1)		Year_ 83	0	.	.	.
Var(1)		Year_ 84	4.0134	1.6360	2.45	0.0071
ARH(1)		Year_ 84	0	.	.	.
Var(1)		Year_ 85	5.4144	2.1356	2.54	0.0056
ARH(1)		Year_ 85	0	.	.	.
Var(1)		Year_ 86	14.2837	5.4923	2.60	0.0047
ARH(1)		Year_ 86	0	.	.	.
Var(1)		Year_ 87	58.2206	23.1442	2.52	0.0059
ARH(1)		Year_ 87	0	.	.	.
Var(1)		Year_ 88	12.8992	5.5662	2.32	0.0102
ARH(1)		Year_ 88	0	.	.	.

Fit Statistics

-2 Res Log Likelihood	891.6
AIC (Smaller is Better)	939.6
AICC (Smaller is Better)	948.2
BIC (Smaller is Better)	949.1

```
/*Nested split-plot with different residual variances and variance
components*/
/*Note: This model is CPU-intensive, prepare for a longer than usual
run time*/
title1 'Nested split-plot with variance components in each year';
Proc Mixed data=wmpotato method=REML covtest;
class Year_ Block_ Wholeplot_ Rotation_ Nitrogen_;
model Yield = Year_|Rotation_|Nitrogen_ / ddfm=kr;
random Block_ Block_*Wholeplot_ / group=Year_ type=VC;
repeated / group=Year_ type=VC;
ods select covparms fitstatistics;
run;
```

Fit Statistics

-2 Res Log Likelihood	854.8
AIC (Smaller is Better)	904.8
AICC (Smaller is Better)	914.2
BIC (Smaller is Better)	872.2

Covariance Parameter Estimates

Cov Parm	Group	Estimate	Standard Error	Z Value	P > Z
Block_	Year_ 77	2.3216	3.3537	0.69	0.2444
Block_	Year_ 79	2.6018	4.1605	0.63	0.2659
Block_	Year_ 80	0.8148	1.5001	0.54	0.2935
Block_	Year_ 81	10.2806	16.8786	0.61	0.2712
Block_	Year_ 82	0	.	.	.
Block_	Year_ 83	0.1311	0.8770	0.15	0.4406

Covariance Parameter Estimates					
Cov Parm	Group	Estimate	Standard Error	*Z* Value	*P* > *Z*
Block_	Year_ 84	3.1495	4.8633	0.65	0.2586
Block_	Year_ 85	0	.	.	.
Block_	Year_ 86	0	.	.	.
Block_	Year_ 87	0	.	.	.
Block_	Year_ 88	0.2564	3.5622	0.07	0.4713
Block_*Wholeplot_	Year_ 77	0	.	.	.
Block_*Wholeplot_	Year_ 79	0.7705	1.2529	0.62	0.2693
Block_*Wholeplot_	Year_ 80	0.7394	0.8696	0.85	0.1976
Block_*Wholeplot_	Year_ 81	7.4135	5.7598	1.29	0.0990
Block_*Wholeplot_	Year_ 82	0.2702	1.3809	0.20	0.4224
Block_*Wholeplot_	Year_ 83	1.6180	1.6054	1.01	0.1568
Block_*Wholeplot_	Year_ 84	0	.	.	.
Block_*Wholeplot_	Year_ 85	0	.	.	.
Block_*Wholeplot_	Year_ 86	0	.	.	.
Block_*Wholeplot_	Year_ 87	58.3238	38.6156	1.51	0.0655
Block_*Wholeplot_	Year_ 88	7.2629	7.3523	0.99	0.1616
Residual	Year_ 77	0.7464	0.2820	2.65	0.0041
Residual	Year_ 79	2.7182	1.2156	2.24	0.0127
Residual	Year_ 80	1.3682	0.6119	2.24	0.0127
Residual	Year_ 81	2.1583	0.9652	2.24	0.0127
Residual	Year_ 82	4.7941	2.1440	2.24	0.0127
Residual	Year_ 83	1.8555	0.8298	2.24	0.0127
Residual	Year_ 84	4.3284	1.6360	2.65	0.0041
Residual	Year_ 85	5.3971	1.9707	2.74	0.0031
Residual	Year_ 86	14.1737	5.1755	2.74	0.0031
Residual	Year_ 87	8.1086	3.6263	2.24	0.0127
Residual	Year_ 88	8.8934	3.9773	2.24	0.0127

```
/*Nested split-plot, different residual variance each year*/
/*Tests for Fixed Effects*/
title1 'Nested split plot with different residual variance in each
year';
title2 'Tests for Fixed Effects';
Proc Mixed data=wmpotato method=REML covtest;
class Year_ Block_ Wholeplot_ Rotation_ Nitrogen_;
model Yield = Year_|Rotation_|Nitrogen_ / ddfm=kr chisq;
random intercept Block_ Block_*Wholeplot_ / subject=Year_;
repeated / group=Year_;
ods select tests3;
run;
```

Type 3 Tests of Fixed Effects						
Effect	Num DF	Den DF	Chi-Square	*F* Value	Pr > ChiSq	Pr > *F*
Year_	10	9.98	851.71	85.52	< 0.0001	< 0.0001
Rotation_	4	44.4	71.00	17.75	< 0.0001	< 0.0001
Year_*Rotation_	40	6.51	154.96	4.49	< 0.0001	0.0269
Nitrogen_	2	35.8	21.19	10.59	< 0.0001	0.0002
Year_*Nitrogen_	20	32.1	29.16	1.24	0.0847	0.2841
Rotation_*Nitrogen_	8	35.8	6.30	0.79	0.6137	0.6167
Year_*Rotati*Nitroge	80	17.5	108.44	1.22	0.0189	0.3278

```
title1 'Nested split plot with different residual variance in each
year';
title2 'Drop 3-way Fixed term';
Proc Mixed data=wmpotato method=REML covtest;
class Year_ Block_ Wholeplot_ Rotation_ Nitrogen_;
model Yield = Year_|Rotation_|Nitrogen_ @2 / ddfm=kr chisq;
random intercept Block_ Block_*Wholeplot_ / subject=Year_;
repeated / group=Year_;
```

```
ods select tests3;
run;
```

Type 3 Tests of Fixed Effects						
Effect	Num DF	Den DF	Chi-Square	F Value	Pr > ChiSq	Pr > F
Year_	10	10.5	796.19	79.81	< 0.0001	< 0.0001
Rotation_	4	52.9	71.73	17.93	< 0.0001	< 0.0001
Year_*Rotation_	40	14.9	117.03	3.02	< 0.0001	0.0122
Nitrogen_	2	68.4	27.46	13.73	< 0.0001	< 0.0001
Year_*Nitrogen_	20	79.5	29.67	1.35	0.0754	0.1747
Rotation_*Nitrogen_	8	126	8.40	1.05	0.3954	0.4025

```
title1 'Nested split plot with different residual variance in each year';
title2 '3-way Fixed term and Rotation*Nitrogen Removed';
Proc Mixed data=wmpotato method=REML covtest;
class Year_ Block_ Wholeplot_ Rotation_ Nitrogen_;
model Yield = Year_ Rotation_ Nitrogen_ Year_*Rotation_
Year_*Nitrogen_ / ddfm=kr chisq;
random intercept Block_ Block_*Wholeplot_ / subject=Year_;
repeated / group=Year_;
ods select tests3;
run;
```

Type 3 Tests of Fixed Effects						
Effect	Num DF	Den DF	Chi-Square	F Value	Pr > ChiSq	Pr > F
Year_	10	10.8	783.24	78.50	< 0.0001	< 0.0001
Rotation_	4	57.2	69.67	17.42	< 0.0001	< 0.0001
Nitrogen_	2	70.7	27.45	13.72	< 0.0001	< 0.0001
Year_*Rotation_	40	18.4	112.77	2.87	< 0.0001	0.0089
Year_*Nitrogen_	20	89.4	29.82	1.37	0.0729	0.1588

```
title1 'Nested split plot with different residual variance in each
year';
title2 '3-way Fixed term, Rotation*Nitrogen, and Year*Nitrogen
Removed';
Proc Mixed data=wmpotato method=REML covtest;
class Year_ Block_ Wholeplot_ Rotation_ Nitrogen_;
model Yield = Year_ Rotation_ Nitrogen_ Year_*Rotation_ / ddfm=kr
chisq;
random intercept Block_ Block_*Wholeplot_ / subject=Year_;
repeated / group=Year_;
ods select tests3;
run;
```

Type 3 Tests of Fixed Effects						
Effect	Num DF	Den DF	Chi-Square	F Value	Pr > ChiSq	Pr > F
Year_	10	10.8	780.44	78.22	< 0.0001	< 0.0001
Rotation_	4	59.7	70.33	17.58	< 0.0001	< 0.0001
Nitrogen_	2	140	70.72	35.36	< 0.0001	< 0.0001
Year_*Rotation_	40	17.4	117.47	2.99	< 0.0001	0.0081

```
/*Get predictions and plot*/
/*Predicted means for Nitrogen*/
/*Predicted means for Year*Rotation*/
title1 'Predicted Means for Nitrogen';
title2 'Predicted Means for Year*Rotation';
Proc Mixed data=wmpotato method=REML covtest;
class Year_ Block_ Wholeplot_ Rotation_ Nitrogen_;
model Yield = Year_ Rotation_ Nitrogen_ Year_*Rotation_ / ddfm=kr
chisq;
random intercept Block_ Block_*Wholeplot_ / subject=Year_;
repeated / group=Year_;
lsmeans Nitrogen_ Year_*Rotation_;
ods select lsmeans;
run;
```

Least Squares Means

Effect	Rotation_	Nitrogen_	Year_	Estimate	Standard Error	DF	*t* Value	*P* > |*t*|
Nitrogen_		N1		43.3676	0.4069	16.2	106.57	< 0.0001
Nitrogen_		N2		45.0254	0.4069	16.2	110.65	< 0.0001
Nitrogen_		N3		45.6490	0.4069	16.2	112.18	< 0.0001
Year_*Rotation_	III		77	21.9185	1.4849	18.3	14.76	< 0.0001
Year_*Rotation_	IIIf		77	21.8704	1.4849	18.3	14.73	< 0.0001
Year_*Rotation_	IIf		77	23.4370	1.4849	18.3	15.78	< 0.0001
Year_*Rotation_	IV		77	25.0148	1.4849	18.3	16.85	< 0.0001
Year_*Rotation_	IVf		77	21.1963	1.4849	18.3	14.27	< 0.0001
Year_*Rotation_	III		79	46.6185	1.5333	20.6	30.40	< 0.0001
Year_*Rotation_	IIIf		79	45.7370	1.5333	20.6	29.83	< 0.0001
Year_*Rotation_	IIf		79	39.4407	1.5333	20.6	25.72	< 0.0001
Year_*Rotation_	IV		79	49.5037	1.5333	20.6	32.29	< 0.0001
Year_*Rotation_	IVf		79	44.3519	1.5333	20.6	28.93	< 0.0001
Year_*Rotation_	III		80	41.1370	1.4485	16.9	28.40	< 0.0001
Year_*Rotation_	IIIf		80	40.8963	1.4485	16.9	28.23	< 0.0001
Year_*Rotation_	IIf		80	40.5259	1.4485	16.9	27.98	< 0.0001
Year_*Rotation_	IV		80	41.5444	1.4485	16.9	28.68	< 0.0001
Year_*Rotation_	IVf		80	42.4815	1.4485	16.9	29.33	< 0.0001
Year_*Rotation_	III		81	38.0444	1.5463	23.7	24.60	< 0.0001
Year_*Rotation_	IIIf		81	39.4519	1.5463	23.7	25.51	< 0.0001
Year_*Rotation_	IIf		81	30.3741	1.5463	23.7	19.64	< 0.0001
Year_*Rotation_	IV		81	39.4296	1.5463	23.7	25.50	< 0.0001
Year_*Rotation_	IVf		81	38.1370	1.5463	23.7	24.66	< 0.0001
Year_*Rotation_	III		82	40.4852	1.5945	23.2	25.39	< 0.0001
Year_*Rotation_	IIIf		82	42.9481	1.5945	23.2	26.94	< 0.0001
Year_*Rotation_	IIf		82	39.7519	1.5945	23.2	24.93	< 0.0001
Year_*Rotation_	IV		82	45.5333	1.5945	23.2	28.56	< 0.0001
Year_*Rotation_	IVf		82	45.6296	1.5945	23.2	28.62	< 0.0001
Year_*Rotation_	III		83	36.5481	1.5354	20.9	23.80	< 0.0001
Year_*Rotation_	IIIf		83	39.5296	1.5354	20.9	25.74	< 0.0001
Year_*Rotation_	IIf		83	31.1444	1.5354	20.9	20.28	< 0.0001
Year_*Rotation_	IV		83	37.9556	1.5354	20.9	24.72	< 0.0001
Year_*Rotation_	IVf		83	37.6148	1.5354	20.9	24.50	< 0.0001
Year_*Rotation_	III		84	57.6741	1.6346	24.9	35.28	< 0.0001
Year_*Rotation_	IIIf		84	56.6333	1.6346	24.9	34.65	< 0.0001
Year_*Rotation_	IIf		84	54.5889	1.6346	24.9	33.40	< 0.0001
Year_*Rotation_	IV		84	58.4000	1.6346	24.9	35.73	< 0.0001
Year_*Rotation_	IVf		84	58.9926	1.6346	24.9	36.09	< 0.0001
Year_*Rotation_	III		85	47.8000	1.7813	29.8	26.84	< 0.0001
Year_*Rotation_	IIIf		85	49.7852	1.7813	29.8	27.95	< 0.0001
Year_*Rotation_	IIf		85	51.4815	1.7813	29.8	28.90	< 0.0001
Year_*Rotation_	IV		85	52.5370	1.7813	29.8	29.49	< 0.0001
Year_*Rotation_	IVf		85	49.7519	1.7813	29.8	27.93	< 0.0001
Year_*Rotation_	III		86	51.0278	2.1328	36.1	23.93	< 0.0001
Year_*Rotation_	IIIf		86	54.5759	2.1328	36.1	25.59	< 0.0001
Year_*Rotation_	IIf		86	48.2093	2.1328	36.1	22.60	< 0.0001
Year_*Rotation_	IV		86	51.4815	2.1328	36.1	24.14	< 0.0001
Year_*Rotation_	IVf		86	52.9870	2.1328	36.1	24.84	< 0.0001
Year_*Rotation_	III		87	48.7556	2.8095	31.5	17.35	< 0.0001
Year_*Rotation_	IIIf		87	41.5556	2.8095	31.5	14.79	< 0.0001
Year_*Rotation_	IIf		87	42.9852	2.8095	31.5	15.30	< 0.0001
Year_*Rotation_	IV		87	45.6741	2.8095	31.5	16.26	< 0.0001
Year_*Rotation_	IVf		87	51.8500	2.8095	31.5	18.46	< 0.0001
Year_*Rotation_	III		88	62.7611	1.9176	35.1	32.73	< 0.0001
Year_*Rotation_	IIIf		88	64.4141	1.9176	35.1	33.59	< 0.0001
Year_*Rotation_	IIf		88	54.2833	1.9176	35.1	28.31	< 0.0001

Least Squares Means									
Effect	Rotation_	Nitrogen_	Year_	Estimate	Standard Error	DF	t Value	P > |t|
Year_*Rotation_	IV		88	58.5241	1.9176	35.1	30.52	< 0.0001
Year_*Rotation_	IVf		88	62.4556	1.9176	35.1	32.57	< 0.0001

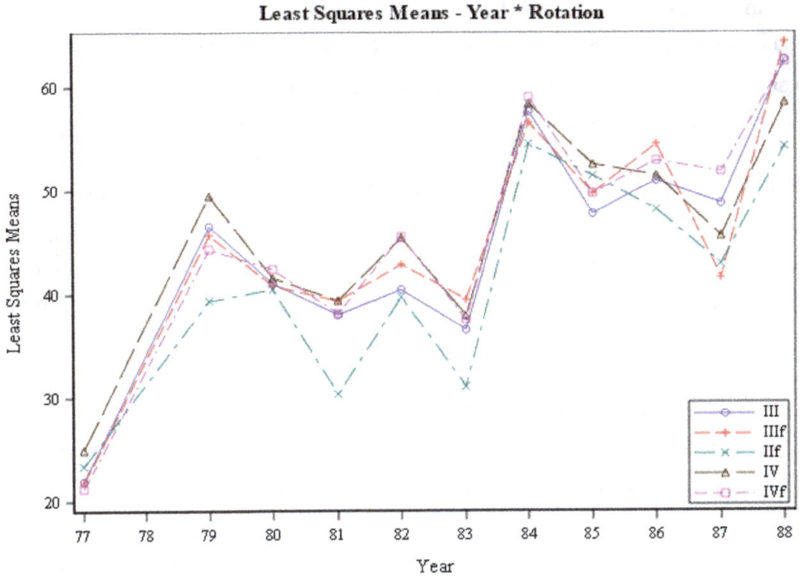


```
/*Plot of year(x) predicted value(y) rotation (group) */

data YR;
set lsmeans;
if Effect = 'Nitrogen_' then delete;
run;

proc template;
define statgraph sgdesign;
dynamic _YEAR_ _ESTIMATE _ROTATION_;
begingraph;
   entrytitle halign=center 'Least Squares Means - Year * Rotation';
   layout lattice / rowdatarange=data columndatarange=data
rowgutter=10 columngutter=10;
      layout overlay / xaxisopts=( label=('Year') linearopts=(
tickvaluesequence=( start=77.0 end=88.0 increment=1.0))) yaxisopts=(
label=('Least Squares Means'));
         seriesplot x=_YEAR_ y=_ESTIMATE / group=_ROTATION_
name='series' display=(markers) clusterwidth=0.5 connectorder=xaxis
grouporder=data;
         discretelegend 'series' / opaque=false border=true
halign=right valign=bottom displayclipped=true across=1 order=rowmajor
location=inside;
      endlayout;
   endlayout;
endgraph;
end;
run;

proc sgrender data=YR template=sgdesign;
dynamic _YEAR_="'YEAR_'n" _ESTIMATE="ESTIMATE" _
ROTATION_="'ROTATION_'n";
run;
```

Appendix 6. Answers to Review Questions

1. Why would you do a long-term rotation experiment instead of several single-year experiments?

   Rotation experiments allow you to study differences between sequences of treatments that are applied over several years. They also allow you to study how the effect of a treatment develops over more than a year.

2. What is the difference between a short-term and a long-term rotation experiment?

   In a short-term rotation experiment the sequences run through one simultaneous cycle, to compare the sequences in the final year. Long-term rotation experiment run through several cycles, and involve analyses of data from more than a year.

3. Why might it be a problem if you ran the rotations over only one series of years?

   The comparisons between the rotations will depend on the specific properties of those years, which may favor one rotation over the others. There is less of a risk of this happening if you run the experiments over more than one series of years.

4. How might you include auxiliary treatments, in addition to the rotation treatments?

   The simplest way to do this is to split the plots into subplots, to form a split-plot design in each year, with the auxiliary treatment factor(s) as the split-plot factor(s).

5. Why might the analysis be more complicated than the analysis of a single-year experiment?

   The results will be recorded from several different years, and these may show different amounts of random variation. The same plot may be observed in several years and, unless these observations are well separated, the results may show a nonuniform correlation structure where the correlations between these observations decline with increasing distance in time.

6. What are the advantages of REML compared to ordinary analysis of variance, and how would exploit these in your analysis?

   REML allows different residual variances to be estimated for the years during the combined analysis of data from several years. It also allows you to fit models to describe the correlations between observations at different times on the same plot.

7. What statistics can you use to decide on the random model?

   You can use the differences between the deviances of two models, if one is a generalization of the other (i.e. if it contains all the random parameters of that model) together with some additional ones. This can be treated can be treated as a chi-square statistic with number of degrees of freedom equal to the number of additional parameters. Otherwise you can use Akaike or Schwarz Bayesian information criteria. The best model is the one with the smallest value of the chosen criterion.

8. How would you assess the fixed terms?

   The standard way to do this is to examine their Wald statistics. These would have exact chi-square distributions if the variance parameters were known but, as those must be estimated, the statistics are only asymptotically distributed as chi-square. In practical situations they are biased (i.e. they tend to give too many significant results). Alternatively, your statistical software may be able to estimate the number of residual degrees of freedom relevant to each term, so that F statistics can be used instead. These should not be subject to the biases of the chi-square statistics.

CHAPTER 12: SPATIAL ANALYSIS OF FIELD EXPERIMENTS

Juan Burgueño

SAS programs for Examples are provided in the electronic supplement.

Answer to review questions

1. Explain with your own words what means "spatial variability analysis"

   In field experimental designs it is related with modelling residual error by considering the spatial distribution of the plots in the field.

2. What are the different models that can be used to model the residual spatial variability?

   Among others, neighbor models, moving average, autoregressive models in row and columns, splines.

3. Why modelling the residual spatial variability usually has larger precision than a standard analysis in which spatial variability is not modeling?

   Because the experimental design is not able to capture all the variability in the field. It is not able to capture variability generated during the experimentation and it is not abel to capture small-scale variability.

4. Is it possible to model spatial variability of an experiment with five treatments in two replicates? Justify your answer.

   It is difficult since there is not so much information, there is a few number of row and columns in the field to adjust most of the model used to model spatial variability.

5. If you are analyzing an experimental design; and you want to perform spatial analysis, are there any changes in the assumptions? If so, mention them.

   Yes, all spatial analysis models assume some degree of relatedness between residuals error compared with the assumption of independence used in standard analyses.

6. What are the advantages of using spatial analysis?

   With spatial analysis it is possible to capture more noise and extract more information about the treatments. Spatial analysis usually is more precise and it adjusts the means of the treatments by the position in the field.

CHAPTER 13: AUGMENTED DESIGNS-EXPERIMENTAL DESIGNS IN WHICH ALL TREATMENTS ARE NOT REPLICATED

Juan Burgueño, José Crossa, Francisco Rodríguez, and Kathleen M. Yeater

Appendix A1. Complete results for Example 1 obtained with the codes in the text. SAS editor file (.sas) provided in electronic supplement.

(P1-f)
The GLIMMIX Procedure

Model Information	
Data Set	WORK.A
Response Variable	y
Response Distribution	Gaussian
Link Function	Identity
Variance Function	Default
Variance Matrix	Diagonal
Estimation Technique	Restricted Maximum Likelihood
Degrees of Freedom Method	Residual

Class Level Information		
Class	Levels	Values
t	8	1 2 3 4 5 6 7 8

Number of Observations Read	12
Number of Observations Used	12

Dimensions	
Covariance Parameters	1
Columns in X	9
Columns in Z	0
Subjects (Blocks in V)	1
Max Obs per Subject	12

Optimization Information	
Optimization Technique	None
Parameters	9
Lower Boundaries	1
Upper Boundaries	0
Fixed Effects	Not Profiled

Fit Statistics	
-2 Res Log Likelihood	16.94
AIC (smaller is better)	34.94
AICC (smaller is better)	214.94
BIC (smaller is better)	29.41
CAIC (smaller is better)	38.41
HQIC (smaller is better)	22.82
Pearson Chi-Square	9.33
Pearson Chi-Square / DF	2.33

Type III Tests of Fixed Effects				
Effect	Num DF	Den DF	F Value	Pr > F
t	7	4	16.06	0.0088

Estimates					
Label	Estimate	Standard Error	DF	*t* Value	*P* > |*t*|
Ch	-5.3333	1.2472	4	-4.28	0.0129
UnT	-4.0000	2.1602	4	-1.85	0.1377
Ch-UnT	-5.6667	1.7638	4	-3.21	0.0325

t Least Squares Means					
t	Estimate	Standard Error	DF	*t* Value	*P* > |*t*|
1	22.3333	0.8819	4	25.32	< 0.0001
2	27.6667	0.8819	4	31.37	< 0.0001
3	28.0000	1.5275	4	18.33	< 0.0001
4	32.0000	1.5275	4	20.95	< 0.0001
5	27.0000	1.5275	4	17.68	< 0.0001
6	38.0000	1.5275	4	24.88	< 0.0001
7	35.0000	1.5275	4	22.91	< 0.0001
8	28.0000	1.5275	4	18.33	< 0.0001

Appendix 1. Complete results for Example 1 obtained with the codes in the text.

(P2-f)

The GLIMMIX Procedure

Model Information	
Data Set	WORK.A
Response Variable	y
Response Distribution	Gaussian
Link Function	Identity
Variance Function	Default
Variance Matrix	Diagonal
Estimation Technique	Restricted Maximum Likelihood
Degrees of Freedom Method	Residual

Class Level Information		
Class	Levels	Values
t	8	1 2 3 4 5 6 7 8
d1	3	1 2 3

Number of Observations Read	12
Number of Observations Used	12

Dimensions	
Covariance Parameters	1
Columns in X	12
Columns in Z	0
Subjects (Blocks in V)	1
Max Obs per Subject	12

Optimization Information	
Optimization Technique	None
Parameters	9
Lower Boundaries	1
Upper Boundaries	0
Fixed Effects	Not Profiled

Fit Statistics	
-2 Res Log Likelihood	16.94
AIC (smaller is better)	34.94
AICC (smaller is better)	214.94

Fit Statistics	
BIC (smaller is better)	29.41
CAIC (smaller is better)	38.41
HQIC (smaller is better)	22.82
Pearson Chi-Square	9.33
Pearson Chi-Square / DF	2.33

Type III Tests of Fixed Effects

Effect	Num DF	Den DF	*F* Value	Pr > *F*
d1	2	4	34.93	0.0029
t(d1)	5	4	8.51	0.0295

Estimates

Label	Estimate	Standard Error	DF	*t* Value	*P* > |*t*|
Ch	-5.3333	1.2472	4	-4.28	0.0129
UnT	-4.0000	2.1602	4	-1.85	0.1377
Ch-UnT	-5.6667	1.7638	4	-3.21	0.0325

d1 Least Squares Means

d1	Estimate	Standard Error	DF	*t* Value	*P* > |*t*|
1	22.3333	0.8819	4	25.32	< 0.0001
2	27.6667	0.8819	4	31.37	< 0.0001
3	31.3333	0.6236	4	50.25	< 0.0001

t(d1) Least Squares Means

t	d1	Estimate	Standard Error	DF	*t* Value	*P* > |*t*|
1	1	22.3333	0.8819	4	25.32	< 0.0001
2	2	27.6667	0.8819	4	31.37	< 0.0001
3	3	28.0000	1.5275	4	18.33	< 0.0001
4	3	32.0000	1.5275	4	20.95	< 0.0001
5	3	27.0000	1.5275	4	17.68	< 0.0001
6	3	38.0000	1.5275	4	24.88	< 0.0001
7	3	35.0000	1.5275	4	22.91	< 0.0001
8	3	28.0000	1.5275	4	18.33	< 0.0001

Appendix A1. Complete results for Example 1 obtained with the codes in the text.

(P3-f)

The GLIMMIX Procedure

Model Information	
Data Set	WORK.A
Response Variable	y
Response Distribution	Gaussian
Link Function	Identity
Variance Function	Default
Variance Matrix	Diagonal
Estimation Technique	Restricted Maximum Likelihood
Degrees of Freedom Method	Residual

Class Level Information

Class	Levels	Values
t	8	1 2 3 4 5 6 7 8
d2	2	1 2

Number of Observations Read	12
Number of Observations Used	12

Dimensions	
Covariance Parameters	1
Columns in X	11
Columns in Z	0

Dimensions	
Subjects (Blocks in V)	1
Max Obs per Subject	12

Optimization Information	
Optimization Technique	None
Parameters	9
Lower Boundaries	1
Upper Boundaries	0
Fixed Effects	Not Profiled

Fit Statistics	
-2 Res Log Likelihood	16.94
AIC (smaller is better)	34.94
AICC (smaller is better)	214.94
BIC (smaller is better)	29.41
CAIC (smaller is better)	38.41
HQIC (smaller is better)	22.82
Pearson Chi-Square	9.33
Pearson Chi-Square / DF	2.33

Type III Tests of Fixed Effects

Effect	Num DF	Den DF	F Value	P > F
d2	1	4	51.57	0.0020
t(d2)	6	4	10.14	0.0210

Estimates

Label	Estimate	Standard Error	DF	*t* Value	P > |*t*|
Ch	-5.3333	1.2472	4	-4.28	0.0129
UnT	-4.0000	2.1602	4	-1.85	0.1377
Ch-UnT	-5.6667	1.7638	4	-3.21	0.0325

d2 Least Squares Means

d2	Estimate	Standard Error	DF	*t* Value	P > |*t*|
1	25.0000	0.6236	4	40.09	< 0.0001
2	31.3333	0.6236	4	50.25	< 0.0001

t(d2) Least Squares Means

t	d2	Estimate	Standard Error	DF	*t* Value	P > |*t*|
1	1	22.3333	0.8819	4	25.32	< 0.0001
2	1	27.6667	0.8819	4	31.37	< 0.0001
3	2	28.0000	1.5275	4	18.33	< 0.0001
4	2	32.0000	1.5275	4	20.95	< 0.0001
5	2	27.0000	1.5275	4	17.68	< 0.0001
6	2	38.0000	1.5275	4	24.88	< 0.0001
7	2	35.0000	1.5275	4	22.91	< 0.0001
8	2	28.0000	1.5275	4	18.33	< 0.0001

Tests of Effect Slices for t(d2) Sliced By d2

d2	Num DF	Den DF	F Value	P > F
1	1	4	18.29	0.0129
2	5	4	8.51	0.0295

Appendix A1. Complete results for Example 1 obtained with the codes in the text.

(P2-r)

The GLIMMIX Procedure

Model Information	
Data Set	WORK.A
Response Variable	Y
Response Distribution	Gaussian
Link Function	Identity

Model Information	
Variance Function	Default
Variance Matrix	Not blocked
Estimation Technique	Restricted Maximum Likelihood
Degrees of Freedom Method	Containment

Class Level Information		
Class	Levels	Values
t	8	1 2 3 4 5 6 7 8
d1	3	1 2 3

Number of Observations Read	12
Number of Observations Used	12

Dimensions	
G-side Cov. Parameters	1
R-side Cov. Parameters	1
Columns in X	4
Columns in Z	8
Subjects (Blocks in V)	1
Max Obs per Subject	12

Optimization Information	
Optimization Technique	Dual Quasi-Newton
Parameters in Optimization	1
Lower Boundaries	1
Upper Boundaries	0
Fixed Effects	Profiled
Residual Variance	Profiled
Starting From	Data

Iteration History					
Iteration	Restarts	Evaluations	Objective Function	Change	Max Gradient
0	**0**	4	47.864285513	.	2.05E-15

Convergence criterion (ABSGCONV=0.00001) satisfied.

Fit Statistics	
-2 Res Log Likelihood	47.86
AIC (smaller is better)	51.86
AICC (smaller is better)	53.86
BIC (smaller is better)	52.02
CAIC (smaller is better)	54.02
HQIC (smaller is better)	50.79
Generalized Chi-Square	21.00
Gener. Chi-Square / DF	2.33

Covariance Parameter Estimates		
Cov Parm	Estimate	Standard Error
t(d1)	17.5333	12.6726
Residual	2.3333	1.6499

Type III Tests of Fixed Effects				
Effect	Num DF	Den DF	*F* Value	*P* > *F*
d1	2	5	2.00	0.2304

Solution for Random Effects							
Effect	t	d1	Estimate	Std Err Pred	DF	*t* Value	*P* > |*t*|
t(d1)	1	1	-626E-16	4.1873	4	-0.00	1.0000
t(d1)	2	2	-4E-14	4.1873	4	-0.00	1.0000
t(d1)	3	3	-2.9418	2.1537	4	-1.37	0.2437
t(d1)	4	3	0.5884	2.1537	4	0.27	0.7982
t(d1)	5	3	-3.8244	2.1537	4	-1.78	0.1504
t(d1)	6	3	5.8837	2.1537	4	2.73	0.0523
t(d1)	7	3	3.2360	2.1537	4	1.50	0.2074

Solution for Random Effects							
Effect	t	d1	Estimate	Std Err Pred	DF	*t* Value	*P* > |*t*|
t(d1)	8	3	-2.9418	2.1537	4	-1.37	0.2437

Estimates					
Label	Estimate	Standard Error	DF	*t* Value	*P* > |*t*|
Ch	-5.3333	6.0516	5	-0.88	0.4185
UnT	-3.5302	2.0294	4	-1.74	0.1569
Ch-UnT	-6.0582	4.5184	5	-1.34	0.2377

d1 Least Squares Means					
d1	Estimate	Standard Error	DF	*t* Value	*P* > |*t*|
1	22.3333	4.2791	5	5.22	0.0034
2	27.6667	4.2791	5	6.47	0.0013
3	31.3333	1.8196	5	17.22	< 0.0001

Tests of Covariance Parameters Based on the Restricted Likelihood					
Label	DF	-2 Res Log Like	ChiSq	P > ChiSq	Note
t(d1=3)	1	51.9494	4.09	0.0216	MI

MI: P-value based on a mixture of chi-squares.

Appendix A1. Complete results for Example 1 obtained with the codes in the text.

(P3-r)

The GLIMMIX Procedure

Model Information	
Data Set	WORK.A
Response Variable	y
Response Distribution	Gaussian
Link Function	Identity
Variance Function	Default
Variance Matrix	Not blocked
Estimation Technique	Restricted Maximum Likelihood
Degrees of Freedom Method	Containment

Class Level Information		
Class	Levels	Values
t	8	1 2 3 4 5 6 7 8
d2	2	1 2

Number of Observations Read	12
Number of Observations Used	12

Dimensions	
G-side Cov. Parameters	2
R-side Cov. Parameters	1
Columns in X	3
Columns in Z	16
Subjects (Blocks in V)	1
Max Obs per Subject	12

Optimization Information	
Optimization Technique	Dual Quasi-Newton
Parameters in Optimization	2
Lower Boundaries	2
Upper Boundaries	0
Fixed Effects	Profiled
Residual Variance	Profiled
Starting From	Data

Iteration History					
Iteration	Restarts	Evaluations	Objective Function	Change	Max Gradient
0	0	4	54.050115446	.	5.27E-16

Convergence criterion (ABSGCONV=0.00001) satisfied.

Fit Statistics	
-2 Res Log Likelihood	54.05
AIC (smaller is better)	60.05
AICC (smaller is better)	64.05
BIC (smaller is better)	60.29
CAIC (smaller is better)	63.29
HQIC (smaller is better)	58.44
Generalized Chi-Square	23.33
Gener. Chi-Square / DF	2.33

Covariance Parameter Estimates			
Cov Parm	Group	Estimate	Standard Error
t(d2)	d2 1	13.4444	20.1208
t(d2)	d2 2	17.5333	12.6726
Residual		2.3333	1.6499

Type III Tests of Fixed Effects				
Effect	Num DF	Den DF	F Value	P > F
d2	1	6	3.85	0.0975

Solution for Random Effects								
Effect	t	d2	Group	Estimate	Std Err Pred	DF	t Value	P > |t|
t(d2)	1	1	d2 1	-2.5208	2.6627	4	-0.95	0.3974
t(d2)	2	1	d2 1	2.5208	2.6627	4	0.95	0.3974
t(d2)	3	2	d2 1	0	3.6667	4	0.00	1.0000
t(d2)	4	2	d2 1	0	3.6667	4	0.00	1.0000
t(d2)	5	2	d2 1	0	3.6667	4	0.00	1.0000
t(d2)	6	2	d2 1	0	3.6667	4	0.00	1.0000
t(d2)	7	2	d2 1	0	3.6667	4	0.00	1.0000
t(d2)	8	2	d2 1	0	3.6667	4	0.00	1.0000
t(d2)	1	1	d2 2	0	4.1873	4	0.00	1.0000
t(d2)	2	1	d2 2	0	4.1873	4	0.00	1.0000
t(d2)	3	2	d2 2	-2.9418	2.1537	4	-1.37	0.2437
t(d2)	4	2	d2 2	0.5884	2.1537	4	0.27	0.7982
t(d2)	5	2	d2 2	-3.8244	2.1537	4	-1.78	0.1504
t(d2)	6	2	d2 2	5.8837	2.1537	4	2.73	0.0523
t(d2)	7	2	d2 2	3.2360	2.1537	4	1.50	0.2074
t(d2)	8	2	d2 2	-2.9418	2.1537	4	-1.37	0.2437

Estimates					
Label	Estimate	Standard Error	DF	t Value	P > |t|
Ch	-5.0417	6.0446	4	-0.83	0.4512
UnT	3.5302	5.5684	4	0.63	0.5605
Ch-UnT	-5.9123	5.8172	6	-1.02	0.3487

d2 Least Squares Means					
d2	Estimate	Standard Error	DF	t Value	P > |t|
1	25.0000	2.6667	6	9.37	< 0.0001
2	31.3333	1.8196	6	17.22	< 0.0001

Tests of Covariance Parameters Based on the Restricted Likelihood					
Label	DF	-2 Res Log Like	ChiSq	P > ChiSq	Note
T(d2=1)	1	58.6165	4.57	0.0163	MI
T(d2=2)	1	58.1353	4.09	0.0216	MI

MI: P-value based on a mixture of chi-squares.

CHAPTER 14: MULTIVARIATE METHODS FOR AGRICULTURAL RESEARCH

Kathleen M. Yeater and María B. Villamil

Review Questions (True or False)

[*Hint*: Your T/F answers can be located using information in Table 1.]

1. We can explore relationships among variables in a data set with MR, CART, or even with cluster analysis among other tools. (T)
2. The main difference between PCA and FA is that the PCA tries to uncover theoretical constructs underlying the data set. (F)
3. SEM and CART are examples of flexible modern techniques that can integrate categorical and continuous variables. (T)
4. MANOVA, PCA, and DA are examples of techniques that require multivariate normality and homogeneity of variances. (T)
5. When the goal of the MA is to predict group membership, then DA, LR, or CART are NOT good choices. (F)
6. The desired result of applying MANOVA to a data set is to create linear combinations of variables that maximize our group differences. (F)
7. No previous knowledge of group membership is required when we explore the data set with CA. (T)
8. If the dependent variable in the data set only takes two values, a LR is the most appropriate technique to describe differences among the two possible outcomes. (T)
9. If your data set does not meet the requirements of multivariate normality you cannot apply any of the MA techniques. (F)
10. You can use CCPA to explore the distribution of several plant species on different environments characterized by topography, moisture level, aspect, etc. (T)

Exercises:

For the following exercises we will be working with the iris data set available from R by typing:

```
> library(datasets)
> iris
```

Once the data set is available on your workspace,

1) Investigate the structure of the data set and create a scatterplot matrix of the variables

```
Sepal.Length, Sepal.Width, Petal.Length, Petal.Width.
```

What can you infer from these results?

```
> str(iris)
'data.frame':   150 obs. of   5 variables:
```

```
$ Sepal.Length: num   5.1 4.9 4.7 4.6 5 5.4 4.6 5 4.4 4.9 ...
$ Sepal.Width : num   3.5 3 3.2 3.1 3.6 3.9 3.4 3.4 2.9 3.1 ...
$ Petal.Length: num   1.4 1.4 1.3 1.5 1.4 1.7 1.4 1.5 1.4 1.5 ...
$ Petal.Width : num   0.2 0.2 0.2 0.2 0.2 0.4 0.3 0.2 0.2 0.1 ...
$ Species      : Factor w/ 3 levels "setosa","versicolor",..: 1 1 1 1
1 1 1 1 1 1 ...
> pairs(iris[,1:4])
```

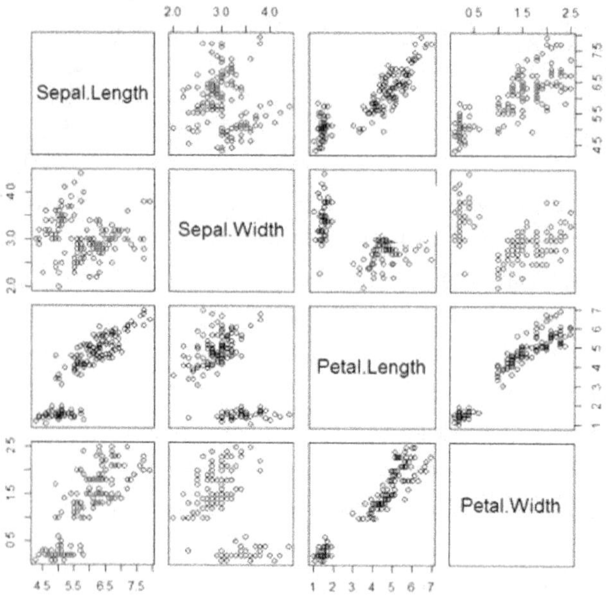

2) Explore the possibility of successfully separating the Iris species based on the sepal and petal variables listed. Hint: You will need the MASS library.

```
> library(MASS)
> iris.lda<-lda(iris [,-5], iris[,5])
> iris.lda
Call:
lda(iris[, -5], iris[, 5])
Prior probabilities of groups:
    setosa versicolor  virginica
 0.3333333  0.3333333  0.3333333

Group means:
           Sepal.Length Sepal.Width Petal.Length Petal.Width
setosa            5.006       3.428        1.462       0.246
versicolor        5.936       2.770        4.260       1.326
virginica         6.588       2.974        5.552       2.026

Coefficients of linear discriminants:
                    LD1         LD2
Sepal.Length  0.8293776  0.02410215
Sepal.Width   1.5344731  2.16452123
Petal.Length -2.2012117 -0.93192121
Petal.Width  -2.8104603  2.83918785

Proportion of trace:
   LD1    LD2
0.9912 0.0088
> plot(iris.lda, abbrev=TRUE, cex=0.8) # Saved as Exercise lda plot.
tiff
```

```
> abline(h=0, v=0, lty=4)
```

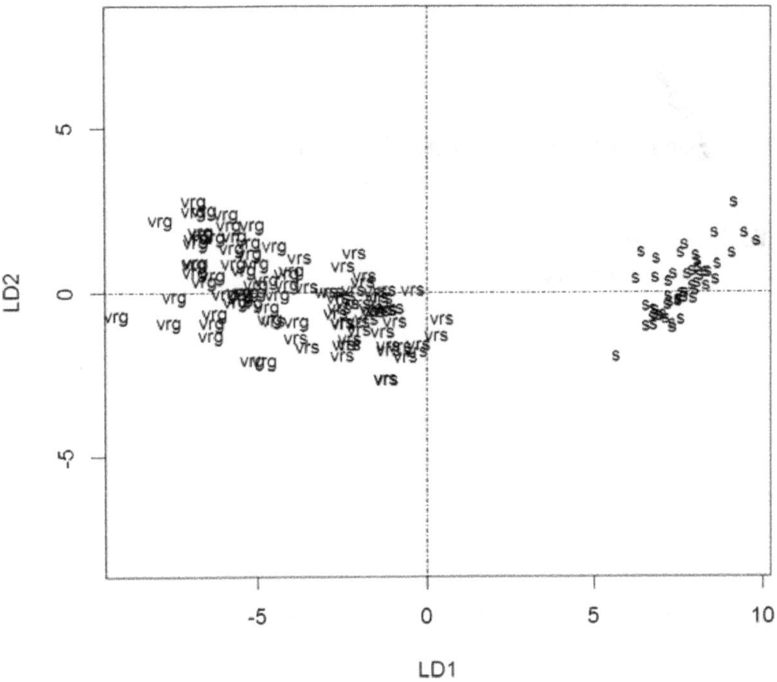

3) Calculate the error rates associated with the classification of each of the Iris species and provide an interpretation of your results.

```
> iris.ld<-predict(iris.lda)
> iris.ldcv<-lda(iris [,-5], iris [,5], CV=T)
> table(true=iris$Species, pred=iris.ldcv$class)
            pred
true         setosa versicolor virginica
  setosa         50          0         0
  versicolor      0         48         2
  virginica       0          1        49
```

Essential website information

R: http://cran.r-project.org/web/views/Multivariate.html

SAS: http://support.sas.com/documentation/cdl/en/statug/63033/HTML/default/viewer.htm

CHAPTER 15: NONLINEAR REGRESSION MODELS AND APPLICATIONS

Fernando Miguez, Sotirios Archontoulis, and Hamze Dokoohaki

Answer To Exercises

Multiple choice correct answers

1. e)
2. d)
3. c)
4. b)
5. b)

Exercise 1

1.a. The simple exponential model is

$$Y = Y_0 \exp(-k)$$

The first partial derivative with respect to k is still

$$\frac{\partial Y}{\partial k} = Y_0 \exp(-kx) \times -x$$

Remember that the derivative of $f(x) = \exp(x)$ is also $\exp(x)$. The second derivative will not be equal to zero, therefore this is a function with a nonlinear parameter k.

1.b. In the model

$$y = \beta_0 + \beta_1 \exp(-\frac{x}{\theta})$$

The partial derivative with respect to the first parameter (β_0) is

$$\frac{\partial y}{\partial \beta_0} = 0$$

Thus, this is a linear parameter. The first partial derivative with respect to the second parameter () is

$$\frac{\partial y}{\partial \beta_1} = \exp\left(-\frac{x}{\theta}\right)$$

The second partial derivative is

$$\frac{\partial y}{\partial^2 \beta_1} = 0$$

Thus, this is a linear parameter. The partial derivative with respect to the parameter q is

$$\frac{\partial y}{\partial \theta} = x \times \frac{[\beta_1 \exp(-\frac{x}{\theta})]}{\theta^2}$$

Thus the second derivative will not be zero and this is not a linear parameter.

Exercise 2

1.a. Virtually any model which has an additive parameter such as

$$y = \beta_0 + \ldots$$

Will have the intercept model only as a subset. One less trivial example would be the exponential decay. Setting k=0 results in a model with a single parameter (Y_0). Another is the Michaelis–Menten equation in the following form

$$y = \frac{ax}{b+x}$$

where setting $b = 0$ results in a $y = a$

2.b.

It is tricky to show strictly that something cannot happen. A simple example could be

$$y = \exp(-kx)$$

In this case we do not have the simple intercept model as a subset of the full model. When $k = 0$, y = 1. As k increases, then y gets closer to zero. Again this is not an intercept model. Another example would be the one parameter logistic (Eq. 3.10, Table 3).

CHAPTER 16: ANALYSIS OF NON-GAUSSIAN DATA

Walter W. Stroup

Answers to Review Questions. Data sets provided as .csv file in electronic supplement.

1. a. Read "plot(block) | variety" as "plot within block after accounting for variety."

b. Before applying treatments there are 2 plots and hence 1 df in each of 10 blocks. Therefore 10 df for plot(block). Accounting for variety removes the df for variety(in this case one. Therefore "plot(block) | variety" has $10-1=9$ df.

2. False. Leaving block*variety out of the RANDOM statement will result in overdispersion, a common form of poorly specified model.
3. False. Unless you include PLOT in the data set, the term PLOT(BLOCK*VARIETY) will be unintelligible to SAS. If you do include it, the algorithm GLIMMIX uses to determine denominator DF may not work properly, so you should check the listing. In any event, BLOCK*VARIETY uniquely identifies "plot(block) | variety" and avoids unintended consequences.
4. a. Y/N

b. FALSE (!!!!)

c. Make absolutely sure you have *both* Y *and* N in the data set for *every* experimental unit!

5. False. If you answered "true" Reread section on notation conventions for mixed models.
6. False.

   delete IRRIG from CLASS statement
7. cumulative logit
8. (b) leaf shape is a "nominal" multinomial variable (i.e. no obvious ranking or categories)
9. False. Science should drive statistics, not vice-versa. If science calls for unequal spacing, statistics can deal with it.
10. No
11. False. Using RANDOM...RESIDUAL will make this impossible.
12. True
13. False
14. False. (correct answer is Poisson)
15. True
16. True
17. True
18. False. If the distribution is binomial, The data scale estimate of the LSMEAN is the probability. (or proportion if the distribution is Beta)
19. True
20. False. Negative binomial is used for *count* data, not proportions. Use Beta for continuous proportion data.

www.ingramcontent.com/pod-product-compliance
Lightning Source LLC
Chambersburg PA
CBHW080834310825
31883CB00004B/13